ELECTRON EMISSION SPECTROSCOPY

ELECTRON EMISSION SPECTROSCOPY

PROCEEDINGS OF THE NATO SUMMER INSTITUTE
HELD AT THE UNIVERSITY OF GENT,
AUGUST 28–SEPTEMBER 7, 1972

Edited by

W. DEKEYSER, L. FIERMANS, G. VANDERKELEN and J. VENNIK

State University, Gent

D. REIDEL PUBLISHING COMPANY

DORDRECHT-HOLLAND / BOSTON-U.S.A.

First printing: December 1973

Library of Congress Catalog Card Number 73–83559

ISBN-13: 978-94-010-2632-1   e-ISBN-13:978-94-010-2630-7
DOI: 10.1007/978-94-010-2630-7

Published by D. Reidel Publishing Company,
P.O. Box 17, Dordrecht, Holland

Sold and distributed in the U.S.A., Canada, and Mexico
by D. Reidel Publishing Company, Inc.
306 Dartmouth Street, Boston,
Mass. 02116, U.S.A.

TABLE OF CONTENTS

PREFACE

Electron emission spectroscopy became recently a major tool for the study of molecules and solids. These volumes contain a rather complete review of the state of the art in this field. Both the physical and chemical aspects are covered extensively by well-known specialists. Different modes of excitation are used in electron emission spectroscopy.

The electron-solid scattering is covered in detail by C. B. Duke, from a theoretical point of view. Elastic and inelastic low energy electron diffraction are extensively discussed in relation to the geometrical, electronic and vibronic structure of solid surfaces.

Auger electron emission spectroscopy (AES) is covered by J. C. Tracy. The technique is discussed from the point of view of surface research. This part also contains a complete literature list concerning the application of AES up to the middle of 1972.

Electron emission produced by X-ray impact, is covered by C. S. Fadley, D. T. Clark, R. P. Gupta and S. K. Sen.

The contribution by C. S. Fadley, entitled 'Theoretical Aspects of X-Ray Photoelectron Spectroscopy', is an up to date discussion of core electron binding energies, valence electron binding energies, multiplet splittings and multi-electron processes. R. P. Gupta and S. K. Sen's contribution provides an introduction to crystal field theory and its application to electron energy level determination. D. T. Clark deals with the more chemical aspects of X-ray photoelectron spectroscopy, i.e. the study of chemical shifts and the relation to the bonding characteristics in molecules.

Finally, E. Lindholm considers the UV excitation in electron emission spectroscopy, a branch known as 'Molecular Photoelectron Spectroscopy'.

These surveys are the rearranged and extended versions of six out of the nine sets of lecture notes of the NATO Advanced Summer Institute held at Gent from August 28 to September 7, 1972. The editors are grateful to the lecturers of this school for providing extended lecture notes and particularly to those who accepted to publish them. In some cases this involved a major rewriting task in order to cover the subject as completely as possible. The support of the NATO Science Committee is gratefully acknowledged.

Our gratitude also goes to the following editors of books and periodicals and authors who granted permission to reproduce figures, diagrams or other material, i.e. Accounts of Chemical Research; American Chemical Society; American Institute of Physics; Analytical Chemical Research; Applied Spectroscopy; Bulletin des Sociétés Chimiques Belges; Chemical Physics Letters; Faraday Division of the Chemical Society; Institute of Physics (London); Journal of the American Chemical Society; Journal of Chemical Physics; Journal of the Chemical Society; Journal of Electron

Spectroscopy; Journal of Physical Chemistry; Journal of Polymers Science; North-Holland Publishing Co.; Nova Royal Society Uppsala; Physical Society of Japan; Plenum Publishing Corporation; Review of Modern Physics; Robert A. Welch Foundation Research Bulletin; Royal Society (London); Royal Society of Sciences of Uppsala; Royal Swedish Academy of Sciences; Solid State Communications; Zeitschrift für Naturforschung; Zeitschrift für Physik; J. Wiley and Sons Publishing Co; I. Adams; S. Aksela; D. A. Allison; A. Bagchi; M. Barber; T. Bergmark; S. A. L. Bergström; G. K. Bohn; H. P. Bonzel; G. Broden; C. R. Brundle; J. M. Burkstrand; C. W. Caldwell Jr; T. A. Carlson; J. C. Carver; P. H. Cutler; D. W. Davis; R. Ditchfield; A. R. Du Charme; C. B. Duke, D. E. Eastman; C. S. Fadley; A. Fahlman; R. W. Finck; C. T. Foxon; T. E. Gallon; U. Gelius; R. L. Gerlach; J. Gerstner; A. M. Gibbons; E. J. McGuire; C. Glupe; K. Hamrin; S. B. M. Hagström; E. Hasilbach; D. M. Hercules; K. Hirabayshi; P. O. Heden; J. Hedman; B. W. Hilland; J. M. Hollander; J. E. Houston; D. A. Huchital; L. D. Hulett; D. W. Jespen; G. Johansson; W. L. Jolly; E. R. Jones; J. Jones; R. C. Jopson; B. A. Joyce; S. Karlsson; J. T. McKinney; M. O. Krausse; A. B. Kunz; G. E. Laramore; U. Landman; M. G. Lagally; B. Lindberg; I. Lindgren; H. Lofgren; P. M. Marcus; H. Mark; N. Mehlhorn; W. E. Moddeman; J. H. Neave; T. C. Ngoe; R. Nordberg; C. Nordling; C. Norris; T. Novakov; G. A. Olah; P. W. Palmberg; R. L. Park; L. G. Parrat; M. Pelavin; F. M. Propst; E. G. McRae; H. Rosencwaig; D. A. Shirley; K. Siegbahn; D. L. Smith; L. C. Snydes; W. E. Swartz; C. D. Swift; N. J. Taylor; T. D. Thomas; J. C. Tracy; C. W. Tucker; M. B. Webb; G. W. Wertheim.

ELECTRON SCATTERING BY SOLIDS:
DETERMINATION OF THE CHEMICAL, GEOMETRICAL,
ELECTRONIC AND VIBRATIONAL STRUCTURE OF SURFACES

C. B. DUKE*

*Dept. of Physics, Materials Research Laboratory,
and Coordinated Science Laboratory University of Illinois Urbana, Ill., U.S.A.*

TABLE OF CONTENTS

* Present address: Xerox Research Laboratories, Phillips Road, Webster, N.Y. 14580, U.S.A.

W. Dekeyser et al. (eds.), Electron Emission Spectroscopy, 1–149. All Rights Reserved
Copyright © 1973 by D. Reidel Publishing Company, Dordrecht-Holland

1. The Structure of Solid Surfaces

A. INTRODUCTION

In these notes we examine the use of various experiments, primarily those associated with generic label low-energy-electron diffraction ('LEED'), to determine the atomic identity, positions, vibrational amplitudes, and electronic structure of 'atoms' (i.e., ion-cores immersed in a sea of valence electrons) in the uppermost 1–5 layers of a single-crystal solid. A general survey of mathematical models of the specifically surface microscopic properties of solids is available elsewhere. [1] Furthermore, the construction of an adequate theory [2–6] of the electron-solid scattering phenomena of primary interest to us requires extensive use of modern quantum-field-theory methods with their concommitant algebraic complexity. Thus we focus our attention on the applications of theories of particle-solid scattering to characterize the structure of solid surfaces. We indicate only briefly the ingredients of models of the electronic, geometrical, and vibrational structure of surfaces and the construction of theories of the scattering process.

To specify the microscopic 'structure' of a solid surface, one must provide four types of information: the atomic identity of ion cores in the uppermost few layers; the positions of these ion cores; their vibrational amplitudes; and the ground-state properties and excitation spectrum of the valence electron fluid in which they are immersed. Much of this information can be determined in more than one way. Thus the choice of which experiments to perform to obtain it often is based on considerations of experimental convenience and/or ease in theoretical interpretation, rather than on a fundamental limitation of the experimental methods not used. However, two general features of sample preparation are common to all methods. First, the samples must be prepared in high vacuum ($p \sim 10^{-10}$–10^{-12} torr) in order that the surface conditions remain constant for the duration ($t \sim 1$ h) of the experiment. Second, single-crystal specimens are almost a necessity because of the difficulties inherent in trying to characterize the surface conditions on polycrystalline or powder samples. For some applications (e.g., the study of impurity segregation at grain boundaries [7]), the second criterion may be relaxed, but in these lectures I shall presume it to be satisfied.

Given the requirements of high-vacuum environment and a single-crystal sample, a further important distinction between types of experiments is based on the geometry of the sample. In the 'tip' geometry, characteristic of field-emission and field ionization microscopy, [8] multiple atomic planes are exposed on a more-or-less spherical tip and electron emission from or ion formation near these planes is used to project a greatly expanded image of the tip on a spherical screen. The area of each planar surface is small ($A \sim (100 \text{ Å})^2$), the electric fringing fields at the surface are large, [9] and the samples generally are confined to the mechanically-strong transition metals. In the 'planar' geometry, characteristic of electron-diffraction and photo-emission experiments, a macroscopic area ($A \sim (1 \text{ cm})^2$) of a single atomic plane is sought by combinations of polishing, etching, ion-bombardment, and heat treatment. Perhaps the most serious shortcoming of this type of experiment is that usually no independent measurement of the surface topography is possible. [10] As, however, essentially all theoretical models of electron-solid and photon-solid interactions either implicitly or explicitly assume the planar geometry, we concentrate our attention on this latter class of experiments. A synopsis of these general experimental design criteria is given in Figures 1 and 2.

Characterization of Solid Surfaces I

A. Maintenance of Constant Surface Condition
   During Measurement ($\tau \sim 1 \text{min}$):
   UHV ($p \sim 10^{-10}$ Torr)

B. Classes of Experiment
   1. Tip-Geometry
      a. $V > 0$ FEM
      b. $V < 0$ FIM

Fig. 1.   Schematic indication of some of the design criteria for experiments utilized to characterize the structure of solid surfaces. The use of ultra-high vacuum (UHV) field-emission microscopy (FEM) and field-ion microscopy (FIM) is discussed in the text.

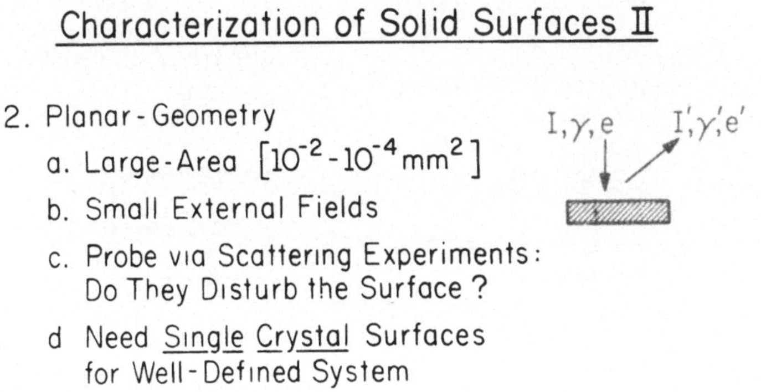

Characterization of Solid Surfaces II

2. Planar-Geometry
   a. Large-Area $[10^{-2} - 10^{-4} \text{mm}^2]$
   b. Small External Fields
   c. Probe via Scattering Experiments:
      Do They Disturb the Surface?
   d Need <u>Single</u> <u>Crystal</u> Surfaces
      for Well-Defined System

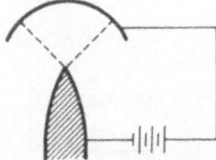

Fig. 2.   General features of experiments used to characterize planar solid surfaces. The probes used in the scattering experiments can be atomic species (I), photons (γ), or electrons (e).

B. THE CHEMICAL STRUCTURE OF PLANAR SURFACES

The nuclei of the atoms in the uppermost layers of the solid are characterized by two numbers; their charge and mass. Their charge is measured by experiments in which a deep 'core' electron is excited by an incident electron or photon. Their mass is most easily and reliably measured by the recoil momentum given a 'high-energy', i.e. keV-MeV noble-gas ion scattering elastically from the surface layer. In the determination of the nuclear charge by core-level excitation, electrons and X-rays usually are the sources of the excitation and the particles detected in the final state. Therefore, one studies the reactions

$$\gamma + S \rightarrow S^* \tag{1a}$$

$$e + S \rightarrow S^* + e' \tag{1b}$$

$$S^* \rightarrow S + \gamma' \tag{1c}$$

$$S^* \rightarrow S + e''. \tag{1d}$$

In Equations (1), S designates the initial ground state of the solid and S^* designates its excited state in which a core electron is excited into the conduction band as shown in Figures 3a and b respectively. The electronic processes involved in the decay of the excited states indicated in Equations (1c) and (1d) are indicated in Figures 3c and d. Actually, although Equations (1) specify the most commonly studied reactions,

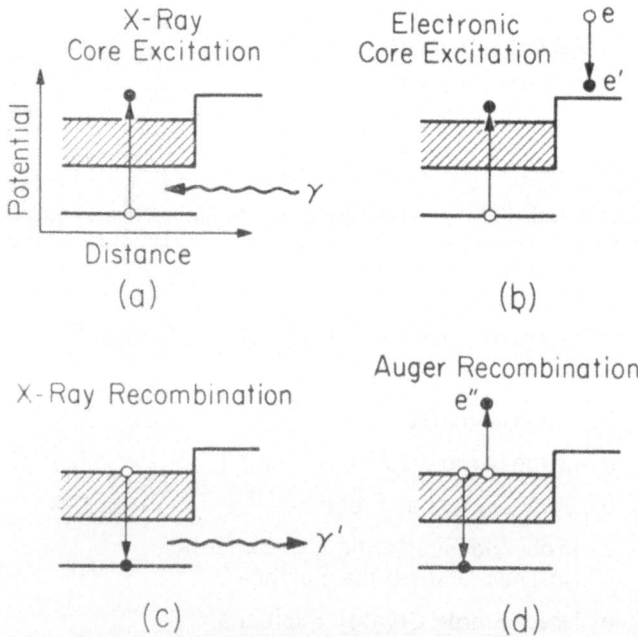

Fig. 3. Schematic diagram of the electronic transitions involved in the reactions designated in Equations (1) in the text.

they fail to include the X-ray Raman effect

$$\gamma + S \rightarrow S^* + \gamma' \tag{2}$$

and the possibility of electron cascades, e.g.

$$e' + S^* \rightarrow S^{**} + ne''' \tag{3a}$$

$$e'' + S \rightarrow S^{***} + me''''. \tag{3b}$$

We neglect both processes here because they have not proven particularly useful for the chemical analysis of surface species.

For our purposes, the important feature of the reactions indicated in Equations (1) is that the energies of the incident X-ray absorption edge (E_γ), of the characteristic energy loss of an incident keV electron ($E_e - E_{e'} = \Delta E'$), of the X-ray fluorescence (E_γ), and of the Auger electrons ($E_{e''}$) all are characteristic of the atomic number of the ion-core in which the core electronic state is excited or filled. Therefore by measuring any of these energies, an indication of the identity of the elements in the target is obtained.

Three questions immediately arise. Which measurements are the most desirable to make in a given case? How is the electronic energy resolution characteristic of these measurements related to the spatial position of the nucleus whose atomic number we are attempting to determine? How does the chemical state of the valence electrons associated with this nucleus affect the measurements? (E.g., can we distinguish between molecular CO and 'atomic' C and O adsorbed on a metal surface?) An adequate response to these questions would carry us too far from our main topic. Besides, reviews of these topics may be found in the literature [11–13] as well as in other lectures at this school by Drs Fadley, Tracy, and Spicer. Therefore, let me just sketch for you some of the issues involved in answering these questions and note some of the current experimental 'best practice' techniques.

The most important experimental design criterion in surface spectroscopy is that of assembling a set of measurements which (hopefully completely) characterizes all of the properties of a (single-crystal) surface in the same high-vacuum system. Thus one must choose a particular configuration of electron, ion, and photon sources and detectors for the particular system under investigation. Most early 'ESCA' (Electron-Spectroscopy-for-Chemical Analysis) experiments are characterized by nearly mono-energetic X-ray photon sources and high-energy-resolution secondary-electron detectors. Other, more recent photoemission experiments, in the visible, UV, and soft-X-ray regions, generally involve broad-band photon sources and moderate energy-resolution ($\Delta E \sim 1$ eV) electron detectors. Both types of experiments are being discussed extensively by other lecturers at this school. However, photoemission experiments are not well-suited for determining the geometrical and vibrational structure of (planar) solid surfaces. For these tasks, electron-scattering experiments, both elastic and inelastic, are utilized.

Important advantages of electron sources and detectors are their relative simplicity, and the general adequacy of electron scattering experiments to determine the important characteristics of solid surfaces. The identification of the elemental charge of the surface atoms is accomplished by Auger spectroscopy in which an incoming electron of energy $100 \, eV \lesssim E \lesssim 10^4 \, eV$ is used to excite a core hole [Equation (1b); Figure 3b] and the electron ejected when the hole is filled [Equation (1d); Figure 3d] is detected. This process yields electrons emitted from ions in the upper 1–4 layers of the solid. [13] The surface sensitivity of the method can be enhanced substantially by using glancing-angle incident electrons. [14] Also, both the Auger [15] and other types of the high-energy-electron-diffraction [16] (HEED) scattering experiments can be operated in a scanning mode to provide spatial resolution along the solid surface. Auger spectroscopy is quite satisfactory for the identification of the elemental charge of surface ion cores because appropriate transitions can be detected for all elements but H and He. However, it is not yet very useful for *chemical* analysis [e.g., is a carbon nucleus in adsorbed CO or an adsorbed graphite form] because of the complicated line shapes which are observed. Photoemission spectroscopy and inelastic electron-solid scattering appear to be the best candidates for determining the electronic-chemical structure of surface compounds.

Ion-scattering experiments, the primary means of measuring the *mass* of surface atoms, [17] exhibit two disadvantages relative to electron-scattering experiments. First, the ion-beam erodes the surface, a fact that forms the basis for determining impurity depth profiles via secondary ion mass spectrometry [18] (SIMS). Second, they are not readily compatible with the electron-scattering experiments needed to obtain the geometrical, electronic, and vibrational properties of solid surfaces. Therefore, although ion scattering techniques are quite surface sensitive [capable of detecting atoms in the top monolayer] they have not yet been incorporated in complete arrays of experiments whose objective is the complete characterization of solid surfaces. A similar result is true for the scattering of very-low-energy ($E \lesssim 10 \, eV$) neutral atoms and molecules from solid surfaces.

Summarizing, from the point of view of practical equipment design, electrons are by far the most convenient and useful probe of the surface region. Photons are less convenient and furthermore are not intrinsically surface sensitive. Finally, low-energy atoms and ions, while highly surface-sensitive, are the least convenient to work with, and are not as versatile as electrons in probing a wide variety of surface properties.

Turning to the issue of the relation between spatial and energy resolution, let us recall the distance scales probed by various 'surface' experiments. [1] The complicated inhomogeneous boundary layer of a solid extends for about $d_{bl} \sim 1$–10 Å from an ideal surface. Electromagnetic radiation in and above the visible frequency range penetrates for distances $\lambda_{em} \sim 10^4$ Å. Low-energy ($5 \, eV \lesssim E \lesssim 500 \, eV$) electrons exhibit electron-electron inelastic collision mean free paths $\lambda_{ee} \sim 5$ Å. Low-energy ($\kappa T \lesssim E \lesssim 10 \, eV$) atoms do not penetrate the solid at all. Static electric fields penetrate metals by $\lambda_m \sim 1$ Å and semiconductors by $\lambda_{sc} \sim 10^3$–10^6 Å. Thus we immediately see that as $\lambda_{em}, \lambda_{sc} \gg d_{bl}$, the optical properties of materials and surface-induced space charge effects in semiconductors are sensitive to the surface properties of the solid only by virtue of boundary conditions. Atom and ion-solid scattering experiments are intrinsically the most sensitive to the specific properties of the surface layer, but, as we have just noted, are fraught with practical problems of versatility and, in the case of atom (molecule) solid scattering, of energy resolution. Thus electrons in the 'low-energy', $25\,eV \lesssim E \lesssim 250\,eV$, range seem to be the best compromise surface-sensitive probes of solids. They can be generated outside the solid and used in elastic-low-energy-electron diffraction (ELEED) and inelastic low-energy-electron diffraction (ILEED) experiments. They can be generated inside the solid by incident higher energy (e.g., keV) electrons as in Auger spectroscopy [13] (AES) and the reflection [13] scanning electron-microscope operated in the secondary-electron mode [16] (SEM). Finally they also can be generated in the solid by energetic photons in the various types of photoemission spectroscopy [11] (PES). In all cases the surface-sensitivity of a particular technique rests on the fact that electrons in this 'low-energy' range rapidly lose energy by inelastic collisions and hence those which escape with little or no energy loss must have originated in the surface layers of the solid.

These considerations establish three results of central importance in surface spectroscopy. First, of the three types of surface-sensitive probes ['low-energy' electrons, keV-MeV ions, slow neutrals] we focus on electrons for reasons of experimental convenience, of utility in determining a wide range of surface properties, and of the relatively non-destructive nature of the scattering processes. Second, low-energy electrons can be used to measure *surface* properties only because they exhibit strong electron-solid interactions, which cause them to undergo energy loss processes in which they create electronic elementary excitations in the solid with a mean free path $\lambda_{ee} \sim 5$-$10\,\text{Å}$. Third, because of the second result, the spatial resolution of an electron-scattering experiment depends on the ability to distinguish experimentally between elastically and inelastically scattered (emitted) electrons. Elastic scattering (emission) originates from the uppermost 1-3 atomic layers, single-loss inelastic scattering from the uppermost 1-6 layers, and multiple inelastic events, in particular secondary-electron emission of electrons of energy $E \lesssim 25\,eV$ independent of the primary electron energy, originate within 100-200 Å of the surface. Thus in the case of electron scattering experiments (ELEED, ILEED) good energy resolution ($\Delta E \sim 1$ eV) implies good spatial resolution ($\Delta d \sim 5\,\text{Å}$) normal to the surface. The resolution parallel to the surface depends predominately on instrument design. [10] In the case of internally-generated electrons (AES, PES, SEM), energy resolution of the emitted electrons can be related to the elastic-inelastic nature of the emission process only via the intermediary of a theory of the creation-energy-loss-emission sequence. Therefore energy resolution of $\Delta E \lesssim 1$ eV in the detector is a necessary, but not sufficient, condition for establishing a spatial resolution $\Delta d \lesssim 5\,\text{Å}$ normal to the surface. In general, crude lineshape analyses are used to distinguish between elastic and inelastic emission of these electrons. [19]

We now turn to our final topic in this subsection, the determination of the chemical state as well as the charge and depth of the nucleus of a 'surface atom'. In principle, PES, AES, and ILEED all measure the excitation spectra of valence electrons in the vicinity of the surface and, consequently, should provide a measure of the local electronic (and hence chemical) structure. A similar result holds for the case of ion-neutralization spectroscopy [20] (INS) in which an externally-generated charged ion provides the core-hole which is filled by the Auger process specified by Equation (1d) and Figure 3d. Although observed spectra sometimes can be correlated with simple theoretical expectations, [21] at the present time the lack of an adequate theory of PES, AES, and INS, and the failure of experimenters to characterize adequately the surfaces on which their measurements are performed conspire to render the extraction of surface chemical structure information from experimental data qualitative at best. This situation is expected to improve rapidly.

Summarizing, current experimental techniques and their theoretical interpretation permit the determination of the elemental charge, Z, and mass, M, of most nuclei in the upper layers of a solid to within a depth resolution of $\Delta d \sim 5$ Å. The chemical structure of the valence electrons associated with these nuclei cannot be determined reliably at the present time. The intrinsic limit of sensitivity of the charge determinations is roughly $\Delta N \sim 10^{10}$–10^{12} nuclei cm^{-2} in the upper five layers of the solid. [13] This limit is not known accurately in the case of the mass determination by ion scattering, but it is thought to be $\Delta N \gtrsim 10^{13}$ nuclei cm^{-2} in a thickness $\Delta d \sim 5$ Å for atoms of heavier mass than those in the substrate. [17, 22, 23]

C. THE GEOMETRICAL STRUCTURE OF PLANAR SURFACES

When discussing the geometry of clean, single-crystal surfaces, it is convenient to

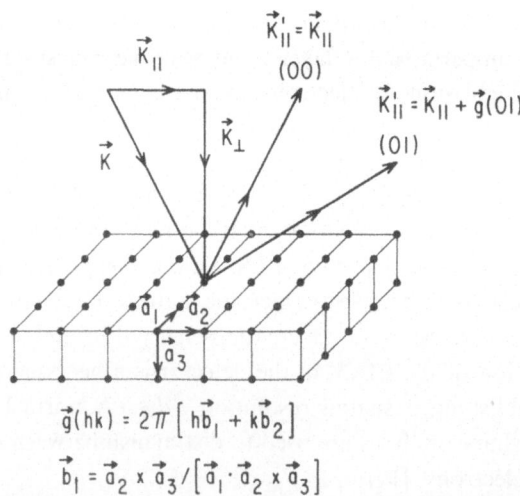

Fig. 4. Schematic illustration of an incident electron beam of wave-vector $\mathbf{k} = \mathbf{k}_\perp + \mathbf{k}_\parallel$, scattered elastically from a single crystal into a state characterized by the wave-vector $\mathbf{k}' = \mathbf{k}'_\parallel + \mathbf{k}'_\perp$. The construction of the reciprocal lattice associated with the single-crystal surface also is shown.

regard the solid as composed of identical two-dimensional planar diffraction gratings parallel to the surface, composed of atomic scatterers, and stacked on the top of each other to form the crystal. In this simple case, exemplified by simple metals, the geometrical structure consists of determining two things: the location of the atomic scatterers within each diffraction grating and the stacking sequence of these gratings to form the crystal.

Because of the periodic nature of the two-dimensional gratings, the use of a diffraction technique is appropriate to determine the positions of the scatterers in each unit cell of the lattice. The atomic spacings between the scatterers, $a \sim 1$ Å, implies that X-ray photons, low-energy ($E \gtrsim 5$ eV) electrons, or thermal neutrons would diffract from the lattice. However, we already have noted that the weak photon-solid interaction permits X-rays to penetrate deeply into the solid ($\lambda_{em} \sim 10^4$ Å) so that they measure *bulk*, not surface, properties of the sample. A similar conclusion applies to thermal neutrons, leaving electrons or thermal H or He atoms as the appropriate diffractive probes of the geometry of clean, single-crystal surfaces. Although diffractive atom-solid scattering has been observed, [24, 25] it is not common, [1] so we focus our attention on electron-solid scattering.

A schematic diagram of electron reflection from a single-crystal solid is shown in Figure 4. The appropriate conservation laws are those of energy (E) conservation and momentum conservation for the component of momentum parallel to the surface

Elastic Electron-Solid Scattering: A Surface-Sensitive Spectroscopy

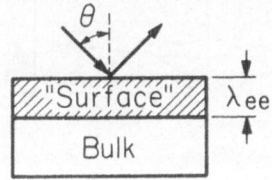

1. Surface Sensitive because of Inelastic-Collision Damping $4\text{Å} \lesssim \lambda_{ee} \lesssim 8\text{Å}$.

2. Spectroscopy for <u>Periodic</u> Geometries because
$$k = (2mE/\hbar^2)^{1/2} \gtrsim 1/(\text{Atomic Dimensions}).$$

Fig. 5.   Schematic summary of the physical reasons why elastic electron-solid scattering is an appropriate diffraction spectroscopy of surface atomic geometry. The values of the inelastic-collision pentration depth, λ_{ee}, are typical ones for electrons whose energy normal to the surface lies in the range $5\,\text{eV} \lesssim E_\perp \lesssim 500\,\text{eV}$. Electrons whose energy normal to the surface is smaller also may probe the surface properties of a solid if these energies occur within an energy band gap of the substrate (a common occurrence on the (100) face of fcc and bcc metals for $E \sim 1$–10 eV). A value of the energy for motion normal to the surface in the requisite range may be achieved by various experimental arrangements, as noted in Figure 6.

modulo the reciprocal lattice vectors **g** associated with the space-group symmetry of the two-dimensional diffraction gratings consisting of planes of atoms parallel to the surface which, when stacked together, comprise the crystal. The geometry of the surface atoms is determined by an analysis of the elastic scattering cross sections of the incident beam of electrons scattered from the solid surface. The reasons that elastic electron diffraction is a probe of the surface rather than bulk atomic positions at a planar solid-vacuum interface are summarized in Figure 5. It seems worth noting at this point that the strong electron-solid interaction, necessary in order to achieve the desired sensitivity to *surface* properties, has the additional consequence that elementary Born-approximation ('kinematical') scattering theories do not suffice to describe the electron-solid scattering process. Therefore the two central problems in the theory of electron-solid scattering are the specification of the (complicated) force law, and the separation of the effects of the surface atomic geometry from those of other (often unknown) features of this force law. [1, 6, 26]

Two 'standard' experimental arrangements are used to extract geometrical structure information from elastic electron diffraction as indicated schematically in Figure 6. The glancing-angle diffraction of keV electrons, referred at as reflection high-energy electron diffraction (RHEED) embodies the arrangement noted in the top panel of

Surface Crystallography

1. Reflection High-Energy Electron Diffraction (RHEED)

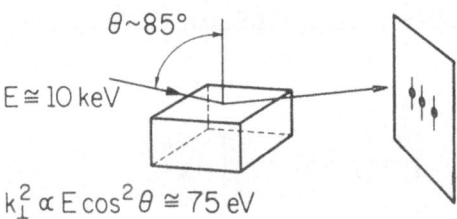

$\theta \sim 85°$

$E \cong 10\,keV$

$k_\perp^2 \propto E\cos^2\theta \cong 75\,eV$

2. Elastic Low-Energy Electron Diffraction (ELEED)

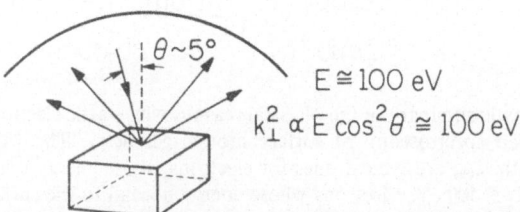

$\theta \sim 5°$

$E \cong 100\,eV$

$k_\perp^2 \propto E\cos^2\theta \cong 100\,eV$

Fig. 6. Schematic indication of the two electron-diffraction techniques used to determine the geometry of single-crystal surfaces: reflection high-energy-electron diffraction (RHEED) and elastic low-energy-electron diffraction (ELEED).

the figure. Its high vacuum use on single-crystal surfaces thus far has been limited to determining the space-group symmetry of clean surfaces and simple adsorbed monolayers. As noted in a recent review, [27] its customary applications have been the examination of roughened crystal surfaces in a poor vacuum ($p \sim 10^{-4}$–10^{-5} torr). A more common technique for single-crystal surface crystallography is that of elastic low-energy electron diffraction (ELEED) indicated in the lower panel of Figure 6. This technique is one of our primary concerns in these lectures. It also has been reviewed recently [6, 28, 29]. As noted in the caption to Figure 5, both techniques yield electron energies for motion normal to the surface in the range characterized by strong inelastic-collision damping.

No other methods have been utilized extensively for the determination of the positions of atomic scatterers in the uppermost 1–5 layers of a solid. The use of double-alignment channeling experiments involving the scattering of MeV rare-gas ions has been proposed, [22, 23] but not widely applied, to determine the deviation of the positions of surface scatterers from those expected in a truncated, but otherwise ideal bulk lattice. Therefore, the current 'best practice' techniques for determining the atomic composition and geometry of a planar, single-crystal solid surface are those summarized in Figure 7.

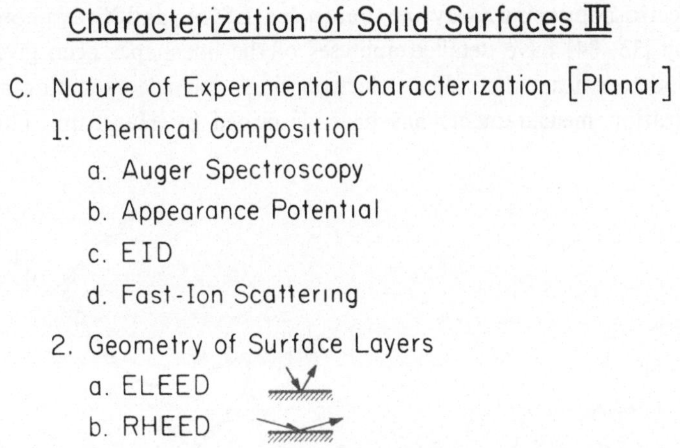

Characterization of Solid Surfaces III

C. Nature of Experimental Characterization [Planar]

1. Chemical Composition

   a. Auger Spectroscopy

   b. Appearance Potential

   c. EID

   d. Fast-Ion Scattering

2. Geometry of Surface Layers

   a. ELEED

   b. RHEED

Fig. 7.   Synopsis of current (1971–72) experimental techniques used for the characterization of the atomic composition and geometry of single-crystal solid surfaces. Auger spectroscopy and fast-ion scattering are discussed in Section 1C. Appearance potential spectroscopy (APS) and electron-ion-desorption (EID) are described in the literature. [12]

D. THE ELECTRONIC STRUCTURE OF PLANAR SURFACES

The two most obvious types of surface-sensitive electronic structure information obtainable from analyses of particle-solid scattering experiments are the particle-hole excitation spectra of the solid (especially 'surface' branches of this spectra like surface plasmons), and the changes in such spectra characteristic of a 'clean' surface upon the adsorption of other atomic species. In our classification scheme we regard modifications of the ground-state charge density and excitation spectra of the valence-electron fluid upon adsorption as elements of the 'electronic' structure of a surface, although such changes carry what often is called 'chemical' information [e.g. whether adsorbed C and O species exist as 'CO' or atomic 'C' and 'O']. In this context, we already have discussed in Section 1B the fact that the common core-electron spectroscopic techniques [Auger-electron spectroscopy (AES), photoemission spectroscopy (PES, ESCA), appearance-potential spectroscopy (APS), and ion-neutralization spectroscopy (INS)] can, in principle, provide insight into the valence-electron charge density by virtue of two phenomena. First, the energy-levels of core electrons depend on the eigenstates of the valence-electrons. Energy shifts (called 'chemical shifts') in these core levels of the order of several electron volts are created by the valence-electron environment of the core. Second, in the core-hole recombination processes shown in Figures 3c and d, the electron which recombines with the hole originates from the valence electron charge density. Therefore, the shape and intensity of the recombination line is determined in part by the excitation spectrum of the valence electrons. Analysis of these lineshapes is complicated, however, by lifetime broadening of the core hole, by dynamic screening of the (localized) core hole by the valence-electron fluid prior to the recombination event, [30–32] and by the occurrence of loss events experienced by incident or exit electrons. [31, 32] Only in the cases of photoemission spectroscopy using X-ray line sources (i.e. ESCA) and X-ray photoabsorption spectroscopy [33, 34] have detailed analyses of the lineshapes been given. A qualitative method to extract the valence-electron single-particle excitation spectra from ion-neutralization measurements has been described by Hagstrum. [20] However,

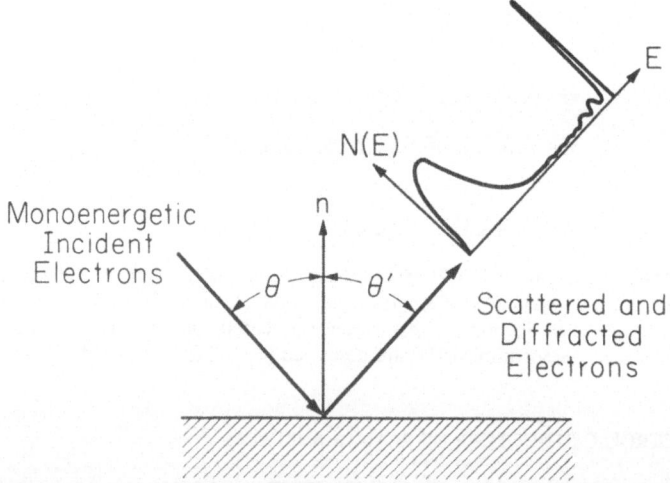

Fig. 8.   Schematic diagram of the experimental arrangement for an inelastic low-energy electron diffraction (ILEED) measurement. The energies, E, of the incident electrons lie in the range 5 eV $\lesssim$ $\lesssim E_0 \lesssim$ 500 eV. The scattering cross sections are determined by the number of electrons with energy $E \leqslant E_0$, i.e., $N(E)$, scattered into a given solid angle $d\Omega'$ at the final-state polar (θ') and azimuthal (ψ') angles. A curve for $N(E)$ when $\theta' \simeq \theta$ is shown in the figure.

attempting to achieve a quantitative characterization of the properties of valence electrons by analyzing fine structure on the lineshapes associated with core-electron transitions is not yet a very rewarding enterprise.

Our main interest in these lectures is the direct determination of the excitation spectra of the valence electrons, rather than their indirect determination via lineshape analyses of core-electron transitions. As an example of a spectroscopic technique in which both types of determination are possible, let us consider inelastic low-energy electron diffraction (ILEED). A schematic diagram of the ILEED experiment is shown in Figure 8. For a given position of the gun generating incident electrons of fixed (and controllable) energy, the scattered electrons are detected by a movable target which contains a retarding grid and phase-sensitive detection apparatus. The measured quantity is the number, $N(E)$, of electrons with final energy E (less than the initial energy, E_0), scattered into a particular final angle. A typical $N(E)$ curve for scattering angles near the elastic diffraction condition is indicated in Figure 8 and, in more detail, in Figure 9. Since typical instrumental energy resolutions are

$$\Delta E \sim 0.5 \, \text{eV} , \tag{4a}$$

it is not possible to resolve elastic scattering from the inelastic scattering 'assisted' by lattice vibrations (phonons). Therefore, the instrument measures the quasi-eleastic scattering cross sections

$$\left(\frac{d\sigma}{d\Omega}\right)_{qe} = \int_{E_0 - \Delta E/2}^{E_0 + \Delta E/2} d\varepsilon \left(\frac{d^2\sigma}{d\Omega \, d\varepsilon}\right) \tag{4b}$$

for $E \cong E_0$, as noted in association with the large peak in $N(E)$ for $E \cong E_0$ in Figure 9. However, the resolution is adequate to measure the cross sections for electrons which have undergone a small number (usually one) of large-energy, $1 \, \text{eV} \lesssim w \lesssim 20 \, \text{eV}$, electronic loss processes. These electrons form the 'discrete' or 'characteristic'-loss peaks in the $N(E)$ curve noted in Figure 9. The 'true secondary' region of the curve consists of electrons with final energies $E \lesssim 30 \, \text{eV}$ which either are inelastically-scattered incident ('primary') electrons or, more likely, 'secondary' electrons from the solid excited by inelastic collisions of the primary electrons [see, e.g., Figure 3b]. The Auger electrons, whose origin is indicated in Figure 3d, are shown in Figure 9 superposed on the high-energy tail of the secondary electron distribution. As we have seen, fine-structure in the Auger peaks and the characteristic-loss peaks both are caused by excitations of the valence electron fluid. Both can be measured by ILEED, but analysis of the characteristic-loss lineshapes provide a simpler and more precise characterization of the excitation spectrum of the valence electrons.

A synopsis of the various techniques for determining the electronic structure of solid surfaces is presented in Figure 10. We already have decided that the use of the more direct analysis of characteristic energy losses in ILEED is preferable to the (indirect) one of core-excitation lineshapes. What we now examine is the distinction between the results obtained from analyzing ILEED and those extracted from visible and UV photoemission, (i.e., PES). We also must indicate why these techniques yield information about the *surface* (as opposed to bulk) properties of the solid.

Visible and UV photoelectron (and presumably, absorption) spectroscopy in principle permit the determination of the particle-hole excitation spectra in those cases in which the electron and hole have nearly the same momentum. These transitions, called 'vertical', 'direct', or 'exciton' interband transitions, are indicated schematically

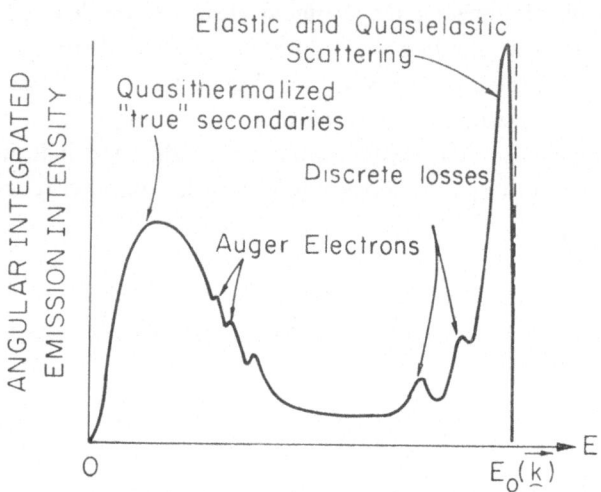

Fig. 9. Schematic indication of the energy-loss profile associated with a measurement of inelastic low-energy electron diffraction (ILEED) intensities for a detector position near that of an elastically-diffracted beam. The physical origin of the various structures in the loss profile are noted in the figure. The angular distributions associated with electrons scattered by the various mechanisms are quite different, so that a loss profile taken at other positions of the detector could appear rather unlike that shown in the figure.

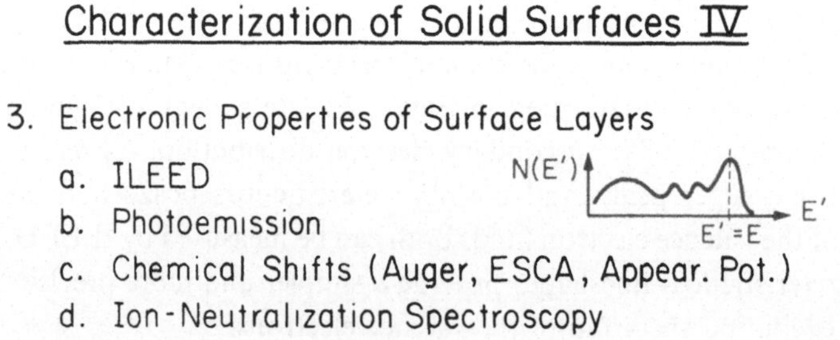

Characterization of Solid Surfaces IV

3. Electronic Properties of Surface Layers

    a. ILEED

    b. Photoemission

    c. Chemical Shifts (Auger, ESCA, Appear. Pot.)

    d. Ion-Neutralization Spectroscopy

Fig. 10. Synopsis of current (1971–72) experimental techniques for the charactization of the electronic structure of the surface layers of a single crystal solid. Many of these techniques, all of which are discussed in the text, also convey information about the bulk properties of the target. A major task of the data analysis is the separation of these two types of information which are both unavoidably contained in the measured quantities.

in Figure 11. In a bulk solid they are characterized by energy and momentum conservation for a photon of energy hv

$$E'(\mathbf{k}') - E(\mathbf{k}) = hv \tag{5a}$$

$$\mathbf{k}' - \mathbf{k} = \mathbf{k}_{ph}. \tag{5b}$$

As for a photon $k_{ph} = 2\pi v/c \sim 10^{-3}$ Å, we see that $k_{ph} \ll k$, $k' \sim 10^{-1}$ Å. Therefore, $\mathbf{k} \cong \mathbf{k}'$ so the transition is 'vertical'. For low-energy electrons, however, the momentum transfer is $\Delta k \sim 1$ Å^{-1}. Consequently large momentum transfers from the incident electron to the valence electron fluid accompany the loss of energy of the incident electron via incoherent particle-hole excitations as indicated for an intraband transition in Figure 11. The different ranges of momentum transfer associated with photon absorption and inelastic low-energy-electron scattering result in the fact that photoemission spectroscopy (PES) and inelastic low-energy electron diffraction (ILEED) probe separate features of the (*bulk*) valence-electron excitation spectrum. Also, the distinctions between the selection rules for electron-induced and photon-induced transitions render PES and ILEED sensitive to different electronic excitations.

In practice, three 'surface' phenomena complicate greatly the simple model of bulk loss processes in PES and ILEED implicit in our discussion of Figure 11. First,

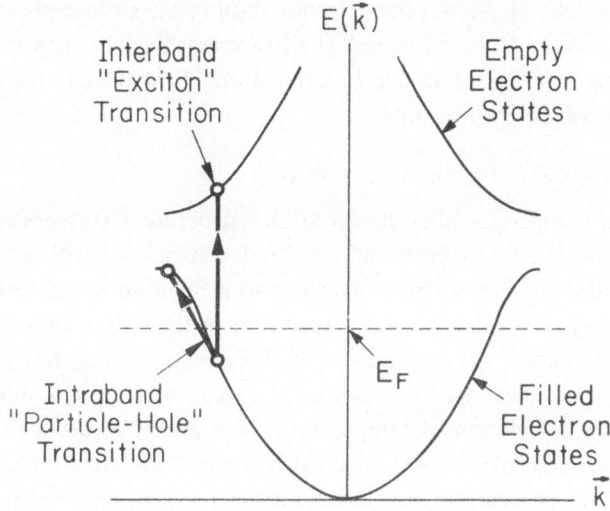

Single-Particle Excitations

Fig. 11.   Illustration of exciton ('vertical', 'direct') interband particle-hole transitions of the valence electron fluid caused by the absorption of a photon and of intraband particle-hole transitions caused by the inelastic scattering of an incident electron. An incident electron also can cause large-momentum-transfer ('indirect', 'diagonal') interband transitions. The diagram is drawn for a bulk solid. The presence of a planar surface breaks the momentum conservation law for motion normal to the surface as, e.g., in the 'surface' photoelectric effect in metals [36].

at a planar surface the component of momentum normal to the surface is not conserved in either photon absorption or inelastic electron scattering. In PES, this fact gives rise to the 'surface-photoelectric effect' which is important for photon energies near the photoemission threshold. [36] In ILEED (PES) it gives rise to structure in the cross sections (energy distribution curves, 'EDC's') associated with the precise shape of the potential near the solid-vacuum interface. [1, 4, 36] Second, as already discussed in Section 1c, the inelastic collisions of the emitted photoelectrons and of both the initial and final electrons in ILEED render the EDC's and ILEED cross sections highly surface sensitive. However, even the most recently used models of this effect in PES are quite primitive, [36] being comparable to the earliest models [37] of inelastic effects in LEED, rather than current 'best-practice' ones. [4, 6] Third, the presence of disordered adsorbates on the surface breaks the momentum conservation law even for the components of momentum parallel to the surface. Thus the interpretation of both LEED [6] and PES [38, 39] is complicated still further in this important case.

These three features of the electron solid interaction, while rendering PES and ILEED sensitive to the surface properties of solids, also require that a detailed theory be used to extract from raw experimental data any quantitative characterization of surface electronic properties. This data-analysis procedure is far more extensively developed for ILEED [3, 4, 6, 40–43] than for PES. [36, 38, 39] Therefore, in these lectures I shall concentrate on the case most thoroughly studied [40–43] using the most sophisticated models: [4, 42, 43] the determination of the surface-plasmon dispersion relation in aluminum from measured ILEED intensities. In principle, [6, 36, 38] analogous analyses of PES data can be carried out. Probably, in the next few years several of these will be forthcoming.

E. THE VIBRATIONAL STRUCTURE OF PLANAR SURFACES

Two types of question concerning the vibrational structure of surfaces are of particular interest. First, are the rms vibrational amplitudes associated with a layer of atoms, labeled by n, parallel to the surface layer ($n \equiv 0$) a function of the distance from the surface? Evidently, if the normal mode spectrum of the lattice vibrations (phonons) contains surface modes, [1] the answer to this question must be 'yes'. Second, if foreign atoms (molecules) are adsorbed on the clean surface of a substrate, what is their behavior as a function of temperature? For example, do such atoms diffuse across the surface and experience chemical reactions? Do they form ordered structures in certain temperature ranges which undergo abrupt phase changes (e.g., 'surface' melting [44])? Or do they remain in relatively immobile disordered arrays whenever they 'stick' to the surface? A synopsis of these issues and an indication of the techniques used to examine periodic planer atomic arrays are given in Figure 12.

In comparison with the analyses of structural and electronic properties of planar surfaces, those of atomistic motion along a planar surface are in an embryonic state. Although an adequate theory of both ELEED and ILEED from a vibrating lattice exists, [2, 6] it has been applied only in a few simple cases. [5, 45–49] The main reason

for the lack of detailed analyses is that any such undertaking confronts three fundamental difficulties. First, as the energy resolutions, ΔE, of all existing instruments
satisfy $\Delta E \geqslant 30$ meV, they are larger than acoustical phonon energies. Thus the only
direct ILEED measurements of the phonon spectra which have been made are those
of local modes of light adsorbates [50] and optical phonons. [51, 52] The latter are
more nearly macroscopic than surface properties of the solid. [1] Being required by
the above considerations to abandon direct ILEED determination of most surface
phonon spectra, the only remaining good probe of the surface vibrational properties
is the temperature dependence of ELEED intensities. Although, photoemission energy
distributions exhibit some temperature dependence, separation of surface from bulk

Characterization of Solid Surfaces $\underline{\text{V}}$

4   Vibronic Properties of Surface Layers

$$[<u_S^2> \cong 2 <u_B^2>]$$

a.   ELEED Temperature Dependence $[<u^2>]$

b   ILEED Phonon Spectra

5   Trace Mechanisms of Chemical Reactions
(Oxidation, Chemisorption-Desorption,
Catalytic Reactions)

Fig. 12.   Synopsis of the two major aspects of the characterization of atomistic motion at solid
surfaces. The two commonly-used techniques (elastic low-energy electron diffraction, ELEED, and
inelastic low-energy electron diffraction, ILEED) for determining the vibrational properties of
surface atoms in planer periodic arrays are indicated. Techniques for studying disordered arrays
and chemical reaction paths are not yet well-developed. The main conclusion of the ELEED
studies is that the rms vibrational amplitudes of surface 'atoms', $\langle u_S^2 \rangle$, are about twice those of
bulk 'atoms' $\langle u_B^2 \rangle$.

effects is likely to prove more difficult than in the case of ELEED and, in any event, no
theory of the temperature dependence has yet been given. (A combination of Duke
and Laramore's LEED theory [2] and Mahan's photoemission theory [53] is straightforward and would be adequate.) However, studies of the temperature dependence
of ELEED intensities exhibit two further fundamental difficulties. First, carefully
taken data on any but a few special systems (e.g., xenon [54]) indicate an extremely
complicated dependence of the ELEED intensities on temperature. [55] Second,
substantial ambiguities occur in theoretical analyses of these data. [45, 48] In the
case of clean metal surfaces, the one generally-accepted conclusion, [6, 45, 55, 56]
arrived at on the basis of such analyses, is that the rms vibrational amplitudes of
surface ion-cores, $\langle u_0^2 \rangle_T$, are about twice those of the bulk ion-cores, $\langle u_{n \to \infty}^2 \rangle_T$; i.e.
$\langle u_0^2 \rangle_T \cong 2 \langle u_\infty^2 \rangle_T$ as indicated in Figure 12. However, the exact values of the $\langle u_n^2 \rangle_T$
extracted from a given data analysis depend sensitively upon both the nature of the
inelastic-collision damping [45] and the detailed way in which the $\langle u_n^2 \rangle_T$ vary with the
layer index, [56] n. In addition, the temperature dependence of a specified maximum
in the ELEED intensities also depends on its dynamical origin. [45, 49] Therefore,

the analysis of the temperature dependence of ELEED intensities, although quite feasible, is not a precise measure of the $\langle u_n^2 \rangle_T$ because the results of such an analysis are influenced strongly by poorly-understood [4, 6] features of the electron-solid interaction.

Summarizing, high-resolution ILEED experiments and studies of the temperature dependence of ELEED intensities are the only techniques currently available for studying the atomistic motion of surface atoms in planar periodic arrays, although the field-ion microscope is useful for such studies on samples in the 'tip' geometry. However, the necessary ILEED experiments generally are not feasible, and the extraction of a quantitative characterization of the vibrational amplitudes of surface atoms via an analysis of ELEED intensities is unreliable because of uncertainties in the electron-solid force law. Therefore, in these lectures, I shall not deal with the atomic vibrations of surface ion-cores in detail. So little is known about surface chemical reactions and adsorbate diffusion that a systematic theoretical discussion of these topics is impossible at the present time.

F. ORGANIZATION OF THE REMAINING LECTURES

In the preceding five sections we have seen that a variety of techniques exist for characterizing individual aspects of the structure of the uppermost 1–5 layers of a single-crystal solid surface. Two features of these techniques are of paramount significance for us. First, no single technique suffices to completely characterize a surface. Thus in any thorough surface study, a variety of techniques must be used, simultaneously, to examine a given sample. Second, quantitative information about the surface structure can be extracted from the data taken using these techniques only via the intermediary of a theoretical model. Thus the minimal adequate facility for surface spectroscopy is an operational high-vacuum system containing instrumentation for several different measurements, and a related theoretical effort providing the model calculations required for extracting from the data a quantitative characterization of the state of the sample.

The simplest minimal instrumental arrangement consists of experiments involving electrons as both the initial and final probe particles. Two electron guns are required, a glancing-incidence keV electron gun to provide incident electrons for AES and RHEED, together with a movable, lower energy gun to provide them for ELEED and ILEED. And ion gun to clean the surface is needed in any case, and a mass spectrometer is highly desirable for adsorption studies. Several detection schemes for the electrons are possible, of which movable electrostatic deflectors provide a versatile detector capable of the best energy resolution ($\Delta E \sim 30$–50 meV) for ILEED experiments. The sample holder must include temperature control and a manipulator capable of executing complete three-dimensional rigid-body rotations.

My intent in the remainder of these lectures is to outline for you the ingredients and current status of the theoretical models available to analyze the data taken using such a minimal experimental arrangement. However, as little detailed theory is required to use AES for chemical analysis and little is available for RHEED, I refer the reader to the literature [13, 15, 16, 27] for discussions of these topics. An outline of the items which we shall consider explicitly is given in Figure 13. The remainder of these notes is organized in accordance with this outline.

Finally, it should be emphasized that the minimal configuration which I am proposing is not necessarily the most desirable one. PES, which will be discussed by other lecturers at this school, would be a welcome addition and is, in fact, quite compatible experimentally with our minimum package. However, in Frank Propst's and my experimental group at Illinois, we did not have infinite resources and used ILEED in lieu of PES for experimental design reasons. It also is true that the theory of surface effects in ILEED is much more thoroughly developed [2–4, 6, 43] than the comparable theory of PES. [19, 36, 38, 53] This fact is a consequence of the experimental design considerations, however, because the ILEED theory was developed explicitly to analyze the experiments planned in the Illinois experimental program.

Structure Determination by LEED

A. Elastic LEED

    1. Electron-Solid Force Law: Microscopic Theory

    2. Geometrical Structure

    3. Vibronic Structure: Temperature Dependence

B. Inelastic LEED

    1. Chemical Condition of the Surface

    2. Electronic Structure [e.g. Surface-Plasmon Dispersion]

Fig. 13.   Schematic indication of the nature of the information about solid surfaces which can be extracted from a theoretical analysis of appropriate low-energy-electron-diffraction (LEED) data.

2. Elastic Electron-Solid Scattering: General Features

In this part of the lecture notes we discuss three aspects of the electron-solid scattering problem. We consider only elastic scattering because the inelastic processes are treated in Section 5. First, we recall the precise mathematical definition of the problem and the nature of its solution. Second, we examine the general features of the electron-solid force law. Finally we discuss how these features are incorporated into theoretical models used to calculate the electron-solid elastic-scattering differential cross sections. Throughout the section, we concentrate our attention on salient features of the calculations for incident electrons in the energy range $5\,\mathrm{eV} \leqslant E \leqslant 500\,\mathrm{eV}$ characteristic of low-energy electron diffraction (LEED).

A. THE ELASTIC SCATTERING PROBLEM

Our first task in the analysis of electron scattering experiments is the formulation of the model calculation of the scattered electron intensity as a function of the parameters characterizing the incident beam, those characterizing the surface of the solid from which it is scattered, and the temperature. A schematic diagram of the elastic scattering process was shown in Figure 4. The general features of the scattered intensity are determined by the conservationlaws describing the scattering process. Initially we envisage this process as occurring between two plane-wave states

$$\varphi_{\mathbf{k}}(\mathbf{r}) = \exp[i\mathbf{k}\cdot\mathbf{r}] \qquad (6)$$

$$\mathscr{H}_0 \varphi_{\mathbf{k}} = E(k)\,\varphi_{\mathbf{k}} = [\hbar^2 k^2/2m]\,\varphi_{\mathbf{k}} \qquad (7)$$

$$\mathscr{H}_0 = -\hbar^2 \nabla^2/2m \qquad (8)$$

which are the electron eigenstates in the absence of the crystal. The elastic scattering from the crystal by definition is described by energy conservation between the initial and final states in the scattering process, $\varphi_{\mathbf{k}}$ and $\varphi_{\mathbf{k}'}$ respectively:

$$E(\mathbf{k}) = E(\mathbf{k}'). \qquad (9)$$

In addition, the assumed periodicity of the crystal parallel to the surface requires that the component of $\mathbf{k}$ parallel to the surface, which we denote by $\mathbf{k}_\parallel$, also be conserved modulus a reciprocal lattice vector, $\mathbf{g}$, of the surface Bravais net

$$\mathbf{k}'_\parallel = \mathbf{k}_\parallel + \mathbf{g}. \qquad (10)$$

The combined effect of these two conservation laws is that the scattered electrons emerge from the solid only at a few discrete angles as noted in Figure 4. The electron emissions in these directions are referred to as 'beams' and labeled by the Miller indices, (hk), of the associated reciprocal lattice vector:

$$\mathbf{g} = 2\pi[h\mathbf{b}_1 + k\mathbf{b}_2] \qquad (11a)$$

$$\mathbf{b}_1 = \mathbf{a}_2 \times \mathbf{a}_3/[\mathbf{a}_1 \cdot \mathbf{a}_2 \times \mathbf{a}_3], \; c.p. \qquad (11b)$$

$$\mathbf{a}_3 = \text{interior surface normal}. \qquad (11c)$$

As implied by Equations (11), the surface unit cell, described by the primitive translation vectors $\mathbf{a}_1$ and $\mathbf{a}_2$, always is chosen to be as small as possible compatible with the requirement that it describe the unit cell of a layer of atoms perpendicular to the surface normal, $\mathbf{a}_3$.

The case of inelastic scattering is more complicated because both the energy and the momentum conservation laws must be extended to include the excitation(s) of the crystal which cause the electrons'energy loss. We return to a discussion of this situation latter in Section 5.

In brief, the theoretical problem of calculating the differential elastic scattering cross section, $d\sigma/d\Omega$, from the surface reduces to that of calculating the angular position and intensity of each of the final ('emerging') beams. The quantities on which the final results depend are noted in Figure 14. Evidently, the detailed predictions of the theory can depend substantially on the model used to describe the target. One of the most important outputs of any theory must be a discussion of how sensitive these predictions are to the dynamical details of the model [e.g., ion-core-scattering] as opposed to the atomic geometry of the lattice. LEED is a useful technique for structure analysis if and only if the dependence of the beam intensities on the dynamical parameters either can be accounted for in detail or can be eliminated.

LEED INTENSITIES FROM (CLEAN) SURFACES II: THEORETICAL PROBLEM:

EVALUATE THE INTENSITY OF THE (hK) BEAM AS A FUNCTION OF

(1) THE CONSERVED BEAM QUANTITIES

(a) ENERGY: $E(\vec{K}') = E(\vec{K})$

(b) PARALLEL MOMENTUM: $\vec{K}'_{\parallel} = \vec{K}_{\parallel} + \vec{g}$

(Bragg condition in surface planes)

(2) MATERIAL PARAMETERS OF THE TARGET

(a) ATOMIC GEOMETRY

(b) TEMPERATURE

(c) SCATTERING POWER OF ION CORES

(d) SCATTERING POWER OF VALENCE ELECTRONS

Fig. 14.   Schematic indication of the parameters on which the intensities of the emerging electron beams depend.

For later convenience we introduce some nomenclature associated with the experimentally measured quantities. The pattern of spots on a (real or hypothetical) fluorescent screen [28] due to the emerging beams is called an *intensity pattern*. The intensity of a given (hk) beam as a function of the incident electron energy E is denoted by $I_{hk}(E)$ and referred to as an *intensity profile*. Both definitions are illustrated in Figure 15. We shall see that one might expect (but rarely observes for $E \lesssim 100$ eV) large primary Bragg peaks in the intensity profile due to the diffraction of the incident electrons from the layers of atoms perpendicular to the surface. These are indicated in Figure 15 for a normally-incident beam. Their absence at low energies, and the occurrence of other more complicated 'secondary' structure in the intensity profiles, are the main experimental observations which motivate and require the somewhat complicated dynamical theory of LEED whose development is traced in the second lecture of this series. The simplest type of model for electron-solid scattering is obtained by considering the atoms in the solid to be rigid, static potentials situated on a periodic lattice. Such a model is referred to as a *static potential model*. It constitutes the basis for band structure calculations for a bulk solid as well as for electron-solid scattering. The Hamiltonian which defines the model is

$$\mathscr{H}_0 = \mathscr{H}_{00} + V \tag{12a}$$

$$\mathscr{H}_{00} = -\hbar^2 \nabla^2 / 2m \tag{12b}$$

$$V = V(\mathbf{r}) = \sum_n v_n(\mathbf{r} - \mathbf{R}_n) \tag{12c}$$

EXPERIMENTALLY MEASURED QUANTITIES IN LEED

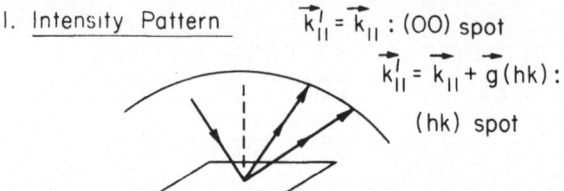

1. <u>Intensity Pattern</u> $\vec{k}'_{\|} = \vec{k}_{\|}$: (OO) spot

$\vec{k}'_{\|} = \vec{k}_{\|} + \vec{g}(hk)$:

(hk) spot

At any fixed energy of the incident beam , get an "intensity pattern" of spots due to diffracted beams.

2 <u>Intensity Profile</u> : The energy dependence of the intensity of a given spot.

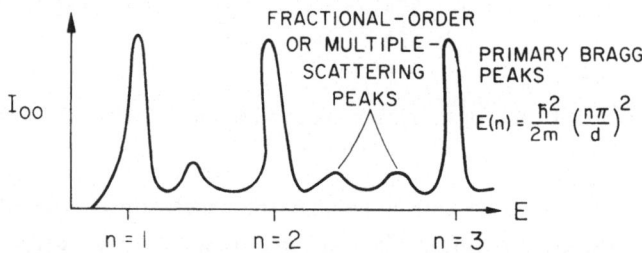

Fig. 15. Illustration of the distinction between an intensity pattern and an intensity profile. In both cases, the detector is biased relative to the gun in such a fashion that only the electrons in the quasi-elastic peaks, indicated in Figures 8 and 9, are collected.

in which $\mathscr{H}_0$ is the kinetic energy of the incident electron and $v_n(\mathbf{r} - \mathbf{R}_n)$ is the static potential of an atom whose nucleus is at the position $\mathbf{R}_n$. In Equations (12) magnetic interactions [e.g., due to the spin-orbit coupling of the incident electron with the 'atomic' potential] are neglected. An equivalent form of the potential, Equation (12c), for a periodic lattice is

$$V(\mathbf{r}) = \sum_{\mathbf{G}} V_{\mathbf{G}} e^{-i\mathbf{G} \cdot \mathbf{r}} \tag{13}$$

in which $\mathbf{G}$ is a reciprocal lattice vector of the bulk (i.e., 3-dimensional) lattice. This is the form of the potential used in 'multiple-beam' analyses of LEED intensities which have been reviewed recently by Stern *et al.* [57] Note that Equation (13) implies that *all* of the atomic potentials are identical. However, the potentials near the surface are *not* necessarily identical to those in the bulk. Hence, Equation (13) provides a good model for the analysis of LEED intensities only if the incident beam penetrates many (e.g., $\gtrsim 10$) atomic layer spacings. This restricts the applicability of the multiple-beam analysis to very-low $(E \lesssim 5 \, \text{eV})$ and very-high $(E \gtrsim 1 \, \text{keV})$ energy incident electrons.

A major problem which any analysis of Equations (12) entails is that of deciding what to use for the $V_n(\mathbf{r} - \mathbf{R}_n)$ [or $V_\mathbf{G}$ in Equation (13)]. In principle, the potential in a solid is not of the form (12c) but depends, in effect, on the eigenvalue E of $\mathcal{H}_\mathrm{o}$ as well on $\mathbf{r}$. The reason for this is the fact that the $V(\mathbf{r} - \mathbf{R}_n)$ are the *self-consistent* ion-core potentials of nuclei immersed in a sea of mobile electrons. The nuclear coulomb potential is screened differently in the solid than in the isolated atoms. This screening is determined by the interactions between the valence electrons in the solid. As these electrons are not considered in Equations (12), the form of $V(\mathbf{r} - \mathbf{R}_n)$ must, in effect, be assumed and not calculated, unless we wish to extend Equations (12) into a more complicated model Hamiltonian. Thus we see that a calculation of elastic-low-energy-electron-diffraction (ELEED) intensities may be divided naturally into two parts: the evaluation of the electron-solid interaction [e.g., $V(\mathbf{r})$ in Equations (12) and (13)] and the solution of the Schrödinger Equation for the motion of the incident electron given this interaction. We now turn to a discussion of each of these topics in turn.

B. THE ELECTRON-SOLID FORCE LAW

In this section, we first discuss the main physical features of the electron-solid interaction. Then we indicate how these features are incorporated into model Hamiltonians describing electron diffraction. Finally, we describe the uncertainties in our present knowledge of the details of the electron-solid force law and their consequences for the use of ELEED for surface crystallography.

The three important parts of the electron-solid interaction for electrons in the energy range of interest in LEED, i.e., $5 \, \text{eV} \lesssim E \lesssim 500 \, \text{eV}$, are indicated in Figure 16. The

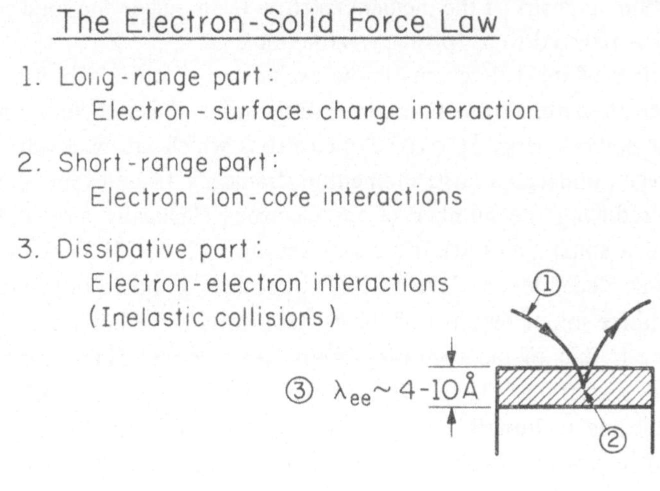

Fig. 16. Schematic indication of the three parts of the electron-solid force law for electrons in the energy range, $5 \, \text{eV} \leqslant E \leqslant 500 \, \text{eV}$, of interest in low-energy-electron diffraction (LEED).

long-range contributions due to the interaction of the incident electron with the charge it induces on the surface cause the image force (even for a fast electron) if it is far from the surface outside of the solid. [4] However, as an electron approaches within a few angstroms of the surface, the emission of surface plasmons becomes likely, leading to the description of elastic electron-solid scattering in terms of a complex, highly non-local, effective one-electron 'optical' potential. [4] Such inelastic excitation of plasmons occurs because of the interaction of the incident electron with the valence electrons in the solid. Thus it is the electron-electron interactions which give rise to the major dissipative contributions to the electron-solid force law by virtue of an incident electron's energy loss to bulk as well as surface plasmons, and to other types of excitation of the valence electron fluid like those indicated in Figure 11.

The electron-electron interactions produce predominately small-momentum-transfer (i.e., forward-scattering) events. Therefore, in a large-scattering-angle reflection spectroscopy like LEED, another mechanism must exist to 'turn the electron around'. The back-scattering of the electron from the ion-cores in the solid performs this function. For modest electron energies, $500 \, \text{eV} \gtrsim E \gtrsim 5 \, \text{eV}$, the inelastic collisions mediated by the electron-electron interaction require that an elastically-scattered electron originate from the upper 1–5 layers of the solid. Thus, of necessity, ELEED intensities in this energy range are sensitive functions of the positions and electronic structure of ion-cores in the surface region of the solid. Only in the (highly-unlikely) event that these surface ion-cores are identical to those in the bulk does ELEED reflect the bulk properties of the solid (e.g., its energy-band structure) directly. However, at lower energies, i.e., $E \lesssim 5$–$10 \, \text{eV}$, the inelastic damping can be small (e.g., $\lambda_{ee} \to 100 \, \text{Å}$). In this case, the ELEED intensities are sentitive to surface properties only if the motion of the electron normal to the surface is restricted by diffraction from the ion-core lattice itself. This occurs, for example, when its energy for motion normal to the surface, $E_\perp$, lies within a bandgap for such motion in the bulk solid. The decay of the intensity of the incident electron beam inside the solid in this special circumstance is referred to as 'primary extinction'.

A final source of inelastic events is the excitation of lattice vibrations (phonons) when the electron scatters from an ion-core in the solid. This is a common occurrence, especially for higher-energy ($E \gtrsim 100 \, \text{eV}$) electrons which are backscattered through large angles (i.e., undergo a large momentum transfer). These events play an important role in reducing the number of the electrons elastically and quasi-elastically scattered into a small-aperature (e.g., slit) detector. [2, 6, 56] They also determine the temperature dependence of the ELEED intensities as noted in Section 1E.

Given the three major features of the electron-solid force law noted in Figure 16, the next issue is that of incorporating them into a model Hamiltonian describing electron-solid scattering. The electron-ion-core interactions are included into the model as indicated in Equations (12) and (13) and summarized in Figure 17. They are described by a static, effective one-electron potential which acts on the incident electron. The consequences of electron-electron interactions are summarized in Figures 18–20. Some of these consequences for *elastic* electron-solid scattering can be simulated in a static-potential model by the insertion of a complex 'external' optical potential designated by $V_{ext}(\mathbf{r})$ in Figure 17. One should keep in mind, however, that static-potential models constructed to analyze LEED intensities are intrinsically more complicated than analogous models used to describe bulk band structures due to the inequivalence of the surface and bulk atomic potentials, and to the high energy of the electrons used in LEED studies, $\varphi < (E - E_F) \lesssim 10^3$ eV, on an atomic scale.

STATIC – POTENTIAL MODELS

$$H = -\frac{\hbar^2 \nabla^2}{2m} + \sum_n V(\vec{r} - \vec{R}_n) + V_{ext}(\vec{r})$$

1. Material Parameters: Ion–core positions $\langle \vec{R}_n \rangle$ and potentials, $V(\vec{r} - \vec{R}_n)$

2. Adequate for Band–Structure Calculations because the Exclusion principle Limits the effect of Electron–Electron Interactions

3. Types of damping

    A. Primary extinction (Bragg Energies)

    B. Phenomenological

        I. Ion–core excitation [complex phase shifts]

        2. "Bulk Losses" [complex $V_{ext}(\vec{r})$]

4. Additional technical approximations [e.g. finite beam analysis]

Fig. 17.   Summary of the main ingredients in static-potential models of electron-solid scattering.

ROLE OF INTERACTIONS WITH VALENCE ELECTRONS

$$H_{ee} = \frac{-\hbar^2 \nabla^2}{2m} + \sum_{i=1}^{N} \left(\frac{-\hbar^2 \nabla_i^2}{2m} \right) + \frac{1}{2} \sum_{i \neq j = 1}^{N+1} \frac{e^2}{|\vec{r}_i - \vec{r}_j|}$$

(1) Screening of "Bare" ion–core potentials

(2) Energy dependent complex inner– potential correction, $\Sigma(\vec{K}, E) \equiv V + i\Gamma$ due to scattering of injected electron from Valence electrons via coulomb interaction

Fig. 18.   Schematic presentation of the model Hamiltonian associated with the coulomb interactions among the valence and incident electrons, and synopsis of the two major consequences of these interactions with regard to an incident electron of energy, E, and momentum $\mathbf{k}$.

THE REAL ELECTRON-ELECTRON INDUCED INNER POTENTIAL (I.e. "PROPER SELF ENERGY")

1. Due to the same mechanism as the cohesive energy of Metals [Wigner and Bardeen, 1935]

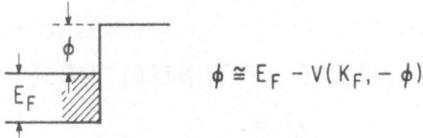

$$\phi \cong E_F - V(K_F, -\phi)$$

2. Decreases with increasing Electron Energy
   Eg. in Born Approximation

$$V(\vec{K}, E) \cong -\pi e^2 n / K^2$$

Fig. 19. Synopsis of the physical origin and main consequences of the real part of the effective one-electron ('inner' or 'optical') potential acting on an elastically-scattered incident electron and caused by the interaction of this electron with the valence electrons in the solid.

THE IMAGINARY ELECTRON-ELECTRON INDUCED INNER POTENTIAL

1. Describes an Energy Loss of the incident electron due to the excitation of a Valence electron:

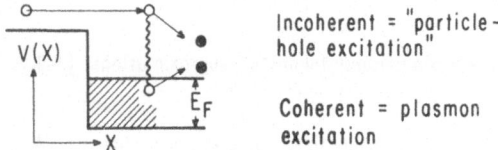

Incoherent = "particle-hole excitation"

Coherent = plasmon excitation

2. Mean Free paths of a 50 Volt electron in Jellium at the density of Aluminum

$$\lambda_{incoh} \cong 12 \, A^\circ \quad ; \quad \lambda_{P\ell} \cong 10 \, A^\circ$$

$$\lambda_{total} = (\lambda_{P\ell}^{-1} + \lambda_{incoh}^{-1})^{-1} \cong 5 \, A^\circ$$

3. Surface Plasmon emisssion probability $P \cong 80\%$

Fig. 20. Synopsis of the physical origin and main consequences of the imaginary (dissipative) part of the effective one-electron ('inner' or 'optical') potential acting on an elastically-scattered incident electron and caused by the interaction of this electron with the valence electrons in the solid.

It does not seem profitable to develop in detail the construction of the model Hamiltonian both because in these lectures we shall not use it directly and because the topic is discussed quite adequately in the literature. [2–4, 6, 37, 56] The static-potential model easily can be extended to describe the influence of lattice vibrations by writing, in 'second-quantized' notation [58] for a monatomic lattice

$$\mathbf{R}_n(t) = \mathbf{R}_n^{(0)} + \mathbf{u}_n(t) \tag{14}$$

$$\mathbf{u}_n(t) = \sum_{\mathbf{p}, \lambda} \mathbf{\varepsilon}_{\mathbf{P}\lambda} \left(\frac{\hbar}{2MN\omega(\mathbf{p})} \right)^{1/2} [b_{\mathbf{p}\lambda}^+ e^{-i\mathbf{p} \cdot \mathbf{R}_n^{(0)}} + b_{\mathbf{p}\lambda} e^{i\mathbf{p} \cdot \mathbf{R}_n^{(0)}}] \tag{15a}$$

$$\mathscr{H}_L = \sum_{\mathbf{p}\lambda} \hbar\omega_\lambda(\mathbf{p}) [b_{\mathbf{p}\lambda}^+ b_{\mathbf{p}\lambda} + \tfrac{1}{2}]. \tag{15b}$$

The $b_{\mathbf{p}\lambda}$ are the annihilation operators of phonons of quasi-wave vector $\mathbf{p}$, polarization index λ, and energy $\hbar\omega_\lambda(\mathbf{p})$. They satisfy the commutation relations

$$[b_{\mathbf{p}\lambda}, b_{\mathbf{p}'\lambda'}] = [b_{\mathbf{p}\lambda}^+, b_{\mathbf{p}'\lambda'}^+] = 0 \tag{16a}$$

$$[b_{\mathbf{p}\lambda}, b_{\mathbf{p}'\lambda'}^+] = \delta_{\lambda\lambda'} \delta_{\mathbf{p}, \mathbf{p}'}. \tag{16b}$$

For convenience, we usually suppress the polarization index, λ. The electronic Hamiltonian, specified by Equations (12), also can be written in this notation as

$$\mathscr{H}_0 = \mathscr{H}_{00} + V \tag{17a}$$

$$\mathscr{H}_{00} = \sum_{\mathbf{k}} \varepsilon(\mathbf{k}) c_{\mathbf{k}}^+ c_{\mathbf{k}} \tag{17b}$$

$$\varepsilon(\mathbf{k}) = \hbar^2 k^2 / 2m \tag{17c}$$

$$V = \sum_{\mathbf{q}, \mathbf{k}} \langle \mathbf{k} + \mathbf{q} | V | \mathbf{k} \rangle c_{\mathbf{k}+\mathbf{q}}^+ c_{\mathbf{k}} = \sum_{\mathbf{q}, \mathbf{k}, n} v_{\mathbf{q}} e^{i\mathbf{q} \cdot \mathbf{R}_n} c_{\mathbf{k}+\mathbf{q}}^+ c_{\mathbf{k}} \tag{17d}$$

$$v(\mathbf{r}) \equiv \sum_{\mathbf{q}} v_{\mathbf{q}} e^{i\mathbf{q} \cdot \mathbf{r}} \tag{18}$$

$$[c_{\mathbf{k}}, c_{\mathbf{k}'}]_+ = [c_{\mathbf{k}}^+, c_{\mathbf{k}'}^+]_+ = 0 \tag{19a}$$

$$[c_{\mathbf{k}}, c_{\mathbf{k}'}^+]_+ = \delta_{\mathbf{k}\mathbf{k}'}. \tag{19b}$$

Thus the Hamiltonian including electron-phonon interactions may be written as

$$\mathscr{H}_1 = \mathscr{H}_0 + \mathscr{H}_L = \mathscr{H}_{00} + \mathscr{H}_L + V \tag{20}$$

with $\mathscr{H}_0$ given by Equations (17) and $\mathscr{H}_L$ given by Equations (15). Equation (20) is referred to as the rigid-ion approximation because the potential, $v(\mathbf{r} - \mathbf{R}_n(t))$, associated with a given ion-position, $\mathbf{R}_n(t)$, rigidly moves as the ion vibrates. This model is very poor for some purposes [e.g., the description of superconductivity] but its refinements require consideration of electron-electron interactions. [1, 2, 6]

The inclusion of electron-electron interactions in the model Hamiltonian is rather complicated because careful distinctions must be drawn between the consequences of real and induced charge densities. [3, 4] Therefore, this aspect of analysis almost always is treated phenomenologically (with only a single exception [4]). The consequences of electron-electron interactions on elastic electron-solid scattering are simulated by use of empirical optical potentials [59] or an equivalent propagator-renormalization procedure [3, 4, 37] embodying a description of the physical processes indicated in Figures 19 and 20. In the case of inelastic electron-solid scattering, the incident electrons' coupling to the loss-modes of the solid is incorporated explicitly into the Hamiltonian by means of an interaction term

$$U_{\text{el}} = \sum_{\mathbf{q}, \mathbf{k}, n} c_{\mathbf{k}+\mathbf{q}}^+ c_{\mathbf{k}} T(n; \mathbf{k} + \mathbf{q}, \mathbf{k}) \tag{21a}$$

$$T(n; \mathbf{k} + \mathbf{q}, \mathbf{k}) = \exp[-i\mathbf{q} \cdot \mathbf{R}_n] \, h(n) \tag{21b}$$

$$h(n) = \sum_{\mathbf{p}} \exp[-i\mathbf{p} \cdot \mathbf{R}_n] \, t(\mathbf{p}) [b_{\mathbf{p}}^\dagger + b_{-\mathbf{p}}] \tag{21c}$$

in which the $b_{\mathbf{p}}$ are the annihilation operators of the *electronic* excitations of the valence electron fluid [e.g., bulk or surface plasmons]. The Hamiltonian of these loss modes themselves is, in principle, of the form given in Equation (15b). However, in practice a more-refined spectral density quantum-field theory formalism is used [3, 4, 43] in which the interactions of these loss-mode excitations result in their damping.

In summary, the inclusion of the electron-electron interactions into the model Hamiltonian yields the formally-complete Hamiltonian

$$\mathscr{H} = \mathscr{H}_{00} + \mathscr{H}_L + V + U_{el} + V_{el\text{-}el} \tag{22}$$

in which we use $V_{el\text{-}el}$ to designate those terms in the electron-electron interaction not included in U_{el} [see, e.g., Figure 18]. The important aspect of our development of LEED theory is that the approximations which work well to describe the ground and low-lying excited states (in the bulk solid) *do not suffice* to provide even a qualitative theory of LEED. The big difference in LEED and low-temperature transport or optical properties is that the excitation of collective and single-particle excitations of the valence-electron-ion-core plasma cause strong 'inelastic-collision' damping of the incident elastic wave field. This damping must be incorporated into the theory *ab initio* in order to achieve even a qualitative description of experimental results. It results in the important property of LEED experiments that they are sensitive to the *surface*, not bulk, properties of the solid.

Finally, the difficulty in achieving a correct microscopic description of the consequences of the electron-electron interactions results in some significant theoretical uncertainties in the use of LEED to characterize quantitatively the properties of solid surfaces. Similar uncertainties arise from incomplete experimental characterization of the nature of the actual surface employed in a given measurement [see

The Major Uncertainties
in Electron-Solid Force Law

1   Long range electron-induced-charge
    interaction (the image force?)

2   Atomic positions in upper 1-4 layers
    rearrangement? expansion?

3   Vibronic amplitudes in upper 1-4 layers
    $\langle u_S^2 \rangle_T \sim 2 \langle u_B^2 \rangle_T$

4   Spatial dependence of the inelastic-
    collision damping

Fig. 21.   Summary of the major uncertainties in the electron-solid force law even in the best-characterized case of low-energy electron diffraction from 'clean' low-index faces of simple metals. These uncertainties are augmented by a lack of detailed knowledge of the electron-ion-core potential in the cases of reactive gases chemisorbed on metals and of ionic and covalent crystals.

Part I]. The best-known feature of the model force law is the electron-ion-core interaction. In the case of the low index faces of clean simple metals, a bulk lattice potential seems to work fairly adequately in those cases for which such a model has been tested. [5, 46, 60–63] However, this situation does not always persist for chemisorbed gases or non-metallic solids. In addition, even for clean metals, the combined theoretical and experimental difficulties in the characterization of the electron-solid scattering system lead to the uncertainties noted in Figure 21. We shall return to a discussion of these uncertainties in our treatment of surface crystallography because we must be able to argue that in spite of the other uncertainties, the positions of atomic scatterers in the surface region can be determined.

C. MODELS OF ELASTIC ELECTRON-SOLID SCATTERING

Any adequate model of electron-solid scattering must describe two phenomena: the response of the valence electrons in the solid to the charge of an incident electron and interaction of this electron with the short-range ion-core potentials which 'turn it around' and, thereby, make reflection spectroscopy (either LEED or RHEED) possible. The interactions of the valence electrons of the solid with an incident electron are dynamic: i.e., they depend on its momentum parallel to the planer surface, its energy, and its distance from the surface. Therefore only a propagator formalism describes this interaction properly. [3, 4, 6] The basic ideas underlying such a formalism are indicated in Figure 22 under the heading 'Inelastic Collision Model' which is the name given to this approach by its initial proponents. [37] Simply stated, one first solves the Schrödinger equation describing the response of the valence electrons in the solid to the incident electron and uses the solutions thus

THE INELASTIC COLLISION MODEL

$$H = H_{ee} + \sum_n V(\vec{r} - \vec{R}_n)$$

Main Concept: Scattering of incident electron from valence electrons as strong or stronger than from lattice potential. Therefore

(1) Use a propagator Formalism

(2) Evaluate propagators relevant for H_{ee} alone

(3) Use these propagators, which describe damped electron waves, to evaluate the electronic (multiple) scattering from the ion cores

Fig. 22.   Schematic indication of the model Hamiltonian and calculational procedure used in the first propagator formalism [37] embodying the electron-electron interactions and called the 'inelastic-collision model'. The detailed form of the original proposal subsequently has been shown [3] to be correct for the interaction of the incident electron with bulk excitations of the valence electrons in the solid. However, the description of its interaction with surface excitations requires an extension [4] of the originally-proposed computational steps indicated in the figure. The latter often are referred to as the ingredients of 'distorted-wave' scattering theory because the electron-electron interactions 'distor' the plane-wave eigenfunction of an incident electron.

obtained as the basis states mixed by the ion cores in a multiple-scattering treatment of the consequences of their short-range potentials. Pictorially stated, the electron-electron interactions 'renormalize' the propagation of the electron 'between' its scatterings from the ion-cores.

The mathematical details of this theory need not concern us directly because their specification is not required for the comprehension of either the physical interpretation or the conceptual content of the predictions of the theory, and because they are displayed explicitly in a series of papers by Duke and co-workers. [2–5, 37, 43] However, for the theorists in the audience I will complete this section by outlining the major steps in the full theory and indicate the *ad hoc* assumptions required to reduce it to the various simple models treated in the literature. [6, 29, 56] The section is concluded with a few remarks about why none of these models live up to the great expectations of their proponents.

We begin by displaying the distorted-wave scattering equations. First, define an electronic propagator $G_0(\mathbf{r}, \mathbf{r}', E)$ associated with the distorted-wave basis satisfying the integral equation [4]

$$G_0(\mathbf{r}, \mathbf{r}', E) = G_{00}(\mathbf{r}, \mathbf{r}', E) + \int d^3r_1 \, d^3r_2 G_{00}(\mathbf{r}, \mathbf{r}_1, E) \, \Sigma(\mathbf{r}_1, \mathbf{r}_2, E) \times$$
$$\times G_0(\mathbf{r}_2, \mathbf{r}', E) \qquad (23)$$

$$\left(E - \frac{\hbar^2 \nabla^2}{2m}\right) G_{00}(\mathbf{r}, \mathbf{r}', E) - \int d^3r_1 V_0(\mathbf{r}, \mathbf{r}_1) \, G_{00}(\mathbf{r}_1, \mathbf{r}', E) = \delta(\mathbf{r} - \mathbf{r}')$$
$$(24)$$

in which $\Sigma(\mathbf{r}', \mathbf{r}, E)$ is the retarded proper self-energy or 'optical potential' due to electron-electron interactions and $V_0(\mathbf{r}, \mathbf{r}')$ is any prescribed static potential. In principle, $V_0(\mathbf{r}, \mathbf{r}')$ is taken to be the Hartree and Hartree-Fock contributions to $\Sigma(\mathbf{r}, \mathbf{r}', E)$, although in practice such a choice is not feasible. [64] The complete propagator of the system, $G(\mathbf{r}, \mathbf{r}', E)$ satisfies the integral equation

$$G(\mathbf{r}, \mathbf{r}', E) = G_0(\mathbf{r}, \mathbf{r}', E) + \int d^3r_1 \, d^3r_2 G_0(\mathbf{r}, \mathbf{r}_1, E) \, V_L(\mathbf{r}_1, \mathbf{r}_2) \, G(\mathbf{r}_2, \mathbf{r}'_, E)$$
$$(25)$$

in which $V_L(\mathbf{r}, \mathbf{r}')$ is the change in potential caused by decomposing a uniform positive 'jellium' background into positive ions of the appropriate charge and position. A multiple-scattering analysis of Equation (25) can be performed if $V_L(\mathbf{r}, \mathbf{r}')$ is of the form

$$V_L(\mathbf{r}, \mathbf{r}') = \delta(\mathbf{r} - \mathbf{r}') \sum_{\mathbf{R}} V_{\mathbf{R}}(\mathbf{r} - \mathbf{R}). \qquad (26)$$

The elastic electron-solid cross section is obtained by examining the asymptotic form of

$$\psi_{\text{scatt}}(\mathbf{r}, E) - \varphi_{\mathbf{k}}(\mathbf{r}) = \int d^3r_1 \, d^3r_2 G(\mathbf{r}, \mathbf{r}_1, E) \, V_L(\mathbf{r}_1, \mathbf{r}_2) \, \varphi_{\mathbf{k}}(\mathbf{r}_2) \qquad (27a)$$

$$\varphi_{\mathbf{k}}(\mathbf{r}) = e^{i\mathbf{k}\cdot\mathbf{r}} + \int d^3r_1 \, d^3r_2 G_{00}(\mathbf{r}, \mathbf{r}_1, E) \, \Sigma(\mathbf{r}_1, \mathbf{r}_2, E) \, \varphi_{\mathbf{k}}(\mathbf{r}_2) \qquad (27b)$$

in which **k** is the wave vector of the incident electron. Therefore in this formulation of the calculation of the elastic electron-solid cross section, all effects of both the induced charge on the solid's surface and the inelastic loss processes are described by the coordinate representation of the electronic proper self-energy, $\Sigma(\mathbf{r}, \mathbf{r}', E)$, for semi-infinite jellium. Details of evaluating this self-energy are displayed by Feibelman *et al.* [4]

In many recent model calculations of elastic low-energy-electron diffraction intensities [61–63, 65] (ELEED) a local complex potential is employed to simulate the consequences of the electron-electron interactions on electron-solid scattering. This model is defined by specifying, *a priori*, the retarded proper self-energy $\Sigma(\mathbf{r}, \mathbf{r}', E)$ to be of the form

$$\Sigma(\mathbf{r}, \mathbf{r}', E) \equiv \delta(\mathbf{r} - \mathbf{r}') \Sigma(\mathbf{r}, E). \tag{28}$$

In practice [61–63, 65] the further assumption is made that

$$\Sigma(\mathbf{r}, E) = V(E)\, \theta(z) \tag{29}$$

in which $V(E)$ [or at least its imaginary part] is taken to be the energy-shell one-electron self-energy in bulk jellium. [66, 67]

Although it cannot be derived from our microscopic theory in any well-defined limit, the local-complex-potential model exhibits the interesting feature that the distorted-wave basis states, $\varphi_\mathbf{k}(r)$ defined by Equation (27b), and propagator, $G_0(\mathbf{r}, \mathbf{r}', E)$ defined by Equation (24), can be evaluated in closed form. We find that if ϱ designates the coordinate parallel to the surface and z that normal to the surface we obtain [4]

$$G_0(\mathbf{r}, \mathbf{r}', E) = \int \frac{d^2 k_\parallel}{(2\pi)^2}\, e^{i\mathbf{k}_\parallel \cdot (\varrho - \varrho')} G_0(z, z', \mathbf{k}_\parallel, E), \tag{30}$$

$$G_0(z, z', \mathbf{k}_\parallel, E) = -\frac{2mi}{\hbar^2} \begin{cases} \dfrac{1}{2k_\perp}\left[e^{ik_\perp|z-z'|} - \left(\dfrac{\tilde{k}_\perp - k_\perp}{\tilde{k}_\perp + k_\perp}\right) e^{ik_\perp(z+z')} \right]; & z, z' < 0 \\[2ex] e^{-i(k_\perp z - \tilde{k}_\perp z')}/(k_\perp + \tilde{k}_\perp); & z < 0 < z' \\[1ex] e^{i(\tilde{k}_\perp z - k_\perp z')}/(k_\perp + \tilde{k}_\perp); & z' < 0 < z \\[1ex] \dfrac{1}{2\tilde{k}_\perp}\left[e^{i\tilde{k}_\perp|z-z'|} + \left(\dfrac{\tilde{k}_\perp - k_\perp}{\tilde{k}_\perp + k_\perp}\right) e^{i\tilde{k}_\perp(z+z')} \right] & 0 < z, z' \end{cases} \tag{31a}$$

$$k_\perp^2(\mathbf{g}, E) = (2mE/\hbar^2) - (\mathbf{k}_\parallel + \mathbf{g})^2 \tag{31b}$$

$$\tilde{k}_\perp^2(\mathbf{g}, E) = (2m[E + V(E)]/\hbar^2) - (\mathbf{k}_\parallel + \mathbf{g})^2 \tag{31c}$$

in which **g** is a reciprocal lattice vector of the lattice potential layers parallel to the surface and

$$\varphi_k(\mathbf{r}) = e^{i\mathbf{k}_\parallel \cdot \varrho}\varphi_k(z) \tag{32a}$$

$$\varphi_k(z) = \begin{cases} e^{ik_\perp z} + \left[\dfrac{k_\perp - \tilde{k}_\perp}{k_\perp + \tilde{k}_\perp}\right] e^{-ik_\perp z}; & z < 0 \\[2ex] 2k_\perp e^{i\tilde{k}_\perp z}/(k_\perp + \tilde{k}_\perp); & z > 0 \end{cases}. \tag{32b}$$

By use of Equations (31) and (32) in the formal scattering theory Equations (25) and (27) we illustrate the type of effects predicted by the distorted-wave scatter-theory. For simplicity, we consider only the distorted-wave Born approximation and use the delta-potential model of the lattice potential

$$v(\mathbf{r} - \mathbf{R}_n) = v_n \, \delta(\mathbf{r} - \mathbf{R}_n). \tag{33}$$

After some algebra we obtain for the asymptotic form of the wave function

$$\psi_{\text{scatt}}(\mathbf{r}) \xrightarrow{z \to -\infty} e^{i\mathbf{k}_\| \cdot \varrho} \psi_{k_\perp}(z) \tag{34a}$$

$$\psi_{k_\perp}(z) = e^{ik_\perp(0, E)z} + \sum_{\mathbf{g}} R(\mathbf{g}, E) e^{-ik_\perp(\mathbf{g}, E)z} \tag{34b}$$

$$R(\mathbf{g}, E) = R_0(0, E) \, \delta_{\mathbf{g}, 0} +$$
$$+ \sum S(0, E) S(\mathbf{g}, E) \frac{miv_n \exp\{i[k_\perp(0, E) + k_\perp(\mathbf{g}, E)] R_{n\perp}\}}{\hbar^2 \tilde{k}_\perp(\mathbf{g}, E)}. \tag{34c}$$

The first term in Equation (34c), i.e.,

$$R_0(0, E) \equiv [k_\perp(0, E) - \tilde{k}_\perp(0, E)]/[k_\perp(0, E) + \tilde{k}_\perp(0, E)] \tag{35a}$$

gives the contribution to the scattered wave from the jellium-vacuum surface alone. The second term in Equation (34c) is the result obtained from the Born approximation of scattering from the lattice potential, Equation (33), alone multiplied by the product of

$$S(\mathbf{g}, E) \equiv 2k_\perp(\mathbf{g}, E)/[k_\perp(\mathbf{g}, E) + \tilde{k}_\perp(\mathbf{g}, E)] \tag{35b}$$

for the incident ($\mathbf{g} \equiv 0$) and final beams, i.e., the product of the transmission coefficients of jellium-vacuum interface in the absence of the lattice potential.

We present the results of these elementary calculations to illustrate three important features of the distorted-wave scattering theory. First, the spatial dependence of the optical potential at solid-vacuum interface not only contributes its own reflection coefficient, $R_0(0, E)$, to the total electron-solid cross-section but also influences the electron's scattering from the short-range electron-ion-core potential. The latter influence is always strong for glancing initial or final beams independent of the energy of the electrons. This fact has the important consequence that the ratios of the absolute intensities of different beams predicted by a model potential will depend on the spatial dependence and magnitude of $\Sigma(\mathbf{r}, \mathbf{r}', E)$. Second, as $\Sigma(\mathbf{r}', \mathbf{r}, E)$ diminishes as E increases, [4] the extent of the dependence of the relative beam intensities on the shape of the optical potential diminishes with increasing energy of the incident electron. For more general models of the optical potential than Equations (28) the magnitude of the 'apparent' real part of $V(E)$ depends on the beam index, $\mathbf{g}$, as well as E, so that the energies as well as the intensities of maxima in the intensity-versus-energy profiles depend on the shape of the optical potential. Third, and finally, we see that $G_0(\mathbf{r}, \mathbf{r}', E)$ does not depend solely on $\mathbf{R} = \mathbf{r} - \mathbf{r}'$. Therefore, simple multiple-scattering theories [37] no longer suffice to solve Equation (25) and more general planer scattering theories [61] must be used for this task. In fact, no adequate solutions to the general scattering equations have been given yet. [4, 6]

Another common model described in the literature is the inelastic collision model introduced via Figure 22. As proposed heuristically by Duke and Tucker [37] and derived microscopically from a quantum-field theory by Duke and Laramore, [3] it consists of avoiding the solution of Equations (23) and (24) by use of the ansatz

$$G_0(\mathbf{r}, \mathbf{r}', E) = \int \frac{d^3k}{(2\pi)^3} G_0(\mathbf{k}, E) e^{i\mathbf{k} \cdot (\mathbf{r} - \mathbf{r}')} \tag{36a}$$

$$G_0^{-1}(\mathbf{k}, E) = E - \hbar^2 k^2 / 2m - \Sigma(\mathbf{k}, E) \tag{36b}$$

in which $\Sigma(\mathbf{k}, E)$ is the proper self-energy of the incident electron in bulk jellium. [66, 67] Equation (36a) exhibits the important analytical feature that $G_0(\mathbf{r}, \mathbf{r}', E)$ depends *only* on $\mathbf{R} = \mathbf{r} - \mathbf{r}'$. Therefore, Equation (25) for the full electron-solid Green's function can be solved by standard multiple-scattering methods in either the configuration [68] or momentum [2, 5] representations. Use of the momentum representation exhibits the important advantage that the consequences of the thermal vibrations of the short-range ion-core potentials can be incorporated into the analysis. [2] The non-locality of the self energy,

$$\Sigma(\mathbf{r}, \mathbf{r}', E) \equiv \int \frac{d^3k}{(2\pi)^3} e^{i\mathbf{k} \cdot (\mathbf{r} - \mathbf{r}')} \Sigma(\mathbf{k}, E) \tag{37}$$

is neglected in all of the numerical applications of Equations (36) because the energy-shell self energy, $\Sigma(E) \equiv \Sigma(k(E), E)$, is used in Equation (36b) where $k(E)$ is the value of k for which $G^{-1}[k(E), E] = 0$.

Although complicated in detail, the solution to Equations (27) by the multiple-scattering method is conceptually simple. It proceeds in five steps. First, Equation (27a) is expanded as a perturbation series with summations over scattering sites indicated explicitly. In practice, this step is accomplished by writing Equation (27a) in terms of a 'T' matrix, $T(\mathbf{r}, \mathbf{r}')$, via

$$\psi_\mathbf{k}(\mathbf{r}) = \varphi_\mathbf{k}(\mathbf{r}) + \int d^3r' \, d^3r'' G_0(\mathbf{r}, \mathbf{r}', E) T(\mathbf{r}', \mathbf{r}'') \varphi_\mathbf{k}(\mathbf{r}'') \tag{38}$$

and subsequently expanding the equation for T which is analogous to Equation (25), i.e.,

$$\begin{aligned} T(\mathbf{r}, \mathbf{r}') &= \delta(\mathbf{r} - \mathbf{r}') \sum_\mathbf{R} v_\mathbf{R}(\mathbf{r} - \mathbf{R}) + \\ &+ \int d^3r'' \sum_\mathbf{R} v_\mathbf{R}(\mathbf{r} - \mathbf{R}) G_0(\mathbf{r}, \mathbf{r}'', E) T(\mathbf{r}'', \mathbf{r}') \\ &= \delta(\mathbf{r} - \mathbf{r}') \sum_\mathbf{R} v_\mathbf{R}(\mathbf{r} - \mathbf{R}) + \\ &\quad \sum_{\mathbf{R}, \mathbf{R}'} v_{\mathbf{R}'}(\mathbf{r} - \mathbf{R}') G_0(\mathbf{r}, \mathbf{r}', E) v_\mathbf{R}(\mathbf{r}' - \mathbf{R}) + \cdots . \end{aligned} \tag{39}$$

The second step consists of defining a 'single-site' t matrix which accounts for all of the diagonal terms in Equation (39). It satisfies

$$\langle \mathbf{k}' | t_{\mathbf{R}}(E) | \mathbf{k} \rangle = \int d^3 r' \, d^3 r \, e^{-i\mathbf{k}' \cdot \mathbf{r}'} t_{\mathbf{R}}(\mathbf{r}' - \mathbf{R}, \mathbf{r} - \mathbf{R}) \, e^{i\mathbf{k} \cdot \mathbf{r}} \equiv$$

$$\equiv e^{-i(\mathbf{k}' - \mathbf{k}) \cdot \mathbf{R}} t_{\mathbf{R}}(\mathbf{k}', \mathbf{k}; E) \tag{40a}$$

$$t_{\mathbf{R}}(\mathbf{k}', \mathbf{k}; E) = v_{\mathbf{R}}(\mathbf{k}' - \mathbf{k}) + \int \frac{d^3 k''}{(2\pi)^3} v_{\mathbf{R}}(\mathbf{k}' - \mathbf{k}'') \, G_0(\mathbf{k}'', E) \, t_{\mathbf{R}}(\mathbf{k}'', \mathbf{k}; E) \tag{40b}$$

$$v_{\mathbf{R}}(\mathbf{q}) = \int d^3 r \, v_{\mathbf{R}}(\mathbf{r}) \, e^{-i\mathbf{q} \cdot \mathbf{r}}. \tag{40c}$$

The third step consists of inserting these expressions for the single-site t matrix back into the integral Equation (39) for T and converting that integral equation into an algebraic equation by use of partial-wave expansion techniques. It is worth noting that the *major* effects of lattice vibrations for *low-energy* ($E \lesssim 100$ eV) electrons are inserted into the theory [2] via replacing the t matrix in Equation (40b) by

$$t_{\mathbf{R}}(\mathbf{k}', \mathbf{k}, E) \to t_{\mathbf{R}}(\mathbf{k}', \mathbf{k}, E) \exp[-W_{\mathbf{R}}(\mathbf{k}' - \mathbf{k})] \tag{41a}$$

$$W_{\mathbf{R}}(\mathbf{q}) = \tfrac{1}{2} \mathbf{q} \cdot \langle \mathbf{u}_{\mathbf{R}}(0) \, \mathbf{u}_{\mathbf{R}}(0) \rangle_T \cdot \mathbf{q} \tag{41b}$$

in which $\exp[-W_{\mathbf{R}}(\mathbf{q})]$ is the Debye-Waller factor associated with the rms lattice displacements $\mathbf{u}_{\mathbf{R}}(t)$ at the site $\mathbf{R}$. Therefore the temperature effects are put into the theory prior to the execution of the third step in the cross-section calculation. Their presence merely changes some of the details of the partial-wave expansions performed during this step.

The fourth step in the solution to Equation (27a) is the partial performance of the sums which appear in the algebraic equations at the completion of the third step. In particular, the lattice is divided into repeating layers of geometrically identical 'subplanes' of scattering centers. All of the multiple scattering in each subplane is summed analytically with the result that the atomic scattering matrices, $t_{\mathbf{R}}$, are replaced by subplane scattering matrices τ_ν of the generic form

$$\tau_\nu = t_{\mathbf{R} = (\mathbf{P}, \nu)} [1 - G^{sp} t_{\mathbf{R} = (\mathbf{P}, \nu)}]^{-1} \tag{42}$$

in which $\mathbf{P}$ labels the sites of the scatterers in a given subplane, ν labels the subplanes, and G^{sp} is an appropriate propagator defined in terms of $G_0(\mathbf{k}, E)$. [2,37]

The complete layer scattering amplitudes, T_ν, satisfy a set of coupled linear matrix equations. The final step in the calculation is the solution of these equations and the insertion of their solutions back into Equation (38) for the wave function. The Fourier transform of $T(\mathbf{r}, \mathbf{r}')$ is simply related [5] to the contribution to the electron-solid reflection coefficient analogous to the last term in Equation (34b). In actual applications of the inelastic-collision model the leading term in Equation (34b), which describes the electron reflection from jellium in the absence of the short-range potential, usually is neglected.

The above outline is our brief aside 'for theorists only' on calculational details. However, it seems appropriate to conclude this section by noting a few features of both the inelastic-collision and local-complex-potential models as they are analyzed in the literature. The only difference between the models lies in their expressions for G_0 and $\varphi_{\mathbf{k}}(\mathbf{r})$ given by Equations (23) and (27b) respectively. This difference enters the final expressions for the cross sections in three places: Equations (40) for the site scattering amplitudes, expressions for the propagators like G^{sp} in the final algebraic equations for the T_v, and in the boundary conditions implicit in Equation (27a). All existing microscopic model calculations [5, 26, 37, 45–49, 60–63, 65, 66] use free-particle propagators in Equation (40b) for the site scattering amplitudes. To this extent they all are identical, semi-empirical, and perhaps inadequate [26] models of the electron-solid scattering process. The local-complex-potential model calculations [61–63, 65] are performed using the distorted-wave propagators, (31), rather than the plane-wave (inelastic-collision-model) propagators, (36), in solving the algebraic equations for the layer scattering amplitudes T. However, most users of the local-complex-potential models [61, 63, 65] ultimately revert back to the boundary conditions of the inelastic-collision model in order to avoid (unwanted) fine structure in their predicted intensity profiles caused by quantum interference effects associated with the sharp change in the optical potential, Equation (29), at $z=0$. This strictly empirical and arbitrary, change in boundary conditions is accomplished by various *ad hoc* prescriptions in the three most recent major calculations. [61, 63, 65] Earlier calculations are reviewed by Estrup and McRae. [29] Thus, contrary to allegations by their proponents, none of the microscopic calculations are, in fact, actually microscopic. They all involve semi-empirical treatments of the short-range electron-ion-core interactions, of both the spatial and energy dependence of the inelastic-collision optical potential, and of the boundary conditions. This fact, of course, does not diminish their utility for the qualitative description of ELEED intensities or for the illustration of particular dynamic effects. It should be emphasized, however, that none of these calculations actually displays the consequences of the models of which they nominally are supposed to be an analysis, and that in addition, none of the existing models embodies an adequate electron-electron-induced optical potential.

3. Elastic Low-Energy Electron Diffraction (ELEED)

In this part of the lecture notes we describe the elastic scattering of 'low-energy', $5 \text{ eV} \leqslant E \leqslant 500 \text{ eV}$, electrons from presumably planar surfaces of single-crystal solids. We proceed in four steps. First, we examine the simplest approach to ELEED; the Born-approximation calculation of the ELEED intensities for scattering from a rigid periodic lattice. Next, recalling from Equations (4) that the energy resolution of most electron spectrometers does not permit phonon-assisted diffraction to be distinguished from elastic diffraction, we discuss the Born-approximation (i.e., linear-response) calculation of the quasielastic scattering cross sections of low-energy electrons from a periodic lattice of vibrating atomic potentials. Having discovered the general features

of ELEED from these two Born-approximation analyses, we demonstrate that because of the strong electron-solid interactions, such analyses must be extended along
the lines described in Section 2C in order to achieve a meaningful description of
experimentally-measured ELEED intensities. Finally, we present a brief survey of
the description of the existing experimental data by current model calculations.

A. SCATTERING FROM A RIGID LATTICE: THE BORN APPROXIMATION

In order to achieve some insight into the nature of electron-solid scattering, we begin
by investigating the simplest possible model: the Born-approximation analysis of
electron scattering from a static array of ion-core scatterers at positions $\{\mathbf{R}_n\}$. From
Equations (23)–(27), we find that the integral form of the Schrödinger equation for
the wave function of an electron scattered from a local static potential $V(\mathbf{r})$ is

$$\psi_\mathbf{k}(\mathbf{r}) = \varphi_\mathbf{k}(\mathbf{r}) + \int d^3r' G_0(\mathbf{r}, \mathbf{r}', E)\, V(\mathbf{r}')\, \psi_\mathbf{k}(\mathbf{r}') \tag{43a}$$

$$V(\mathbf{r}) = \sum_n v_n(\mathbf{q})\, e^{i\mathbf{q} \cdot (\mathbf{r} - \mathbf{R}_n)} \tag{43b}$$

$$G_0(\mathbf{r}, \mathbf{r}', E) = \lim_{\delta \to 0^+} \frac{1}{(2\pi^3)} \int d^3k\, \frac{e^{i\mathbf{k} \cdot (\mathbf{r} - \mathbf{r}')}}{E - \hbar^2 k^2/2m + i\delta}$$
$$= \frac{-m\, e^{ik(E)|\mathbf{r} - \mathbf{r}'|}}{2\pi\hbar^2\, |\mathbf{r} - \mathbf{r}'|} \tag{44a}$$

$$k(E) \equiv (2mE/\hbar^2)^{1/2}. \tag{44b}$$

The $v_n(\mathbf{q})$ are the Fourier transforms of the ion-core potentials at the sites $\{\mathbf{R}_n\}$. If
we use scattering-theory boundary conditions, we take $|\mathbf{r}| \gg |\mathbf{r}'|$ so that

$$G_0(\mathbf{r}, \mathbf{r}', E) \xrightarrow{|\mathbf{r}| \gg |\mathbf{r}'|} \frac{-m\, e^{ik(E)r}}{2\pi\hbar^2\, r}\, e^{-ik(E)\hat{\mathbf{r}} \cdot \mathbf{r}'} \tag{45}$$

where $\hat{\mathbf{r}}$ is a unit vector directed along $\mathbf{r}$. The scattered wave emerges along $\hat{\mathbf{r}}$ so that
we define the final wave vector by

$$\mathbf{k}' \equiv k(E)\, \hat{\mathbf{r}}. \tag{46}$$

Therefore in the asymptotic limit that $|\mathbf{r}| \gg |\mathbf{r}'|$ Equations (43) give

$$\psi_\mathbf{k}(\mathbf{r}) = \varphi_\mathbf{k}(\mathbf{r}) + f(\mathbf{k}', \mathbf{k})\, \frac{e^{ik(E)r}}{r}. \tag{47}$$

In the Born Approximation the scattering amplitude, $f(\mathbf{k}', \mathbf{k})$ is written as

$$f(\mathbf{k}', \mathbf{k}) = \frac{-m}{2\pi\hbar^2} \int d^3r'\, e^{-i\mathbf{k}' \cdot \mathbf{r}'} V(\mathbf{r}')\, e^{i\mathbf{k} \cdot \mathbf{r}'} =$$
$$= \frac{-m}{2\pi\hbar^2} \sum_n v_n(\mathbf{k}' - \mathbf{k})\, e^{-i(\mathbf{k}' - \mathbf{k}) \cdot \mathbf{R}_n}. \tag{48}$$

In the case that all of the scattering potentials are identical [i.e., $v_n(\mathbf{q}) = v(\mathbf{q})$] the Born approximation to the elastic scattering cross section is given by

$$d\sigma/d\Omega = |f(\mathbf{k}', \mathbf{k})|^2 = \left| \frac{-m}{2\pi\hbar^2} \sum_n v(\mathbf{k}' - \mathbf{k}) e^{-i(\mathbf{k}' - \mathbf{k}) \cdot \mathbf{R}_n} \right|^2, \tag{49}$$

which is the usual text-book [69] expression for scattering from a periodic potential. In fact, one often defines an interference function (sometimes called a structure factor in the solid-state literature [70]) $S(\mathbf{q})$ via

$$S(\mathbf{q}) = \frac{1}{N} \sum_n e^{-i\mathbf{q} \cdot \mathbf{R}_n}. \tag{50}$$

Writing $\mathbf{k}' = \mathbf{k} + \mathbf{q}$ in Equation (48) we obtain

$$\frac{d\sigma}{d\Omega} = \left(\frac{mN}{2\pi\hbar^2} \right)^2 |v(\mathbf{q})|^2 |S(\mathbf{q})|^2 \tag{51}$$

in which N denotes the number of atoms in the monatomic lattice. Thus the scattering amplitude factors into the interference function, $S(\mathbf{q})$, which depends only on the lattice geometry, and a form factor, $v(\mathbf{q})$, which depends only on the dynamics of the electron-ion-core interaction.

A calculation of the electron-solid cross sections using the Born-approximation is referred to as a 'kinematical' analysis. As an example, we consider the scattering of an electron beam from the (110) face of an fcc material (e.g., aluminum). The original model of scattering from such a crystal face was proposed by Davisson and Germer [71] to be diffraction from a single layer of atoms, i.e., the surface layer. In this model, the interference function, $S(\mathbf{q})$, given by Equation (50) becomes

$$S(\mathbf{q}) = \frac{1}{N_\parallel} \sum_n e^{-i\mathbf{q} \cdot (\mathbf{R}_n)_\parallel} = \delta_{\mathbf{q}, \mathbf{g}}. \tag{52}$$

We have defined the parallel and perpendicular components $\mathbf{k}$ via

$$\mathbf{k}_\parallel \cdot \mathbf{R}_\perp = 0 \tag{53a}$$
$$\mathbf{k}_\perp \cdot \mathbf{R}_\parallel = 0 \tag{53b}$$

where the magnitude of $\mathbf{k}_\parallel$ is specified by boundary conditions [Figure 4] and that of $k_\perp$ is determined by energy conservation

$$k_\perp^2(\mathbf{g}, E) \equiv \frac{2mE}{\hbar^2} - (\mathbf{k}_\parallel + \mathbf{g})^2. \tag{54}$$

As anticipated, the scattered electrons emerge from the crystal in a series of beams, the intensity of each of which is determined by

$$|v(\mathbf{q})|^2 \rightarrow |v[\mathbf{q}_\parallel = \mathbf{g}, q_\perp = k_\perp(0, E) + k_\perp(\mathbf{g}, E)]|^2 \tag{55}$$

which is a slowly decreasing function of energy for a given beam index $\mathbf{g} = 2\pi (h\mathbf{b}_1 + k\mathbf{b}_2)$. Thus we see that for a two-dimensional diffraction grating, the kinematical approximation predicts an intensity profile for each beam which is a monotonically decreasing function of the primary beam energy. This prediction is in strong disagreement with measured profiles such as the one indicated schematically in Figure 15. Equations (51), (52), and (55) also reveal the major 'physical' features of the Born-approximation analysis of this model. The geometry of the scattering potential determines the spot pattern whereas the details of the electron-solid interaction determine the intensities of the various scattered beams as a function of the incident beam parameters and the beam index.

Having noted that electron-diffraction from a single atomic layer leads to monotonically decreasing intensity profiles (in disagreement with experiment), it is natural to ask whether multilayer scattering in the first Born-approximation can improve the description of experimental data. In order to examine this question, we recall from Equations (31) and (36) that the occurrence of inelastic collisions by the incident electron gives $k_\perp(\mathbf{g}, E)$ a small imaginary component. I.e., if we focus our attention solely on the elastic scattering of an electron by the solid, inelastic collisions have the effect of 'absorbing' particles from the incident beam. Therefore, the electron's momentum normal to the surface is determined by

$$[k_{\perp 1}(\mathbf{g}, E) + ik_{\perp 2}(\mathbf{g}, E)]^2 = \frac{2m}{\hbar^2}[E + \Sigma(E)] - (\mathbf{k}_\| + \mathbf{g})^2 \tag{56}$$

in which $\Sigma(E)$ is the complex potential describing the effects of both elastic and inelastic electron-electron collisions of the incident electron with the valence electrons of the solid [see Sections 2B and 2C]. Because of these interactions, Equation (56) replaces Equation (54) for electronic motion 'inside' the solid. Using Equation (56) and noting that $\mathbf{k}_\perp'$ has the opposite direction to $\mathbf{k}_\perp$, we obtain for the interference function [Equation (50)]:

$$S(\mathbf{k}' - \mathbf{k}) = \delta_{\mathbf{k}_\|', \mathbf{k}_\| + \mathbf{g}} \sum_{v=0}^{\infty} e^{-i\mathbf{g} \cdot \mathbf{d}_v} e^{i[k_\perp(0, E) + k_\perp(\mathbf{g}, E)] d_{v\perp}} \tag{57}$$

in which $\mathbf{d}_v$ is the radius vector from the origin of the central cell on the top atomic layer ($v=0$) to that on the vth layer. Note, for example, that in the case of a (110) face of an fcc lattice, if a is the lattice spacing of the simple cubic basis, the primitive Bravais net of the surface layer is specified by

$$\mathbf{a}_1 = a(1/\sqrt{2}, 0) \tag{58a}$$

$$\mathbf{a}_2 = a(0, 1) \tag{58b}$$

and

$$\mathbf{d}_v = va\left(\frac{1}{2\sqrt{2}}, \frac{1}{2}, \frac{1}{2\sqrt{2}}\right) \equiv v\mathbf{d}_0, \tag{59}$$

i.e., the origin of the central cell is shifted by $[(\mathbf{a}_1 + \mathbf{a}_2)/2]$ from layer to layer, and the layer spacing is $d = a/2\sqrt{2}$. Because of the inelastic-collision damping, the series in Equation (57) is a simple geometrical one with the sum:

$$S(\mathbf{k'} - \mathbf{k}) = \delta_{\mathbf{k}_{\parallel}', \, \mathbf{k}_{\parallel} + \mathbf{g}} [1 - R(0, \mathbf{g}, E]^{-1} \tag{60}$$

$$R(\mathbf{g}, \mathbf{g'}, E) = \exp\{i [k_{\perp 1}(\mathbf{g}, E) + k_{\perp 1}(\mathbf{g'}, E)] d\} \times$$
$$\times \exp\{- [k_{\perp 2}(\mathbf{g}, E) + k_{\perp 2}(\mathbf{g'}, E)] d\} \times$$
$$\times \exp\{- i(\mathbf{g} + \mathbf{g'}) \cdot \mathbf{d_0}\}. \tag{61}$$

Equation (58) predicts an interference function which exhibits maxima near energies which satisfy

$$[k_{\perp}(0, E) + k_{\perp}(\mathbf{g}, E)] d - \mathbf{g} \cdot \mathbf{d_0} = 2\pi n \tag{62}$$

which are the nth order 'primary' Bragg peaks. In geometrical terms, this is just the condition that

$$\mathbf{k} - \mathbf{k'} = \mathbf{G} \tag{63}$$

where $\mathbf{G}$ is a reciprocal lattice vector of the full three-dimensional lattice.

Summarizing, from Equations (51), (55), and (60) we obtain the expression for the Born-approximation to the elastic differential electron-solid cross-section:

$$\frac{d\sigma}{d\Omega} = \left| \frac{m}{2\pi\hbar^2} v[\mathbf{g}, k_{\perp}^+(0, \mathbf{g})] \right|^2 [N \delta_{\mathbf{k}_{\parallel}' - \mathbf{k}_{\parallel}, \, \mathbf{g}}]^2 \frac{1}{|1 - R(0, \mathbf{g}, E)|^2} \tag{64a}$$

$$v(\mathbf{g}, k_{\perp}) = \int d^3 r \, e^{-i\mathbf{g} \cdot \boldsymbol{\varrho}} e^{-ik_{\perp} z} v(\boldsymbol{\varrho}, z) \tag{64b}$$

$$k_{\perp}^+(\mathbf{g}, \mathbf{g'}) \equiv k_{\perp}(\mathbf{g}, E) + k_{\perp}(\mathbf{g'}, E) \tag{64c}$$

and $R(\mathbf{g}, \mathbf{g'}, E)$ is given by Equation (61). Two main features are evident from Equations (64). The existence of discrete final beams at fixed values of $\mathbf{k}_{\parallel}'$ are the cause of the discrete spot pattern illustrated in Figure 15. The occurrence of the $|1 - R|^{-2}$ factor in Equation (64a) is the origin of the primary-Bragg peaks noted in the intensity profile shown in Figure 15. The widths of these peaks are determined in the Born-approximation solely by the imaginary part of the $k_{\perp}(\mathbf{g}, E)$. A proper multiple-scattering analysis predicts considerably larger widths. [72]

B. SCATTERING FROM A VIBRATING LATTICE: LINEAR-RESPONSE THEORY

Our discussion in this section is developed in three steps. First, we examine the nature and consequences of the general conservation laws describing inelastic electron-solid scattering from planar periodic solids. Then we develop the linear-response analysis of electron scattering from a periodic array of vibrating atomic scatterers. In this analysis, the electron-scatterer interaction is treated in the Born-approximation just as in the previous section. However, the consequences of the vibrations of these scatterers are described 'exactly' (i.e., without further approximation) within this framework. We conclude the section with a brief survey of the applications of the Born-approximation analysis to determine the lattice displacements of surface scatterers.

Turning to our discussion of the consequences of the appropriate conservation laws, we see in Figure 23 a schematic diagram of an inelastic electron reflection process. An incident electron of energy E and component of momentum parallel to the surface, $\mathbf{k}_{\parallel}$, is reflected back from the surface with energy

$$E' = E - w \tag{65a}$$

and momentum parallel to the surface

$$\mathbf{k}'_{\parallel} = \mathbf{k}_{\parallel} - \mathbf{p}_{\parallel} + \mathbf{g}. \tag{65b}$$

In Equations (65) $\mathbf{p}_{\parallel}$ is the momentum parallel to the surface of the excitation(s) of the solid responsible for the electron's energy loss, w, and $\mathbf{g}$ is a reciprocal lattice vector associated with the (presumed identical) planes of ion cores in the target solid parallel to its surface.

If we assume that both w and $\mathbf{k}'_{\parallel} - \mathbf{k}_{\parallel}$ are known precisely, and that only a single excitation of the solid is created by the incident electron, then we can draw some

INELASTIC LEED FROM CLEAN SURFACES: KINEMATICS

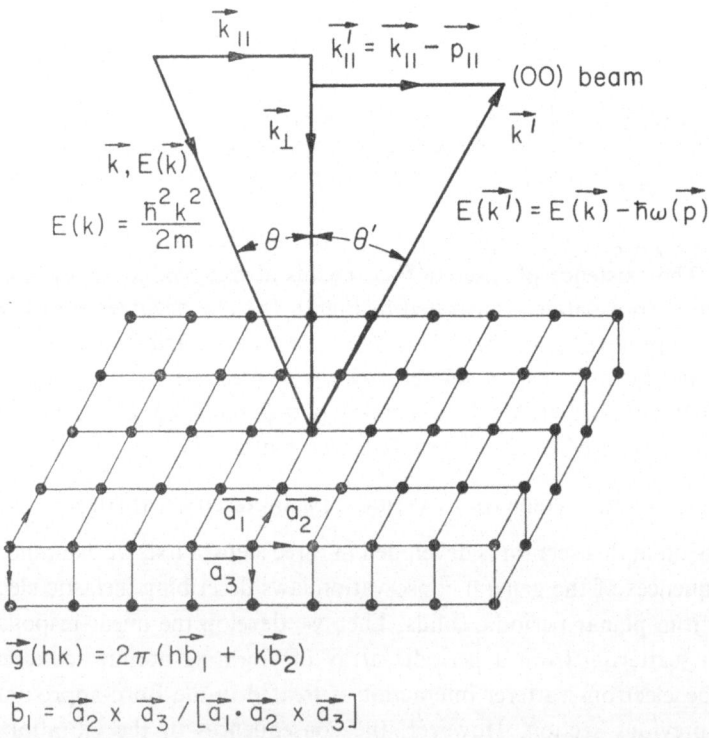

Fig. 23.   Schematic diagram of the kinematics for an electron loss process in which an excitation of parallel momentum, $\mathbf{p}_{\parallel}$, and energy, $\hbar\omega(\mathbf{p})$, is excited.

general conclusions from the conservation laws, Equations (65), independent of the detailed dynamics of the electron-solid interaction. Consider first the case of a bulk excitation of the solid characterized by the dispersion relation:

$$w = \hbar\omega_b(\mathbf{p}_{\parallel}, p_{\perp}). \tag{66}$$

The excitation's momentum parallel to the surface, $\mathbf{p}_{\parallel}$, is determined from momentum conservation. Thus, in general, if $w > \hbar\omega_b(0)$ there exists an excitation of energy w whose momentum normal to the surface, $p_{\perp}$, is specified by Equation (66). The conservation of momentum normal to the surface plays no obvious role in this process because the symmetry of the solid in this direction is broken by the surface. This fact suggests that the excitation of bulk loss processes [e.g., in the quasi-elastic or characteristic-loss regions of the energy-loss profile shown in Figure 9] merely broadens the angular width of scattered beam relative to that width which is characteristic of elastically-scattered electrons. [This latter width is determined by the structural disorder of the surface layers [6]]

In the case of surface excitations, the loss-mode dispersion relation is of the form

$$w = \hbar\omega_s(\mathbf{p}_{\parallel}). \tag{67}$$

The quantity $\mathbf{p}_{\parallel}$ is determined by momentum conservation. If w is fixed and the detector angle is varied, a surface loss will be observed only at those angles for which Equations (65) and (67) simultaneously are satisfied. Therefore, the excitation of surface modes causes a conelike distribution of inelastically scattered electrons such that the axis of the cone is the 'inelastic-specular' direction defined by

$$\mathbf{k}'_{\parallel} = \mathbf{k}_{\parallel} + \mathbf{g} \tag{68a}$$

$$E' = E - \hbar\omega_s(0). \tag{68b}$$

The above considerations presume a measuring instrument which exhibits infinitely good angular and energy resolution. In practice, the angular resolution $(\Delta\theta \sim 1°)$ of existing spectrometers is adequate for our purposes. However, for phonon excitations, $w \cong \hbar\omega_{ph} \sim 10$ meV whereas in most cases the energy resolution of slow-electron spectrometers is $\Delta E \sim 500$ meV $\gg \hbar\omega_{ph}$. (There are, of course, exceptions [50–52].) A typical measurement is that of the total secondary electron energy distribution for an incident beam of energy E_0 and given direction. A schematic diagram of such a distribution is given in Figure 9 as discussed in Section 1D.

If for a given instrument of energy resolution ΔE, $\hbar\omega_{ph} \ll \Delta E$, the energy-loss of electrons to phonons cannot be measured directly. In this case we refer to phonon-assisted electron scattering as 'quasi-elastic'. Such scattering is included in the 'elastic' peak evident in Figure 9. As we shall see, momentum transfer to the electrons by the phonons is observable with existing spectrometers. Therefore phonon-assisted diffraction is significant in determining the angular shape of a given 'elastic' diffracted beam. If, however, one examines the dependence of the scattered intensity in a given beam on the energy of the incident beam, the phonon-assisted diffraction primarily leads to the introduction of a temperature dependence of the 'elastically' scattered intensities. The physical origin of this temperature dependence is the phonon-assisted scattering of the incident electrons outside the *angular aperature* of the detector. If both the angular aperature and energy resolution are of such magnitudes that phonon-assisted scattering from wide regions of the Brillouin zone are observed as 'elastic' events, then the measured intensity profiles become almost independent of temperature and the angular dependence of the beam intensity is no longer a meaningful observation. [73]

We now turn to the main topic of this section: the quasielastic scattering of electrons from a vibrating lattice as described using the Born-approximation. In terms of our Hamiltonian specified by Equations (14)–(20), the Born-approximation to the differential inelastic scattering cross section for scattering by an angle (θ, φ) with energy-loss w is given by [69]

$$\left(\frac{d^2\sigma}{d\Omega \, d\varepsilon}\right)_{fi} = \left(\frac{E - w}{E}\right)^{1/2} \left(\frac{m}{2\pi\hbar^2}\right)^2 |\langle \mathbf{k} + \mathbf{q}, f| V |\mathbf{k}, i\rangle|^2 \, \delta \left(w - E_i + E_f\right).$$

(69)

The electron makes a transition from a state $(\mathbf{k}, E)$ to the state $(\mathbf{k}+\mathbf{q}, E-w)$ while the lattice makes a transition from the state $|i\rangle$ to the state $|f\rangle$. Thus the initial and final wave functions are

$$\psi_i = \exp(i\mathbf{k} \cdot \mathbf{r}) \, \varphi_i \{\mathbf{R}\} \tag{70a}$$

$$\psi_f = \exp[i(\mathbf{k} + \mathbf{q}) \cdot \mathbf{r}] \, \varphi_f \{\mathbf{R}\}. \tag{70b}$$

The $\{\mathbf{R}\}$ denote a complete set of variables describing the crystal. In the present case, they consist of a set of equilibrium positions for the ion-cores $\{\mathbf{R}_n^{(0)}\}$, and occupation numbers of the phonons associated with the lattice displacements, $\mathbf{u}_n$, as defined by Equations (14) and (15).

In terms of the electron-lattice interaction, Equations (12c) and (17), and the lattice-displacement operators, Equations (15), we obtain the expression for the matrix elements of V in a plane-wave basis:

$$\langle \mathbf{k} + \mathbf{q}, f| V |\mathbf{k}, i\rangle = \int d\{\mathbf{R}\} \int d\mathbf{r} e^{-i(\mathbf{k}+\mathbf{q}) \cdot \mathbf{r}} \bar{\varphi}_f \{\mathbf{R}\} \times$$

$$\times \sum_n v(\mathbf{r} - \mathbf{R}_n) \, e^{i\mathbf{k} \cdot \mathbf{r}} \varphi_i \{\mathbf{R}\}$$

$$= v(\mathbf{q}) \int d\{\mathbf{R}\} \, \bar{\varphi}_f \{\mathbf{R}\} \sum_n e^{-i\mathbf{q} \cdot \mathbf{R}_n} \varphi_i \{\mathbf{R}\}$$

$$\equiv v(\mathbf{q}) \, \langle f| \sum_n e^{-i\mathbf{q} \cdot \mathbf{R}_n} |i\rangle. \tag{71}$$

In Equation (71) we have assumed that all of the ion-cores have the same potential form factor $v(\mathbf{q})$. More generally we must use form factors which depend on the site index:

$$\langle \mathbf{k} + \mathbf{q}, f| V |\mathbf{k}, i\rangle = \sum_n v_n(\mathbf{q}) \, \langle f| e^{-i\mathbf{q} \cdot \mathbf{R}_n} |i\rangle. \tag{72}$$

Equation (71) does not specify the experimentally-measured differential inelastic electron-scattering cross section for two reasons. First, the final state of only the electron is observed in the experiment. Therefore the lattice may be in any final state consistent with a momentum-energy transfer of $(\mathbf{q}, w)$ to the electron. Consequently, we must sum Equation (69) over the final state 'f' indices of the lattice. Second, the lattice was not in a given state prior to the scattering event but was in an incoherent superposition of such states characteristic of its temperature, T. Thus in Equation (69) we must average over the initial state indices, in accordance with

$$\langle M \rangle_T \equiv \sum_i \langle i| M |i \rangle e^{-E_i/\kappa T}/Z \tag{73a}$$

$$Z = \sum_i e^{-E_i/\kappa T}. \tag{73b}$$

Performing these two operations leads to our expression for the observed differential inelastic electron-scattering cross section

$$\frac{d^2\sigma}{d\Omega\,d\varepsilon} = \left[\frac{E - w}{E}\right]^{1/2} \left[\frac{m}{2\pi\hbar^2}\right]^2 \sum_{fi} e^{-E_i/\kappa T} \delta\left(w - E_i + E_f\right) \times$$

$$\times \left|\langle f| \sum_n v_n(\mathbf{q}) e^{-i\mathbf{q}\cdot\mathbf{R}_n} |i \rangle\right|^2 Z^{-1}. \tag{74}$$

If we could remove the delta function in energy, we could use the final-state completeness relation to remove the sum over 'f' in Equation (74). We achieve this objective by using the integral representation for the delta function

$$\delta(E) = \frac{1}{2\pi\hbar} \int_{-\infty}^{\infty} dt\, e^{-iEt/\hbar} \tag{75}$$

and noting that the E_i and E_f are eigenvalues of the lattice Hamiltonian, Equation (15b), i.e.,

$$\mathcal{H}_L \Phi_\alpha\{\mathbf{R}\} = E_\alpha \Phi\{\mathbf{R}\}. \tag{76}$$

Consequently, we write Equation (74) as

$$\frac{d^2\sigma}{d\Omega\,d\varepsilon} = \left[\frac{E - w}{E}\right]^{1/2} \left[\frac{1}{2\pi\hbar^2}\right]^2 \frac{1}{2\pi\hbar} \int_{-\infty}^{\infty} dt\, e^{-iwt/\hbar} \sum_{fi} \times$$

$$\times \langle i| e^{iE_it/\hbar} \sum_m \bar{v}_m(\mathbf{q}) e^{i\mathbf{q}\cdot\mathbf{R}_m} e^{-iE_ft/\hbar} |f \rangle \times$$

$$\times \langle f| \sum_n v_n(\mathbf{q}) e^{-i\mathbf{q}\cdot\mathbf{R}_n} |i \rangle e^{-E_i/\kappa T}/Z$$

$$= \left[\frac{E - w}{E}\right]^{1/2} \left[\frac{m}{2\pi\hbar^2}\right]^2 \frac{1}{2\pi\hbar} \int_{-\infty}^{\infty} dt\, e^{-iwt/\hbar} \sum_{m,n} \bar{v}_m(\mathbf{q}) v_n(\mathbf{q}) \times$$

$$\times \langle e^{i\mathbf{q}\cdot\mathbf{R}_m(t)} e^{-i\mathbf{q}\cdot\mathbf{R}_n(0)} \rangle_T \tag{77}$$

in which we used the fact that

$$\langle i|\, e^{iE_i t/\hbar} e^{i\mathbf{q}\cdot\mathbf{R}_m} e^{-iE_f t/\hbar}|f\rangle = \langle i|\, e^{i\mathcal{H}_{Lt}/\hbar} e^{i\mathbf{q}\cdot\mathbf{R}_m} e^{-i\mathcal{H}_{Lt}/\hbar}|f\rangle$$

$$\equiv \langle i|\, e^{-i\mathbf{q}\cdot\mathbf{R}_m(t)}|f\rangle \tag{78a}$$

$$\mathbf{R}_m(t) = \mathbf{R}_m + \mathbf{u}_m(t) \tag{78b}$$

$$\mathbf{u}_m(t) = \sum_{\mathbf{p},\lambda} \varepsilon_{\mathbf{p}\lambda}\left(\frac{\hbar}{2MN\omega_\lambda(\mathbf{p})}\right)^{1/2} [b^\dagger_{\mathbf{p},\lambda} e^{-i\mathbf{p}\cdot\mathbf{R}_m + i\omega_\lambda(\mathbf{p})t} + b_{\mathbf{p},\lambda} e^{i\mathbf{p}\cdot\mathbf{R}_m - i\omega_\lambda(\mathbf{p})t}]. \tag{78c}$$

I.e., we have introduced the Heisenberg representation for the lattice position operators:

$$A(t) \equiv e^{i\mathcal{H}_{Lt}/\hbar} A e^{-i\mathcal{H}_{Lt}/\hbar}. \tag{79}$$

We also have suppressed the superscript (0) on the equilibrium positions of the ion cores.

The reason that it is convenient to write the cross section in the form of Equation (77) is that for harmonic oscillater systems, the thermal average can be evaluated in closed form. [69] In the case of a monatomic lattice the final result is given by

$$\frac{d^2\sigma}{d\Omega\, d\varepsilon} = \left(\frac{E-w}{E}\right)^{1/2}\left(\frac{m}{2\pi\hbar^2}\right)^2\left(\frac{1}{2\pi\hbar}\right) \times$$

$$\times \sum_{m,n} \bar{v}_m(\mathbf{q})\, v_n(\mathbf{q})\, e^{i\mathbf{q}\cdot(\mathbf{R}_m - \mathbf{R}_n)} \int dt\, e^{-iwt/\hbar} e^{-Q_{mn}(t)} \tag{80a}$$

$$Q_{mn}(t) = \frac{\hbar^2 q^2}{2MN}\sum_{\mathbf{p},\lambda} [\hbar\omega_\lambda(\mathbf{p})]^{-1}[-i\sin[\theta_{mn}(t)] +$$

$$+ \{1 - \cos[\theta_{mn}(t)]\}\{2N[\omega_\lambda(\mathbf{p})] + 1\}] \tag{80b}$$

$$\theta_{mn}(\mathbf{p},t) = \omega_\lambda(\mathbf{p})\, t + \mathbf{p}\cdot(\mathbf{R}_m - \mathbf{R}_n) \tag{80c}$$

$$N(\omega) = [\exp(\hbar\omega/\kappa T) - 1]^{-1}. \tag{80d}$$

This expression gives the inelastic electron-scattering cross section accurate to second-order in the $v(\mathbf{q})$ but to all orders in the electron-phonon interaction as described by the rigid-ion model.

It is appropriate to note at this point that if the potential form-factors, $v_m(\mathbf{q})$, are independent of the atomic position index, the electron-scattering properties factor out of the expression for the cross section giving

$$\frac{d^2\sigma}{d\Omega\, d\varepsilon} = \left(\frac{E-w}{E}\right)^{1/2}\left(\frac{m}{2\pi\hbar^2}\right)^2 \frac{1}{\hbar}\, |v(\mathbf{q})|^2\, S(\mathbf{q}, w) \tag{81a}$$

$$S(\mathbf{q}, w) = \frac{1}{2\pi}\int_{-\infty}^{\infty} dt\, e^{-iwt/\hbar}\sum_{n,n}\langle e^{i\mathbf{q}\cdot\mathbf{R}_m(t)} e^{-i\mathbf{q}\cdot\mathbf{R}_n(0)}\rangle_T$$

$$= \sum_{n,m} e^{i\mathbf{q}\cdot(\mathbf{R}_m - \mathbf{R}_n)}\int_{-\infty}^{\infty}\frac{dt}{2\pi}\, e^{-iwt/\hbar} e^{-Q_{mn}(t)}. \tag{81b}$$

The quantity $S(\mathbf{q}, w)$ is the well-known pair-correlation function. [69] It describes the linear response of the atomic lattice to a time-dependent potential of wave-vector $\mathbf{q}$ and frequency $\omega = w/\hbar$.

In order to extract from Equations (81) the expression for the elastic scattering cross section, we decompose $Q_{mn}(t)$ into its time-independent and time-dependent terms:

$$Q_{mn}(t) = 2W(\mathbf{q}) - F_{mn}(t) \tag{82a}$$

$$W(\mathbf{q}) = \left(\frac{\hbar^2 q^2}{4M}\right) \frac{1}{N} \sum_{\mathbf{p}, \lambda} \{2N[\omega_\lambda(\mathbf{p})] + 1\}/\hbar\omega_\lambda(\mathbf{p}) \tag{82b}$$

$$F_{mn}(t) = \left(\frac{\hbar^2 q^2}{2M}\right) \frac{1}{N} \sum_{\mathbf{p}, \lambda} [\hbar\omega_\lambda(\mathbf{p})]^{-1} \{[1 + N(\hbar\omega_\lambda(\mathbf{p}))] \times$$

$$\times \exp[i\theta_{mn}(t)] + N[\hbar\omega_\lambda(\mathbf{p})] \exp[-i\theta_{mn}(t)]\}. \tag{82c}$$

The classification of contributions to the cross-section as 's'-phonon terms arises from expanding

$$\exp[iQ_{mn}(t)] = \exp[-2W(\mathbf{q})] \sum_{s=0}^{\infty} \frac{1}{s!} [F_{mn}(\mathbf{q}, t)]^s \tag{83}$$

and identifying the sth-order term with processes associated with s-phonons. For example, the elastic-scattering cross section arises from the $s=0$ term in Equation (83):

$$\left(\frac{d^2\sigma}{d\Omega \, d\varepsilon}\right)_{elastic} = \left(\frac{mN}{2\pi\hbar^2}\right)^2 |v(\mathbf{q}) \, e^{-W(\mathbf{q})}|^2 \, |S(\mathbf{q})|^2 \, \delta(w) \tag{84a}$$

$$S(\mathbf{q}) = \frac{1}{N} \sum_n e^{-i\mathbf{q} \cdot \mathbf{R}_n} \tag{84b}$$

which is identical to the result which would be derived for a rigid lattice, except for the vertex-renormalization factor, $\exp[-W(\mathbf{q})]$, that measures the reduction in the total intensity of the elastic scattering due to the occurrence of inelastic scattering. This factor, $\exp[-W(\mathbf{q})]$, is called the Debye-Waller factor.

The single-phonon emission and absorption processes result from the $s=1$ term in Equation (83). The t-integral gives energy conservation and the n, m sums give momentum conservation. We obtain prior to the m, n sums:

$$\left(\frac{d^2\sigma}{d\Omega \, d\varepsilon}\right)_{1 \, phonon} = \left[\frac{E-w}{E}\right]^{1/2} \left[\frac{mN}{2\pi\hbar^2}\right]^2 |v(\mathbf{q}) \, e^{-W(\mathbf{q})}|^2 \frac{\hbar^2 q^2}{2M} \times$$

$$\times \frac{1}{N} \sum_{\mathbf{p}, \lambda} \frac{1}{\hbar\omega_\lambda(\mathbf{p})} \{[1 + N(\omega_\lambda(\mathbf{p}))] \, \delta[w - \hbar\omega_\lambda(\mathbf{p})] \, |S(\mathbf{p} + \mathbf{q})|^2 +$$

$$+ |S(-\mathbf{p} + \mathbf{q})|^2 \, N[\omega_\lambda(\mathbf{p})] \, \delta[w + \hbar\omega_\lambda(\mathbf{p})]\}. \tag{85}$$

The m, n sums are absorbed in the 'phonon-assisted' interference functions

$$S(\pm \mathbf{p} + \mathbf{q}) = \frac{1}{N} \sum_n \exp[-i(\mathbf{q} \pm \mathbf{p}) \cdot \mathbf{R}_n]. \tag{86}$$

The first term in Equation (85) describes phonon emission by the electron and the second term describes phonon absorption.

Equations (85) and (86) often are applied to described slow neutron scattering in bulk crystals. [69] In that case, the interference functions are taken to be delta functions, i.e.,

$$S(\mathbf{q} \pm \mathbf{p}) \rightarrow \delta_{\mathbf{q}, \pm \mathbf{p}}. \tag{87}$$

Therefore the sum over $\mathbf{p}$ disappears in Equation (85) so that each neutron scattering event leads to the adsorption of a prescribed lattice phonon whose momentum and energy are determined from:

$$\mathbf{G} - \mathbf{p} = \mathbf{k}' - \mathbf{k} = \text{observed momentum transfer} \tag{88a}$$

$$\hbar\omega(\mathbf{p}) = w = \text{observed energy gain} \tag{88b}$$

for phonon absorption by low-energy neutrons (which cannot emit phonons). The occurrence of $\mathbf{G} \neq 0$ in Equation (88a) is referred to as an 'Umklapp process' in which the neutron 'simultaneously' scatters from the rigid lattice and emits a phonon. These processes play a key role in the actual experimental determination of phonon spectra by slow neutron scattering.

In the case of electron scattering from surfaces, the momentum conservation law for the *electronic* momentum applies only to the component of momentum parallel to the surface. Thus, for the geometry shown in Figure 23, we see that the angular position of the detector determines the phonon's parallel momentum via

$$\mathbf{p}_{\|} = \mathbf{k}_{\|} - \mathbf{k}'_{\|} + \mathbf{g} \tag{89a}$$

for phonon emission in the diffracted beam labeled by $\mathbf{g}$. As noted earlier, the energy of the phonon is determined, in principle, by the loss-energy observed with the detector:

$$w = \hbar\omega(\mathbf{p} = \mathbf{p}_{\|} + \mathbf{p}_{\perp}). \tag{89b}$$

The component of the phonon's momentum perpendicular to the surface, $p_{\perp}$, is determined from Equations (89) rather than a diffraction condition like Equation (87). (In the case of surface phonons, $\mathbf{p}_{\|}$ determines both ω and the magnitude of the decay length $p_{\perp} = i|p_{\perp}|$. Therefore Equations (89) and our subsequent discussion directly apply only to bulk phonons.) The residual effects of the conservation of the component of the electron's momentum perpendicular to the surface are revealed by a maximum in $S(\mathbf{q} + \mathbf{p})$ near the appropriate values of $p_{\perp}$. If we take the phonons to be 'undamped' ($p_{\perp}$ is real), the structure factors are given by:

$$S(\mathbf{k}' - \mathbf{k} \pm \mathbf{p}) = \delta_{\pm \mathbf{p}_{\|}, \mathbf{k}_{\|}' - \mathbf{k}_{\|} - \mathbf{g}'} [1 - R(0, \mathbf{g}', E; \pm p_{\perp}, w)]^{-1} \tag{90a}$$

$$R(\mathbf{g}, \mathbf{g}', E; p_\perp, w) = \exp\{-i(\mathbf{g} + \mathbf{g}')\cdot\mathbf{d}\} \times$$
$$\times \exp\{i[k_{11}(\mathbf{g}, E) + k'_{11}(\mathbf{g}', E - w) + p_\perp] d_\perp\} \times$$
$$\times \exp\{-[k_{12}(\mathbf{g}, E) + k'_{12}(\mathbf{g}', E - w)] d_\perp\}. \qquad (90b)$$

In Equations (90) the value of $p_\perp$ is determined from Equations (89) and $\mathbf{g}'$ is the label that the $\mathbf{k}'_\parallel$ beam would have in the absence of phonon emission. The quantities $k_{11}(\mathbf{g}, E)$ and $k_{12}(\mathbf{g}, E)$ are the components of the complex momentum of the electron as determined from Equation (56), and the factor $[1 - R]^{-1}$ in Equation (90a) is determined as described in Equations (57)–(61).

When Equations (90) are inserted into Equation (85) for the cross section we obtain for the phonon-emission cross section associated with the (00) beam [$\mathbf{g}' = 0$]:

$$\left(\frac{d^2\sigma}{d\Omega\, d\varepsilon}\right)_{\substack{(00)\ \text{beam} \\ 1\ \text{phonon emission}}} = \left(\frac{E - w}{E}\right)^{1/2} \left|\frac{mNv(\mathbf{q})}{2\pi\hbar^2} e^{-W(\mathbf{q})}\right|^2 \frac{\hbar^2 q^2}{2M} \times$$

$$\times \frac{6\pi^2}{p_m^3} \sum_\lambda \int \frac{dp_\perp\, d^2 p_\parallel}{2\pi} \frac{(2\pi)^2}{(2\pi)^2 A} \delta(\mathbf{p}_\parallel - \mathbf{k}'_\parallel - \mathbf{k}_\parallel) |1 - R(0, 0, E; p_\perp, w)|^{-1} \times$$

$$\times \delta[w - \hbar\omega_\lambda(\mathbf{p})] \{1 + N[\hbar\omega_\lambda(\mathbf{p})]\}/\hbar\omega_\lambda(\mathbf{p}). \qquad (91)$$

By writing the prefactor of p_m^{-3} in Equation (91) we implicitly assume a Debye spectrum for the phonons so that $N = V(p_m^3/6\pi^2)$. In performing the integral over $p_\perp$, it is convenient to note that for a given $\hbar\omega_\lambda(p) = w$ and value of $\mathbf{p}_\parallel$, there usually are two $p_\perp$ values satisfying

$$\hbar\omega(\mathbf{p}_\parallel, p_\perp) = w$$

giving, for spherically-symmetric phonon dispersion relations:

$$p_{\perp\lambda} = \pm p_{\perp\lambda}(w, \mathbf{p}_\parallel). \qquad (92)$$

Thus if we introduce the density of states

$$\varrho_{\perp\lambda}(w, \mathbf{p}_\parallel) = \left[\frac{\partial[\hbar\omega_\lambda(\mathbf{p})]}{\partial p_\perp}\right]^{-1}_{w, \mathbf{p}_\parallel}. \qquad (93)$$

Equation (91) gives

$$\left(\frac{d^2\sigma}{d\Omega\, d\varepsilon}\right)_{\substack{(00) \\ 1\ \text{phonon} \\ \text{emission}}} = \left(\frac{E - w}{E}\right)^{1/2} \left|\frac{mNv(\mathbf{q})}{2\pi\hbar^2} e^{-W(\mathbf{q})}\right|^2 \left(\frac{\hbar^2 q^2}{2m}\right) \frac{3\pi}{p_m^3 A} \times$$

$$\times \left[\frac{1 + N(w/\hbar)}{w}\right] \sum_{\lambda=1}^{3} \varrho_{\perp\lambda}(w, \mathbf{k}'_\parallel - \mathbf{k}_\parallel) \left\{\left|\frac{1}{1 - R(0, 0, E, p_\perp, w)}\right|^2 + \right.$$

$$\left. + \left|\frac{1}{1 - R(0, 0, E, -p_\perp, w)}\right|^2\right\}_{w, \mathbf{p}_\parallel}. \qquad (94)$$

For acoustical phonons described by the Debye model, $\hbar\omega_\lambda(\mathbf{p}) = \hbar v_\lambda p$, we get

$$\varrho_{\perp\lambda}(w, \mathbf{p}_{\parallel}) = \hbar^{-1}\left[\frac{\partial\omega_\lambda(p)}{\partial p_\perp}\right]_{w,\,p_{\parallel}}^{-1} =$$

$$= \left(\frac{p}{\hbar v_\lambda p_\perp}\right)_{w,\,p_{\parallel}} = \frac{w}{\hbar v_\lambda[w^2 - (\hbar v_\lambda p_{\parallel})^2]^{1/2}}. \tag{95}$$

Equations (94) and (95) give our final result for a Debye model of acoustical phonons:

$$\left(\frac{d^2\sigma}{d\Omega\,d\varepsilon}\right)_{\substack{(00)\\ \text{1 phonon}\\ \text{emission}}} = \left[\frac{E-w}{E}\right]^{1/2}\frac{mNv(\mathbf{q})\,e^{-W(\mathbf{q})}2}{2\pi\hbar^2}\left|[1+N(w/\hbar)]\frac{3\pi}{p_m^2 A}\right. \times$$

$$\times \sum_\lambda\left(\frac{\hbar^2 q^2}{2M}\right)\left(\frac{1}{\hbar v_\lambda p_m}\right)\frac{w}{[w^2 - \hbar v_\lambda p_{\parallel})^2]^{1/2}} \times$$

$$\times \{|1 - R(0, 0, E; p_{\perp\lambda}, w)|^{-2} + |1 - R(0, 0, E; -p_{\perp\lambda}, w|^{-2}\}_{w,\,\mathbf{p}_{\parallel}}. \tag{96}$$

The external momentum transfer is determined from the beam parameters via

$$k(E) \equiv [2mE/\hbar^2]^{1/2} \tag{97a}$$

$$k'(E) = [2m(E-w)/\hbar^2]^{1/2} \tag{97b}$$

$$k_{\parallel} = k(E)\sin\theta \tag{98a}$$

$$k'_{\parallel} = k'(E)\sin\theta' \tag{98b}$$

$$k_\perp(E) = [k^2(E) - k_{\parallel}^2]^{1/2} \tag{99a}$$

$$k_\perp(E-w) = [k'^2(E) - k_{\parallel}'^2]^{1/2} \tag{99b}$$

where θ and θ' are defined in Figure 23. However, the value of $\mathbf{q}$ to be used in $v(\mathbf{q})$ $\exp[-W(\mathbf{q})]$ in Equation (96) is [for a $\mathbf{k}'$ beam of index $\mathbf{g}'$]:

$$\mathbf{q}_{\parallel} = \mathbf{k}'_{\parallel} - \mathbf{k}_{\parallel} = \mathbf{g}' - \mathbf{p}_{\parallel} \tag{100a}$$

$$q_\perp = [k_{\perp 1}(0, E) + k_{\perp 1}(\mathbf{g}', E - w)] + i[k_{\perp 2}(0, E) + k_{\perp 2}(\mathbf{g}', E - w)]. \tag{100b}$$

Equation (96) has been derived in detail because it illustrates the influence of the large-energy discrete loss processes on the low-energy quasielastic loss cross sections. The modified conservation laws of E and $\mathbf{k}_{\parallel}$ (rather than three dimensional momentum conservation) have the interesting consequence that momentum conservation normal to the surface is observable via 'sideband-diffraction' resonances in the inelastic cross section. [3, 47, 74] These resonances occur only for bulk excitations and require a precise definition of the loss-energy w for their observation. The inelastic scattering cross-section exhibits maxima as a function of either E or θ' when the Bragg form factors, $|1 - R|^{-2}$, pass through a resonance, i.e., when

$$\mathbf{g}'\cdot\mathbf{d} + [k_{\perp 1}(0, E) + k'_{\perp 1}(\mathbf{g}', E - w) \pm p_\perp(w, \mathbf{p}_{\parallel})]d_\perp = 2n\pi. \tag{101}$$

This maximum is due to the diffraction of the combined wave fields of the incident electron, final electron, and (bulk) phonon from the lattice. Thus it is literally the diffraction from the lattice of the wave fields associated with the phonon emission (absorption) sidebands to the elastic-scattering event.

The limited instrumental energy resolution of most modern detectors renders neither the elastic nor the inelastic differential cross sections directly measurable experimentally. Thus, as noted in Section 1D, the measured quantity is the quasielastic scattering cross section defined by

$$\left(\frac{d\sigma}{d\Omega}\right)_{qe} \equiv \int_{E-\frac{1}{2}\Delta E}^{E+\frac{1}{2}\Delta E} d\varepsilon \left(\frac{d^2\sigma}{d\Omega \, d\varepsilon}\right) \tag{4b}$$

where ΔE is the experimental energy resolution. If we assume $\Delta E \gg \hbar\omega_\lambda(p_m)$, then from Equations (80) and (82) we obtain

$$\left(\frac{d\sigma}{d\Omega}\right)_{qe} \cong \left(\frac{m}{2\pi\hbar^2}\right)^2 \sum_{m,n} e^{i\mathbf{q}\cdot(\mathbf{R}_m-\mathbf{R}_n)} \bar{v}_m(\mathbf{q}) \, v_n(\mathbf{q}) \, e^{-2W(\mathbf{q})} \exp\left[F_{mn}(t=0)\right]. \tag{102a}$$

$$F_{mn}(0) = \frac{\hbar^2 q^2}{2M} \frac{1}{N} \sum_{\mathbf{p},\lambda} [\hbar\omega_\lambda(\mathbf{p})]^{-1} \{(1 + N[\omega_\lambda(\mathbf{p})]) \, e^{i\mathbf{p}\cdot(\mathbf{R}_m-\mathbf{R}_n)}$$
$$+ N[\omega_\lambda(\mathbf{p})] \, e^{-i\mathbf{p}\cdot(\mathbf{R}_m-\mathbf{R}_n)}\}. \tag{102b}$$

By expanding $\exp[F_{mn}(0)]$ we recover the usually-quoted [75, 76] expressions for the multiphonon scattering of low-energy electrons. The conventional procedure [75, 76] is to expand $\exp[F_{mn}(t=0)]$ and perform the multiple integrals over $(\mathbf{p}, \lambda)$ and (m, n) term by term. The constribution to the quasielastic scattering cross section of the term linear in $F_{mn}(0)$ is called the *thermal-diffuse* scattering cross-section. A more complete discussion of sum rules and comparisons with experiment may be found in the literature. [56]

Perhaps the most important feature of phonon-assisted quasielastic LEED is its determination of the angular width of the individual scattered beams in the absence of surface disorder. The commonly-considered case [76] is that in which only the top layer of the crystal scatters the electrons. In this case Equations (102) give

$$\left(\frac{d\sigma}{d\Omega}\right)_{qe \atop \text{1-phonon}} = \left|\frac{mv_0(\mathbf{q})}{2\pi\hbar^2} e^{-W(\mathbf{q})}\right|^2 \frac{\hbar^2 q^2}{2M} \frac{N_0^2}{N} \sum_{\mathbf{p},\lambda} \frac{1}{\hbar\omega_\lambda(\mathbf{p})} \times$$
$$\times \{(1 + N[\omega_\lambda(\mathbf{p})]) \, \delta_{\mathbf{p}_\parallel'-\mathbf{q}_\parallel} + N[\omega_\lambda(\mathbf{p})] \, \delta_{\mathbf{p}_\parallel', \mathbf{q}_\parallel}\}. \tag{103a}$$

Taking the high-temperature limit and using a Debye model for the phonon spectrum leads to the simple result:

$$\left(\frac{d\sigma}{d\Omega}\right)_{qe \atop 1\text{-phonon}} = N_0^2 \left|\frac{mv_0(\mathbf{q})}{2\pi\hbar^2} e^{-W(\mathbf{q})}\right|^2 \sum_{\lambda=1}^{3} \frac{3\pi q^2 \kappa T}{p_m^3 M v_\lambda^2} \int_{-p_m}^{p_m} \frac{dp_\perp}{[q_\parallel^2 + p_\perp^2]}$$

$$= N_0 \left(\frac{mv_0(\mathbf{q})}{2\pi\hbar^2} e^{-W(\mathbf{q})}\right)^2 \frac{6\pi q^2 \kappa T \tan^{-1}(p_m/q_\parallel)}{M p_m^3 q_\parallel} \sum_{\lambda=1}^{3} \frac{1}{v_\lambda^2}$$

(103b)

where N_0 is the number of ion cores in the top layer of the solid. The $q_\parallel^{-1}$ dependence, where $\mathbf{q}_\parallel = \mathbf{k}_\parallel' - \mathbf{k}_\parallel$ is determined by the position of the detector as shown in Figure 23, is a predicted and experimentally confirmed feature of the angular profile of a given LEED beam. [76] The existing experimental data are not sufficiently accurate to extract precise information about either the surface-phonon modes or the in elastic-collision electron penetration depth from the angular lineshapes, although McKinney et al. [76] attempt to estimate the consequence of these effects in their data analysis. They also verified both the κT and q^2 dependence of $(d\sigma/d\Omega_{qe})$ predicted by Equation (103b).

Our final topic in this section is the application of the linear-response theory of electron-solid scattering to extract values of the rms values of the thermal displacements, $\langle u_v^2 \rangle_T$, from an analysis of the temperature dependence of ELEED intensities. Although such analyses [54, 55, 77–93] exhibit several systematic difficulties because they neglect the consequences of multiple electron-scatterer collisions [2, 48, 49] they have uncovered the systematic feature of surface scatterers that $\langle u_0^2 \rangle_T$ is about twice the value for corresponding bulk scatterers (see, e.g., Section 1E).

In order to write our formulas in a form compatible with the analyses in the literature, [54, 55, 77–93] we re-examine Equation (77) written in the form

$$\frac{d^2\sigma}{d\Omega\,d\varepsilon} = \left(\frac{E-w}{E}\right)^{1/2} \left(\frac{m}{2\pi\hbar^2}\right)^2 \frac{1}{2\pi\hbar} \sum_{m,n} e^{i\mathbf{q}\cdot(\mathbf{R}_m - \mathbf{R}_n)} \bar{v}_w(\mathbf{q})\, v_n(\mathbf{q})$$

$$\times \int_{-\infty}^{\infty} dt\, e^{iwt/\hbar} \langle e^{i\mathbf{q}\cdot\mathbf{u}_m(t)} e^{-i\mathbf{q}\cdot\mathbf{u}_n(0)} \rangle_T.$$

(104)

Using standard combinatorial techniques, [2] we can reduce the thermal average in Equation (104) to a form equivalent to Equations (80):

$$\langle e^{i\mathbf{q}\cdot\mathbf{u}_m(t)} e^{-i\mathbf{q}\cdot\mathbf{u}_n(0)} \rangle_T = e^{-W_m(\mathbf{q},t)} e^{-W_n(\mathbf{q},0)} \exp[-i\mathbf{q}\cdot\mathbf{D}(m,n,t)\cdot\mathbf{q}]$$

(105a)

$$W_m(\mathbf{q},t) = \tfrac{1}{2}\langle[\mathbf{q}\cdot\mathbf{u}_m(t)]^2\rangle_T$$

(105b)

$$\mathbf{D}(m,n,t) = -i\langle\mathbf{u}_m(t)\,\mathbf{u}_n(0)\rangle_T.$$

(105c)

The expression, (105c) for $W_m(\mathbf{q})$ is independent of m for bulk phonons thereby reducing to the Debye-Waller factor given by Equation (82b). In the harmonic approximation [i.e., $\mathcal{H}_L$ is given by Equation (15b) or its generalization to include surface modes] $W_m(\mathbf{q}, t)$ is independent of t, so we shall suppress this variable via setting $W_m(\mathbf{q}, t) \to W_m(\mathbf{q})$. The quasi-elastic scattering cross-section is found from Equations (102), (104), and (105) to be given by

$$\left(\frac{d\sigma}{d\Omega}\right)_{qe} = \left(\frac{m}{2\pi\hbar^2}\right)^2 \sum_{m,n} e^{i\mathbf{q}\cdot(\mathbf{R}_m - \mathbf{R}_n)} \bar{v}_m(\mathbf{q}) \, e^{-\overline{W}_m(\mathbf{q},\,0)} v_n(\mathbf{q}) \times$$

$$\times \, e^{-W_n(\mathbf{q},\,0)} \exp\left[i\mathbf{q}\cdot D(m, n, 0)\cdot\mathbf{q}\right]. \tag{106}$$

Equation (106) is the generalization of Equations (102) which permits a discussion of the effects of a layer-index dependence of *both* the atomic form factors, $v_m(\mathbf{q})$ and the atomic displacements, $\mathbf{u}_m(0)$. A standard observation is that for the (00) beam $\mathbf{q}$ is normal to the surface so that

$$W_m(\mathbf{q}) \to 2\,|k_\perp(0, E)|^2 \, \langle u_{m\perp}^2 \rangle_T \tag{107}$$

where we used the fact in the presence of inelastic-collision damping

$$W_m(\mathbf{q}) = \tfrac{1}{2}\mathbf{q}\cdot\langle \mathbf{u}_m(0)\,\mathbf{u}_m(0) \rangle_T \cdot \bar{\mathbf{q}} \tag{108a}$$

$$F_{mn}(t) = -\,i\mathbf{q}\cdot D(m, n, t)\cdot\bar{\mathbf{q}}. \tag{108b}$$

From Equations (108) we see that if we pick a simple form for $\langle u_m^2 \rangle$, e.g.,

$$\langle u_m^2 \rangle_T = \langle u^2 \rangle_T \left[1 + A F_m\right], \tag{109}$$

the sums in Equation (106) can be performed numerically if we neglect thermal diffuse and multiphonon contributions to the quasielastic cross section. The resulting expression is just the differential elastic scattering cross section

$$\left(\frac{d\sigma}{d\Omega}\right)_{qe}^{(0)} = \left(\frac{m}{2\pi\hbar^2}\right)^2 \left| \sum_{m=0}^{\infty} e^{-i\mathbf{q}\cdot\mathbf{R}_m} v_m(\mathbf{q}) \, e^{-W_m(\mathbf{q})} \right|^2. \tag{110}$$

Jones *et al.* [79] performed this calculation for various values of F_n using a form for $k_{12}(0, E)$ in which k_{12} linearly decreased with increasing energy, E, of the incident electrons. The comparison of these calculations with some of their experimental data is shown in Figure 24. In this figure, the effective value of $\langle u_{m\perp}^2 \rangle$ has been converted into an effective value of the Debye Θ parameter via the use of the Debye spectrum in the high temperature limit to set [see, e.g., Equations (103)]

$$\langle u_{\perp\infty}^2 \rangle = \hbar^2 T / M\kappa\Theta_\infty^2 \tag{111a}$$

$$\Theta_\infty = \hbar\bar{v}_\lambda p_w / \kappa \tag{111b}$$

in which $\bar{v}_\lambda$ is an average speed of sound. Thus Equation (109) becomes

$$\frac{1}{\Theta_m^2} = \frac{1}{\Theta_\infty^2} \left[1 + A F_m\right]. \tag{112}$$

which describes the calculated curves shown in Figure 24.

The interpretation of plots, like that shown in Figure 24, of the values of either $\langle u_{\perp}^2 \rangle_{\text{eff}}$ or Θ as a function of the energy of the associated peak in the elastic intensity profile usually involved the three concepts indicated in Figure 25. First, each point on the plot is obtained by examining the temperature dependence of a particular peak in the intensity profile as indicated by item (1) in Figure 25. It is found universally [54, 55, 77–93] that the higher-energy peaks yield smaller values of $\langle u^2 \rangle_{\text{eff}}$ (larger values of Θ). Figure 24 provides an example of this result. The conventional [54, 77–93]

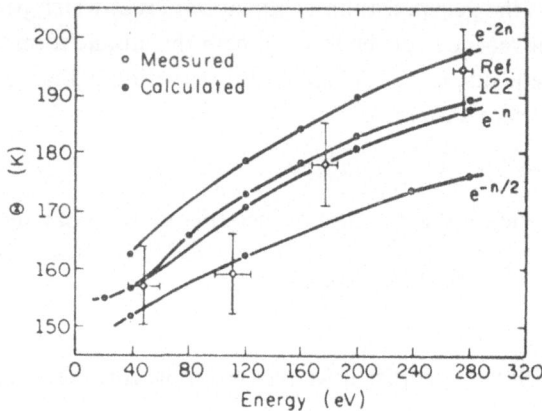

Fig. 24. Comparison of effective Debye temperatures calculated from Equation (112) in the text with the experimental values (denoted by open circles with error bars) extracted from data by using a model in which all atomic form factors are identical, all atoms have the same $\langle u_{\perp}^2 \rangle_T = \hbar^2 T / m\kappa\Theta^2$, and $k_{\perp 2}(0, E)$ decreases linearly with increasing electron energy. The selection of the F_n and evaluation of $k_{\perp}(0, E)$ is described in detail in [79]. (After Jones *et al.* [79]).

Vibronic Properties of Surface Atoms
via ELEED Temperature Dependence

1. Each peak in the Intensity Profile exhibits a different $d[\ln I]/dt$

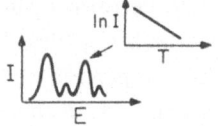

2. Beam Penetration (λ_{ee}) depends upon beam energy

3. $\langle u_n^2 \rangle$ depends on the layer-index n. $[\langle u_s^2 \rangle \cong 2 \langle u_B^2 \rangle]$

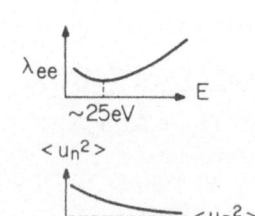

Fig. 25. Schematic indication of the three central concepts used in interpreting plots of the exponential temperature dependence of maxima in ELEED intensity profiles as a function of the energy of the peak.

rationalization of these observations is that since with increasing energy the elastic beam of electrons penetrates further into the lattice (item (2), Figure 25), they are scattered increasingly by bulk rather than surface ion-cores. Hence, the decrease in $\langle u^2 \rangle_{\mathrm{eff}}$ with increasing beam energy reflects a decrease in $\langle u_v^2 \rangle_T$ with increasing v (i.e., distance from the surface). This conclusion is indicated by item (3) of Figure 25. Thus if one knows λ_{ee} and the form factors $v_n(q)$ as a function of energy, then by using the appropriate complex optical potential in Equation (56), e.g. [66]

$$\Sigma(E) = V(E) - i\Gamma(E) \tag{113a}$$

$$\Gamma(E) = \hbar \{2[E + V(E)]/m\}^{1/2}/\lambda_{ee} \tag{113b}$$

$$V(E) \equiv -V_0, \tag{113c}$$

one can extract the $\langle u_v^2 \rangle_T$ from the data by using Equations (100) and (110). The solid curves in Figure 24 are generated in a similar fashion by taking $\lambda_{ee}(E)$ from empirical considerations and the $\langle u_v^2 \rangle_T$ from model calculations. [79] The fact that this type of analysis is qualitatively but not quantitatively correct is evident from Figure 24. It also has been found by numerous other workers. [55, 77–93] In particular, the analysis suffers from the fatal flaw that multiple-scattering effects, ignored in linear-response calculations, are of critical importance in achieving a quantitative description of ELEED intensities. [5, 6] An example of the consequences of multiple scattering is the fact that an increase in the surface $\langle u_0^2 \rangle_T$ relative to bulk $\langle u^2 \rangle_T$ leads to reduced scattering by the surface atoms and hence increased penetration of the elastic beam inside the solid. In a multiple-scattering analysis, this effect alone provides a decrease in $\langle u^2 \rangle_{\mathrm{eff}}$ for peaks in the ELEED intensities at increasingly larger energy. [45, 48] Therefore the energy dependence of λ_{ee} serves only to complicate, not to cause, the observed trend for the observed $\langle u^2 \rangle_{\mathrm{eff}}$ to decrease with increasing peak energy. Since such complicating effects of multiple scattering are evident in many features of the LEED intensities, we next turn to a discussion of them.

C. BEYOND THE BORN APPROXIMATION

In previous sections we have emphasized repeatedly that although the Born approximation is useful as a simple illustration of the qualitative features of LEED (both elastic and inelastic), the strong electron-solid interactions render it inadequate for a quantitative description of the diffracted intensities. This section constitutes a brief interlude in the main stream of our presentation in which a few examples of this inadequacy are displayed.

Two analytical indications of difficulty with the Born approximation already are evident from electron-atom scattering (i.e., in gases) for electrons in the energy range, $5 \,\mathrm{eV} \lesssim E \lesssim 500 \,\mathrm{eV}$ of interest in LEED. The elastic electron-atom cross sections are of the order of $(1 \,\text{Å})^2$, and are peaked in the forward direction. Furthermore, the inelastic cross sections are comparable with the elastic ones. Although the details evidently depend both on the atomic species comprising the target and on the incident electron's energy, the general features of the physical situation are clear. A solid appears to an incident low-energy electron as an array of scatterers whose cross-sections are comparable to their spacing (especially for (penetrating) forward scattering). Thus both elastic and (initial) inelastic scattering events are expected to occur close to the surface, and multiple scattering processes should be relatively common. Furthermore, the occurrence of those coherent multiple-elastic-scattering processes characteristic of the bulk energy-band structure of solids is not expected for most of the LEED energy range because of the combined influence of the inelastic scattering cross sections [37] (which limits elastic scattering to the surface region of the solid) and the geometrical, vibrational, and electronic inequivalence [1, 26] between the surface scatterers and those in the bulk. Therefore, we anticipate that the calculations of the LEED intensities will be complicated, must be based on a distorted wave scattering theory such as that described in Section 2C, and, for suitably general models, may lead to diffracted intensities bearing little detailed relation to the bulk energy-band structure of the solid.

An explicit demonstration of the inadequacy of the Born approximation to describe ELEED even from light elements (i.e., weak scatterers exhibiting small electron-atom cross sections) is given in Figure 26. In this figure experimental elastic intensity profiles

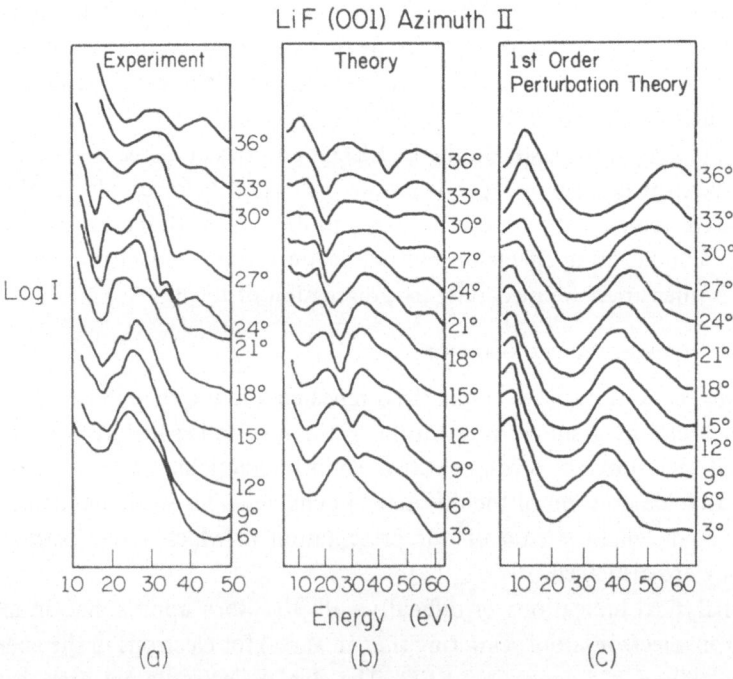

Fig. 26. Analysis of experimental [94] LEED intensities from LiF [panel (a)] using a complete multiple scattering theory [panel (b)] and the Born approximation [panel (c)]. All panels show the logarithm of the specularly reflected intensity as a function of the incident electron's energy. The polar angles of incidence are indicated in the figure. The azimuthal angles are specified in [94]. (After Holland et al. [95].)

for the (00) [i.e., specular] beam of electrons diffracted from the (001) face of LiF are displayed [94] in panel (a). The occurrence of complicated structure in these profiles, not obviously related to the kinematical Bragg peaks calculated in Section 3A., is evident from the figure. However, such Bragg peaks are evident in the results of the model calculations [95] shown in panel (c). The intensity profiles shown in this panel were calculated using a model in which multiple elastic scattering *within* planer layers of ion-cores was accounted for, but the electron was presumed to scatter once

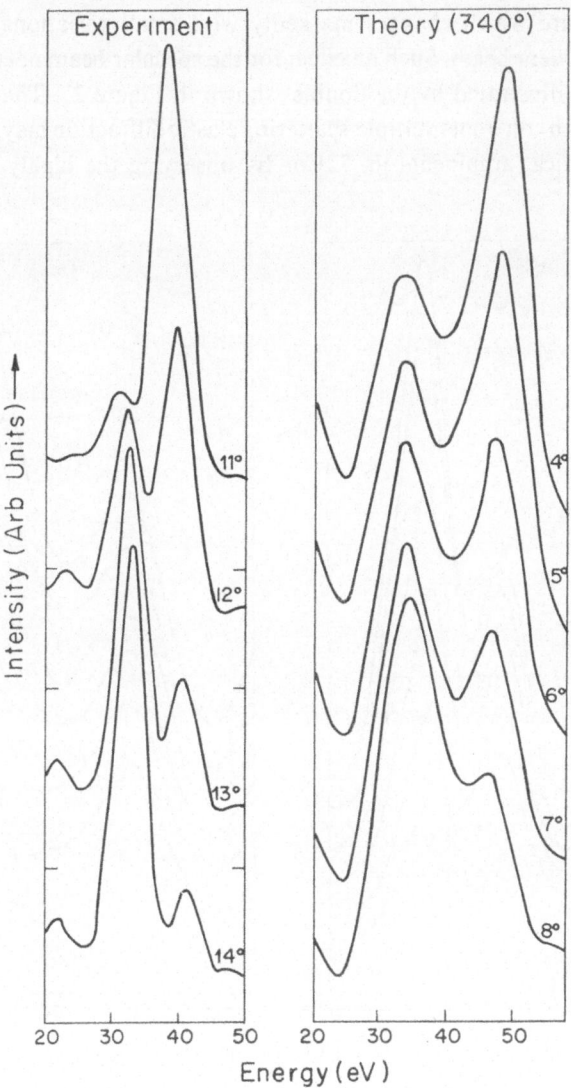

Fig. 27. The intensity-voltage profiles for the (00) beam from Cu(001) for a series of angles of incidence. The zeros for different curves are displaced arbitrarily. The experimental results were taken at room temperature. (After Holland [49].)

and only once from each layer. Evidently, the calculated profiles shown in panel (c) bear little resemblance to the observed ones displayed in panel (a). The dramatic consequences of including in the model multiple scattering between layers are revealed by comparing panel (b) with panel (c). The complete calculations shown in panel (b), obtained using the inelastic-collision model [Section 2C], exhibit a much improved correspondence with the measurements. However, quantitative descriptions of the data require a still more accurate description of the electron-solid force law.

In addition to the general lack of resemblance between observed intensity profiles and those predicted by the Born approximation, the measured intensities often exhibit fine structure which changes markedly with small variations in the angle or energy of the incident beam. Such an effect for the specular beam of electrons reflected from Cu(100) is illustrated by the doublet shown in Figure 27. That such structures are associated with coherent multiple-scattering elastic diffraction may be demonstrated either by theoretical arguments [6, 72] or by observing the highly non-kinematical

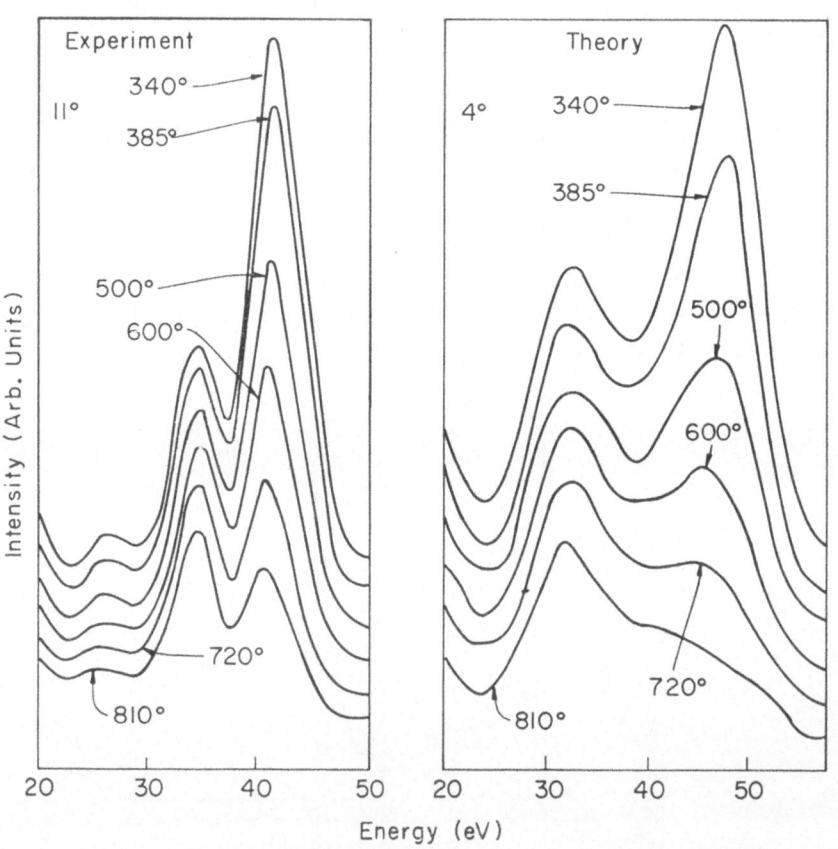

Fig. 28. The effect of temperature changes on an intensity-voltage profile for the (00) beam from Cu(001). Corresponding experimental and theoretical curves refer to the same temperature and have the same base-line, but curves for different temperatures have slightly different base lines.
(After Holland [49].)

temperature behavior exhibited by this doublet [48, 55] as shown in Figure 28. The results of the model calculations displayed in the right-hand panels of Figures 27 and 28 reveal that a theory [2] which incorporates multiple-scattering phenomena indeed can describe the qualitative features of such observations. However, as found in Figure 26, their quantitative description requires a more refined characterization of the electron-solid force law. For our present purposes, the important point illustrated by Figures 26–28 is necessity of incorporating a description of multiple-elastic scattering phenomena in any theory of isothermal ELEED intensities or of their temperature dependence. The theory described in Section 2C provides a convenient and popular vehicle for this purpose.

Having convinced ourselves that a consideration of multiple scattering effects is essential to any microscopic theory of LEED, we must ask another, in many respects more important, question. "Given this centrality of multiple scattering for describing the detailed behavior of ELEED intensities, are there residual manifestations of the kinematic model [Born approximation] which can be utilized in a study of surface crystallography?" Stated more pragmatically but less precisely, "can we discern systematic features of LEED intensities which are sensitive to the positions of the atomic scatterers but insensitive to the host of uncertain features [Section 2B] in the electron-solid force law?" Fortunately for the use of ELEED as a technique to determine surface crystallography, the answer to these questions is 'yes' provided the inelastic-collision damping is sufficiently strong (e.g., λ_{ee} in Equations (113) satisfies $\lambda_{ee} \lesssim 8$ Å).

The fundamental residual manifestation of the kinematical model in a dynamical theory lies in the fact that although the structure in the dynamical intensity profiles is complicated, the intensities are largest near those energies at which a primary Bragg peak occurs in the Born approximation. Implicit in the early calculations of Morse, [96] this feature of dynamical theories was rediscovered and its significance for surface crystallography identified by Tucker and Duke, [72] who gave it the short-hand label 'the Bragg envelope concept'. The essential features of Tucker and Duke's argument are illustrated in Figures 29 and 30.

We see in Figure 29 the consequences of increasing the scattering amplitude of atomic scatterers described by the t-matrix [Equation (40b)]

$$t_{\mathbf{R}}(\mathbf{k}', \mathbf{k}, E) = -\frac{2\pi h^2}{m} \left[\frac{\exp(2i\delta) - 1}{2i |k(E)|} \right] \tag{114}$$

in which

$$k(E) = \{2m [E + \Sigma(E)]/\hbar^2\}^{1/2} \tag{115}$$

in accordance with Equation (56). For small values of δ (i.e., $\delta \lesssim \pi/5$ for the other parameters use to construct Figure 29), the dynamical theory produces a simple kinematic-like peak in the intensity profile, although this peak may be shifted from the kinematical Bragg energy [Equation (62)]. For larger values of δ, we see in the figure a more complex structure, consisting of multiple overlapping peaks. However, this structure occurs for incident-electron energies in the vicinity of the kinematical primary Bragg energy.

The important feature of the complicated multiple-peak structure occurring for strong-atomic scatterers is its dependence on the parameters characterizing the incoming beam of electrons (Figure 14). No matter what detailed shape the structure near the maximum in the diffracted intensity assumes, this maximum always occurs for values of the beam parameters near those characteristic of the kinematical Bragg condition. This fact is illustrated explicitly in Figure 30. In this figure we show the dependence of this structure on the polar angle of incidence, θ, of the electron. Similar results are obtained for other azimuthal directions (ψ) of the incident beam. We see that although the detailed shape of the structure does indeed depend on θ, its energy centroid follows the predictions of the kinematical model for the Bragg peaks as θ is increased.

Evidently, the energy centroid of the multiple-scattering peaks does not occur at exactly the Bragg energies for a variety of reasons discussed by Tucker and Duke. [72] However, the general behavior expected of a large block of ELEED intensity profiles taken at different (θ, ψ) combinations is clear. Complicated fine structure should occur at all energies, but should become most prominent in the vicinity of the kinematical Bragg energies. Thus the intensity profiles may be visualized as a dense array of randomly positioned individual multiple-scattering peaks modulated by a kinematical 'Bragg envelope' characteristic of the geometry of scatterers in the uppermost layers of the target. This concept is intrinsically appealing because it permits the systematic

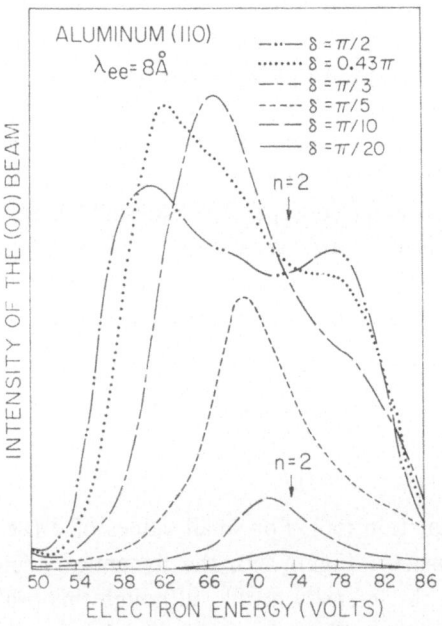

Fig. 29.   Calculated intensity of the (00) diffraction beam with normal incidence on the aluminum (110) surface for various phase shifts from $\pi/20$ to $\pi/2$. The calculations were performed for electron energies in the vicinity of the $n = 2$ Bragg peak position with $\lambda_{ee} = 8$ Å.
(After Tucker and Duke [72].)

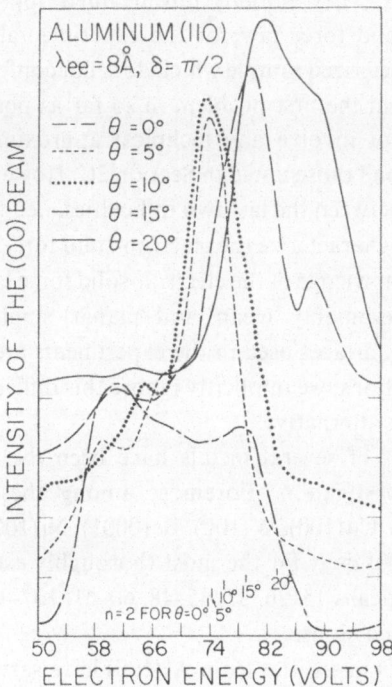

Fig. 30. Calculated intensity of the (00) diffracted beam from the aluminum (110) surface as the angle of inclination is increased in 5° steps from normal incidence. The short vertical lines near the bottom of the figure mark the kinematical position of the $n = 2$ Bragg peak to be associated with the various curves. The phase shift is $\pi/2$ and λ_{ee} is 8 Å for all curves. The azimuth is along the short axis of the rectangular unit cell. (After Tucker and Duke [72].)

interpretation of a common general feature exhibited by a great mass of otherwise intractable ELEED intensity data [6, 72]. Of more importance to us, however, is the fact that this Bragg-envelope concept underlies a variety of data-averaging procedures used in surface crystallography. [97] We return to this topic in the next lecture [Part IV of these notes].

D. DYNAMICAL MODELS OF ELEED: COMPARISON OF THEORETICAL PREDICTIONS WITH EXPERIMENTAL DATA

1 *Historical Introduction*

In the preceding sections we have established two central results. First, the electron-solid interaction is strong. Second, certain features of this interaction are poorly characterized (e.g., inelastic-collision damping caused by surface-plasmon excitation). In this section we try to achieve some perspective on current models of ELEED by examining the sensitivity of their predictions to the various features of the assumed electron-solid interaction, and by comparing these predictions to some of the existing experimental data. The model calculations can fail to describe the observations adequately for three reasons. They embody unwarranted approximations; they utilize an inaccurate electron-solid force law; or the experimental measurements are performed on a poorly-characterized sample which does not conform to the experimenter's conception of it. We avoid the first problem in so far as possible by discussing only calculations which do not involve any technical approximations (e.g., the use of perturbation theory) beyond those noted in Section 2C. However, we have no adequate means of distinguishing between the last two difficulties, i.e., between 'theoretical' and 'experimental' failures to characterize the electron-solid force law. We try to minimize the uncertainties in our knowledge of the electron-solid force law by examining electron diffraction only from presumably 'clean' and 'planar' single-crystal metal surfaces. Undoubtedly the sample surfaces used in the experiments are neither. In most of our discussions of the calculations, we implicitly ignore this difficulty, not from ignorance, but from lack of a viable alternative.

The close-packed faces of several metals have been the subject of relatively extensive experimental investigation. Foremost among these are [6, 56] Ag(111), Al(100), Al(110), Al(111), Cu(100), W(100), Be(0001), Ni(100), Ni(110), and Ni(111). Of these, however, Al(100) is by far the most thoroughly examined both experimentally [98–106] and theoretically [5, 26, 37, 45–48, 60–61, 107–110] Thus we concentrate on this system in our discussions.

Calculations of ELEED intensities from Al(100) have served as illustrations of the consequences of dynamical models of electron diffraction for over five years. Listed in chronological order, such calculations have been given by Hofmann and Smith, [98, 99] Marcus et al. [100] Duke and Tucker, [26, 37, 72] Hoffstein and Boudreaux, [110] Duke et al [47, 48, 66] Tong and Rhodin, [46] Laramore et al. [60] Jepsen et al. [61] Laramore and Duke [5] and Hirabayashi. [108, 109]

The calculations of Hoffmann and Smith [98, 99] and of Hoffstein and Boudreaux [110] were performed using the 'wave-function-matching' method in which the wave functions inside the solid first are calculated using a pseudopotential and then are joined to wave functions outside the solid satisfying suitable boundary conditions. Thus a calculation consists of two steps: the evaluation of the pseudopotential and the subsequent imposition of the electron-diffraction boundary conditions on the wave functions. The various calculations differ both in the selection of the pseudopotential (which is evaluated for a bulk rigid lattice) and the technical details of the wave-function joining procedure at the solid-vacuum interface.

Hofmann and Smith [98, 99] used a crude pseudopotential, technical approximations of unspecified accuracy, and the local-complex-potential model [Equation (29)] of inelastic-collision damping. They were among the first to recognize the need for a substantial amount of (empirically-determined) inelastic-collision damping, and achieved a tolerable description of the limited data (normal incidence, $20\,\text{eV} \lesssim E \lesssim 70\,\text{eV}$, (00), (01), (02) beams) available to them using $\text{Im}\,[V(E)] = 2.5$ eV in Equation (29).

A major difficulty with any model analysis of modest amounts of experimental data lies in the fact that most calculations contain enough parameters (either explicit or hidden) that they always can be brought into qualitative agreement with the data. This result is a consequence of the 'Bragg-envelope' principle discussed in the previous section. All models and most data exhibit prominent structure near the kinematical primary Bragg energies appropriately shifted by the 'inner-potential' $\text{Re}\,[V(E)]$ in Equation (29). Therefore although the comparison of the predictions of these calculations with experimental data over narrow ranges of incident energy, angles, and beam indices usually appears to be satisfactory, this result is not an adequate test of a microscopic model because such qualitative agreement between model predictions and experimental data can be achieved for a variety of quite different models. [26, 60]

Recognizing the ambiguities inherent in the analysis of small amounts of data using models containing adjustable parameters, the more recent model calculations (i.e., later than those of Hofman and Smith) are based on an analysis of larger blocks of data, a specification of all model parameters *a priori*, or both. The former approach became feasible for aluminum upon the publication of an extensive experimental study by Jona [101] of ELEED intensity profiles from the (100), (110), and (111) faces. The calculations of both Marcus *et al.* [100] and Hoffstein and Boudreaux [110] constitute attempts to refine the model for the short-range electron-ion-core potential, neglecting any inelastic-collision damping. A comparison of the results of calculations by Hoffstein and Boudreaux [110] with various experimental data [101, 111] for the intensities of electrons specularly reflected from Al(100) is shown in Figure 31. It is evident from the figure that although some correspondence exists between clusters of peaks in the model predictions and maxima in the experimental data, models in which inelastic-collision damping is neglected predict intensities which are several orders of magnitude too large and excessive fine structure in the intensity profiles.

From the upper panel of Figure 31 we see, however, that the insertion into the model of inelastic-collision damping of the incident electron's elastic wave field both reduces the magnitude of the predicted scattered intensities, eliminates the unobserved fine structure and produces peak-widths of the correct order of magnitude. Indeed, from Figure 31 we might infer that the prominent maxima in the predictions of the inelastic-collision model are comprised of clusters of maxima predicted by Hoffstein and Boudreaux's static-potential model. The sense in which such an inference is warrented was described in Figures 29 and 30, and the discussion thereof. It subsequently has been verified in the context of still another model by Jepsen *et al.* (see Figure 3 of [61].

The calculations shown in Figure 31 are based on models which describe well either the electron-ion-core potential (lower panel) or inelastic-collision damping (upper panel) but not both. It seems clear from the figure that these models, although of historical interest, require further refinement. Therefore, we next inquire into what are the ingredients of an adequate model of the electron-solid force law, and how distinguishable are the ELEED intensities predicted by different models thereof. The various models differ from each other in numerous ways. We organize the remainder of our discussion by examining, in turn, the selection of the electron-ion-core potential and the description of electron-scattering from it; the consequences of using different boundary conditions at the vacuum-solid interface [Section 2C]; the inclusion of a description of lattice vibrations in the model; the sensitivity of the predicted ELEED intensities to the choice of the electron-ion-core potential in the case that the actual solid is presumed to be a truncated but otherwise ideal bulk solid; and, finally, the consequences on the ELEED intensities of specifically surface phenomena. Although

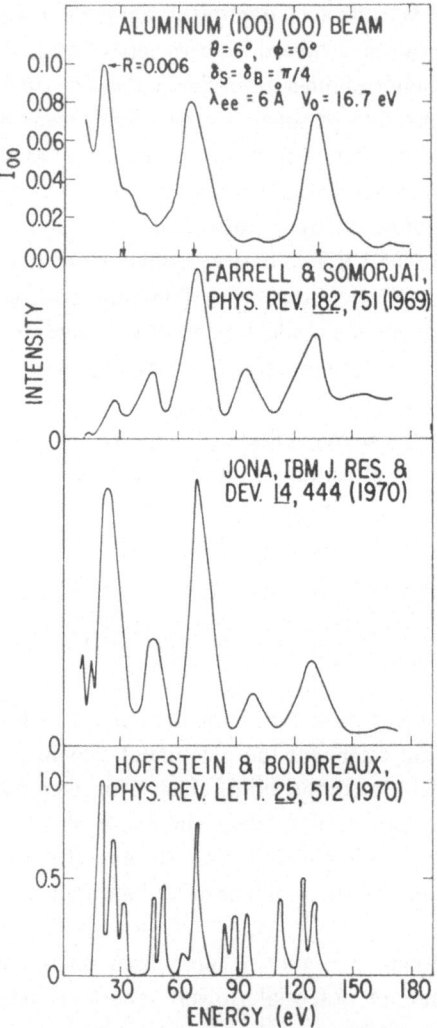

Fig. 31. Comparison of experimental and calculated intensity profiles for the (00) beam of electrons scattered from Al(001) for a 6° angle of incidence in the plane determined by the normal to the surface and a vector between two nearest-neighbor surface atoms. The lower three panels are Figure 2 in [110]. In the upper panel is shown the intensity profile calculated using the matrix inversion analysis of the isotropic-scatterer inelastic-collision model for the parameters indicated in the figure. The azimuthal angle φ is measured relative to the primitive unit cell. Arrows designate the energies of the kinematical primary Bragg maxima. The magnitude of the maximum reflection coefficient (i.e., 0.6%) is indicated in the figure for comparison with the calculations of Hoffstein and Boudreaux shown in the bottom panel. (After Duke and Tucker [26].)

some comparisons of the model predictions with experimental data will be made during the course of these discussions, we conclude this section with a summary of the model descriptions of ELEED intensities from the low-index faces of aluminum, and a synopsis of our main conclusions.

2. The Electron-Ion-Core Potential

We recall from part II of the lecture notes that the model evaluation of ELEED intensities can be divided into two steps: selection of a model electron-solid interaction and its subsequent utilization in an appropriate theory of ELEED. In this subsection we deal with two aspects of the first step. First we survey the nature and construction of model electron-ion-core potentials which have been used to describe ELEED from aluminum. Then we examine how many phase shifts must be used to describe elastic scattering from these potentials.

The most recent calculations [5, 60, 61] have been based on a model embodying both inelastic-collision damping [via Equations (29) or (36) and (113)] and variants of the muffin-tin electron-ion-core potential obtained by Snow [112] for *bulk* aluminum. Jepsen *et al.* [61] used Snow's potential with an arbitrary rigid shift of the energy zero chosen to maximize agreement between their calculated ELEED intensities and Jona's data. Laramore and Duke, [5, 60] however, always require that the electronic inner potential used in the propagator renormalization, i.e., $V(E)$ in Equation (113), be equal to the constant part of the muffin-tin potential between the ion-cores. Any other convention causes an internal inconsistency in the propagator renormalization defined via Equation (23). They achieved this result by using Snow's 'corrected' value of $V_0 = -1.225$ ryd, defined by Equation (113c), between the ion-cores, but his potential as specified in Table I of [112] inside the ion-cores. This choice also makes $V_0 = \zeta + \Phi$ where ζ is the Fermi-energy of aluminum and Φ is its work function. The distinction between these two 'Snow' potentials is conceptual as well as procedural in nature, because they represent opposite treatments of the spatial dependence of the exchange-correlation contributions to the 'effective' potential acting on a high-energy electron. The Laramore-Duke model is based on the supposition that the polarization of the valence electrons by the incident electron occurs primarily in the regions between the muffin-tin ion cores, whereas the Marcus-Jepsen model is based on the supposition that this polarization is spatially uniform. The Laramore-Duke procedure corresponds directly to Snow's APW method of including exchange-correlation contributions to the one-electron energies because, as noted in Section 2C, Snow and others use G_{00}, (given by Equation (24) in which $V_0 \equiv 0$), rather than G_0 (given by (23)) in the APW Equations analogous to Equation (40b) for the electronic motion inside the ion-core radius. In Snow's calculations, [113] correlation contributions to the energy are neglected; exchange inside the core-radius is described by Slater's (energy-independent) free-electron local-density approximation; and exchange outside this core radius is treated by shifting the value of the constant part of the muffin-tin potential by the exchange energy of a uniform free-electron fluid. Only this latter term can readily be made energy dependent by use of a propagator renormalization in the region between ion cores, in which case the value of the exchange-correlation contributions to the constant muffin-tin potential depends naturally on the state of the incident electron. Hence, in this model, the spatial dependence of the muffin-tin potential depends explicitly on the energy of the incident electron if a model embodying an energy-dependent $V(E)$ in Equation (113a) is used. Laramore and Duke chose to set $V(E)$ equal to its value at the Fermi energy in order to avoid this additional complication in their initial calculations because the functional form of $V(E)$ is not well-known for actual metallic densities.

A third electron-ion-core potential was used by Tong and Rhodin [46] in their analysis of ELEED from Al(100). They attribute the construction of this potential to Pendry and Capart who, by an unspecified technique, evaluated the potential explicitly 'for the LEED energy region'. As we shall see later, Laramore et al. [60] demonstrated that the ELEED intensities calculated using the Tong-Rhodin potential are essentially indistinguishable from those calculated using Laramore and Duke's version of Snow's potential.

The consequences of all three electron-ion-core potentials on electron-solid scattering are incorporated into models of the scattering process via the intermediary of

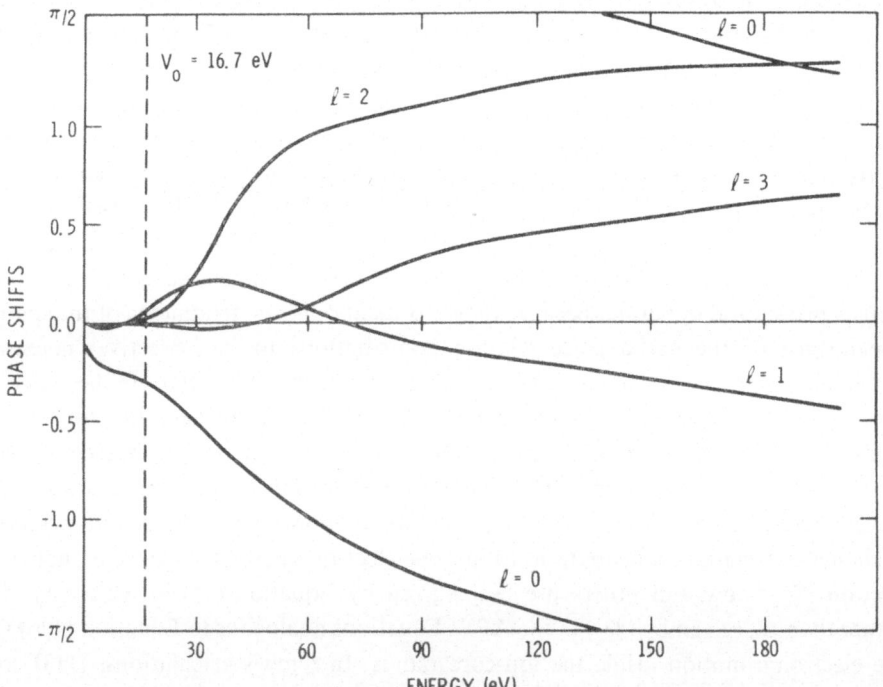

Fig. 32. Phase shifts obtained from the APW potential of Snow. [112] The energy scale is measured relative to the constant value of the potential between the muffin-tin potentials. This zero level is 16.7 eV below the vacuum level and so a zero-energy electron outside the crystal will have $V_0 = 16.7$ eV energy when it has entered the solid. The dotted line at 16.7 eV indicates the zero level for external electron energies measured relative to the vacuum. (After Laramore and Duke [5].)

electron-ion-core phase shifts. The differential elastic-scattering cross section, $d\sigma/d\Omega$, for electron scattering from an individual ion core is specified in terms of these phase shifts by the familiar formulae [114]

$$\frac{d\sigma}{d\Omega} = |f(\theta)|^2 \tag{116a}$$

$$f(\theta) = [2ik(E)]^{-1} \sum_l (2l+1) [e^{2i\delta_l} - 1] P_l(\cos\theta) \tag{116b}$$

in which the $\delta_l(E)$ are the energy-dependent phase shifts in the lth partial wave and $k(E)$ is defined by Equation (115). As an example, the phase shifts associated with the Laramore-Duke-Snow potential are shown in Figure 32. Those characterizing the Tong-Rhodin-Pendry potential are essentially identical, [46, 60] whereas the Marcus-Jepsen-Snow potential exhibits larger s and p wave phase shifts at lower energies ($E \lesssim 60$ eV). [61]

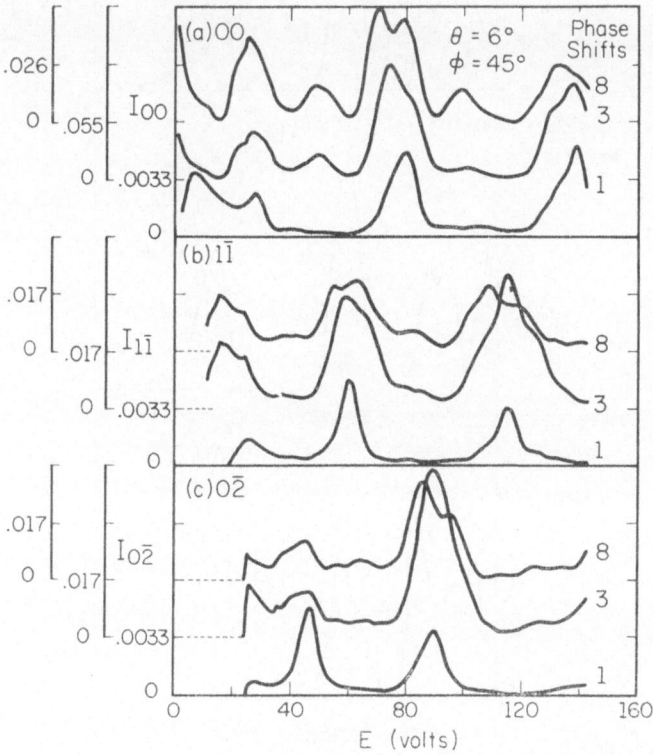

Fig. 33.   Effect on the LEED spectra of Al(001) of using different numbers of phase shifts from the Marcus-Jepsen-Snow potential for Al in the theoretical calculation. The three curves in each box were calculated with one, three, and eight phase shifts as indicated. (a) 00 beam, (b) $1\bar{1}$ beam, and (c) $0\bar{2}$ beam. For all spectra $\theta = 6°$, $\varphi = 45°$, see [101]. Note that 0_1, 0_3, 0_8 on the left refer to the origins of the three scales of intensities for the three curves in each box. The theoretical spectra, calculated with zero real potential between atomic scatterers but a constant imaginary potential of 0.3 Ryd, are shifted by -7.5 eV to align the peaks with Jona's data. [101] (After Jepsen et al. [61].)

Given the use of a phase-shift description of electron-ion-core scattering, the question immediately arises of "how many phase shifts are required to describe this process adequately?" For the purpose of describing ELEED from 'rigid-lattice' aluminum for $E \lesssim 150$ eV, three seem to suffice. The effect of introducing higher phase shifts is a slight modification of the fine structure in the ELEED intensities as shown in Figure 33. Comparable and larger modifications easily can be caused by the uncertainties in the electron-solid force law described in Section 2B.

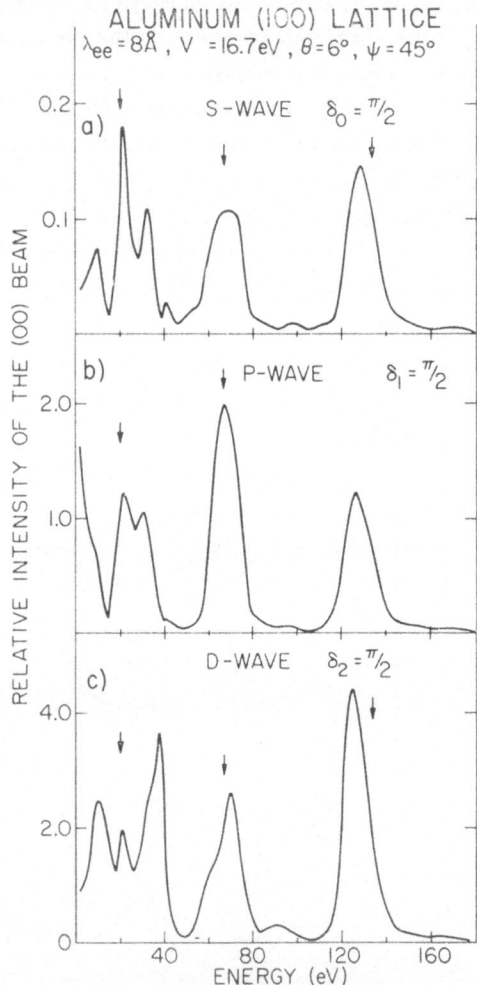

Fig. 34. Effect of higher partial waves upon the calculated intensity profiles. The panels of the figure show intensity profiles calculated for s, p, and d wave scatterers placed in an Al(100) lattice structure. The strength of the scatterers is at the unitary limit in each case. The polar angle is denoted by θ and the azimuthal angle by ψ. ψ is measured relative to the non-primitive cubic unit cell axes. The parameters used in the calculation are shown in the figure. The arrows mark the kinematical positions of the Bragg peaks for an inner potential of 16.7 eV. The scales on the ordinates indicate the relative values of the scattering cross section. (After Laramore and Duke [5].)

Finally, it seems relevant to note that the occurrence of large higher-order phase shifts does not invalidate the 'Bragg-envelope concept' introduced in Section 3C. This fact is illustrated by Figure 34. It is an important result for surface crystallography because it indicates that even for strong electron-ion-core scattering the geometrical arrangement of the scatterers exerts a dominant influence on the systematic features (although not the finer details) of the ELEED intensity profiles.

3. Boundary Conditions

In Section 2C we described the application of various boundary conditions at the solid-vacuum interface used in models of electron-solid scattering. The consequences of using these boundary conditions has been examined by Jepsen et al. [61] Some of

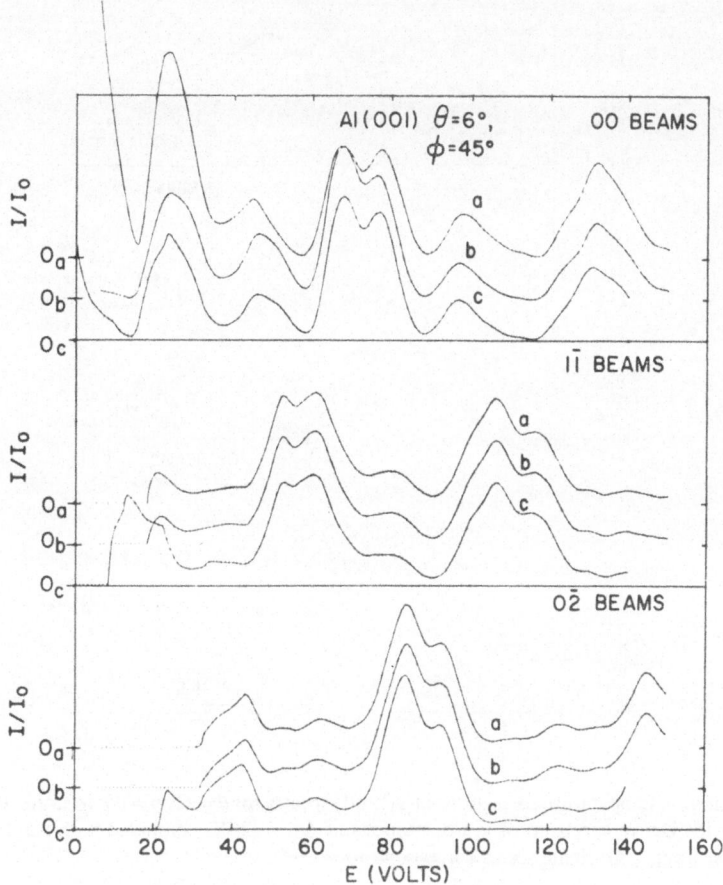

Fig. 35. Effects of different matching procedures on calculated Al(001) spectra at $\theta = 6°$, $\varphi = 45°$, $\mathrm{Im}[V(E)] = 4.1$ eV for (00), (11), (02) beams. Plots a: abrupt step at surface in real and imaginary part of potential $-7.5 - 4.1$ eV. Plots b: no reflection matching (the inelastic-collision model). Plots c: $\mathrm{Im}[V(E)]$ reflection matching (abrupt step only in imaginary part of potential, spectrum calculated and shifted -7.5 eV). (After Jepsen et al. [61].)

their results are shown in Figure 35. Evidently, the use of different boundary conditions usually introduces rather small alterations in the predicted ELEED intensities. Only for rather low energies, $E \lesssim 50$ eV, in the specular beam are pronounced changes in the intensity profiles caused by varying the boundary conditions.

4. *Lattice Vibrations*

The incorporation of a description of lattice vibrations into models of electron-solid scattering was discussed in Section 3C in association with Equations (41). These

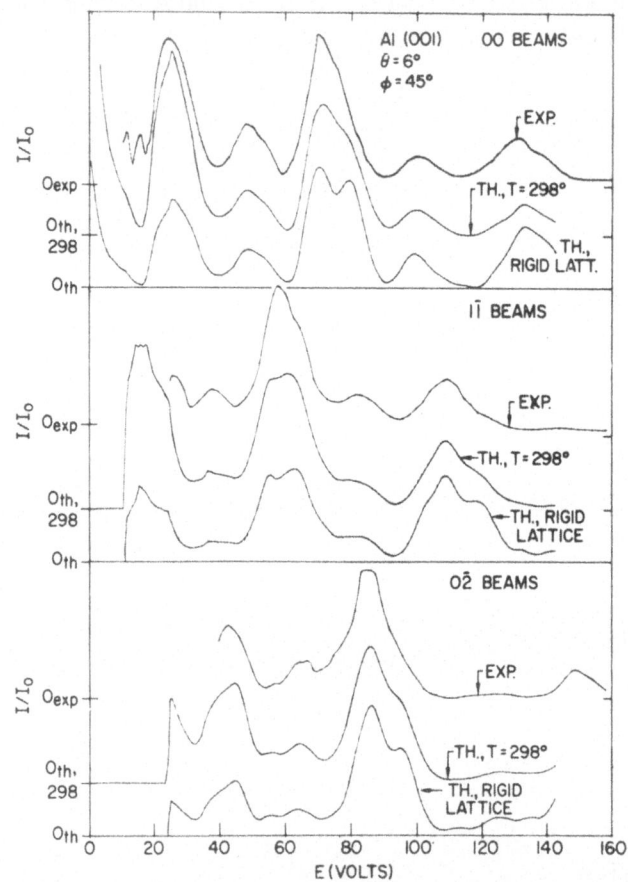

Fig. 36.   Effect of temperature corrections on Al(001) spectra for $\theta = 6°$, $\varphi = 45°$; spectra at $T = 298$ K computed with Debye spectrum of lattice motions at $\Theta_D = 426$ K, compared with rigid lattice and experimental spectra. Ordinate scales correspond to:

beams	peak energy	(I/I_0) rigid lattice	(I/I_0) $T = 298$ K
00	70.5	0.053	0.0190
1Ī	62.5	0.020	0.0092
0Ī	86.5	0.028	0.0085

(After Jepsen *et al.* [61]).

equations describe the model used in all vibrating-lattice model calculations performed thus far. The effects on the ELEED intensities of this extension of the model are illustrated in Figure 36 for the case in which the bulk and surface atoms are taken to exhibit *idential* vibrational motion and electronic structure. Evidently, the lattice vibrations create large changes in both the magnitude and shape of the intensity profiles. Moreover, in examining Figure 36 we should recall that three vibrating-lattice phenomena have not been incorporated into the model calculation: virtual-phonon vertex corrections, [2] the finite energy and angular resolution of the spectrometer [Section 3B]; and the inequivalence of the vibrational motion of the surface and bulk scatterers. [2, 6, 45, 47] A consideration of any of these introduces still further deviations between the rigid- and vibrating-lattice models. [6, 56]

5. *Consequences of Different 'Bulk' Potentials*

Having discussed in Section 3D2 the construction of three different potentials to describe ELEED from aluminum, it seems appropriate to indicate the extent to which they lead to differences in the ELEED intensities which may be distinguished by comparison with the available experimental data. A comparison of some of the predictions of the Tong-Rhodin-Pendry and Laramore-Duke-Snow potentials is shown in Figure 37. Both calculations were performed using a model in which only three (s, p, d) phase shifts were used to describe the electron-ion-core scattering. The important conclusion which I draw from this figure is that the two model potentials

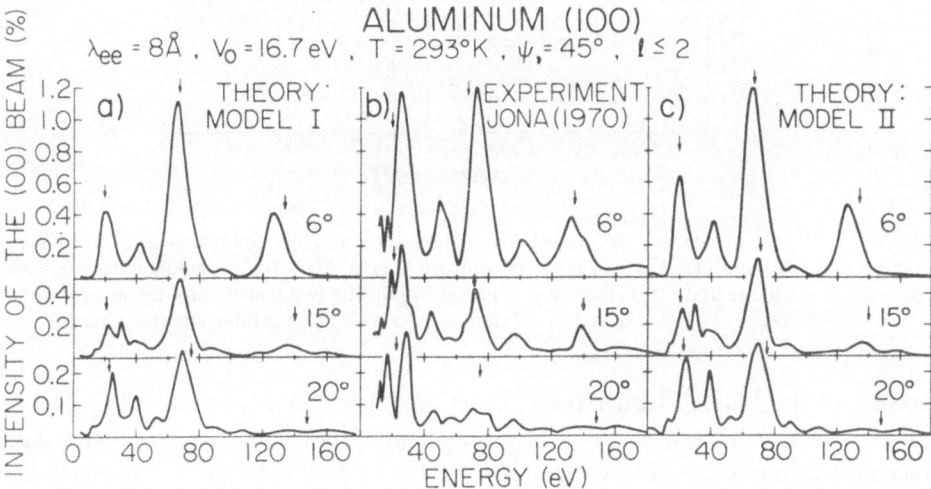

Fig. 37. Comparison between model predictions and experimental data for the specular beam of electrons scattered from Al(100). The parameters describing the incident beam and model calculations are noted in the figure. Both 'theory' calculations were performed using the matrix-inversion analysis described in [60] and a Debye temperature of $\Theta_D = 426$ K for aluminum. Solid arrows indicate the kinematical Bragg energies for an inner potential of $V_0 = 16.7$ eV. The experimental data are those of Jona [101]. Model I refers to the Tong-Rhodin-Pendry potential and Model II to the Laramore-Duke-Snow potential. (After Laramore *et al.* [60].)

yield ELEED intensities which resemble each other more closely than either resembles Jona's data.

A superficial improvement in the comparison of the model calculations with Jona's data can be achieved by expanding the horizontal (energy) scale and contracting the vertical (intensity) scale. This is illustrated by Figure 38 in which a more extensive comparison between the predictions of the Laramore-Duke-Snow model and Jona's data is presented. The results of comparable calculations based on a rigid-lattice

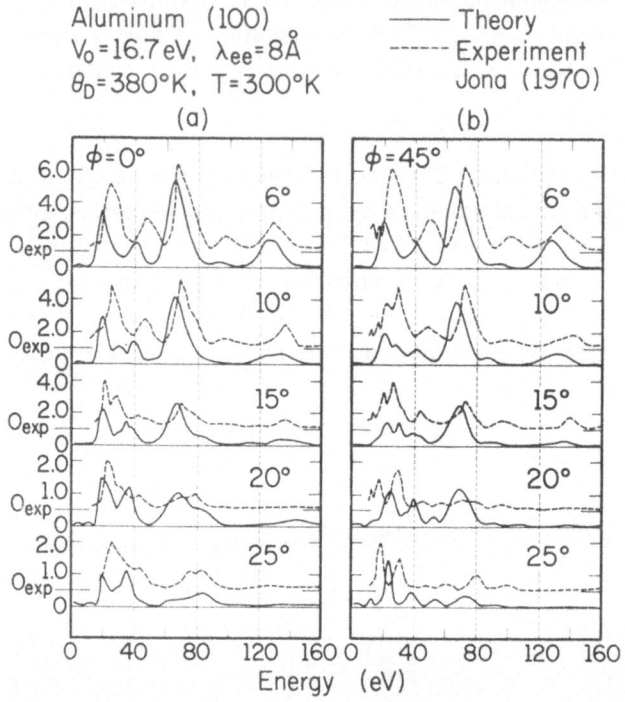

Fig. 38. Comparison between the predictions of the Laramore-Duke-Snow model [5] and Jona's data [101] for the specular beam of electrons diffracted from Al(100). The model parameters are indicated in the figure and the method of calculation in [5]. The two panels show the intensities for an incident beam along a diagonal ($\varphi = 0$) and side ($\varphi = 45°$) of a primitive unit surface cell.

version of the Marcus-Jepsen-Snow model are shown in Figure 39. The peaks in these calculated intensities are in better correspondence with those in Jona's data by construction: an effective value of $\mathrm{Re}\, V(E) = -7.5$ eV is assigned in the inelastic-collision boundary condition. This value is manifestly inconsistent with that of $V_c = -12.4$ eV used as the constant muffin-tin potential between the ion-core potentials and the value of zero used in their evaluation of the G_0 propagators defined in Equation (23). Shifting the Laramore-Duke-Snow intensities by 6–7 eV to higher energies [i.e., using $V_0 \cong 10$ eV in Equation (113c)] brings their predicted intensity maxima also into essentially exact correspondence with those of Jona. Thus we see

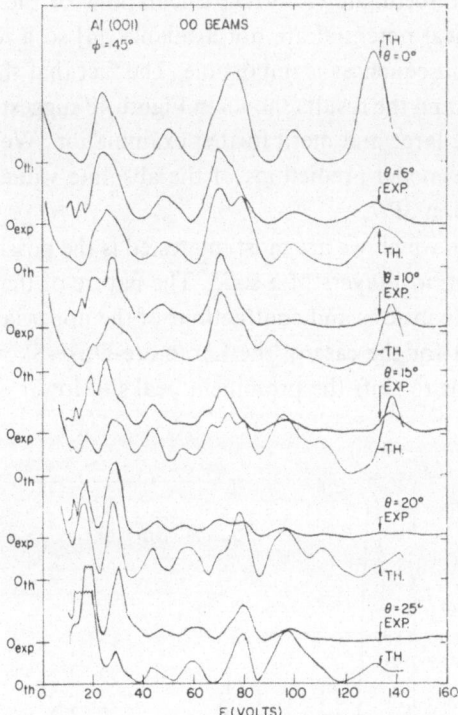

Fig. 39. Al(100), (00) intensity profiles measured and calculated, as functions of polar angle θ at $\varphi = 45°$. Calculated with Marcus-Jepsen-Snow potential and absorption ($\text{Im}[V(E)] = \beta = 4.1 \text{ V} = 0.30$ ryd., step of $-i\beta$ in potential at surface, spectra shifted -7.5 eV. Zero levels of experimental and theoretical curves marked for each angle. (After Jepsen et al. [61].)

again that two model potentials cannot be distinguished on the basis of comparison with Jona's data.

6. Specifically Surface Effects

We now come to the final topic in our discussion of the sensitivity of the predicted ELEED intensities to the various ingredients of the model electron-solid force law: How large are the consequences of specifically surface phenomena like lattice expansion (contraction), $\langle u_S^2 \rangle \cong 2 \langle u_B^2 \rangle$, and surface-plasmon-induced inelastic-collision damping? Unfortunately, the changes in the ELEED intensities due to any one of these phenomena depends on the magnitude of *all* the parameters in the model force-law, not just on the particular ones under consideration. Therefore, we proceed by giving estimates of the size, nature, and interdependence of the consequences of the various effects.

Of all the possibilities, the consequences of surface-plasmon losses on the ELEED intensities is the least well-understood. The effect of using model optical potentials of different shapes can be substantial [115] as indicated in Figure 40. A similar sensitivity to the shape of the imaginary (absorptive) part of the optical potential has been

documented. [116] Unfortunately, correct calculations of the surface-plasmon contributions to the optical potential are not available, [4] so a reliable estimate of the
magnitude of their consequences is impossible. The fact that these contributions lead
to the image force [4] and the results shown in Figure 40 suggest that surface-plasmon-
induced effects can be large and merit further examination. We return to this topic in
our discussion of the model predictions of the absolute values of LEED intensities
from Al(100) in Section 3D7.

The surface effect in which we are most interested is the possibility of altered lattice
geometry in the uppermost layers of a solid. The nature of the effect on the ELEED
intensity profiles of expansions and contractions of the upper layer spacing of Al(100)
is shown in Figure 41 for the case of the Laramore-Duke-Snow potential. Evidently
lattice expansion tends to shift the prominent peaks to lower energy and lattice con-

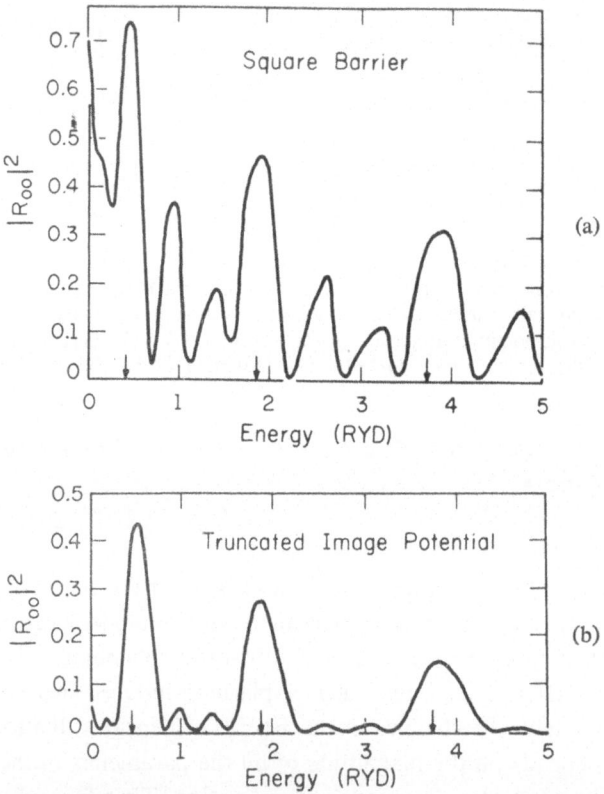

Fig. 40.   The reflection coefficient of the (00) beam scattered from a one-dimensional potential
consisting of four delta function 'ion-core' potentials inside a step-well of height 10 eV [Panel (a)]
and a truncated image potential of this height [Panel (b)]. The finite number of delta-function 'ion-
cores' rather than a complex potential is used to simulate inelastic-collision damping. The lattice
parameters were chosen to simulate $W(100)$. The calculation is described in Ref. 115. The arrows
designate the energies of the kinematical primary Bragg peaks for a semi-infinite medium. (After
Gerstner and Cutler [115].)

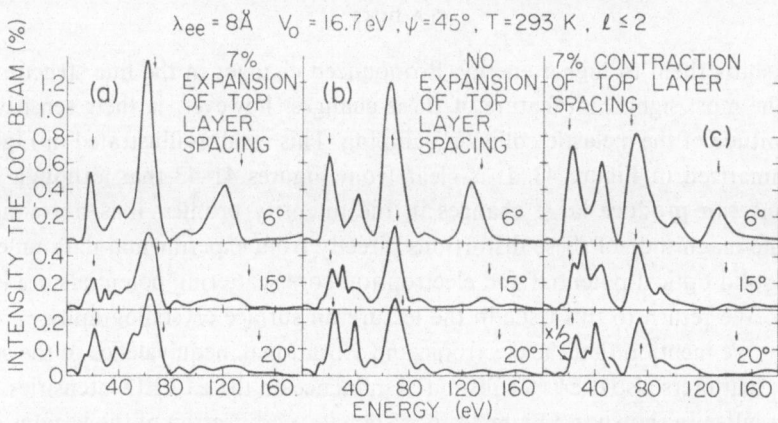

Fig. 41. Effect of a distortion of the upper layer spacing on the absolute reflectivity of the (00) beam from Al(100). The calculations are based on the Laramore-Duke-Snow model using the parameters indicated in the figure. The arrows mark the kinematical positions of the Bragg peaks using $V_0 = 16.7$ eV. (After Laramore and Duke [5].)

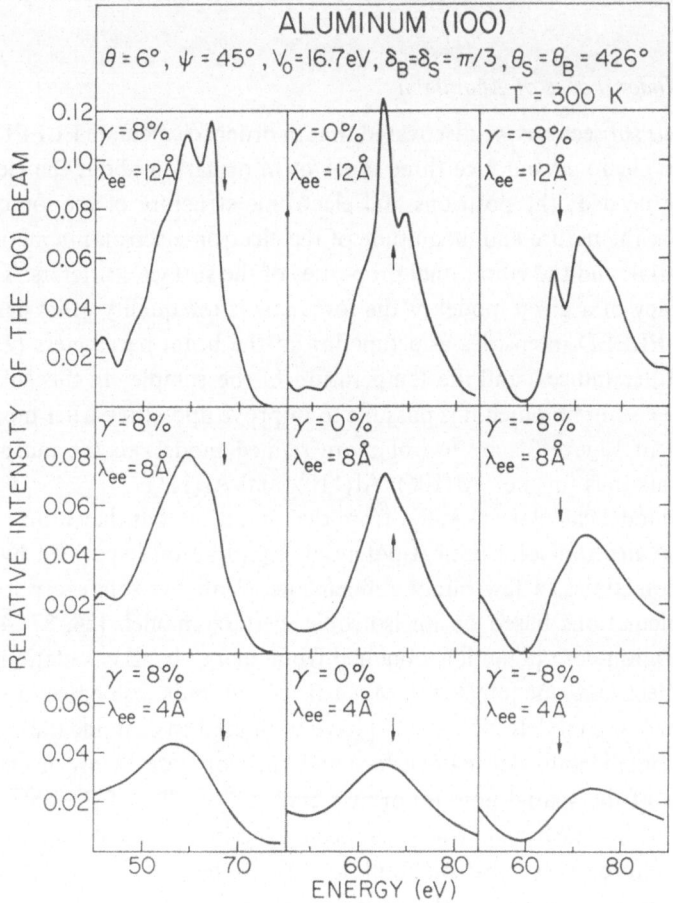

Fig. 42. Illustration of the dependence of the energy profile on the electron mean-free path ($\lambda_{ee} = 2\lambda_{mfp}$) for expanded ($\gamma > 0$), unexpanded ($\gamma = 0$) and contracted ($\gamma < 0$) lattices. The spacing of the 'bulk' layers is taken to be $d = 2.0201$ Å and that of the upper two layers is taken to be $d' = d(1 + \gamma)$. The calculations were performed using the double-diffraction analysis of the finite-temperature isotropic-scatterer inelastic-collision model [48] for the parameters shown in the figure. The arrows mark the kinematical positions of the Bragg peaks for an undistorted lattice. The azimuthal angle, ψ, is measured relative to the non-primitive cubic unit cell. (After Duke et al. [48].)

traction shifts them to higher energy. Pronounced changes in the fine structure also occur. The most significant feature of these changes, however, is their sensitivity to the magnitude of the inelastic-collision damping. This result is illustrated in Figure 42 and summarized in Figure 43. It is clear from Figures 41–43 that although lattice distortions can produce large changes in the intensity profiles, it is not simple to extract the magnitude of these distortions directly from experimental data unless the electron solid optical potential and electron-ion-core scattering potentials are known precisely. We return to this issue in the lecture on surface crystallography.

Finally, we mention that the electronic and vibrational inequivalence of the surface and bulk scatterers also exerts a substantial influence on the ELEED intensities. Some typical results are shown in Figure 44. A systematic codification of the various effects is complicated, and may be found in papers by Duke and coworkers. [6, 48, 56] For our present purposes, the important aspect of these results is the large magnitude of the modifications of the ELEED intensities caused by these phenomena, and the necessity to distinguish such modifications from those due to alterations in surface geometry.

7. *The Low-Index Faces of Aluminum*

In the previous subsections we discovered that in order to calculate ELEED intensities from a given 'clean' crystal face three types of information about the electron-solid force law are needed: the positions and electronic structure of the ion-cores in the surface region, the nature and magnitude of the electron-electron-interaction-induced optical potential, and the vibrational properties of the surface scatterers. The measure of the adequacy of a given model of the force law is the quality of its description of the absolute ELEED intensities as a function of the beam parameters (E, θ, ψ, and exit-beam Miller indices) and the temperature of the sample. In this subsection we examine the existant to which it is possible to improve upon the earlier model calculations, shown in Figure 31, by use of more refined models in the most-extensively studied systems thus far, i.e., Al(100), Al(110), and Al(111).

All extant model calculations suffer from the limitation that they utilize quite crude models of the electron-solid optical potential, e.g., like that specified by Equations (113) with, sometimes, a few minor refinements. With the sole exception of a few schematic calculations based on an isotropic-scatterer model, [26, 37, 48, 66] they all utilize bulk ion-core potentials obtained from energy-band calculations. I.e., they ignore the electronic inequivalence of surface and bulk ion-cores. Furthermore, primarily schematic models [45, 48, 107] have been used to examine the consequences of the vibrational inequivalence of surface and bulk ion-cores. Only recently have the consequences of the spatial non-uniformity and non-locality of the optical potential and of specifically surface phenomena been examined via calculations based on models using more-or-less 'realistic' bulk ion core potentials.

The initial question posed by theoretical workers was "how well does a model of the solid as a truncated but otherwise ideal bulk material describe the observed ELEED intensities?" We already have seen in Figures 36–39 the extent of the description of

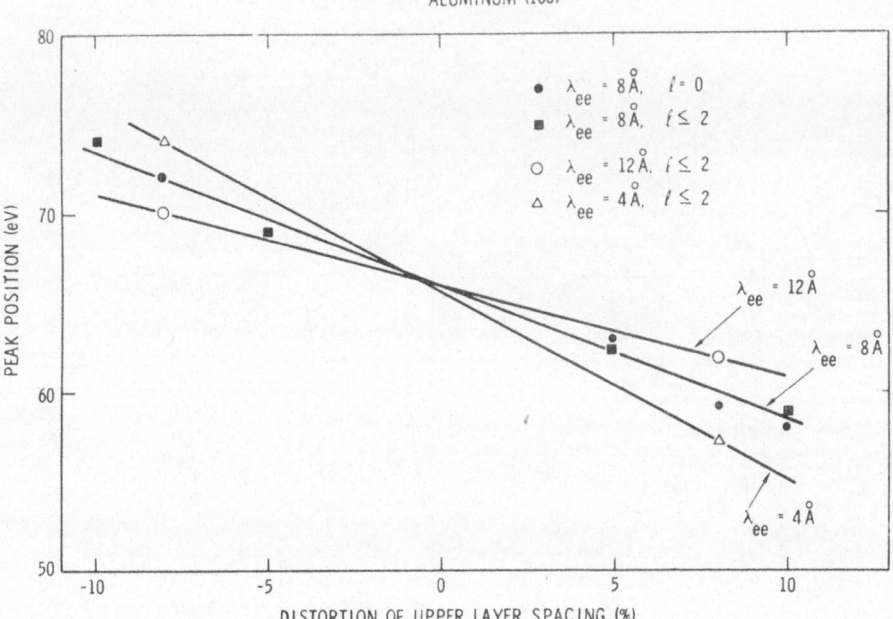

ALUMINUM (100)

Fig. 43. Position of the $n = 3$ 'Bragg' peak of the specular beam from Al(100) as a function of both the distortion of the upper layer spacing and the effective electron mean-free-path ($\lambda_{ee} = 2\lambda_{mfp}$). The parameters used in the calculations were $V_0 = 16.7$ eV, $\theta = 6°$, $\psi = 45°$, $\Theta_D = 426$K with the results of both isotropic-scatterer calculations [48] and calculations involving the higher partial waves predicted by the Laramore-Duke-Snow potential being shown in the figure. The upper layer spacing is expressed in terms of the 'bulk' layer spacing as indicated in the caption to Figure 42. There are points for all three curves at $\gamma = 0$, $E = 66$ eV. For purposes of clarity these points are not shown in the figure. (After Laramore and Duke [5].)

Jona's Al(100) data [101] by the most-refined calculations [5, 60, 61] of this type which have been performed thus far. Generally speaking, the energies of large peaks can be predicted *a priori* to within about ± 7 eV and 'fit' by a model over a wide range of data to within ± 4 eV. In the data surveyed by Laramore and Duke, [5] different experimental measurements of these peak positions differed by as much as ± 9 eV. Therefore, the model calculations displayed in Figures 36–39 agree with Jona's data about as well as Jona's data agree with those of other workers. [106, 111]

Jona [101] did not measure the absolute magnitude of the scattered intensities, but Burkstrand subsequently has. [106] A comparison of the results of various model calculations with Burkstrand's data is presented in Figure 45. This figure illustrates the important result that the predicted absolute intensities depend on two poorly-known features of the electron-solid force law: the optical potential and the vibrational amplitudes of surface scatterers. The absolute magnitude of LEED intensities from Al(100) is reproduced correctly by a constant local optical potential if and only if this potential is taken to extend about one full lattice spacing beyond the geo-

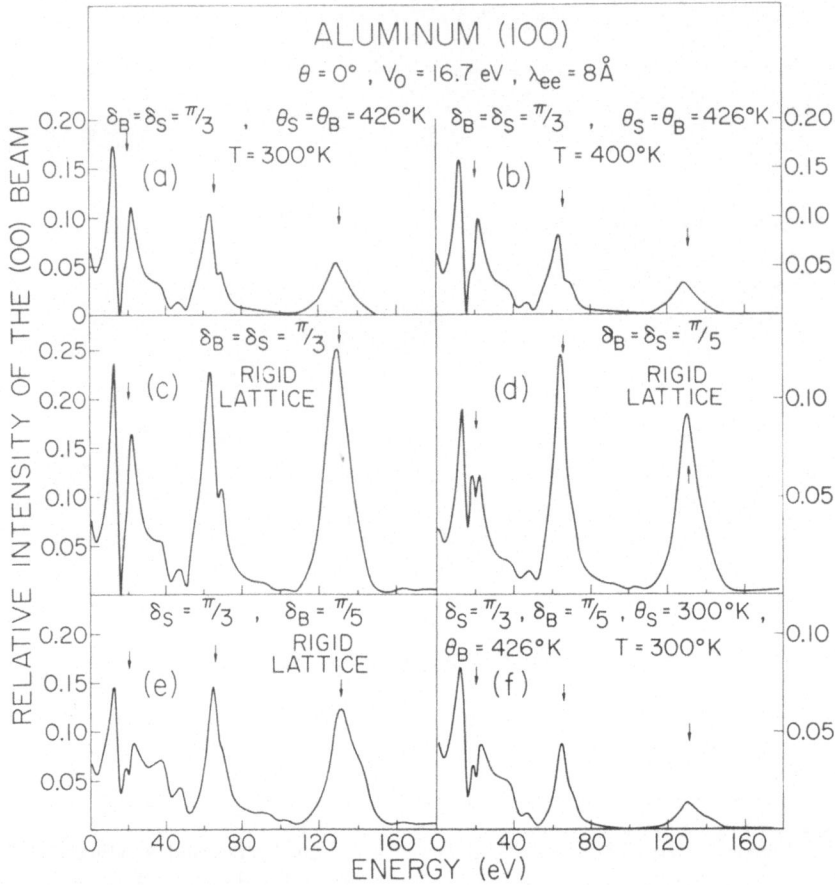

Fig. 44. Effect of the electronic and the vibrational properties of the ion cores on the LEED intensity profiles. The model calculations are made in the double-diffraction approximation [48] and use the isotropic-scatterer inelastic collision model with the parameters indicated in the figure. The parameters of panels (a) and (b) were chosen such as to make the effective scattering cross section of a vibrating ion-core at $E = 50$ eV for a scattering angle of 180° be the same as (a) or less than (b) the corresponding cross section for a rigid ion-core of phase shift $\delta = \pi/5$. The parameters of panel (f) were chosen to make the effective scattering cross section of the vibrating ion-core at an energy $E = 50$ eV for a scattering angle of 180° be the same for the bulk and surface ions and equivalent to the corresponding cross section of a rigid ion-core of phase shift $\delta = \pi/5$. The arrows indicate the kinematical positions of the Bragg peaks. (After Duke et al. [48].)

metrical boundary of a truncated bulk solid (the $d = 3.03$ Å curves in Figure 45). In this case, taking the surface vibrations to be larger than those in the bulk by an amount consistent with the temperature dependence of the large peaks in the intensity profile [107], an acceptable description of the magnitude as well as energies of maxima in the intensity profile is obtained. We interpret this result to be a consequence of strong surface-plasmon induced damping in aluminum.

Another mode of presenting ELEED intensities is provided by 'rotation diagrams' for which the energy (E) and polar angle (θ) of the incident beam are held fixed, while the scattered intensity is plotted as a function of the azimuthal angle (ψ) relative to an axis in the crystal face. Such diagrams presumably provide good measures of the adequacy of a dynamical model because in the kinematical model [Section 3B] they would be horizontal straight lines (i.e., $d\sigma/d\Omega$ in Equations (64) is a function only of E and θ). A comparison of the predictions of the Laramore-Duke-Snow

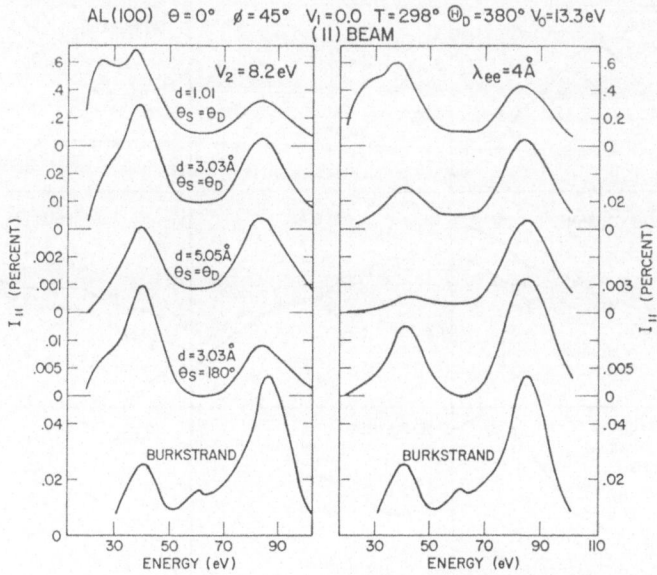

Fig. 45. Description of the absolute intensities of electrons diffracted from Al(100) by several model potentials. The left-hand panels shows results obtained using the optical potential $\Sigma(E) = V_1 - iV_2$ in the inelastic-collision model. The right-hand panel shows results obtained using $\Sigma(E) = V_1 - i\hbar \times \times \{2(E+V_1)/m\}^{1/2}/\lambda_{ee}$. Both optical potentials begin a distance d outside a plane passing through the first row of atoms. The value $d = 1.10$ Å corresponds to a truncated bulk solid. The values of the dynamical parameters are indicated in the figure. An overlapping atomic potential was used to obtain the $l \leqslant 2$ phase shifts employed in the calculation. The measured absolute intensities (bottom panels) were obtained by Burkstrand [106].

model with the data of de Bersuder et al. [103] on Al(100) is shown in Figure 46. The main results evident in the figure are the sensitivity of the model predictions to the value of the energy of the incident electron and the generally adequate description of the data to within the accuracy with which this energy is known. The agreement between the model calculations and experimental data is less satisfactory at larger angles of incidence and lower energies as indicated by Figure 47. The calculations shown in this figure were undertaken to examine the consequences of the boundary conditions. In particular, since the inelastic-collision model boundary conditions do not lead to 'surface state resonances' of the sort discussed by Lauzier et al. [102]

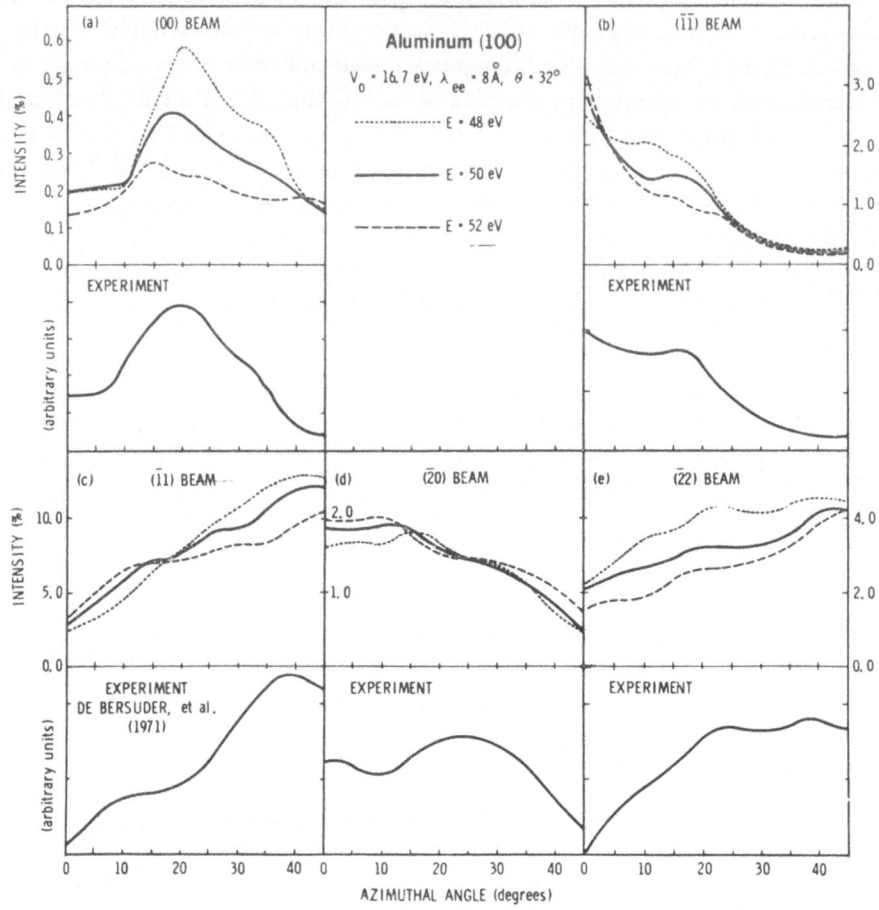

Fig. 46.   Comparison between theoretical and experimental [103] LEED rotation diagrams for Al(100) for an incident angle of 32°. The theoretical curves are shown for $E = 48$ eV (dotted curve), 50 eV (solid curve), and 52 eV (dashed curve) and the experimental curves are shown for $E = 50$ eV. The units of the theoretical curves are in percent reflectivity while the units of the experimental curves are arbitrary. The parameters used in the calculation are shown in the figure. The beams are indexed according to the non-primitive cubic two-dimensional unit cell. The electron-solid force law used in the calculation is that of the Laramore-Duke-Snow model described in the text. (After Laramore [117].)

we see that in contrast to their claims, [102] their data, shown in Figure 47, does not support the necessity of the existence of such states.

Additional calculations for Al(100) have predicted the temperature dependence of the ELEED intensities [45, 47, 48, 107] (including the consequences of vibrationally inequivalent surface and bulk ion-core scatterers) and the effects of various types of distortion of the surface geometry relative to that in the bulk. [5, 48, 97, 118, 119] However, analyses of the temperature dependence of ELEED have been hampered by lack of data and the need to use more than three phase shifts to achieve an adequate description of the data for incident electron energies $E \gtrsim 75$ eV. Discussions of modified surface geometries are inconclusive because, as noted above, the uncertainties in the presently-available data exceed the discrepancies between predictions of the truncated-bulk-solid model and the data themselves.

In addition to those described above for Al(100), fairly extensive model calculations also have been performed for Al(110) [5, 37, 48, 60, 66, 72, 120, 121] and Al(111). [5, 37, 60, 66, 72, 120, 121] The major result emanating from these analyses is the fact that model potentials which describe adequately Jona's data [101] on Al(100), do not provide as satisfactory a description of his data on Al(110). [5, 48, 60, 121] This conclusion is illustrated in Figure 48 for the Tong-Rhodin-Pendry model, although it also is true for the Laramore-Duke-Snow [5] and Marcus-Jepsen-Snow [121] model potentials. The principal difficulty, evident in Figure 48, is the fact that the value of the real part of the optical potential (called the 'inner' potential) which correctly locates the energies of the major peaks on one crystal face fails to do so on the other faces. Although inner-potential changes of the order of 1–2 eV could be rationalized from work-function considerations [122], the observed

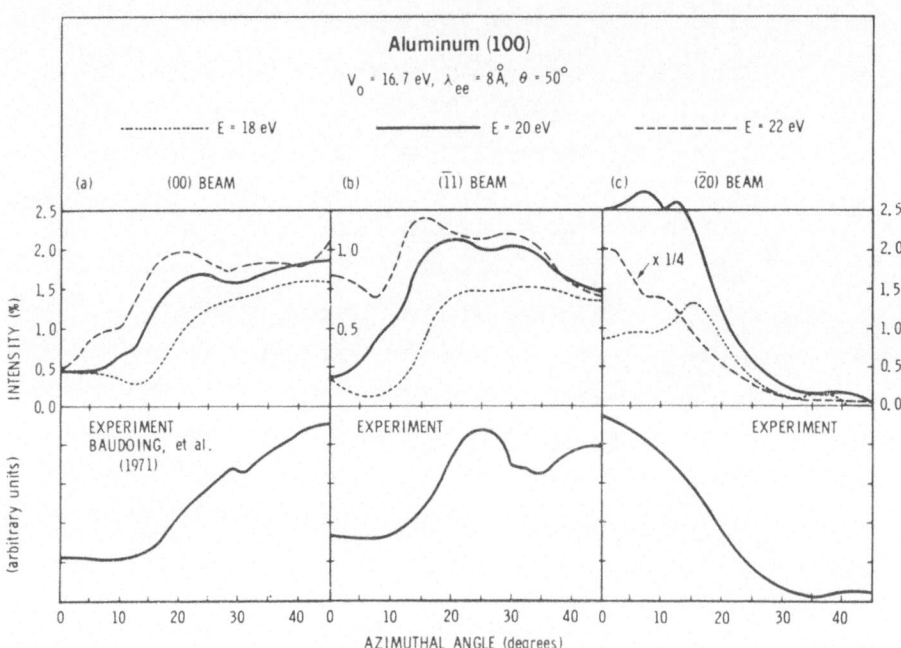

Fig. 47. Comparison between theoretical and experimental [102, 104] LEED rotation diagrams for Al (100) for an incident angle of 50°. The theoretical curves are shown for $E = 18$ eV (dotted curve), 20 eV (solid curve), and 22 eV (dashed curve), and the experimental curves are shown for $E = 20$ eV. The units of the theoretical curves are in percent reflectivity, while the units of the experimental curves are arbitrary. The parameters used in the calculation are shown in the figure. The beams are indexed according to the non-primitive cubic two-dimensional unit cell. The electron-solid force law used in the calculation is the Laramore-Duke-Snow model described in the text. (After Laramore [117].)

discrepancies [5, 48] are of the order of 5–10 eV. Such large shifts in the energies of the maxima in the ELEED intensity profiles must be caused by substantial alterations in the chemical composition or geometry (including typography) of the surface layers. [5, 48]

8. *Synopsis*

It seems appropriate to conclude our discussion of microscopic models with a summary of the major conclusions derived thus far and an indication of future directions. Truth, like beauty, lies in the eyes of the beholder. Workers in the field of ELEED are an individualistic and emotional lot, some of whom register strong (if unsubstantiated) objections to the remarks made below. Thus the reader is duly cautioned

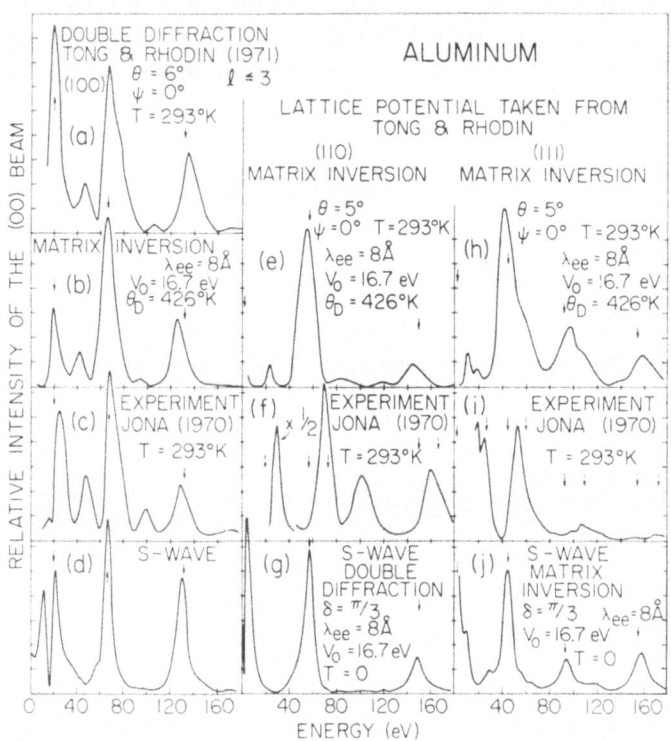

Fig. 48. Comparison between model predictions and experimental data for the specular beam of electrons scattered from Al(100), (110), and (111). The phase shifts for the potential were taken from Tong and Rhodin [46] and the experimental data from Jona. [2] The angle of incidence θ and azimuthal angle ψ (defined in [101]) are noted in the figure, as are the parameters used in the calculation. The lower left-hand panel was calculated using the double-diffraction approximation and the parameters indicated in the lower central panel. Solid arrows indicate the kinematical Bragg energies for an inner potential of $V_0 = 16.7$ eV which is equal to the sum of the Fermi energy and work function of Al(100). Dashed arrows designate the kinematical Bragg energies for an inner potential $V_0 \equiv 0$.
(After Laramore *et al.* [10].)

that while I regard these conclusions as both elementary and rather obvious, they are not universally accepted.

Three major conclusions can be derived from the considerations given earlier in this section. First, it is evident from Figures 36–39, and 45–48 that microscopic 'truncated-bulk-solid' models predict ELEED intensities from 'clean' metals in semi-quantitative agreement with the observations *provided* one describes the electron-solid optical potential via adjustable parameters [e.g., V_0 and λ_{ee} in Equations (113)]. Second, the model descriptions of the data are better at smaller angles of incidence (see, e.g., Figures 38, 39, 46, 47). Third, the discrepancies between these model calculations and existing data are comparable to those between the data of different workers (i.e., peak energies commonly vary by ± 4 eV and peak heights by $\pm 100\%$). These three conclusions are our 'primary' ones derived directly from comparison of the model calculations of three separate groups (Cornell, [46] IBM, [61, 62] and Illinois [5, 48, 60, 117]) with data taken by four distinct groups (California, [111] Grenoble, [102–104] IBM, [101] and Illinois [105, 106]). Therefore, I regard these results as characteristic of the current 'state-of-the-art' in both their theoretical and experimental aspects.

Two additional conclusions are 'derived' consequences of the three primary ones. First, the electron-solid optical potential is the weak link in our fundamental understanding of ELEED. Its spatial dependence is essentially unknown; its apparent variations from one crystal face to another are thus far unexplained by the obvious considerations; and the magnitudes of both its real and imaginary parts are inextricably interwoven with the degree of vibrational and geometrical inequivalence of the surface ion-cores from those in the bulk. Second, the current accuracy of the experimental characterization of solid-vacuum surfaces (even of 'clean' metals) simply does not provide sufficiently reproducible and well-defined intensity data to be useful in refining further the microscopic calculations. A wide variety of models predict peaks within ± 4 eV of each other whose intensity is uncertain to within a factor of two.

Finally, what is the direction of current activity in this area? I think that for a few years any major advances must be predominately experimental in nature. The reproducibility of ELEED intensity data must be improved considerably, as must the accuracy with which energies, angles, and sample temperatures are measured. Workers at different laboratories must learn to characterize their samples with sufficient precision to reduce the peak-position uncertainties in the ELEED intensities to about ± 1 eV and the peak-intensity uncertainties to about $\pm 20\%$. Such numbers imply substantial advances in characterizing the chemical and mechanical cleanliness of the surface, in the control of sample alignment and temperature, and in the measurement and control of the absolute energies and directions of the incident and scattered electrons. The microscopic theoretical description of ELEED has improved substantially in the past three years as can be appreciated by comparing the lower panel in Figure 31 with Figures 38, 39, and 45. However, further improvements require much more precise data in order to resolve some of the uncertainties in our understanding of the electron-solid force law.

4. Surface Crystallography

A. DEFINITION OF THE PROBLEM

As one of the main applications envisaged for elastic low-energy-electron-diffraction (ELEED) is the determination of the geometrical structure of single-crystal solid surfaces, in this lecture we review the nature of problem of structure-determination, and the current status of attempts at its solution. As in the case of the X-ray crystallography of bulk solids, [123, 124] it is convenient to decompose the problem into two parts: the determination of the space-group symmetry (Bravais space-lattice) of the lattice and the determination of the structure of the unit cells which occupy the individual points of the Bravais lattice. In X-ray diffraction, the former can be obtained directly from the *geometry* (i.e., directions) of the X-ray beams diffracted from the crystal. No information about the *intensities* of the beams is required for this determination. An analysis of the intensities is required only for the examination of the structure of a unit cell. Thus, for example, the direction of the diffracted beams alone suffices to determine the structure of monatomic solids with one atom per unit cell (e.g., most elemental metals).

The surface structure problem is more complicated geometrically because the presence of the surface destroys the translational symmetry of the solid normal to the surface. Thus the solid must be viewed as unit cells of length normal to the surface equal to the thickness of the sample. The periodicity of the solid parallel to the surface is reflected in the conservation law [Equation (10); Figure 4]:

$$\mathbf{k}'_\parallel = \mathbf{k}_\parallel + \mathbf{g} \tag{10}$$

for the component of momentum, $\mathbf{k}_\parallel$, of the incident electron parallel to the presumed planer surface. The vector, $\mathbf{g}$, belongs to the reciprocal lattice of the Bravais net characterizing the translational symmetry of the solid parallel to the surface. The intensity pattern or 'spot-pattern' [see Figure 15] resulting from Equation (10) determines the space-group symmetry of the lattice parallel to the surface. [125, 126] Five Bravais 'nets' characterize the space-group symmetries of surfaces of ideal crystals as described by Wood. [125, 126] For example, if the incident beam is normal to the surface ($\mathbf{k}_\parallel = 0$), the directions of the exit beams directly correspond to the various directions in the reciprocal lattice in accordance with Equation (10). Using Wood's notation $\mathbf{a}$ and $\mathbf{b}$ for the surface unit mesh, we find

$$\mathbf{g} = 2\pi(h\mathbf{a}^* + k\mathbf{b}^*) \tag{117a}$$

$$\mathbf{a}^* = \mathbf{a} \times \mathbf{n}/(\mathbf{a} \cdot \mathbf{b}) \tag{117b}$$

$$\mathbf{b}^* = \mathbf{b} \times \mathbf{n}/(\mathbf{a} \cdot \mathbf{b}) \tag{117c}$$

in which $\mathbf{n}$ is the interior surface normal. Combining the definition of elastic scattering, (i.e., $k'_\parallel = (2mE/\hbar^2) \sin \theta'$ in which E is the energy of the incident beam and θ' is the exit angle) with Equation (10) gives an expression for the exit angles as a function of the incident beam energy and spacings of the unit cells parallel to the surface. For example, considering the exit beams along b^* for a square lattice gives the expressions for the exit angles

$$\sin \theta_{0k} = \frac{2\pi k}{a} \left[\frac{\hbar^2}{2mE} \right]^{1/2} . \tag{118}$$

The lattice spacing a is determined by using the version of Equation (118)

$$E = 150.4k^2/a^2 \sin^2 \theta_{0k} \tag{119}$$

for a given (integral) value of k measuring E in eV. A plot of E versus $\sin^{-2} \theta_{0k}$ gives a line of slope $150.4k^2/a^2$ from which a is extracted. For example, Andersson and Kasemo [127] have demonstrated that such a plot for the (100) face of nickel gives a result $(a=2.46 \text{ Å})$ essentially identical to that obtained for a bulk sample by X-ray diffraction $(a=2.49 \text{ Å})$.

Knowing that electrons exhibit wavelike behavior which leads them to diffract from the lattice in the energy range $E \geqslant 0$, it is interesting to recall why they do not simply exhibit Laue patterns as observed in a conventional X-ray transmission geometry experiment. The basic reason, described in Sections 1C and 2B is that their strong interactions with the valence electrons of the solid cause them to experience *inelastic* collisions with these electrons in the upper 2–5 layers of the crystal. This contrasts with the weak interaction of X-rays with solids which, in turn, leads most X-rays to experience only one elastic collision with the ion-cores in the solid as they traverse the sample. On the one hand, the high probability of inelastic electron-solid collisions is desirable because it renders the observation of elastic electron-solid cross sections sensitive to the properties of the upper 2–5 layers of the solid [see Section 1C]. On the other hand, it is undesirable because the large *inelastic* electron-electron cross-sections are accompanied by large *elastic* electron-ion-core cross sections which render multiple elastic scattering phenomena important in determining the elastic diffraction intensities. Thus we recall [Section 1C] that the central problem of surface crystallography is the separation of the geometrical structure information from the dynamical effects associated with the combined phenomena of strong elastic and inelastic electron-solid scattering as indicated in Figure 49.

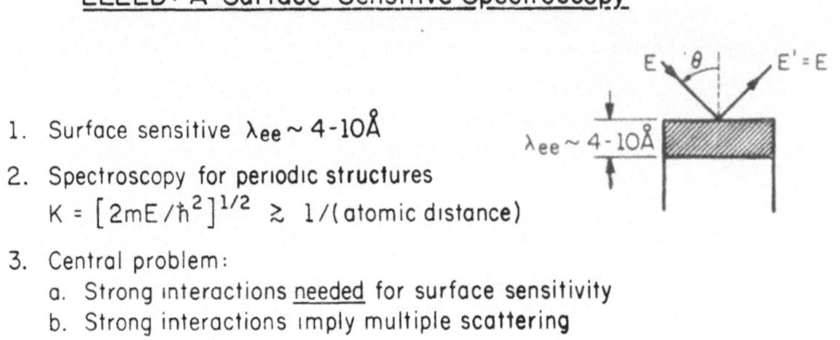

ELEED: A Surface-Sensitive Spectroscopy

1. Surface sensitive $\lambda_{ee} \sim 4\text{-}10\text{Å}$

2. Spectroscopy for periodic structures
   $$K = \left[2mE/\hbar^2 \right]^{1/2} \gtrsim 1/(\text{atomic distance})$$

3. Central problem:
   a. Strong interactions <u>needed</u> for surface sensitivity
   b. Strong interactions imply multiple scattering

Fig. 49. A summary of the reasons why elastic low-energy-electron diffraction (ELEED) is an appropriate probe of the geometrical arrangement of the surface atoms in a solid. The concomitant theoretical problem implied by the strong electron-solid interaction also is indicated.

Combining the observations in the previous two paragraphs, we see that although a study of elastic low-energy-electron diffraction (ELEED) can provide, in principle, a determination of the packing sequence and layer spacings of the top 2–5 layers of a single-crystal surface, the measurement and analysis of the *intensities* of the scattered electron beams is required for this purpose even in the case of monatomic lattices whose bulk unit cell contains only one atom. This additional complexity relative to the X-ray case is caused by the strong electron-solid interaction which restricts the elastic beam to within the upper few layers of the solid. However, the strong interaction also causes an added 'dynamical' complexity relative to the X-ray case: i.e., as discussed in Section 3C it renders inadequate the use of a single-scattering ('Born-approximation' or 'Linear-response') model to calculate the scattered beam intensities. Therefore the simple expression relating the energy in the diffracted beams to the interference function [Equation (51)], which forms the theoretical backbone of X-ray crystallography [123, 124] is not an appropriate basis for a surface crystallography based on ELEED. Thus we are confronted with the dual problems of needing to analyze the scattered intensities to extract even elementary geometrical structure information (like layer packing sequences for simple metals) and not being able to rely on the conventional single-scattering model to provide such an analysis.

As might be expected, different theoretical procedures are useful in attacking the various specific aspects of the general problem of determining the positions of surface ion-cores. Consequently, we summarize in Figure 50 and 51 the precise geometrical quantities sought and the data from which they may be extracted in the cases of clean surfaces and surfaces covered by a presumably distinct adsorbed monolayer, respectively.

Whereas Figures 50 and 51 summarize the structure-determination problems associated with the geometry of defect-free planar surfaces, even chemically clean macroscopic surfaces contain both defects and steps. In addition, adsorbed monolayers of lower symmetry than the substrate often exhibit several distinct but equivalent orientations relative to the substrate. Consequently, adsorbed monolayers generally form in localized islands separated by antiphase domains. [6, 28] This type of surface

Objectives of Surface Crystallography I

Clean Surfaces

a. Surface Bravais net :
   the intensity pattern

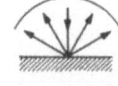

b Unit - cell geometry
   (e.g. layer spacing and
   packing sequence) :
   the intensity profile

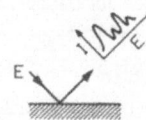

Fig. 50.   Schematic summary of the specific information sought about the geometry of surface ion-cores at a presumably clean, planar surface of a single-crystal solid. The types of data required to provide this information also are indicated.

disorder can produce marked effects on the intensity patterns. For example, the observed pattern in the presence of a low-symmetry adsorbed monolayer often is a superposition of the patterns predicted for the various orientations of the domains. The various types of surface disorder also can broaden or 'split' the scattered beams of electrons, depending on the physical nature of the disorder. [6, 128] Thus we see that the intensities of the scattered beams convey information not only about the unit-cell geometry of an ideal planar single-crystal surface, but also about the statistical distribution of defects at a non-ideal surface containing steps, domains, and/or other types of disorder (e.g., that due to the thermal motion of surface scatterers).

Objectives of Surface Crystallography II

Chemisorbed Monolayers

a. Surface Bravais net:
   the intensity pattern

b Surface unit cell:
   monolayer-beam intensities

c. Monolayer-substrate registry:
   substrate-beam intensities

Fig. 51. Schematic summary of the specific information sought about the geometry of a presumably distinct, planar, domain-free monolayer adsorbed on a defect-free, planar single-crystal surface. The types of data required to provide this information also are indicated. The dashed scattered beams in the figure indicate the additional beams which occur when the monolayer geometry exhibits a lower symmetry than the planer layers of the single-crystal substrate. These new beams are referred to as 'monolayer' beams whereas those present in the absence of the adsorbed layer are called 'substrate' beams.

These considerations reveal that in order to analyze 'practical' ELEED intensity data, it is necessary to distinguish the consequences of a strong electron-solid interaction *per se* from those associated with non-ideal surface morphology. At the present time, such a distinction usually cannot be made with confidence. Analyses of ELEED from disordered surfaces have, been given, thus far, only for one [128, 129] and two-[130] dimensional models. They have been confined to the use of the Born approximation which, as we saw in Sections 3C and 3D, is manifestly inadequate to describe electron solid scattering. An important prediction of these models, however, is that a distribution of steps on an surface can introduce significant modulations of the intensity profile [Figure 15] as well as create related changes in the angular pattern of the scattered beams. Such a surface typography may well be the origin [6] of the discrepancies between the planer-surface predictions of Laramore and Duke and the data of Jona as shown in Figure 48. Therefore the rudimentary nature of current models of electron-scattering from disordered surfaces should not blind us to the potentially important consequences of surface defects on observed ELEED intensities.

The importance of surface typography notwithstanding, however, in these lectures we confine our attention to the case of planar, defect-free surfaces. The central topic of our discussion is the relationship between kinematical and dynamical model analyses of ELEED intensities as applied for the purpose of determining the crystallographic structure of the uppermost layers of a single-crystal solid. The pervasive, indeed almost exclusive use of kinematical analyses in the experimental literature on surface crystallography is evident upon inspection of any recent review of the topic. [28, 29, 131–134] Thus the examination of their relationship to the 'dynamical' models described in Section 3D and the determination of the unit-cell geometry (in addition to the Bravais net) of both clean surfaces and adsorbed species are the two 'outstanding' problems in modern surface crystallography.

B. SURVEY OF THEORETICAL APPROACHES

Confronted with the problem of determining the geometry of the 'surface' unit-cells (of essentially infinite extent into the crystal), theoretical workers have proposed two rather different approaches to its solution. We refer to these as the microscopic-model and data-reduction methods of surface crystallography, and discuss them in turn.

The objective of the microscopic-model approach is the development of a model of electron-solid scattering which describes, in detail, measured ELEED intensities from solid surfaces. This approach was described in Sections 2C and 3D. Obviously, if such a model is available, the geometry of the solid surface can be extracted from a data analysis. In practive, however, the application of this approach to surface crystallography encounters two major obstacles. First, for its unequivocal success it requires a precise characterization of the experimental surface and a detailed knowledge of the electron-solid force law, neither of which currently is available even for simple systems like elemental metals. Second, from the point-of-view of crystallography, it is excessively ambitious in that it characterizes the electronic and vibronic structure of the surface as well as its geometrical structure. Because of the wealth of dynamical information about the force-law required by (or, in principle, extracted from) a model analysis, certain ambiguities concerning the geometrical structure *per se* can arise in data analyses. These sorts of ambiguities are illustrated well by our discussion of the analysis of ELEED intensities on the low index faces of aluminum [Section 4D7]. On the one hand, if one examines a small set of data, ambiguities almost always arise between the values of dynamical parameters (e.g., values of the inner-potential between muffin-tin ion-core scatterers) and geometrical parameters (like the spacing of the upper two layers of the solid). On the other hand, as the analysis is extended to more data, it inevitably works adequately for some parts of the data but poorly for other parts. Consequently, one is confronted with making subjective choices between alternative sets of parameters. In many cases, all of the satisfactory alternatives are not displayed explicitly because the 'dynamical' parameters are chosen (to some extent arbitrarily) at the beginning of the calculation and only the geometrical parameters are varied. Such a pre-selection of the dynamical parameters is particularly unsuited for the study of chemisorbed species and rearranged surfaces for which the electronic potential is not well-known. Thus we anticipate that the major conceptual difficulties encountered with the microscopic-model approach to surface crystallography will be those of the uniqueness and intrinsic accuracy of the values of the geometrical parameters extracted from data analyses. Technical difficulties (e.g., the requirement of extensive computer computations) also occur, but probably can be overcome just as similar ones in X-ray crystallography have been. [123, 124]

The data-reduction method of surface crystallography is based on abstracting from model calculations and experimental ELEED data those features which are sensitive to the geometrical structure parameters of the solid force law. [72] It has evolved as a set of prescriptions for extracting from observed ELEED intensities certain features which can be analyzed directly by a single-scattering (kinematical) model analogous to that used in the analysis of X-ray diffraction. [97, 118, 119, 134–136] Thus one seeks to 'reduce' the experimental data, which exhibit numerous dynamical manifestations of the strong electron-solid interactions, to forms from which structural information can be extracted directly and unambiguously via kinematical analyses. The price paid in effecting such a reduction is an intrinsic uncertainty of about $\Delta a \sim 0.1$ Å in the resulting structural parameters. In Section 3D we argued that, at the present time, uncertainties in the electron-solid force law render this uncertainty of no significance, i.e., the microscopic-model approach is comparably or less accurate once the consequences of ambiguities in the force law are incorporated into the error analysis. However, for simple systems microscopic-model analysis are expected ultimately to provide a refinement of structural parameter values extracted initially using the data-reduction method. Indeed, it seems most sensible to regard the two methods as complementary. The data-reduction method requires less dynamical input data and consequently produces intrinsically less accurate structural information. The microscopic-model method, while capable of producing structural information with high accuracy, requires a more precise statement of the electron-solid force law to achieve this accuracy. Thus we anticipate that the application of the data-reduction methods can provide useful initial values of structural and electronic parameters which, if desired, can be refined further by microscopic-model calculations.

A synopsis of the major characteristics of the two theoretical approaches to surface crystallography is presented in Figure 52. In the remaining sections in this part of

Theoretical Approaches to Surface Crystallography via ELEED

1. Microscopic - Model Methods
   a. Uncertainties in the force law
   b. Lack of clear decision criteria

2. Data - Reduction Methods
   a. Intrinsically approximate hence inaccurate
   b. Several different types

Fig. 52. Synopsis of the major characteristics of the two theoretical approaches to the application of elastic low-energy electron diffraction (ELEED) for the determination of the geometry of the scatterers in the upper-most 2–5 layers of a single-crystal solid. A more detailed discussion is given in Section 4B of the text.

the lecture notes we concentrate on the data-reduction methods because of their simplicity and their direct contact with the descriptions of ELEED data given historically by experimentalists. [131–134] The vehicle used in these methods to 'reduce' the data to a form which is to be compared with the predictions of a kinematical analysis is that of averaging the observed ELEED intensities over values of one or more of the incident beam parameters (E, θ, φ). An important characteristic of these data-reduction or 'averaging' methods, noted in Figure 52, is that the choice of both the ELEED intensities themselves and the beam variables over which to average depends upon the type of structural information which we seek to determine. I.e., there are several different types of averaging procedures, each with their own unique characteristics and merits. By way of providing a 'table of contents' for the following

Data-Averaging Methods

1. "Clean Surfaces"
   a Average over ϕ
   b Average over θ for
      Fixed Momentum Transfer

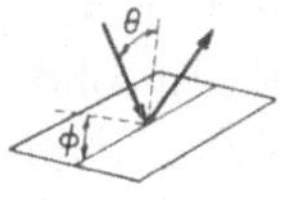

2 Chemisorbed Monolayers
   a Monolayer Unit Cell Geometry: Energy-
      Average the "Extra" Monolayer Beams.
   b. Monolayer-Substrate Registry: Use
      High-Energy Region or θ-ϕ Averages.

Fig. 53. Synopsis of the features of current (1972) data-averaging methods to reduce experimental ELEED intensities to a form which can be analyzed by a kinematical analysis. The details of these methods are described in Sections 4C and 4D of the text.

two sections, we outline the characteristics of current averaging methods in Figure 53. We next turn to a description of these methods in the cases of clean surfaces [Section 4C] and adsorbed monolayers [Section 4D] respectively.

C. CLEAN SURFACES

It is convenient to decompose the problem of determining the atomic geometry of a clean, defect-free planar single-crystal surface into three parts. First, one must construct the Bravais net of the two-dimensional atomic planes parallel to the surface out of which the solid is composed by stacking these planes on top of each other. Second, the chemical identities, electron-ion-core scattering factors, and positions of the atomic species in each layer must be evaluated. Finally, the registry (i.e., packing sequence and spacing) of the various layers must be determined. For simplicity, I shall discuss each of these parts in the limiting case of a monatomic metal. Surprisingly enough, even this elementary limit has not been treated in detail in the literature for any materials except   the low-index faces of Al, [5, 48, 97] Ag(111), [135] Cu(100), [127] Ni(100), [127] and Ni(111). [136]

The simplest model of the atomic structure of clean metal surfaces is that obtained by considering the surface to be obtained by the truncation of an otherwise ideal bulk single-crystal. The intensity patterns associated with the low-index faces of most metals reveal that the geometry and spacings within the two-dimensional atomic layers parallel to the surface are identical to those in the bulk to within experimental error. [1] The nature of these analyses of the intensity patterns was indicated in Section 4A.

In the elementary case of a clean metal surface, the chemical identities and positions of the atomic scatterers within an individual two-dimensional layer are regarded as 'known' *a priori* (once, of course, that the cleanliness of the surface has been established by, e.g., Auger spectroscopy). This fact provides a valuable experimental test of model electron-ion-core scattering factors used to determine the layer spacing and packing sequence. [118] In general, these scattering factors must be obtained initially from a model ion-core potential [1, 5, 6, 46, 60–63, 137] and subsequently used to determine the geometry of the basis in each two-dimensional atomic layer. [97] If the basis geometry is known, however, we can invert the procedure for its determination to evaluate the atomic scattering factors of its constituents. We next describe this process, as proposed by Duke and Laramore, [118] in the limit of a rigid-lattice model of the solid. The consequences of lattice vibrations can be incorporated into the model as described in Equations (41) *et seq.*

One central concept provides the foundation for the averaging methods [97, 118, 138, 139] designed to distinguish properties of the top layer of a solid from those of underlying layers. In the limit of strong inelastic-collision damping the multiple scattering from the underlying layers primarily redistributes the 'kinematic' intensity as a function of incident beam energy *within* a given scattered beam, rather than redistribute appreciable intensity from one beam to another. [97] Consequently, Tucker and Duke proposed [97, 138, 139] that by averaging out this fine structure in the dynamical intensities, $I_{hk}(E)$, over a range of energy $(E_0 + \Delta/2) \geqslant E \geqslant (E_0 - \Delta/2)$, they could construct a set of quantities

$$M_{hk} = \int_{E_0 - \Delta/2}^{E_0 + \Delta/2} I_{hk}(E) \, \mathrm{d}E \tag{120}$$

directly proportional to the average kinematical intensities

$$I_{hk}^{\text{kin}} = \int_{E_0 - \Delta/2}^{E_0 + \Delta/2} I_{hk}^{\text{kin}}(E) \, \mathrm{d}E \tag{121}$$

$$I_{hk}^{\text{kin}}(E) = |F_{hk}(E)|^2 \tag{122a}$$

$$F_{hk}(E) = \sum_{\alpha} f_{\alpha}(E, \theta_{hk}) \exp\left[2\pi i (h x_{\alpha} + k y_{\alpha})\right]. \tag{122b}$$

In Equations (121) and (122) the quantity $f_{\alpha}(E, \theta)$ is the atomic scattering amplitude as a function of energy and scattering angle for the ion-core scatterer labeled by α; h and k are the Miller indices corresponding to the reciprocal lattice vector g which designates the scattered electron beam; and x_α and y_α are the fractional coordinates of the atoms in the two-dimensional monolayer unit cells. If we designate the electron's scattering angle by θ_{if}, the general form of the monolayer structure factor is given in terms of the electron-ion core scattering phase shifts, $\delta_\alpha^l(E)$, via:

$$F_{hk}(E) = [2ik(E)]^{-1} \sum_l (2l + 1) P_l(\cos\theta_{if}) \times$$

$$\times \sum_\alpha \{\exp[2i\delta_\alpha^l(E)] - 1\} \exp[2\pi i (hx_\alpha + ky_\alpha)] \times$$

$$\times \exp\{[k_\perp(0, E) + k_\perp(\mathbf{g}_{hk}, E)] z_\alpha\}, \tag{123a}$$

for a rigid lattice of scatterers with vertical positions z_α measured with respect to the mean value

$$\bar{z} = \sum_\alpha z_\alpha = 0. \tag{123b}$$

An ambiguity exists in the choice of $k(E)$ and $k_\perp(\mathbf{g}, E)$ because they could be taken to be the values characteristic of the electron either outside [Equations (97)–(99)] or inside, e.g.,

$$k^2(E) \equiv 2m[E - \Sigma(k, E)]/\hbar^2, \tag{124}$$

the crystal. The electron's proper self energy, $\Sigma(k, E)$, in Equation (124) usually is evaluated using a simplified model such as that given by Equations (113). Duke and Laramore resolve this ambiguity by using the real part of $k(E)$ inside the solid and reflectivity rather than scattering boundary conditions with the set of conventions:

$$I_{hk}^{\mathrm{kin}}(E) = \frac{k_{0\perp}(\mathbf{g}_{hk}, E)}{k_{0\perp}(0, E)} \left| \frac{2\pi i F_{hk}(E)}{k_\perp(\mathbf{g}_{kh}, E)} \right|^2 \tag{125}$$

$$k_{0\perp}^2(\mathbf{g}, E) = 2mE/\hbar^2 - (\mathbf{k}_\parallel + \mathbf{g})^2 \tag{126a}$$

$$k_\perp^2(\mathbf{g}, E) = 2m[E - \Sigma(E)]/\hbar^2 - (\mathbf{k}_\parallel + \mathbf{g})^2 \tag{126b}$$

$$\Sigma(E) = -V_0 - \hbar^2[2m(E + V_0)/\hbar^2]^{1/2}/m\lambda_{\mathrm{ee}} \tag{127}$$

$$\mathbf{g}_{hk} = 2\pi(h\mathbf{b}_1 + k\mathbf{b}_2). \tag{128}$$

In Equations (126) the symbol $\mathbf{k}_\parallel$ designates the momentum parallel to the surface of the incident electron and $\mathbf{g}$ is a reciprocal lattice vector associated with the Bravais net of the two-dimensional, planar atomic diffraction gratings out of which the solid is constructed. The real part of the proper self-energy, i.e., V_0, is chosen to be the sum of the work function and Fermi-energy for simple metals. [5]

In the 'usual' case that we presume to know the atomic scattering factors [i.e., the $f_\alpha^l(E, \theta)$ or $\delta_\alpha^l(E)$] the monolayer unit-cell geometry is determined by the 'energy-averaging' method in which we proceed in four steps. First, from atomic scattering potentials (or some other source) a selection of the phase shifts, $\delta_\alpha^l(E)$, associated with the monolayer atoms is made for insertion into Equations (123). Second, the kine-matical intensities are calculated from Equations (125–128) and averaging according to Equation (121). Third, these averages are compared with the M_{hk} obtained via

Equation (120) from the observed intensity profiles, $I_{hk}(E)$, for the hk beam. Fourth, steps 2–4 are iterated for different atomic positions (e.g., $x_\alpha, y_\alpha, z_\alpha$) until the kinematical averages I_{hk}^{kin} agree with the M_{hk} to within a predetermined accuracy. For example, Duke and Laramore [118] require that the quantity

$$R \equiv \sum_{hk} |M_{hk} - I_{hk}^{kin}| / \sum_{hk} M_{hk} \tag{129}$$

be less than $R_m = 0.1$. When $R \le R_m$, the positions of the scatterers in the monolayer are taken to be determined to within an accuracy $\Delta a \cong 0.1$ Å provided $\lambda_{ee} \lesssim 4$ Å.

In our present case of a clean simple-metal monolayer, the $(x_\alpha, y_\alpha, z_\alpha)$ all are known *a priori* from the analysis of the intensity pattern. Therefore in principle the above procedure can be inverted to determine the electron-ion-core phase shifts, $\delta_\alpha^l(E) = \delta^l(E)$ for the identical metallic ion-cores in the upper layer of the solid. That such a procedure is sufficiently accurate to be useful is shown by Figure 54. To generate

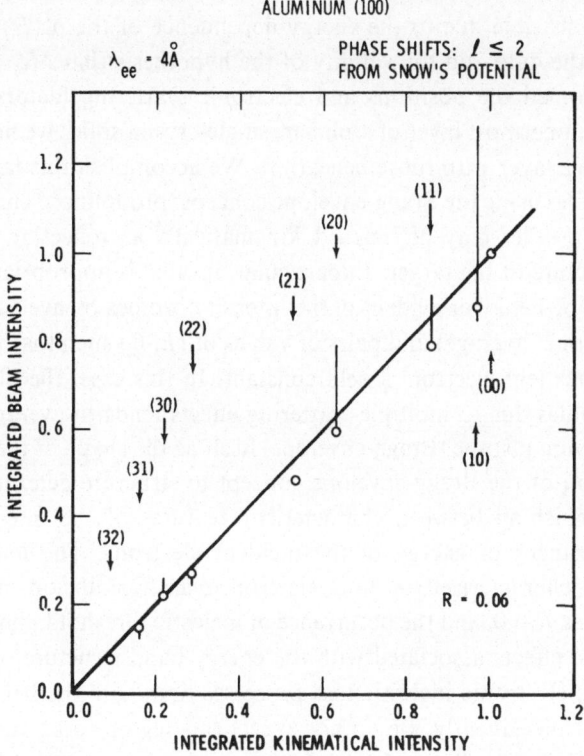

Fig. 54. Normalized integrated dynamical intensities versus the normalized kinematical intensities for the beams characteristic of normally-incident electrons scattered from clean Al(100). The phase shifts describing the aluminum ion cores were obtained from the Laramore-Duke-Snow model potential. The beams are labeled by the Miller indices characteristic of the clean aluminum 'substrate'. The energy interval used to define the open-circle values is 283 eV $\le E \le$ 383 eV. The vertical lines passing through the circles indicate the range of ratios M/I^{kin} obtained using also the intervals 303 eV $\le E \le$ 383 eV, 243 eV $\le E \le$ 383 eV, and for the (00), (10), (11), and (20) beams, 85 eV $\le E \le$ $\le$ 165 eV. (After Duke and Laramore [118].)

this figure, the Laramore-Duke-Snow model was used to compute the $\delta^l(E)$ [see e.g. Figure 32], these in turn were utilized to evaluate the 'dynamical' $I_{hk}(E)$ required to obtain the M_{hk} via Equation (120). The M_{hk} generated in this fashion provide the ordinates in Figure 54 whereas the I_{hk}^{kin} calculated from Equations (123)–(128) constitute the abscissa. The open circles give the values of this M_{hk}/I_{hk}^{kin} ratio for the beams indicated. If they all fell on the solid line, then $M_{hk} \equiv I_{hk}^{kin}$ so the dynamical and kinematical averages would be identical. In fact, the two averages, while not precisely equal, are nearly so for most beams. Therefore if the M_{hk} are obtained from experimental ELEED intensity data, plots of the form of Figure 54 provide an excellent check on the model used to determine the $\delta^l(E)$. In particular, if $\lambda_{ee} \lesssim 4$ Å any dynamical model calculation of the intensity profiles must use an electron-ion-core interaction which predicts phase shifts consistent with the energy-averaged beam intensities regardless of how well (or poorly) it describes the individual intensity profiles over a limited range of beam parameters (i.e., incident energy and angle). [118] A direct determination of the $\delta^l(E)$ from the M_{hk} obtained using experimental ELEED data does not appear feasible due to the energy dependence of the $\delta^l(E)$ and the uncertainties in both the data and the validity of the hypothesis that $M_{hk} = I_{hk}^{kin}$.

Having determined the positions and electronic scattering factors of the atomic scatterers in the uppermost layer of a planar, single-crystal solid, we now can evaluate the registry of this layer with those beneath it. We accomplish this task by analyzing the intensity profiles using the Bragg-envelope concept: pronounced clusters of maxima in the intensity profiles may be indexed kinematically as reflecting the underlying geometrical structure of the target. In particular, it often is appropriate to accentuate the Bragg-envelope behavior evident in the intensity profiles by averaging over values of φ (the 'azimuthal' average) and pairs of values of (E, θ) such that the momentum-transfer to the incident electron is held constant. In this case, the 'fine-structure' in the intensity profiles due to multiple-scattering effects tends to average to a constant background, leaving just the 'Bragg-envelope' itself as the shape of the averaged data.

The application of the Bragg-envelope concept to structure determination is facilitated by distinguishing between characteristic features of the intensity profiles in three separate ranges of energy of the incident electrons. The *low-energy region*, $0 \lesssim E \lesssim 25$ eV, is characterized by long electron inelastic-collision mean free paths, e.g., $\lambda_{ee} \sim 200$ Å as $E \to 0$, and the occurrence of inelastic threshold phenomena. Therefore the dynamic effects associated with the energy-band structure of the bulk solid and the discrete, electronic inelastic loss processes dominate the behavior of the intensity profiles. Consequently, we do not expect this region to be very useful for the study of the surface crystallography of clean crystals.

The *intermediate-energy region*, 25 eV $\lesssim E \lesssim 100$–300 eV, is characterized by strong inelastic-collision damping ($\lambda_{ee} \sim 2$–8 Å), and strong elastic electron-ion-core scattering. It is in this energy region that ELEED from clean crystals is most surface sensitive. However, in this region the intensity profiles generally appear highly 'dynamical' in character as may be seen from Figures 31, 36–39, and 45–48. (I.e., the fine structure in between the clusters of peaks near the kinematical Bragg energies is comparable in intensity to these 'kinematical' clusters.) Also, as initially noted by Farnsworth, [140] the intensity profiles from geometrically similar but electronically distinct materials can differ substantially. The key to utilizing this region of the intensity profiles to determine surface crystallography lies in the application of data-averaging to eliminate the unwanted dynamical fine structure while preserving maxima at the energies of the Bragg-envelope clusters.

A number of simplifications in the intensity profiles occur in the *high-energy region* $E \gtrsim 100$–500 eV. The precise definition of this region varies from one crystal face and

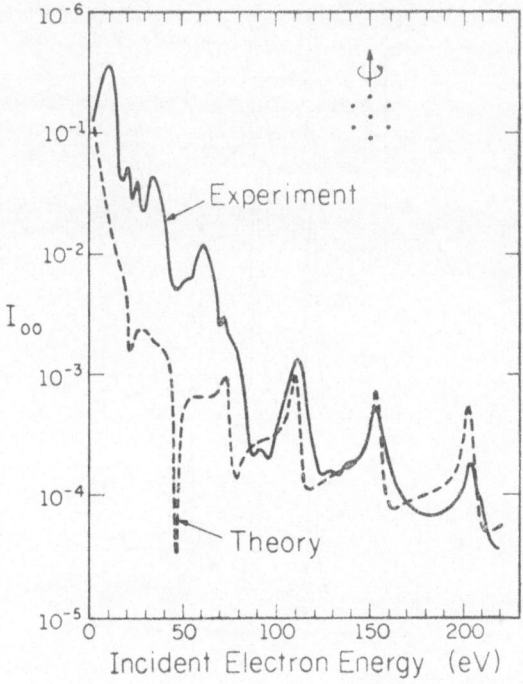

Fig. 55. Comparison of theoretical [142] and experimental [141] intensity profiles for the specular beam of electrons scattered from graphite. The azimuth is noted in the figure and the angle of incidence is 4°. The calculation is described in [142]. (After Hirabayashi. [142]).

material to the next, occurring for energies above about 100 eV in carbon but not until about 450 eV for $W(100)$. In this region the inelastic-collision mean free path is increasing so that ELEED is becoming less sensitive to the specifically surface properties of the solid. This effect is enhanced by the large vibrational amplitudes of surface scatterers which, by virtue of Equations (41), reduce the scattering from the surface atoms relative to that from bulk atoms by an amount roughly proportional to $\exp[-\beta ET]$, in which β is a constant for a given material and T is greater than the Debye temperature of the material. Thus the elastic as well as inelastic scattering of the electron is diminishing in this energy region leading to intensity profiles which 'appear' kinematical. This result is illustrated by Figure 55 in which we see some early data taken by Lander and Morrison [141] as analyzed by Hirabayashi. [142] (Hirabayashi's analyses [141, 142] of this data, one of which is shown in the figure, were the first to systematically incorporate (a phenomenological description of) inelastic-collision damping and hence achieve a sensible description of experimental measurements.) It is evident from the figure that although the theoretical model of the electron-solid interaction is not adequate to describe the data in the medium energy region $25 \text{ eV} \lesssim E \lesssim 100 \text{ eV}$, in the high energy region only kinematical-like structures near the

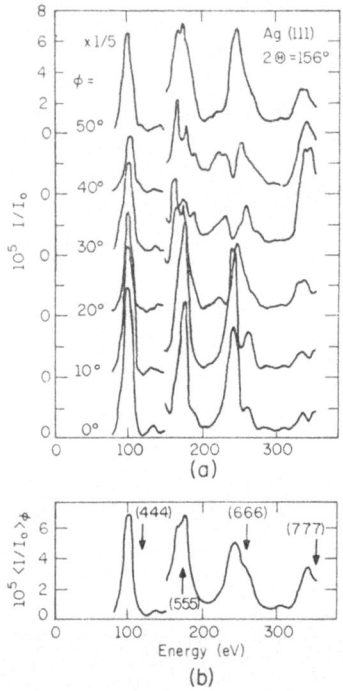

Fig. 56.   Intensity profiles for electrons scattered from Ag(111) at incident and exit angles of $\theta = 12°$. The azimuthal angles, φ, are indicated in the figure. The original raw data are shown in panel (a). In panel (b) is illustrated the angular average over azimuthal angles, φ, of the data shown in panel (a). (After Lagally et al. [135]).

Bragg energies occur. Their prediction does not require a comparably accurate description of this interaction. This observation is a general result, independent of the model [72, 131] which renders the high-energy region of the intensity profiles directly useful for surface crystallography without the need for data averaging. [72, 131] The disadvantage of using this energy region for surface crystallography is the concommitant loss of sensitivity to specifically surface effects caused by the increasing values of λ_{ee} and the enhanced lattice vibrations of surface scatterers.

    Anticipating that the intermediate-energy region will prove most useful for applications in which surface-sensitivity is important (e.g., the study of adsorbed monolayers or altered upper layer spacings of clean crystals), we conclude this section by a discussion of the constant-momentum-transfer data averages [97, 119, 134–136] and their comparison with the predictions of kinematical model analyses.

The constant-momentum-transfer average initially was proposed by Lagally *et al.* [135] who applied it to analyze data on Ag(111). The fact that averaging the intensity profiles of the specular beam over the azimuthal angle, φ, smooths out fine structure in the observed intensity is illustrated by Figure 56. The manifestations of the Bragg-envelope concept are evident in the original intensity profiles [panel (a)] as well as in the azimuthal average [panel (b)].

An additional average over pairs of (E, θ) values can be performed by recalling that the kinematical intensities of the (hk) beam, given by Equations (64), depend only on the momentum transfer:

$$\mathbf{S} = \mathbf{g}_{hk} + [k_\perp(0, E) + k_\perp(\mathbf{g}_{hk}, E)]\,\mathbf{n}, \tag{130}$$

in which $\mathbf{n}$ is the exterior surface normal and the $k_\perp(\mathbf{g}, E)$ are given by Equation (56) or Equation (126b). The momentum transfer $\mathbf{S}$ often is measured in units of

$$S_0 = 2\pi/d \tag{131a}$$

in which d designates the spacing between two identical layers parallel to the surface. Thus for an ideal clean crystal surface, the kinematical Bragg peaks occur at

$$S/S_0 = m \tag{131b}$$

in the specular $(\mathbf{g} \equiv 0)$ beam if m is an integer. The important feature of Equations (131) for our present purposes is that $\mathbf{S}$ is not only independent of φ for $\mathbf{g}=0$, but in general is constant along certain trajectories in the E-θ plane. Therefore, the most complete average possible is a combined 'grand' average over both φ and some appropriate measure of arc length along these trajectories. In the particular case of the specular beam Equations (56) and (130) give

$$\mathbf{S} = 2\{2m[E\cos^2\theta - \Sigma(E)]/\hbar^2\}^{1/2}\,\mathbf{n}. \tag{132}$$

This is a complex vector because of the inelastic-collision damping described by the imaginary part of $\Sigma(E)$ [see e.g. Equations (113)]. If we neglect the imaginary part of $\mathbf{S}$, we can use Equations (113) to obtain the scalar

$$S \cong 2[2m(E\cos^2\theta + V_0)/\hbar^2]^{1/2} \tag{133}$$

as the quantity to be held constant along the E-θ trajectories. From Equation (133) we see that $E^{1/2}$ and $\cos\theta$ are the appropriate parametric variables in terms of which we should describe the intensities. In particular, we should present intensity plots by fixing a minimal value, E_m, of E for $\cos\theta=1$ at a given S, and then using $\cos\theta$ as the abscissa of the plot, with the understanding that for each value of $\cos\theta$, the value of $E \geqslant E_m$ is chosen so that S as specified by Equation (133) remains fixed. We note, however, that Equation (133) differs in form from those usually employed in the literature. [131] Typically an 'inner-potential' V_i is taken to be related to S via

$$S = 2[2m(E + V_i)\cos^2\theta/\hbar^2]^{1/2}. \tag{134a}$$

This form results from the implicit assumption that the inner-potential correction applies to $\mathbf{k}_{\parallel}$ as well as $k_{\perp}(\mathbf{g}, E)$; an assumption in conflict with the conservation law on $\mathbf{k}_{\parallel}$. Its use should give rise to an 'angular dependent inner potential', i.e.,

$$V_i = V_0/\cos^2\theta \tag{134b}$$

which, in fact, has been observed experimentally. [144] For purposes of compatibility with the literature, however, we use Equation (134a) with $V_0 \equiv V_i$ for the remainder of our discussion. If, for a set of azimuthal averages at fixed values of θ, one further averages over E-θ pairs using Equation (133), Equation (134a), or $V_0 \equiv 0$, he smooths out stil further data like that shown in panel (b) of Figure 56 for both Ag(111) [135] and Ni(111). [136]

The central question [134–136] concerning the constant-momentum average is not whether this procedure smooths the data, but rather whether the resulting average is described adequately by kinematical model calculations. Duke and Smith [119] examined this question by comparing the appropriate averages of (exact) model calculations of LEED intensities with kinematical intensities evaluated using the same model electron-solid force law. This procedure exhibits the advantage that it contains no empirical ingredients: the model potential is specified *a priori* and is identical *by definition* for the dynamical (i.e., exact) and kinematical (i.e., approximate) intensities. In this respect their considerations differ from any analysis of experimental data because in the latter certain empirical assumptions about the electron-solid interaction always must be introduced in order to define a 'kinematical' intensity for comparison with the averaged data. Examples of such assumptions by Lagally *et al.* [135, 136] include their use of a particular theoretical model for the atomic scattering factor, a simple sharp-junction, semiclassical model for surface plasmon losses, a heuristic model with an adjustable parameter to describe the (bulk?) inelastic-collision damping of the incident-electrons' wave function, and the presumption of an ideal bulk geometry right up to the surface. The use of these assumptions, while necessary in actual data analyses, preclude a quantitative examination of validity of the analytical technique *per se* by means of the analyses.

Representative results of Duke and Smith's calculations for Ag(111), in which they reproduced the analysis of Lagally *et al.* [135] using simple isotropic-scatterer model calculations of the intensity of the (00) beam in lieu of the experimental data, are shown in Figures 57–59. Their construction of the azimuthal average is indicated in Figure 57. The smoothing effect of this average is evident from the figure. The crucial issue, we recall, is the relationship between this average and the kinematical (single-scattering) intensities calculated *using the same electron-solid force law*. The azimuthal average, kinematical intensity profile, and ratio between them are shown in Figure 58. The energies of the primary Bragg maxima in the kinematical intensities are evident from the figure. If the azimuthal-averaged intensities, $\langle I \rangle_\varphi$, were proportional to the kinematical intensities, I_{kin}, then the ratio $\langle I \rangle_\varphi / I_{kin}$ shown in the lower panel of Figure 58 would be a constant independent of the incident electron's energy, E. Obviously, such is not the case. For energies near the Bragg maxima the averaged intensities are systematically smaller than the kinematical intensities whereas for energies between the Bragg maxima the kinematical intensities are the smaller of

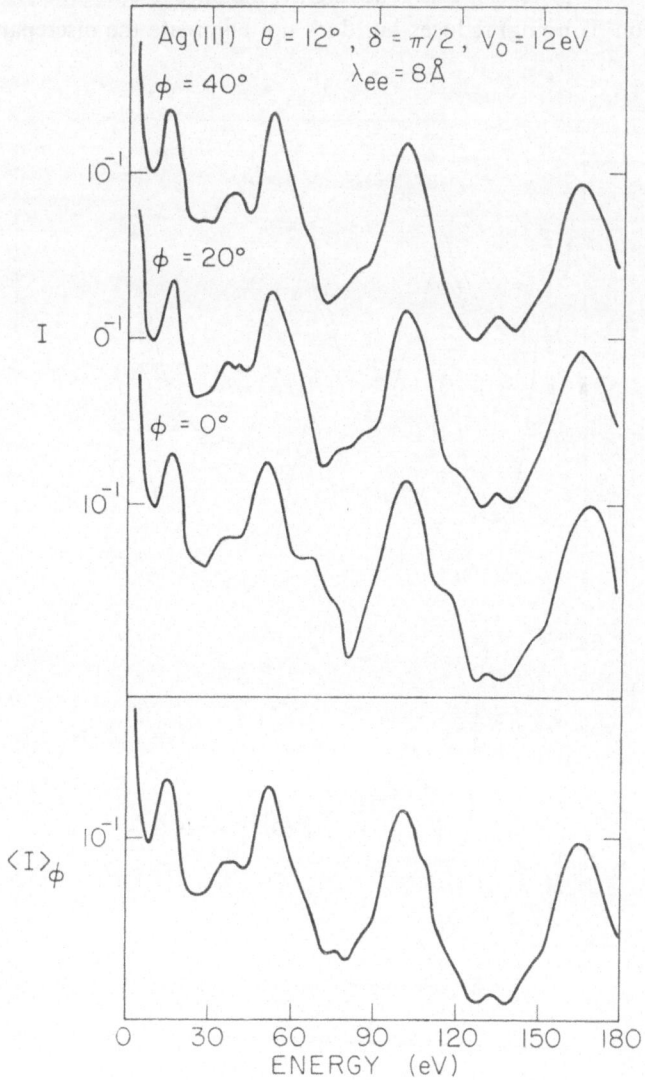

Fig. 57. Intensity of the (00) beam as a function of energy for three values of the azimuthal angle, φ, and the φ-averaged intensity. Intensity profiles for $\varphi = 0$, $10°$, $20°$, $30°$, $40°$, and $50°$, were used in the average. Three of these are shown in the figure. The $\varphi = 20°$ and $40°$ intensity profiles should be identical by virtue of symmetry. The values of the dynamical parameters are indicated in the figure. (After Duke and Smith [119].)

the two. In order-of-magnitude of the maximum discrepancy between the two sets of intensities is 10–100% for the parameters used in constructing Figures 57 and 58 (i.e., $\delta = \pi/2$, $\lambda_{ee} = 8$ Å). These values of the parameters are representative of the scattering of medium-energy ($50 \lesssim E \lesssim 400$ eV) electrons from monatomic metals. The estimated uncertainty indicated by error bars in the lower panel of Figure 58 stems mainly from the use of a finite number of atomic layers (5) in the model calculations and secondarily from the large ($\Delta\varphi = 10°$) mesh size in the azimuthal-average grid. Inspection of Figure 59 reveals that reducing the magnitude of λ_{ee} (i.e., increasing the inelastic-collision damping) reduces but does not eliminate the discrepancy between

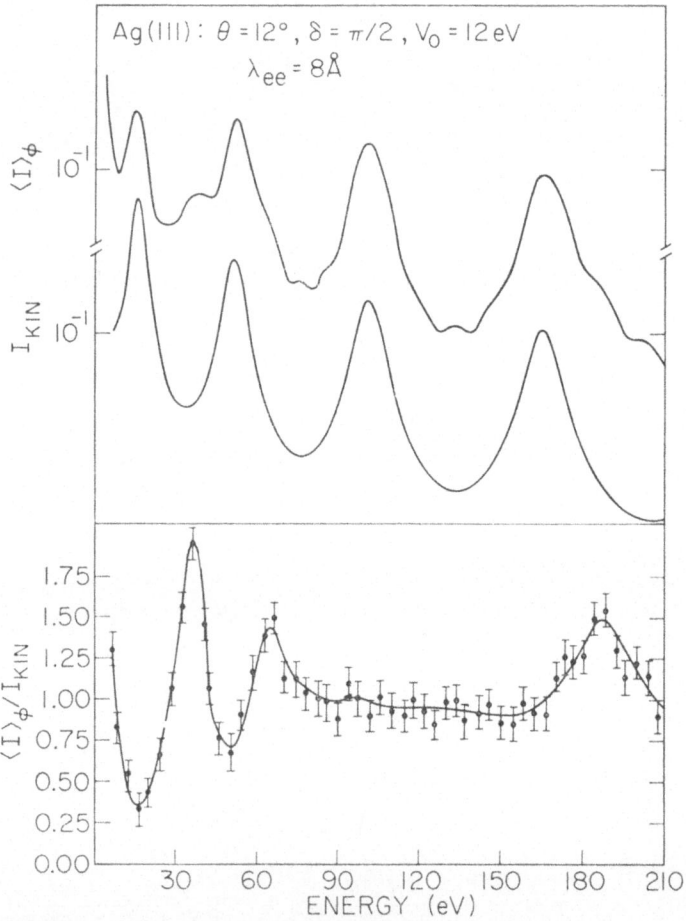

Fig. 58.   Comparison of the azimuthal (φ) averaged intensity of the (00) beam with the kinematic intensity (upper panel) and the ratio of φ averaged intensity to kinematic intensity (lower panel). Intensity profiles for $\varphi = 0$, 10°, 20°, 30°, 40°, and 50° were used in the average. The error bars in the lower panel indicate the estimated uncertainty in performing the average from the given set of six intensity profiles. The values of the dynamical parameters are indicated in the figure.
(After Duke and Smith [119].)

the average and kinematical intensities. A value of $\lambda_{ee} \sim 2\text{–}4$ Å describes the maximum damping expected from a study of the electron-electron interaction induced collision mechanisms. [37]

The physical mechanism for the systematic discrepancy between the kinematic and dynamic intensities was described by Tucker and Duke [72] (see, e.g. Figures 29 and 30). The 'sharing' of the scattered electrons' intensity between several nearly-degenerate electronic states in the solid causes broad peaks in the dynamical intensity profiles composed of several overlapping multiple-scattering-induced resonance maxima. Thus the apparent energy width of a maximum in the dynamical intensity profiles easily can exceed the 'intrinsic' inelastic-collision induced width of an isolated multiple-scattering resonance peak. The energy width of the kinematic Bragg peaks, however, is at most that of an isolated resonance. (In the case of weak inelastic-collision damping the kinematical peaks possess smaller energy widths than isolated multiple-scattering resonance maxima.) The azimuthal average simply smooths out the fine structure in the overlapping dynamical resonances. Therefore the averaged intensity profiles

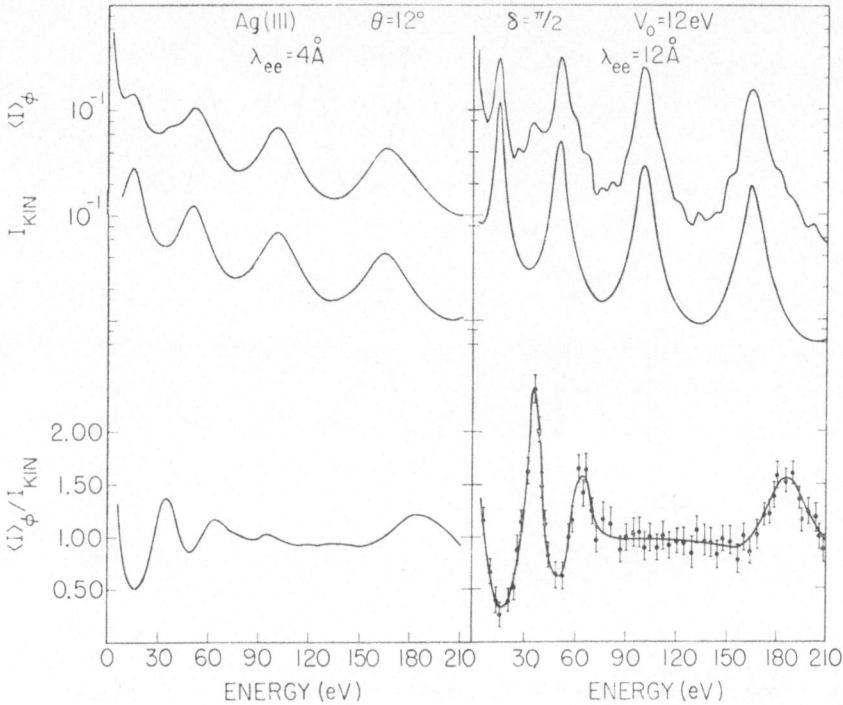

Fig. 59. Comparison of the azimuthal (φ) averaged intensity of the (00) beam with the kinematic intensity (upper panel) and the ratio of φ averaged intensity to kinematic intensity for two values of λ_{ee} (lower panel). For each λ_{ee} the six values of $\varphi = 0$, 10°, 20°, 30°, 40°, and 50° were used in constructing the average. The error bars in the ratio at $\lambda_{ee} = 12$ Å indicate the estimated uncertainty in performing the average from the given set of six intensity profiles. This uncertainty is much smaller for $\lambda_{ee} = 4$ Å so error bars are not indicated for that case. The values of the dynamical parameters are indicated in the figure. (After Duke and Smith [119].)

are expected to exhibit flatter, broader maxima than the kinematical intensity profiles: precisely the behavior evident in Figures 58 and 59.

Because the origin of the discrepancy between the averaged and kinematical intensities resides in a fundamental aspect of any multiple-scattering process, it cannot be eliminated merely by introducing additional averages to smooth further the dynamical intensities. We have noted, however, Lagally *et al.*'s argument that an additional average over energy-incident-angle $(E\text{-}\theta)$ pairs of parameters leading to a constant momentum transfer improves substantially the agreement between the average and kinematical intensities. Following their prescription [135] for Ag(111) leads to the

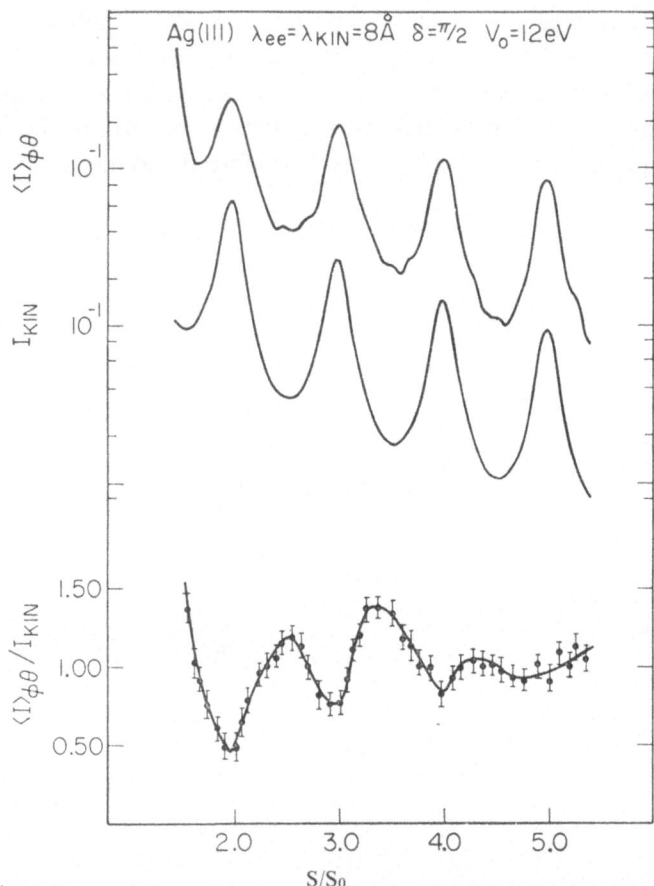

Fig. 60. Comparison of the grand averaged intensity of the (00) beam with the kinematic intensity (upper panel) and the ratio of the grand averaged intensity to the kinematic intensity (lower panel). The six values of $\varphi = 0$, 10°, 20°, 30°, 40°, and 50° were used in constructing the azimuthal averages. Then six azimuthal averages for angles of incidence $\theta = 12°$, 14°, 16°, 18°, 20°, and 22° were in turn averaged by selecting energies and incidence angles corresponding to fixed momentum transfers. The intensities are plotted as a function of a normalized momentum transfer, S/S_0, chosen so that the Bragg peaks occur at integral values of S/S_0. The error bars in the ratio indicate the estimated uncertainty in performing the average from the given set of thirty-six intensity profiles. The values of the dynamical parameters are indicated in the figure. (After Duke and Smith [119].)

'grand' average over both φ and $(E\text{-}\theta)$ indicated in Figure 60. Evidently, the discrepancy between the averaged and kinematical intensities has been reduced somewhat relative to that evident in Figure 58. It remains a factor of about 10–50%, however, even for the case of the grand average.

Duke and Smith proceeded from their analysis of Ag(111) to one of Al(100) with the objective of determining those functions and variables describing ELEED intensities such that by averaging over a range of values of the incident beam parameters (E, θ, φ) the results of the kinematical analysis (first-Born approximation) are recovered as closely as possible. In a general formulation of this problem, they used as the distinguishing characteristic of the kinematical model the fact that it predicts scattered intensities which depend *only* on the momentum transfer, **S**. Therefore, the incident beam parameters (E, θ, φ) may be regarded as determining a three-dimensional vector space in which the equation

$$\mathbf{S}(E, \theta, \varphi) = \text{const} \tag{135}$$

determines a family of two-dimensional surfaces over which the kinematical intensities are constant but the 'exact' dynamical scattered intensities are not. In these terms the correspondence between the averaged data and kinematical model predictions would be ideal if a suitable measure of the area, dA, along these surfaces and suitable functions, F, of the LEED intensities, I, could be constructed such that

$$\int_{\mathbf{S}=\text{const}} dA\,[F(I) - F(I^{\text{kin}})] = 0, \tag{136}$$

in which the exact (I) and kinematical (I^{kin}) intensities both depend parametrically upon the beam parameters (E, θ, φ).

Lagally *et al.*'s [135] proposed solution to this problem consisting of setting

$$F(I) = I \tag{137a}$$

$$dA = d\varphi\,d\theta. \tag{137b}$$

We already have seen that this proposed solution satisfies Equation (136) only to within an error:

$$\int dA\,[F(I) - F(I^{\text{kin}})] \Big/ \int dA\,F(I^{\text{kin}}) \sim 0.25 \tag{138}$$

for physically reasonable parameters describing the electron-solid interaction. Moreover, this error is a sentitive function of the inelastic collision damping, diminishing as the damping increases.

Duke and Smith's analysis of Al(100) established that the appropriate measure of the area along the surfaces of constant S is not given by Equation (137b) but rather by

$$dA = d\varphi\,d\,[\cos\theta], \tag{139}$$

(i.e., the angular measure on the unit sphere of the direction of the incident beam) and that a maximum mesh size for the numerical evaluation of the integral in Equation (136) is

$$\Delta \varphi \lesssim 5° \tag{140a}$$

$$\Delta [\cos \theta] \lesssim 0.01 . \tag{140b}$$

They concluded, however, that there exists no choice of F for which Equation (136) is a theorem. Their study of the discrepancies intrinsic in the comparison between averaged dynamical intensities and kinematical model's predictions permitted the verification that error estimates of the order specified by Equation (138) could be expected *a priori* provided one used

$$F(I) = I/\cos^2 \theta \tag{141a}$$

for the isotropic-scatterer model (probably

$$F(I) = I \tag{141b}$$

more generally) and *large regions of integration*, e.g.,

$$0 \leqslant \varphi \leqslant 180° \tag{142a}$$

$$0.5 \leqslant \cos \theta \leqslant 1 . \tag{142b}$$

For small integration areas along the constant-S surface the errors in the sense of Equation (138) easily can exceed unity.

A final question of interest is that of "given the intrinsic accuracy of the averaging method noted above, what applications can we envisage for this method?" An issue of importance for 'clean' single-crystal solid surfaces is whether or not the layer spacings of the uppermost few layers are expanded (or contracted) relative to those in the bulk solid. As discussed earlier, [1] such alterations of the layer spacing can range from 1–10%. We show in Figure 61 a comparison of the averaged dynamical and kinematical interference functions for the cases of a truncated bulk solid (second panel from the top), an upperlayer spacing expanded by 10% (top panel), and an upper layer spacing contracted by 10% (third panel from the top). It is evident from the figure that in this case (of strong scattering and strong damping), the data averaging method easily could be used to detect an expansion or contraction by 10% of the upper layer spacing of the solid provided a sufficiently refined grid [e.g., that specified by Equations (142)] is used in constructing the average. However, neither of the clean-surfaces studied thus far by the averaging method (Ag(111) [135] and Ni(111) [136]) have been found to exhibit upper layer spacings or registries detectably different from those characteristic of the bulk metal.

In summary, an energy-averaging method permits comparison of the dynamical ELEED intensities with the predictions of a kinematical calculation for the purpose of experimentally verifying the atomic scattering factors predicted by microscopic models of the short-range, electron-ion-core potentials in simple solids whose two-dimensional unit-cell geometry (parallel to the surface) can be determined directly from the intensity pattern. If $\lambda_{ee} \lesssim 4$ Å constant-momentum-transfer averages agree sufficiently closely with the predictions of the kinematical model to be useful in determining shifts in the upper-layer spacing of a solid by 5–10%. To date, however, no shifts of the upper layer spacings from their bulk values have been documented definitively, although Al(110) clearly exhibits either a distorted surface geometry or typography. [5] We conclude that although the applicability of data-averaging methods

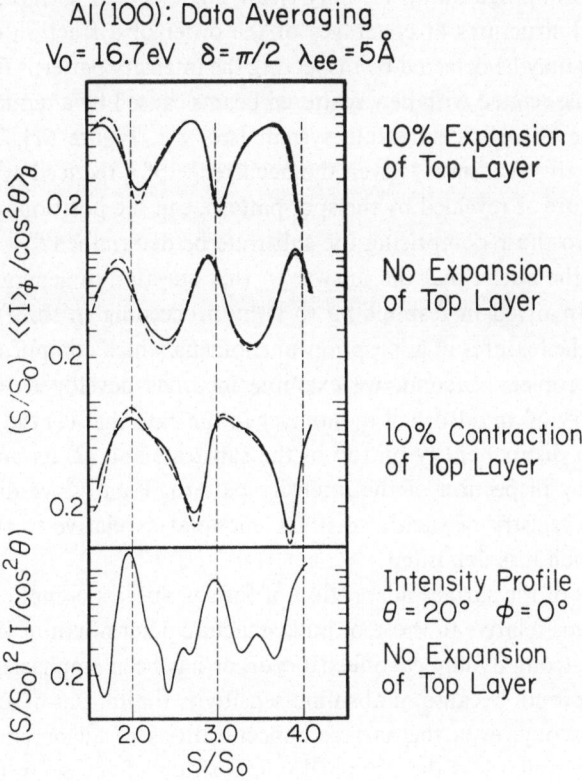

Fig. 61.   Averaged dynamical intensities (solid lines) and their kinematical analogs (dashed lines) for electrons reflected in the (00) beam from Al(100). Both the averaged dynamical intensities and the kinematical averages have been multiplied by $(S/S_0)^2$ in order to obtain a function analogous to the kinematical interference function. The lower panel shows an intensity profile for a particular value of θ and φ in order to illustrate how extensively the averages shown in the upper three panels smooth out the individual intensity profiles. The top panel shows the dynamical and kinematical averages for a crystal in which the upper layer spacing is expanded 10% relative to the bulk. The second panel shows the averages for a truncated bulk solid. The third panel shows the averages for the case in which the upper layer spacing is contracted 10% relative to the bulk value. The dynamical averages were evaluated using the azimuthal (φ) grid 0(9°) 45° and the polar grid for $\cos \theta = 0.6 \times$ $\times (0.025)$ 1.0. The kinematical averages were performed using the $\cos \theta$ grid 0.6(0.01) 1.0. A check on the mesh size indicated an accuracy of a few percent relative to the use of smaller grid meshes. The dynamical parameters are indicated in the figure. The peaks in $\langle\langle I \rangle_\varphi / \cos^2 \theta \rangle_\theta$ for the truncated bulk lattice do not fall at integral values of S/S_0 because the range of values for $\cos \theta$ used in the polar-angle average leads to a large discrepancy between Equation (134a) (which was used in constructing the average [135]) and Equation (133) (which is required by the conservation laws.) (After Duke and Smith [119].)

to clean surfaces has been established theoretically, much more experimental work is required before their applications lead to high-precision characterizations of the geometry of such surfaces.

D. ADSORBED MONOLAYERS

One of the most significant discoveries [131–133] to emerge from studies of low energy electron diffraction is that under certain circumstances if foreign atoms or molecules are permitted to impinge on an initially-clean single-crystal surface, they stick to it forming ordered structures at coverages of the order of a fraction of a monolayer. These structures may be detected by inspecting the intensity patterns for the formation of extra 'spots' associated with new scattered beams caused by a reduced translational symmetry of the adsorbate-substrate system [see, e.g. Figure 51]. The question of interest to us in this section is "given the occurrence of a reduced-symmetry ordered adsorbate structure as revealed by the spot pattern, can the positions of the adsorbate species relative to those comprising the substrate be determined?"

Admitting at the outset that the answer to this question in general is 'no', we discuss the issues involved in responding to it by proceeding in three steps. First, we survey some of the features of adsorption-phenomena which complicate the structure-determination problem. Second, we examine methods developed to determine the unit-cell geometry of an adsorbed monolayer in the case that (1) the monolayer may be regarded as a distinct entity on top of the substrate and (2) its unit-cell geometry is not obvious by inspection of the intensity pattern. Finally, we discuss the determination of the registry of such a 'distinct' monolayer relative to the single-crystal substrate on which it is deposited.

Three features of the surface adsorption of foreign species complicate the resulting structure problems relative to those of bulk structure determination via X-ray diffraction. [97] First, strong dynamical effects occur, as has been emphasized repeatedly in these lectures. Second, because of absolute-sensitivity limitations of auger and photo-electron spectroscopy, even the surface concentrations of adsorbed species are not known precisely, much less the chemical composition of the semi-infinite unit cells normal to the surface. This problem is compounded further if reactive gases are used as the adsorbed species. They undoubtedly displace 'impurity' gases from the walls of the vacuum system as well as adsorb on the surface of the target. Finally, defect and domain structures, common in ordered adsorbed systems, are difficult to detect if the adsorbate-induced unit cell is comparable in size to that of the substrate.

The uncertainties resulting from these complicating features have spawned a bewildering variety of 'microscopic' models of the structure of individual adsorption systems. [131–133] A distinction usually is introduced between monolayer formation ('chemisorption on top' [132]) in which case the adsorbed species do not diffuse into the substrate and 'reconstruction' in which case they do. Yet no definitive empirical test that a given system exists in one state or the other has been proposed. In principle, the various possibilities can be distinguished by comparison of the observed ELEED intensities with the predictions of microscopic models embodying the different geometrical arrangements of the atomic scatterers. Such an analysis of a $C(2 \times 2)$ sodium structure on Ni(100) has been undertaken by Andersson and Pendry with the results shown in Figure 62. As anticipated in Section IV.B, some of the data (e.g., the $(\frac{1}{2}\frac{1}{2})$ beam intensities) never are described adequately while others (e.g., the (00) beam intensities) can be fit quite well by adjusting the spacing between the Na monolayer and the Ni(100) substrate. Although an intuitive discrimination between the model potentials employed in the theoretical study can be obtained, the method seems ill-suited for quantitative structure determination in the absence of a sensitivity analysis defining the uncertainties in the structure induced by those in the electronic potential and precise criteria for distinguishing the 'best' structure [such as the criterion given

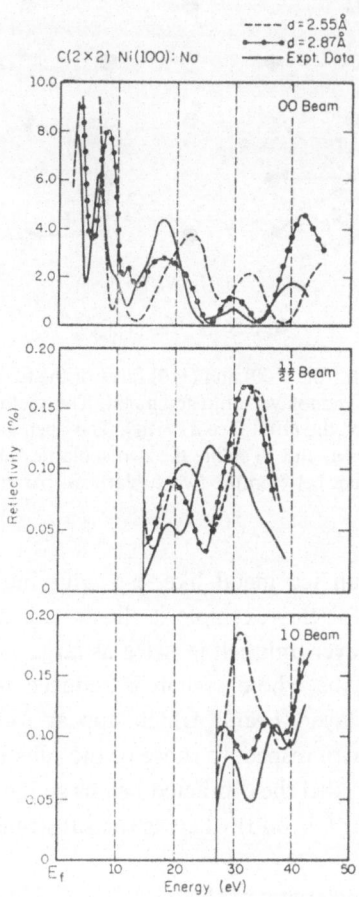

Fig. 62.   LEED spectra taken at normal incidence from a $C(2 \times 2)$ sodium structure on a nickel 100 surface. ——experiment; - - -theory, $d = 2.55$ Å; -.-. theory, $d = 2.87$ Å. Experimental (00) beam × 3 above 25 eV; theoretical $(\frac{1}{2}\frac{1}{2})$ beam × 4; theoretical (10) beam × 2. The theory curves were obtained from an unspecified electron-solid potential as described in [145]. (After Andersson and Pendry [145].)

in Equation (129)]. Figure 62 illustrates well the interpretational difficulties which confront structure analysis via the microscopic-model method even in simple cases.

Confronted with earlier manifestations of this situation, Tucker and Duke [97, 138, 139] noted that if all one requires is the unit-cell geometry of the upper layer *alone*, it could be obtained via the energy-averaging method described by Equations (120)–(129). This type of analysis is most useful in those cases for which monolayer formation occurs 'on top' of the substrate and the monolayer unit cell is large. (These cases occur when the interactions between adsorbed species are strong relative to their interactions with the substrate atoms.) To illustrate an application of the energy-averaging method, let us consider the simple example of a C (2×2) structure on the

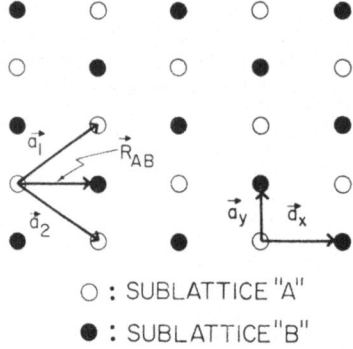

O : SUBLATTICE "A"

● : SUBLATTICE "B"

Fig. 63.   Choice of sublattices for the (100) and (110) faces of the fcc structure for the two sublattice model of a C (2×2) adsorbed monolayer on a fcc metal. The vectors $\mathbf{a}_x$ and $\mathbf{a}_y$ on the right side define the primitive unit cell (for the (100) face $a_x = a_y$). The open and filled circles denote the two sublattices A and B. The vectors $\mathbf{a}_1$ and $\mathbf{a}_2$ define the two sublattice unit cell used in the calculations and $\mathbf{R}_{AB}$ is the displacement between the two sublattices. (After Tucker and Duke [97].)

(100) and (110) faces of an fcc metal like, e.g. aluminum. The geometry of the monolayer and substrate for this example is illustrated in Figure 63. As described earlier, because the monolayer unit cell is twice as large as that of the substrate, the space-group symmetry of the whole system is reduced to that of the monolayer. Therefore the new 'monolayer' beams which appear following adsorption of the monolayer are half-order with respect to those of the substrate. We use the reciprocal lattice of the monolayer to label the diffracted beams so the monolayer beams contain a single odd index (i.e., h or k is odd) whereas the substrate beams contain both even or both odd indices.

The first important characteristic of the energy-averaging method is its elimination of any substrate dependence of the averaged intensities of the (new) monolayer beams. This result is illustrated in Figures 64 and 65 which show calculated intensity profiles and their energy-averages respectively for a C(2×2) monolayer on the Al(100), Al (110), and Al (111) substrates. The marked dependence of the individual intensity profiles on the substrate is evident in Figure 64. From Figure 65, however, we see that the M_{hk}, defined by Equation (120), obtained for the various substrate geometries agree well both with each other and with the average kinematical intensities given by Equations (121) and (122). Tucker and Duke [97] demonstrated the validity of this conclusion for a wide variety of model parameters using the isotropic-scatterer version of the inelastic-collision model.

The second important characteristic of the energy-averaging method is its applicability despite major modifications in the electron-solid force law which alter dramatically the appearance of the individual intensity profiles. Figures 66 and 67 illustrate the enormous changes in the substrate and monolayer beams respectively of different

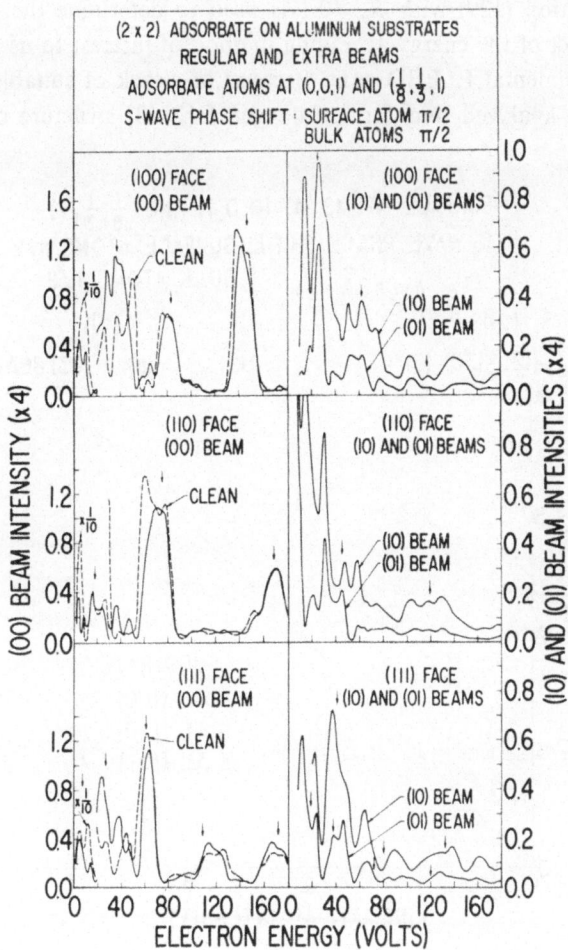

Fig. 64. Intensity profiles of a regular beam (00) and two extra beams (10) and (01) for an adsorbate with atoms at (0, 0, 1) and ($\frac{1}{8}$, $\frac{1}{3}$, 1) placed on aluminum (100), (110), and (111) substrates. For the (00) beam the profile for the clean surface (dashed curve) is shown for comparison. The exponential decay length is 8 Å and the s-wave phase shift is $\pi/2$ for both the adsorbate and substrate atoms. Arrows designate the energies of the maxima of the Bragg envelopes (after Tucker and Duke [97].)

model chemisorbed monolayers on the Laramore-Duke-Snow model of Al(100). For example, the difference between p- and d-wave adsorbates (at identical positions) almost interchanges maxima and minima in the intensity profiles. (Recall our comments in connection with Figures 62 about the consequences of uncertainties in the dynamical as well as geometrical parameters.) As shown in Figure 68, however, if $\lambda_{ee} \lesssim 4$ Å, the energy-averaging method still reduces the energy integrals, M_{hk}, of these highly-dynamical intensity profiles to their kinematical counterparts as defined in Equations (121) and (122). That the results shown in Figure 68 for mixed p- and d-wave adsorbates are valid in general was demonstrated by Duke and Laramore. [118] They also showed that the uncertainty in the positions of the surface-layer scatterers imposed by the non-exact nature of the energy-averaging method is $\Delta a \sim$ ~ 0.1 Å if Equation (129) with $R_m = 0.1$ is used to determine the 'final' geometry.

The final aspect of the energy-averaging method of interest to us is its application to analyze experimental ELEED data. Because of a lack of suitable data, only one system has been analyzed thus far: an oxygen $C(2 \times 8)$ structure on Rh(100). The

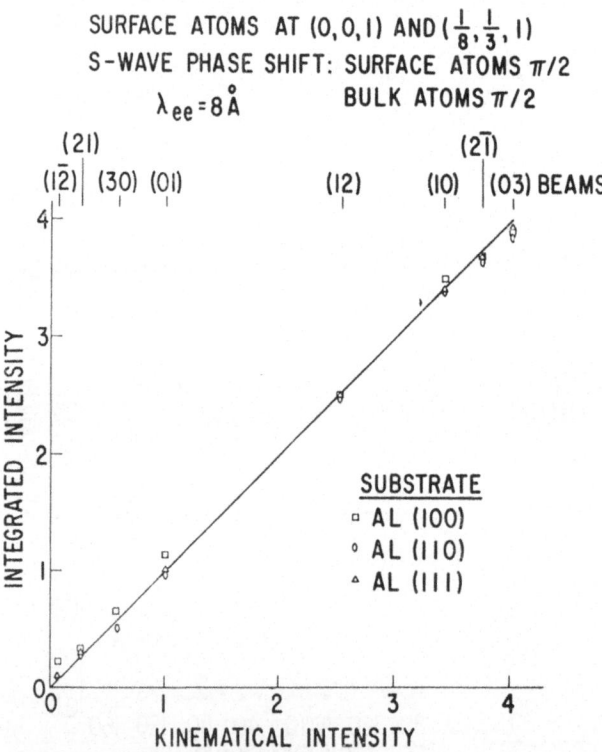

Fig. 65. Comparison of calculated integrated dynamical and kinematical intensities for normally incident electrons scattered from an adsorbate with atoms at $(0, 0, 1)$ and $(\frac{1}{8}, \frac{1}{3}, 1)$ placed on aluminum (100), (110), and (111) substrates at a height of one substrate layer spacing. The exponential decay length is 8 Å and the s-wave phase shift is $\pi/2$ for both the adsorbate and substrate atoms. The integrated dynamical intensities are normalized to the kinematical (12) beam intensity.
(After Tucker and Duke [97].)

results of Tucker and Duke's analysis [97] of this case is summarized in Figure 69. The value of 0.28 for the relibility index [defined in Figure 69 not by Equation (129)] was obtained by optimizing the vertical position of the oxygen atom at (3/7,0). As a value of $R \leqslant 0.5$ indicates a structure which is basically correct but needs refinement it is clear that the energy-averaging method constitutes a *practical* method of determining the unit-cell structure of the top layer of a solid. Obviously, a more accurate structure could be obtained in this particular case if a more accurate form factor for electron scattering from the adsorbed oxygen and more extensive data were available.

We now turn to our final topic in this subsection, the determination of the monolayer-substrate registry for cases in which an adsorbed species forms a distinct monolayer

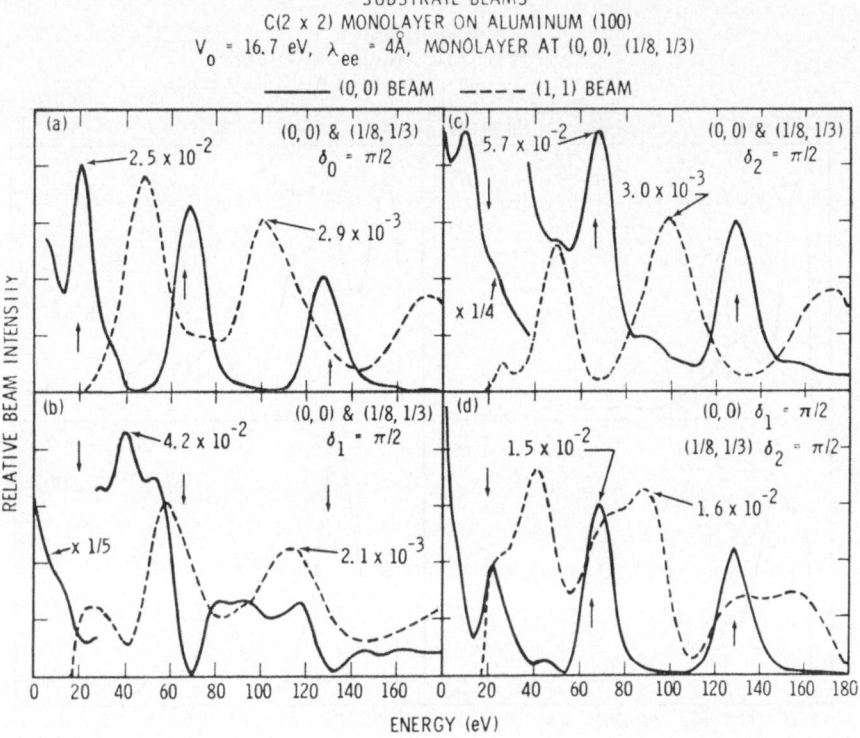

Fig. 66.   Intensity profiles for two substrate beams for the case of $C(2 \times 2)$ monolayer on the (100) face of aluminum. The beams are labeled according to the reciprocal lattice vectors of the combined monolayer-substrate system. There are two atoms per unit cell of the monolayer at the horizontal positions (0,0) and $(\frac{1}{2}, \frac{1}{2})$ in terms of the unit vectors of the monolayer Bravais net and at a distance 2.0201 Å from the center of the top layer of substrate scatterers. The monolayer phase shifts are indicated in the figure. The phase shifts describing the aluminum ion-cores of the substrate were obtained from Laramore-Duke-Snow APW potential. The other parameters used in the calculation are indicated in the figure as are the reflectivities corresponding to some of the peaks in the intensity profiles. The calculations were performed for the case of a rigid lattice and a normally incident beam. Arrows designate the kinematical primary Bragg energies in the (00) beam.
(After Duke and Laramore [118].)

'on top' of the substrate. As in the case of the internal monolayer geometry, little
work has been done on this topic primarily because of lack of data on suitably characterized experimental systems. Tucker and Duke [97] proposed that this information
could be extracted from the *substrate* beams either by microscopic-model calculations
or by constant-momentum-transfer averaging. The utility of the latter proposal
depends on the magnitude of the inelastic-collision damping. Using the model self-
energy of Equation (113), the constant-momentum-transfer-average method is ex-
pected to be useful only if $(\lambda_{ee}/a) \lesssim 1$, where a is the monolayer-substrate spacing.
In this limit, scattering from the monolayer can be as strong as that from the substrate.
Therefore the predominant diffraction phenomena reflect the interference of electrons
scattered from the adsorbed layer with those scattered from the substrate as a whole

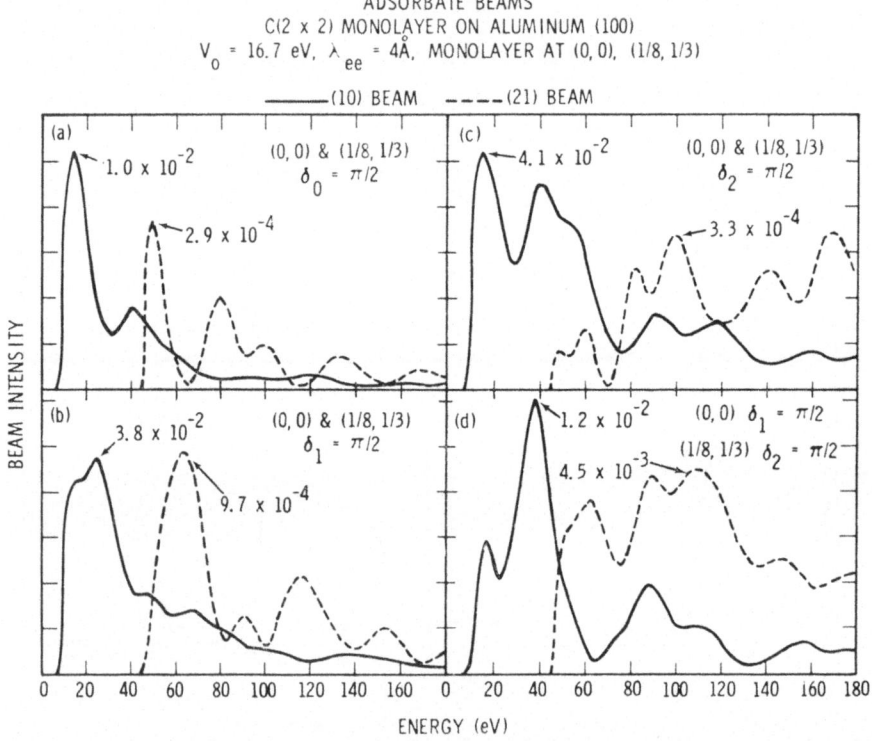

Fig. 67.   Intensity profiles for two monolayer ('adsorbate') beams associated with a C(2 × 2) monolayer on the (100) face of aluminum. The beams are labeled according to the reciprocal lattice vectors
of the monolayer. There are two atoms per unit cell of the monolayer at the horizontal positions
$(0, 0)$ and $(\frac{1}{8}, \frac{1}{3})$ in terms of the monolayer Bravais net and at a distance 2.0201 Å above the substrate.
The monolayer phase shifts are indicated in the figure. The phase shifts describing the aluminum
ion-cores of the substrate were obtained from the Laramore-Duke-Snow APW potential. The other
parameters used in the calculation are indicated in the figure as are the reflectivities corresponding
to the peaks in the intensity profiles. The calculations were performed for the case of a rigid lattice
and a normally incident beam. (After Duke and Laramore [118].)

and hence are particularly sensitive to the monolayer-substrate registry. A similar conclusion holds for very low energy electrons, $E \lesssim 10$ eV, which confront a band-gap in a low-index face of the pure substrate. In this case, the substrate acts like an almost totally reflecting mirror so that observed interference phenomena again reflect the monolayer-substrate registry. [146, 147] This concept has been applied within the context of a simple model to determine the monolayer-substrate lattice spacings of disordered monolayers of alkali metals (N, Ba, K, and Cs) on Ni(100). [147]

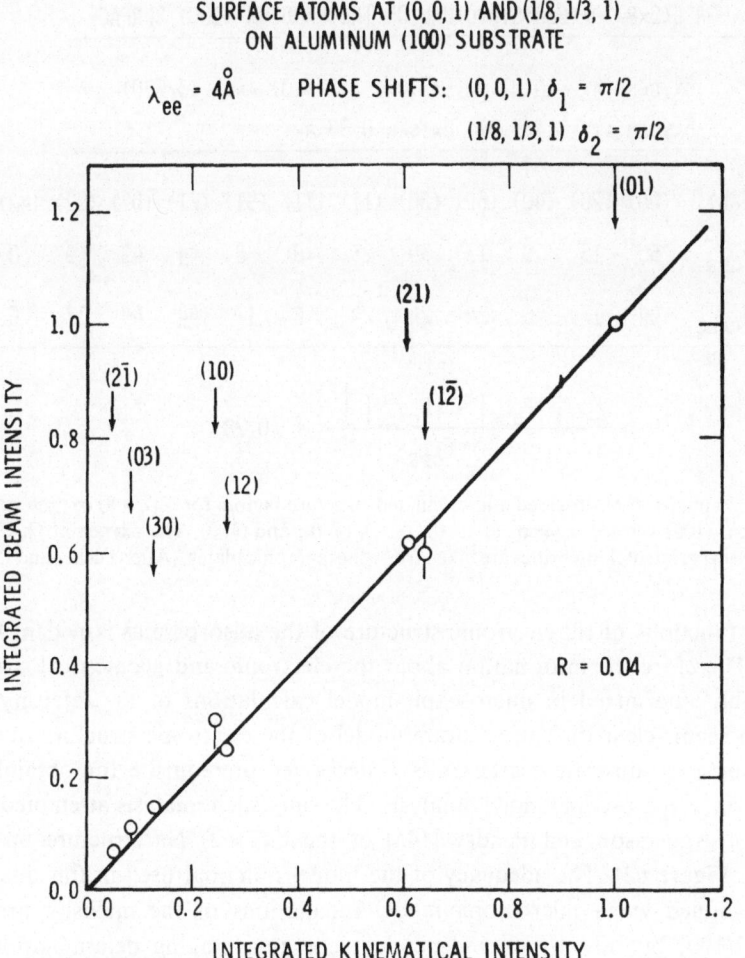

Fig. 68. Normalized integrated dynamical intensities versus the normalized integrated kinematical intensities for the case of a C(2 × 2) monolayer on Al(100). One atom in the monolayer unit cell is a p-wave scatterer and the other atom in the monolayer unit cell is a d-wave scatterer. The positions of the adsorbate atoms are expressed in terms of the lattice vectors describing the Bravais net of the monolayer. The open circles indicate the results for eight adsorbate beams obtained using the energy interval 85 eV $\leq E \leq 165$ eV. R is a reliability index indicating the goodness of fit as defined in Equation (129) of the text. The vertical lines passing through the open circles indicate the range of values of the ratio M/I^{kin} obtained using also the intervals 100 eV $\leq E \leq 160$ eV and 85 eV $\leq E \leq 180$ eV. (After Duke and Laramore [118].)

The sensitivity of the intensity profiles to the monolayer-substrate registry, initially documented by Tucker and Duke, [97] may be seen by comparing the profiles shown in Figures 70 and 71 with those in Figures 66 and 67. The structure used to obtain Figures 70 and 71 is related to that used for Figures 66 and 67 by a translation of the monolayer by one-half of a primitive lattice vector across the surface and about 0.5 Å toward the surface. Dramatic changes in the intensity profiles of the (00) and mono-layer beams are visible in these four figures. However, these intensity profiles also are

COMPARISON OF OBSERVED AND CALCULATED STRUCTURE FACTORS
FOR C(2x8) OXYGEN STRUCTURE ON THE RHODIUM (100) SURFACE

OXYGENS AT (0,0), (1/7,0), (2/7,0) AND (3/7,0)

OXYGEN AT (3/7,0) RAISED 0.35 Å

(HK)	(00)	(20)	(40)	(60)	(80)	(11)	(31)	(51)	(71)	(02)	(22)	(42)
F_{OBS}	50	15	0	13	39	23	0	8	44	49	9	0
F_{CALC}	44	17	6	24	24	24	6	17	44	44	17	6

$$R = \frac{\sum \left| |F_{OBS}| - |F_{CALC}| \right|}{\sum F_{OBS}} = 0.28$$

Fig. 69.   Comparison of observed and calculated structure factors for $C(2 \times 8)$ oxygen structure on the rhodium (100) surface oxygens at $(0, 0)$, $(\frac{1}{7}, 0)$, $(\frac{2}{7}, 0)$, and $(\frac{3}{7}, 0)$. The oxygen at $(\frac{3}{7}, 0)$ is raised 0.35 Å. The experimental intensities are taken from photographic plates. (After Tucker and Duke [97].)

sensitive functions of the electronic structure of the adsorbate as is evident from the figures. Therefore the information about the electronic and geometrical structure is inextricably intermixed in microscopic-model calculations of the intensity profiles per se. It seems clear that an accurate model of the electronic structure of both the monolayer and substrate scatterers is a necessary prerequisite for obtaining their registry via a microscopic-model analysis. The only such analysis attempted thus far is that of Andersson and Pendry [145] of the $C(2 \times 2)$ Na structure on Ni(100) [see, e.g. Figure 62]. The adequacy of the model potential used in this analysis has been examined via a microscopic-model calculations of the intensity profiles for clean Ni(100) but not by an independent energy-averaging determination of the atomic scattering factors of the Na atoms in the (presumed ideal) $C(2 \times 2)$ mono-layer. However, our discussion of Figures 62, 66, 67, 70, and 71 suggests that con-siderable additional work is required before reliable values of the geometrical param-eters can be extracted from such an analysis. Since Andersson and Pendry's work is the most complete consideration of actual experimental data available at the present time, we see that the monolayer-substrate registry problem remains un-solved even in the simple case of a monolayer adsorbed 'on top' of the substrate.

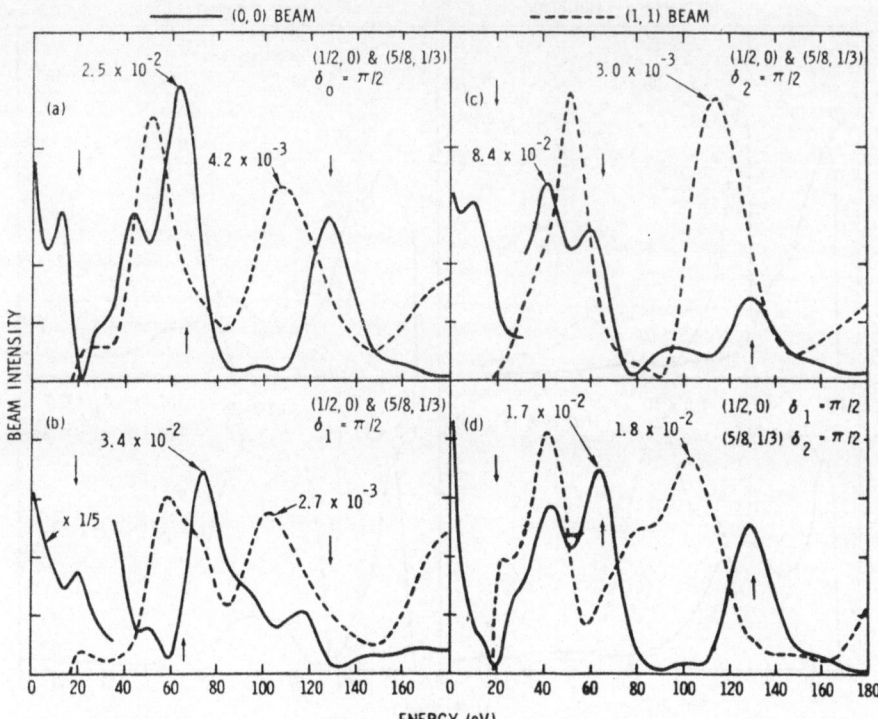

SUBSTRATE BEAMS

C(2 × 2) MONOLAYER ON ALUMINUM (100)

V_o = 16.7 eV, λ_{ee} = 4Å, MONOLAYER AT (1/2, 0), (5/8, 1/3)

Fig. 70.   Intensity profiles for two substrate beams for a $C(2 \times 2)$ monolayer on the (100) face of aluminum. The beams are labeled according to the reciprocal lattice vectors of the monolayer which contains two atoms per unit cell at the horizontal positions $(\frac{1}{2}, 0)$ and $(\frac{5}{8}, \frac{1}{3})$ in terms of the unit vectors of the monolayer Bravais net, and 1.515 Å above the substrate. The monolayer phase shifts are indicated in the panels of the figure. The phase shifts describing the aluminum ion cores of the substrate were obtained from the Laramore-Duke-Snow APW potential. The other parameters used in the calculation are indicated in the figure, as are the reflectivities corresponding to some of the peaks in the intensity profiles. The calculation is for the case of a rigid lattice and a normally incident beam. Arrows indicated the kinematical Bragg energies associated with the substrate (i.e., $d = 2.0201$ Å) although the monolayer is only 1.515 Å above the substrate. (After Duke and Laramore [118].)

Determination of the geometrical structure of reconstructed surfaces lies is only now beginning to be examined [6, 148].

E. SUMMARY AND PERSPECTIVE

Summarizing, we have seen that a principle application envisaged for elastic low-energy-electron diffraction (ELEED) is the determination of the atomic geometry of the uppermost few layers of a single-crystal solid. This anticipation was based on observations, originating with those by Davisson and Germer, [71] of marked changes

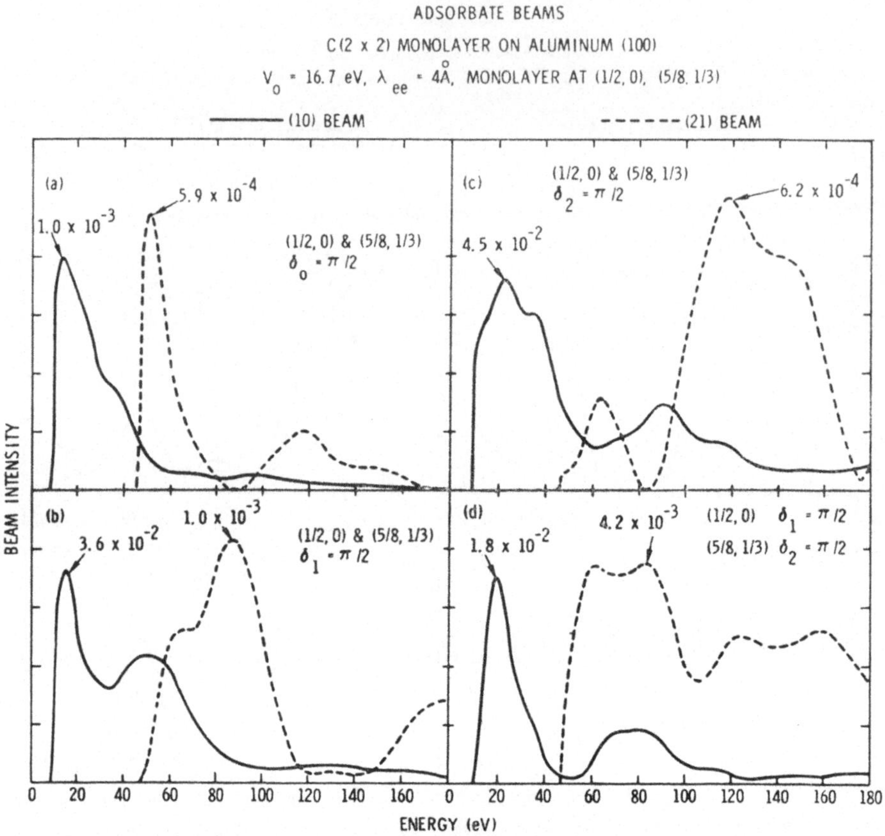

Fig. 71. Intensity profiles for two monolayer ('adsorbate') beams for a C(2 × 2) monolayer in the (100) face of aluminum. The beams are labeled according to the reciprocal lattice vectors of the mono-layer which contains two atoms per unit cell at the horizontal positions $(\frac{1}{2}, 0)$ and $(\frac{5}{8}, \frac{1}{3})$ in terms of the monolayer Bravais net, and 1.515 Å above the substrate. The monolayer phase shifts are indicated in the panels of the figure. The phase shifts describing the aluminum ion cores of the substrate were obtained from the Laramore-Duke-Snow APW potential. The other parameters used in the calcula-tion are indicated in the figure, as are the reflectivities corresponding to some of the peaks in the in-tensity profiles. The calculation is for the case of a rigid lattice and a normally incident beam.
(After Duke and Laramore [118].)

in the intensity pattern upon the adsorption of foreign species. From measurements of the intensity patterns alone, however, one can infer only the space-group symmetry for translations along a planar surface. Measurements of the actual positions of the atomic species in either the adsorbed layers or the substrate have been delayed over four decades by the absence of a suitable theory of ELEED intensities, and a suitable standard of experimental surface characterization.

Modern theoretical work on ELEED has eliminated uncertainties in the theory *per se* as a major source of delay in the applications of ELEED to surface crystallography. In this section we have reviewed this work with the objective of answering four questions: What is the extent and accuracy of the structural information which, in principle, it is possible to extract from ELEED intensities? What are the currently-proposed practical schemes for surface-crystallography via ELEED? How much does the extent and accuracy of the information obtainable from these schemes differ from that which, in principle, is possible? What systems have been examined using the modern theoretical methods?

We found that although, in principle, highly precise information about planar, periodic structures is obtainable, in practice uncertainties in the surface typography, the vibrational properties of surface atoms, the inelastic-collision damping, and even the electronic structure of chemisorbed species greatly reduce the accuracy with which structural parameters *per se* can be extracted from an analyses of observed ELEED intensities. Faced with these uncertainties, we concentrated our attention on data-reduction methods which, at the expense of providing structural parameters with an 'intrinsic' inaccuracy of $\Delta a \sim 0.1$ Å, provide a relatively simple analytical framework for data analysis analogous to the kinematical theory used in the X-ray [123, 124] and experimental ELEED [131–134] literature. Of equal if not greater importance, the "input" dynamical electron-ion-core form factors can be tested *directly in the experimental system* by application of the energy-averaging method to systems of known (e.g., trivial) unit-cell geometry in the upper layer of the solid [see, e.g. Section 4C]. Such data-reduction methods have been applied to very few systems [Al(100), [97] Ag(111), [135] Ni(111), [136] and Rh(100)–C(2 × 8)–O [97]] because of the lack of available measurements of the diffracted intensities. Hopefully, the detailed specification and verification of the bounds of validity of these methods, such as we have indicated above, will stimulate their application to a wider variety of systems.

Application of microscopic-model methods to determine surface geometry, although in principle more accurate than the data-reduction methods, in practice are plagued by substantial ambiguities between various combinations of the electronic, vibrational, and geometrcal parameters. This observation gives rise to questions concerning these methods' widespread practical utility. For example, the bulk solid-state analogue of the requirement that the electron-ion-core form factors be well-known is the requirement that the energy-band structure of a solid be established before its crystal structure could be determined. Yet, for most solids the energy-band structures are still disputed. Even in simple, technologically-important solids, materials with the same crystal structure have widely different band structures [e.g., magnetic metals, 3–5 compound semi-conductors]. Therefore by analogy, if not logic, one can argue that requiring a description of the electronic (and consequently vibronic) structure of surface atoms as a prerequisite for determining their geometry renders their atomic geometry unobservable except possibly in a few simple cases. Indeed, one of the disturbing features of our results is the current reliance on such methods to determine monolayer-substrate registries. Only a few systems have been studied by these methods [Al(100), [5, 48, 97] Al(110), [5, 48] and Ni(100):Na [145]]. The results on all of these systems have proven inconclusive because of uncertainties in the model potential, the ELEED intensity data, or both. Perhaps the future holds more promise [6, 148].

We conclude, therefore, that in the case of surface crystallography (as in that of the determination of other features of the electron-solid force law; see Section 3D8) the major advances in the near future must be experimental in nature. The analytical schemes described above must be tested using measured intensities in a wider range of systems, the electronic and vibrational properties of whose surfaces are characterized independently. Only after the various techniques for structure determination have been tested on several simple, well-characterized experimental systems can their merits and limitations be assessed accurately.

5. Inelastic Low-Energy Electron Diffraction (ILEED)

In this concluding part of the lecture notes we describe the inelastic scattering of 'low-energy' $5\,\mathrm{eV} \lesssim E \lesssim 500\,\mathrm{eV}$ electrons from the presumably planar surfaces of single-crystal solids. We recall from Sections 1B, 1D, and 3B that many types of inelastic scattering events are possible. The incident electron can excite lattice vibrations, low-energy intraband particle-hole pairs, interband particle-hole pairs, plasmons, and core electrons in the target. Having discussed in Section 1B the incident electron's energy loss via its creation of ('deep') core holes and in Section 3B its energy loss via phonon creation, in this part of the lectures we examine its energy loss to electronic excitations of the valence-electron fluid, especially plasmons. In particular, we evaluate the inelastic scattering differential cross sections in the discrete or 'characteristic' loss region of secondary electron energy spectra such as those shown in Figures 8 and 9. The 'true' secondary peak in these spectra results from the emission from the solid of valence-band electrons excited by the incident electron as it loses energy by a cascade of multiple characteristic energy losses.

As emphasized in Sections 1B and 1D, the major applications envisaged for inelastic electron scattering are chemical analysis of the surface region via study of the Auger-electron emission [see Figure 9] and determination of surface electronic structure of the valence electron fluid by studying the surface branches of its elementary excitation spectrum (e.g., surface plasmons, resonant electronic excitations of chemisorbed complexes). Of the various applications, the measurement of the surface-plasmon dispersion relation at clean surfaces is simplest both theoretically and experimentally. Consequently, we proceed in three steps. First, we outline the methods of presenting ILEED data and the qualitative concepts required to interpret the general features of these data. Second, we discuss the level of theoretical sophistication required for an adequate description of the various modes of data presentation. Finally we describe the application of the microscopic quantum-field theory of ILEED [3, 4, 6, 43] to determine the surface plasmon dispersion relation on Al(111) from measured ILEED intensities. [40-43]

A. DEFINITIONS AND GENERAL PRINCIPLES

In this section we begin by reviewing the conservation laws appropriate for a discussion of inelastic-low-energy-electron diffraction (ILEED) as already described for the case of phonon-excitation in Section 3B. We then describe the various types of data displays and conclude with a discussion of the concepts required to describe the major qualitative features of the observed ILEED intensities.

The characteristic distinguishing inelastic *diffraction* from inelastic *scattering* is the satisfaction of the momentum conservation law

$$\mathbf{k}'_{\|} = \mathbf{k}_{\|} - \mathbf{p}_{\|} - \mathbf{g} \tag{143}$$

for the component of momentum parallel to a single-crystal, planer solid surface. Primed quantities characterize the final states, and unprimed ones the initial states. The vectors $\mathbf{k}$ designate electron momenta, $\mathbf{p}$ momenta of the excitations of the solid, and $\mathbf{g}$ the reciprocal lattice vectors of the Bravais net associated with the presumed-identical planes of scatterers parallel to the surface. The important aspect of experimental observations of inelastic diffraction is the existence of a spot pattern in the nelastic cross sections, characterized by a fixed energy loss

$$w \equiv E - E', \tag{144}$$

as well as in the elastic cross sections ($w \equiv 0$). The objective of a theory of inelastic electron diffraction is the description of the diffracted intensities in these ILEED spot patterns as a function of the beam parameters of the incident and exit electrons, and of the dynamical parameters (lattice spacing, electronic structure, etc.) characterizing the target solid. The Hamiltonian on which the theory discussed in this section is based already has been described in Section 2B.

The state of the incident electrons can be specified by the incident beam energy E and incident beam direction (θ, ψ), where the polar axis is taken perpendicular to the solid surface, and the azimuthal angle ψ is measured relative to an axis in the crystal face. (We use ψ rather than φ to describe the azimuthal angle in this part of the lecture notes for compatibility with the literature. [3, 43]) Similar considerations hold for the scattered electrons with energy E' and direction given by (θ', ψ'). However, instead of E' it is generally more convenient to use w, the energy lost by the electrons, defined by Equation (144), as the independent variable. The results of both experiments and calculations are presented for a fixed direction of the incident beam $(\theta, \psi$ fixed) and scattering in a plane $(\psi' = \psi + \pi)$. This procedure reduces to three, E, w, and θ', the number of independent variables specifying the inelastic cross sections. By holding two of them fixed and varying the third, we generate 'energy (intensity) profiles', 'loss-profiles', and 'angular profiles' respectively as indicated in Figure 72. In detail, these three modes of presenting the data are defined as follows: (a) In the case of an *energy profile* we hold fixed the direction (θ, ψ) of the incident beam, the direction of the scattered beam (θ', ψ'), and the loss energy w. We calculate the scattered intensity as a function of incident beam energy E. (b) In the case of a *loss profile* we hold fixed the direction (θ, ψ) and the energy E of the incident beam and the direction (θ', ψ') of the scattered beam. We calculate the scattered intensity as a function of the loss energy w. (c) In the case of an *angular profile*, we hold fixed the direction (θ, ψ) and energy E of the incident beam as well as the loss energy w. We calculate the scattered intensity as a function of the final angle θ' (with $\psi'=\psi+\pi$).

The general consequences of the conservation laws have been indicated in Figure 23. They were described in Section 3B in conjunction with this figure. We recall that for values of w above but near the threshold excitation energy, the conservation laws predict a single complicated structure peaked about the inelastic specular direction [defined by $\mathbf{p}_{\parallel}=0$, $w=\hbar\omega_b(0)$] for bulk plasmons and a doublet structure with its central minimum in the corresponding direction for surface plasmons. Unfortunately, such simple lineshapes are almost never observed. [40–42, 149–151] Consequently, we must examine microscopic models of the inelastic-diffraction process in order to describe the observations.

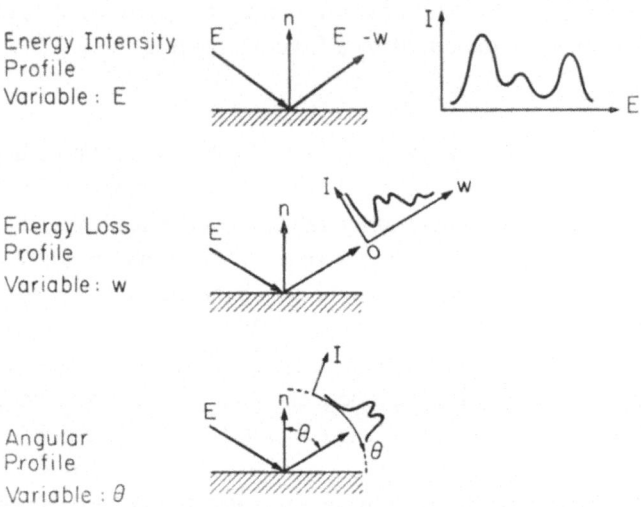

Fig. 72. Schematic diagram of the three types of display of inelastic low-energy-electron diffraction (ILEED) intensity data. The 'loss profile' display is called the secondary-electron-emission spectrum if all final angles (θ', ψ') are integrated over.

Another distinguishing feature of ILEED (as opposed to 'inelastic scattering') is the existence of simple relationships between resonance maxima in the elastic cross sections and those in the inelastic cross sections. [71, 149–154] These relationships are direct manifestations of the 'two-step' nature of plasmon-assisted diffraction as elastic-diffraction prior to energy-loss [71] or vice-versa. [152, 153] I.e., the electron loses energy by a predominantly forward-scattering loss event $(p_{\parallel}\ll k)$ and consequently must scatter elastically from the lattice in order to get 'turned around'. A schematic indication of the consequences of the 'two-step' diffraction mechanism is given in Figure 73. In a kinematical model the maximum in the elastic energy profile at $E=E_B$ would be a primary Bragg peak caused by the coherent elastic scattering of the incident beam of electron from the periodic planes of scatterers in the solid. The two maxima in the inelastic energy profile at $E=E_B$, E_B+w, evidently are analogous diffraction phenomena. The occurrence of such phenomena in experimental data, [149–153] and our emphasis in the model calculations on describing them, [40–43, 154] provides further motivation for us to refer to our analysis of inelastic low-energy electron-solid scattering as one of inelastic low-energy-electron *diffraction*. This feature of the cross-sections is not predicted by single-step (i.e., Born-approximation) models of inelastic scattering because such models do not permit the correlation of resonance maxima in the elastic and inelastic channels. The semiclassical theory of Lucas and Sunjić, [155] for example, says nothing about the fact that resonant elastic

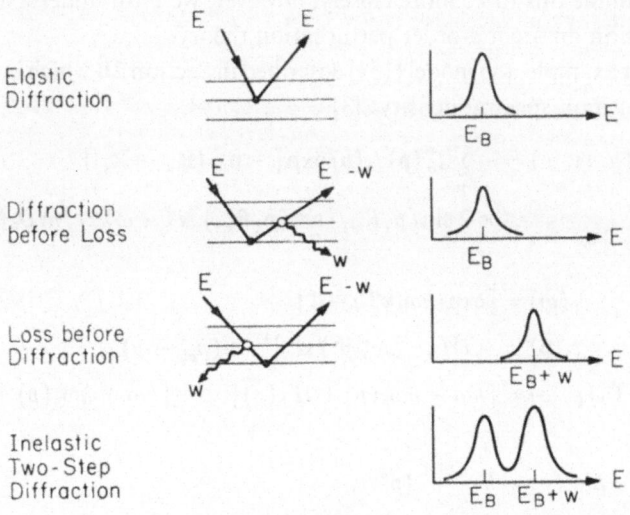

Fig. 73. Schematic indication of the relationship between a diffraction resonance in the elastic energy ('intensity') profile (top panel), the diffraction-before-loss (DL) energy profile (second panel), the loss-before-diffraction (LD) profile (third panel), and total inelastic energy profile (bottom panel) predicted by a two-step diffraction model. The profiles are illustrated for a bulk plasmon emission process in the direction of the specular beam ($k'_{\parallel} = k_{\parallel}$).

reflection makes inelastic processes appear prominent at certain energies which are simply related to the energies of maxima in the elastic-scattering intensities. Also their model describes only the 'total' inelastic cross-section integrated over angles. Therefore, it is not directly useful for studying plasmon dispersion relations because the angular dependence of the intensity of the inelastically scattered electrons is the quantity which provides information about the dispersion of plasmons. [40–43]

Since features associated with kinematical Bragg resonances are going to interest us, and since these resonances are consequences of the vestiges in ELEED of momentum conservation for motion normal to the surface, it is appropriate to inquire more generally into the consequences of the vestiges of this conservation law in ILEED. We already have noted in our discussion of phonon-assisted diffraction (Section 3B) that in the Born approximation, a 'Bragg' resonance in an inelastic channel associated with the scattered beam labeled by **g** occurs if

$$\mathbf{g} \cdot \mathbf{d} + [k_{\perp 1}(0, E) + k'_{\perp 1}(\mathbf{g}, E - w) \pm p_{\perp}(w, p_{\parallel})] d_{\perp} = 2n\pi \qquad (145a)$$

$$\hbar\omega_b [p_{\perp}(w, p_{\parallel}), \mathbf{p}_{\parallel}] = w \qquad (145b)$$

for the emission of bulk phonons of momentum $p_{\perp}$ normal to the surface. For plasmons, however, the forward-scattering nature of the inelastic loss vertex prevents the first Born approximation from contributing anything but background to the reflected intensity. [3] Nevertheless, an analogous 'sideband diffraction' resonance also occurs in the two-step (i.e., second-order perturbation-theory) calculation. [3, 47, 74, 154] In order to examine this topic more closely, however, we must undertake a somewhat lengthy digression on second-order perturbation theory.

Consider, for example, the model [154] described in Section 2B which is characterized by the bulk plasmon spectral density: [3]

$$\Lambda_b(n, m, \omega) = - \sum_{\mathbf{p}} t_b^*(\mathbf{p}) \, t_b(\mathbf{p}) \exp[- \mathbf{p}_{\parallel} \cdot (\mathbf{R}_w - \mathbf{R}_n)]$$

$$\times \, 2 \sin(p_{\perp} R_{m\perp}) \sin(p_{\perp} R_{n\perp}) \, N(-\omega) \, 2i \, \mathrm{Im} \, D_b(\mathbf{p}, \omega), \qquad (146a)$$

$$N(\omega) = [\exp(\hbar\omega/\kappa T) - 1]^{-1} \qquad (146b)$$

$$t_b(\mathbf{p}) = - i [(\pi e^2 \hbar\omega_b/p^2) \, \Omega^2]^{1/2} \, \theta(p_{cb} - p), \qquad (146c)$$

$$D_b(\mathbf{p}, \omega) = [\hbar\omega - \hbar\omega_b(p) + i\Gamma_b(p)]^{-1} - [\hbar\omega + \hbar\omega_b(p) + i\Gamma_b(p)]^{-1}, \qquad (146d)$$

$$\hbar\omega_b(p) = \hbar\omega_b + A p^2 \qquad (147a)$$

$$\Gamma_b(p) = \Gamma_b + B_1 p^2 + B_2 p^4. \qquad (147b)$$

In Equations (146) p_{cb} denotes a cut-off momentum and $\Omega \equiv A \cdot d_{\perp}$ is the volume of a unit cell of area A and depth $d_{\perp}$. Using this model, the second-order perturbation-theory expression for the differential inelastic cross section is obtained by squaring the sum of the amplitudes for the two processes indicated in Figure 74. The result is given by [42]

$$\left(\frac{d^2\sigma}{d\varepsilon \, d\Omega} \right)^{(4)}_{bp} = \left(\frac{E - w}{E} \right)^{1/2} \left(\frac{2\pi m}{h^2} \right) \frac{m\pi e^2 \hbar\omega_b}{h^2} \frac{\Omega^2}{A} \sum_{\mathbf{g}} \int \frac{dp_{\perp}}{\pi}$$

$$\times \, \frac{1}{[p_{\parallel}^2 + p_{\perp}^2 + (\Gamma_b/A)]} \, \frac{\Gamma_b(\mathbf{p}_{\parallel}, p_{\perp})}{[w - \hbar\omega_b(\mathbf{p}_{\parallel}, p_{\perp})]^2 + [\Gamma_b(\mathbf{p}_{\parallel}, p_{\perp})]^2}$$

$$\times \, M[k_{\perp}(0, E), k'_{\perp}(0, E - w), p_{\perp}, \mathbf{g}] \, |A_b(\mathbf{g}, E) + A_c(\mathbf{g}, E - w)|^2, \qquad (148a)$$

$$M(k_{\perp}, k'_{\perp}, p_{\perp}, \mathbf{g}) = |[1 - R(k_{\perp}, k'_{\perp}, p_{\perp}, \mathbf{g})]^{-1}$$

$$- [1 - R(k_{\perp}, k'_{\perp}, - p_{\perp}, \mathbf{g})]^{-1}|^2 \qquad (148b)$$

$$R(k'_{\perp}, k_{\perp}, p_{\perp}, \mathbf{g}) = \exp[i(k_{\perp} + k'_{\perp} + p_{\perp}) d_{\perp} - i\mathbf{g} \cdot \mathbf{a}] \qquad (148c)$$

$$A_b(\mathbf{g}, E) = - \frac{mi\tau(E)}{\hbar^2 A k_\perp(\mathbf{g}, E)} \frac{1}{\{1 - R[k_\perp(0, E), k_\perp(\mathbf{g}, E), \mathbf{g}]\}} \quad (148d)$$

$$R(k'_\perp, k_\perp, \mathbf{g}) = \exp[i(k'_\perp + k_\perp)d_\perp - i\mathbf{g}\cdot\mathbf{a}] \quad (148e)$$

$$A_c(\mathbf{g}, E - w) = - \frac{mi\tau(E - w)}{\hbar^2 A k'_\perp(-\mathbf{g}, E - w)}$$

$$\times \frac{1}{\{1 - R[k'_\perp(-\mathbf{g}, E - w), k'_\perp(0, E - w), \mathbf{g}]\}}. \quad (148f)$$

$$\tau(E) = t(E)[1 - G^{sp}(\mathbf{k}_\parallel, E)t(E)]^{-1}. \quad (148g)$$

In Equations (148) A is the area of a unit cell, $d_\perp$ the spacing between atomic layers, **a** the vector displacement of a unit cell from one layer to the next, $t(E)$ the electron-ion-core scattering amplitude for the (presumed-identical) scatterers, and $G^{sp}(\mathbf{k}_\parallel, E)$ is a propagator defined by Duke et al. [66] Finally, the value of $\mathbf{p}_\parallel$ is obtained from Equation (143), and we use the symbol $k_\perp(\mathbf{g}, E)$ to denote the perpendicular component of momentum in the gth beam inside the crystal as given by Equation (126b).

Although Equations (148) may appear lengthy and unwieldy, they illustrate five important predictions of second-order perturbation theory. [3, 154] First, the factor $M(k_\perp, k'_\perp, p_\perp, \mathbf{g})$ in Equations (148a) and (148b) yields the sideband-diffraction resonances described by Equations (145). Thus it serves the function in second-order perturbation theory that the expression in braces in Equations (94) and (96) serve in first-order perturbation theory. Second, the *elastic* scattering vertices for diffraction before loss (DL, see, e.g. Figure 74a) and loss-before-diffraction (LD, see, e.g. Figure 74b) are given by the factors $A_b(\mathbf{g}, E)$ and $A_c(\mathbf{g}, E-w)$ respectively in Equations

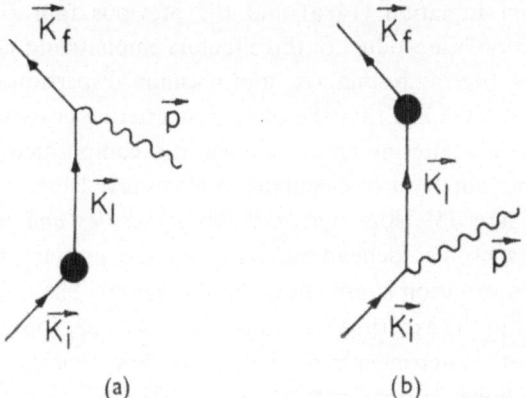

(a)   (b)

Fig. 74. Perturbation-theory diagrams for the scattering amplitudes for diffraction followed by loss, (a), the loss followed by diffraction, (b). Solid lines indicate renormalized electron propagators, wavy lines indicate boson propagators, and shaded dots indicate 'bare' electron-lattice vertices. Energy, but not momentum, is conserved at each vertex. Conservation of momentum parallel to the surface results from performing the sums over $\mathbf{R}_n$ associated with Equations (21) in the text.
(After Duke et al. [74].)

(148a), (148d), and (148f). These factors are responsible for the maxima at E_B and E_B+w respectively shown in Figure 73. Their contributions to the cross-section add *coherently* because they describe a transition of the incident electron between the same initial and final states. Third, the coherent sum of the two elastic-diffraction vertices *multiplies* the sideband-diffraction $M(k_\perp, k'_\perp, p_\perp, \mathbf{g})$ term. Therefore, resonances in either of these terms merely *modulate* those in the other term. Strong maxima occur when the two sets of resonances overlap. Doublets or pairs of doublets occur when they do not. Fourth, the direct consequences of the dispersion of the bulk plasmons are described via the term

$$\text{Im}\, D_b(\mathbf{p}, w) = \frac{2\Gamma_b(\mathbf{p}_\|, p_\perp)}{[w - \hbar\omega_b(\mathbf{p}_\|, p_\perp)]^2 + \Gamma_b^2(\mathbf{p}_\|, p_\perp)} \tag{150}$$

in Equation (148a). For bulk plasmons, this term tends to project out a single value of $p_\perp$ in the integral in Equation (148a) given w and $\mathbf{p}_\|$ in terms of the initial and final electron beam parameters via Equations (143) and (144). [See, e.g. the discussion at the beginning of Section 3B.] For surface plasmons, however, the analogous term

$$\text{Im}\, D_s(\mathbf{p}_\|, w) = \frac{2\Gamma_s(\mathbf{p}_\|)}{[w - \hbar\omega_s(\mathbf{p}_\|)]^2 + [\Gamma_s(\mathbf{p}_\|)]^2}, \tag{151}$$

$$\hbar\omega_s(\mathbf{p}_\|) = \hbar\omega_s + C_1 p_\| + C_2 p_\|^2, \tag{152a}$$

$$\Gamma_s(\mathbf{p}_\|) = \Gamma_s + D_1 p_\| + D_2 p_\|^2, \tag{152b}$$

is large only for values of w and $p_\|$ such that the equation $w = \hbar\omega_s(p_\|)$ is satisfied. Since $p_\|$ is known from Equation (143) and w is known from the retarding voltage in the collector of the LEED spectrometer, the two-step model offers a potential technique for determining the surface-plasmon dispersion relation. Fifth and finally, we see from Equation (148a) and the previous four comments that the calculated cross section is a product of three factors emphasizing sideband-diffraction resonances, elastic-scattering resonances, and plasmon-dispersion resonances respectively. Therefore we anticipate that the inelastic-scattering cross sections will reflect structure in the elastic-scattering cross sections in a complicated fashion. That this anticipation is borne out both in calculated and observed loss profiles for Al(100) is illustrated in Figure 75. Note that *both the intensities and the energies* of the maxima in the loss profiles depend explicitly on the primary beam energy (and angles although this variation is not shown in the figure). The origin of these effects is clearly mis-stated in the experimental literature. [149–153, 156–158] Moreover, the systematic feateres of the dependence on the primary beam energy of the experimental ILEED intensities shown in this figure are amazingly well-reproduced by the simple two-step model calculations in which *all parameters were obtained independently of the ILEED intensities shown in the figure.*

We finally have accumulated enough information about the two-step model to return to the topic which originally motivated our discussion of its details: the role of the vestiges of momentum-conservation for motion normal to the surface on the predicted ILEED intensities. First, we emphasize that the concept of normal-momentum conservation is defined *only* in perturbation theory. We saw in Sections 3A and 3C that 'Bragg' peaks occur in low-order perturbation theory, but *not* in a complete multiple scattering theory. As, however, the two-step model is simply a version of second-order perturbation theory, we expect the concept to have validity within the model even if not in practice. The residue of perpendicular momentum conservation at the elastic-scattering vertices is manifested as energy-tuned 'primary' Bragg peaks in A_b and A_c when the appropriate $[I-R]^{-1}$ factors in Equations (148d) and (148f) become large. This aspect of the model's predictions was described in the discussion of Figure 73. The location of these peaks is independent of the nature and quantum numbers of loss-modes excited by the incident electron. For bulk loss modes, however, the sideband-diffraction phenomenon associated with kinematical resonances in the $M(k_\perp, k'_\perp, p_\perp, \mathbf{g})$ factors in Equations (148a) and (148b) provides another 'momentum-

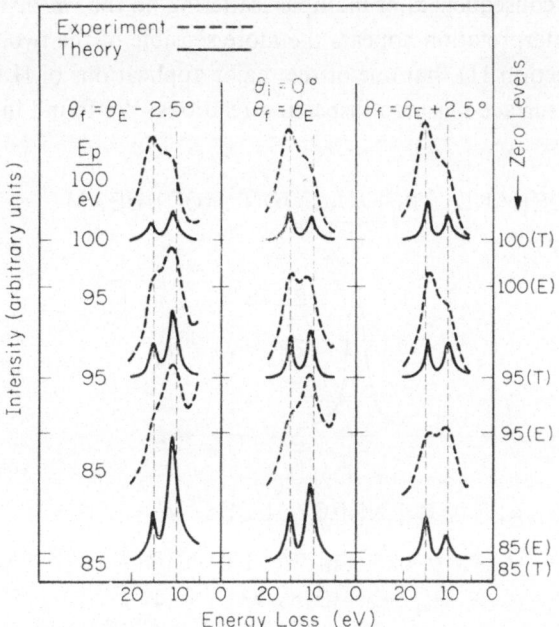

Fig. 75. A comparison of theoretical (full lines) and experimental (dashed lines) loss profiles of electrons incident normally on Al(100) and scattered inelastically in the (11) beam. The primary beam energy is denoted by E_p which also labels the zero levels on the right-hand side of the graph. The notation $T(E)$ designates the theoretical (experimental) curve. The exit angles in the experiment are denoted by θ_f while θ_E defines the direction of the elastically diffracted beam. The theoretical curves in each vertical panel are computed for an emergence angle $1°$ above the measured exit angles. (See [42]. The absolute value of this angle may be in error by as much as $\pm 4°$.) The parameters characterizing the surface plasmon dispersion and damping used in the theory are given by $\hbar\omega_s = 10.1 - 0.7p_\| + 10p_\|^2$ and $\Gamma_s = 0.9 + 0.94p_\|$. Those characterizing the bulk plasmon are taken from keV electron transmission measurements. [42] Elastic electron-ion-core scattering is described by the s-wave inelastic-collision model with $\lambda_{ee} = 8$ Å, $V_0 = 16.7$ eV and $\delta = \pi/4$. For these parameters, the kinematically calculated (11) beam intensities exhibit a primary Bragg peak at $E_B = 86$ eV. A large peak occurs in the experimental (11) intensity profile at this energy. (After Bagchi and Duke [42].)

(i.e. $p_\perp$) tuned' vestige of perpendicular momentum conservation. Its manifestations in the energy profiles are indicated schematically in Figure 76. Each of the peaks shown in Figure 73 can split into a doublet. A detailed discussion of the kinematical conditions needed to observe these resonances is given by Laramore and Duke. [154] The main criterion is evident from Equations (148): w must be substantially greater than $\hbar\omega_b$ and $p_\parallel$ must be small so that the maximum in $\operatorname{Im} D_b(p_\parallel, p_\perp, w)$ occurs at sufficiently large values of $p_\perp$ for the resonances in M to occur at rather different values of $(\mathbf{k}_\parallel, E)$ than those in A_b or A_c. The predictions of a numerical calculation illustrating the possibility of the resulting quartet rather than doublet structure in the energy profile is shown in Figure 77. As shown in Figure 78, experimental measurements on Al(100) reveal the predicted transition from a doublet to a quartet structure with increasing w in the energy profiles. As we shall see in the next section, however, such an observation could be caused by multiple-scattering effects as well as by the vestiges of normal momentum conservation in the two-step model. Given our present knowledge of the consequences of multiple scattering on the *elastic* intensity profiles, [72] the former interpretation appears the more sensible of the two.

We noted in Section 1D that one of the major applications of ILEED was for the determination of surface-plasmon dispersion relations. We found in Section 3D that

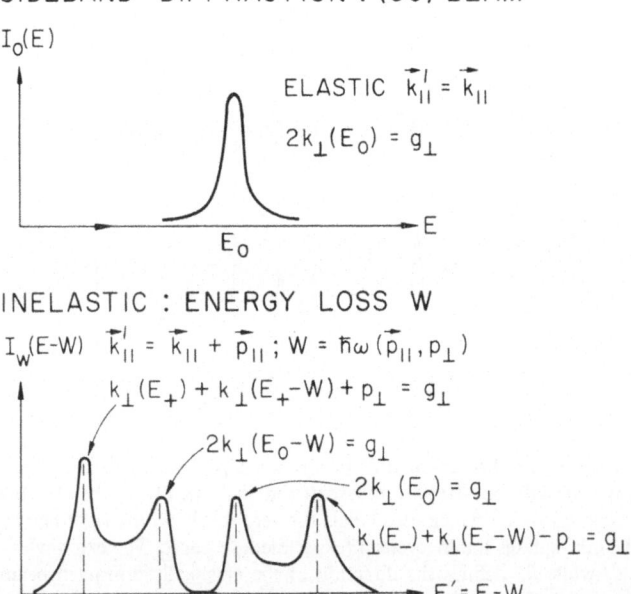

Fig. 76.   Schematic illustration of the sideband-diffraction splitting of the doublet-structure in the inelastic energy profile. This splitting occurs only for bulk plasmons, and even then only if $w \gg \hbar\omega_b$ as defined in Equation (147a) in the text so that the resonance conditions in the figure hold for modest values of $p_\perp$. The quantity $g_\perp$ is $2\pi m/d$ where d is the layer spacing normal to the surface and m is an integer.

the ELEED intensities may be sensitive functions of poorly-known features of the electron-solid force law and that, consequently, their quantitative theoretical description is rather difficult although their qualitative description is quite feasible if one uses a sufficiently refined model. Finally, we have just seen that the detailed behavior of the ILEED energy profiles depends on that of the ELEED intensity profiles. Combining these three observations we are confronted with the question "if one cannot

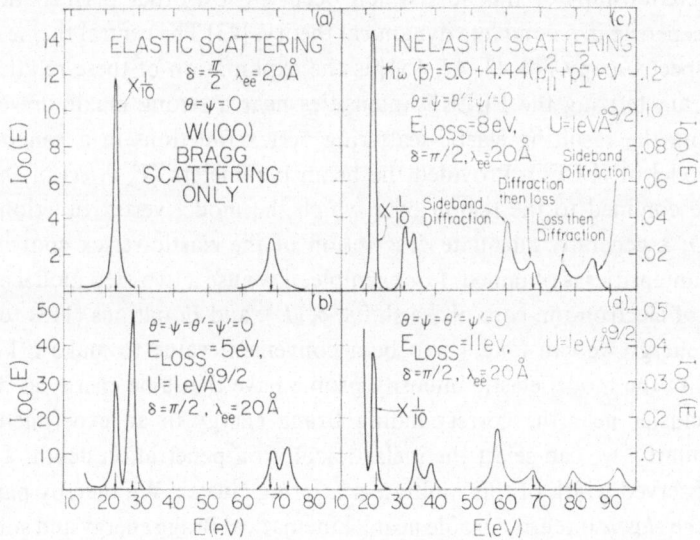

Fig. 77. Intensity profiles obtained for elastic diffraction, (a), and inelastic diffraction, (b)–(d), from a model embodying only primary Bragg scattering of electrons from a rigid lattice. The elastic profile was obtained using the Born approximation. The inelastic profiles were calculated using Equations (2.13), (4.17a), and (4.17b) in [3]. Damping of the plasmons was neglected (i.e. $\Gamma_b(p) \equiv 0$).
(After Laramore and Duke [154].)

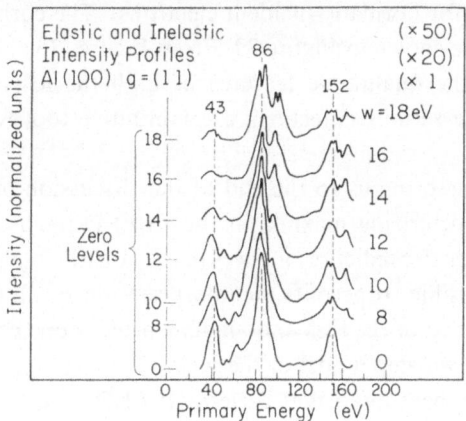

Fig. 78. Elastic and inelastic energy profiles for the (11) beam associated with electrons normally incident on Al(100). The angle of the outgoing beam is taken to be that for which the elastic intensity exhibits a maximum (for each value of the primary beam energy). (After Burkstrand and Propst [105]).

guarantee a quantitative microscopic description of ELEED intensities, how can he hope to analyze ILEED intensities to extract quantitative measures of surface plasmon dispersion (or anything else)?" We conclude our introductory section on general principles by responding to this crucial inquiry.

The essential feature of ILEED on which the prospects for high-precision excitation spectroscopy depend is the existence of direct relationships between resonance-maxima in the ILEED intensities with those in the ELEED intensities. We have just displayed relationships of this sort which occur in low-order perturbation theory. Similar ones persist in a complete dynamical theory. [43] The central theme of ILEED excitation spectroscopy [40–43, 47, 154] is the exploitation of these relationships by initially parameterizing the ELEED intensities near a strong maximum and subsequently using the resulting elastic-scattering vertex functions in a generalized [43] two-step model of ILEED. Provided the beam parameters (E, θ, ψ) of the incident electron are confined to the region over which the model vertex function describes the ELEED, a generally adequate description of the elastic-vertex contributions to the ILEED intensities is obtained. For example, if we use a two-step model embodying a given set of electron-ion-core phase shifts, $\delta_\alpha^l(E)$, and Equations (113) for the electronic self energy we can take V_0 to be a convenient value to make a kinematical Bragg peak in the model elastic intensity profile have the same energy as a multiple-scattering cluster near the corresponding Bragg energy in an experimental elastic profile. Similarly, we can select the inelastic-collision penetration depth, $\lambda_{ee} \sim 2$–8 Å, to fit the observed width of this multiple-scattering cluster. We thereby parameterize accurately the *elastic* intensity profile near a kinematical Bragg energy and subsequently use this parameterization to *predict* the inelastic cross-sections associated with the contribution to the two-step diffraction process of elastic scattering near this energy. We have seen in Figure 75 an example of the accuracy of this procedure. That figure illustrates the prediction of the systematic features of the loss profiles associated with the (11) beam of electrons scattered from Al(100) near a prominent multiple-scattering cluster at $E_B = 86$ eV (for normally-incident electrons). The corresponding results for the angular profiles are shown in Figure 79. From Figures 75 and 79 we infer that the procedure describes the qualitative features of both the loss and angular profiles but may exhibit some serious defects as a quantitative tool for the description of angular profiles.

These considerations bring us to the end of our discussion of 'general' principles. The concept of parameterizing maxima in the ELEED intensities and utilizing the resulting elastic vertices to predict resonances in the ILEED intensities has been shown to be qualitatively sensible. It permits the interpretation of all the major features of the observed dependence of the loss and angular profiles on the incident beam parameters in terms of a 'two-step' model of the inelastic scattering process. We now turn to a discussion of the next important issue in ILEED excitation spectroscopy: can we convert this qualitative insight based on a two-step picture into a high-precision determination of the dispersion and damping of surface excitations of the valence-electron fluid of the target?

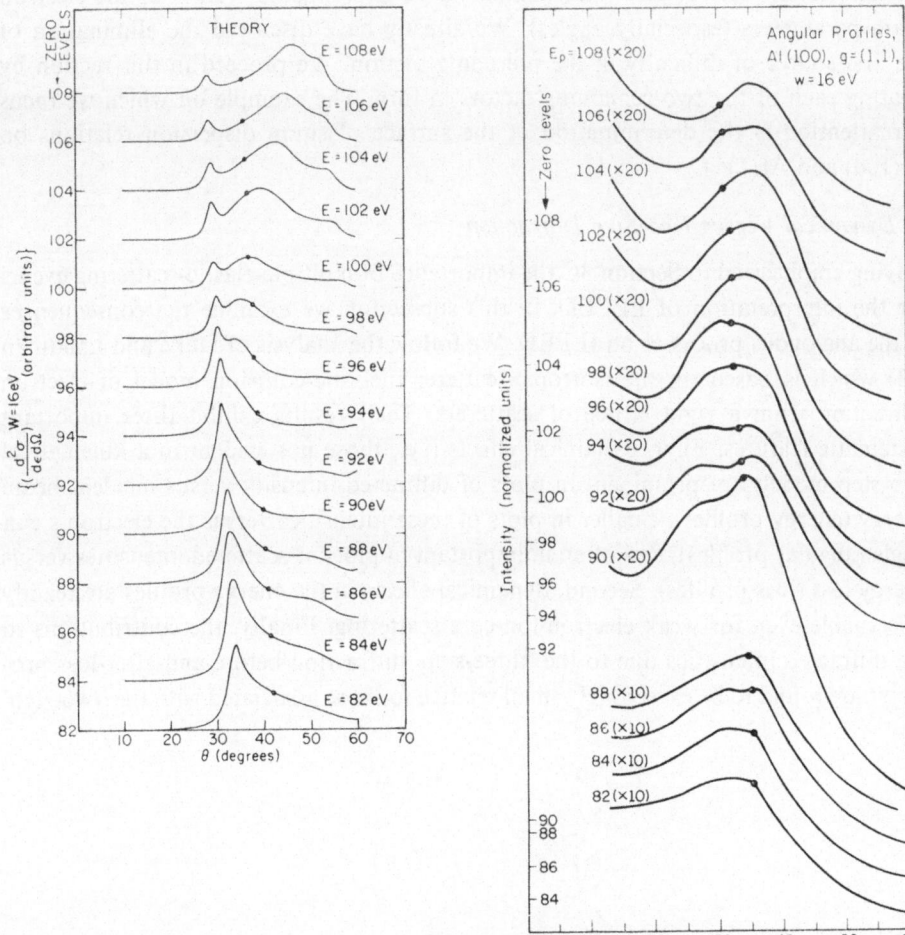

Fig. 79. Theoretical and experimental curves for the angular profile of electrons incident normally on Al(100) and scattered inelastically in the (11) beam. The loss energy is fixed at 16 eV and the primary beam energy E is varied. Dots indicate the direction of emergence of the elastically scattered (11) beam. All parameters that are used in the model calculations are indicated in the caption to Figure 75. The experimental collector angle is uncertain to within $\pm 4°$ because of uncertainties in target and beam alignment and the angular spread of the incident beam. (After Bagchi and Duke [42].)

B. SURFACE-EXCITATION SPECTROSCOPY VIA ILEED

The major problem arising in the analysis of ILEED data to examine the electronic structure of surfaces is that of uniqueness. Consequently, we must inquire "how accurately and uniquely may the surface-plasmon dispersion relation characteristic of clean metal surfaces be extracted from an analysis of ILEED intensities from these surfaces?" This accuracy is limited predominately by three factors: lack of a detailed specification of the electron-solid force law, dynamical (as opposed to 'two-step') effects in inelastic diffraction, and uncertainties in the experimental values of the electron beam parameters (especially angles). We already have discussed the elimination of the first source of difficulty in the preceding section. We proceed in this section by treating each of the two remaining factors in turn. The example on which we focus our attention is the determination of the surface plasmon dispersion relations on Al(100) and Al(111).

1. *Dynamical Versus Two-Step Diffraction*

Having emphasized in Section 3C the importance of multiple-elastic-scattering events for the interpretation of ELEED, in this subsection we examine the consequences of the analogous processes on ILEED. We follow the analysis of Duke and Landman [43] which is based on the isotropic-scatterer inelastic-collision model of electron diffraction from a rigid lattice of scatterers. Their results exhibit three important systematic features. First, dynamical effects (i.e., those not evident in a kinematical two-step model) are prominent in plots of diffracted intensity versus incident beam energy (energy profiles), smaller in plots of these intensities versus the electron's exit angle (angular profiles), and often unimportant in plots of scattered intensities versus energy loss (loss profiles). Second, dynamical effects in the energy profiles are readily discernable even for weak electron-ion-core scattering. Finally, the contributions to the diffracted intensities due to the 'three-step' diffraction-before-and-after-loss processes are found to be exceedingly small relative to those associated with the 'two-step'

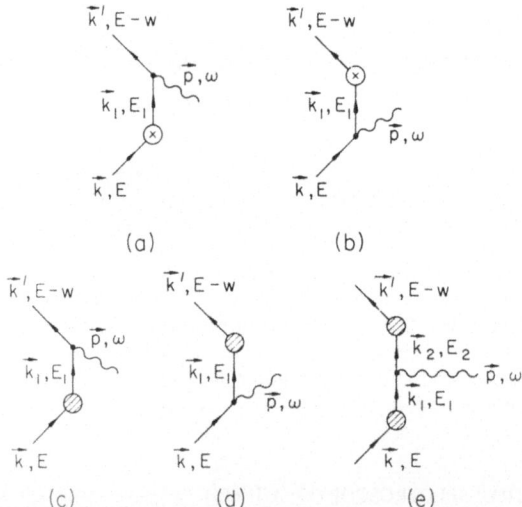

Fig. 80.   Diagrams contributing to the scattering amplitudes of ILEED: (a) 2-step, diffraction before loss; (b) 2-step, loss followed by diffraction; (c) multiple elastic diffraction before loss; (d) loss followed by multiple elastic diffraction; (e) multiple elastic diffraction both before and after loss. The perturbation-theory definition of the expressions for the cross sections associated with these diagrams is given by Duke and Laramore. [3] The shaded circle indicates the summation of an arbitrary number of individual elastic scattering events [designated by the circled cross in Figures 1a and 1b]. (After Duke and Landman [43].)

processes of diffraction before or after loss. These features, moreover, permit us to draw two important conclusions from the analysis. First, the extraction from experimental ILEED intensities of surface plasmon dispersion relations may be based on a kinematical two-step model provided the analysis is confined to a consideration of loss profiles. Second, the consequences of the vestiges of momentum conservation normal to the surface in a kinematical model calculation of the excitation of bulk plasmons cannot be distinguished clearly from those of multiple-elastic scattering. Thus kinematical momentum conservation conditions for motion normal to the surface seem to be entirely irrelevant in the interpretation of ILEED intensities.

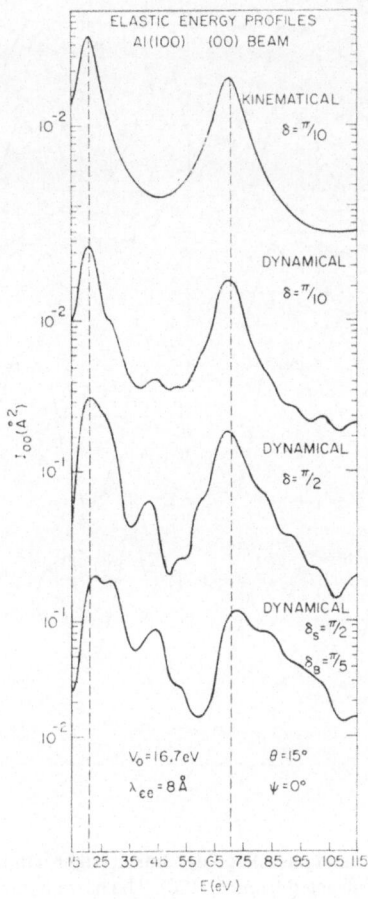

Fig. 81. Elastic energy profiles for the (00) beam of electrons scattered from Al(100). The primary and final polar angles of incidence and exit respectively are 15°. Both incident and exit beams lie in a plane containing a unit vector of the primitive unit surface cell. The dynamical elastic scattering parameters are indicated in the figure. All 'dynamical' calculations were performed for a rigid-lattice model using the isotropic-scatterer, inelastic-collision-model analysis of Tucker and Duke. [72] The 'kinematical' calculations are performed using the Born approximation. Vertical dashed lines designate the energies of the kinematical primary Bragg peaks. (After Duke and Landman [43].)

The 'dynamical' analysis given by Duke and Landman [43] extends the 'kinematical' two-step model calculations, which encompass the two contributions to the electron-solid scattering amplitude initially shown in Figure 74 and reviewed in panels (a) and (b) of Figure 80, by summation of all diagrammatic contributions to the cross section in which the electron has scattered elastically from the lattice an arbitrary number of times but has undergone only a single loss-event. It is convenient to divide these diagrams into three classes. First, one sums all of the diagrams in which

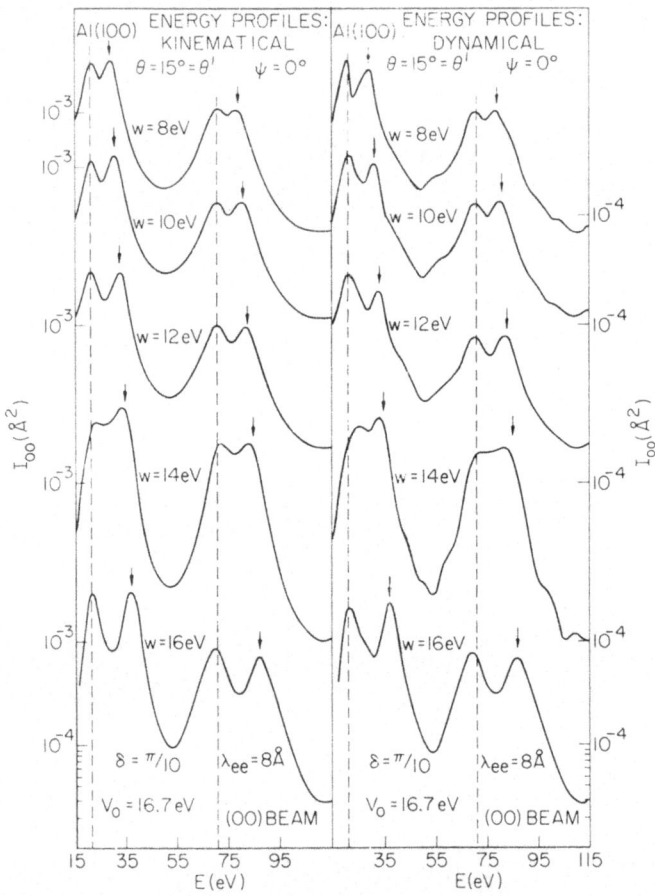

Fig. 82. The kinematical (left-hand panel) and dynamical (right-hand panel) inelastic energy profiles for the (00) beam of electrons diffracted from Al(100). The beam parameters and the parameters used to describe the elastic-electron-solid scattering are indicated in the figure. Vertical dashed lines indicate the energies, E_B, of the kinematical Bragg peaks in the elastic energy profiles. Downward-pointing arrows designate the energies $E_B + w$. The 'kinematical' profiles were evaluated using their two-step model of Bagchi and Duke. [41, 42] The 'dynamical' profiles were evaluated according to Equations (23)–(41) in the first paper of Duke and Landman. [43] The plasmon dispersion relations were taken to be those of Bagchi and Duke: [41] $\hbar\omega_s(p) = 10.1 - 0.7p_{\parallel} + 10p_{\parallel}^2$; $\Gamma_s(p_{\parallel}) = 0.9 + 0.74p_{\parallel}$, $\hbar\omega_b(p) = 14.2 + 3.048p^2$; and $\Gamma_b(p) = 0.53 + 0.103p^2 + 1.052p^4$. All energies are measured in eV and momenta in reciprocal angstroms. (After Duke and Landman [43].)

one or more elastic scattering events occur before the loss process. The sum of these diagrams defines a generalized diffraction-before-loss (D-L) process, indicated diagrammatically in Figure 80c. Second, one sums all of the diagrams in which one more elastic scattering occur after the loss event. This sum defines a generalized loss-before-diffraction (L-D) process indicated diagrammatically in Figure 80d. Finally, one sums all of the diagrams in which the incident electron experiences one or more elastic scatterings *both* before and after the loss event. The sum of these diagrams defines a generalized diffraction-followed-by-loss-followed-by-diffraction

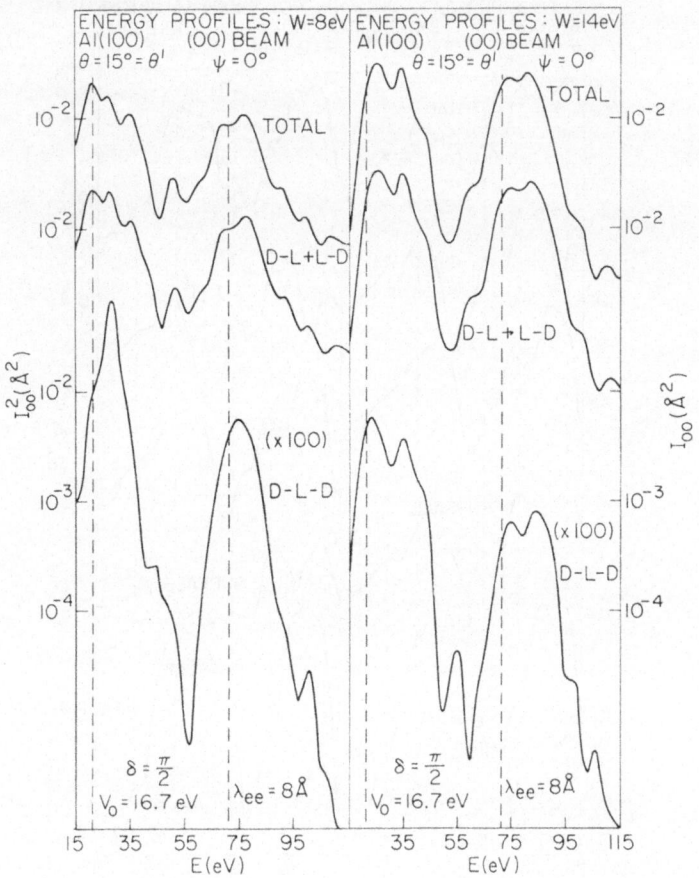

Fig. 83. Various contributions to the dynamical inelastic energy profiles at loss energies $w = 8$ eV and $w = 14$ eV for the (00) beam of electrons scattered from Al(100). The upper curves designate the cross section obtained by taking into account diffraction before-loss (D-L), Loss-before-diffraction (L-D) and diffraction before-and-after-loss (D-L-D) as defined in Equations (20)–(23) in Duke and Landman. [43] The center curves give the renormalized two-step contribution [D-L and L-D] to the cross sections. The lower curves describe the 'three-step' contributions [D-L-D] alone. Vertical dashed lines indicate the energies of prominent maxima in the elastic intensity profiles [see Figure 81].
  The plasmon dispersion relations used in the analysis are indicated in the caption to Figure 82.
(After Duke and Landman [43].)

(D-L-D) process indicated in Figure 80e. Rather than reproduce the analytical results of Duke and Landman, we simply indicate the main conclusions of their analysis
of the consequences of these three summations. In particular, we proceed by examining
the energy, angular, and loss profiles in turn.

To illustrate the effects of multiple-elastic scattering on the ILEED energy profiles,
we examine the case of electrons specularly reflected at an angle of $\theta = 15°$ from Al(100).
For reference, we present in Figure 81 the elastic ($w=0$) energy profiles for several
different models of the electron-ion-core interactions. In the case of 'weak' elastic

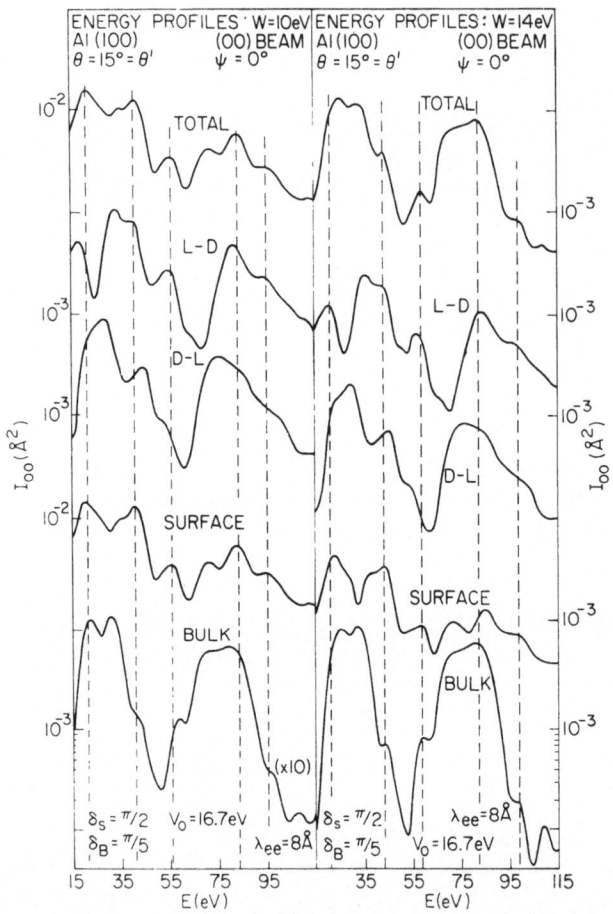

Fig. 84. Various contributions to the dynamical inelastic energy profiles for the (00) beam of electrons scattered from Al(100), for loss-energies $w = 12$ eV and 14 eV. The beam parameters and those
used to describe the elastic electron-solid diffraction are indicated in the figure. The plasmon dispersion
relations used in the analysis are given in the caption to Figure 82. The labels L-D and D-L designate
the loss-before-diffraction and diffraction-before-loss contributions to the cross section, respectively
as specified by Equations (28), (30), (31), (36), (37) and (38) in Duke and Landman. [43] The labels
'surface' and 'bulk' indicate the calculated intensities using only surface and bulk plasmon emission
respectively. Vertical dashed lines are solely to guide the eye to identify the energies of prominent
structures in the intensities [see also Figure 81]. (After Duke and Landman [43].)

scattering, i.e., $\delta = \pi/10$, the kinematical primary Bragg peaks occurring at energies of 21 eV and 71 eV persist when going from the kinematical to the dynamical description. However, a secondary peak at 45 eV, missing in the kinematical curve, occurs in the dynamical results. Additional dynamical structure also is evident for energies $E > 75$ eV. The dynamical structure is enhanced by going to a stronger electron-ion-core scattering $(\delta = \pi/2)$ for a truncated bulk solid or increased scattering of the surface (δ_S) relative to the bulk (δ_B) ion cores.

In Figure 82 we present the kinematical and dynamical inelastic energy profiles in the weak-scattering limit for a series of loss energies (w). In this limit, the diffraction before loss (D-L) and the loss followed by diffraction (L-D) peaks occur at the same energies in both the 'dynamical' and 'kinematical' intensities. However, dynamical effects are evident in the predicted line shapes, especially for those values (10 eV and 14 eV) of the loss energy near the surface and bulk plasmon excitation thresholds (10.1 eV and 14.2 eV respectively).

As the electron-ion-core scattering becomes stronger (i.e., $\delta > \pi/10$ in this model) dynamical effects become more prominent in the energy profiles and, in fact, simulate the consequences of sideband diffraction. [43] Moreover, two additional questions arise: What is the relative importance of surface and bulk plasmon excitations in causing a given 'observed' structure? How large are the contributions of the three mechanisms [diffraction-before-loss (D-L); loss-before-diffraction (L-D); and diffraction-before-and-after-loss (D-L-D)] of inelastic diffraction?

Turning first to the latter question, the contributions to the inelastic cross sections from a renormalized two step mechanism are compared to the contributions arising from the three step processes in Figure 83. It is evident that even in the 'strong-scattering' limit $(\delta = \pi/2)$ the contribution of the three-step process is nearly negligible for both values of w. Consequently, we identify the dynamical structures, which are clearly evident in this limit, with the renormalization of the elastic scattering vertices in the 'two-step' mechanism.

We still must examine the magnitudes of the D-L and L-D mechanisms as well as the relative significance of bulk and surface plasmon losses for a given value of the loss-energy, w. To do this, we consider the case of enhanced elastic scattering from the surface layer $[\delta_S > \delta_B]$ in order to accentuate the dynamical fine structure in the energy profiles [see, e.g., Figure 81]. Typical inelastic energy profiles for this case are shown in Figure 84 in which the separate D-L and L-D as well as surface and bulk-plasmon contributions also are displayed explicitly. Comparison of the contributions of the D-L and L-D mechanisms with the complete profile [top panel, Figure 84] illustrates the persistence in the total scattered intensity of fine structure associated with the loss-before-diffraction process. The total intensity is not a simple sum of the L-D and D-L intensities because these two processes are coherent (i.e., they lead to the same final state). However, the surface and bulk plasmon contributions [lower two panels in Figure 84] do add to give the total intensity because they are incoherent. As expected, for $w \leqslant 12$ eV the surface-plasmon-emission contribution dominates that associated with bulk plasmon emission. Note, however, the substantial importance of the surface-plasmon process even at $w = 14$ eV for 40 eV $\lesssim E \lesssim 60$ eV. Also, the surface plasmon contribution to the intensity exhibits considerably more dynamical fine structure than that of the bulk plasmons because of the absence of the integration over components of momenta normal to the surface.

Turning to an examination of the angular profiles, the central issue for us is whether or not prominent surface-plasmon peaks in these profiles are predicted to occur at

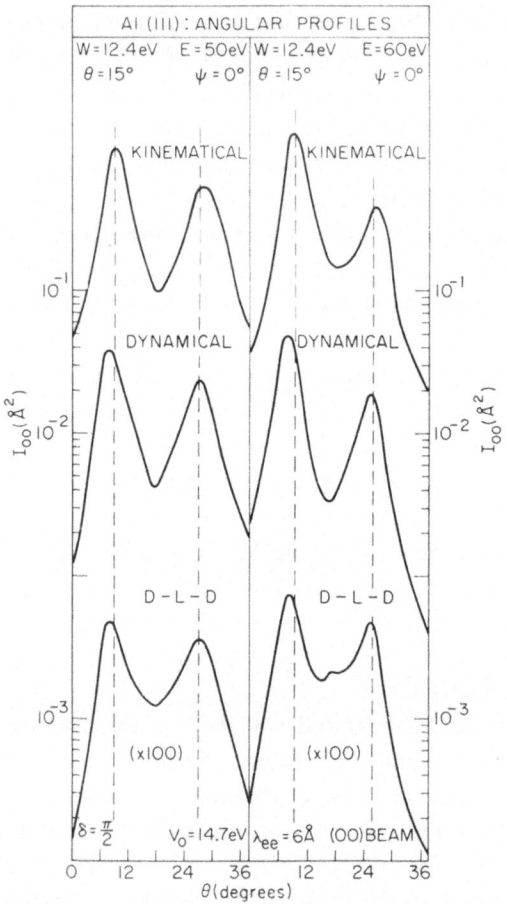

Fig. 85. Kinematical and dynamical angular profiles for electrons diffracted in the (00) beam from Al(111), for primary beam energies of 60 eV (right-hand panel) and 50 eV (left-hand panel). The above energies correspond to those of a prominent peak in the kinematical elastic energy profile and to a surface-plasmon energy above the aforementioned peak energy respectively [see e.g. Figure 86]. The loss of $w = 12.4$ eV (above the threshold for surface plasmon excitation but below the one corresponding to bulk plasmon excitation) is the one analyzed by Bagchi and Duke. [42] We also use Bagchi and Duke's dispersion relations for the bulk and surface plasmons (see caption to Figure 82). The notation D-L-D designates the contribution from diffraction before-and-after loss as defined in Equation (22) in Duke and Landman. [43] The parameters used in the model of the elastic scattering cross section are noted in the figure. Vertical dashed lines are included for convenience in visualization. (After Duke and Landman [43].)

the same angles by the kinematical and dynamical model calculations. We see in Figure 85 that, unfortunately, they may be displaced by 1–2° relative to their kinematically-predicted angles by dynamical (i.e., multiple-elastic-scattering) effects. However, the 'three-step' D-L-D contribution to the angular profile is negligible in the dynamical model, leading to a simple 'two-step' interpretation in this as well as the kinematical model. The two incident beam energies, $E = 50$ eV, 60 eV, were chosen to be the energy of a Bragg peak in the kinematical energy profile [see, e.g., pane

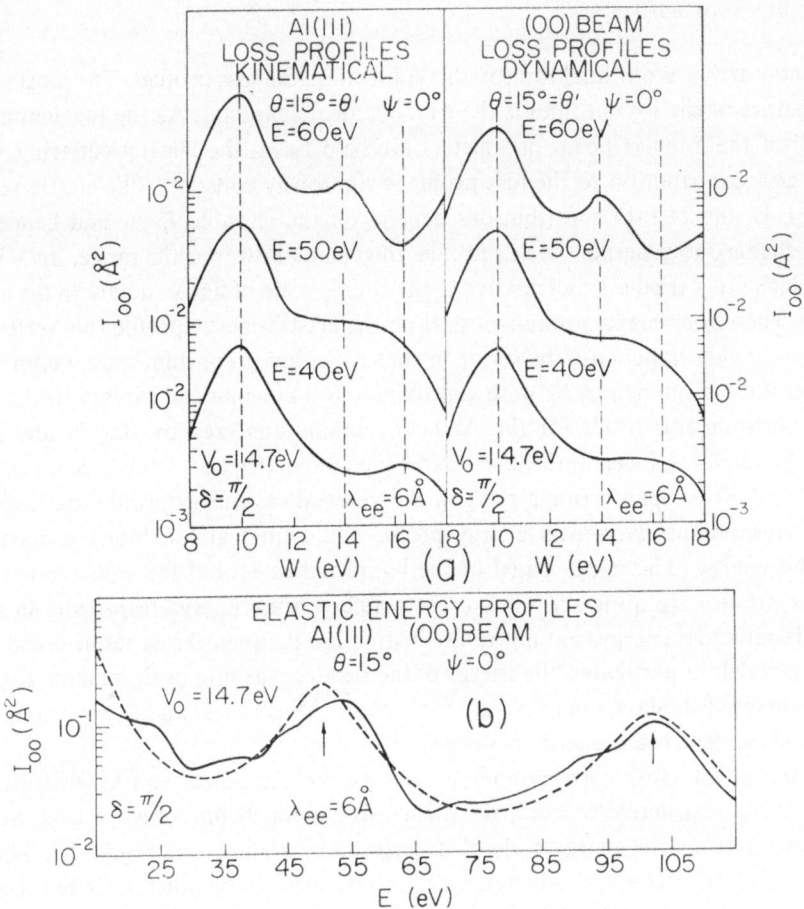

Fig. 86. Panel (a) displays the kinematical and dynamical loss profiles for electrons diffracted in the (00) beam from Al(111). The profiles are shown for various energies above and below the kinematical bragg energy [$E_B = 50$ eV] in the elastic energy profile of the (00) beam. The plasmon dispersion relations used in constructing the figure are those used by Bagchi and Duke [42] and indicated in the caption to Figure 82. The parameters used in describing the elastic electron-solid scattering are indicated in the figure. Vertical dashed lines are included for convenience in visualization. Panel (b) contains a plot of the kinematical (dashed line) and dynamical (solid line) elastic intensity profiles obtained using these parameters (see also Bagchi and Duke [42]). Arrows indicate the kinematical primary Bragg energies. All beam parameters are noted in the figure. (After Duke and Landman [43].)

(b) of Figure 86] and that of this Bragg peak plus a surface-plasmon energy. Therefore the D-L process dominates the profiles at $E = 50$ eV whereas the L-D process is the largest at $E = 60$ eV. The examples of dynamical effects on the angular and loss profiles are given for Al(111) in order to be directly comparable with the analyses given in the following section of these lecture notes. Finally, it is informative to recall that the doublet structure evident in Figure 85 is a direct consequence of the conservation laws, i.e.,

$$\mathbf{k}'_{\parallel} = \mathbf{k}_{\parallel} \pm \mathbf{p}_{\parallel}(w) \tag{153a}$$

$$w = \hbar\omega_s[\mathbf{p}_{\parallel}(w)]. \tag{153b}$$

We now arrive at our final topic in this subsection, the loss profiles. The most significant feature of the loss profiles is their kinematical character. As the incident energy and all of the angular beam parameters are held fixed, the elastic-scattering vertex in the D-L contribution to the loss profile is rigorously constant. The elastic vertices in the L-D and D-L-D contributions depend on the variable $E - w$ and hence vary as w is changed to generate the loss profile. In general, however, the range, $\Delta w \sim 10$ eV, over which w is varied is small relative to the energy scale of fine structure in the energy profile. Therefore, near a prominent peak in the elastic energy profile this vertex also exhibits roughly 'kinematic' behavior in that to within a constant scale factor it behaves as if the prominent peak were approximately a kinematical primary Bragg peak.

We illustrate this result for the Al(111) example analyzed by Bagchi and Duke [42] in Figure 86. As seen in the lower panel [i.e., panel (b)] the incident beam energies 40, 50, and 60 eV span a Bragg peak in the kinematical energy profile (dashed line). The corresponding dynamical energy profile (solid line) also exhibits a maximum near this energy. The upper panel of the figure indicates that for some values of w the loss profiles are quite sensitive to the primary beam energy (especially near the bulk plasmon loss energy) but not to the distinction between the dynamical and kinematical model. In particular, the energy of the surface plasmon peak remains constant on the energy scale shown in the figure for both the kinematical and dynamical models when the incident beam energy is varied.

Because of the close correspondence between the dynamical and kinematical loss profiles, it is instructive to compare them using a much finer energy grid. Such a comparison, using the energy scale of $\Delta w = 50$ meV, is shown in Figure 87. The important features evident in this figure are the occurrence of differences between the peaks in the kinematical and dynamical loss profiles on the energy scale $\Delta w \lesssim 50$ meV and the dependence of the surface-plasmon peak on the energy of the incident beam (see also Figure 75). The former result establishes a loss-energy scale, $\Delta w \sim 50$ meV, above which differences in the peak positions in the kinematical and dynamical loss profiles can be neglected. The latter result indicates that because of the 'two-step' nature of ILEED, a complete model analysis of the scattered intensities is required to evaluate the dispersion of surface excitations. In particular, one *cannot* extract the dispersion relation from the observed loss profiles by a direct application of the con-

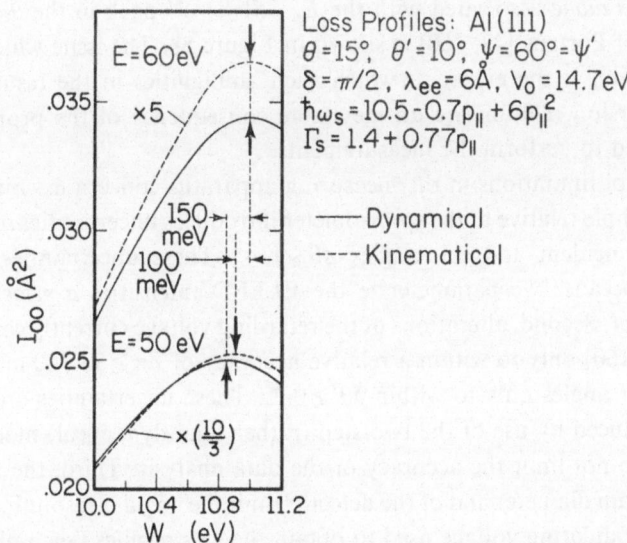

Fig. 87. Comparison between kinematical and dynamical loss profiles for the (00) beam of electrons scattered from Al(111) at two incident energies, $E = 60$ eV (top) and $E = 50$ eV (bottom). The calculations were performed using the s-wave version of the two-step ILEED model [43] with the parameters indicated in the figure. Vertical arrows indicate the positions of the peaks. Vertical dashed lines indicate the shifts in peaks' positions both for a given incident beam energy (0 and 50 meV at 60 eV and 50 eV respectively) and for different incident beam energies.
(After Duke and Landman [159].)

servation laws given, e.g., in Equations (153). The factors M and $|A_b + A_c|^2$ in the surface plasmon analogue of Equations (148) impose an important modulation on the contribution to the scattered intensities from $\operatorname{Im} D_s(\mathbf{p}_\parallel, w)$ alone.

In summary, we have established that dynamical effects may be regarded as irrelevant for extracting surface plasmon dispersion relations from experimental data *provided one selects the proper subset of data (i.e., loss profiles) to analyze*. However, the extraction must be based on a complete two-step model analysis as described in Section 5A rather than on a simple application of the energy-momentum conservation laws given by Equations (143) and (144).

2. Consequences of Instrumental Uncertainties

The objective of surface-plasmon spectroscopy via ILEED is the determination of the quantities $\{\hbar \omega_s, C_1, C_2, \Gamma_s, D_1,$ and $D_2\}$ in Equations (152) from observed ILEED intensities subject to the limitations noted in the previous subsection. To achieve this goal, we first select a given maximum in the ELEED intensity profiles as the basis for our analysis. We next parameterize the ELEED intensities as described in Section 5A and proceed to calculate the ILEED intensities using the two-step model for various values of $\{\hbar \omega_s, C_1, C_2, \Gamma_s, D_1, D_2\}$. We then discover [42, 43] to our dismay that a variety of values of these parameters provide comparable descriptions of any given set of experimental data. For example, the range of values of C_1 and C_2 describing the *loss profiles alone* associated with the $E_B = 51 \pm 1$ eV peak in the $\theta = 15°$, $\psi = 60°$ Al(111) data of Porteus [42, 160] is shown in Figure 88. The issue which we discuss in this subsection is the extent to which such ambiguities in the resulting surface-plasmon dispersion relation are an inevitable consequence of the properties of the instrument used to perform the measurements.

Three types of limitations in the measuring apparatus concern us. First, the alignment of the sample relative to the spectrometer introduces uncertainties in the absolute values of the incident and exit angles $\Delta\theta \sim 1$–$4°$. These uncertainties are of little interest here because we parameterize the ELEED intensities *a priori* for a given target alignment. Second, alterations of the retarding voltage currently can be achieved [40, 105, 106, 160] only to within a relative accuracy of $\delta w \gtrsim 50$–100 meV and those of the collector angles only to within $\delta\theta' \gtrsim 0.2°$. These uncertainties are comparable to those introduced by use of the two-step (rather than dynamical) model, however, so they also do not limit the accuracy of the data analysis. Third, the finite sizes of the incident beam diameter and of the detector limit the angular resolution to $\Delta\theta' \sim 2°$, whereas the modulating voltage used to obtain the loss profiles (via a phase-sensitive detection derivative method [12, 40]) limits the loss-energy resolution to $\Delta w \sim 1$ eV. These potential sources of uncertainty are large relative to those intrinsic to the theoretical model, and consequently must be examined more closely.

In order to investigate the consequences of finite angular and energy-loss resolution in a typical instrument, Duke and Landman [159] calculated the average value of the loss-profiles over a range of values for $\bar{w} - \Delta w \leqslant w \leqslant \bar{w} + \Delta w$ and $\bar{\theta}' - \Delta\theta' \leqslant \theta' \leqslant \bar{\theta}' +$

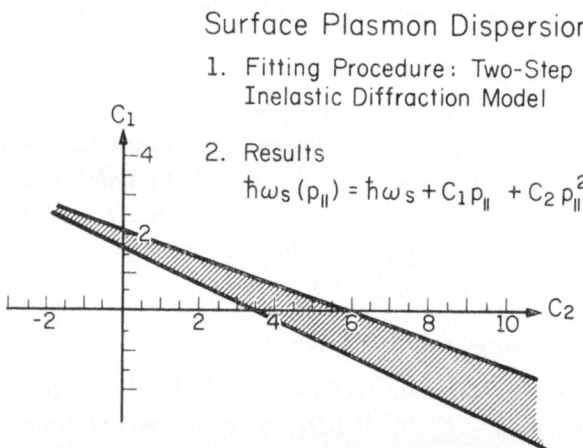

Fig. 88. Region of ambiguity in the surface plasmon dispersion relation of Al(111), (see e.g. Equations (152) in the text). Any dispersion relation whose coefficients lie in the shaded region yields in a good fit between the theoretical and observed loss and angular profiles. The calculations were performed using the two-step model with 4-phase-shifts derived from the Laramore-Duke-Snow APW potential, for the (00) beam of electrons with an inner potential $V_0 = 14.7$ eV, mean free path $\lambda_{ee} = 6$ Å and damping $\Gamma_s = 1.4 + 0.74p_{\parallel}$. The theoretical loss and angular profiles were fitted to experimental ones for incident energies $E = 50$ and 60 eV with $\theta = 15°$ and $\psi = 60°$. (After Duke and Landman [159].)

$+\Delta\theta'$. As noted above, typical values of Δw are 1.0 eV and of $\Delta\theta'$ are 2°. A comparison of the averaged theoretical loss profiles (as a function of $\bar{w}$) with the loss profiles obtained by setting $w \equiv \bar{w}$ at each value of $\bar{w}$ is shown in Figure 89. The significant result evident from the figure is the failure of the averaging procedure to alter the energy of the peak in the loss profile to within the resolution ($d\bar{w} = 25$ meV) of the calculation. This result is valid for a variety of different elastic-diffraction conditions and surface-plasmon dispersion relations within the region of ambiguity shown in Figure 88. [159, 160] Therefore two exceedingly important results have been established. First, it is both sensible and important to take experimental loss profiles using a (w, θ') grid which is *much smaller* than the instrumental resolution (i.e., $dw \sim 50$ meV $\ll \Delta w \sim 500$ meV; $d\theta' \sim 0.5° < \Delta\theta' \sim 1°$). Second, if we are interested in peak energies rather than peak shapes, we can analyze the experimental loss profiles using the ideal rather than the averaged theoretical profiles. This latter result permits a substantial saving in the computer time required for an analysis.

Kinematical Loss Profiles: Al(111)   Snow APW
$\theta = 15°$, $\theta' = 10°$, $\psi = 60° = \psi'$, $\Delta W = 1.0$ eV, $\Delta\theta' = 2°$
$\Gamma_s = 1.4 + 0.77 p_{\parallel}$, $V_0 = 14.7$ eV, $\lambda_{ee} = 6$ Å

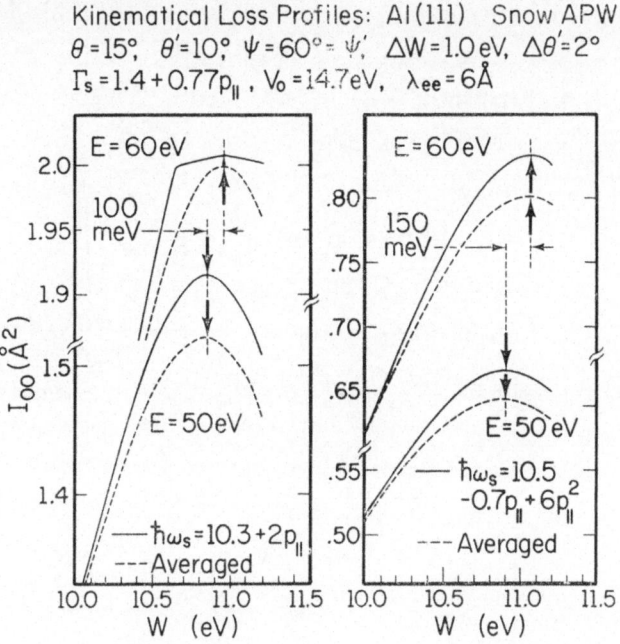

Fig. 89.   Comparison of infinite-resolution (solid curves) and finite-resolution (dashed curves) loss profiles for the (00) beam of electrons scattered from Al(111). Results of calculations using a 4-phase-shift two-step model with two different dispersion relations (left and right panels) and for two primary energies, $E = 60$ eV (top) and $E = 50$ eV (bottom) are shown. The values of the parameters used are given in the figure. Vertical arrows indicate the positions of the peaks, and vertical dashed lines are drawn for ease in determining shifts in peaks' positions. As indicated, only shifts due to changes in diffraction conditions is observed. The averaging was performed using a procedure corresponding to the experimental one. For each value of w the intensity was calculated using a 5 point (equally spaced) Simpson rule in the interval $[w - 0.5$ eV; $w + 0.5$ eV$]$ which was then divided by the integration interval. The same averaging procedure was performed for the angular parameter in the interval $[\theta' - 1°, \theta' + 1°]$. (After Duke and Landman [159].)

C. SURFACE-PLASMON DISPERSION ON Al(111)

In this section we conclude our discussion of ILEED by describing the application of the analytical procedure outlined in the preceding two sections to the determination of the surface-plasmon dispersion relation for nominally-clean Al(111). This is the only existing example [40-43, 159, 160] of the use of ILEED as a precision spectroscopic probe of the electronic structure of surfaces. In fact, even in this case the analysis is not yet complete because the analytical technique itself has been refined on a continuing basis as the analysis has proceeded.

The analysis was begun by Duke and Bagchi [41] who performed a series of calculations of ILEED from Al(111) and Al(100) using surface plasmon dispersion relations, Equations (152), which are predicted by RPA calculations appropriate [4, 161, 162] to semi-infinite 'jellium': a hypothetical material in which the ion-core charge density has been 'smeared' out into a uniform positive background. Comparing the predicted ILEED intensities with the data of Porteus [40, 42] on Al(111) and Burkstrand [105, 106] on Al(100), Duke and Bagchi concluded [41] that the RPA

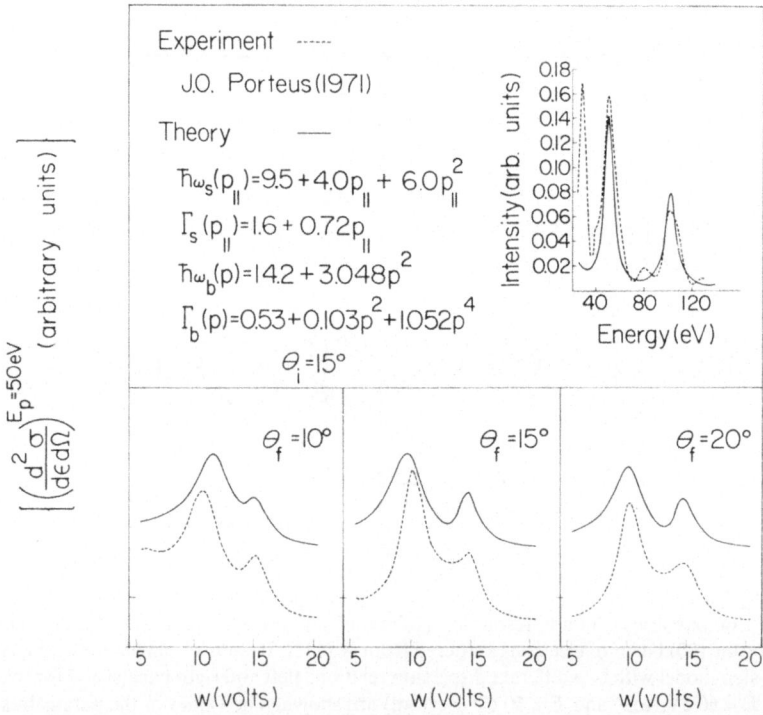

Fig. 90. Comparison of theoretical (full lines) and experimental (dotted lines) loss profiles for the (00) beam on Al(111). The primary beam energy is 40 eV and θ_f is the exit angle. A comparison of the experimental elastic intensity profile measured by Porteus with the theoretical curve calculated using the model of s-wave scatterers, [66] with $\lambda_{ee} = 6$ Å, $V_0 = 14.7$ eV, and $\delta = \pi/4$ is noted in the inset. The parameters used to describe the plasmon losses are taken to be as close to their RPA values as feasible and are indicated in the figure. (After Duke and Bagchi [41].)

dispersion relation was clearly inconsistent with the measurements. Sample calculations of this type (but with Γ_s and C_2 adjusted to provide some resemblance between
the calculations and the data) are shown in Figure 90. Thus the question arose of
whether a surface plasmon dispersion relation could be determined which was consistent with all of the available data.

In order to respond to this question, Bagchi and Duke [42] undertook a detailed
analysis of Porteus' original data on Al(111). This data was rather limited in extent,
consisting of ELEED and ILEED intensities associated with the approximately
specular scattering of electrons incident on an epitaxially grown film of Al(111) at
polar and azimuthal angles [40] $\theta = 15°$ and $\psi = 60°$. The elastic intensity profile
associated with the scattered electrons is shown as the dashed line in the inset to
Figure 90. Following the procedure outlined in the preceding sections, Bagchi and
Duke began by fitting the prominent peaks in the experimental ELEED intensity
profile by a kinematical model calculation. As can be seen from the inset to Figure 90,
this procedure is quite reasonable for the ($\theta = 15°$, $\psi = 60°$) specular beam from Al(111).
The subsequent steps in the analytical procedure adopted by Bagchi and Duke de-
pended, however, on two historical accidents. First, most of the available data at
that time was presented in the form of angular profiles associated with a single
maximum ($E_B = 50$ eV, $\theta = 15°$, $\psi = 60°$) in the ELEED intensity profile. Second, the
dynamical calculations described in Section 5B1 had not yet been performed. Therefore the difficulties inherent in using a two-step model to analyze angular profiles
were unrecognized. Consequently, Bagchi and Duke analyzed the existing angular-
and loss-profile data and concluded that a surface-plasmon dispersion relation consistent with all of it was [42]

$$\hbar\omega_s(p_\parallel) = 10.1\,(\pm\,0.1) + 0\,(\pm\,1)\,p_\parallel + 8\,(\pm\,2)\,p_\parallel^2. \tag{154}$$

In Equation (154) and the figure captions we give the dispersion relations for energies
measured in eV and momenta measured in Å^{-1}. For reference, the original RPA
calculations predicted [161, 162]

$$\hbar\omega_s^{RPA}(p_\parallel) \cong 11 + 4p_\parallel + 3p_\parallel^2. \tag{155}$$

Two features of Equation (154) require further discussion. First, Bagchi and Duke
discovered that the coefficient of the term linear in $p_\parallel$ [i.e., C_1 in Equation (152a)]
depends on the value of the coefficient of the quadratic term [i.e., C_2 in Equation
(152a)]. The error bars in Equations (154) reflect the combined uncertainty in the two
as found in Bagchi and Duke's analysis. If, however, C_2 were known accurately, the
error in C_1 would be much smaller. For example, if $C_2 = 10$ eV Å^2, then Bagchi
and Duke find $C_1 = -0.7\,(\pm 0.3)$ eV Å. Second, Bagchi and Duke evaluated C_2
from an analysis of the angular profiles: a procedure subsequently shown [43] to
be a potential source of error because of dynamical effects (recall Figure 85 and
the discussion thereof). Therefore the error for C_2 shown in Equation (154) is
probably too small because of this systematic flaw in the original analytical procedure. [40, 42]

A more recent analysis [159, 160] of ILEED data for Al(111) taken by Porteus constitutes an improvement on that of Bagchi and Duke in three ways. First, much more extensive loss-profile data was taken and used as the basis for the analysis, thereby removing the uncertainties resulting from consideration of the angular profiles. Second, the ILEED intensities associated with five separate maxima in the ELEED intensities (obtained for different incident beam (E, θ, ψ) parameters) were analyzed for internal consistency. Third and finally, a finer mesh $(d\bar{w}=100 \text{ meV}, d\bar{\theta}=1°)$ than that $(d\bar{w}=400 \text{ meV}, d\bar{\theta}'=2°)$ used in the original analysis [40–42] was employed in parts of the more recent one. [160]

Although the revised analysis is still in progress at the time of writing these lecture notes, we can illustrate some of the general features of the preliminary results. Bagchi and Duke used the isotropic-scatterer model to parameterize the ELEED intensities associated with a prominent peak in the intensity profile at $E_B=50$ eV for $\theta=15°$ and $\psi=60°$ on Al(111). The elastic intensity profile used by them is shown by the dashed line in panel (b) of Figure 86 and the solid line in Figure 90. In the recent analysis of Duke and Landman, [160] the Laramore-Duke-Snow potential is used to describe the ELEED intensities. A resulting kinematical ELEED intensity profile is shown in Figure 91 together with three 'two-step' ILEED energy profiles associated with surface plasmon dispersion relations all of which describe the $d\bar{w}=0.4$ eV, $d\bar{\theta}=2°$ data for the $(E_B=50 \text{ eV}, \theta=15°, \psi=60°)$ peak in the ELEED intensity profile. The

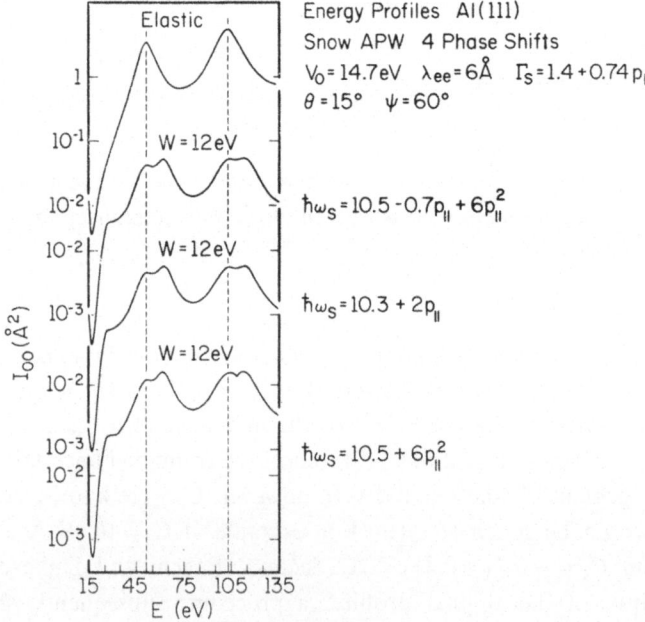

Fig. 91. Elastic and inelastic energy profiles predicted by the kinematic analyses of the Laramore-Duke-Snow potential for electrons incident at $\theta=15°$ and $\psi=60°$ on Al(111). The parameters used in the electronic proper self energy [Equations (113)] and the surface-plasmon dispersion relation [Equations (152)] are noted in the figure.

complete range of ambiguity in (C_1, C_2) associated with the analysis of this set of data was shown in Figure 88. Comparisons of the angular and loss profiles associated with these three surface-plasmon dispersion relations (all of which lie in the region of uncertainty in Figure 88) with Porteus' data are shown in Figures 92–94. The lesson to be learned from these figures is that the data are insufficiently refined to be used as the basis for discriminating between the three dispersion relations. Furthermore, it is not feasible to increase the sensitivity of the analysis by examining loss profiles for larger values of $p_{\parallel}$ (i.e., $\theta' < 8°$ or $\theta' > 26°$) because the combined effects of increased plasmon damping and experimental noise renders the observed peak energies indistinct.

Two avenues exist for further refining the analysis to reduce the region of ambiguity shown in Figure 88. If we confine our attention to a given maximum in the ELEED intensity (e.g., that at $E_B = 50$ eV, $\theta = 15°$, $\psi = 60°$), then we must analyze data taken on a finer $d\bar{w}$, $d\bar{\theta}'$ grid. For example, it is evident from Figure 89 that loss profiles which establish the peak energy to within ± 50 meV can distinguish between the linear and quadratic dispersion relations used to construct Figures 91–94. On the other hand, we also could examine the loss profiles associated with various ELEED maxima at $(E_{Bi}, \theta_i, \psi_i)$ and require internal consistency between the dispersion relations resulting from these analyses. This procedure has been carried out [160] for

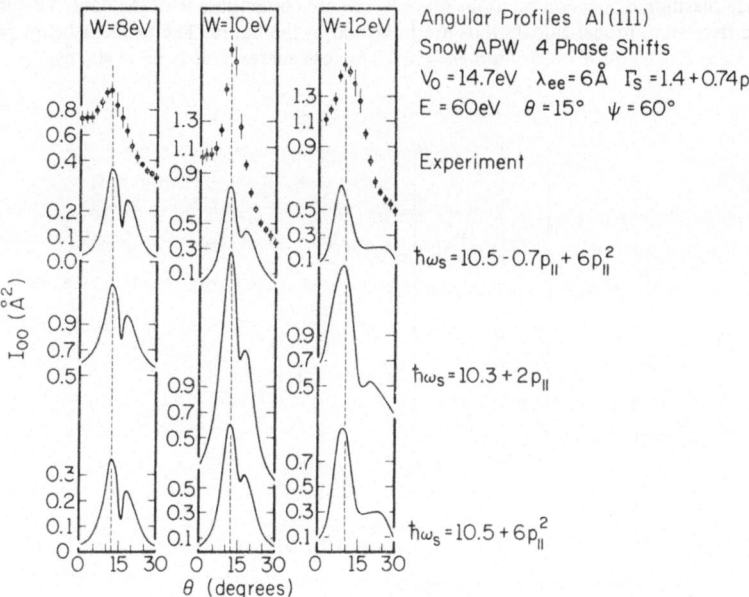

Fig. 92.   Comparision with Porteus' data on Al(111) of angular profiles associated with three surface-plasmon dispersion relations all of which are compatible with the data. The parameters used in the model calculations are noted in the figure. The two-step model provided the basis for these calculations. The angular-profile grid is $d\bar{\theta} = 2°$. The incident beam parameters are given in the figure.

five ELEED maxima on Al(111). For each maximum, the analogue of Figure 88 is constructed. The intersection of the various regions of ambiguity occurs near $C_1 \cong 2$ and $C_2 \cong 0$. Therefore if we believe that the electron-surface-plasmon coupling is independent of the incident electron's energy (as predicted by all existing theories

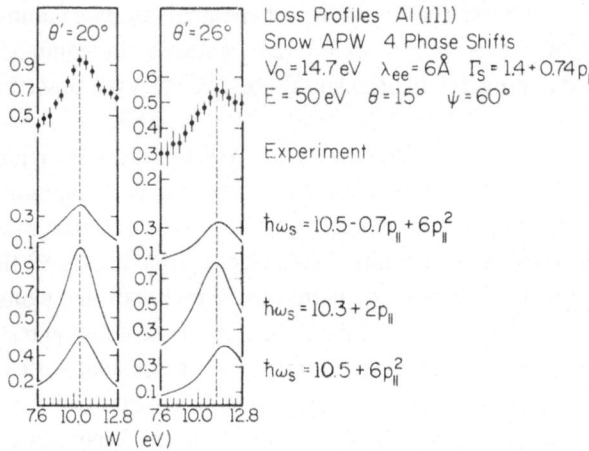

Fig. 93. Comparison with Porteus' data on Al(111) of super-specular loss profiles associated with three surface-plasmon dispersion relations all of which are compatible with the data. The parameters used in the (two-step) model calculations are indicated in the figure. The incident-beam parameters are $E = 50$ eV, $\theta = 15°$, and $\psi = 60°$. The loss-energy grid is $d\bar{w} = 400$ meV.

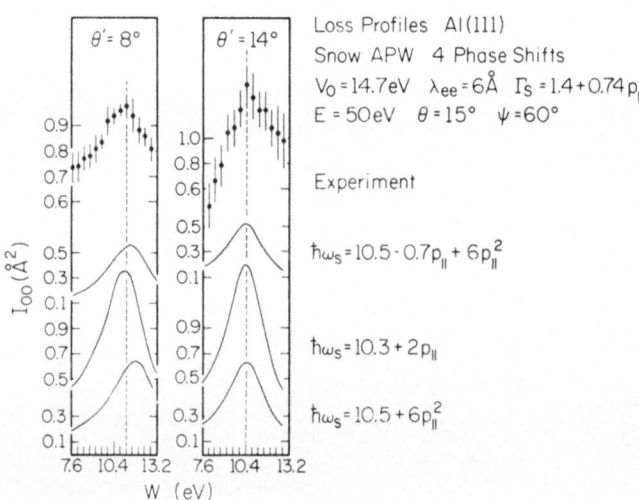

Fig. 94. Comparison with Porteus' data on Al(111) of sub-specular loss profiles associated with three surface-plasmon dispersion relations all of which are compatible with the data. The beam parameters and dynamical parameters used in the (two-step) model calculations are indicated in the figure. The loss-energy grid is $d\bar{w} = 400$ meV.

of this coupling [4]), then we conclude that the best currently-available measure of the surface-plasmon dispersion is

$$\hbar\omega_s(p_{\parallel}) = 10.5\,(\pm 0.1) + 2.0\,(\pm 1)\,p_{\parallel} + 0\,(\pm 2)\,p_{\parallel}^2. \tag{156}$$

The failure of the range of values of the constant and quadratic terms in Equations (154) and (156) to overlap is a direct consequence of using loss-profile data associated with different ELEED maxima to obtain Equation (156) but angular-profile data associated with the $(E_B = 50 \text{ eV}, \theta = 15°, \psi = 60°)$ maximum to obtain Equation (154). Such sensitivity of the results to the analytical procedure is uncomfortable even if explicable on the basis of the insight gained in Section 5B1. Further investigation into the nature, magnitude, and origin of the resulting uncertainties in the surface plasmon dispersion relation is currently in progress. It already is clear from Equations (155) and (156), however, that early microscopic models of surface-plasmon dispersion exhibit unsuspected flaws, and that ILEED seems capable of being used as a high-precision quantitative probe of the electronic surface excitation spectra of solids.

6. Synopsis

My theme in these notes has been the construction of quantitative spectroscopic probes of the properties of the upper 2–4 layers of a single crystal solid in a high-vacuum environment. Considerations of surface-sensitivity, convenience, non-destructiveness, and quantitativeness led us to settle on electrons-in-electrons-out as the most satisfactory single set of experiments to perform this surface characterization. The chemical (i.e., 'elemental') composition of the surface may be determined by electron-stimulated Auger spectroscopy. The atomic geometry of surface scatterers may be examined by elastic-low-energy-electron diffraction. The vibrational structure of the surface scatterers is reflected in the temperature dependence of the elastic electron-diffraction intensities. Finally, the electronic structure of the ground state determines some features of the elastic scattering whereas the electronic excitation spectrum may be explored via inelastic low-energy electron diffraction.

Because many of the experimental techniques used to achieve the desired quantitative surface spectroscopy are over four decades old, I emphasized the recent developments which are rendering them useful as well as historical. Foremost among these are the routinization of the achievement of a high-vacuum environment, [12] the invention of Auger electron spectroscopy, [12, 13] the construction of an adequate microscopic theory of low-energy ($5 \text{ eV} \lesssim E \lesssim 500 \text{ eV}$) electron-solid scattering [1–6] [Part III], and the selective application of features of this theory to extract from experimental data quantitative measures of the condition of the surface [Parts IV and V]. During the course of our survey, we discovered that the four-decades-old theoretical barriers to the rapid development of surface spectroscopy have been removed. Therefore in the immediate future the major advances probably will come from the experimental tests and subsequent analytical refinement of models within the framework of the existing general theory. [1–6]

References

1. Duke, C. B., *Ann. Rev. Mater. Sci.* **1**, 165 (1971).
2. Duke, C. B. and Laramore, G. E., *Phys. Rev.* **B2**, 4765 (1970).
3. Duke, C. B. and Laramore, G. E., *Phys. Rev.* **B3**, 3183 (1971).
4. Feibelman, P. J., Duke, C. B., and Bagchi, A., *Phys. Rev.* **B5**, 2436 (1972).
5. Laramore, G. E. and Duke, C. B., *Phys. Rev.* **B5**, 267 (1972).
6. Duke, C. B., *Adv. Chem. Phys.* (in press).
7. Phillips, V. A. and Lifshin, E., *Ann. Rev. Mater. Sci.* **1**, 1 (1971).
8. Gomer, R., *Field Emission and Field Ionization*, Harvard University Press, Boston (1961).
9. Young, R. D. and Clark, H. E., *Phys. Rev. Letters* **17**, 351 (1966).
10. Young, R. D., *Physics Today* **24**, 42 (November 1971).
11. Carlson, T. A., *Physics Today* **25**, 30 (January 1972).
12. Duke, C. B. and Park, R. L., *Physics Today* **25** (August 1972).
13. Chang, C. C., *Surface Sci.* **25**, 53 (1971).
14. Ertl, G., in *Electron Spectroscopy* (ed. by D. A. Shirley), North Holland, Amsterdam (1972), p. 47.
15. McDonald, N. C. *Appl. Phys. Letters* **16**, 76 (1970).
16. Kammlott, G. W., *Surface Sci.* **25**, 120 (1971).
17. Smith, D. P., *Surface Sci.* **25**, 171 (1971).
18. Castaing, R. and Hennequin, J. F., *Adv. Mass. Spectrometry* **5**, 419 (1971).
19. Eastman, D., *Abstracts of the 32nd Physical Electronics Conference*, Sandia Laboratories, Albuquerque (1972), p. El.
20. Hagstrum, H., *J. Research, NBS* **74A**, 433 (1970).
21. Grimley, T. B., *J. Vac. Sci. Technol.* **8**, 31 (1971).
22. Davies, J. A., *J. Vac. Sci. Technol.* **8**, 487 (1971).
23. Bøgh, E., *Abstracts of the 32nd Physical Electronics Conference*, Sandia Laboratories, Albuquerque (1972), p. D1.
24. Frisch, R. and Stern, O., *Z. Physik* **84**, 430 (1933).
25. Weinberg, W. H. and Merrill, R. P., *Phys. Rev. Letters* **25**, 1198 (1970).
26. Duke, C. B. and Tucker, C. W., Jr., *Phys. Rev.* **B3**, 3561 (1971).
27. Bauer, E., in *Techniques of Metals Research*, II (ed. by R. F. Bunshah), Interscience, New York (1969), p. 501.
28. Bauer, E., in *Techniques of Metals Research*, II (ed. by R. F. Bunshah), Interscience, New York (1969), p. 559.
29. Estrup, P. J. and McRae, E. G., *Surface Sci.* **25**, 1 (1971).
30. Natta, M. and Joyes, P., *J. Phys. Chem. Solids* **31**, 447 (1970).
31. Langreth, D. C., *Phys. Rev. Letters* **26**, 1229 (1971).
32. Laramore, G. E., *Phys. Rev. Letters* **27**, 1050 (1971): *Solid State Commun.* **10**, 85 (1972).
33. Gähwiller, C., Brown, F. C., and Fujita, H., *Rev. Sci. Inst.* **41**, 1275 (1970).
34. Gähwiller, C. and Brown, F. C., *Phys. Rev.* **B2**, 1918 (1970).
35. Brown, F. C., Gähwiller, C., Fujita, H., Kunz, A. B., Scheifley, W., and Carrera, N., *Phys. Rev.* **B2**, 2126 (1970).
36. Schaich, W. L. and Ashcroft, N. W., *Phys. Rev.* **B3**, 2452 (1971).
37. Duke, C. B. and Tucker, C. W., Jr., *Surface Sci.* **15**. 231 (1969); *Phys. Rev. Letters* **23**, 1163 (1969).
38. Penn, D. R., *Phys. Rev. Letters* **28**, 1041 (1972).
39. Eastman, D. E. and Cashion, J. K., *Phys. Rev. Letters* **27**, 1520 (1971).
40. Bagchi, A., Duke, C. B., Feibelman, P. J., and Porteus, J. O., *Phys. Rev. Letters* **27**, 998 (1971).
41. Duke, C. B. and Bagchi, A., *J. Vac. Sci. Technol.* **9**, 738 (1971).
42. Bagchi, A. and Duke, C. B., *Phys. Rev.* **B5**, 2784 (1972).
43. Duke, C. B. and Landman, U., *Phys. Rev.* **B6**, 2956 (1972); **B6**, 2968 (1972).
44. Henrion, J. and Rhead, G. E., *Surface Sci.* **29**, 20 (1972).
45. Laramore, G. E. and Duke, C. B., *Phys. Rev.* **B2**, 4783 (1970).
46. Tong, S. Y. and Rhodin, T. N., *Phys. Rev. Letters* **16**, 711 (1971).
47. Duke, C. B., Howsmon, A. J., and Laramore, G. E., *J. Vac. Sci. Technol.* **8**, 10 (1971).
48. Duke, C. B., Laramore, G. E., Holland, B. W., and Gibbons, A. M., *Surface Sci.* **27**, 523 (1971).

49. Holland, B. W., *Surface Sci.* **28**, 258 (1971).
50. Propst, F. M., and Piper, T. C., *J. Vac. Sci. Technol.* **4**, 53 (1967).
51. Ibach, H., *Phys. Rev. Letters* **24**, 1416 (1970).
52. Ibach, H., *Phys. Rev. Letters* **27**, 253 (1971).
53. Mahan, G. D., *Phys. Rev.* **B2**, 4334 (1970).
54. Ignatjevs, A., Rhodin, T. N., Tong, S. Y., Lundqvist, B. I., and Pendry, J. B., *Solid State Commun.* **9**, 1851 (1971).
55. Reid, R. J., *Surface Sci.* **29**, 623 (1972).
56. Duke C. B., in *LEED: Surface Structure of Solids* (ed. by M. Láznička), JCMF, Prague, (1972).
57. Stern, R. M., Perry, J. J. and Boudreaux, D. S., *Rev. Mod. Phys.* **41**, 275 (1969).
58. Schweber, S. S., *An Introduction to Relativistic Quantum Field Theory*, Row, Peterson, and Company, Evanston (1961), Chapter 6.
59. Slater, J. C., *Phys. Rev.* **51**, 840 (1937).
60. Laramore, G. E., Duke, C. B., Bagchi, A., and Kunz, A. B., *Phys. Rev.* **B4**, 2058 (1971).
61. Jepsen, D. W., Marcus, P. M., and Jona, F., *Phys. Rev. Letters* **26**, 1365 (1971); *Phys. Rev.* **B5**, 3933 (1972).
62. Capart, G., *Surface Sci.* **26**, 479 (1971).
63. Pendry, J. B., *J. Phys.* **C4**, 2514 (1971).
64. Duke, C. B., *J. Vac. Sci. Technol.* **6**, 152 (1969).
65. Strozier, J. A., Jr. and Jones, R. O., *Phys. Rev.* **B3**, 3228 (1971).
66. Duke, C. B., Anderson, J. R., and Tucker, C. W. Jr., *Surface Sci.* **19**, 117 (1970).
67. Hedin, L. and Lundqvist, S., *Solid State Phys.* **23**, 2 (1969).
68. Beeby, J. L., *J. Phys.* **C1**, 82 (1968).
69. Kittel, C., *Quantum Theory of Solids*, Wiley, New York (1967), Chapter 19.
70. Harrison, W. A., *Pseudopotentials in the Theory of Metals*, Benjamin, New York (1963).
71. Davisson, C. J. and Germer, L. H., *Phys. Rev.* **30**, 705 (1927).
72. Tucker, C. W., Jr. and Duke, C. B., *Surface Sci.* **24**, 31 (1971).
73. Weber, W. H. and Webb, M. B., *Phys. Rev.* **177**, 1103 (1969).
74. Duke, C. B., Laramore, G. E., and Metze, V., *Solid State Commun.* **8**, 1189 (1970).
75. Barnes, R. F., Lagally, M. G., and Webb, M. B., *Phys. Rev.* **171**, 627 (1968).
76. McKinney, J. T., Jones, E. R., and Webb, M. B., *Phys. Rev.* **160**, 523 (1967).
77. MacRae, A. U. and Germer, L. H., *Phys. Rev. Letters* **8**, 489 (1962).
78. MacRae, A. U., *Surface Sci.* **2**, 522 (1964).
79. Jones, E. R., McKinney, J. T., and Webb, M. B., *Phys. Rev.* **151**, 146 (1966).
80. Lyon, H. B. and Somorjai, G. A., *J. Chem. Phys.* **44**, 3707 (1966).
81. Goodman, R. M., Farrell, H. H., and Somorjai, G. A., *J. Chem. Phys.* **48**, 1046 (1968).
82. Morabito, J. M. Jr., Steiger, R. F., and Somorjai, G. A., *Phys. Rev.* **176**, 638 (1969).
83. Estrup, P. J., in *The Structure and Chemistry of Solid Surfaces* (ed. by G. A. Somorjai), Wiley, New York (1969).
84. Woodruff, D. P. and Seah, M. P., *Phys. Letters* **30A**, 263 (1969).
85. Woodruff, D. P. and Seah, M. P., *Phys. Stat. Sol.* **(a)1**, 471 (1970).
86. Andersson, S. and Kasemo, B., *Solid State Commun.* **8**, 1885 (1970).
87. Reid, R. J., *Phys. Stat. Sol.* **(a)2**, K109 (1970).
88. Tabor, D. and Wilson, J., *Surface Sci.* **20**, 203 (1970).
89. Baudoing, R., Corotte, C., and Mascall, A., *J. Phys.* **31**, C1–21 (1970).
90. Theeten, J. B., Dommange, J. L., and Bonnerot, J., *Solid State. Commun.* **8**, 643 (1970).
91. Theeten, J. B., Bonnerot, J., Dommange, J. L., and Hurault, J. P., *Solid State Commun.* **9**, 1121 (1971).
92. Reid, R. J., *Phys. Stat. Sol.* **(a)4**, K211 (1971).
93. Tabor, D., Wilson, J. M., and Bastow, J. T., *Surface Sci.* **26**, 471 (1971).
94. McRae, E. G. and Caldwell, C. W., Jr., *Surface Sci.* **7**, 41 (1967).
95. Holland, B. W., Hannum, R. W., and Gibbons, A. M., *Surface Sci.* **25**, 567 (1971).
96. Morse, P. M., *Phys. Rev.* **35**, 1310 (1930).
97. Tucker, C. W., Jr. and Duke, C. B., *Surface Sci.* **29**, 237 (1972).
98. Hoffman, F. and Smith, H. P., Jr., in *The Structure and Chemistry of Solid Surfaces* (ed. by G. A. Somorjai), Wiley, New York (1969).
99. Hoffman, F. and Smith, H. P., Jr., *Phys. Rev.* **B1**, 2811 (1970).

100. Marcus, P. M., Jepsen, D. W., and Jona, F., *Surface Sci.* **17**, 442 (1969).
101. Jona, F., *IBM J. Res. Develop.* **14**, 444 (1970).
102. Lauzier, G., de Bersuder, L., and Hoffstein, V., *Phys. Rev. Letters* **27**, 735 (1971).
103. de Bersuder, L., Hoffstein, V., and Lauzier, J., *Surface Sci.* **27**, 338 (1971).
104. Baudoing, R., de Bersuder, L., Gaubert, C., Hoffstein, V., Lauzier, J., and Taub, H., *J. Vac. Sci. Technol.* **9**, 634 (1972).
105. Burkstrand, J. M. and Propst, F. M., *J. Vac. Sci. Technol.* **9**, 731 (1972).
106. Burkstrand, J. M., *Phys. Rev.* (to be published).
107. Laramore, G. E., *Phys. Rev.* **B6**, 1097 (1972).
108. Hirabayashi, K., *J. Phys. Soc. Japan* **30**, 211 (1971).
109. Hirabayashi, K., *Surface Sci.* **28**, 621 (1971).
110. Hoffstein, V. and Boudreaux, D. S., *Phys. Rev. Letters* **25**, 512 (1970).
111. Farrell, H. H. and Somorjai, G. A., *Phys. Rev.* **182**, 751 (1969).
112. Snow, E. C., *Phys. Rev.* **158**, 683 (1967).
113. Snow, E. C. and Weber, J. T., *Phys. Rev.* **157**, 570 (1967).
114. Schiff, L. I., *Quantum Mechanics*, Second Edition, McGraw-Hill, New York (1955), p. 105.
115. Gerstner, J. and Cutler, P. H., *Surface Sci.* **9**, 198 (1968).
116. Gersten, J. I., *Phys. Rev.* **B2**, 3457 (1970).
117. Laramore, G. E., *Phys. Rev.* **B6**, 2950 (1972).
118. Duke, C. B. and Laramore, G. E., *Surface Sci.* **31**, 659 (1972).
119. Duke, C. B. and Smith, D. L., *Phys. Rev.* **B5**, 4730 (1972).
120. Hoffstein, V. and Boudreaux, D. S., *Phys. Rev.* **B3**, 2447 (1971).
121. Jepsen, D. W., Marcus, P. M., and Jona, F., *Phys. Rev.* **B6**, 3684 (1972).
122. Duke, C. B., *J. Vac. Sci. Technol.* **6**, 152 (1969).
123. Buerger, M. J., *Crystal Structure Analyses*, Wiley, New York (1960).
124. Buerger, M. J., *Contemporary Crystallography*, McGraw-Hill, New York (1970).
125. Wood, E. A., *Bell System Tech. J.* **43**, 541 (1964).
126. Wood, E. A., *J. Appl. Phys.* **35**, 1306 (1964).
127. Andersson, S. and Kasemo, B., *Surface Sci.* **25**, 273 (1971).
128. Houston, J. E. and Park, R. L., *Surface Sci.* **21**, 209 (1970); **26**, 269 (1971).
129. Park, R. L. and Houston, J. E., *Surface Sci.* **18**, 213 (1969).
130. Ertl, G. and Küppers, J., *Surface Sci.* **21**, 61 (1970).
131. Lander, J. J., *Adv. Solid-State Chem.* **2**, 26 (1965).
132. Somorjai, G. A. and Farrell, H. H., *Adv. Chem. Phys.* **20**, 215 (1971).
133. May, J. W., *Adv. Cataly.* **21**, 151 (1970).
134. Robertson, W. D., *J. Vac. Sci. Technol.* **8**, 403 (1971).
135. Lagally, M. G., Ngoc, T. C., and Webb, M. B., *Phys. Rev. Letters* **26**, 1557 (1971).
136. Lagally, M. G., Ngoc, T. C., and Webb, M. B., *J. Vac. Sci. Technol.* **9**, 645 (1972).
137. Fink, M., Martin, M. R., and Somorjai, G. A., *Surface Sci.* **29**, 303 (1972).
138. Duke, C. B. and Tucker, C. W., Jr., *J. Vac. Sci. Technol.* **8**, 5 (1971).
139. Tucker, C. W., Jr. and Duke, C. B., *Surface Sci.* **23**, 411 (1970).
140. Farnsworth, H. E., *Phys. Rev.* **43**, 900 (1933).
141. Lander, J. J. and Morrison, J., *J. Appl. Phys.* **35**, 3593 (1964).
142. Hirabayashi, K., *J. Phys. Soc. Japan* **24**, 846 (1968).
143. Hirabayashi, K. and Takeishi, Y., *Surface Sci.* **4**, 150 (1966).
144. Stern, R. M. and Gervais, A., *Surface Sci.* **17**, 273 (1969).
145. Andersson, S. and Pendry, J. B., *J. Phys.* **C6**, 601 (1973).
146. Watts, C. M. K., *Surface Sci.* **23**, 453 (1970).
147. Andersson, S. and Kasemo, B., *Surface Sci.* **32**, 78 (1972).
148. Duke, C. B., Lipari, N. O., Laramore, G. E., and Theeten, J. B., *Solid State Commun.* (to be published).
149. Weber, W. H. and Webb, M. B., *Phys. Rev.* **177**, 1103 (1969).
150. Seah, M. P., *Surface Sci.* **17**, 161 (1969); **24**, 357 (1971).
151. Porteus, J. O. and Faith, W. N., *Phys. Rev.* **B2**, 1532 (1970); *J. Vac. Sci. Technol.* **9**, 1062 (1972).
152. Turnbull, J. C. and Farnsworth, H. E., *Phys. Rev.* **54**, 507 (1938).
153. Reichertz, P. P. and Farnsworth, H. E., *Phys. Rev.* **75**, 1902 (1949).
154. Laramore, G. E. and Duke, C. B., *Phys. Rev.* **B3**, 3198 (1971).

155. Lucas, A. A. and Sunjić, M., *Phys. Rev. Letters* **26**, 229 (1971).
156. Wei, P. S. P. and Smith, A. W., *Surface Sci.* **27**, 675 (1971).
157. Steinrisser, F. and Sickafus, E. N., *Phys. Rev. Letters* **27**, 992 (1971).
158. Tharp, L. N. and Scheibner, E. J., *J. Appl. Phys.* **38**, 3320 (1967).
159. Duke, C. B. and Landman, U., *Phys. Rev.* **B7**, 1368 (1973); **B8**, July 15 (1973).
160. Duke, C. B., Landman, U., and Porteus, J. O., *J. Vac. Sci. Technol.* **10**, 183 (1973).
161. Beck, D. E., *Phys. Rev.* **B4**, 1555 (1971).
162. Heger, Ch. and Wagner, D., *Phys. Letters* **34A**, 448 (1971); *Z. Physik* **244**, 449 (1971).

THEORETICAL ASPECTS OF
X-RAY PHOTOELECTRON SPECTROSCOPY

CHARLES S. FADLEY

Department of Chemistry, University of Hawaii,
Honolulu, H.I. 96822, U.S.A.

1. Introduction

In these notes, we shall review several important aspects of the basic theory necessary
for interpreting and understanding experiments based on X-ray photoelectron spectroscopy (*XPS* or *ESCA*). In view of the breadth of material to be covered, only an
outline of the various theoretical models will be presented, including discussions of
the basic assumptions involved, examples of the agreement obtained between theory
and experiment, and statements concerning the remaining unsolved theoretical or
experimental problems. At several points, references will be made to inherent similarities and differences between XPS and the closely related ultraviolet photoelectron
spectroscopy (*UPS*). The aim of these notes will thus be to provide a basic background
for more specialized study. The bibliography presented here is not intended to be
an exhaustive one, but rather to contain the most important review papers, or in
some cases, original papers, on a given topic, as well as certain illustrative examples
of the use of various theoretical models.

The fundamental experiment in either X-ray- or ultraviolet-photoelectron spectroscopy involves exposing the specimen to be studied to a flux of nearly mono-
energetic photons with mean energy hv. Photoelectrons, as well as Auger electrons and
secondary electrons, will be emitted from the specimen. There are basically three
parameters characterizing each emitted photoelectron: its kinetic energy, its direction
of emission, and, for certain special experimental situations, the orientation of its
electron spin. If it is assumed that we can somehow distinguish photoelectrons from
Auger- and secondary-electrons, then these three parameters give rise to three fundamental measurements that are possible on the photoelectron flux [1, 2]:

(1) *The distribution of photoelectron intensity with kinetic energy*: This measurement
of course requires an electron spectrometer, of which several types are being utilized.
In the most common dispersive spectrometers, an individual kinetic energy distribution
is usually measured at a fixed angle of electron emission (or small range of emission
angles) relative to both the photon source and the specimen. In non-dispersive retarding grid spectrometers, the range of emission angles accepted by the spectrometer
may be very large.

(2) *The distribution of photoelectron intensity with angle of emission*: Such angular
distribution measurements can be made relative to either the photon propagation
direction or to axes fixed with respect to the specimen. Generally, these measurements
require kinetic energy distribution measurements at each of several angles of emission.

W. Dekeyser et al. (eds.), Electron Emission Spectroscopy, 151–224. All Rights Reserved
Copyright © 1973 by D. Reidel Publishing Company, Dordrecht-Holland

(3) *The spin polarization or spin distribution of the photoelectron intensity*: Such measurements require a specimen that has somehow been magnetically polarized, usually by an external field, so that more photoelectrons may be emitted with one of the two possible spin orientations than with the other. Then the relative numbers of spin-up and spin-down photoelectrons are measured [3]. Such spin polarization measurements have so far only been made with ultraviolet radiation for excitation.

The additional time and experimental complexity required for angular distribution or spin polarization measurements have resulted in the fact that most work in XPS or UPS up to the present time has involved only kinetic energy distribution measurements with a fixed geometry of the photon source, specimen, and spectrometer. However, measurements of both types (2) and (3) seem fruitful from several points of view [2–6], and their expansion seems both desirable and inevitable.

In a given kinetic energy distribution, we wish to be able to predict both the positions of the various photoelectron peaks in kinetic energy and also their relative intensities. Also, the variation of the intensity of each peak with angle of emission is an important theoretical problem. We shall treat the three most important aspects of the theory pertaining to photoelectron kinetic energies and shall review in a fourth section the theory of photoelectron peak intensities and their angular distributions. The subject outline is as follows:

(a) Core electron binding energies
(b) Valence electron binding energies
(c) Multiplet splittings and multi-electron processes
(d) Relative intensities and angular distributions.

Before proceeding to these topics, however, it is well to discuss the fundamental conservation equations applying to photoelectron spectroscopy. The energy conservation equation can be written for a free atom or molecule as

Total initial energy $=$ Total final energy

$$hv + E^i = E_{kin} + E^f(k)' \tag{1}$$

where hv is the photon energy, E^i is the *total* initial energy of the atom or molecule, E_{kin} is the photoelectron kinetic energy, and $E^f(k)'$ is the total final energy of the atom or molecule after ejection of an electron from the kth orbital as a photoelectron. For atoms the index k thus stands for the quantum numbers nl or nlj (for example, $3s$ or $4f_{7/2}$). For molecules, k is determined by the overall symmetry of the various molecular orbitals (for example, 1σ or 4π). E^i and $E^f(k)'$ may include contributions from electronic, vibrational, rotational, and translational motion. In view of the present instrumental contributions to linewidth in an XPS spectrum of approximately $0.5-1.0$ eV, it is reasonably accurate to assume that the initial state energy E^i is unique, or equivalently, that the various sources of excitation from the ground states of all types of motion are negligible with respect to ~ 0.5 eV. Such sources would be thermal excitation, photon bombardment, and collisions with photo-, Auger-, and secondary-electrons. Because of the disruptive character of the photoemission process, this same simplicity need not hold for the final state energy $E^f(k)'$, however.

The most obvious source of additional energy in the final state arises because of momentum conservation. In order to conserve total momentum between the states represented by the left and right sides of Equation (1), the atom or molecule must in general recoil in a direction opposite to that of the electron emission. That is, the momentum conservation equation analogous to Equation (1) is

$$\text{Total initial momentum} = \text{Total final momentum}$$
$$\mathbf{P}_{ph} + 0 = \mathbf{P}_e + \mathbf{P}_r, \tag{2}$$

where $\mathbf{P}_{ph}$ is the photon momentum and has a magnitude of $h\nu/c$, the momentum associated with E^i is taken to be zero, $\mathbf{P}_e$ is the photoelectron momentum, and $\mathbf{P}_r$ is the recoil momentum of the atom or molecule, treated as a center-of-mass translation. The photoelectron velocities typically encountered in XPS can be considered to a good approximation to be non-relativistic. For $E_{kin} = 500$ eV, $v/c = 0.044$ and for $E_{kin} = 1500$ eV, $v/c = 0.076$, where $v =$ photoelectron velocity. In this approximation, it is a simple matter to show that for photoelectrons originating from valence electronic levels (for which $E_{kin} \approx h\nu$) $|\mathbf{P}_{ph}| \approx v/2c |\mathbf{P}_e|$. Therefore, in general $|\mathbf{P}_{ph}| \ll |\mathbf{P}_e|$ and $\mathbf{P}_e \approx -\mathbf{P}_r$. $E^f(k)'$ can thus be written as a sum of a recoil energy E_r plus a term containing the energies due to all other modes of motion $E^f(k)$:

$$E^f(k)' = E^f(k) + E_r. \tag{3}$$

By means of Equation (2) it can be shown that for a given $h\nu$ and E_{kin}, E_r increases with decreasing atomic or molecular mass. For excitation of valence shell photoelectrons with AlKα radiation ($h\nu = 1486$ eV), Siegbahn et al. [7] have calculated the following recoil energies for different atoms: H – 0.9 eV, Li – 0.1 eV, Na – 0.04 eV, K – 0.02 eV, and Rb – 0.01 eV. It is thus clear that only for the lightest atoms H, He, and Li does the recoil energy have a significant magnitude in comparison with the present 0.5–1.0 eV instrumental line-widths in XPS spectra. For almost all cases, we can thus neglect E_r and write Equation (1) simply as

$$h\nu + E^i = E_{kin} + E^f(k), \tag{4}$$

for which there is no net translational energy in the state described by $E^f(k)$. $E^f(k)$ may still contain various electronic, vibrational, or rotational excitations relative to the simple ground state E^i, however. Vibrational and rotational structure in $E^f(k)$ are routinely observed in high resolution UPS spectra from gases, for example [8, 9]. However, the present resolution limitations in XPS have so far prevented observation of other than various electronic states in $E^f(k)$. The origins of these electronic states and the complex spectra they give rise to we discuss in Section 4.

The binding energy of a given electron is defined as the positive energy required to remove it to infinity with zero kinetic energy. If we denote the binding energy of the kth electron by $E_b(k)$, then this definition requires that

$$E_b(k) \equiv E^f(k) - E^i \tag{5}$$

and Equation (4) can be rewritten as

$$hv = E_{kin} + E_b(k).$$ (6)

Equation (6) refers to an atom or molecule in free space and so implicitly utilizes the vacuum level as the reference level for binding energies. Thus, we can rewrite this equation, which should hold in the limit of dilute gases, as

$$hv = E_{kin} + E_b^V(k) \quad \text{(gases)}$$ (7)

where the superscript V indicates a vacuum level reference. The vacuum level reference also arises naturally in theoretical calculations on atoms and molecules.

For the case of solid specimens, however, we are more or less obliged to connect the specimen electrically to the spectrometer in an attempt to maintain it at a known fixed potential during photoemission. For the simplest possible case of a metallic specimen in a metallic spectrometer, the energy levels and kinetic energies which result are as shown in Figure 1. In a metal at absolute zero, the Fermi level E_F or electron chemical potential has the simple interpretation of being the highest occupied level, as shown in Figure 1. This interpretation of E_F is also very nearly true at normal

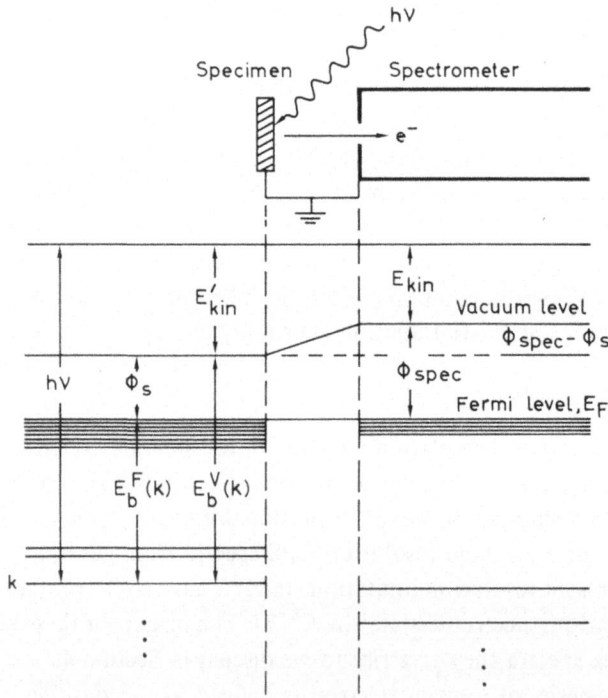

Fig. 1. Energy level diagram for a metallic specimen electrically connected to an electron spectrometer. The closely spaced levels near the Fermi level E_F represent the filled portion of the valence bands. The deeper levels are core levels. An analogous diagram also applies in principle to semiconducting or insulating specimens, with the only difference being that E_F lies somewhere between the filled valence bands and the empty conduction bands above.

experimental temperatures. The work function ϕ for a solid is defined to be the energy separation between the vacuum level and the Fermi level. When connected as shown in Figure 1, thermodynamic equilibrium between specimen and spectrometer results, and the electron chemical potentials (Fermi levels) will be identical. The respective vacuum levels for specimen and spectrometer need not be equal however, so that in passing from the surface of the specimen into the spectrometer, a photoelectron will feel an accelerating or retarding potential equal to $\phi_s - \phi_{spec}$, where ϕ_s is the specimen work function and ϕ_{spec} is the spectrometer work function. Thus, an initial kinetic energy E'_{kin} at the surface of the specimen becomes E_{kin} inside the spectrometer, and

$$E_{kin} = E'_{kin} + \phi_s - \phi_{spec}. \tag{8}$$

From Figure 1 it is thus clear that binding energies in a metallic solid can be measured quite easily relative to the identical Fermi levels of specimen and spectrometer. The pertinent equation is

$$h\nu = E_{kin} + E_b^F(k) + \phi_{spec} \quad \text{(solids)}, \tag{9}$$

where the superscript F indicates a Fermi level reference. (Note that a superscript f refers to final state quantities and an F to the Fermi level.) Provided that we could also determine the specimen work function ϕ_s, vacuum-referenced binding energies could be obtained from

$$E_b^V(k) = E_b^F(k) + \phi_s. \tag{10}$$

ϕ_s has not been measured in connection with XPS experiments, however, and in fact, very careful measurements would be necessary to include the strong dependence of ϕ_s on specimen surface condition. Thus, it is virtually always Equation (9) that is used in analyzing XPS data on metals, as well as other solid specimens.

From this discussion, it is clear that Fermi-referenced binding energies are operationally very convenient for metallic specimens. For semiconductors and insulators, however, it is not so easy to locate the Fermi level, which lies somewhere between the predominantly filled valence bands and the predominantly empty conduction bands [10]. Also, the assumption of thermodynamic equilibrium during photoemission becomes somewhat doubtful if the resistance of the specimen is too high in comparison with the currents necessary to replenish the electrons lost from the specimen surface due to X-ray excitation. A further problem is that theoretical calculations on semiconductors and insulators seldom use the Fermi level as a reference level, nor is it possible to easily calculate its position. More often, the vacuum level emerges as a natural reference level in calculations on such systems, as well as in certain types of core electron chemical shift calculations in metals. Thus, the present state of affairs is that Fermi-referenced binding energies or chemical shifts for solids are often compared with vacuum-referenced theoretical calculations of the same quantities. In most cases, such comparisons have yielded fair to good agreement [11], but further study of this problem is certainly called for.

An effect closely connected with the reference level for binding energies is that of positive charging of the specimen due to electron emission. Such charging effects result in a net retardation of the photoelectrons before they enter the spectrometer, and thus spuriously large binding energies are measured. The magnitudes of these effects can be of the order of 1–100 eV and have been observed in both solids [1] and gases [9]. The extent of charging depends upon many factors, including photon flux, electron flux from surrounding surfaces, temperature, and the surface and bulk conductivities of solid specimens. If the specimen is charged uniformly with a positive potential V_c relative to the spectrometer, then the energy conservation equations for gases and solids become

$$hv = E_{kin} + E_b^V(k) + V_c \quad \text{(gases)} \tag{11}$$
$$hv = E_{kin} + E_b^F(k) + V_c + \phi_{spec} \quad \text{(solids)}. \tag{12}$$

If a non-zero V_c is through ignorance assumed to be zero, we thus measure binding energies $E_b^V(k)$ and $E_b^F(k)$ greater than the actual values by an amount V_c. If V_c varies throughout the volume of the specimen that is active in producing photoelectrons, a broadening of photoelectron peaks will also result. Such charging effects are also in need of further study.

A final important effect to be considered in the analysis of any photoelectron spectrum is that of inelastic scattering. In addition to producing a background of inelastically scattered photoelectrons on the low kinetic energy side of each peak in a spectrum, such effects limit the depths within the specimen from which photoelectrons that have not suffered significant inelastic scattering can be emitted. It is only for such photoelectrons that Equations (7) and (9) hold. As the inherent instrumental contribution to linewidth in XPS is of the order of 1 eV, a significant amount of inelastic scattering can at present be defined as that which decreases the photoelectron kinetic energy by ~0.5 eV or more. For solids, the depth limitations imposed by inelastic scattering may be rather stringent, and present estimates range from approximately 10 Å to 100 Å [7, 12–14]. Further investigations of these escape depths are certainly called for, but at the present time we can definitely conclude that caution is necessary in preparing the surfaces of solid specimens and in interpreting spectra obtained from them, if surface effects are either to be avoided or specifically studied. In fact, the magnitudes of these penetration depths raise the definite question as to whether photoelectron spectroscopy deserves to be considered a bulk measurement, a surface measurement, or something intermediate between the two. In Section 5, we discuss an experimental procedure involving angular distribution measurements which may make it possible to more clearly delineate the bulk and surface aspects of photoelectron spectroscopy.

2. Core Electron Binding Energies

There seems little doubt that the majority of XPS studies to date has been involved primarily with the precise measurement of core electron binding energies, and in particular, with the measurement of chemical shifts in these binding energies. XPS is rather uniquely qualified for such studies, as the usual X-rays utilized (MgKα = 1254 eV and $AlK\alpha = 1486$ eV) can penetrate to levels well below the vacuum level. The more common ultraviolet radiation sources presently limit UPS to valence levels and weakly bound core levels within ~ 40 eV of the vacuum level, on the other hand.

The core levels of any atom can by definition be considered to represent filled sub-shells, and are found in XPS spectra to be relatively sharp in energy, with typical experimental widths of approximately 1–10 eV. The width observed for a core photo-electron peak depends upon several factors of both inherent and instrumental type. The most important inherent sources of width are:

(1) the lifetime of the core hole created by photoemission,

(2) various possible values for the final state energy $E^f(k)$, as represented for example by multiplet splittings or multi-electron effects (see Section 4), and

(3) unresolvable chemically-shifted peaks.

The most important instrumental sources and their typical magnitudes are:

(1) the exciting X-ray linewidth (approximately 1.0 eV for $AlK\alpha$ without mono-chromatization and approximately 0.5 eV with),

(2) the finite resolving power of the electron spectrometer (0.3 eV for 0.03% re-solution at $E_{kin} = 1000$ eV), and

(3) non-uniform charging of the specimen (variable magnitude).

The minimum linewidths observed to date have been very close to 0.5 eV, and are presumably determined in large measure by instrumental contributions. Thus, pro-vided that the various inherent sources of linewidth and also non-uniform charging are not too large, it is possible in principle to measure chemical shifts of the order of 0.1 eV between two or more photoelectron peaks resulting from emission from the same subshell.

If we consider the same atom A either in two chemically inequivalent sites in the same compound labelled 1 and 2 or in two different compounds which can be simi-larly labelled 1 and 2, then the chemical shift ΔE_b of the k electron binding energy can be written simply as the difference of two binding energies. For gaseous specimens, this means that

$$\Delta E_b^V(A, k, 1\text{--}2) = \left(E_b^V(k)\right)_1 - \left(E_b^V(k)\right)_2$$
$$= (E_{kin})_2 - (E_{kin})_1 \quad \text{(gases)} \tag{13}$$

where A, k, 1–2 represent the minimum number of parameters required to specify a chemical shift, that is, the atom and level, and the two chemical sites or compounds involved. Here, we have neglected charging effects. For solids, the corresponding equation is

$$\Delta E_b^F(A, k, 1\text{--}2) = \left(E_b^F(k)\right)_1 - \left(E_b^F(k)\right)_2$$
$$= (E_{kin})_2 - (E_{kin})_1 + (\phi_{spec})_2 - (\phi_{spec})_1 + (V_c)_2 - (V_c)_1, \tag{14}$$

where possible effects due to spectrometer work function changes or differences in charging potential have been included. Provided that both of the latter effects are negligible, Equation (14) simplifies to a form identical to that of Equation (13),

$$\Delta E_b^F (A, k, 1\text{--}2) = (E_b^F (k))_1 - (E_b^F (k))_2$$
$$= (E_{kin})_2 - (E_{kin})_1 \quad \text{(solids)}. \tag{15}$$

As we have noted previously, many if not all theoretical calculations of chemical shifts have an implicit vacuum reference level. This is quite satisfactory for gas phase work, but not necessarily for work on solids. For the latter case, the relationship between vacuum-referenced and Fermi-referenced chemical shifts is from Equation (10):

$$\Delta E_b^V (A, k, 1\text{--}2) = \Delta E_b^F (A, k, 1\text{--}2) + (\phi_s)_1 - (\phi_s)_2. \tag{16}$$

Thus, in directly comparing vacuum-referenced theoretical calculations and Fermi-referenced experimental values, it is required to neglect the work function difference between the two solids, $(\phi_s)_1 - (\phi_s)_2$. In most work to date, no serious effects of work function differences have been observed, although there is generally more scatter on a plot of measured chemical shifts against calculated chemical shifts for solids than on a corresponding plot for gases [7, 9, 11, 15]. This additional scatter could be connected with reference level effects or specimen charging or both. Both of these effects deserve further study.

The various theoretical procedures for calculating and interpreting chemical shifts span a very broad range in both rigor and complexity. A quite considerable literature has built up on such procedures and we can only hope to present a very brief review here. For further details, several reviews of various aspects of the subject are available [7, 9, 11, 16–18]. In a recent review, Shirley [11] has presented the various procedures in approximate order of descending rigor (although not necessarily descending utility). We shall adhere to a similar, although not identical, outline. In such an ordered list, the most exact procedure for calculating a binding energy is to carry out accurate self-consistent field calculations on both the initial state of the system to obtain the total energy E^i and also on the final hole state of the system to obtain the total energy $E^f(k)$ (cf. Equation (5)). In both of these calculations, relativistic effects and electron correlation effects should be considered. Chemical shifts can then be obtained as a difference of two such binding energies, so that two initial state calculations and two final state calculations are involved. Such calculations are obviously complicated and time consuming and they have been carried out on relatively few atoms and small molecules [11]. At the other end of this scale of rigor can be placed procedures such as the correlation of binding energy shifts with Pauling charges derived from electronegativity arguments [7]. There are several procedures intermediate between these two approximate limits and we shall review these briefly below. Each of these methods has its own advantages and disadvantages, and the selection of a given one depends upon the particular compounds under study and the accuracy with which interpretation is desired.

The self-consistent field (SCF) calculations necessary for total-energy-difference binding energy determinations can be performed by means of either the non-relativistic Hartree-Fock method or the relativistic Dirac-Fock method. As the Hartree-Fock method has been widely used in calculations on both atoms and molecules at different levels of exactness and also serves as a reference method for several more accurate and less accurate methods of computing electronic energy levels, we shall outline it here in simplest form. The wave function Ψ for an N-electron system is represented by a single Slater determinant of N one-electron orbitals. Each one-electron orbital is composed of a product of a spatial part φ_i and a spin part denoted by α or β. We assume for simplicity that each spatial orbital φ_i is doubly occupied and thus paired with both possible spin functions α and β, although this need not be required in general. Thus, N must be an even number and we can set $N=2n$, where n is the number of orbitals φ_i $(i=1,...,n)$. Then Ψ can be written as

$$\Psi = \frac{1}{\sqrt{N!}} \begin{vmatrix} \varphi_1(1)\alpha(1) & \varphi_1(1)\beta(1)...\varphi_n(1)\alpha(1) & \varphi_n(1)\beta(1) \\ \varphi_1(2)\alpha(2) & \varphi_1(2)\beta(2)...\varphi_n(2)\alpha(2) & \varphi_n(2)\beta(2) \\ \vdots & \vdots & \vdots & \vdots \\ \varphi_1(N)\alpha(N) & \varphi_1(N)\beta(N)...\varphi_n(N)\alpha(N) & \varphi_n(N)\beta(N) \end{vmatrix}$$

(17)

where the integers $1...N$ label electron coordinates for both spatial and spin parts of each orbital. The N-electron Hamiltonian for this system, which we assume to have P nuclei, will be given in electrostatic units by

$$\mathcal{H} = -\frac{\hbar^2}{2m} \sum_{i=1}^{N} \nabla_i^2 - \sum_{i=1}^{N}\sum_{l=1}^{P} \frac{z_l e^2}{r_{il}} + \sum_{i>j}^{N} \frac{e^2}{r_{ij}} + \sum_{l>w}^{P} \frac{z_l z_w e^2}{r_{lw}}$$

(18)

$$\underset{\substack{\text{Kinetic} \\ \text{energy}}}{} \qquad \underset{\substack{\text{Electron-nuclear} \\ \text{attraction}}}{} \qquad \underset{\substack{\text{Electron-} \\ \text{electron} \\ \text{repulsion}}}{} \qquad \underset{\substack{\text{Nuclear-nuclear} \\ \text{repulsion}}}{}$$

If such a Hamiltonian is used together with the variational principle to determine that Ψ for which the total energy $E=\langle\Psi|\mathcal{H}|\Psi\rangle$ is a minimum, the Hartree-Fock equations are obtained. These n equations can be used to determine a self-consistent set of orbitals φ_i, as well as to calculate the total energy E of the state described by Ψ. For a system with fixed nuclear positions, the nuclear-nuclear repulsion term represents an additive constant in the expression for total energy. Thus it is generally disregarded. In atomic units (1 AU $=27.21$ eV, 1 Bohr radius $=a_0=0.529$ Å), the Hartree-Fock equations in diagonal form are

$$\left[-\frac{1}{2}\nabla_1^2 - \sum_{l=1}^{P} \frac{z_l}{r_{1l}} \right]\varphi_i(1) + 2\left[\sum_{j=1}^{n} \int \varphi_j^*(2)\frac{1}{r_{12}}\varphi_j(2)\,d\tau_2 \right]\varphi_i(1)$$

$$\underset{\substack{\text{Kinetic} \\ \text{energy}}}{} \quad \underset{\substack{\text{Electron-} \\ \text{nuclear} \\ \text{attraction}}}{} \qquad\qquad \underset{\substack{\text{Electron-electron} \\ \text{repulsion}}}{}$$

$$-\sum_{j=1}^{n}\left[\int \varphi_j^*(2)\frac{1}{r_{12}}\varphi_i(2)\,d\tau_2 \right]\varphi_j(1) = \varepsilon_i\varphi_i(1),$$

(19)

$$\underset{\substack{\text{Exchange} \\ \text{interaction}}}{}$$

where the ε_i's are termed energy eigenvalues, one-electron energies, or orbital energies. The origins of the individual terms are labelled. Note the factor of two multiplying the electron-electron repulsion term because of the double occupancy of each orbital. Also, as the exchange interaction is only possible between electrons with parallel spins, there are only n possible terms for each orbital φ_i, including the interaction between $\varphi_i\alpha$ and $\varphi_i\beta$. These equations can be rewritten by dividing and multiplying the exchange terms by $\varphi_i^*(1)\,\varphi_i(1)$ to give

$$\left[-\tfrac{1}{2}\nabla_1^2 - \sum_{l=1}^{P}\frac{z_l}{r_{1l}}\right]\varphi_i(1) + 2\left[\sum_{j=1}^{n}\int \varphi_j^*(2)\,\frac{1}{r_{12}}\,\varphi_j(2)\,d\tau_2\right]\varphi_i(1)$$

$$-\left[\frac{\displaystyle\sum_{j=1}^{n}\int \varphi_i^*(1)\,\varphi_j^*(2)\,\frac{1}{r_{12}}\,\varphi_i(2)\,\varphi_j(1)\,d\tau_2}{\varphi_i^*(1)\,\varphi_i(1)}\right]\varphi_i(1) = \varepsilon_i\varphi_i(1). \qquad (20)$$

This equation has the form of a pseudo-eigenvalue equation $\mathscr{F}\varphi_i = \varepsilon_i\varphi_i$ where $\mathscr{F}$ is termed the Fock operator and depends upon the form of all the φ_i's. It is convenient to express this simply as

$$\left[-\tfrac{1}{2}\nabla_1^2 - \sum_{l=1}^{P}\frac{z_l}{r_{1l}}\right]\varphi_i(1) + \sum_{j=1}^{n}[2J_j - K_j]\,\varphi_i(1) = \varepsilon_i\varphi_i(1) \qquad (21)$$

by defining the Coulomb and exchange operators J_j and K_j such that

$$J_j\varphi_i(1) = \int \varphi_j^*(2)\,\frac{1}{r_{12}}\,\varphi_i(1)\,\varphi_j(2)\,d\tau_2 \qquad (22)$$

$$K_j\varphi_i(1) = \int \varphi_j^*(2)\,\frac{1}{r_{12}}\,\varphi_i(2)\,\varphi_j(1)\,d\tau_2. \qquad (23)$$

Thus, the matrix elements of these operators are the two-electron Coulomb integrals J_{ij} and exchange integrals K_{ij}:

$$J_{ij} \equiv \langle \varphi_i(1)|\,J_j\,|\varphi_i(1)\rangle = \iint \varphi_i^*(1)\,\varphi_j^*(2)\,\frac{1}{r_{12}}\,\varphi_i(1)\,\varphi_j(2)\,d\tau_1\,d\tau_2 \qquad (24)$$

$$K_{ij} \equiv \langle \varphi_i(1)|\,K_j\,|\varphi_i(1)\rangle = \iint \varphi_i^*(1)\,\varphi_j^*(2)\,\frac{1}{r_{12}}\,\varphi_i(2)\,\varphi_j(1)\,d\tau_1\,d\tau_2. \qquad (25)$$

From these definitions, it is clear that $J_{ij}=J_{ji}$, $K_{ij}=K_{ji}$, and $J_{ii}=K_{ii}$. Once the Hartree-Fock equations have been solved to the desired self-consistency, the orbital energies ε_i can be obtained from

$$\varepsilon_i = \langle \varphi_i(1)|\,\mathscr{F}\,|\varphi_i(1)\rangle = \varepsilon_i^0 + \sum_{j=1}^{n}(2J_{ij} - K_{ij}) \qquad (26)$$

where ε_i^0 is the expectation value of the one-electron operator for kinetic energy and electron-nuclear attraction

$$\varepsilon_i^0 = \langle \varphi_i(1) | -\tfrac{1}{2}\nabla_1^2 - \sum_{l=1}^{P} \frac{z_l}{r_{1l}} | \varphi_i(1) \rangle .$$

By comparison, the total energy of the state described by Ψ is given by

$$E = \langle \Psi | \mathcal{H} | \Psi \rangle = 2 \sum_{i=1}^{n} \varepsilon_i^0 + \sum_{i,j=1}^{n} (2J_{ij} - K_{ij}). \qquad (27)$$

Note that this is not simply the sum of all the one-electron energies for the $2n$ electrons in the system, as the sum of the Coulomb and exchange terms is multiplied by one instead of two in the total energy expression to avoid counting these terms twice. This means that measured binding energies (which we shall show to be very close to the ε_i's in value) cannot be directly used to determine total energies and hence such quantities as reaction energies.

In using the difference method for computing binding energies, it is necessary to compute both E^i and $E^f(k)$ corresponding to the wave-functions Ψ^i and $\Psi^f(k)$, respectively. The final state wave function will be characterized by having a hole in the kth subshell and thus only $N-1$ electrons. As the photo-emission process occurs on a time scale very short compared to that of nuclear motion ($\sim 10^{-16}$ s compared to $\sim 10^{-13}$ s), the nuclear positions in $\Psi^f(k)$ can be assumed to be identical to those in Ψ^i. However, in general, the $N-1$ passive electrons in $\Psi^f(k)$ will not have the same spatial distribution as those in Ψ^i due to relaxation around the k hole. Such hole-state calculations have been carried out on both atoms [11, 17, 19, 20] and small molecules [11, 21–24]. If binding energies determined in this way are corrected for relativistic effects where necessary, very good agreement with experimental core electron binding energies has been obtained. For example, an agreement of approximately 0.2% is found between theoretical and experimental $1s$ binding energies of Ne ($E_b^V(1s)=870$ eV) and Ar ($E_b^V(1s)=3205$ eV) [11, 19].

The relativistic correction generally increases core electron binding energies and its magnitude depends on the ratio of the characteristic orbital velocity to the velocity of light. Shirley [11] estimates that $1s$ binding energies increase by a factor of approximately $\tfrac{3}{4}(v/c)^2$ as a rough guide in determining the size of relativistic effects. For example, the correction for $C1s$ is only about 0.2 eV out of 290 eV ($\sim 0.06\%$), whereas for the deeper core level $Ar1s$, it is about 24 eV out of 3205 eV ($\sim 0.75\%$).

An additional correction which should in principle be made to such binding energy calculations is that dealing with electron-electron correlation [11, 19, 25, 26]. The intuitive expectation for such corrections might be that because the initial-state SCF calculation does not include favorable correlation between a given core electron and the other $N-1$ electrons, the calculated E^i value would be too large and thus that the binding energy $E_b^V(k)=E^f(k)-E^i$ would be too small. However, in comparing relativistically corrected hole-state calculations on several small atoms and ions with experimental binding energies, the remaining error due to correlation has been found to change sign from level to level within the same system [11, 19, 25]. Such deviations from simple expectations appear to have their origins primarily in the different types of correlation possible for final hole states in different core or valence levels. For example, $E_b^V(1s)$ for Ne shows a correlation correction δE_{corr} in the expected direction (that is, so as to increase E_b) of approximately 0.6 eV out of 870.2 eV ($\sim +0.07\%$) [26], whereas δE_{corr} for $E_b^V(2s)$ acts in the opposite direction by approximately 0.9 eV out of 48.3 eV ($\sim -1.8\%$) [25]. For core levels in closed shell systems such as Ne, it appears that such corrections can be computed approximately from a sum of electron pair correlation energies $\varepsilon(i, j)$ calculated for the ground state of the system [25]. For example, in computing the $1s$ binding energy in Ne, the correction has the form of a sum over pair correlation energies between the $1s$ electron and all other electrons in the atom. Such correlation energies are dependent upon both overlap and spin orientation, as the exchange interaction partially accounts for correlation of electrons with parallel spin. For Ne $1s$, this sum is thus:

$$\delta E_{corr} = \varepsilon(1s\alpha, 1s\beta) + \varepsilon(1s\alpha, 2s\alpha) + \varepsilon(1s\alpha, 2s\beta)$$
$$+ 3\varepsilon(1s\alpha, 2p\alpha) + 3\varepsilon(1s\alpha, 2p\beta), \tag{28}$$

with values of $\varepsilon(1s\alpha, 1s\beta) = +1.09$ eV, $\varepsilon(1s\alpha, 2s\alpha) = +0.01$ eV, $\varepsilon(1s\alpha, 2s\beta) = +0.06$ eV, $\varepsilon(1s\alpha, 2p\alpha) = +0.11$ eV, $\varepsilon(1s\alpha, 2p\beta) = +0.15$ eV. Note the smaller magnitudes of $\varepsilon(i, j)$ for electrons with parallel spins. Also, it is clear that most of the correlation correction arises from the strongly overlapping $1s$ electrons. Equation (28) is only a first approximation, however, and more exact calculations involving explicit estimates of all types of correlation in both Ne and Ne$^+$ with a $1s$ hole give better agreement with the experimental $1s$ binding energy [26]. The experimental value is $E_b^V(1s) = 870.2$ eV, in comparison to $\delta E_{corr} = 1.4$ eV, $E_b^V(1s) = 870.8$ eV based on Equation (28) [25] and $\delta E_{corr} = 0.6$ eV, $E_b^V(1s) = 870.0$ eV based on the more accurate calculation [26]. δE_{corr} is decreased in the latter calculation primarily because of correlation terms present in Ne$^+$ but not present in Ne. The sum of pair correlation energies $\varepsilon(i, j)$ in Ne$^+$ is larger than that in Ne by about 30%, and other terms not describable as pair interactions are present in Ne$^+$ but not Ne.

Aside from verifying that hole-state energy difference calculations can yield very accurate values for core electron binding energies in atoms and molecules, such investigations also lead to at least two important conclusions concerning the final hole state formed by photoelectron emission. First, the formation of the positive hole causes a relaxation (also referred to as polarization or rearrangement) of all the electrons in $\Psi^f(k)$ as compared to those in Ψ^i. Thus, the one-electron orbitals φ_i used to make up Ψ^i will not be the same as the one-electron orbitals φ_i' used to make up $\Psi^f(k)$, although in many cases the approximation $\varphi_i \approx \varphi_i'$ can be made. A further point relevant to molecules is that the core electron holes formed will tend to be localized on one atomic center, as opposed to being distributed over all centers as might be expected from an LCAO Hartree-Fock calculation including all electrons [22–24]. In the simple example of O_2, a hole in the $1\sigma_g$ or $1\sigma_u$ molecular orbitals (which can be considered in first approximation to be made up primarily of a sum or difference of $1s$ orbitals on the two oxygen atoms, respectively) might be expected to result in a charge of $+\frac{1}{2}e$ on each oxygen atom in the molecule. However, such a state does not minimize the total energy associated with the final state Hamiltonian. The lowest energy state localizes the $1s$ core hole entirely on either oxygen atom. These pairs of equivalent final states yield the correct values of $E^f(k)$ for computing binding energies. For O_2, the localized hole states yield a value of $E_b^V(1s)=542$ eV, in comparison with an experimental value of 543 eV, and a delocalized hole state value of 554 eV. Thus, localizing the hole represents a correction of 12 eV ($\sim 2.2\%$).

In order to avoid the difficulties associated with hole state calculations in determining binding energies, a very often used approximation is to assume that Koopmans' Theorem well describes the relationship between initial and final state total energies [27]. The basis of this theorem is the assumption that the initial one-electron orbitals φ_i making up Ψ^i are precisely equal to the final orbitals φ_i' making up $\Psi^f(k)$. The final state total energy $E^f(k)$ can then be calculated from the formula for E^i (cf. Equation (27)) simply by eliminating those terms dealing with one of the two electrons occupying the kth orbital initially. This procedure leaves as the Koopmans' Theorem value for $E^f(k)$,

$$E^f(k)^{KT} = \varepsilon_k^0 + 2\sum_{i \neq k}^{n} \varepsilon_i^0 + \sum_{i,\, j \neq k}^{n} (2J_{ij} - K_{ij}) + \sum_{i \neq k}^{n} (2J_{ik} - K_{ik})$$

or

$$E^f(k)^{KT} = \varepsilon_k^0 + 2\sum_{i \neq k}^{n} \varepsilon_i^0 + \sum_{i,\, j = 1}^{n} (2J_{ij} - K_{ij}) - \sum_{i = 1}^{n} (2J_{ik} - K_{ik}). \qquad (29)$$

The Koopmans' Theorem binding energy of the kth electron is then by the difference method (cf. Equation (27)),

$$E_b^V(k)^{KT} = E^f(k)^{KT} - E^i = -\varepsilon_k^0 - \sum_{i = 1}^{n} (2J_{ik} - K_{ik})$$

or, making use of Equation (26) for the orbital energy ε_k,

$$E_b^V(k)^{KT} = -\varepsilon_k. \qquad (30)$$

Thus, the binding energy of the kth electron in this approximation is equal to the negative of the orbital energy ε_k. This result is Koopmans' Theorem, as we have indicated by the superscript KT. In reality, the relaxation of the $N-1$ passive orbitals about the k hole will lower $E^f(k)$ relative to $E^f(k)^{KT}$, and thus binding energies estimated with Koopmans' Theorem should always be too large (barring fortuitous cancellation with possible relativistic and correlation corrections). If the error due to such electronic relaxation is denoted by $\delta E_{\text{relax}} > 0$, then we can write

$$\begin{aligned} E_b^V(k) &= E_b^V(k)^{KT} - \delta E_{\text{relax}} \\ E_b^V(k) &= -\varepsilon_k - \delta E_{\text{relax}}. \end{aligned} \qquad (31)$$

It should be noted, however, that Koopmans' Theorem as we have derived it applies only to closed-shell systems (that is, systems that are adequately represented by a single Slater determinant with doubly-occupied one electron orbitals). For any other case, there will in general be several possible couplings of spin and orbital angular momenta in the open shell or shells, and each distinct coupling will give rise to a different initial or final state energy. These states might, for example, be described in terms of Russell-Saunders (L-S) coupling, and would in general be represented by a linear combination of Slater determinants [28]. Although each of these determinants would have the same gross configuration (for example, $3d^5$), various possible combinations of $m_s = \pm \frac{1}{2}$ and m_l would be possible within the open shells. Slater [28] has pointed out that, provided final state relaxation is neglected, a binding energy $E_b^V(k)^{KT}$ computed as the difference between the average total energy for all states within the final configuration and the average total energy for all states within the initial configuration is equal to the one-electron energy ε_k computed from an initial-state Hartree-Fock calculation utilizing Coulomb and exchange potentials averaged over all states possible within the initial configuration. This we can write as

$$\overline{E_b^V(k)^{KT}} = \overline{E^f(k)^{KT}} - E^i = -\varepsilon_k \tag{30a}$$

and it represents a generalization of Koopmans' Theorem to open-shell systems. The various final states discussed here are the cause of the multiplet splittings to be considered in Section 4.

The most direct way of calculating δE_{relax} is of course to carry out SCF Hartree Fock calculations on both the initial and final states and to compare $E_b^V(k)$ as calculated by a total energy difference method with $E_b^V(k)^{KT} = -\varepsilon_k$. Such calculations have been carried out by various authors on both atoms and molecules [7, 11, 17, 19, 20, 29]. As representative examples of the magnitudes of these effects, for the neon atom, $E_b^V(1s) = 868.6$ and $E_b^V(1s)^{KT} = 891.7$, giving $\delta E_{relax} \approx 23$ eV ($\sim 2.6\%$), and $E_b^V(2s) = 49.3$ eV and $E_b^V(2s)^{KT} = 52.5$ eV, giving $\delta E_{relax} \approx 3$ eV ($\sim 6.0\%$). Effects of similar magnitude are found in the $1s$ levels of molecules containing first-row atoms [21]. Also, in certain cases, the presence of a localized hole may also cause considerable valence electron polarization relative to the initial state [21–23]. Thus δE_{relax} appears to be in the range of 1–10% of the binding energy involved, with greater *relative* values for more weakly bound electrons [17, 19]. Several procedures have also been advanced for estimating δE_{relax} [11, 17, 30, 31]. It has also been pointed out by Manne and Åberg [32] that a Koopmans' Theorem binding energy should represent an average binding energy as measured over both one-electron- and multi-electron-transitions in photoemission. This analysis we discuss in detail in Section 4B. The binding energies referred to in all of the above discussion have been those derived from one-electron transitions, as this is the dominant mode by which photoemission occurs.

Up to this point, we have been considering the calculation of total core electron binding energies rather than chemical shifts in these energies. In terms of a non-relativistic Koopmans' Theorem calculation of a binding energy as $E_b(k)^{KT} = -\varepsilon_k$, we have noted that the three most important types of correction are for relaxation (polarization, rearrangement) effects (δE_{relax}), relativistic effects (δE_{relat}), and correlation effects (δE_{corr}). Thus, an accurate vacuum-referenced binding energy can be calculated as

$$E_b^V(k) = E_b^V(k)^{KT} - \delta E_{\text{relax}} + \delta E_{\text{relat}} + \delta E_{\text{corr}}$$
$$= -\varepsilon_k - \delta E_{\text{relax}} + \delta E_{\text{relat}} + \delta E_{\text{corr}}. \tag{32}$$

A chemical shift in such binding energies between two chemically inequivalent sites or compounds labelled 1 and 2 is thus (cf. Equation (13)),

$$\Delta E_b^V(A, k, 1\text{–}2) = \left(E_b^V(k)\right)_1 - \left(E_b^V(k)\right)_2$$
$$= -(\varepsilon_k)_1 + (\varepsilon_k)_2 - (\delta E_{\text{relax}})_1 + (\delta E_{\text{relax}})_2$$
$$+ (\delta E_{\text{relat}})_1 - (\delta E_{\text{relat}})_2 + (\delta E_{\text{corr}})_1 - (\delta E_{\text{corr}})_2$$
$$\Delta E_b^V(A, k, 1\text{–}2) = -\Delta\varepsilon_k - \Delta(\delta E_{\text{relax}}) + \Delta(\delta E_{\text{relat}}) + \Delta(\delta E_{\text{corr}}). \tag{33}$$

In view of the physical origins of the relaxation, relativistic, and correlation corrections for a given core level, they will tend to have values of approximately the same magnitude from one site or compound to another. Thus, in many cases, it would be

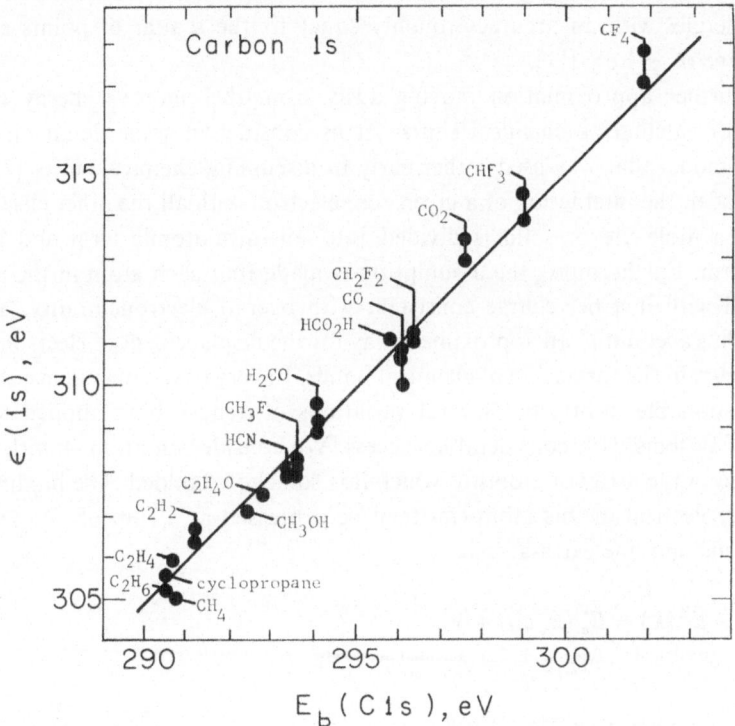

Fig. 2. Plot of calculated carbon 1s orbital energies against experimental binding energies for several carbon-containing gaseous molecules. For some molecules, more than one calculated value is presented. The slope of the straight line is unity. All of the theoretical calculations were of roughly double-zeta accuracy or better. The original references are given in [11], from which this figure is taken.

expected that $\Delta(\delta E_{\text{relax}})$, $\Delta(\delta E_{\text{relat}})$, and $\Delta(\delta E_{\text{corr}})$ would be considerably smaller in magnitude than the individual corrections to either $[E_b^V(k)]_1$ or $[E_b^V(k)]_2$, and therefore that $-\Delta\varepsilon_k$ would represent a quite good approximation to the chemical shift $\Delta E_b^V(A, k, 1\text{–}2)$ [17]. For similar reasons, the quality of the wave function utilized in obtaining ε_k is often not as critical as might be imagined. That is, approximate wave functions with the same degree of self-consistency for both systems 1 and 2 may yield a reasonably accurate value of $\Delta\varepsilon_k$ (which is, after all, a small perturbation primarily due to changes in valence electron charge distribution). For these reasons, the use of Koopmans' Theorem in conjunction with various approximate calculation procedures such as minimal-basis-set- or double-zeta-basis-set- Hartree-Fock calculations has met with success in analyzing much chemical shift data [11]. It appears that molecular wave functions of double-zeta quality can be utilized to predict chemical shifts from $-\Delta\varepsilon_k$ which agree with experiment to within $\sim \pm 1$ eV, in spite of the fact that the orbital energies tend to be as much as 10–20 eV higher than the experimental binding energies for such levels as C1s and N1s. In Figure 2, experimental C1s binding energies for different gaseous molecules are compared to 1s orbital energies from various theoretical calculations of roughly double-zeta occupancy. Although the two scales are shifted relative to one another by about 15 eV, the points lie very near to a straight line of slope unity. Thus, chemical shifts should be calculable from these orbital energies with an accuracy roughly equal to the scatter of points about the straight line or ± 1 eV [11].

As a further approximation moving away from the rigorous energy difference method for calculating chemical shifts, let us consider in some detail one type of potential model that was used rather early in discussing chemical shifts [7, 16, 17]. In this model, the interaction of a given core electron with all the other electrons and nuclei in a molecule or solid is divided into an intra-atomic term and an extra-atomic term. Furthermore, the assumption is made that each atom in the array has associated with it a net charge consistent with overall electroneutrality. These net changes thus account in an approximate way for the displacement of electronic charge which occurs in the formation of chemical bonds. In very covalent systems, this model is of questionable utility, but several variations of it have been applied to a wide variety of systems with considerable success. We consider an atom A with a charge q_A situated in the array of atoms to which it is somehow bonded. The binding energy of the kth electron in this atom can then be expressed as a sum of two terms, one intra-atomic and one extra-atomic:

$$E_b^V(k) = E_b^V(k, q_A) + V.$$

(34)

$$\underset{\text{compound}}{} = \underset{\substack{\text{free ion of} \\ \text{charge } q_A}}{} + \underset{\substack{\text{potential due to all} \\ \text{other atoms}}}{}$$

The first term is a binding energy for the kth electron in a free ion of charge q_A and the second term is the total potential due to all other atoms in the array. The first term can be evaluated by means of a free-ion Hartree-Fock calculation, for example. The simplest way to calculate the second term is to assume that the other atoms behave as point charges in creating the potential V. Thus,

$$V = e^2 \sum_{i \neq A} \frac{q_i}{r_{iA}} \tag{35}$$

where the summation is over all atoms in the array. If the array is a crystal, then V represents a convergent infinite sum that is closely related to the Madelung energy of the crystal [17]. Thus, both terms in Equation (34) may be relatively easy to obtain for a variety of systems. Calculating a chemical shift using Equation (34) gives

$$\Delta E_b^V (A, k, 1\text{-}2) = E_b^V (k, q_{A,1}) - E_b^V (k, q_{A,2}) + V_1 - V_2, \tag{36}$$

where $q_{A,1}$ and $q_{A,2}$ are the net charges on atom A in the sites 1 and 2, respectively. It is instructive to consider the predictions of this model for several simple systems, as it is found to explain qualitatively and semi-quantitatively several basic features of chemical shifts.

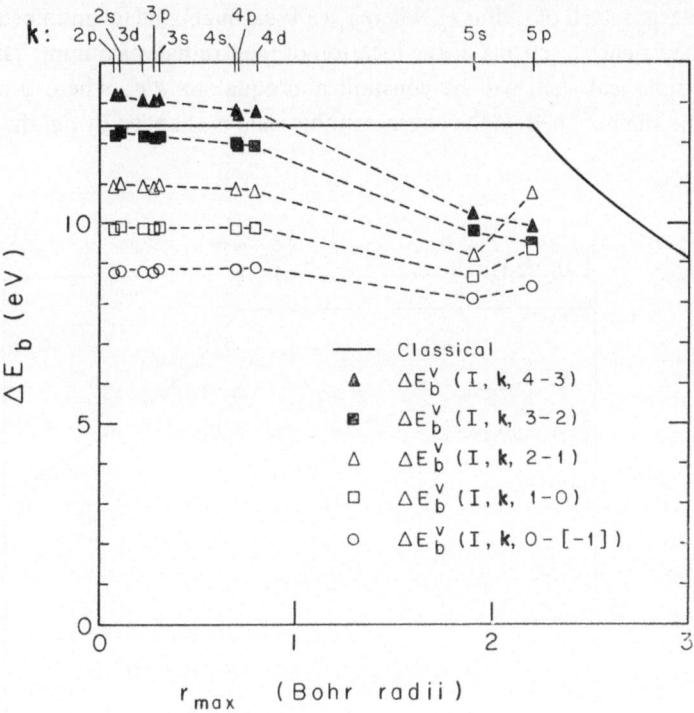

Fig. 3. Calculated free-ion binding energy shifts caused by the removal of a valence $5p$ electron from various configurations of iodine, plotted against the location of the radial maxima for the various orbitals. Koopmans' Theorem was used in obtaining the shifts so they represent changes in orbital energies. The configurations are: $+4 = 5s^2 5p$, $+3 = 5s^2 5p^2$, $+2 = 5s^2 5p^3$, $+1 = 5s^2 5p^4$, $0 = 5s^2 5p^5$, and $-1 = 5s^2 5p^6$. The solid curve shows the classical shift resulting from the removal of an electron from a thin spherical shell of charge with the radius of the $5p$ maximum. (From [17].)

The free ion term in Equation (36) represents a change in binding energy concomitant with a change in the valence electron orbital occupation of the atom. Fadley *et al.* [17] calculated such changes for removal of successive valence electrons from various ionic states of I, Br, Cl, F and Eu, using a minimum-basis-set Hartree-Fock calculation and Koopmans' Theorem. These results are presented in Figures 3–7, where the shifts are plotted against the location of the radial maximum for each orbital. Several systematic features of these results can be noted. For iodine, all core levels shift by very nearly the same amount. This is basically true also for Br and Cl, although as the atomic size decreases, there is less constancy in the core shifts. In all of the halogens, the p valence electrons are largely external to the core, as is evidenced by the location of the core and valence orbital radial maxima. For Eu, which by contrast has valence $4f$ electrons overlapping considerably with the core electrons, the core shifts are not at all constant, and can be about twice as large per unit change in valence shell occupancy as for the halogens. All of these results are qualitatively consistent with a very simple classical model of the interaction between core and valence electrons. The valence electron charge distribution is approximated by a spherical charged shell of radius r_v, where r_v can reasonably be taken to be the average radius of the valence electrons or the location of their radial maximum. The potential inside this spherical shell will be constant and equal to q/r_v, where q is the total charge in the valence shell. If the charge on this shell is changed by δq, the potentials,

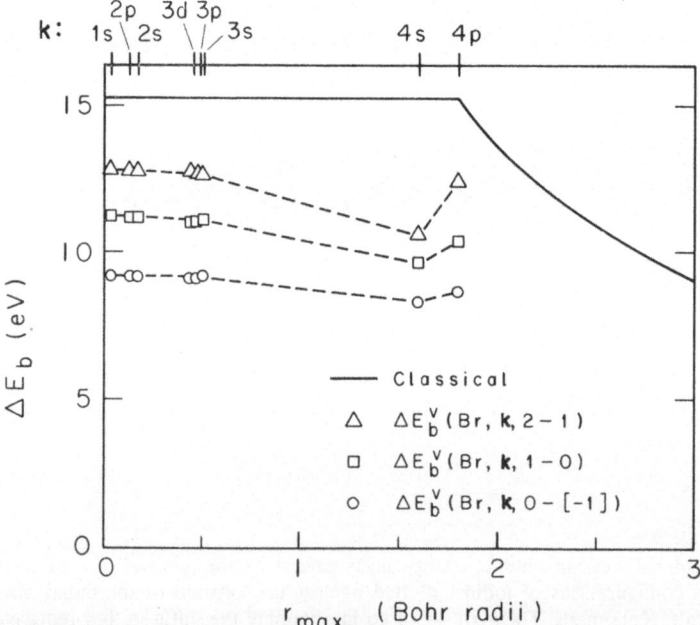

Fig. 4. Calculated free-ion binding energy shifts caused by the removal of a valence $4p$ electron from bromine. The configurations are: $+2 = 4s^24p^3$, $+1 = 4s^24p^4$, $0 = 4s^24p^5$, and $-1 = 4s^24p^6$. (From [17].)

and thus binding energies, of all the core electrons located well inside the shell will shift by an amount $\delta E_b^V = \delta q / r_v$. Such classical calculations are shown as the solid lines in Figures 3–7 and are found to give results in qualitative and semi-quantitative agreement with the more accurate Hartree-Fock calculations. In general then, all core electrons which overlap relatively little with the valence shell should shift by approximately the same amount. The magnitude of the shift per unit change in charge should also increase as the valence shell radius r_v decreases. Thus, if an approximate indicator of the sensitivity of an atom to chemical shifts is taken to be $\delta E_b^V / \delta q$, then the more electronegative elements in the periodic table should be more sensitive [17]. A more accurate estimate of $\delta E_b^V / \delta q$ for any atom is given by the change in ε_k upon removal of one valence electron. From Equation (26), this will be given by $J_{kv} - K_{kv}$ (spins parallel) or J_{kv} (spins anti-parallel). As the core-valence exchange integral K_{kv} will be of significant magnitude only if there is appreciable overlap between the core and valence orbitals, we can neglect K_{kv} in comparison to J_{kv}. Thus, $\delta E_b^V / \delta q$ should be approximately equal to J_{kv}, the core-valence Coulomb integral. The magnitudes of such Coulomb integrals are in good agreement with the shifts calculated in

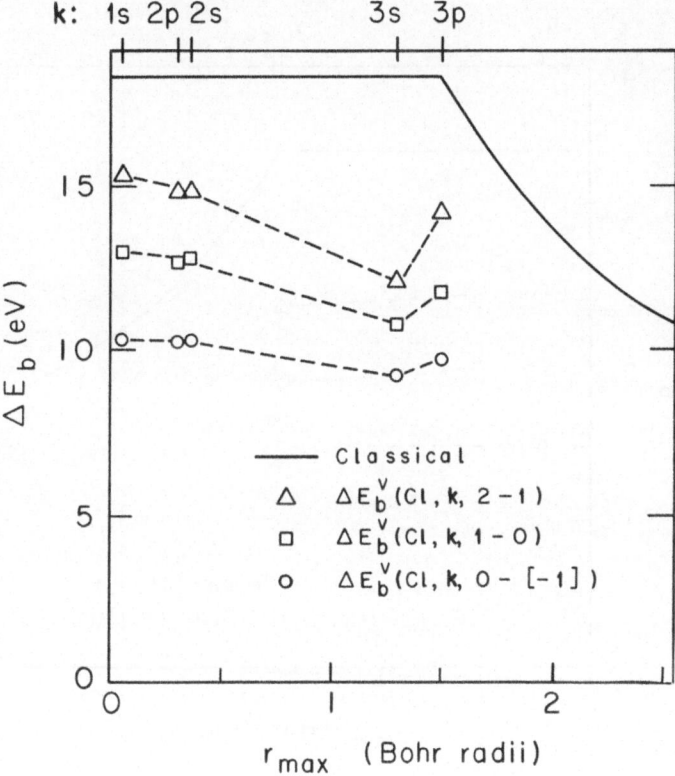

Fig. 5. Calculated free-ion binding energy shifts caused by the removal of a valence $3p$ electron from chlorine. The configurations are: $+2 = 3s^2 3p^3$, $+1 = 3s^2 3p^4$, $0 = 3s^2 3p^5$, and $-1 = 3s^2 3p^6$. (From [17].)

Figures 3–5 for I, Br, and Cl. As a final point, we note that the free ion term $\delta E_b^V/\delta q$ is of the order of 10 eV/electron charge for essentially all elements.

If we now consider the potential term V in Equation (34), it is found that its value also will be of the order of 10 eV for a transfer of unit electron charge from one atom to its nearest neighbors [7, 16, 17], as, for example, in a highly ionic alkali halide crystal. Furthermore, for a given molecule or solid the free-ion term $(\delta E_b^V/\delta q)\cdot\delta q$ will be opposite in sign to V, as V must account for the fact that charge is not displaced to infinity, but only to adjacent atoms during chemical bond formation. Thus, both the free-ion- and potential-terms in Equation (34) must be calculated with high accuracy if the resultant binding energy (or chemical shift) value is to have acceptable accuracy. This represents one of the possible drawbacks of such potential models.

Several other models based essentially on Equation (34) have been utilized in analysing core electron chemical shifts [11], and the detailed theoretical justifications for such models have been discussed by Basch [33] and Schwartz [21]. For example, Siegbahn et al. [9], and Gelius et al. [15] have been able to describe the 1s binding energy shifts for a variety of compounds of C, N, O, F, and S with the following

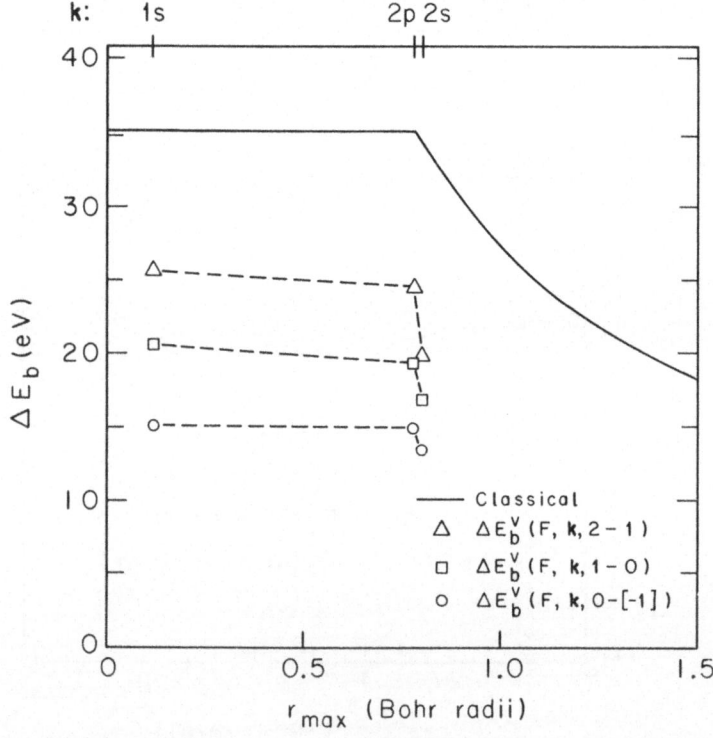

Fig. 6. Calculated free-ion binding energy shifts caused by the removal of a valence 2p electron from fluorine. The configurations are: $+2 = 2s^22p^3$, $+1 = 2s^22p^4$, $0 = 2s^22p^5$, and $-1 = 2s^22p^6$. (From [17].)

equation:

$$\Delta E_b^V (A, k, 1\text{-}2) = C_A q_A + V + l, \tag{37}$$

where 2 represents a fixed reference compound. The various atomic charges q_i in each molecule were estimated using CNDO theory, and these charges used in Equation (35) to compute V. Then the constants C_A and l were determined empirically by a least squares fit to the experimental data. Such fits give a reasonably consistent description of the data, as is shown in Figure 8, and in particular the parameters C_A are found to be fairly close to the $1s$-valence Coulomb integral $J_{1s\text{-valence}}$ computed for atom A. Thus, Equation (37) as utilized in this semi-empirical way is consistent with a somewhat more exact theoretical model. Note, however, that all molecules are not adequately described by this model and that the points for CO and CS_2 lie far from the straight line predicted by Equation (37). As might be expected, if an orbital energy difference based on near Hartree-Fock wave functions is used for the calculated shift of CO, much better agreement with experiment is obtained, as is shown in Figure 2.

In another potential model proposed by Davis *et al.* [34], a series of chemical shift measurements on core levels in all the atoms of several related molecules are used to derive a self-consistent set of atomic charges. For each atom in each molecule, the measured chemical shift is written in terms of undetermined atomic charges as

$$\Delta E_b^V (A, k, 1\text{-}2) = C_A' q_A + e^2 \sum_{i \neq A} \frac{q_i}{r_{Ai}}, \tag{38}$$

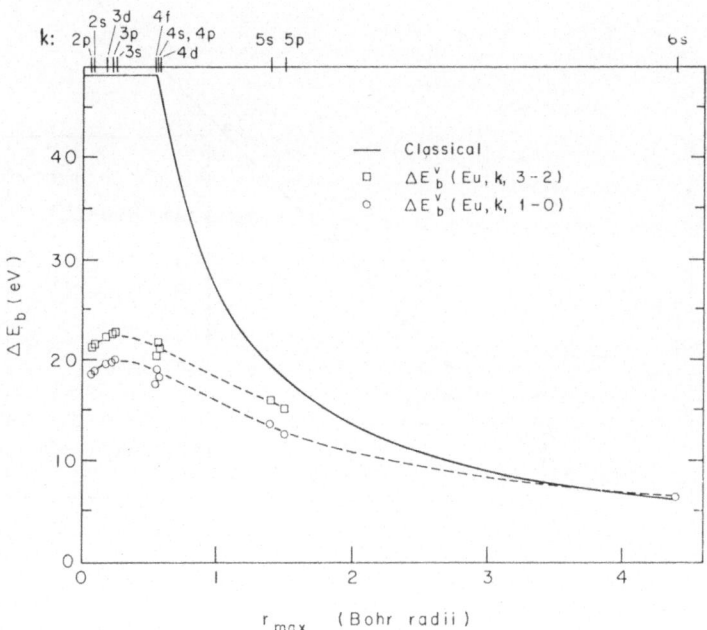

Fig. 7. Calculated free-ion binding energy shifts caused by the removal of a valence $4f$ electron from europium. The configurations are: $+3 = 4f^6$, $+2 = 4f^7$, $+1 = 4f^6 6s^2$, and $0 = 4f^7 6s^2$. (From [17].)

where C'_A is set equal to $J_{1s\text{-valence}}$ for atom A. The resultant set of equations is solved self-consistently for the q_A values on each atom. Such calculations on a series of fluorinated benzenes [34] give charges which agree rather well with charges obtained from CNDO calculations, as is apparent in Figure 9.

Another procedure for analyzing chemical shift data that can be at least indirectly related to the various potential models is based on summing empirically determined shifts associated with each of the groups bonded to the atom of interest. Each group shift is assumed to be constant and independent of the other groups present and is determined from a series of chemical shift measurements on compounds representing suitable combinations of the groups. The chemical shift associated with atom A in a given compound is thus written as

$$\Delta E_b(A, k, 1\text{--}2) = \sum_{\text{groups}} \Delta E_b(\text{group}), \qquad (39)$$

where 2 constitutes some reference compound against which all of the group shifts are determined. The applicability of this procedure has been demonstrated on a large number of carbon- and phosphorus-containing compounds [15, 35], and some results obtained for phosphorus compounds are shown in Figure 10. The relationship of this

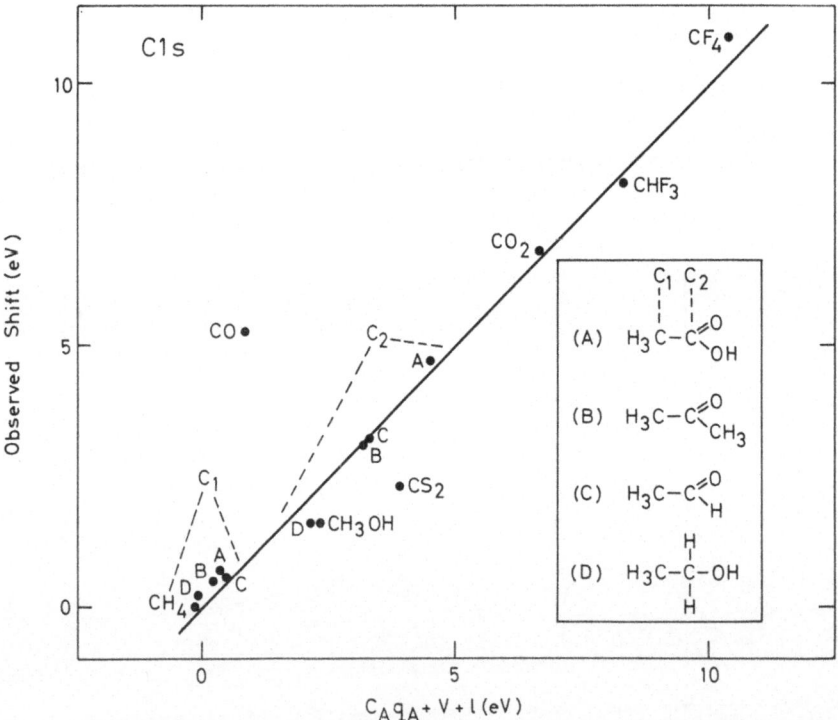

Fig. 8. A comparison of experimental carbon $1s$ chemical shift values for several molecules with shifts calculated using the potential model of Equation (37). The shifts were measured relative to CH₄. The parameters of the straight line were $C_A = 21.9$ eV/unit charge and $1 = 0.80$ eV. (From [9].)

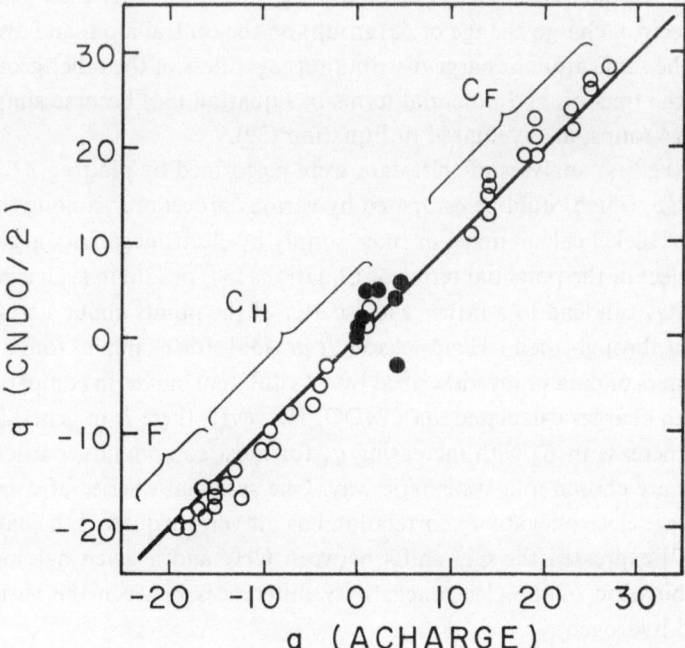

Figure 9. Atomic charges for the various fluorinated benzenes as calculated by the CNDO/2 method and as derived experimentally ('ACHARGE') from chemical shift measurements on carbon and fluorine and Equation (38). Charges are in units of 1/100 of an electronic charge. The filled circles represent average hydrogen charges. (From [34].)

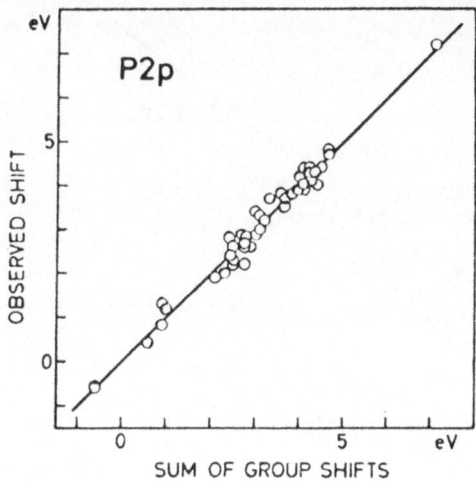

Fig. 10. A comparison of measured phosphorous $2p$ chemical shifts with shifts calculated using the group shift model of Equation (39). Compounds studied as solids. (From [35].)

procedure to a potential model is possible if it is assumed that each group induces a valence electron charge change of δq(group) on the central atom and also possesses essentially the same atomic charge distribution regardless of the other groups present. Then both the free-ion and potential terms in Equation (34) become simply additive for different groups, as is required in Equation (39).

Some of the first analyses of shift data were performed by plotting ΔE_b against an atomic charge, which could be estimated by various procedures, among them CNDO or extended Hückel calculations, or most simply by electronegativity arguments. The implicit neglect of the potential terms of Equations (34) or (36) in such a corelation of ΔE_b against q_A can lead to a rather wide scatter of the points about a straight line or curve drawn through them. Hendrickson et al. [36], for example, found two rather distinct clusters of data points described by two different curves in comparing nitrogen $1s$ shifts with charges calculated via CNDO. However, there is in general observed a systematic increase in E_b with increasing q_A for most compounds, particularly if the compounds are chosen in a systematic way. One systematic series of compounds for which a simple electronegativity correlation has proven adequate is the halomethanes. Thomas [37] expressed the C$1s$ shifts between CH_4 and a given halomethane as a linear combination of the electronegativity differences between the various ligands present and hydrogen:

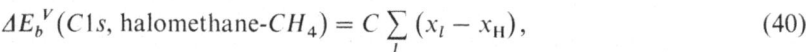

$$\Delta E_b^V (C1s, \text{halomethane-}CH_4) = C \sum_l (x_l - x_H), \tag{40}$$

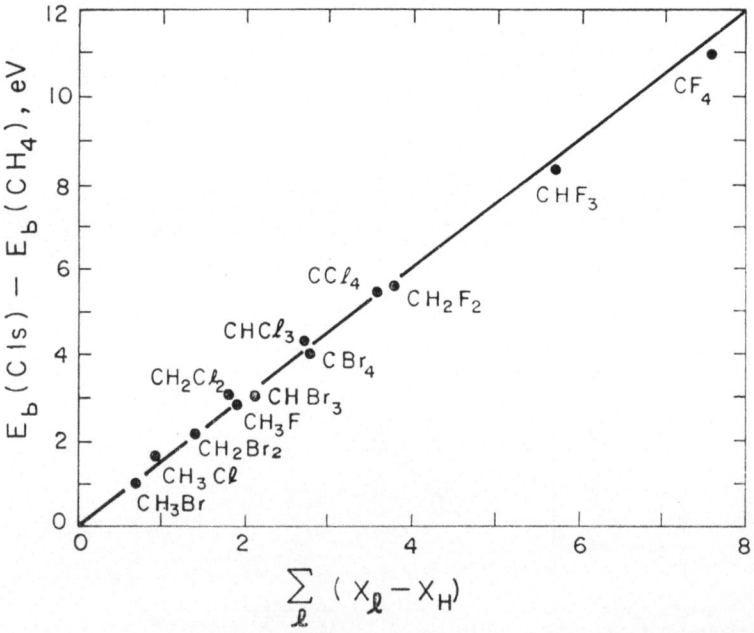

Fig. 11.   Carbon $1s$ chemical shifts for halogenated methanes relative to CH_4 plotted against shifts calculated on the basis of a sum of ligand-hydrogen electronegativity differences, as in Equation (40). (From [11, 37].)

where C is a constant, X_l is the ligand electronegativity and X_H is the electronegativity of hydrogen. Such a correlation is shown in Figure 11 [11, 37]. The explanation for the success of this correlation would seem to be as a simplification of the group shift approach, in which each monatomic ligand induces a charge transfer δq_l proportional to $X_l - X_H$, and the potential term involved is also simply proportional to δq_l for a nearly constant carbon-ligand bond length. Thus, the potential model Equation (36) reduces to the form of Equation (40).

Among the other methods utilized to analyze chemical shift data, mention should be made of a procedure introduced by Jolly and Hendrickson [38] for relating chemical shifts to thermochemical data (see also [39]). In this procedure, it is assumed that to a good approximation the atomic core of an atom with nuclear charge Z and a single core electron hole acts on any surrounding valence electrons in an equivalent way to the filled core of an atom with nuclear charge $Z + 1$. If the core electron overlap with the valence electrons is small, then the nuclear shielding should be nearly complete and this assumption is reasonable. Thus, for example the O^{6+} core can be exchanged for the N^{6+*} core, where the asterisk denotes a $1s$ hole on the nitrogen. Such core exchanges can be utilized to write binding energy shifts in terms of thermodynamic heats of reaction, and hence to either predict shifts from thermodynamic data or thermodynamic data from shifts. As one example of the application of this procedure, let us consider $1s$ photoelectron emission from gaseous NH_3 and N_2 as chemical reactions in which the electron is assumed to be formed with no kinetic energy:

$$NH_3 \rightarrow NH_3^{+*} + e^- \quad \Delta E_1 = E_b^V(N1s, NH_3)$$
$$N_2 \rightarrow N_2^{+*} + e^- \quad \Delta E_2 = E_b^V(N1s, N_2),$$

where the * denotes the presence of a $1s$ hole. These reactions are endothermic with internal energy changes ΔE_1 and ΔE_2 given by the $1s$ binding energies in NH_3 and N_2. Subtracting the second reaction from the first gives

$$NH_3 + N_2^{+*} \rightarrow NH_3^{+*} + N_2 \quad \Delta E = \Delta E_1 - \Delta E_2$$
$$= E_b^V(N1s, NH_3) - E_b^V(N1s, N_2)$$
$$= \Delta E_b^V(N1s, NH_3 - N_2),$$

with an internal energy change precisely equal to the $N1s$ chemical shift between NH_3 and N_2. However, this reaction involves the somewhat unusual species N_2^{+*} and NH_3^{+*}. Now, it is assumed that the N^{6+*} core can be replaced by the O^{6+} core in either N_2^{+*} or NH_3^{+*} with only a small gain or loss of energy that we shall term the core exchange energy ΔE_{ce}. As long as the core exchange energy is very nearly the same in both N_2^{+*} and NH_3^{+*}, then the overall energy change associated with the reaction is not affected by core exchange. Thus, we have a final reaction of

$$NH_3 + NO^+ \rightarrow OH_3^+ + N_2 \quad \Delta E = \Delta E_b^V(N1s, NH_3 - N_2) + \Delta E_{ce} - \Delta E_{ce}$$
$$= \Delta E_b^V(N1s, NH_3 - N_2).$$

Thus, the chemical shift is equal to a thermodynamic heat of reaction involving well known species. This procedure has been applied to 1s shifts in compounds of N, C, O, B, and Xe, and very good agreement is obtained between experimental ΔE_b values and thermochemical estimates of these shifts. Such a comparison is shown in Figure 12. This analysis is also closely related to the isodesmic processes discussed by Clark [40].

Finally, let us briefly mention a few other procedures that have been utilized to interpret core electron binding energy shifts. An attempt has been made to correlate binding energy shifts for different levels in the same atom [41]. From Figures 3–7 it is clear that the outer core and valence levels of a given atom need not shift by the same amount as inner core levels. Such *relative* shifts of different levels can for certain cases be simply related to the basic Coulomb and exchange integrals involved, and then utilized to determine certain properties of the valence electron charge distribu-

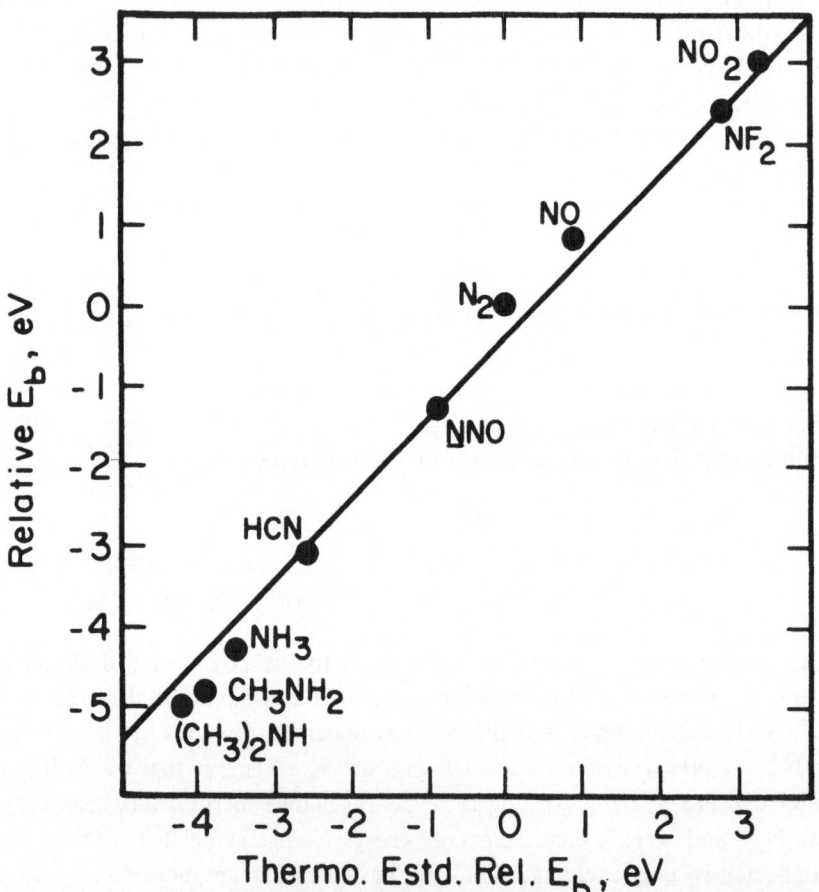

Fig. 12. Plot of experimental N1s binding energies relative to N_2 for several molecules versus values calculated using equivalent core exchange and thermodynamic data. The slope of the line is unity. (From [39].)

tion. In particular, the relative shifts of the inner core $3d_{5/2}$ and valence $5p_{1/2}$ levels have been measured for iodine in various alkali iodides and HI, and these relative shifts are found to be consistent with a simple bonding model of the compounds involved [41]. Also, attempts have been made to correlate binding energy shifts with diamagnetic shielding constants derived from NMR measurements [11, 33]. However, the difficulty of separating out diamagnetic and paramagnetic contributions to shielding have prevented extensive application of this correlation to date [11]. Also, binding energy shifts for a closely related set of tin compounds correlate reasonably well with Mössbauer chemical shift values [42], although no detailed theoretical justification for this correlation has been presented. Atomic charges on iodine in several compounds derived from Mössbauer measurements are also in approximate agreement with charges obtained from a potential model analysis of binding energy shift data [17].

It is clear that the theoretical interpretation of core electron binding energies or chemical shifts in these energies can be attempted in several ways at varying levels of sophistication. When binding energies are calculated by the most rigorous total-energy-difference method, including perhaps corrections for relativistic effects and electron-electron correlation, values in very good agreement with experiment have been obtained for several atoms and small molecules. This agreement verifies that all of the basic physical effects involved have been recognized and can be accounted for quantitatively. If binding energies are calculated from orbital energies via Koopman's Theorem, an error due to neglect of final state relaxation is incurred. This error can be from 1–10% of the total binding energy, and can be estimated in several ways. In calculating chemical shifts of binding energies between two different sites or compounds by means of Koopmans' Theorem, however, a fortuitous cancellation of a large fraction of the relativistic, correlation, and relaxation corrections occurs. Thus, orbital energies can be used with reasonable success in predicting shifts, although anomalously large final state relaxation around a localized hole represents an ever-present source of error in such analyses. The interaction of a core electron with its environment can be simplified even further, giving rise to several socalled potential models with varying degrees of quantum-mechanical and/or empirical input. All of these models can be useful in interpreting shifts, although it may be necessary to restrict attention to a systematic set of compounds for the most approximate of these. The direct connection of chemical shifts with thermochemical heats of reaction is also possible. Finally, it is worthwhile to note that one of the primary reasons that chemical shifts can be analyzed by such a wide variety of methods is that their origin is so simply and directly connected to the molecular charge distribution. In turn, it is very often this charge distribution that is of primary interest in a given chemical or physical investigation.

3. Valence Electron Binding Energies

The theoretical calculation of valence electron binding energies for atoms, molecules, and solids is a distinct and highly developed subject which considerably predates the

present upsurge of interest in electron spectroscopy. Therefore, it is not appropriate
to review the various techniques involved here. Rather we shall restrict ourselves to
a discussion of the way in which useful information about valence electrons can be
extracted from X-ray photoelectron spectra. Also, as the companion technique UPS
has been extensively utilized in studying valence electrons, several comments are in
order as to the relative merits of XPS and UPS for such studies, and the ways in which
they complement one another.

XPS has been applied to the study of the valence levels of both gaseous molecules
[9] and solids [43]. Representative spectra are shown for gaseous O_2 in Figure 13
[9] and for the valence bands of several solids containing transition metal atoms in
Figure 14 [44]. In studying molecules, the quantities of primary interest are the energies

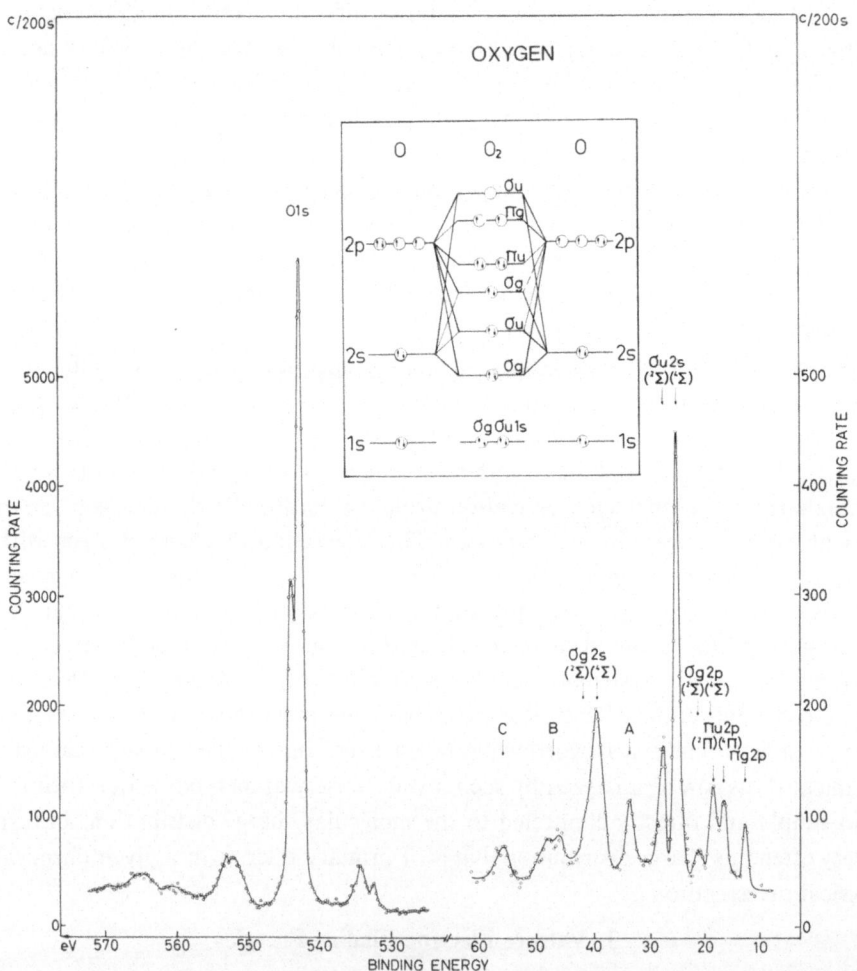

Fig. 13. XPS spectra of the core and valence electrons in gaseous oxygen, together with assignments
of the various peaks to certain molecular orbitals and final states. (From [9].)

of the orbitals and their make-up, usually expressed in terms of some combination of atomic orbitals (cf. the inset in Figure 13). For certain ionic solids, a similar molecular orbital structure can be observed [45], but for semiconductors and metals one is primarily interested in determining the densities of states of the various valence bands as a function of energy [2, 44]. The total density of states for all bands is denoted by $\varrho(E)$, with units of number eV^{-1}.

It has already been amply demonstrated in work on both gases [46, 47], and solids [43] that the relative intensities of various valence electron peaks can change quite drastically with changes in photon energy, particularly between the X-ray and uv range. Such variations in relative intensity are presumably connected with changes in the relative photoelectric cross sections σ_k of the various electronic states involved. It has been suggested [47] that such variations could be profitably utilized for characterizing the spatial character of valence states. Certainly such studies represent an area worth further development and one in which XPS and UPS complement one another quite well. The exploration of the photon energy range of approximately 50 eV to a few hundred eV is also a very intriguing possibility in this respect [48]. Gelius [49] has also recently discussed a method for utilizing empirically determined *atomic* valence shell relative cross sections in the analysis of molecular orbital spectra. The agreement between theoretical calculations based on this method and experiment is very good for several small molecules [49]. The use of such relative intensity analyses in the determination of valence state character will be discussed further in Section 5.

In discussing XPS work on solids that is aimed at determining the density of states

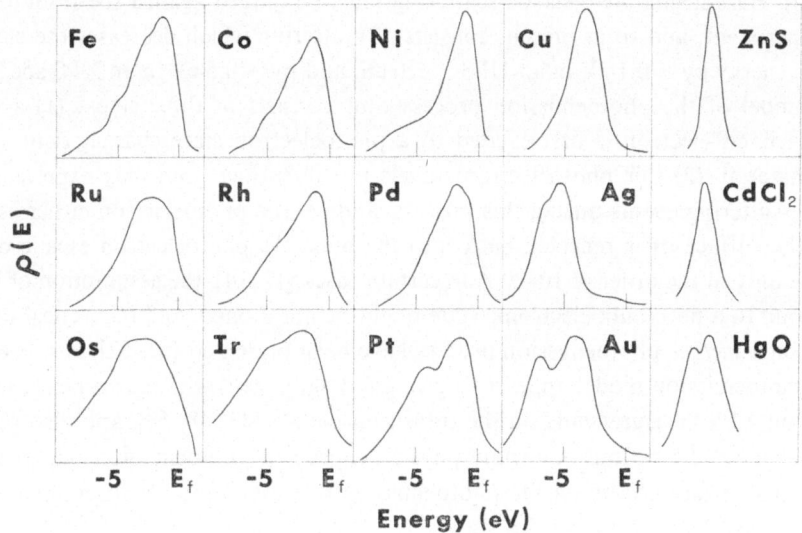

Fig. 14. XPS spectra from the valence bands of fifteen solids containing 3d (first row), 4d (second row), and 5d (third row) transition metals. Each row is sequential in atomic number, so that the filling of the d bands can be followed from left to right. The peaks for Zn, CdCl₂, and HgO lie at -13 eV, -14 eV, and -12 eV, respectively. (From [44].)

$\varrho(E)$ in a crystalline solid, it is useful to consider the various assumptions and procedures required to derive a quantity closely related to $\varrho(E)$ from a measured photoelectron spectrum. This also permits certain statements to be made about the inherent differences between XPS and UPS.

Each valence electron Bloch state in a crystalline solid is specified completely by giving its energy E and wavevector $\mathbf{k}$. In our previous notation 'k' thus corresponds to $E, \mathbf{k}$ rather than to nl, nlj, etc. Here we have also followed standard notation in setting $E \equiv$ the negative of the Fermi-referenced binding energy $\equiv -E_b^F(k)$. For a given state $E, \mathbf{k}$ which is irradiated with photons of energy $h\nu$, photoelectric transitions are possible to states at an energy situated a distance $h\nu$ above E. Since we are using the Fermi level as a reference level, the state at $E + h\nu$ corresponds to a kinetic energy of $h\nu - E - \phi_{\text{spec}}$ (cf. Equation (9)). These transitions often seem to be controlled by a selection rule which states that the final wave vector must equal the initial wave vector [48]. The probability for a given transition is proportional to the square of a one-electron dipole matrix element between the initial state $E, \mathbf{k}$ and the final state $E', \mathbf{k}'$. The photoelectric cross section $\sigma_{E, \mathbf{k}}(h\nu)$ is in turn proportional to this matrix element squared. Alternatively, operator commutation relations can be used to replace the dipole matrix element $\langle E, \mathbf{k} | \mathbf{r} | E', \mathbf{k}' \rangle$ by either a momentum matrix element $\langle E, \mathbf{k} | \mathbf{p} | E', \mathbf{k}' \rangle$ or a matrix element of the gradient of the total potential V in the one-electron Hamiltonian for the solid $\langle E, \mathbf{k} | \nabla V | E', \mathbf{k}' \rangle$ [48, 50]. The energy and wave vector selection rules require that $E' = E + h\nu$ and $\mathbf{k}' = \mathbf{k}$. A further parameter that can be utilized in discussing such a photoemission event is an escape function $T(E', \mathbf{k}')$, which is equal to the probability that an electron in state $E', \mathbf{k}'$ will escape from the surface of the solid with negligible inelastic scattering [48, 50]. Here negligible inelastic scattering must be judged against the resolution of the experiment, and so at present consists of scattering which decreases the electron kinetic energy by ~ 0.1 eV in an UPS spectrum and by ~ 0.5 eV in an XPS spectrum. This model of the photoemission process thus consists of three steps: (1) a single valence band electron is first excited to a photoelectron state characteristic of the *bulk* material, (2) this photoelectron travels to the surface, and may experience inelastic scattering events during this process, and (3) the photoelectron either escapes from the surface or is reflected back into the bulk. As photoelectron escape depths may be only of the order of 10–20 Å in certain cases [12–14], the assumption of initial excitation to a pure bulk electronic state might be questioned, and theoretical models based on a *surface* photoemission process have been presented [51, 52]. However, the bulk photoemission model appears to give good agreement with experiment for both UPS and XPS measurements on the transition metals [43, 48, 53], and also permits separating out the various important physical aspects of photoemission. The presence of distinct surface effects on the photoemission process cannot be discounted, however.

Based on the bulk photoemission model and a neglect of inelastic scattering effects, the probability that a given transition $E, \mathbf{k} \to E', \mathbf{k}'$ will occur and also result in a photoelectron whose kinetic energy bears a simple relationship to E via Equation (9) can be written as the following proportionality relation

$$P_{E,\mathbf{k}\to E',\mathbf{k}'} \propto |\langle E, \mathbf{k}| \mathbf{r} |E', \mathbf{k}'\rangle|^2 \, T(E', \mathbf{k}') \propto \sigma_{E,\mathbf{k}}(hv) \, T(E', \mathbf{k}'), \qquad (41)$$

or, if the energy and wave vector selection rules are included,

$$P_{E,\mathbf{k}\to E+hv,\mathbf{k}} \propto |\langle E, \mathbf{k}| \mathbf{r} |E + hv, \mathbf{k}\rangle|^2 \, T(E + hv, \mathbf{k})$$
$$\propto \sigma_{E,\mathbf{k}}(hv) \, T(E + hv, \mathbf{k}). \qquad (42)$$

An ideal observed kinetic energy distribution $N_i(E_{kin}) = N_i(hv - E - \phi_{spec})$ containing no inelastically scattered electrons can be calculated by integrating $P_{E,\mathbf{k}\to E+hv,\mathbf{k}}$ over all allowed transitions corresponding to an initial energy E [48]. This integration must be over all occupied initial valence band states and all final unoccupied states and therefore must span all wave vector values. $N_i(E_{kin})$ can thus be written as [48]:

$$N_i(E_{kin}) = N_i(hv - E - \phi_{spec})$$
$$\propto \sum_{\substack{\text{all} \\ \text{bands}}} \int \sigma_{E,\mathbf{k}}(hv) \, T(E', \mathbf{k}') \, \delta(E' - E - hv) \, \delta(\mathbf{k}' - \mathbf{k}) d^3 k \qquad (43)$$

where the summation is over all occupied and unoccupied bands and each transition must be from an occupied state to an unoccupied state. The delta functions simply account for conservation of energy and wave vector, so that each transition corresponds to $E' = E + hv$, $\mathbf{k}' = \mathbf{k}$. Such calculations of $N_i(E_{kin})$ have been performed in the analysis of UPS spectra of metals [48, 54, 55], including also the effects of inelastic scattering, and the results are in good agreement with experiment, especially as to certain changes in the shape of $N_i(E_{kin})$ with changes in hv.

In discussing XPS measurements on solids, as well as certain essential differences between XPS and UPS, it appears to be a useful approximation to assume that the $\mathbf{k}$ integration involved in determining $N_i(E_{kin})$ can be performed in such a way that the final result is expressed as a product of an average photoelectric cross-section for initial states at energy E, $\bar{\sigma}_E(hv)$; an initial occupied density of states at energy E; a final unoccupied density of states at energy $E + hv$, and an average escape function at energy $E + hv$, $\bar{T}(E + hv)$. Furthermore, we should multiply the initial density of states $\varrho(E)$ by the Fermi function

$$F(E) = \frac{1}{e^{(E - E_F)/kT} + 1},$$

to allow for thermal excitation of electrons near the Fermi level. It is adequate to assume that all final states described by $\varrho(E + hv)$ are unoccupied, as $hv - E \gg kT$. Thus $N_i(E_{kin})$ is given in this approximation by

$$N_i(E_{kin}) = N_i(h_v - E - \phi_{spec})$$
$$\propto \bar{\sigma}_E(hv) \, \varrho(E) \, F(E) \, \varrho(E + hv) \, \bar{T}(E + hv). \qquad (44)$$

However, it should be noted that the separation of factors resulting from the $\mathbf{k}$ integration of Equation (43) may not be rigorously possible in terms of $\mathbf{k}$-conserving transitions unless the average cross-section over all states at energy E is replaced with a less easily calculable average cross-section over only those states at energy E from which allowed **k**-conserving transitions can be made to states an energy hv above. Either surface effects [51, 52] or many body effects [56, 57] could tend to alter the **k**-conservation requirement, however. Also, this separation of factors permits isolating the several basic assumptions that are presently utilized in analyzing XPS data. From the comparisons between theory and experiment that are presently available, these assumptions appear to be reasonably accurate, at least in first approximation.

We can allow for certain instrumental and inherent linewidth contributions to the observed energy distribution by integrating $N_i(E_{kin})$ as given in Equation (44) over a linewidth function $L(E'-E)$. $L(E'-E)$ can thus include several sources of linewidth, among these being: (1) the linewidth of the exciting radiation, (2) the electron spectrometer resolution, (3) lifetime broadening by the $E, \mathbf{k}$ hole, and (4) thermal broadening of the initial state. Thus, we have finally

$$N_i(E_{kin}) = N_i(hv - E - \phi_{spec})$$

$$\propto \int_{-\infty}^{\infty} \bar{\sigma}_{E'}(hv)\,\varrho(E')\,F(E')\,\varrho(E'+hv)\,\bar{T}(E'+hv)\,L(E'-E)\,\mathrm{d}E'$$

(45)

which should represent an approximation to an observed spectrum $N(E_{kin})$ with primary exception that the effects of inelastic scattering have not been included [44].

In dealing with X-ray photoelectron spectra, there are several simplifications that appear to be possible in Equation (45), thereby permitting a more direct comparison to be made of $N_i(E_{kin})$ and the occupied density of states $\varrho(E)$.

First, the unoccupied final density of states $\varrho(E+hv)$ can be considered constant over the energy range pertinent to excitation from the valence bands by soft X-rays, as the final state photoelectrons are situated at more than 1000 eV into the continuum and therefore should behave very much like free electrons [58, 59]. That is, the appropriate final density of states will be proportional to $E_{kin}^{\frac{1}{2}}$. This function is negligibly smaller for electrons ejected from the bottom of the valence bands than for those emitted from the top of the valence bands (for example, $E_{kin}(bottom) \approx 1240$ eV and $E_{kin}(top) \approx 1250$ eV for excitation with MgKα radiation). Such constancy of $\varrho(E+hv)$ cannot be assumed in UPS work however, and considerable modulation of UPS spectra by the final density of states has been observed [48]. For example, Figure 15 shows the changes in shape of the energy distribution curves of crystalline gold which occur as photon energy is varied between 10.2 eV and 26.9 eV [48]. These changes are believed to be caused by final density of states modulation, as expressed most precisely by the integration over **k** in Equation (43). As further justification for the effective constancy of the final density of states in XPS measurements, it was found by Eastman and Cashion [60] that UPS spectra from the valence bands of Au obtained at high photon energies of approximately 40 eV tend to have very nearly the same appearance as corresponding XPS spectra.

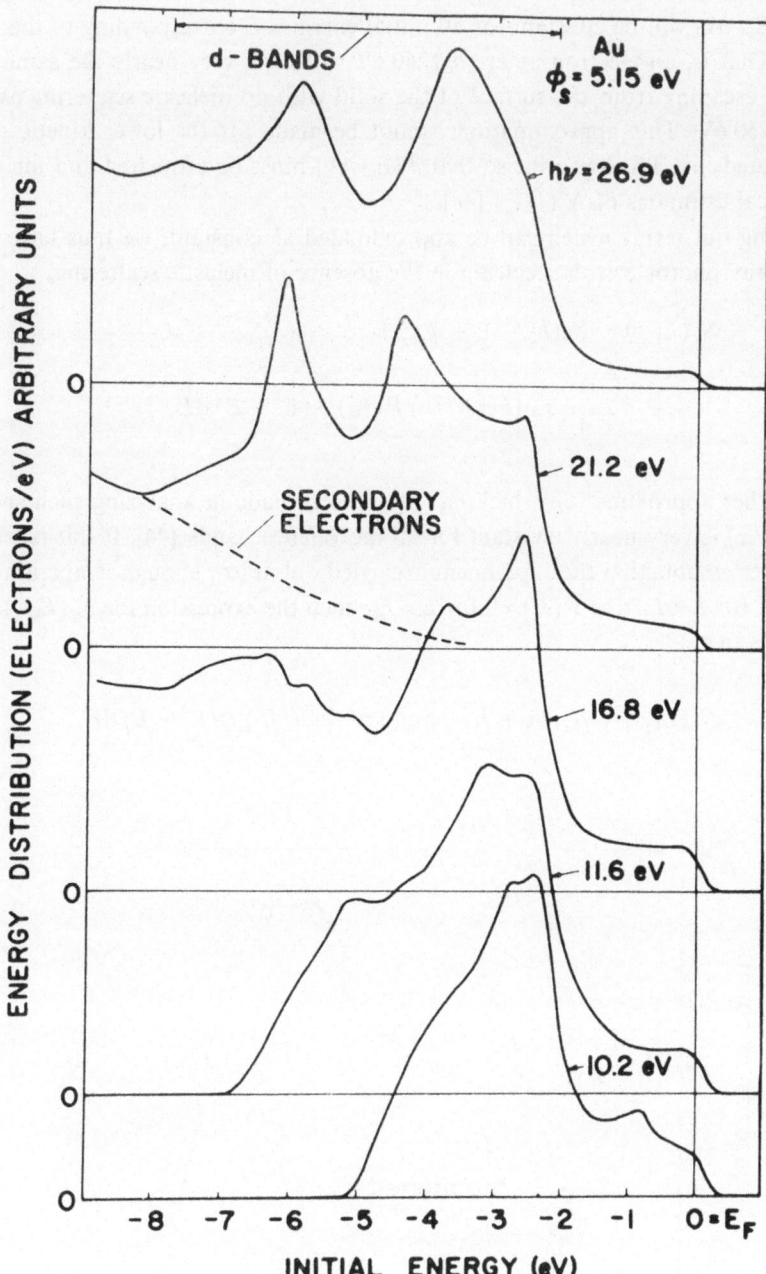

Fig. 15. Valence band UPS spectra for crystalline gold at various photon energies. The work function $\phi_s = 5.15$ eV, so that each spectrum has a minimum initial energy of $hv - 5.15$. (From [48].)

The high kinetic energy of X-ray photoelectrons also means that the escape function $\bar{T}(E+h\nu)$ will be constant for all initial energies E corresponding to the valence bands. That is, an electron at $E_{kin} \approx 1240$ eV will have very nearly the same probability of escaping from the surface of the solid with no inelastic scattering as one at $E_{kin} \approx 1250$ eV. This approximation cannot be made for the lower kinetic energies investigated in UPS, however, so that $\bar{T}(E+h\nu)$ must be estimated and included in theoretical estimates of $N_i(E_{kin})$ [48].

Striking out terms which can be approximated as constant, we thus have for the ideal X-ray photoelectron spectrum in the absence of inelastic scattering,

$$N_i(E_{kin}) = N_i(h\nu - E - \phi_{spec})$$

$$\propto \int_{-\infty}^{\infty} \bar{\sigma}_{E'}(h\nu)\, \varrho(E')\, F(E')\, L(E' - E)\, dE'. \tag{46}$$

A further approximation which has often been made in analyzing such spectra is that $\bar{\sigma}_E(h\nu)$ is very nearly constant for all the valence bands [44]. If this is true, and we further assume that the experiment is carried out at low enough temperatures that $F(E) \approx 1$ for $E < E_F$ and $F(E) \approx 0$ for $E > E_F$, then the expression for $N_i(E_{kin})$ further simplifies to

$$N_i(E_{kin}) = N_i(h\nu - E - \phi_{spec}) \propto \int_{-\infty}^{\infty} \varrho(E')\, L(E' - E)\, dE'. \tag{47}$$

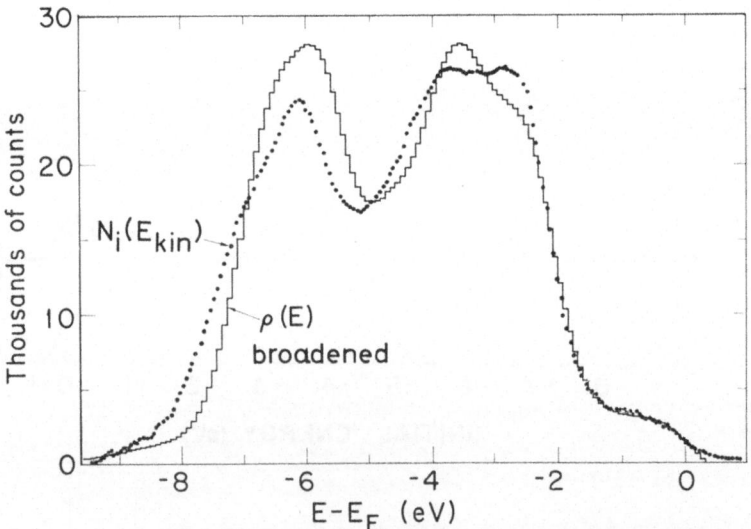

Fig. 16. An XPS spectrum for the valence bands of Au obtained with monochromatized AlKα radiation and corrected for inelastic scattering is compared with a broadened relativistic theoretical density of states curve. The experimental results are shown as points. (From [53].)

That is, in the simplest possible interpretation, $N_i(E_{kin})$ is a broadened version of the occupied density of states curve for temperatures near absolute zero. No quantitative calculations or experiments are available for assessing how valid the approximation of constant $\bar{\sigma}_E(h\nu)$ might be for various systems. For a set of bands of the same character, as for example, transition metal d bands, there is at least qualitative justification for the assumption, although even here it is known that the spatial character of the d bands changes from the 'bonding' bottom to the 'non-bonding' top of the bands [61]. Between such different states as are represented by s and d bands, a considerable difference in $\bar{\sigma}_E(h\nu)$ might exist. For the d bands of a transition metal such as gold, the present experimental evidence does not appear to indicate an appreciable change in $\bar{\sigma}_E(h\nu)$ over the observed valence band spectrum. In support of this statement, Figure 16 directly compares a high resolution XPS spectrum obtained with monochromatized A1Kα radiation with a relativistic theoretical calculation of the density of states for gold [53]. The experimental spectrum has been corrected for the effects of inelastic scattering, and so could correspond to Equation (46) or perhaps to Equation (47). The theoretical $\varrho(E)$ has been broadened by integrating over an appropriate line shape function $L(E'-E)$. The agreement is good as to overall structure, shape, and width. The constancy of $\bar{\sigma}_E(h\nu)$ over the various bands cannot in general be taken for granted, however, and further theoretical and experimental study of this question is called for.

It is also possible to correct XPS valence band spectra in a straightforward way for the effects of inelastic scattering. Very often such a correction is not absolutely necessary, as scattering is usually manifested as a rather flat background on the low kinetic energy side of the valence band peaks that distorts spectrum fine structure and relative intensities very little. The effects of inelastic scattering are often more pronounced in UPS spectra, however, as is shown in Figure 15. Correction of XPS spectra is, however, possible provided that a relatively sharp core photoelectron peak is present at a kinetic energy not too much lower than those of the valence band photoelectrons [44]. These reference core photoelectrons and the valence photoelectrons should thus be affected in a nearly identical way by inelastic scattering. A further condition on the selection of this core reference peak is that its sources of inherent width are essentially the same as those included in $L(E'-E)$. For example, the presence of multiplet splittings or multielectron effects in a core spectrum that are not found in the $L(E'-E)$ of the valence spectrum could render the core spectrum unsuitable for an inelastic scattering correction. Provided that such a peak can be found, however, the correction procedure is as outlined below [1].

Let us denote the measured core and valence photoelectron spectra by $N^c(E_{kin})$ and $N(E_{kin})$, respectively. Each spectrum can be considered to be a vector whose elements are the number of counts in a given kinetic energy channel. A simple form is assumed for the ideal core spectrum in the absence of inelastic scattering, $N_i^c(E_{kin})$. This could be of Lorentzian or Gaussian shape, for example, with a mean position very nearly equal to that of the observed photoelectron peak and a width slightly less than that of the observed peak. Choosing relative widths in this way permits correcting for just inelastic scattering and not for the instrumental and inherent contributions to linewidth expressed by $L(E'-E)$. The latter correction would involve an attempt at basic resolution enhancement and thus in the limit setting $N_i^c(E_{kin})$ equal to a delta function. $N_i^c(E_{kin})$ and $N^c(E_{kin})$ can be connected by a response matrix R which must approximately represent the effects of inelastic scattering. This we denote symbolically as

$$N^c(E_{kin}) = RN_i^c(E_{kin}). \tag{48}$$

R can be determined by making physically reasonable constraints on its mathematical form, and solving the resulting set of equations [1]. It is then a relatively simple matter to calculate the inverse of this matrix and apply it to the observed valence spectrum $N(E_{kin})$ to derive a corrected spectrum $N_i(E_{kin})$,

$$N_i(E_{kin}) = R^{-1}N(E_{kin}). \tag{49}$$

This procedure also has the advantage of incidentally correcting for the $\alpha_{3,4}$ satellites arising in an ordinary achromatic X-ray source. Figure 17 shows the application of this procedure to a Cu valence band spectrum. All of the spectra presented in Figure 14 have been corrected in this way [43].

In summary, for studies of densities of states in solids, both UPS and XPS exhibit certain unique characteristics and advantages. Somewhat better resolution is possible in a UPS measurement, primarily due to the narrower radiation sources presently available. Also, UPS spectra contain in principle information on both the initial and final density of states functions, together with certain k-dependent aspects of these functions. The interpretation of an XPS spectrum in terms of the initial density of

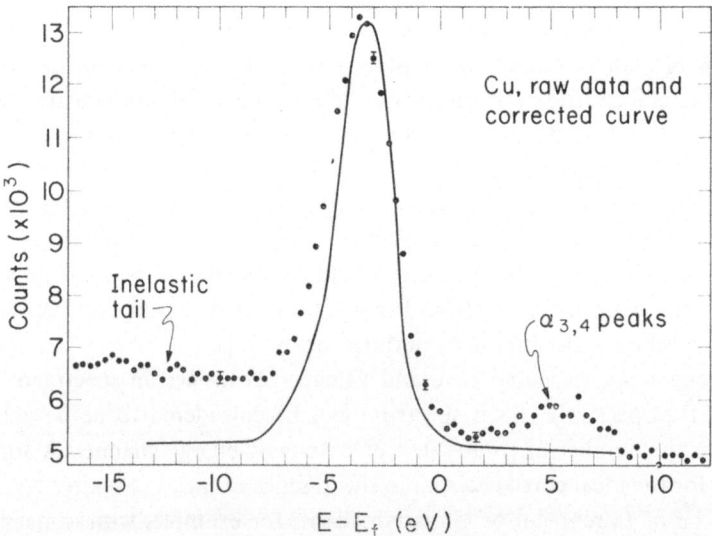

Fig. 17. XPS spectrum for the valencs bands of Cu, showing the effects of a correction for inelastic scattering according to Equation (49). (From [1, 43, 61].)

states appears to be more direct, however. Also, the effects of inelastic scattering are more easily corrected for in an XPS spectrum. Finally, the two techniques are very complementary in the sense that they are controlled by cross sections which may have different relative values for different bands, thereby providing information on the types of states involved.

4. Multiplet Splittings and Multi-Electron Processes

In this section, we shall discuss two distinct, but closely related, sources of additional structure in photoelectron spectra. Multiplet splittings of electron binding energies can be described in terms of a one-electron-transition model of the photoemission process. On the other hand, multi-electron processes involve the excitation of two or more electrons during photoemission and represent a fundamentally different effect.

The fundamental energy conservation relation governing the photoemission process has been stated in Equation 1. As we have noted, the binding energy of an electron from the kth subshell is by definition given by $E_b(k) = E^f(k) - E^i$. Thus, if more than one energy $E^f(k)$ is possible for the final state of the system with a hole in the kth subshell, the binding energy will be split into components. Each of these components can then be specified by the quantum numbers of the different final states. If E^i is assumed to be a unique energy for all measurements on a given system, the separations of the different binding energies will be independent of it and will be given simply by some difference $E^f(k)_1 - E^f(k)_2$. Thus, the reference level for absolute binding energies is not critically important in the analysis of structure due to various final-state energies. In what we shall call multiplet splittings, the different $E^f(k)$ values arise from a detailed consideration of the potential experienced by the electron ejected, including perhaps the coupling of the final state hole to the valence electrons. The various $E^f(k)$ values occuring in connection with multi-electron transitions involve the excitation of one or more additional electrons to different bound subshells or perhaps to free continuum states; in this case, therefore, the final states may differ in the quantum numbers of the additional excited electrons, plus any further differences introduced by multiplet effects. Thus, it is clear that the two types of structure are closely related. If a multi-electron process involves the excitation of additional electrons into the continuum, $E^f(k)$ will also include the kinetic energies of these electrons.

Before proceeding to discuss these two effects in detail, we note several other sources of structure in photoelectron spectra which may be erroneously attributed to binding energy splittings or multi-electron processes [62, 63]: (1) Auger electron peaks: These are easily distinguished because they have a constant kinetic energy, regardless of excitation source. (2) Photoelectron peaks arising from weak satellite or impurity X-rays: These will usually have fixed relative intensities and positions in comparison to the intense X-ray components and can also be detected by utilizing different X-ray sources (e.g., Al and Mg). (3) Inelastic scattering processes occurring during photoelectron escape from the sample: These will cause almost identical low-kinetic-energy tails on each photoelectron peak, particularly for peaks close together in kinetic energy. (4) Chemical effects on electron binding energies caused by spurious chemical reaction in the sample (as, for example, at the surface of a solid sample): Such chemical effects are more difficult to distinguish, but observations of the positions, shapes, and relative intensities of core and valence photoelectron peaks for all the expected constituents of a sample, including possible contaminants, will generally give a very good indication of its chemical state. Thus, with careful analysis of photoelectron spectra, it is generally possible to distinguish structure due to multiplet splittings and multi-electron processes. It may not be possible, however, to completely separate structure due to the two effects, as multi-electron processes often exhibit their own multiplet splittings.

A. MULTIPLET SPLITTINGS OF ELECTRON BINDING ENERGIES

We shall here review the experimental and theoretical aspects of electron binding energy *splittings*, as contrasted with the core-electron binding energy *shifts* discussed in Section 2. A binding energy shift can be considered to be the result of a change in the spatially-averaged Coulomb and exchange potentials exerted upon an electron by its environment. Because the potential has been averaged over the region of the electron, the initial states in the photoemission process can be considered in first approximation to be closed-shell systems with one hole present in the final state. In this model, therefore, the resulting binding energies for different core electrons are fully characterized by the usual one-electron quantum numbers for free atoms: n, l, and j. In describing valence-electron binding energies, on the other hand, the various electron-electron and electron-nuclear interactions must be considered in detail, with the result that quantum numbers appropriate to the symmetry of the total atomic ensemble must be used. For example, an atomic state may be specified by $d^5 \, ^6S$, a homonuclear diatomic molecular state by $(\sigma_g)^2 \, ^1\Sigma_g$, or a state in an octahedral environment by $(t_{2g})^2 T_1$. Thus, multiplet splittings form an integral part of the study of photoelectron spectra from valence electrons [8]. If similar considerations are applied to the description of core electron binding energies, or, in particular, the final states of binding energy measurements with single holes in different core levels, the binding energies are found to be split into several components relative to a description in terms of n, l, and j alone. Such splittings we shall term multiplet splittings, as they arise from the various electronic multiplet states occuring in atoms, molecules, and solids. Multiplet effects involving core electron holes are very commonly used in discussing structure observed in X-ray emission spectra [64–68] and Auger electron spectra [69–71]. However, it is only recently that such effects were first discussed in connection with X-ray photoelectron spectra emanating from core levels [9, 62, 72–75]. The measurement and interpretation of such splittings have recently been reviewed in more detail [63, 76].

As a simple illustration of one type of multiplet splitting found in XPS studies [62, 75] we consider photoemission from the $3s$ level of a Mn^{2+} free ion, as shown on the left-hand side of Figure 18. The ground state of this ion can be described in Russell-

Saunders (L-S) coupling as $3d^5\,{}^6S$ (that is, $S=\frac{5}{2}$, $L=0$). In this state, the five $3d$ spins are coupled parallel. Upon ejecting a $3s$ electron, however, two final states may result: $3s3d^5\,{}^5S(S=2, L=0)$ or $3s3d^5\,{}^7S(S=3, L=0)$. The difference between these two states is that in the ${}^5S$ state, the spin of the remaining $3s$ electron is coupled anti-parallel to those of the five $3d$ electrons, whereas in the ${}^7S$ state the $3s$ and $3d$ spins are coupled parallel. Because the exchange interaction acts only between electrons with parallel spins, the ${}^7S$ energy will be lowered relative to the ${}^5S$ energy because of the favorable effects of $3s$–$3d$ exchange. The magnitude of this energy separation will be proportional to the $3s$–$3d$ exchange integral $K_{3s,\,3d}$, and will be given by [28]:

$$\Delta\left[E_b(3s)\right] = E^f(3s3d^5\,{}^5S) - E^f(3s3d^5\,{}^7S) = \Delta E^f(3s3d^5)$$

$$= 6K_{3s,\,3d}$$

$$= \frac{6e^2}{5}\int_0^\infty\int_0^\infty \frac{r_<^2}{r_>^3}P_{3s}(r_1)\,P_{3d}(r_2)\,P_{3s}(r_2)\,P_{3d}(r_1)\,dr_1\,dr_2,$$

$$(50)$$

where e is the electronic charge, $r_<$ and $r_>$ are chosen to be the smaller and larger of r_1 and r_2 in performing the integrations, and $P_{3s}(r)/r$ and $P_{3d}(r)/r$ are the radial

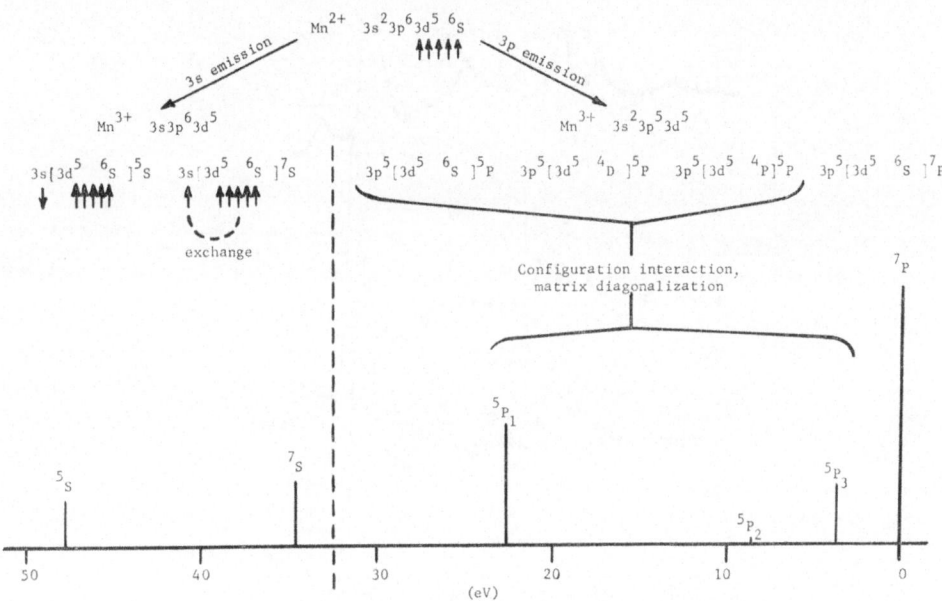

Fig. 18. The various final state multiplets arising from $3s$ and $3p$ photoemission from a Mn^{2+} ion. Within the S and P manifolds, separations and relative intensities have been computed using simple multiplet theory as discussed in the text. The separation and relative intensity of the ${}^7S$ and ${}^7P$ multiplets were fixed at the values observed for $3s(1)$ and $3p(1)$ in the MnF_2 spectrum of Figure 19, to facilitate comparison with experiment. (From [63].)

wave functions for $3s$ and $3d$ electrons. The factor $\frac{1}{5}$ results from angular integrations involved in computing $K_{3s,\,3d}$. A Hartree-Fock calculation of the energy splitting in Equation (50) for Mn^{3+} gives a value of $\Delta E^{J}(3s3d^{5}) \approx 13$ eV [62, 75]. As this predicted splitting is considerably larger than typical XPS linewidths, it is not surprising that rather large $3s$ binding energy splittings have been observed in solid compounds containing Mn^{2+} [62, 75]. Such splittings are clearly evident in the $3s$ regions of the spectra from MnF_2 and MnO in Figure 19. Roughly the left half of each of these spectra represents $3s$ emission, and the splittings observed are approximately one-half of those predicted from Equation (50). The reasons for this discrepancy we discuss below. The peaks denoted by '$\alpha_{3,\,4}$' are due to $K\alpha_3$ and $K\alpha_4$ satellite X-rays.

In discussing such core binding energy splittings, it is worthwhile to present a general model of the photoemission process, including the relevant selection rules [63]. If the photoelectron is ejected from a filled nl subshell containing q electrons, and an unfilled $n'l'$ valence subshell containing p electrons is present, we can write the overall

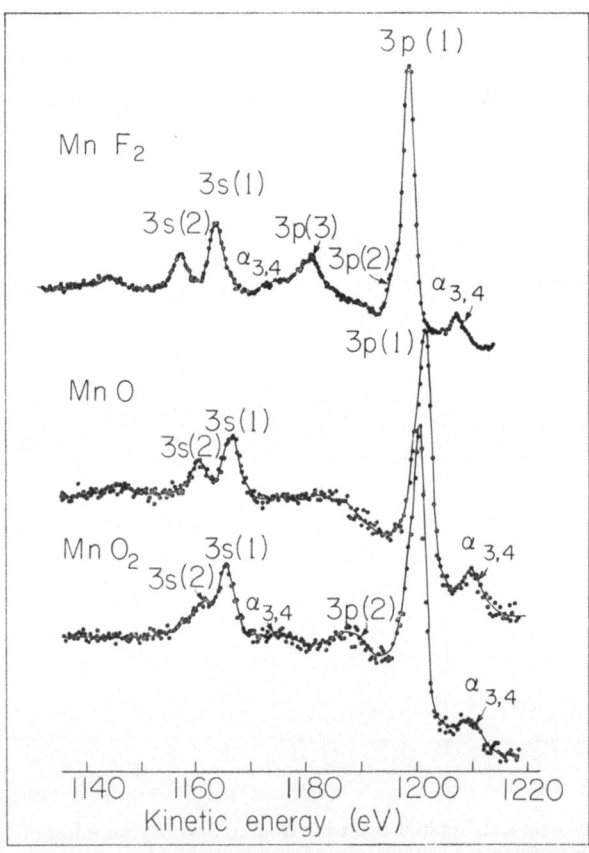

Fig. 19. XPS spectrum from three solid compounds containing Mn, in the kinetic energy region corresponding to emission of Mn3s and 3p electrons. Peaks arising from $\alpha_{3,\,4}$ satellite X-rays are so labelled. (From [62].)

photoemission process as:

$$(nl)^q (n'l')^p \overset{h\nu}{\rightarrow} \underbrace{(nl)^{q-1} (n'l')^p}_{} + \text{photoelectron} \tag{51}$$

$$\text{(filled)} (L, S) \qquad (L', S')$$

L and S denote the total orbital and spin angular momenta of the initial state and L' and S' represent the same quantities for the final state. As $(nl)^q$ is a filled subshell, its orbital and spin angular momenta must both be zero and therefore L and S correspond to the orbital and spin momenta of the valence subshell $(n'l')^p$. In the final state, L' and S' represent momenta resulting from the coupling of $(nl)^{q-1}$ (or, equivalently, a single core electron hole) with $(n'l')^p$. The transition probability per unit time for photoelectron excitation is proportional to the square of a dipole matrix element between the initial and final state wave functions (see Section 5 for a detailed discussion). In a purely one-electron model of photoemission, this matrix element simplifies to the one-electron form $\langle \varphi_{nl} | \mathbf{r} | \varphi_{ph} \rangle$, where φ_{nl} and φ_{ph} are one-electron orbitals for the initial subshell and the final photoelectron state, respectively. The one-electron selection rule on angular momentum is thus

$$l_{ph} = l \pm 1. \tag{52}$$

Conservation of total spin and total orbital angular momenta also requires that

$$\Delta S = S' - S = \pm \tfrac{1}{2} \tag{53}$$

and

$$\Delta L = L' - L = 0, \pm 1, \pm 2, ..., \pm l. \tag{54}$$

Also, in a strict one-electron interpretation of the transition shown in relation (51) in which only one electron is assumed to change its orbital with all other orbitals remaining unrelaxed in the final state, an additional selection rule results. This rule states that the coupling of the unfilled valence subshell $(n'l')^p$ in the final state must be the same as that in the initial state: that is to total spin and orbital angular momenta of L and S. Finally, any coupling scheme for $(nl)^{q-1}$ and $(n'l')^p$ must be consistent with the Pauli exclusion principle. Since $(nl)^{q-1}$ is assumed to represent a single hole in an otherwise filled subshell, it must therefore couple to a total spin of $\tfrac{1}{2}$ and a total orbital angular momentum of l. Within this model, it has also been shown by Cox and Orchard [77] that the total intensity of a given final state specified by L', S' will be proportional to its total degeneracy, as well as to the one-electron matrix element squared. Thus, in Russell-Saunders coupling

$$I_{tot}(L', S') \propto (2S' + 1)(2L' + 1). \tag{55}$$

In the case of s-electron binding energy splittings, $l=0$ and Equations (52)–(54) reduce to

$$\Delta S = \pm \tfrac{1}{2} \tag{56}$$

$$\Delta L = 0 \tag{57}$$

$$I_{tot}(L', S') \propto 2S' + 1. \tag{58}$$

Thus, only two final states are possible corresponding to $S'=S\pm\frac{1}{2}$, and the relative intensities of these will be given by the ratio of their multiplicuties, or

$$\frac{I_{tot}(L, S + \frac{1}{2})}{I_{tot}(L, S - \frac{1}{2})} = \frac{2S + 2}{2S}. \tag{59}$$

The energy separation of these two states can be calculated from simple atomic multiplet theory and is given by [28]:

$$\Delta[E_b(ns)] = E^f(L, S - \frac{1}{2}) - E^f(L, S + \frac{1}{2}) \tag{60}$$

$$\Delta[E_b(ns)] = (2S + 1) K_{ns, n'l'} \quad \text{for} \quad S \neq 0 \tag{61}$$

$$\Delta[E_b(ns)] = \qquad 0 \qquad \text{for} \quad S = 0. \tag{62}$$

Here $K_{ns, n'l'}$ is the $ns–n'l'$ exchange integral and can calculated from

$$K_{ns, n'l'} = \frac{e^2}{2l' + 1} \int_0^\infty \int_0^\infty \frac{r_<^{l'}}{r_<^{l'+1}} \times$$

$$\times P_{ns}(r_1) P_{n'l'}(r_2) P_{ns}(r_2) P_{n'l'}(r_1) dr_1 \, dr_2, \tag{63}$$

where we have used the same notation as that in Equation (50). Thus, Equations (59)–(63) indicate that such s-electron binding energy splittings should yield a doublet with a more intense component at lower binding energy (corresponding to an exchange-favored final state of $S'=S+\frac{1}{2}$) and a component separation that is directly connected to both the initial state spin and the spatial distributions of the core and valence electrons as reflected in the exchange integral. Thus, the potential for extracting useful and unique information from such splittings exists.

The analysis of non-s binding energy splittings of this type is not so straightforward, however, primarily due to the fact that the core electron hole represented by $(nl)^{q-1}$ (which has associated with it a spin of $\frac{1}{2}$ and an orbital angular momentum of l) can couple in various ways with the valence subshell $(n'l')^p$ (which can have various spins S'' and orbital angular momenta L'', including the initial values S and L) to form a final state with a given total spin S' and total orbital angular momentum L'. Thus, the number of allowed final states increases and their energy separations will in general be determined by both Coulomb and exchange integrals through different coupling schemes. The simplest procedure for calculating these energy separations is again to use atomic multiplet theory [28, 75].

As an example of a non-s splitting, let us consider 3p electron emission from Mn^{2+}, as indicated in the right portion of Figure 18 [62, 75]. For this case, $(nl)^{q-1}=3p^5$, $(n'l')^p=3d^5$ and the initial state, as before, is ^6S ($S=\frac{5}{2}$, and $L=0$). The previously stated selection rules imply that the allowed final states correspond to ^7P ($S=3$, $L=1$) and ^5P ($S=2, L=1$). Although a $^5S(S=2, L=0)$ final state would be consistent with selection rule 54, it requires changing the coupling of $3d^5$ from its initial ^6S and so is not allowed. There is only one way for $3p^5$ to couple with $3d^5$ to form a ^7P state, that being with $3p^5$ (always coupled to total spin$=s=\frac{1}{2}$ and total orbital angular momentum $= l = 1$) coupled with $3d^5$ in its initial state coupling of ^6S ($S = \frac{5}{2}$, $L = 0$). However, there are three ways to form the allowed ^5P final state:

$$3p^5 (s = \tfrac{1}{2}, l = 1) \quad \text{with} \quad 3d^5 \, {}^6S \, (S'' = \tfrac{5}{2}, L'' = 0)$$
$$3p^5 (s = \tfrac{1}{2}, l = 1) \quad \text{with} \quad 3d^5 \, {}^4D \, (S'' = \tfrac{3}{2}, L'' = 2)$$
$$3p^5 (s = \tfrac{1}{2}, l = 1) \quad \text{with} \quad 3d^5 \, {}^4P \, (S'' = \tfrac{3}{2}, L'' = 1).$$

Thus, four distinct final states are possible for $3p$ emission from Mn^{2+}, one ^7P and three ^5P. As there are off-diagonal matrix elements of the Hamiltonian between the various ^5P coupling schemes [28], they do not individually represent eigenfunctions. The eigenfunctions describing the ^5P final states will thus be linear combinations of the three schemes:

$$\Psi\,({}^5P_1) = C_{11}\psi\,({}^6S) + C_{12}\psi\,({}^4D) + C_{13}\psi\,({}^4P)$$
$$\Psi\,({}^5P_2) = C_{21}\psi\,({}^6S) + C_{22}\psi\,({}^4D) + C_{23}\psi\,({}^4P) \qquad (64)$$
$$\Psi\,({}^5P_3) = C_{31}\psi\,({}^6S) + C_{32}\psi\,({}^4D) + C_{33}\psi\,({}^4P),$$

where we have labelled each ^5P configuration by the $3d^5$ coupling involved and the C_{ij}'s are expansion coefficients. However, because a purely one-electron transition only reaches final states with the initial state coupling ^6S for $3d^5$, only those components of the ^5P states represented by $C_{i1}\Psi\,({}^6S)$ are accessible.

The energy eigenvalues corresponding to these eigenfunctions will then give the separations between the ^5P states. Such eigenfunctions and eigenvalues can most easily be determined by diagonalizing the 3×3 Hamiltonian matrix for the ^5P states, where each matrix element is expressed as some linear combination of J_{3d-3d}, K_{3d-3d}, J_{3p-3d}, and K_{3p-3d} [1, 28]. If Coulomb and exchange integrals from a Hartree-Fock calculation on Mn^{2+} are used, such matrix diagonalization calculations yield the relative separations indicated in Figure 18 [62, 75].

In determining the *total* intensity ratios for the ^5P and ^7P states, Equation (55) can be used to give:

$$I_{tot}\,({}^5P):I_{tot}\,({}^7P) = [I\,({}^5P_1) + I\,({}^5P_2) + I\,({}^5P_3)]:I_{tot}\,({}^7P) = 5:7,$$

and similarly for the two peaks resulting from $3s$ electron emission:

$$I_{tot}\,({}^5S):I_{tot}\,({}^7S) = 5:7.$$

The individual intensities of $^5P_1$, $^5P_2$, and $^5P_3$ can be computed by noting that only that portion of each ^5P state corresponding to the initial $3d^5$ coupling of ^6S can be reached. Thus, the intensity of each state is proportional to $|C_{i1}|^2$. The relative peak heights in Figure 18 have been calculated in this way, and the experimental $3s(1)$–$3p(1)$ separation and relative intensity for MnF_2 were used to empirically fix the separation between the $3s$ and $3p$ regions.

The spectral structure shown in Figure 18 thus represents the simplest free-ion theoretical estimate. Somewhat more sophisticated calculations utilizing separate Hartree-Fock total energies for each final state yield results very close to these for energy separations [75]. The predicted structure can be compared to the $3s$–$3p$ experimental spectrum from solid MnF_2 shown in Figure 19, as manganese exists approximately as Mn^{2+} $3d^5$ ^6S in this compound. There is at least qualitative agreement between theory and experiment if the following identifications are made: $^5S = 3s(2)$, $^7S = 3s(1)$, $^5P_3 = 3p(3)$, $^5P_2$ too weak to observe, $^5P_3 = 3p(1)$ [62, 75]. However, if the clearly resolved $3s$ doublet is analyzed in detail, we find the following discrepancies in separations and intensities:

<table>
<tr><td></td><td>Separation</td><td>Relative intensity</td></tr>
<tr><td>Experiment:</td><td>$\Delta E_b(3s) = 6.5$ eV,</td><td>$3s(2):3s(1) = 1.0:2.0$</td></tr>
<tr><td>(MnF_2)</td><td></td><td></td></tr>
<tr><td>Theory:</td><td>$\Delta E_b(3s) = 13.0$ eV,</td><td>$^5S:^7S = 1.0:1.4 = 5:7$.</td></tr>
<tr><td>(free-ion)</td><td></td><td></td></tr>
</table>

The origins of these discrepancies have been discussed previously [62, 63, 75]. As to separation, the effects of valence electron delocalization and covalency occurring due to chemical bond formation, as well as the neglect of favorable anti-parallel electron correlation in the ^5S final state would tend to lead to too large a theoretical estimate. The correlation argument is roughly equivalent to saying that $\varepsilon(3s\beta, 3d\alpha) > \varepsilon(3s\alpha, 3d\alpha)$ in the nomenclature of Equation (28). In fact, recent multi-configuration Hartree-Fock calculations on free-ion Mn^{3+} by Bagus et al. [78] which allow for correlation to a certain extent give a ^5S–^7S separation of only 8.2 eV, in much better agreement with experiment. These calculations would appear to indicate that neglect of anti-parallel correlation is a major contributor to the large discrepancies found between experiment and simple free-ion calculations as outlined above. They also predict weak multiplet peaks for E_{kin} below $3s(2)$ [78]. One such peak appears at $E_{kin} \approx 1144$ eV in MnF_2 (Figure 19). Also, these peaks are derived from ^5S states, thereby explaining the high $3s(2):3s(1)$ ratio [78].

Any detailed theoretical analysis of the MnF_2 $3p$ spectrum must also include the effects of crystal field splitting and spin-orbit coupling, both of which will be of the order of 1 eV, as well as the possible sources of inaccuracy already noted for the $3s$ doublet. However, the separations and relative intensities of the peaks observed seem at least semi-quantitatively understood in terms of a simple L-S coupling model. Ekstig et al. [68] have carried out matrix diagonalization calculations like those described here but for more complex sets of final $3p$-hole states in $3d$ transition metal atoms in an attempt to interpret soft X-ray emission spectra from solids. The theoretical aspects of calculating such multiplet splittings have recently been reviewed by Freeman et al. [76].

An examination of the $3s$ regions of the spectra for several Mn and Fe compounds in Figures 19 and 20 shows splittings in qualitative agreement with the predictions of Equations (56)–(62). Doublets are observed for all systems containing unpaired electrons (MnF_2, MnO, MnO_2, FeF_3, and Fe metal). The diamagnetic compounds

$K_4Fe(CN)_6$ and $Na_4Fe(CN)_6$ exhibit no splitting of the $3s$ photoelectron peaks, in consistency with possessing zero spin. In addition to the structure observed for Fe, $3s$ binding energy splittings also appear to be present in the other ferromagnetic transition metals Co and Ni [62].

Deeper core levels in $3d$ transition metal atoms should also exhibit similar splittings, although the magnitudes will be reduced because of the decreased overlap between

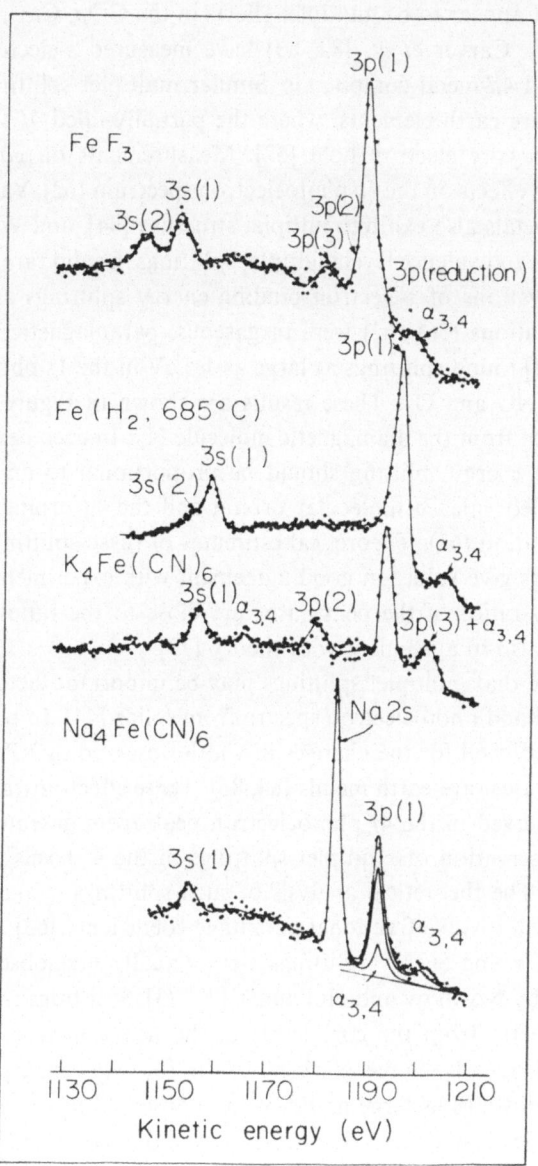

Fig. 20.   XPS spectra from four solids containing Fe in the Fe3s, 3p region. (From [61].)

the core-electron orbital and the $3d$ orbital. For example, the $2p$ levels in MnF_2 are found to be broadened by $\sim 1.5\,eV$ relative to low-spin compounds, and this broadening is probably connected to multiplet effects [62]. The effects of correlation should be less important in the analysis of such deeper level splittings, and simple free-ion calculations agree rather well with experiment [76, 78].

Multiplet splitting measurements have by now been made for a number of solid compounds containing atoms with unfilled valence shells. Novakov [79] and Wertheim et al. [80] have measured $3s$ splittings in several 3d-metal compounds. Helmer [81] has observed what appear to be multiplet effects in the $Cr2p$, $Cr3s$, and $Cr3p$ binding energies in Cr_2O_3. Carver et al. [82, 83] have measured s-electron splittings in a number of 3d- and 4d-metal compounds. Similar multiplet splittings should also be observed in the rare earth elements, where the partially-filled $4f$ shell can couple in various ways to a core electron hole [62]. Measurements on *gaseous* Eu indicate probable multiplet effects on the $4d$ photoelectron spectrum [62]. Valence band spectra from rare earth metals also exhibit multiplet structure [84], and Wertheim et al. [80] have reported core and valence level multiplet splittings in solid rare earth compounds.

The first observations of s-electron binding energy splittings analogous to those described by Equations (56)–(62) were in gaseous, paramagnetic molecules [9, 74]. Hedman et al. [74] found splittings as large as $1.5\,eV$ in the 1s photoelectron spectra of the molecules NO and O_2. These results are shown in Figure 21 along with an unsplit 1s spectrum from the diamagnetic molecule N_2. In each case, it can be shown that the observed energy splitting should be proportional to an exchange integral between the unfilled valence molecular orbital and the $1s$ orbital of N or O [9], in analogy with Equation (61). Theoretical estimates of these splittings from molecular orbital calculations give values in good agreement with experiment [22, 23, 74]. The observed intensity ratios of the peaks are very close to the ratios of the final state degeneracies [9], also in agreement with theory [77].

Finally, we note that multiplet splittings may be important factors in the interpretation of valence-band photoelectron spectra from solids [44]. In particular, multiplet effects account very well for the changes in width observed in XPS spectra from the $4f$ electrons in various rare-earth metals [84, 85]. These effects are also consistent with a broadening observed in the $4f$ photoelectron peak from gaseous Eu [62], as well as with direct observation of multiplet splittings in the $4f$ levels of solid rare earth compounds [80]. The theoretical analysis of such splittings is analogous to that for core levels, and will involve fractional parentage coefficients [63].

Core-electron binding energy splittings were actually first observed in somewhat different systems by Novakov and Hollander [72, 73]. Splittings were found in X-ray photoelectron spectra from the core levels of the heavy metals Th, U, and Au in various solid compounds. Some of their results for Au compounds are shown in Figure 22 [73]. Splittings as large as $10\,eV$ were observed in some solids [72]. However, these splittings cannot be explained in terms of free-atom-like multiplets such as those discussed above. It has been proposed that they may be due to crystal-field-like splittings of the core electronic levels [72, 86]. However, calculations based on

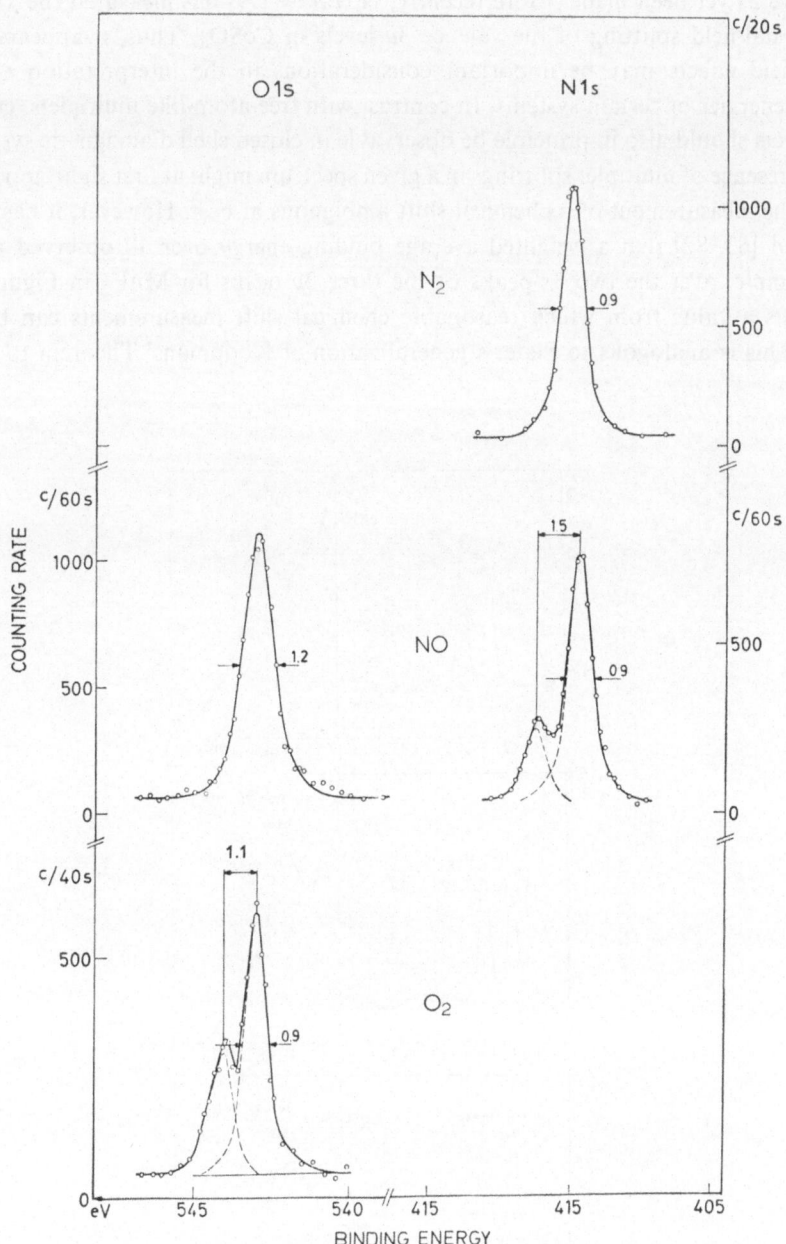

Fig. 21. XPS spectra from the 1s core electrons of the gaseous molecules N_2, NO, and O_2. The 1s peaks from the paramagnetic molecules NO and O_2 are split due to final state multiplets. (From [9].)

1st-order perturbation theory do not seem adequate to quantitatively explain the observed results [86]. A theoretical formulation for interpreting such splittings has been proposed by Gupta *et al.* [87], but no detailed analyses of the experimental data have as yet been made. More recently, Novakov [79] has measured the known 2 eV crystal-field splitting of the valence $3d$ levels in $CoSO_4$. Thus, it appears that crystal-field effects may be important considerations in the interpretation of the binding energies of certain systems. In contrast with free-atom-like multiplets, crystal field effects should also in principle be observable in closed shell diamagnetic systems.

The presence of multiplet splittings in a given spectrum might at first sight appear to render the measurement of a chemical shift ambiguous at best. However, it has been suggested [63, 82] that a weighted average binding energy over all observed peaks (for example, over the two $3s$ peaks or the three $3p$ peaks for MnF_2 in Figure 19) represents a value from which reasonable chemical shift measurements can be inferred. This is analogous to Slater's generalization of Koopmans' Theorem to open

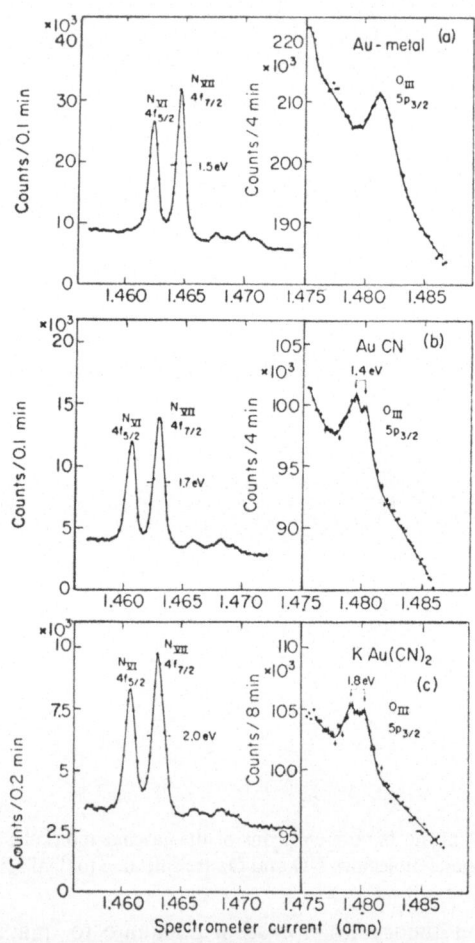

Fig. 22. XPS spectra from several solids containing Au. Note the apparent splittings of the Au$5p_{3/2}$ peaks in AuCN and KAu(CN)$_2$, and the corresponding broadenings of the Au$4f_{7/2}$ peaks for those compounds. (From [73, 86].)

shell systems (Equation (30a)). This suggestion has received little quantitative verification as yet, however.

In comparison to chemical shifts of core-electron binding energies, multiplet splittings of these energies thus represent higher-order effects yielding a different type of information. Chemical shift measurements detect a change in the average potential experienced by an electron, whereas multiplet effects appear to have the capability of determining the valence electron configuration of an atom or the detailed potential around it. The two types of measurements are thus complementary. Numerous applications of multiplet splittings measurements seem possible in the study of the transition series metals, the rare earths, the transuranium elements, and open-shell systems in general. Crystal-field like effects on core levels also deserve further study.

B. MULTI-ELECTRON PROCESSES

Multi-electron processes in connection with X-ray photoemission were first studied in detail by Krause, Carlson, and coworkers [88–94]. In these studies, gaseous neon and argon were exposed to X-rays with energies in a range from 270 eV to 1.5 keV. Measurements were then made of both the charge distributions of the resulting ions and the kinetic energy distributions of the ejected photoelectrons. From these measurements, it was concluded that two-electron and even three-electron transitions occur in photo-absorption, with total probabilities which may be as high as 20% for each absorbed photon. By far the most likely multi-electron process is a two-electron transition, which is approximately ten times more probable than a three-electron transition. Two types of two-electron transitions can be distinguished, depending upon whether the second electron is excited to a higher bound state ('shake-up') or to an unbound continuum state ('shake-off'). These are indicated in the transitions below (cf. the one-electron transition in relation (51)):

Shake-up:

$$(nl)^q \, (n'l')^p \xrightarrow{h\nu} \underbrace{(nl)^{q-1} \, (n'l')^{p-1} \, (n''l'')^1}_{(L', \, S')} + \text{photoelectron} \qquad (65)$$
$$(L, \, S)$$

Shake-off:

$$(nl)^q \, (n'l')^p \xrightarrow{h\nu} (nl)^{q-1} \, (n'l')^{p-1} \, (E''_{\text{kin}} l'')^1 + \text{photoelectron}. \qquad (66)$$

Here $(n'l')^p$ represents some outer subshell from which the second electron is excited; it can be filled or partially filled. Either shake-up or shake-off requires energy that will lower the kinetic energy of the primary photoelectron. Thus, multi-electron processes lead to satellite structure on the low-kinetic energy side of the one-electron photoelectron peak. It was also pointed out in these early studies that the probability for shake-off processes should increase from zero for a value of $h\nu$ equal to the one-electron-transition binding energy to a constant value for $h\nu$ equal to a few times this value. Thus, X-ray photoemission experiments with $h\nu \approx 1.5$ keV should exhibit a maximum effect from shake-off processes for electrons whose binding energies are less than approximately 500 eV.

Higher resolution XPS spectra have been obtained recently for neon by Siegbahn *et al.* [9] and for neon and helium by Carlson *et al.* [95]. A spectrum obtained in a recent study on neon [95] is shown on a semi-logarithmic plot in Figure 23. The notations α_3, α_4, etc., indicate photoelectron peaks due to various satellite X-rays of Mg (themselves the result of multielectron excitations in the Mg anode of the X-ray tube). The dotted curve represents intensity due to inelastic scattering. The solid curve is thus the net spectrum due to elastically scattered photoelectrons. The two-electron transitions responsible for the observed spectrum intensity at relative energies less than ~ -36 eV are diagrammatically indicated in Figure 23. The total two-electron intensity in this spectrum is estimated to be approximately 10% of that of the one-electron peak [95].

The theoretical calculation of peak positions and relative intensities in multi-electron spectra has usually been done with a model based on the 'sudden approximation' [9, 32, 88–96]. This model has also been used in the explanation of satellite X-rays in terms of multi-electron excitations [96]. A concise review of this model has been presented by Manne and Åberg [32]. A fundamental assumption is that the primary photoabsorption process which excited a core electron from the nlth subshell into a continuum photoelectron state occurs so rapidly that the valence electrons do not have time to readjust to the change in the potential due to the core. The primary excitation is considered in this sense to be 'sudden'. A theoretical criterion for the

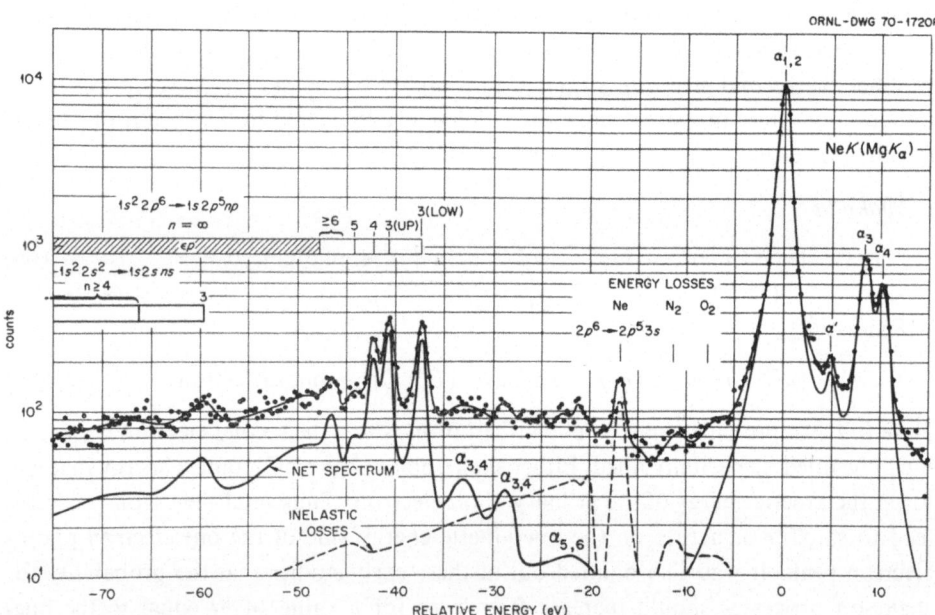

Fig. 23.   Photoelectron spectrum of Ne1s electrons excited by MgKα X-rays. Intensity due to inelastic scattering has been determined experimentally and subtracted to yield the net spectrum. The net spectrum contains peaks due to shake-up and shake-off processes, as well as those due to Mg satellite X-rays. The kinetic energy scale is relative to the intense one-electron peak at $E_{kin} = 616$ eV. Note the logarithmic intensity scale. (From [95].)

validity of the sudden approximation is that [96]

$$[E^f(nl, n'l') - E^f(nl)]\, \tau/h \ll 1, \tag{67}$$

where $E^f(nl, n'l')$ is the total final state energy resulting from a two-electron transition of the shake-up type, $E^f(nl)$ is the total final state energy for a one-electron transition involving electron nl and τ is the time of transit of the nl photoelectron past the $n'l'$ subshell in leaving the atom. The energy difference in Equation (67) is thus just the separation of the intense one-electron photoelectron peak from a certain two-electron satellite at lower kinetic energy (cf. Figure 23). As an indication of the orders of magnitude occurring in this inequality, for a typical X-ray photoelectron of $E_{kin} = 1000$ eV, $v/c \approx 0.06$ or $v \approx 2 \times 10^9$ cm s^{-1}. For a typical atomic diameter of 2 Å, the transit time can be roughly estimated as $\tau \approx 2 \times 10^{-8}/(2 \times 10^9) \approx 10^{-17}$ s. Thus, $\tau/\hbar \approx \frac{1}{65}$ eV^{-1}, and for satellite separations as large as 65 eV, the sudden approximation would appear to be violated. However, this criterion for the applicability of the sudden approximation has proven in practice to be overly restrictive [96], and calculations using the approximation have given reasonable agreement with experiment for several systems for which Equation (67) was not fully satisfied.

A further assumption implicit in this treatment is that the one-electron orbitals describing the initial and final states do not change appreciably. The initial wave function of the N-electron system is denoted by $\Psi^i(N)$, and is expressed as an anti-symmetrized product of the nlth one-electron orbital and a remainder $\Psi_R(N-1)$. That is,

$$\Psi^i(N) = A\varphi_{nl}\psi_R(N-1) \tag{68}$$

where A is an anti-symmetrizing operator, and

$$\mathscr{H}^i(N)\, \Psi^i(N) = E^i\Psi^i(N), \tag{69}$$

where $\mathscr{H}^i(N)$ is the N-electron Hamiltonian. ψ_R is thus *not* a wave function for the $N-1$ electron system. The actual final states will be described by a complete set of orthonormal wave functions $\Psi^f(N-1)_j \neq \psi_R(N-1)$, where $j = 0, 1, 2\ldots$. The index j is used to distinguish different final states connected with different two-electron transitions of the $n'l' \rightarrow n''l''$ type and/or different multiplet states associated with one-electron or two-electron transition final states. We can arbitrarily choose $\Psi^f(N-1)_0$ to represent a one-electron-transition final state, that is, one in which simple relaxation has occurred in an electron configuration identical to that in $\psi_R(N-1)$. Thus, if $H^f(N-1)$ is the final-state $N-1$ electron Hamiltonian, these wave functions must by definition satisfy

$$\mathscr{H}^f(N-1)\, \Psi^f(N-1)_0 = E^f(nl)_0\, \Psi^f(N-1)_0 \quad (j=0) \tag{70}$$

$$\mathscr{H}^f(N-1)\, \Psi^f(N-1)_j = E^f(nl, n'l')_j\, \Psi^f(N-1)_j \quad (j=1,2,3\ldots). \tag{71}$$

We have here neglected the effects of possible multiplet splittings in $\Psi^f(N-1)_0$, or, equivalently, have taken the average energy of all multiplets corresponding to a single nl hole to be $E^f(nl)_0$.

The sudden approximation implies that the $N-1$ passive electrons not directly involved in the primary photoelectron emission event do not change their overall spatial distribution appreciably during the emission process. However, the initial function $\psi_R(N-1)$ representing these electrons clearly is not a valid wave function for the final state Hamiltonian, and thus can be described as some mixture of true final state wave functions. This can be expressed mathematically by writing $\psi_R(N-1)$ as a linear combination of the orthonormal set of final state wave functions $\Psi^f(N-1)_j$:

$$
\psi_R(N-1) = \sum_{j=0}^{\infty} C_j \Psi^f(N-1)_j =
$$
$$
= \sum_{j=0}^{\infty} \langle \Psi^f(N-1)_j \,|\, \psi_R(N-1) \rangle \, \Psi^f(N-1)_j. \tag{72}
$$

The total final state wave function, including the photoelectron in an orbital φ_{ph}, will be given by

$$
\Psi^f(N)_j = A \varphi_{ph} \Psi^f(N-1)_j \tag{73}
$$

and the transition probability from the initial state to this state will be simply

$$
P_j \propto |\langle \Psi^i(N)| \, \mathbf{r} \, | \Psi^f(N)_j \rangle|^2
$$
$$
P_j \propto |\langle A\varphi_{nl}\psi_R(N-1)| \, \mathbf{r} \, | A\varphi_{ph} \Psi^f(N-1)_j \rangle|^2, \tag{73a}
$$

where $\mathbf{r}$ in this relationship represents an N-electron operator. If we assume that the one-electron matrix element $\langle \varphi_{nl} | \mathbf{r} | \varphi_{ph} \rangle$ controlling the primary photoelectron emission event changes little with changes in the photoelectron kinetic energy corresponding to different final states $\Psi^f(N-1)_j$, this transition probability reduces to

$$
P_j \propto |\langle \psi_R(N-1) \,|\, \Psi^f(N-1)_j \rangle|^2
$$
$$
P_j \propto \left| \left\langle \sum_{j'=0}^{\infty} \langle \Psi^f(N-1)_{j'} \,|\, \psi_R(N-1) \rangle \, \Psi^f(N-1)_{j'} \,|\, \Psi^f(N-1)_j \right\rangle \right|^2
$$
$$
P_j \propto |\langle \Psi^f(N-1)_j \,|\, \psi_R(N-1) \rangle|^2. \tag{74}
$$

The transition probability to a given final state $\Psi^f(N-1)_j$ is thus proportional to the square of the overlap of that state with $\psi_R(N-1)$ (i.e., the expansion coefficient C_j in Equation (72)).

The arguments used in reaching Equation (74) are completely analogous to those used in deriving the Franck-Condon factors for molecular vibrational excitations [8]. The Franck-Condon factors represent overlaps between an initial vibrational state and various final vibrational states. The initial and final Hamiltonians differ due to an electronic excitation that occurs on a time scale very short compared to nuclear motion; this is the analog of the sudden approximation.

Equation (74) has been used in some of the most recent theoretical calculations of the energy separations and relative intensities of the two-electron processes shown in Figure 23 [9]. The form of this equation thus implies very strict 'selection rules' between $\Psi^f(N-1)_j$ and the fictitious state $\psi_R(N-1)$ [9]. If $\Delta X = X_R - X'$, then

$$\Delta J = \Delta L = \Delta S = \Delta M_J = \Delta M_S = 0. \tag{75}$$

That is, the multi-electron processes are 'monopole' transitions relative to $\psi_R(N-1)$.

Some previous results for Ne$1s$ emission are summarized in Table I, where calculated two-electron peak separations and relative intensities are compared with experiment. The various final state configurations are noted, and for this case $\psi_R(N-1)$ corresponds to an *unrelaxed* Ne$^+$ $1s\,2s^2 2p^6\ ^2S$. There is reasonable agreement between theoretical and experimental separations, but the theoretical values are uniformly high by about 1.8 eV out of 40 eV. This has been explained as being due to a $2p$–$2p$ correlation error in the Hartree-Fock calculation for the one-electron $2p^6$ final state that is of much lower magnitude in the various $2p^5np$ two-electron final states. Theoretical and experimental relative intensities are also in fair agreement. It should also be noted in connection with this data that the various L–S multiplets formed as

TABLE I

Summary of data relevant to multi-electron transitions accompanying the formation of a $1s$ hole in neon by MgKα or AlKα X-rays.

(1) Shake-off transitions:			*Expt.*	*Theory* (Equation (80))
	Total intensity for shake-off of one electron from Ne $2s$ and $2p$ subshells (Ne $\rightarrow$ Ne^{2+})		16.5%[a]	16.1%[a]
	Total intensity for shake-off of two electrons from Ne $2s$ and $2p$ subshells (Ne$\rightarrow$Ne^{3+})		0.8%[b]	–

(2) Shake-up transitions:		Separation of 1-elec. and 2-elec. peaks (eV)		Relative intensities	
		Experiment[c]	Theory[d]	Experiment[c]	Theory[e]
Initial:	Ne $1s^2 2s^2 2p^6\ ^1S$	–	–	–	–
	$\uparrow$				
	$E_b^V(1s) = 870.2$ eV (expt)[d]				
	$\downarrow$				
1-elec. final:	Ne$^+$ $1s 2s^2 2p^6\ ^2S$	0	0	100%	100%
2-elec. final:	Ne$^+$ $1s 2s^2 2p^5 3p\ ^2S$ (LOW)	37.2	35.6	2.9%	2.3%
	Ne$^+$ $1s 2s^2 2p^5 3p\ ^2S$ (UP)	40.6	39.5	2.9%	2.9%
	Ne$^+$ $1s 2s^2 2p^5 4p\ ^2S$ (LOW)	47.2	40.5	1.7%	–
	Ne$^+$ $1s 2s^2 2p^5 5p\ ^2S$ (LOW)	44.3	42.4	0.6%	–
	Ne$^+$ $1s 2s^2 2p^5 4p\ ^2S$ (UP)	46.4	44.6	0.6%	–
	Partial sum of shake-up intensity:			8.7%	

[a] From [89].
[b] From [91].
[c] Average of experimental data from [9, 95].
[d] From [9].
[e] From [95].

final states must be considered. For example, the peaks indicated as '(LOW)' and '(UP)' in Table I and Figure 23 are due to a multiplet splitting of the type noted on the right-hand side of Figure 18 for the ^5P states of Mn^{3+}. In the case of Ne^+, ^2S states can be formed in two ways from the same total configuration $1s2s^22p^53p$: one in which the $1s$ electron is coupled with $2s^22p^53p$ ^1S and one in which it is coupled with $2s^22p^53p$ ^3S [9, 89]. Thus, there may be considerable interaction between multi-electron processes and multiplet splittings, and a complete specification of $\Psi^f(N-1)_j$ must include possible multiplet effects.

From the data of Table I for total shake-off and partial shake-up intensities, it is clear that $\sim 25\%$ of all events involve a multi-electron process. Thus, in terms of Equation (74), $|\langle \Psi^f(N-1)_0 | \psi_R(N-1)\rangle|^2 \approx 0.75$ or $\langle \Psi^f(N-1)|\psi_R(N-1)\rangle \approx 0.86$ to give a rough idea of the extent of final state relaxation. Note also that all final states in Table I satisfy the selection rules in Equation (75).

Using Equation (74), Manne and Åberg [32] have also pointed out that a weighted-average binding energy over all one-electron and multi-electron processes is precisely equal to a Koopmans' Theorem binding energy based on a one-electron energy from a Hartree-Fock calculation on the initial state (see also Section 2). This can be shown by noting that if $\psi_R(N-1)$ were a valid final state, it would correspond to a Koopmans' Theorem binding energy in which the passive orbitals have not relaxed. That is,

$$E_b^V(nl)^{KT} = \langle \psi_R(N-1)| \mathcal{H}^f(N-1)|\psi_R(N-1)\rangle - E^i. \tag{76}$$

Furthermore, by definition

$$E_b^V(nl) = \langle \Psi^f(N-1)_0| \mathcal{H}^f(N-1)|\Psi^f(N-1)_0\rangle - E^i \tag{77}$$

$$E_b^V(nl, n'l')_j = \langle \Psi^f(N-1)_j| \mathcal{H}^f(N-1)|\Psi^f(N-1)_j\rangle - E^i. \tag{78}$$

Using the expansion of $\psi_R(N-1)$ in terms of the $\Psi^f(N-1)_j$'s represented by Equation (72), we can show that Equation (76) is equivalent to

$$E_b^V(nl)^{KT} = |\langle \Psi^f(N-1)_0 | \psi_R(N-1)\rangle|^2 E_b^V(nl) +$$

$$+ \sum_{j=1}^{\infty} |\langle \Psi^f(N-1)_j | \psi_R(N-1)\rangle|^2 E_b^V(nl, n'l')_j. \tag{79}$$

This is simply a weighted average over the one-electron transition binding energy $E_b^V(nl)$ and the various two-electron transition binding energies $E_b^V(nl, n'l')_j$. The coefficients are just the transition probabilities or, in turn, the experimental peak heights corresponding to each $\Psi^f(N-1)_j$. The validity of Equation (79) has been checked by Manne and Åberg [32] by computing the average binding energy over an experimental Nels spectrum in which both the one-electron peak and two-electron satellites are apparent. The experimental average $1s$ binding energy is 886 eV, as compared to a Koopman's Theorem value of $E_b^V(1s)^{KT} = 892$ eV. The experimental one-electron binding energy is by comparison $E_b^V(1s) = 870$ eV.

It should thus be clear that the sudden approximation is only equivalent to the use of Koopmans' Theorem in an average sense. To say that the photoemission process is sudden in the sense of inequality (67) is not to say that the use of Koopmans' Theorem will give a binding energy very close to the one-electron-transition experimental value. However, the Koopmans' Theorem value does describe in an average way the totality of possible final states, including those involving multi-electron processes. That is, although the relaxation energy δE_{relax} lowers the binding energy corresponding to the one-electron transition relative to the Koopmans' Theorem value (cf. Equation (31)), the binding energies corresponding to the various multi-electron processes will be higher than the one-electron value and generally also higher than the Koopmans' Theorem value because of the extra energy involved in exciting the second, third, etc., electrons.

If a one-electron orbital approximation is used in describing $\Psi^f(N-1)_j$ and $\psi_R(N-1)$, they can be represented as single Slater determinants or sums of these determinants. If it is furthermore assumed that no relaxation of the $N-2$ passive orbitals occurs in the final state of a two-electron transition, than a one-electron analog of Equation (74) results [89],

$$P_{n''l''} = \sum_{\substack{j \\ (n''l'')}} P_j = |\langle \varphi_{n''l''} | \varphi_{n'l'} \rangle|^2, \tag{80}$$

where $P_{n''l''}$ represents the total probability of a two-electron transition of the type $nl \rightarrow$ photoelectron, $n'l' \rightarrow n''l''$, and the sum on j extends over all states $\Psi^f(N-1)_j$ corresponding to such excitations. Associated with Equation (80) is thus a one-electron selection rule of the monopole type

$$l'' - l' = \Delta l = 0. \tag{81}$$

All of the two-electron transitions presented in Table I satisfy this one-electron selection rule. Equations (80) and (81) have been used successfully in the analysis of much experimental data [88–95], although strictly speaking Equations (74) and (75) are more rigorously correct. Some of these one-electron calculations of relative intensities are presented in Table I. The agreement is reasonably good, especially as to the total probability for electron shake-*off* as measured by several experimental methods [89].

In general, theoretical calculations of the intensities of multi-electron processes in rare gases based on the sudden approximation have given reasonable agreement with experiment if the two electrons involved in the transition are in subshells with different principal quantum numbers (e.g., the $1s$ and $2p$ electrons in neon) [9, 89, 95]. However, if the two electrons have the same principal quantum number (e.g., the $2s$ and $2p$ electrons in neon), electron correlation effects appear to increase the experimental multi-electron intensities above those calculated theoretically using the simple sudden approximation models [89, 95].

In addition to the multi-electron processes observed in photoemission from noble gases, such transitions also appear to be present in X-ray photoelectron spectra from gaseous molecules [9, 95] and solids [62, 97, 98]. Spectra from N_2 [9, 95], O_2 [95], and CO_2 [95] exhibit such structure with 10 to 15% probability. Rather intense satellite peaks have also been reported for potassium in inorganic salts [62], for nickel and copper in oxides and halides [97] and for alkali metal and halogen atoms in the alkali halides [98]. Photoelectron spectra from three potassium halides are shown in Figure 24. The effects of inelastic scattering are indicated by the dotted curve. The broad peaks observed at approximately 14 eV below the narrow peaks due to one-electron photoemission of K3s electrons have been explained in terms of two-electron transitions connected with the photoemission of K3p electrons [2, 98]. Novakov [99] has also observed very strong satellite peaks in connection with photoemission from the 2p levels of 3d transition metal atoms in semiconductor compounds. Some of these results for CuS and Cu_2O are shown in Figure 25. Data such as this have been approximately correlated with free-ion energy levels [98] and also the unoccupied

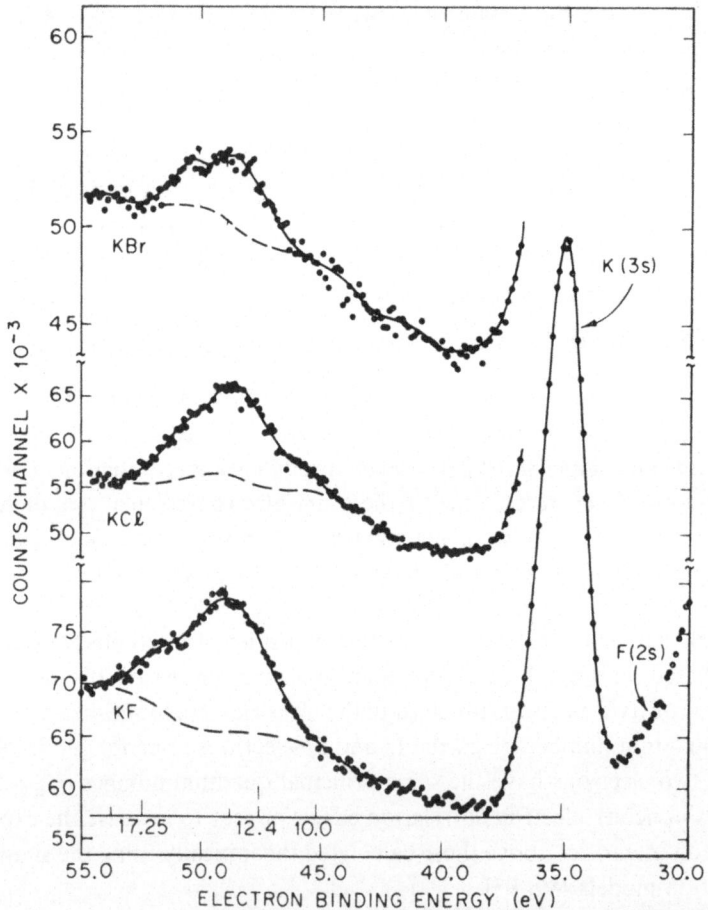

Fig. 24. XPS spectra from solid KBr, KCl, and KF in the region near the K3s peak. The satellites situated at binding energies of approximately 50 eV are thought to be due to shake-up processes. (From [98].)

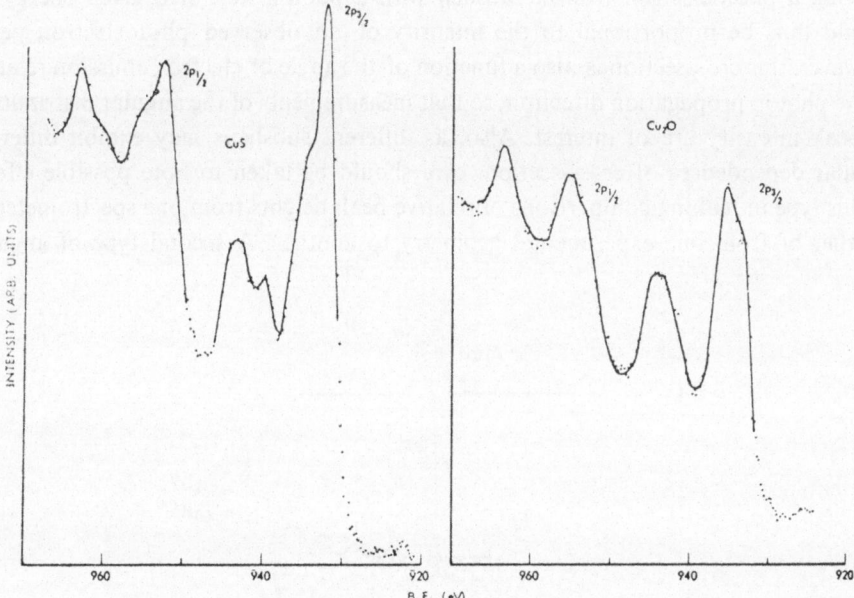

Fig. 25. XPS spectra from Cu2$p_{1/2}$ and Cu2$p_{3/2}$ electrons in solid CuS and Cu$_2$O. Note the intense satellites on the low kinetic energy side of each peak. (From [97]).

valence band structure as measured in optical absorption [99], although a definitive theoretical analysis awaits completion.

Although our discussion has been confined primarily to the observations and interpretations of two-electron transitions, it is entirely possible that three-electron transitions may occur [89, 92]. For example, in neon gas, Table I indicates that triple ionization has been observed in connection with the X-ray excitation of a N$e$ls photoelectron [89]. This triple ionization has a relative intensity of $\sim 1\%$ of all events and is attributed to two shake-off electrons from the 2s and 2p subshells [89]. A relative intensity of $\sim 1\%$ for triple ionization is also consistent with the known ratios of the $\alpha_{1,2}$ and $\alpha_{5,6}$ satellite X-rays of Mg, as can be seen in Figure 23.

Multi-electron effects analogous to those we have been discussing have been predicted to occur also in metals [56, 57]. The multi-electron excitations for a metallic system are more complex and may involve the excitation of plasmons [56]. No definite experimental verifications of such effects have as yet been made, however.

It is thus clear that multi-electron transitions represent an important consideration in the interpretation of any X-ray photoelectron spectrum. Where a detailed analysis of these transitions is possible, they may yield significant information about electronic structure, particularly as to the excited states of valence electrons.

5. Relative Intensities and Angular Distributions

The photoelectric cross section represents the transition probability per unit time for exciting a photoelectron from a subshell with a photon flux of a given energy. It should thus be proportional to the intensity of an observed photoelectron peak. However, this crosssection is also a function of the angle of electron emission relative to the photon propagation direction, so that measurements of the angular distribution of peak intensity are of interest. Also, as different subshells may exhibit different angular dependences of cross-section, care should be taken to note possible effects of this type in making comparisons of relative peak heights from one spectrometer to another or from one experimental geometry to another. A second type of angular

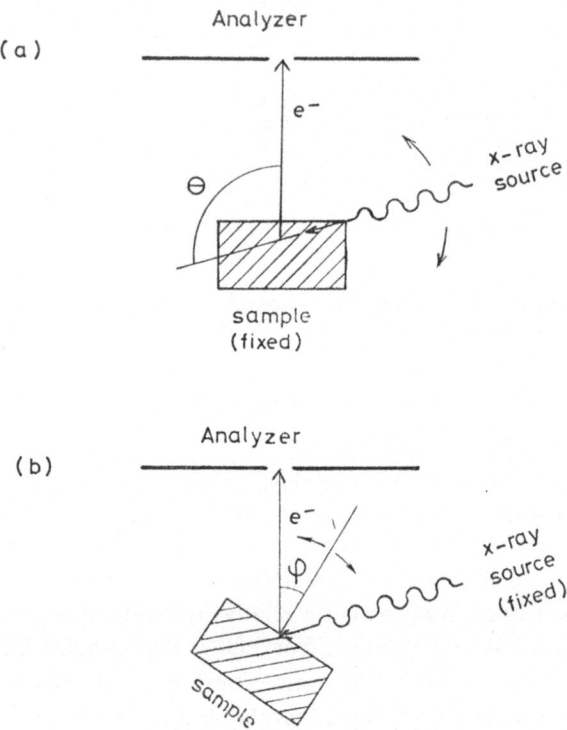

Fig. 26.   Schematic diagram showing the two types of angular distribution measurement. In (a), the angle θ between the incident X-ray and the electron emission direction is varied with the sample position fixed. In (b), a solid sample is rotated relative to a fixed analyzer slit and X-ray source (fixed θ), thereby varying the angle φ.

distribution measurement is also permitted if the sample has a preferred axis of orientation, as, for example, the axis perpendicular to the planar surface of a poly-crystalline solid sample or the crystal axes of a single-crystal solid sample. The experimental arrangements for these two types of angular distribution measurements are indicated in Figure 26. We shall first discuss the basic aspects of atomic and mole-cular photoelectric cross sections and their angular dependence, and then review angular distribution measurements from solids.

A. FREE-ATOMS AND MOLECULES

The basic theory of photon absorption in the one-electron approximation has been reviewed by Bethe and Salpeter [100]. Bates [101] and Cooper [102] have discussed the application of this theory to many-electron atoms. Fano and Cooper [103] have reviewed the experimental and theoretical data bearing on atomic photo-absorption, including a consideration of various multi-electron effects. More recently Cooper and Manson [104] have discussed the one-electron theory of angular distributions in photoemission from atoms. Cooper and Zare [105, 106] and Grimm [107] have reviewed the theory of angular distributions for both atoms and molecules. We shall first review only the one-electron theory for free-atoms, indicating its basic assumptions and predictions.

Several simplifying assumptions are usually made in calculating photoelectric sections in a non-relativistic approximation [100, 102]:

(1) The effect of the photon is treated as a perturbation.

(2) It is assumed that the photon wavelength λ is much larger than the average diameter of the subshell from which emission occurs. This is a reasonably good, although borderline, assumption for AlKα or MgKα X-rays with $\lambda \approx 10$ Å. Such an assumption is termed a neglect of retardation effects and if it is valid, the photon interaction connecting the initial and final states can be described in terms of a dipole operator $\mathbf{r}$. Alternatively, a momentum operator $\mathbf{p} = i\hbar\nabla$ or an operator ∇/V can be used to describe this interaction, provided that appropriate factors involving the energy $h\nu$ are included in the calculation of matrix elements (cf. for example, the discussion of Section 3). The equivalence of the matrix elements of these three operators can be shown by noting the form of the Hamiltonian (Equation (18)), and using simple commutation relations between $\mathcal{H}$, $\mathbf{p}$, and $\mathbf{r}$ [100].

(3) Also, it is assumed that the wave functions of the initial and final states can be described in terms of anti-symmetric products of one-electron orbitals φ_{nl} and φ_{ph} with factors representing the $N-1$ passive orbitals, as shown in Equations (68) and (73). A further approximation is to represent the initial and final wave functions as single Slater determinants or linear combinations of Slater determinants, which automatically satisfy Equations (68) and (73). In either case, it is thus possible to speak of the primary effect of photoelectron emission as the excitation of an electron from the orbital φ_{nl} to a continuum photoelectron orbital φ_{ph}. Both one-electron and multi-electron transitions can in principle be described at this level of approximation.

(4) A further assumption that is often made is that only one electron changes its orbital during photoelectron emission, with all of the passive electron orbitals remaining unchanged or unrelaxed. This is identical to the assumption necessary for deriving Koopmans' Theorem, and of course deals with what appears to be a purely one-electron transition. However, it has been pointed out [102] that such an assumption yields cross sections that represent an average behavior over all final states resulting from nl emission, in much the same way that Equation (79) predicts that $E_b^V(nl)^{KT}$ represents a weighted average binding energy over all possible final states.

(5) It is also assumed that Russell-Saunders $(L-S)$ coupling applies to both initial and final states, and furthermore, that all of the initial state magnetic levels specified by M_L and M_S are equally populated [108].

Subject to the first two assumptions above, the probability per unit time for a transition from an initial state $\Psi^i(N)$ to a final state $\Psi^f(N)_j$ including the orbital φ_{ph} is proportional to the square of a dipole matrix element between them, as stated in Equation (73a). The photoelectric cross section is proportional to the transition probability and is given by

$$\sigma_{nl}(hv) = \sigma_{nl}(E_{kin}) = \frac{4\pi\alpha a_0^2}{3}(hv)\,|\langle\Psi^i(N)|\,\mathbf{r}\,|\Psi^f(N)\rangle|^2 \tag{82}$$

where the dependence of σ_{nl} on hv can be expressed either as $\sigma_{nl}(hv)$ or $\sigma_{nl}(E_{kin}= =hv-E_b(nl))$, α is the fine structure constant, a_0 is the Bohr radius and the operator $\mathbf{r}$ is an N-electron operator

$$\mathbf{r} = \sum_{i=1}^{N} \mathbf{r}_i. \tag{83}$$

If both $\Psi^i(N)$ and $\Psi^f(N)_j$ are written as anti-symmetric products of the forms given in Equations (68) and (73), Equation (82) further simplifies to

$$\sigma_{nl}(E_{kin}) = \frac{4\pi\alpha a_0^2}{3}(hv)\,|\langle\varphi_{nl}|\,\mathbf{r}\,|\varphi_{ph}\rangle\,\langle\psi_R(N-1)\,|\,\Psi^f(N-1)_j\rangle|^2 \tag{84}$$

where $\psi_R(N-1)$ is the $N-1$ electron remainder of the initial state wave function and $\Psi^f(N-1)_j$ is the final state ion core wave function (cf. Equations (73a) and (74)). The cross section is thus proportional to the square of a product of a one-electron dipole matrix element between φ_{nl} and φ_{ph} and an overlap integral between $\psi_R(N-1)$ and $\Psi^f(N-1)_j$. The operator $\mathbf{r}$ in Equation (84) is a one-electron operator such as we have used everywhere in this discussion except in Equations (73a), (82) and (83). The properties of the dipole operator, including its spin-independent character, dictate the following selection rules on the angular momentum associated with the photo-electron and on the changes in the total spin and orbital angular momenta between $\Psi^i(N)$ and $\Psi^f(N-1)_j$ [101, 102]:

$$l_{ph} = l \pm 1 \tag{52}$$
$$\Delta S = S' - S = \pm \tfrac{1}{2} \tag{53}$$
$$\Delta L = L' - L = 0, \pm 1, \pm 2, \dots \pm l. \tag{54}$$

If $\psi_R(N-1)$ and $\Psi^f(N-1)_j$ do not have the same symmetry (for atoms, the same S and L), the overlap factor in Equation (84) will be zero due to orthogonality. This is the origin of the previously stated selection rule that the passive electron coupling must be the same in the initial and final states (that is, that $S_R = S'$ and $L_R = L'$).

If both $\Psi^i(N)$ and $\Psi^f(N)$ are further assumed to be single Slater determinants,

Equation (84) simplifies in first order to [101, 102]:

$$\sigma_{nl}(E_{kin}) = \frac{4\pi\alpha a_0^2}{3}(h\nu)\,|\langle\varphi_{nl}|\,\mathbf{r}\,|\varphi_{ph}\rangle|^2 \cdot \prod_{i\neq nl}|\langle\varphi_i\,|\,\varphi_i'\rangle|^2,\qquad(85)$$

where the $N-1$ electron overlap becomes a product of overlaps between initial and final one-electron orbitals for all of the passive electrons. This equation can be related to Equation (80) provided that all passive orbitals except $n'l' \to n''l''$ are assumed to be unchanged in the final state.

If the fourth basic assumption of no final state relaxation is now made, the overlap integrals all become unity and

$$\sigma_{nl}(E_{kin}) = \frac{4\pi\alpha a_0^2}{3}(h\nu)\,|\langle\varphi_{nl}|\,\mathbf{r}\,|\varphi_{ph}\rangle|^2 =$$

$$= \frac{4\pi\alpha a_0^2}{3}(h\nu)\left|\int \varphi_{nl}^*\mathbf{r}\varphi_{ph}\,d\tau\right|^2.\qquad(86)$$

It is in this simple one-electron form that most theoretical calculations are at present made.

It is common to sum over the possible final state orbitals φ_{ph} which can be reached by excitation of a given φ_{nl} and also to average over the various orbitals φ_{nl} in a given subshell [100, 102]. Such summing and averaging is done over the two possible values of photoelectron angular momentum ($l+1$ and $l-1$), as well as over various values of the initial and final magnetic quantum numbers m and M. The specification of m and M has been eliminated from Equations (84), (85), and (86) for simplicity, but is implied. This summing and averaging yields a total nl subshell cross section of the form

$$\bar{\sigma}_{nl}(E_{kin}) = \frac{4\pi\alpha a_0^2}{3}(h\nu)\left[lR_{E_{kin},\,l-1}^2 + (l+1)R_{E_{kin},\,l+1}^2\right]$$

or equivalently,

$$\bar{\sigma}_{nl}(E_{kin}) = \frac{4\pi\alpha a_0^2}{3}\left[E_{kin} + E_b^V(nl)\right]\left[lR_{E_{kin},\,l-1}^2 + (l+1)R_{E_{kin},\,l+1}^2\right]\quad(87)$$

where the $R_{E_{kin},\,l\pm 1}$ are radial integrals common to all the one-electron dipole matrix elements between φ_{nl} and φ_{ph}, and the factors l and $l+1$ result from the summing and averaging. The radial integrals are given by

$$R_{E_{kin},\,l\pm 1} = \int_0^\infty P_{nl}(r)\,rP_{E_{kin},\,l\pm 1}(r)\,dr,\qquad(88)$$

where $P_{nl}(r)/r = R_{nl}(r)$ is the radial part of the φ_{nl} orbital and $P_{E_{kin},\,l\pm 1}(r)/r$ is the radial part of the continuum φ_{ph} orbital.

It should be noted that for an open shell system, the total subshell cross section calculation outlined above will represent a sum of cross sections for the various allowed final multiplet states [101]. For emission from a closed inner shell, the weight of each of these multiplets is just its total multiplicity, as shown in Equation (55). Thus, for example, the simplest interpretation of the 3s cross section in Mn^{2+} would separate it into parts such that $\frac{5}{12}$ represents a ^5S final state and $\frac{7}{12}$ represents a ^7S final state. For a closed shell system, no such multiplets arise.

Total atomic subshell cross sections for photon energies relevant to XPS have been calculated by Cooper [102], Bearden [109], Rakavy and Ron [110], Brysk and Zerby [111], Manson and Cooper [112], Cooper and Manson [104], Manson and Krause [113] and Kennedy and Manson [114]. These calculations have made use of both the non-relativistic theory outlined above, and also relativistic methods based on the Dirac equation [110, 111]. Comparisons with experiment are often made through the total atomic absorption coefficient for X-rays, which at lower X-ray energies consists primarily of a sum over the several subshell cross sections. Such comparisons yield reasonably good agreement between experiment and theory (~ 5–10%) except near threshold where $hv \approx E_b^V(nl)$ [109, 111]. Cooper and Manson [104] have also calculated relative subshell cross sections which compare favorably with the experimental values of Krause [115].

In general, it is found that for hv well above threshold, as is the case in XPS measurements, transitions to $l_{ph} = l + 1$ are much more probable than those to $l_{ph} = l - 1$ [102, 112]. Thus, the term $(l + 1) \cdot R_{E_{kin}, l+1}$ dominates the term $l R_{E_{kin}, l-1}$ in Equation (87). Also, $\bar{\sigma}_{nl}(E_{kin})$ is generally a decreasing function of E_{kin} for hv well above threshold. However, large oscillations and zeroes in the cross section may occur as hv is increased above threshold [102, 104, 112]. Such oscillations can be explained in terms of the changing overlap character of an oscillatory $P_{nl}(r)$ and an oscillatory $P_{E_{kin}, l \pm 1}(r)$ with changing E_{kin} [102]. As E_{kin} is increased, the effective wavelength of the oscillations in $P_{E_{kin}, l \pm 1}$ decreases and the oscillations penetrate more deeply into the region of non-zero $P_{nl}(r)$ 'within' the atom. The matrix element $R_{E_{kin}, l \pm 1}$ may thus consist of contributions due to the constructive overlap of one or more lobes in $P_{nl}(r)$ and $P_{E_{kin}, l \pm 1}$. If, as E_{kin} is varied, the relative signs of the overlapping lobes change, $R_{E_{kin}, l \pm 1}$ must change sign, and therefore at some kinetic energy intermediate between the sign change, a zero in $R_{E_{kin}, l \pm 1}$ and $\bar{\sigma}_{nl}(E_{kin})$ can result. A corollary of this argument is that atomic orbitals $P_{nl}(r)$ which exhibit no oscillations with r should show cross sections which decrease smoothly with E_{kin} and exhibit no zeroes [102]. Examples of such orbitals would be 1s, 2p, 3d, and 4f.

A further important property of $\bar{\sigma}_{nl}(E_{kin})$ is that it may be much different for different subshells at the same excitation energy hv or for the same subshell at different excitation energies hv. Such effects have already proven very useful in studying the valence electron states in gases [46] and solids [60, 84, 116]. Molecular orbitals with 2s character are found to be emphasized in XPS studies, whereas orbitals with 2p character are more pronounced in UPS studies [46]. Similarly, the 4f electrons in rare earth metals are seen much more clearly in XPS spectra than in UPS spectra as is shown in Figure 27 [43]. The intimate connection of the spatial distribution of an orbital to its cross section has led to the suggestion that photoelectron spectroscopic studies at different photon energies could be utilized as a very direct probe of orbital character [47].

The total subshell cross-section $\bar{\sigma}_{nl}(E_{kin})$ represents an integration over all electron emission angles relative to the photon propagation direction. The detailed dependence of photoelectron emission on angle can be expressed in terms of a differential cross-section, $d\bar{\sigma}_{nl}(E_{kin})/d\Omega$. If retardation is again neglected and the exciting radiation is assumed to be randomly polarized, this differential cross-section is given by [104]

$$\frac{d\bar{\sigma}_{nl}(E_{kin})}{d\Omega} = [\bar{\sigma}_{nl}(E_{kin})/4\pi] \cdot [1 - \tfrac{1}{2}\beta(E_{kin})\,P_2(\cos\theta)]$$

$$= [\bar{\sigma}_{nl}(E_{kin})/4\pi] \cdot [1 + \tfrac{1}{2}\beta(E_{kin})\,(\tfrac{3}{2}\sin^2\theta - 1)], \qquad (89)$$

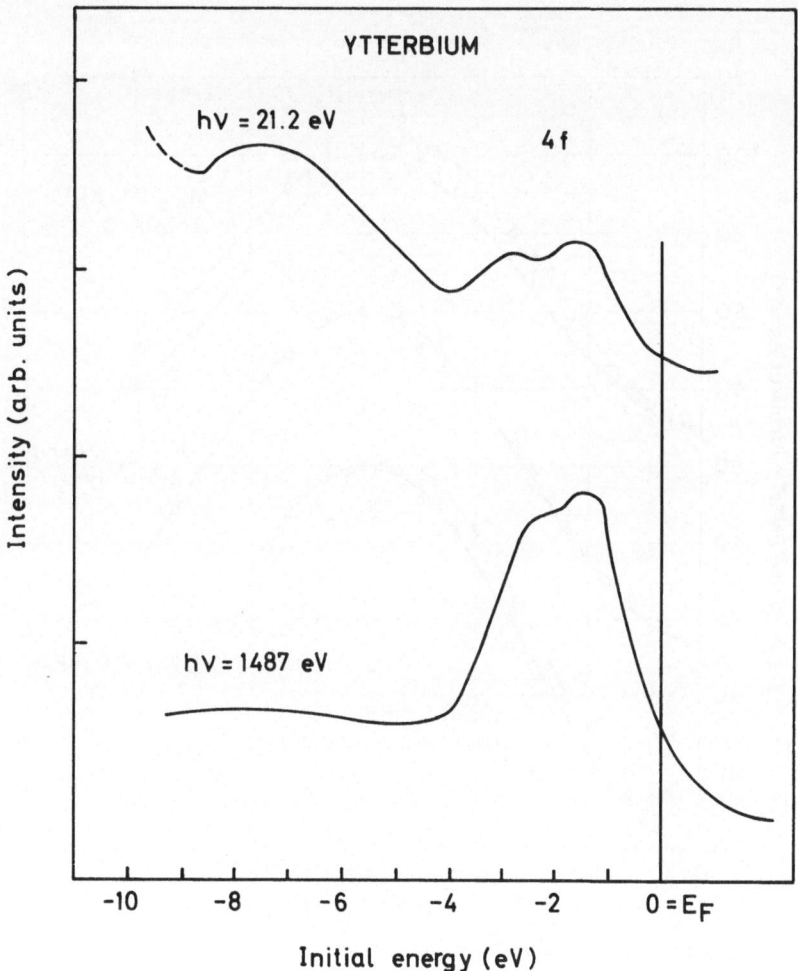

Fig. 27. UPS and XPS spectra from Yb metal illustrating the increased relative sensitivity of the XPS spectrum to the 4f states. (From [116].)

where $\beta(E_{kin})$ is termed the asymmetry parameter, θ is the angle between the photon propagation direction and the photoelectron emission direction (cf. Figure 26), and $P_2(\cos\theta) = \frac{1}{2}(3\cos^2\theta - 1)$. $\beta(E_{kin})$ can be expressed in terms of the radial integrals $R_{E_{kin}, l\pm 1}$ and certain continuum orbital phase shifts $\delta_{l\pm 1}(E_{kin})$ giving the relative phase of the sinusoidal oscillations occurring in $P_{E_{kin}, l\pm 1}(r)$ for very large values of r. This expression is:

$$\beta(E_{kin}) = \frac{\{l(l-1)R^2_{E_{kin}, l-1} + (l+1)(l+2)R^2_{E_{kin}, l+1} \times}{(2l+1)\left[lR^2_{E_{kin}, l-1} + (l+1)R^2_{E_{kin}, l+1}\right]}$$
$$\frac{-6l(l+1)R_{E_{kin}, l+1}R_{E_{kin}, l-1}\cos\left[\delta_{l+1}(E_{kin}) - \delta_{l-1}(E_{kin})\right]\}}{}$$

$$(90)$$

and the term in $\cos\left[\delta_{l+1} - \delta_{l-1}\right]$ represents an interference between outgoing $l+1$ and $l-1$ waves. The range of β is $-1 \leqslant \beta \leqslant +2$. A positive value of β indicates that

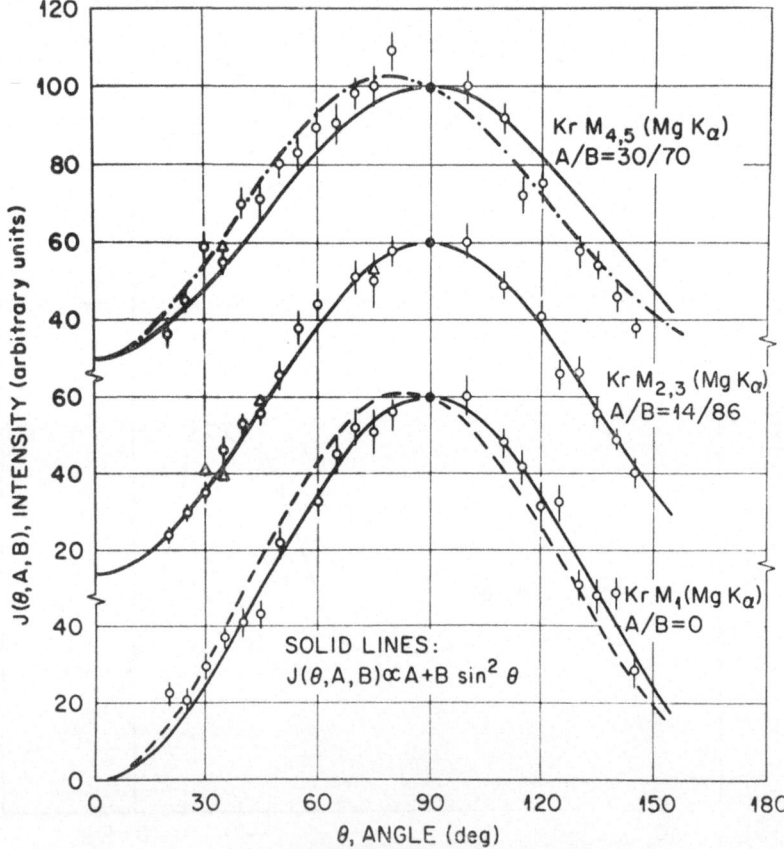

Fig. 28. Experimental angular distributions of $3s$ ($=M_1$), $3p$ ($=M_{2,3}$), and $3d$ ($=M_{4,5}$) photoelectrons excited from gaseous Kr with MgKα X-rays. The curves represent least-squares fits to the data points of a relationship of the form of Equation (93), in which A and B were treated as empirical constants. (From [115].)

photoelectrons are preferentially emitted at angles perpendicular to the photon direction ($\theta=90°$), whereas a negative value indicates preferential emission either parallel or anti-parallel to this direction ($\theta=0°$, $180°$). A value of $\beta=0$ yields an isotropic distribution. For s-electron emission $l=0$, and only transitions to $l=1$ waves are possible. $\beta=+2$ for this case and

$$\frac{d\bar{\sigma}_{ns}(E_{kin})}{d\Omega} = \frac{\bar{\sigma}_{ns}(E_{kin})}{4\pi} \sin^2\theta, \tag{91}$$

with maximum intensity at $\theta=90°$ and zero intensity at $\theta=0°$, $180°$. For the other limiting case of $\beta=-1$,

$$\frac{d\bar{\sigma}_{nl}(E_{kin})}{d\Omega} = \frac{\bar{\sigma}_{nl}(E_{kin})}{4\pi} \cos^2\theta, \tag{92}$$

the photoelectron intensity is zero at $\theta=90°$, and has its maximum value at $\theta=0°$, $180°$. No matter what the value of β is, the form of Equation (89) dictates that the distribution should be symmetric about a perpendicular to the photon propagation direction ($\theta=90°$).

Equation (89) is equivalent to

$$\frac{d\bar{\sigma}_{nl}(E_{kin})}{d\Omega} = A + B \sin^2\theta, \tag{93}$$

where A and B are constants, provided that $A=(\bar{\sigma}_{nl}/4\pi)(1-\beta/2)$ and $B=(\bar{\sigma}_{nl}/4\pi) \cdot 3\beta/4$. From an empirical determination of A and B, β can thus be determined from $\beta=4B/(3A+2B)$. A comparison between the function predicted by Equation (93) and experimental results is shown in Figure 28. The parameters A and B have in this case been empirically adjusted to give the best fit to data obtained for photoemission from Kr3s, Kr3p, and Kr3d levels with MgKα X-rays. The data are reasonably well described by Equation (93). Note that the 3s data are consistent with Equation (91) as expected. Also, a decrease in β with increasing orbital angular momentum is observed, although β is clearly positive for all three cases presented in Figure 28. Calculated values of β are also in reasonable agreement with experiment ($\sim \pm 5\%$) [104]. The small asymmetry of the Kr3d experimental curve about $\theta=90°$ involving a slight displacement toward smaller angles is believed to be due to retardation effects [104, 115]. Manson [117] and Kennedy and Manson [114] have also pointed out that for certain subshells, theory predicts that $\beta(E_{kin})$ may exhibit large oscillations with E_{kin}.

A final point made clear by data such as that shown in Figure 28 is that in order for comparisons of peak intensities in photoelectron spectra to be meaningful, the angular geometry of the experiment must be known and allowed for.

For molecules, equations identical in form to 82, 84, 85, 86, 87, 89, and 90 obtain for the cross section, total cross section and angular distribution due to unpolarized radiation [105, 106, 118]. The only modifications or additional assumptions necessary in deriving these equations for molecules are as follows: (1) Symmetry designations appropriate to the molecular geometry are used in describing initial and final states, as well as valence, and perhaps core, one-electron orbitals. (2) The Born-Oppenheimer approximation is assumed to be valid, so that electronic motion can be treated separately from vibrational or rotational motion. (3) In deriving the angular distribution, it is assumed that the molecules are randomly oriented in space, such as would be the case in gas-phase measurements.

It is clear that even if the initial electronic state of a molecule is assumed to be associated with a single set of vibrational and rotational quantum numbers, each final electronic state may be associated with various vibrational or rotational excitations in the molecular ion. These vibrational excitations are responsible for the bands observed in UPS spectra of molecules, for example [8]. Within the Born-Oppenheimer approximation, the cross section leading to a given final electronic state represents the total probability for all final vibrational and rotational states associated with that state. The total intensity represented by the cross section will be distributed among various final vibrational states according to Franck-Condon factors, for example.

Cox and Orchard [77] have derived the relative probabilities of reaching different final electronic states for emission from both filled and unfilled subshells. A specialization of their results to filled-subshell emission from atoms yields Equation (55).

Although little experimental or theoretical work has been performed on XPS angular distributions from molecules, Carlson et $al.$ have performed such UPS measurements on a series of 27 atoms and small molecules [119]. The results obtained are in agreement with an $A + B \sin^2 \theta$ expression, and the β values for different valence

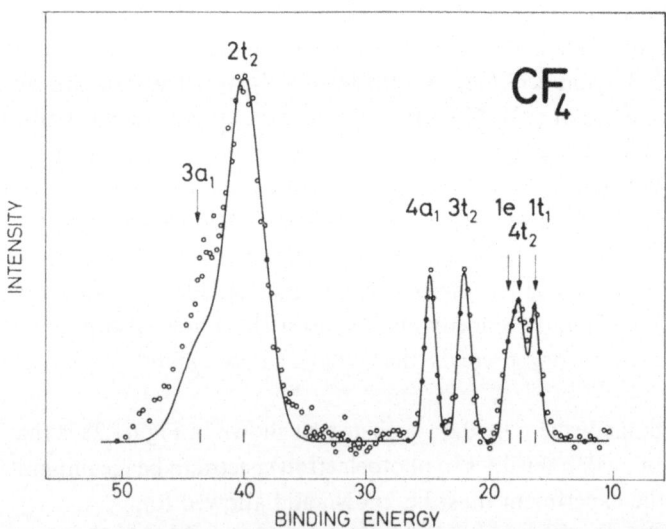

Fig. 29. Experimental and theoretical XPS spectra for the molecular orbitals of CF_4. The theoretical calculations made use of experimentally determined relative atomic subshell cross sections and Equation (98). (From [49].)

photoelectron bands have been used to characterize the electronic states giving rise to them. Also, some evidence was found for a partial breakdown of the Born-Oppenheimer approximation by studying the angular distribution of different vibrational peaks within the same band [119]. If this approximation is valid, each vibrational peak within a band should have the same angular distribution (same β value); for a few of the bands studied, this does not seem to be the case.

Gelius [49] has recently discussed a procedure for relating relative *atomic* valence subshell cross sections to the relative cross sections of molecular orbitals. In applications to several small molecules, theoretical calculations based on this model gave very good agreement with experiment, and in certain cases theoretical ambiguities as to orbital assignments could be resolved by comparing the different possible assignments with experiment. An example of such results is shown in Figure 29 for CF_4. The rationalization of this procedure in its simplest form involves only one assumption beyond those already indicated in discussing atoms and molecules: namely, that the photoelectron state φ_{ph} can be represented by a plane wave $e^{i\mathbf{k}\cdot\mathbf{r}}$ where $\mathbf{k}$ is the wave vector of the photoelectron and $\mathbf{r}$ is its position vector. If a given one-electron molecular orbital is denoted by ϕ_j, than the cross section for a transition from this orbital to φ_{ph} can be shown to be proportional to the square of the overlap of ϕ_j and φ_{ph},

$$\sigma_j \propto |\langle \phi_j | e^{i\mathbf{k}\cdot\mathbf{r}} \rangle|^2. \tag{94}$$

The derivation of this equation involves using the momentum-operator form of the photon interaction. The short de Broglie wavelengths of the valence photoelectron states involved in XPS (~ 0.35 Å) mean that the overlap of the oscillatory plane wave function with any molecular orbital will tend to average to zero unless the orbital is changing rapidly in comparisons with distances of the order of 0.35 Å. Thus, the major contributions to σ_j will come from regions very near the nuclei in which rapid oscillations of ϕ_j can occur. The total cross section for the molecular orbital can thus be written as a sum of terms localized on each atom in the molecule,

$$\sigma_j = \sum_A \sigma_{Aj} \tag{95}$$

where the summation runs over all atoms A in the molecule and σ_{Aj} represents the contribution of atomic orbitals localized on atom A to the total cross-section for the molecular orbital ϕ_j. ϕ_j is next assumed to be expressable in an LCAO format:

$$\phi_j = \sum_{A,\lambda} C_{A\lambda j} \varphi_{A\lambda} \tag{96}$$

where $\varphi_{A\lambda}$ is an atomic orbital on atom A with symmetry $\lambda = nl$. It is then possible to express the σ_{Aj}'s in terms of gross atomic populations $P_{A\lambda j}$ derived from the coefficients $C_{A\lambda j}$ and total atomic subshell cross sections $\sigma_{A\lambda}$ as

$$\sigma_{Aj} = \sum_\lambda P_{A\lambda j} \sigma_{A\lambda}. \tag{97}$$

Combining Equations (95) and (97) yields the total molecular orbital cross section in terms of total atomic subshell cross sections

$$\sigma_j^{(MO)} = \sum_{A,\lambda} P_{A\lambda j} \sigma_{A\lambda}^{(AO)}. \tag{98}$$

where we have specifically indicated those cross sections related to the molecular orbital and to free atoms. Such a theoretical model was utilized together with Hartree-Fock estimates of gross atomic populations to derive the theoretical curve in Figure 29. The relative atomic subshell cross-sections were determined empirically. The extension of such analyses to other systems and to more exact theoretical descriptions has been discussed by Gelius [49].

B. SOLIDS

In solids the fundamental photoemission process is modified in several respects. In a bulk model of photoemission from a crystalline solid, both the final state φ_{ph} and the initial state φ_k must be described in terms of Bloch waves. Also, the periodic potential in a crystalline solid can introduce strong directionality effects on the photoelectron angular distribution. Such effects might be described in terms of either electron diffraction or channeling. In addition, the highly probable inelastic scattering events in solids lead to rather short escape depths for photoelectrons. These short escape depths in turn appear to give rise to distinctly different behavior in the angular distributions of photoelectrons from bulk and surface atoms.

Electron diffraction or channeling effects appear to have been observed in X-ray photoelectron spectra from both sodium chloride single crystals [4] and gold single crystals [5, 6]. These effects are manifested as oscillations in the photoelectron intensity as the electron emission direction is varied relative to some crystalline axis (cf. Figure 26b). No such oscillations are found in randomly oriented polycrystalline specimens. The angular distributions of Au4f electrons which have escaped with minimal inelastic scattering from a gold single crystal are shown in the upper portion of Figure 30, along with the experimental geometry [5, 6]. In this experiment, the single crystal was rotated about an axis perpendicular to the $\langle 111 \rangle$ direction. The pronounced peaks observed in these Au4f angular distributions are believed to be associated with electron diffraction effects, although no detailed theoretical analysis of such data has as yet been performed. In support of this hypothesis is the fact that each peak in Figure 30 is separated from the $\langle 111 \rangle$ direction by an amount very nearly equal to an interplanar angle between the (111) plane and a low index plane in the fcc gold crystal [5, 6]. The low index planes corresponding to each separation are labelled. More detailed studies of such gold crystals show a complex pattern of peaks that changes with the rotation axis selected but which has certain symmetries consistent with the known symmetries of the crystal [6]. Thus, it appears reasonable to think of such angular distribution patterns as analogous to those observed in low energy electron diffraction (LEED) or electron microscopy. The two types of distributions are generated by somewhat different processes, however, in that the scattering electrons in

XPS (or UPS) are generated by a dipole transition from a state with definite spatial character, whereas in LEED or electron diffraction, the scattering electrons are generated by an external beam.

Theoretical treatments of these closely related diffraction or channeling processes indicate that the appearance of the angular distribution should depend on both the

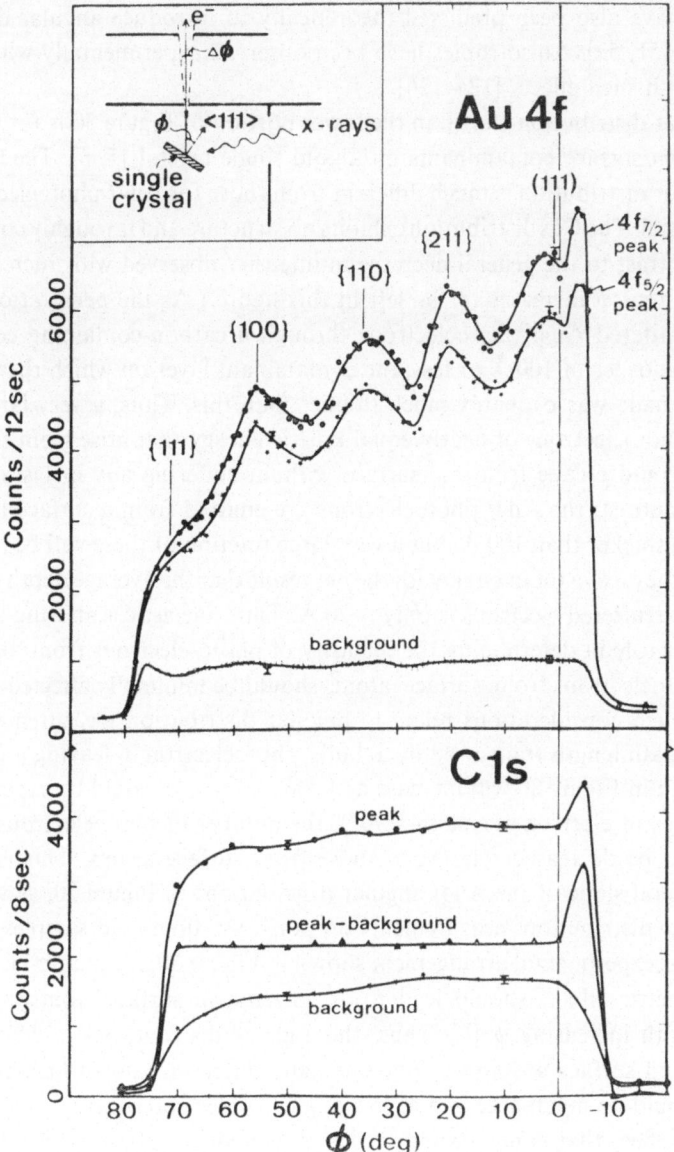

Fig. 30. Angular distributions of Au4*f* and C1s photoelectrons expelled from a gold single crystal by Mg Kα radiation. The interplanar angles between (111) and several low index planes are also indicated above the Au4*f* distribution. The C1s photoelectrons arose from surface contaminants. (From [5].)

spatial distribution and the de Broglie wavelength of the continuum φ_{ph} [120–123]. In XPS or UPS, we might also expect the spatial character of φ_{ph} to depend on that of the initial state φ_k. In fact, some slight differences have been observed between the angular distributions of core and valence photoelectrons with essentially the same de Broglie wavelengths, but differing initial state spatial distributions [4–6]. More detailed theoretical and experimental analyses of such effects are called for.

In connection with UPS experiments, the effects of both crystalline order and the surface have also been predicted theoretically to introduce angular distribution anisotropics [51, 52]. Anisotropics have been observed experimentally which may be connected with such effects [124–126].

The angular distribution shown in the lower portion of Figure 30 is for C1s photoelectrons from surface contaminants on a gold single crystal [5, 6]. The appearance of this angular distribution is much different from those for Au4f photoelectrons from the same crystal. The C1s distribution exhibits no structure and is roughly constant with angle, in contrast to the general decrease in intensity observed with increasing ϕ for Au4f. (Note that ϕ increases to the left in this figure.) As the penetration depth of elastically scattered Au4f photoelectrons through a carbon-containing contaminant layer is of the order of 100 Å or less, the contaminant layer on which these measurements were made was probably much thinner than this. Thus, a very large fraction of the C1s photoelectrons of nearly equal kinetic energy that arise from the contaminant layer could escape from the surface without suffering any inelastic scattering events. By contrast, the Au4f photoelectrons are emitted from a surface layer of the crystal much thicker than 100 Å, but a very large fraction of these will be inelastically scattered as they leave the crystal, with the net result that the average depth of emission of elastically scattered electrons is only ~ 20 Å. Thus, inelastic scattering should play an important role in determining the intensity of photo-electrons from 'bulk' atoms, whereas photoelectrons from 'surface' atoms should be minimally affected by inelastic scattering. Such considerations relate to angular distribution measurements in that the average path length tranversed by a 'bulk' photoelectron in leaving a crystal such as that shown in Figure 30 will increase as $1/\cos\phi$. Thus, it would be expected that at grazing angles of electron escape ($\phi \approx 90°$), the number of photoelectrons from bulk atoms would be decreased relative to those from surface atoms. This is consistent with the general slope of the Au4f angular distributions in Figure 30, as well as with Au4f angular distribution measurements on polycrystalline gold samples [5, 6]. For the particular experimental arrangement shown in Figure 30, it can also be shown that simple geometric effects should lead to an *increase* in surface-atom photoelectron intensities with increasing ϕ [6]. Thus, the angular distributions of photoelectrons from bulk and surface atoms are fundamentally different, and surface-atom photoemission should be much accented at low angles of electron escape [5, 6].

A further effect that is manifested in the angular distributions of Figure 30 is the onset of total X-ray reflection at grazing incidence [6, 127]. This effect, together with the resulting concentration of X-ray energy very close to the surface of the specimen gives rise to the intensity peaks observed at $\phi \approx -4°$ for both Au4f and C1s [6, 127].

In summary, measurements of photoelectron peak intensities as a function of incident photon energy, photoelectron emission angle relative to the photon propagation direction, or photoelectron emission angle relative to the axes of a crystalline sample, appear capable of yielding considerable information about both the detailed dynamics of the photoemission process and the spatial distributions of the wave functions involved. An added benefit from angular distributions measurements on solids is that it may be possible to semi-quantitatively distinguish surface and bulk atoms.

Acknowledgments

The author expresses his gratitude to Prof. Stig Hagström and the Physics Department, Linköping University, Linköping, Sweden, for support during the preparation of these notes.

References

1. Fadley, C. S., Lawrence Radiation Laboratory Report, URCL-19535 (1970) (thesis).
2. Hagström, S. B. M. and Fadley, C. S., in *X-ray Spectroscopy* (ed. by L. V. Azaroff), McGraw-Hill Publ. Co., New York, 1972, in press.
3. Busch, G., Campagna, M., and Siegmann, H. Ch., in *Electron Spectroscopy* (ed. by D. A. Shirley), North-Holland Publ. Co., Amsterdam, 1972, p. 827, plus references therein.
4. Siegbahn, K., Gelius, U., Siegbahn, H., and Olsen, E., *Phys. Letters* **32A**, 221 (1970).
5. Fadley, C. S. and Bergström, S. Å. L., *Phys. Letters* **35A**, 375 (1971).
6. Fadley, C. S. and Bergström, S. Å. L., in *Electron Spectroscopy* (ed. by D. A. Shirley), North-Holland Publ. Co., Amsterdam, 1972, p. 233.
7. Siegbahn, K., Nordling, C., Fahlman, A., Nordberg, R., Hamrin, K., Hedman, J., Johansson, G., Bergmark, T., Karlsson, S.-E., Lindgren, I., and Lindberg, B., *ESCA: Atomic, Molecular and Solid State Structure Studied by Means of Electron Spectroscopy*, Nova Acta Regiae Soc. Sci. Upsaliensis Ser. VI, Vol. 20, 1967. Second Revised edition in preparation, North-Holland Publ. Co., Amsterdam.
8. Turner, D. W., Baker, C., Baker, A. D., and Brundle, C. R., in *Molecular Photoelectron Spectroscopy*, John Wiley and Sons, London, 1970.
9. Siegbahn, K. Nordling, C., Johansson, G., Hedman, J., Hedén, P. F., Hamrin, K., Gelius, U. Bergmark, T., Werme, L. O., Manne, R., and Baer, Y., *ESCA Applied to Free Molecules*, North-Holland Publ. Co., Amsterdam, 1969.
10. Kittel, C., *Introduction to Solid State Physics*, John Wiley & Sons, New York, 1971.
11. Shirley, D. A., in *Adv. Chem. Phys.*, Vol. XXIII (ed. by I. Prigogine and S. A. Rice), John Wiley, New York, 1973.
12. Baer, Y., Hedén, P. F., Hedman, J., Klasson, M., and Nordling, C., *Solid State Comm.* **8**, 1479 (1970).
13. Steinhardt, R. G., Hudis, J., and Perlman, M. L., in *Electron Spectroscopy* (ed. by D. A. Shirley), North-Holland Publ. Co., Amsterdam, 1972, p. 557.
14. Klasson, M., Hedman, J., Berndtsson, A., Nilsson, R., and Nordling, C., *Physica Scripta* **5**, 93 (1972).
15. Gelius, U., Hedén, P. F., Hedman, J., Lindberg, B. J., Manne, R., Nordberg, R., Nordling, C., and Siegbahn, K., *Chem. Phys. Letters* **4**, 471 (1970).
16. Fadley, C. S., Hagström, S. B. M., Hollander, J. M., Klein, M. P., and Shirley, D. A., *Science* **157**, 1571 (1967).
17. Fadley, C. S., Hagström, S. B. M., Klein, M. P., and Shirley, D. A., *J. Chem. Phys.* **48**, 3779 (1968).
18. Shirley, D. A., (ed.), *Electron Spectroscopy*, North-Holland Publ. Co., Amsterdam, 1972.
19. Bagus, P. S., *Phys. Rev.* **139**, A619 (1965).
20. Rosen, A. and Lindgren, I., *Phys. Rev.* **176**, 114 (1968).

21. Schwartz, M. E., *Chem. Phys. Letters* **5**, 50 (1970).
22. Bagus, P. S. and Schaefer, H. F., *J. Chem. Phys.* **55**, 1474 (1971).
23. Bagus, P. S. and Schaefer, H. F., *J. Chem. Phys.* **56**, 224 (1972).
24. Snyder, L. C., *J. Chem. Phys.* **55**, 95 (1972).
25. Verhaegen, G., Berger, J. J., Desclaux, J. P., and Moser, C. M., *Chem. Phys. Letters* **9**, 479 (1971).
26. Moser, C. M., Nesbet, R. K., and Verhaegen, G., *Chem. Phys. Letters* **12**, 230 (1971).
27. Koopmans, T., *Physica* **1**, 104 (1933).
28. Slater, J. C., *Quantum Theory of Atomic Structure*, McGraw-Hill Book Company, New York, 1960, Vol. II.
29. Schwartz, M. E., *Chem. Phys. Letters* **6**, 631 (1970).
30. Liberman, P., *Bull. Am. Phys. Soc. Ser. II* **9**, 731 (1964).
31. Hedin, L. and Johansson, A., *J. Phys.* **B2**, 1336 (1969).
32. Manne, R. and Åberg, T., *Chem. Phys. Letters* **7**, 282 (1970).
33. Basch, H., *Chem. Phys. Letters* **5**, 3371 (1970).
34. Davis, D. W., Shirley, D. A., and Thomas, T. D., *J. Chem. Phys.* **56**, 671 (1972).
35. Hedman, J., Klasson, M., Lindberg, B., and Nordling, C., in *Electron Spectroscopy* (ed. by D. A. Shirley), North-Holland Publ. Co., Amsterdam, 1972, p. 681.
36. Hendrickson, D. N., Hollander, J. M., and Jolly, W. L., *J. Inorg. Chem.* **8**, 2642 (1969).
37. Thomas, T. D., *J. Am. Chem. Soc.* **92**, 4184 (1970).
38. Jolly, W. L. and Hendrickson, D. N., *J. Am. Chem. Soc.* **92**, 1863 (1970).
39. Jolly, W. L. in *Electron Spectroscopy* (ed. by D. A. Shirley), North-Holland Publ. Co., Amsterdam, p. 629.
40. Clark, D. T., these proceedings (1973).
41. Hashmall, J. A., Mills, B. E., Shirley, D. A., and Streitwieser, A., *J.A.C.S.* **94**, 445 (1972).
42. Barber, M., Swift, P., Cunningham, D., and Frazer, M. J., *Chem. Comm.* **1970**, 1338 (1970).
43. Hagström, S. B. M., in *Electron Spectroscopy* (ed. by D. A. Shirley), North-Holland Publ. Co., Amsterdam, 1972, p. 515.
44. Fadley, C. S. and Shirley, D. A., *NBS J. Res.* **74A**, 543 (1970).
45. Prins, R. and Novakov, T., *Chem. Phys. Letters* **9**, 593 (1971).
46. Hamrin, K., Johansson, G., Gelius, U., Fahlman, A., Nordling, C., and Siegbahn, K., *Chem. Phys. Letters* **1**, 613 (1968).
47. Price, W. C., Potts, A. W., and Streets, D. G., in *Electron Spectroscopy*, (ed. by D. A. Shirley), North-Holland Publ. Co., Amsterdam, 1972, p. 187.
48. Eastman, D. E., in *Electron Spectroscopy* (ed. by D. A. Shirley), North-Holland Publ. Co., Amsterdam, 1972, p. 487.
49. Gelius, U., in *Electron Spectroscopy* (ed. by D. A. Shirley), North-Holland Publ. Co., Amsterdam, 1972, p. 311.
50. Berglund, C. N., and Spicer, W. E., *Phys. Rev.* **136**, A1030–A1044 (1964).
51. Mahan, G. D., *Phys. Rev.* **B2**, 4334 (1970).
52. Schaich, W. L. and Ashcroft, N. W., *Phys. Rev.* **B3**, 2452 (1971).
53. Shirley, D. A., in *Electron Spectroscopy* (ed. by D. A. Shirley), North-Holland Publ. Co., Amsterdam, 1972, p. 603.
54. Smith, N. V., *Phys. Rev. Letters* **23**, 1232 (1969).
55. Smith, N. V., *Phys. Rev.* **B3**, 1862 (1971).
56. Hedin, L. and Lundqvist, S., *Solid State Physics* **23**, 1 (1969).
57. Doniach, S., *Phys. Rev.* **B2**, 3898 (1970).
58. Fadley, C. S., and Shirley, D. A., *Phys. Rev. Letters* **21**, 980 (1968).
59. Baer, Y., *Phys. Kondens. Mat.* **9**, 367 (1969).
60. Eastman, D. E. and Cashion, J. K., *Phys. Rev. Letters* **24**, 310 (1970).
61. Cuthill, J. R., McAlister, A. J., Williams, M. L., and Watson, R. E., *Phys. Rev.* **164**, 1006 (1967).
62. Fadley, C. S. and Shirley, D. A., *Phys. Rev.* **A2**, 1109 (1970).
63. Fadley, C. S., in *Electron Spectroscopy* (ed. by D. A. Shirley), North-Holland Publ. Co., Amsterdam, 1972, p. 781.
64. Coster, D. and Druyvesteyn, M. J., *Z. Physik* **40**, 765 (1927).
65. Nefedov, V. I., *Izv. Akad. Nauk SSSR Ser. Fiz.* **28**, 816 (Engl. transl. *Bull. Acad. Sci. USSR, Phys. Ser.* **28**, 724 (1964).

66. Nefedov, V. I., *J. Struct. Chem.* **5**, 603; **5**, 605 (1964).
67. Nefedov, V. I., *J. Struct. Chem.* **7**, 672 (1966).
68. Ekstig, B., Källne, E., Noreland, E., and Manne, R., *Physica Scripta* **2**, 38 (1970).
69. Burhop, E. H. S., *The Auger Effect and other Radiationless Transitions*, Cambridge Univ. Press, London, 1952.
70. Asaad, W. N. and Burhop, E. H. S., *Proc. Phys. Soc.* **71**, 369 (1958).
71. Stalherm, D., Cleff, B., Hillig, H., and Mehlhorn, W., *Z. Naturforsch.* **24a**, 1728 (1969).
72. Novakov, T. and Hollander, J. M., *Phys. Rev. Letters* **21**, 1133 (1968).
73. Novakov, T. and Hollander, J. M., *Bull. Am. Phys. Soc.* **14**, 524 (1969).
74. Hedman, J., Hedén, P. F., Nordling, C. and Siegbahn, K., *Phys. Letters* **29A**, 178 (1969).
75. Fadley, C. S., Shirley, D. A., Freeman, A. J., Bagus, P. S., and Mallow, J. V., *Phys. Rev. Letters* **23**, 1397 (1969).
76. Freeman, A. J., Bagus, P. S., and Mallow, J. V., *Int. J. Magn.* **9**, 35 (1973).
77. Cox, P. A. and Orchard, F. A., *Chem. Phys. Letters* **7**, 273 (1970).
78. Bagus, P. S., Freeman, A. J., and Sasaki, F., *Phys. Rev. Letters* **30**, 850 (1973).
79. Novakov, T., private communication (1971).
80. Wertheim, G. K., Cohen, R. L., Rosencwaig, A., and Guggenheim, H. J., in *Electron Spectroscopy* (ed. by D. A. Shirley), North-Holland Publ. Co., Amsterdam, 1972, p. 813.
81. Helmer, J. C., private communication (1971).
82. Carver, J. C., Carlson, T. A., Cain, L. C., and Schweitzer, G. K., in *Electron Spectroscopy*, (ed. by D. A. Shirley), North-Holland Publ. Co., Amsterdam, 1972, p. 803.
83. Carver, J. C., Schweitzer, G. K., and Carlson, T. A., *J. Chem. Phys.* **57**, 973 (1972).
84. Hedén, P. O., Löfgren, H., and Hagström, S. B. M., *Phys. Rev. Letters* **26**, 432 (1971).
85. Brodén, G., Private communication (1971).
86. Apai, G. R., Delgass, W. N., Hollander, J. M., Novakov, T., and Shirley, D. A., Lawrence Radiation Laboratory Report UCRL–19530, 252 (1969).
87. Gupta, R. P., Rao, B. K., and Sen, S. K., *Phys. Rev.* **A3**, 545 (1971).
88. Krause, M. O., Vestal, M. L., Johnston, W. H., and Carlson, T. A., *Phys. Rev.* **133**, A385 (1964).
89. Krause, M. O., Carlson, T. A., and Dismukes, R. D., *Phys. Rev.* **170**, 37 (1968).
90. Carlson, T. A. and Krause, M. O., *Phys. Rev.* **137**, A1655 (1965), and **140**, A1057 (1965).
91. Carlson, T. A., *Phys. Rev.* **156**, 142 (1967).
92. Krause, M. O. and Carlson, T. A., *Phys. Rev.* **149**, 52 (1966).
93. Krause, M. O. and Carlson, T. A., *Phys. Rev.* **158**, 18 (1967).
94. Carlson, T. A., Hunt, W. E., and Krause, M. O., *Phys. Rev.* **151**, 41 (1966).
95. Carlson, T. A., Krause, M. O., and Moddeman, W. E., *J. Phys.* **C2**, 102 (1971).
96. Aberg, T., *Phys. Rev.* **156**, 35 (1967).
97. Novakov, T., *Phys. Rev.* **B3**, 2693 (1971).
98. Wertheim, G. K. and Rosencwaig, A., *Phys. Rev. Letters* **26**, 1179 (1971).
99. Novakov, T., in *Electron Spectroscopy* (ed. by D. A. Shirley), North-Holland Publ. Co., Amsterdam, 1972, p. 821.
100. Bethe, H. A. and Salpeter, E. E., *Handbuch der Physik* (Springer-Verlag Berlin-Göttingen-Heidelberg) **35**, p. 88.
101. Bates, D. R., *Monthly Notices Roy. Astron. Soc.* **106**, 432 (1946).
102. Cooper, J. W., *Phys. Rev.* **128**, 681 (1962).
103. Fano, U. and Cooper, J. W., *Rev. Mod. Phys.* **40**, 441 (1968).
104. Cooper, J. W. and Manson, S. T., *Phys. Rev.* **177**, 157 (1969).
105. Cooper, J. and Zare, R. N., *J. Chem. Phys.* **48**, 942 (1968).
106. Cooper, J. and Zare, R. N., in *Lectures in Theoretical Physics* (ed. by S. Geltman, K. Mahanthappa and W. Brittlin), Gordon & Breach, New York, 1969, vol. XIC.
107. Grimm, F. A., in *Electron Spectroscopy* (ed. by D. A. Shirley), North-Holland Publ. Co., Amsterdam, 1972, p. 199.
108. Lin, S. H., *Can. J. Phys.* **46**, 2719 (1968).
109. Bearden, A. J., *J. Appl. Phys.* **37**, 1681 (1966).
110. Rakavy, G. and Ron, A., *Phys. Rev.* **159**, 50 (1967).
111. Brysk, H. and Zerby, C. D., *Phys. Rev.* **171**, 292 (1968).
112. Manson, S. T. and Cooper, J. W., *Phys. Rev.* **165**, 126 (1968).
113. Manson, S. T. and Krause, M. O., private communication (1970).

114. Kennedy, D. J. and Manson, S. T., *Phys. Rev.* **A5**, 227 (1972).
115. Krause, M. O., *Phys. Rev.* **177**, 151 (1969).
116. Brodén, G., Hagström, S. B. M., and Norris, C., *Phys. Rev. Letters* **24**, 1173 (1970); Hagström S. B. M., Hedén, P.-O., and Löfgren, H., *Solid State Comm.* **8**, 1245 (1970); Eastman, D. E. and Kusnietz, M., *Phys. Rev. Letters* **26**, 846 (1971).
117. Manson, S. T., *Phys. Rev. Letters* **26**, 219 (1971).
118. Tully, J. C., Berry, R. S., and Dalton, B. J., *Phys. Rev.* **176**, 95 (1968).
119. Carlson, T. A., McGuire, G. E., Jonas, A. E., Cheng, K. L., Anderson, C. P., Lu, C. C., and Pullen, B. P., in *Electron Spectroscopy* (ed. by D. A. Shirley), North-Holland Publ. Co., Amsterdam, 1972, p. 207.
120. Hirsch, P. E. and Howie, A., *Electron Microscopy of Thin Crystals*, Butterworths, London, 1953.
121. Hirsch, P. B., Howie, A., Nicholson, R. B., Pashley, D. W., and Whelan, M. J., *Electron Microscopy of Thin Crystals*, Butterworths, London, 1965.
122. De Wames, R. E. and Hall, W. F., *Acta Cryst.* **A24**, 206 (1968).
123. Duke, C. B., these proceedings (1973).
124. Gerhardt, U. and Dietz, E., *Phys. Rev. Letters* **26**, 1477 (1971).
125. Koyama, R. Y. and Hughey, L. R., *Phys. Rev. Letters* **29**, 1518 (1972).
126. Gustaffson, T., Nilsson, P. O., and Walldén, L., *Phys. Letters* **37A**, 121 (1971).
127. Henke, B., *Phys. Rev.* **A6**, 94 (1972).

CRYSTAL FIELD THEORY AND CALCULATION
OF THE INNER SHELL VACANCY LEVELS

R. P. GUPTA* and S. K. SEN

Department of Physics, University of Manitoba, Winnipeg, Canada

1. Crystal Electric Potential

The study of the energy levels of ions in solids is complicated because of the electric potential due to the rest of the ions. Crystal field theory is a simplified approach for dealing with the complexity. We make the assumption that the ions (ligands) surrounding a given ion (metal ion) are fixed, nonoverlapping, hard spherical charge distributions with a definite arrangement (crystal structure). We then calculate the electrostatic potential in the neighbourhood of the given ion in terms of certain parameters (crystal field parameters) representative of the arrangement and charge of the surrounding ions, and allow the potential to interact with the ion assumed to possess free space Hamiltonian. The energy level structure and the energy separation between the different levels are thus obtained in terms of the crystal potential parameters. The experimental information on the energy levels of the metal ion is then used to calculate the crystal potential parameters. Although *a priori* calculation of the crystal potential parameters is possible in principle, various simplifying assumptions have to be made even in most elaborate calculations. As a result the theoretical values of the parameters are rarely anywhere near to the experimental values. However, the true symmetry of the ligands surrounding the metal ion is not seriously affected by our simplified assumptions and the description of its energy levels remains valid in terms of unknown crystal potential parameters, as long as 'covalence' effects between the metal ions and the ligands are small. In the covalent compounds such as $IrCl_6$, MnF_2, CoF_2 the observed energy levels are not correlated to the free-ion; they are easier to interpret as belonging to molecular orbitals obtained by combining the atomicorbitals. The ligand field theory is a hybridization of the pure crystal field theory and the molecular orbital theory and we shall not consider it here.

In crystal field theory the assumption that the ligand ions do not overlap with the metal ions and have spherically symmetric charge distributions enables us to replace them by point changes situated at their geometric sites. Then the electrostatic potential V at a point (r, θ, ϕ), in a coordinate system, with origin at the nucleus of the metal ion, due to the surrounding point charges q_j at (R_j, θ_j, ϕ_j) is (Figure 1).

$$V(r, \theta, \phi) = \sum_j \frac{q_j}{|\mathbf{R}_j - \mathbf{r}|} = \sum_j q_j \sum_{k=0}^{\infty} \frac{r_<^k}{r_>^{k+1}} P_k(\cos \omega). \tag{1.1}$$

* Visiting scientist from the University of Allahabad, India.

W. Dekeyser et al. (eds.), Electron Emission Spectroscopy, 225–258. All Rights Reserved
Copyright © 1973 by D. Reidel Publishing Company, Dordrecht-Holland

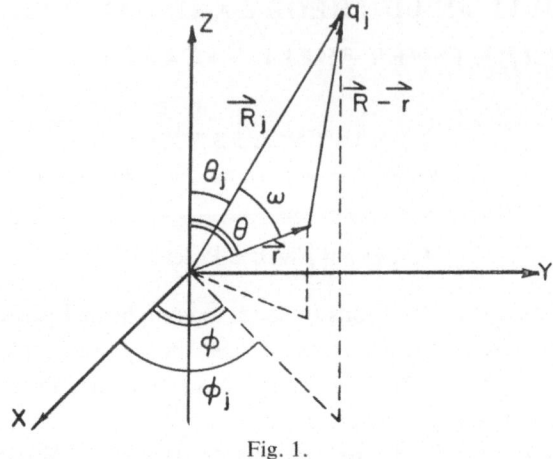

Fig. 1.

Here $r_<$ and $r_>$ are respectively the smaller and the larger of R_j and r, and $P_k(\cos\omega)$ are the Legendre polynomials [1]. The point $\mathbf{r}$ is within the metal ion. Our assumption of non overlapping ions is then equivalent to putting the constraint $r \leqslant R_j$ in Equation (1.1). We take advantage of the spherical harmonic addition theorem,

$$P_k(\cos\omega) = \frac{4\pi}{2k+1} \sum_{q=-k}^{k} (-1)^q Y_k^{-q}(\theta_j, \phi_j) Y_k^q(\theta, \phi), \tag{1.2}$$

to write potential V as

$$V = \sum_j q_j \sum_{k=0}^{\infty} \frac{r^k}{R_j^{k+1}} \frac{4\pi}{2k+1} \sum_{q=-k}^{k} (-1)^q Y_k^{-q}(\theta_j, \phi_j) Y_k^q(\theta, \phi). \tag{1.3}$$

Here spherical harmonics are related to the associated Legendre polynomials by

$$Y_n^m(\theta, \phi) = (-1)^{(m+|m|)/2} \left[\frac{(2n+1)(n-|m|)!}{2(n+|m|)!} \right]^{1/2} \times$$

$$\times \frac{1}{\sqrt{2\pi}} P_n^{|m|}(\cos\theta) \exp(im\phi) \equiv$$

$$\equiv (-1)^{(m+|m|)/2} \Theta_{nm} \frac{1}{\sqrt{2\pi}} \exp(im\phi). \tag{1.4}$$

In our work it is more convenient to use tesseral harmonics which are real quantities defined as [2, 3, 4]:

$$Z_{ko}^c = Y_k^0 = \frac{1}{\sqrt{2\pi}} \Theta_{ko}; \quad Z_{ko}^s = 0;$$

$$Z_{kq}^c = \frac{1}{\sqrt{2}} [Y_k^{-q} + (-1)^q Y_k^q] = \frac{\Theta_{kq}}{\sqrt{\pi}} \cos(q\phi) \tag{1.5}$$

$$q > 0$$

$$Z_{kq}^s = \frac{i}{\sqrt{2}} [Y_k^{-q} - (-1)^q Y_k^q] = \frac{\Theta_{kq}}{\sqrt{\pi}} \sin(q\phi).$$

Then

$$P_k(\cos\omega) = \frac{4\pi}{2k+1} \sum_{q=0}^{k} \sum_{\alpha=c,s} Z_{kq}^{\alpha}(\theta,\phi)\, Z_{kq}^{\alpha}(\theta_j,\phi_j) \qquad (1.6)$$

and the potential,

$$V(r,\theta,\phi) = \sum_j q_j \sum_{k=0}^{\infty} \frac{r^k}{R_j^{k+1}} \frac{4\pi}{2k+1} \sum_{q=0}^{k} \sum_{\alpha=c,s} Z_{kq}^{\alpha}(\theta,\phi)\, Z_{kq}^{\alpha}(\theta_j,\phi_j). \qquad (1.7)$$

If we now define

$$\gamma_{kq}^{\alpha} = \sum_j \frac{4}{2k+1}\, q_j\, \frac{Z_{kq}(\theta_j,\phi_j)}{R_j^{k+1}}, \qquad (1.8)$$

then

$$V(r,\theta,\phi) = \sum_{k=0}^{\infty} r^k \sum_{q=0}^{k} \sum_{\alpha=c,s} \gamma_{kq}^{\alpha} Z_{kq}^{\alpha}(\theta,\phi), \qquad (1.9)$$

It is clear that the problem now is to know γ_{kq}^{α}'s; they can be calculated if the geometric position of the ligand ions and their charges are known. It is generally sufficiant to calculate them for nearest neighbour ligands only but, if required, the lattice sum in

TABLE I [2]

$\Theta_{00} = \sqrt{2}/2$

$\Theta_{10} = (\sqrt{6}/2)\cos\theta$

$\Theta_{11} = (\sqrt{3}/2)\sin\theta$

$\Theta_{20} = (\sqrt{10}/4)(3\cos^2\theta - 1)$

$\Theta_{21} = (\sqrt{15}/2)\sin\theta\cos\theta$

$\Theta_{22} = (\sqrt{15}/4)\sin^2\theta$

$\Theta_{30} = (\sqrt{14}/4)(5\cos^3\theta - 3\cos\theta)$

$\Theta_{31} = (\sqrt{42}/8)(5\cos^2\theta - 1)\sin\theta$

$\Theta_{32} = (\sqrt{105}/4)\sin^2\theta\cos\theta$

$\Theta_{33} = (\sqrt{70}/8)\sin^3\theta$

$\Theta_{40} = (3\sqrt{2}/16)(35\cos^4\theta - 30\cos^2\theta + 3)$

$\Theta_{41} = (3\sqrt{10}/8)(7\cos^3\theta - 3\cos\theta)\sin\theta$

$\Theta_{42} = (3\sqrt{5}/8)(7\cos^2\theta - 1)\sin^2\theta$

$\Theta_{43} = (3\sqrt{70}/8)\cos\theta\sin^3\theta$

$\Theta_{44} = (3\sqrt{35}/16)\sin^4\theta$

$\Theta_{50} = (\sqrt{22}/16)(63\cos^5\theta - 70\cos^3\theta + 15\cos\theta)$

$\Theta_{51} = (\sqrt{165}/16)(21\cos^4\theta - 14\cos^2\theta + 1)\sin\theta$

$\Theta_{52} = (\sqrt{1155}/8)(3\cos^3\theta - \cos\theta)\sin^2\theta$

$\Theta_{53} = (\sqrt{770}/32)(9\cos^2\theta - 1)\sin^3\theta$

$\Theta_{54} = (3\sqrt{385}/16)\cos\theta\sin^4\theta$

$\Theta_{55} = (3\sqrt{154}/32)\sin^5\theta$

$\Theta_{60} = (\sqrt{26}/32)(231\cos^6\theta - 315\cos^4\theta + 105\cos^2\theta - 5)$

$\Theta_{61} = (\sqrt{273}/16)(33\cos^5\theta - 30\cos^3\theta + 5\cos\theta)\sin\theta$

$\Theta_{62} = (\sqrt{2730}/64)(33\cos^4\theta - 18\cos^2\theta + 1)\sin^2\theta$

$\Theta_{63} = (\sqrt{2730}/32)(11\cos^3\theta - 3\cos\theta)\sin^3\theta$

$\Theta_{64} = (3\sqrt{91}/32)(11\cos^2\theta - 1)\sin^4\theta$

$\Theta_{65} = (3\sqrt{2002}/32)\cos\theta\sin^5\theta$

$\Theta_{66} = (\sqrt{6006}/64)\sin^6\theta$

TABLE II [2]

$$Z_{00} = \frac{\sqrt{2}}{2} \frac{1}{\sqrt{2\pi}}$$

$$Z_{10} = \frac{\sqrt{6}}{2} \frac{1}{\sqrt{2\pi}} \cdot \frac{z}{r}$$

$$Z_{11}{}^c = \frac{\sqrt{3}}{2} \frac{1}{\sqrt{\pi}} \cdot \frac{x}{r}$$

$$Z_{11}{}^s = \frac{\sqrt{3}}{2} \frac{1}{\sqrt{\pi}} \cdot \frac{y}{r}$$

$$Z_{20} = \frac{\sqrt{10}}{4} \frac{1}{\sqrt{2\pi}} \frac{3z^2 - r^2}{r^2}$$

$$Z_{21}{}^c = \frac{\sqrt{15}}{2} \frac{1}{\sqrt{\pi}} \cdot \frac{xz}{r^2}$$

$$Z_{21}{}^s = \frac{\sqrt{15}}{2} \frac{1}{\sqrt{\pi}} \cdot \frac{yz}{r^2}$$

$$Z_{22}{}^c = \frac{\sqrt{15}}{4} \frac{1}{\sqrt{\pi}} \frac{x^2 - y^2}{r^2}$$

$$Z_{22}{}^s = \frac{\sqrt{15}}{4} \frac{1}{\sqrt{\pi}} \frac{2xy}{r^2}$$

$$Z_{40} = \frac{3\sqrt{2}}{16} \frac{1}{\sqrt{2\pi}} \frac{(35z^4 - 30z^2r^2 + 3r^4)}{r^4}$$

$$Z_{44}{}^c = \frac{3\sqrt{35}}{16} \frac{1}{\sqrt{\pi}} [(x^4 - 6x^2y^2 + y^4)/r^4]$$

$$Z_{60} = \frac{\sqrt{26}}{32} \frac{1}{\sqrt{2\pi}} \frac{(231z^6 - 315z^4r^2 - 5r^6)}{r^6}$$

$$Z_{64}{}^c = \frac{3\sqrt{91}}{32} \frac{1}{\sqrt{\pi}} \frac{[(11z^2 - r^2)(x^4 - 6x^2y^2 + y^4)]}{r^6}$$

$$Z_{66}{}^c = \frac{\sqrt{6006}}{64} \frac{1}{\sqrt{\pi}} \frac{(x^6 - 15x^4y^2 + 15x^2y^4 - y^6)}{r^6}$$

$$Z_{66}{}^s = \frac{\sqrt{6006}}{64} \frac{1}{\sqrt{\pi}} \frac{[(3x^2 - y^2)(x^2 - 3y^2)(2xy)]}{r^6}$$

Equation (1.8) could also be performed over other neighbours. We shall later see that most of the interest in the crystal potential is to know how it perturbs the multiplets of the ground state configuration of an ion which shall always have $l \leqslant 3$; the matrix elements of $-eV$ between the multiplet states will be non zero only for $k \leqslant 2l$. We shall, therefore, confine ourselves to the evaluation of γ_{kq}^{α}'s for $k \leqslant 6$.

We shall now consider two simple cases to demonstrate how γ_{kq}^{α}'s are calculated and to get a feeling of the form of crystal potential functions. We shall need Θ_{kq} and Z_{kq}^{α}, and some of them which we shall use are given in Tables I and II respectively.

A. OCTAHEDRAL SYMMETRY CRYSTAL POTENTIAL O_h:

(Examples – transition metal Sulphates, Fluosilicates, Oxides)
Six ligand ion positions, (R_j, θ_j, ϕ_j), assuming that Z-axis passes through two of

them and through the metal ion are $R_1(a, 0, 0)$, $R_2(a, \pi, 0)$, $R_3(a, \pi/2, 0)$, $R_4(a, \pi/2, \pi)$, $R_5(a, \pi/2, \pi/2)$, $R_6(a, \pi/2, 3\pi/2)$ (Figure 2). Each ligand has a charge Q.

The potential at a point $r(r, \theta, \phi)$ should be the same as at a point $r'(r, \pi-\theta, \phi)$. This is equivalent to $Z_{kq}^{\alpha}(\theta, \phi) = Z_{kq}^{\alpha}(\pi-\theta, \phi)$ from Equation (1.9). Since $\cos(\pi-\theta) = -\cos\theta$, and since $Z_{kq}^{\alpha}(\theta, \phi)$ through their relation to Θ_{kq} (Equation (1.5), Table I) carry odd powers of $\cos\theta$ for odd $(k+q)$, we find $Z_{kq}^{\alpha}(\pi-\theta, \phi) = -Z_{kq}^{\alpha}(\theta, \phi)$ for odd $(k+q)$. Therefore, the terms with odd $(k+q)$ should be absent from Equation (1.9). Further, every ϕ is equivalent to $\phi+\pi/2$, $\phi+2\pi/2$, $\phi+3\pi/2$. If V is to be the same for such points we should have (Equation (1.5))

$$\cos(q\phi) = \cos(q(\phi + \pi/2)) = \cos(q(\phi + 2\pi/2)) = \cos(q(\phi + 3\pi/2))$$

and/or

$$\sin(q, \phi) = \sin(q(\phi + \pi/2)) = \sin(q(\phi + 2\pi/2)) = \sin(q(\phi + 3\pi/2)).$$

These relations are satisfied by the same condition, $q=0$ or multiple of 4. Our task is reduced to calculating the terms with $k=0, 2, 4, 6$ and $q=0, 4$.

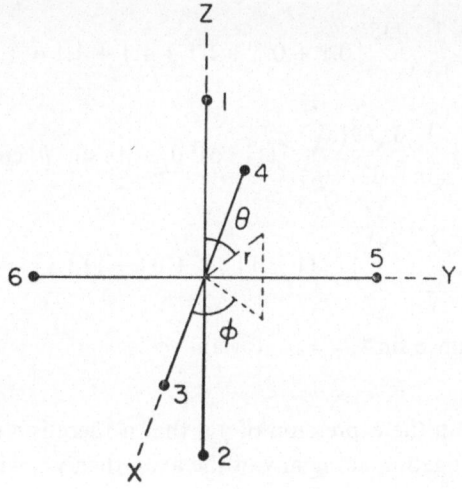

Fig. 2.

The γ_{kq}^{α}'s for $k=0, 2, 4, 6$ and $q=0, 4$, calculated using Equation (1.8), are:

$$\gamma_{00} = \frac{4\pi Q}{a} \sum_{j=1}^{6} Z_{00} = \frac{4\pi q}{a} \frac{1}{\sqrt{2\pi}} 6 \frac{\sqrt{2}}{2} = \frac{12\sqrt{\pi}}{a} Q$$

$$\gamma_{20} = \frac{4\pi Q}{5a^3} \frac{1}{\sqrt{2\pi}} \sum_{j=1}^{6} \frac{\sqrt{10}}{4} (3\cos^2\theta_j - 1) =$$

$$= \frac{4\pi Q}{20a^3} \sqrt{\frac{10}{2\pi}} [2 + 2 - 1 - 1 - 1 - 1]$$

$$= 0$$

$$\gamma_{40} = \frac{4\pi Q}{9a^5} \frac{1}{\sqrt{2\pi}} \frac{3\sqrt{2}}{16} \sum_{j=1}^{6} (35\cos^4\theta_j - 30\cos^2\theta_j + 3)$$

$$= \frac{4\pi Q}{9a^5} \frac{1}{\sqrt{2\pi}} \frac{3\sqrt{2}}{16} [2\cdot(35 - 30 + 3) + 4.3] = \tfrac{7}{3}\sqrt{\pi}\,Q/a^5$$

$$\gamma_{60} = \frac{4\pi Q}{13a^7} \frac{1}{\sqrt{2\pi}} \frac{\sqrt{26}}{32} \left[\sum_{j=1}^{6} (231\cos^6\theta_j - 315\cos^4\theta_j + 105\cos^2\theta_j - 5) \right]$$

$$= \frac{4\pi Q}{13a^7} \frac{1}{32} \sqrt{\frac{13}{\pi}} [2\cdot(231 - 315 + 105 - 5) - 4.5] = \frac{3}{2}\sqrt{\frac{\pi}{13}}\,Q/a^7$$

$$\gamma_{44}^c = \frac{4\pi Q}{9a^5} \frac{1}{\sqrt{\pi}} \frac{3\sqrt{35}}{16} \sum_{j=1}^{6} [\sin^4\theta_j \cos 4\phi_j]$$

$$= \frac{4\pi Q}{9a^5} \frac{3}{16} \sqrt{\frac{35}{\pi}} [0.1 + 0.1 + 1.1 + 1.1 + 1.1 + 1.1] = \frac{\sqrt{35\pi}}{3}\frac{Q}{a^5}$$

$$\gamma_{64}^c = \frac{4\pi Q}{13a^7} \frac{1}{\sqrt{\pi}} \frac{3\sqrt{91}}{32} \sum_{j=1}^{6} [(11\cos^2\theta_j - 1)\sin^4\theta_j \cos 4\phi_j]$$

$$= \frac{4\pi Q}{13a^7} \frac{3}{32} \sqrt{\frac{91}{\pi}} [2(11 - 1)0.1 + 4(-1)1.1] = -\frac{3}{2}\sqrt{\frac{7\pi}{13}}\,Q/a^7$$

$$\gamma_{44}^s = 0$$
$$\gamma_{64}^s = 0 \quad \text{since } \sin 4\phi_j = 0 \quad \text{for all} \quad \phi_j\text{'s}.$$

It should be noted from the expression of γ_{20} that if there is a slight deformation of the octahedron by elongation along any of the axes, then γ_{20} will not be zero and we will have a small quadrupole term in the crystal potential. We therefore write

$$V_{0h}(r, \theta, \phi) = \gamma_{00}Z_{00}(\theta, \phi) + \gamma_{20}r^2 Z_{20}(\theta, \phi) + \gamma_{40}r^4 Z_{40}(\theta, \phi) +$$
$$+ \gamma_{44}^c r^4 Z_{44}^c(\theta, \phi) + \gamma_{60}r^6 Z_{60}(\theta, \phi) + \gamma_{64}^c r^6 Z_{64}^c(\theta, \phi)$$

$$(1.10)$$

as a general representative of octahedral potential.

B. HEXAGONAL SYMMETRY CRYSTAL POTENTIAL C_{3h}, D_{3h}:

(Examples – lanthanide ethyl sulphates, anhydrous trichlorides)
We shall first define C_{3h} and D_{3h} symmetry. We consider (Figure 3) three equilateral triangles, two of them equal and stacked parallel with the same orientation, and the axis joining their centres (z-axis) passing normal to both. We place the third triangle in the middle of the other two, normal to the axis through its centre, with any orientation. Let the origin be the centre of the middle triangle and the metal ion be placed at

the origin and the ligands at the nine corners of the three triangles. We then have z-axis as threefold axis of symmetry C_3 and a reflection symmetry in the horizontal plane (x-y plane through the origin) giving C_{3h} symmetry. For D_{3h} symmetry we should also have three two-fold axes of symmetry perpendicular to the C_3 axis. This is possible only when the middle triangle is oriented either as the other two triangles or at an angle $\pi/3$ with respect to them.

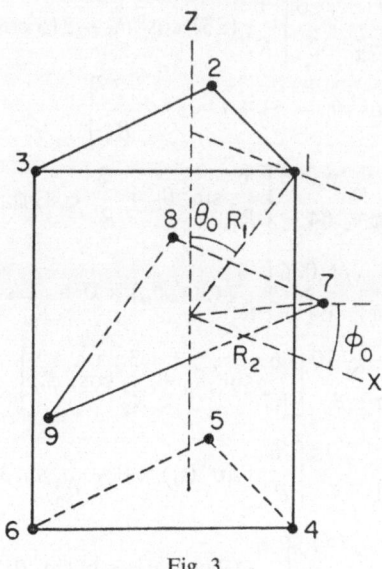

Fig. 3.

The position of the nine ligand ions each with charge Q are $\mathbf{R}_1 (R_1, \theta_0, 0)$, $\mathbf{R}_2 (R_1, \theta_0, 2\pi/3)$, $\mathbf{R}_3 (R_1, \theta_0, 4\pi/3)$, $\mathbf{R}_4 (R_1, \pi-\theta_0, 0)$, $\mathbf{R}_5 (R_1, \pi-\theta_0, 2\pi/3)$, $\mathbf{R}_6 (R_1, \pi-\theta_0, 4\pi/3)$, $\mathbf{R}_7 (R_2, \pi/2, \phi_0)$, $\mathbf{R}_8 (R_2, \pi/2, \phi_0+2\pi/3)$, $\mathbf{R}_9 (R_2, \pi/2, \phi_0+4\pi/3)$, and for D_{3h} symmetry $\phi_0 = \pi/3$ or 0.

Due to the reflection symmetry (θ equivalent to π-θ) here again all terms of the crystal potential, Equation (1.9), with $k+q$ odd are zero. Also due to C_3 rotation symmetry, every ϕ is equivalent to $\phi+2\pi/3$, $\phi+4\pi/3$.

If V is to be the same for such points we should have (Equation (1.5))

$$\cos(q\phi) = \cos(q(\phi + 2\pi/3)) = \cos(q(\phi + 4\pi/3))$$

and/or

$$\sin(q\phi) = \sin(q(\phi + 2\pi/3)) = \sin(q(\phi + 4\pi/3)).$$

These relations are satisfied by the same condition, $q=0$ or multiple of 3. We are thus left with the terms with $k=0, 2, 4, 6$ for $q=0, 6$, and $k=3, 5$ for $q=3$. Corresponding γ_{kq}^α's, using Equation (1.8), are:

$$\gamma_{00} = 4\pi Q \frac{1}{\sqrt{4\pi}} \left[\frac{6}{R_1} + \frac{3}{R_2} \right]$$

$$\gamma_{20} = \frac{4\pi}{5} Q \frac{1}{\sqrt{2\pi}} \frac{\sqrt{10}}{4} \left[\frac{6(3\cos^2\theta_0 - 1)}{R_1^3} + (-1)\frac{3}{R_2^3} \right]$$

$$\gamma_{40} = \frac{4\pi}{9} Q \frac{1}{\sqrt{2\pi}} \frac{3\sqrt{2}}{16} \left[\frac{(35\cos^4\theta_0 - 30\cos^2\theta_0 + 3)6}{R_1^5} + 3\times\frac{3}{R_2^5} \right]$$

$$\gamma_{60} = \frac{4\pi}{13} Q \frac{1}{\sqrt{2\pi}} \frac{\sqrt{26}}{32} \left[\frac{6}{R_1^7} (231\cos^6\theta_0 - 315\cos^4\theta_0 + \right.$$
$$\left. + 105\cos^2\theta_0 - 5) + (-5)\times\frac{3}{R_2^7} \right]$$

$$\gamma_{66}^c = \frac{4\pi}{13} Q \frac{1}{\sqrt{\pi}} \frac{\sqrt{6,006}}{64} \left[\frac{6}{R_1^7} \sin^6\theta_0 + \frac{3}{R_2^7} \cos 6\phi_0 \right]$$

$$\gamma_{66}^s = \frac{4\pi}{13} Q \frac{1}{\sqrt{\pi}} \frac{\sqrt{6,006}}{64} \left[\frac{6}{R_1^7} (\sin^6\theta_0)\times 0 + \frac{3}{R_2^7} \sin 6\phi_0 \right]$$

$$\gamma_{33}^c = \frac{4\pi}{7} Q \frac{1}{\sqrt{\pi}} \frac{\sqrt{70}}{8} \left[\frac{6}{R_1^4} \sin^3\theta_0 + \frac{3}{R_2^4} \cos 3\phi_0 \right]$$

$$\gamma_{33}^s = \frac{4\pi}{7} Q \frac{1}{\sqrt{\pi}} \frac{\sqrt{70}}{8} \left[\frac{6}{R_1^4} (\sin^3\theta_0)\times 0 + \frac{3}{R_2^4} \sin 3\phi_0 \right]$$

$$\gamma_{53}^c = \frac{4\pi}{11} Q \frac{1}{\sqrt{\pi}} \frac{\sqrt{770}}{32} \left[\frac{6}{R_1^6} (9\cos^2\theta_0 - 1)\sin^3\theta_0 + (-1)\frac{3}{R_2^6} \cos 3\phi_0 \right]$$

$$\gamma_{53}^s = \frac{4\pi}{11} Q \frac{1}{\sqrt{\pi}} \frac{\sqrt{770}}{32} \left[\frac{6}{R_1^6} (9\cos^2\theta_0 - 1)(\sin^3\theta_0)\times \right.$$
$$\left. \times 0 + (-1)\frac{3}{R_2^6} \sin 3\phi_0 \right]$$

(For D_{3h} symmetry $\phi_0 = \pi/3$ or 0; thus all γ_{kq}^s above are zero).

$$V_{C_{3h}} = \gamma_{00}Z_{00} + \gamma_{20}r^2Z_{20} + \gamma_{40}r^4Z_{40} + \gamma_{60}r^6Z_{60} + \gamma_{66}^c r^6 Z_{66}^c +$$
$$+ \gamma_{66}^s r^6 Z_{66}^s + [\gamma_{33}^c r^3 Z_{33}^c + \gamma_{33}^s r^3 Z_{33}^s + \gamma_{53}^c r^5 Z_{53}^c +$$
$$+ \gamma_{53}^s r^5 Z_{53}^s]. \qquad (1.11)$$

We shall later limit ourselves to the evaluation of the matrix elements of crystal potential within a configuration i.e. within the states having same orbital angular momentum l. But by the rules of the coupling of angular momenta, $l+k+l$ should be even for non zero matrix elements and therefore k should be even. However, odd k term in the potential may couple different configurations. For most purposes the configuration mixing is ignored, and we shall also ignore it; the square bracketed terms in Equation (1.11) are then irrelevant.

2. Multiplet Splitting in Crystalline Potential

In this section we shall consider the effect of crystalline environment (1) briefly on the multiplet structure of metal ions and on the multiplet structure of their nuclei and (2) in some detail on the structure of the inner-shell vacancy energy levels. We shall first summarize our knowledge of the theory of multiplets.

A. MULTIPLET THEORY

The Hamiltonian of an ion with N-electrons and with an infinite mass nucleus of charge $Z|e|$ in a field free space (considering only the electrostatic coulomb interaction) is generally written as

$$H = -\left(\frac{\hbar^2}{2m_e} \sum_i \nabla_i^2 + \sum_i \frac{Ze^2}{r_i}\right) + \sum_{i>j} \frac{e^2}{r_{ij}} \tag{2.1}$$

where $\hbar$ is the Planck's constant h divided by 2π, m_e is the electronic mass, ∇_i^2 is the kinetic energy operator of the ith electron, r_i is the distance of the ith electron from the nucleus and r_{ij} is the distance between the ith and jth electrons. The solution of the wave equation $H\psi = E\psi$ with H given by Equation (2.1) is complicated because of the electrostatic repulsion between the electrons, $\sum_{i>j} e^2/r_{ij}$, since it couples the ith electrons with each other and prevents separation of variables. We, therefore, seek that $\sum_{i>j} e^2/r_{ij}$ be replaced by some suitably averaged potential and let the electron move in that average potential of other electrons. The Hamiltonian (Equation (2.1)) then takes the form $H = \sum_i H_i$ making it possible to write ψ as a product of single particle wavefunctions u_i, i.e. $\psi = \Pi_i u_i$ and to separate the wave equation $H\psi = E\psi$ into N single electron wave equation,

$$\left[-\left(\frac{\hbar^2}{2m} \nabla_i^2 + \frac{Ze^2}{r_i}\right) + \frac{1}{2} \sum_i \left(\sum_{j \neq i} \left(\frac{e^2}{r_{ij}}\right)_{Av}\right)\right] u_i = \varepsilon_i u_i. \tag{2.2}$$

The solution is hydrogen-like in the sense that the angular dependence of u_i could be taken care of by spherical harmonic $Y_{l_i}^{m_{l_i}}(\theta_i, \phi_i)$ given by

$$Y_n^m(\theta, \phi) = (-1)^{(m+|m|)/2} \left[\frac{(2n+1)(n-|m|)!}{4\pi(n+|m|)!}\right]^{1/2} \times$$
$$\times P_n^{|m|}(\cos\theta) \times \exp(im\phi) \tag{2.3}$$

where P_n^m are the Legendre functions. In the one-electron framework the problem is reduced to determining the radial part $u'_{n_i l_i}(r)$ of the function u_i.

Hartree's prescription of estimating $\sum_{j \neq i}(e^2/r_{ij})_{Av}$ could be mathematically understood as follows: We know

$$\frac{1}{r_{ij}} = \sum_{k=0}^{\infty} \frac{4\pi}{2k+1} \frac{r_<^k}{r_>^{k+1}} \sum_{q=-k}^{k} Y_k^q(\theta_i \phi_i) Y_k^{q*}(\theta_j \phi_j), \tag{2.4}$$

where $r_<$ is the smaller of the r_i and r_j and $r_>$ is the greater of the r_i and r_j. The potential due to the charge distribution $eu_j^*u_j$ of the jth electron at the site of the ith electron is

$$e \int \frac{u_j^*u_j}{r_{ij}} \, d\tau_j .$$

Now if we substitute Equation (2.4) in it, retain only the $k=0$ term, and sum over all $j(\neq i)$ electrons, we get the Hartree potential and Equation (2.2) becomes the Hartree equation. The Pauli principle is then superimposed by allowing maximum two electrons of opposite spin to occupy an orbital u_i.

Fock pointed out that since the electrons are undistinguishable, the interchange of the electron coordinates should not change the solution of the Schrödinger equation. We should, therefore, expect that a linear combination of the ψ's (recalling that $\psi = \Pi_i u_i(i)$) obtained by all possible interchanges of coordinates between the orbitals u_i's should be an appropriate wave function. Electrons obey Fermi-Dirac statistics and, therefore, ψ should be antisymmetric under the interchange of coordinates between any pair of u's. Slater noted that this type of behaviour could be very conveniently represented by a determinantal function,

$$\Psi = \frac{1}{\sqrt{N!}} \begin{vmatrix} u_1(1) & u_2(1) & \ldots & u_N(1) \\ u_1(2) & u_2(2) & \ldots & u_N(2) \\ \ldots & \ldots & \ldots & \ldots \\ u_1(N) & u_2(N) & & u_N(N) \end{vmatrix} \quad (2.5)$$

where $u_i(j)$ includes both spatial and spin parts and is generally called a spin-orbital. Since a determinant vanishes if its two rows or columns are same, the Pauli principle (if $u_i = u_j$, two columns are equal; if $(r_i, s_i) = (r_j, s_j)$, two rows are equal) is automatically satisfied. The use of ψ in the wave equation with Hamiltonian given by Equation (2.1) and the application of variation principle for minimising the energy lead to the so called Hartree-Fock wave equation,

$$-\left(\frac{\hbar^2}{2m} \nabla_i^2 + \frac{Ze^2}{r_i} \right) u_i(i) + \left[\sum_j \int u_j^*(j) \frac{e^2}{r_{ij}} u_j(j) \, d\tau_j \right] u_i(i) -$$

$$-\left[\sum_j \int u_j^*(j) \frac{e^2}{r_{ij}} u_i(j) \, d\tau_j \right] u_j(i) = \varepsilon_i u_i(i) \quad (2.6)$$

where again only $k=0$ term in the expansion of $1/r_{ij}$ (Equation (2.4)) is considered to allow for the separation of variables. (The restriction $j \neq i$ is now redundant since for $j=i$ the two terms involving r_{ij} are the same and cancel each other.) The third term is the so called exchange potential and arises because of the antisymmetry of the wavefunction and vanishes if u_j and u_i have opposite spin parts. The exchange term, has an opposite sign to the direct term, and lowers the energy for the electrons with parallel spins.

The electron, being in a spherically symmetric potential, should have no preference for any direction (except for the spin direction when the exchange potential is considered), and the observable ε_i (or any other observable) should not depend on angular part $Y_{l_i}^{m_{l_i}}(\theta_i, \phi_i)$ of the spin-orbital u_i, or in other words ε_i should be degenerate in l_i and m_{l_i}. In addition, there exists a spin degeneracy if we ignore the exchange potential (in Hartree approximation). For the closed shells there is only one possible determinantal function ψ. In incomplete shells there are not enough electrons to fill all the $(2s+1)(2l_i+1)$ spin orbitals with different $m_{l_i} = l_i, l_i-1, \cdots -l_i$ and $m_s = = +\frac{1}{2}, -\frac{1}{2}$ (but with the same radial part u'_{n,l_i}) and therefore more than one product function ψ (in Hartree approximation) or determinantal function ψ (in Hartree-Fock approximation) could be formed to give the same solution of the wave equation for the ion. The exchange term will favour energetically determinants with maximum parallel spin (within the limitation of the Pauli principle) and the various determinants may be grouped according to the spin arrangements of the electrons. When we include the nonspherical part of electron-electron repulsion term $\Sigma_{i \neq j} e^2/r_{ij}$ ($k > 0$ in Equation (2.4)) as perturbation on spherically symmetric potential, we no longer expect the electrons to enjoy angular independence, i.e. various $l_i s$ do not remain good parameters to describe the quantum mechanics of the ion. However, the ion as a whole being unable to discriminate between various directions in a field free space should still possess the angular independence; all determinantal functions ψ with the same total orbital angular momentum quantum number L, obtained by the vector addition of l_i's of all electrons in an ion, should constitute a distinct class in the solution of the wave equation. We can in that case represent the various states of the ion by the total orbital angular momentum $\mathbf{L}$ and the total spin angular momentum $\mathbf{S}$, provided we assume that there is no electrodynamic interaction (e.g. spin-orbit interaction) involved in the ion. In atomic spectroscopy an energy level specified by $\mathbf{L}$ and $\mathbf{S}$ is called a term (LS). The projection of $\mathbf{L}$ and $\mathbf{S}$ on any conveniently chosen axis, the z-axis, are written as M_L and M_S respectively in analogy with the single-particle functions. As long as there is no preferred direction for the ion the z-axis could be arbitrarily defined and no physical observable can depend on the choice of the z-axis, i.e. an LS term should be degenerate in M_L and M_S. This gives $(2S+1)(2L+1)$ fold degeneracy for a term since $M_L = L, L-1, \cdots, -L$ and $M_S = S, S-1, \cdots, -S$.

We shall now examine with the aid of an example what terms are possible with a given configuration say, $1s^2 2s^2 2p^6 3s^2 3p^6 3d^{10} 4s^2 4p^2$ of Ge. The closed shells, s^2, p^6 and d^{10} contribute nothing to L and S, since the possible orientations of angular momentum vector exactly cancel within a closed shell. The term structure arises due to incomplete shells. We have two electrons available to accommodate in six p-spin-orbitals and this could be done in $^6C_2 = 6!/(6-2)! \times 2! = 15$ ways i.e. there are 15 possible determinants for the two equivalent p-electrons. Since both the electrons have same principal quantum number ($n=4$) and angular momentum quantum numbers ($l=1, s=\frac{1}{2}$) we can distinquish between various determinants by the m_l and m_s values of the electrons. ($m_s = \frac{1}{2}$ is denoted by $+$ and $m_s = -\frac{1}{2}$ by $-$ sign.) Consistent with the exclusion principle the determinants can be classified as in the Table III.

TABLE III

$M_L \backslash M_S$	1	0	-1
2		$(1^+ 1^-)$	
1	$(1^+ 0^+)$	$(1^+ 0^-)(1^- 0^+)$	$(1^- 0^-)$
0	$(1^+ - 1^+)$	$(1^+ - 1^-)(1^- - 1^+)(0^+ 0^-)(1^- - 1^-)$	$(1^- - 1^-)$
-1	$(0^+ - 1^+)$	$(0^+ - 1^-)(0^- - 1^+)$	$(0^- - 1^-)$
-2		$(-1^+ - 1^-)$	

The state $(1^+ 1^-)$ has $M_L = 2$, $M_S = 0$; since we are considering 2 equivalent p-electrons it must belong to $L=2$ and $S=0$, i.e. ^1D term. Its other companions will have $M_L = 1, 0, -1, -2$ and $M_S = 0$, and together with them it forms the singlet D term. The state $(1^+ 0^+)$ has $M_L = 1$ and $M_S = 1$; it must belong to $L=1$ and $S=1$, i.e. ^3P term. The ^3P term will have 9 determinants: $M_L = 1, 0, -1$ and $M_S = 1, 0, -1$. The remaining state has $M_L = 0$, $M_S = 0$ and must belong to $L=0$, $S=0$, i.e. ^1S term. Our example will thus have ^1D, ^3P and ^1S terms with degeneracies 5, 9 and 1 respectively and obey the rule that atom has $(2S+1)(2L+1)$ fold degeneracy. A state $|L, M_L, S, M_S\rangle$ is represented by a single determinant only when a given M_L, M_S pair has one determinant, otherwise it will be a linear combination of all the determinants corresponding to the given M_L, M_S pair. The energy of the ^1D and ^3P terms are

$$E(^1D) = \langle L = 2, S = 0 | H | L = 2, S = 0 \rangle = \langle 1^+ 1^- | H | 1^+ 1^- \rangle$$

$$E(^3P) = \langle L = 1, S = 1 | H | L = 1, S = 1 \rangle = \langle 1^+ 0^+ | H | 1^+ 0^+ \rangle .$$

The ^1S term, $|L=0, S=0\rangle$, as well as the other two terms are represented by linear combination of the three determinants $(1^+ - 1^-)$, $(1^- - 1^+)$, $(0^+ 0^-)$. The energy of the ^1S term is, therefore, one of the three roots of the equation $\langle L=0, S=0| H |L=0, S=0\rangle = E$. The trace of secular determinant formed by the matrix elements of H between the three determinants is equal to the sum of the three roots. Thus we can easily find $E(^1S)$ by calculating only the diagonal matrix elements and substracting $E(^1D)$ and $E(^3P)$ from the sum of the diagonal matrix elements.

Upto here we have assumed that the only force in the free ion governing its mechanics is the electrostatic coulomb force. In fact there are many other interactions of less important magnitude which cause the partial or total removal of the degeneracy of a term. The most important of them in a field free space is the magnetic interaction between electron orbits and spins:

$$\sum_{ij} [a_{ij} \mathbf{l}_i \cdot \mathbf{s}_j + b_{ij} \mathbf{l}_i \cdot \mathbf{l}_j + c_{ij} \mathbf{s}_i \cdot \mathbf{s}_j] .$$

Out of this the diagonal first term is the most important and is the one we shall consider here. We write it explicitly as $\sum_i \xi(r_i) \mathbf{l}_i \cdot \mathbf{s}_i$ with $\xi(r) = (\hbar^2/2m_e^2 c^2)(1/r) \times (\partial U(r) \partial r)$, where c is the speed of light and $U(r)$ is the spherically symmetric part of the ion potential. We shall restrict ourselves to the so called $L - S$ coupling scheme, which assumes that the spin-orbit coupling energy is small compared to the term separation energies. This assumption allows us to neglect the matrix elements of $\Sigma_i \xi(r_i)\, \mathbf{l}_i \cdot \mathbf{s}_i$ between the states belonging to different terms. Now within each term we can use the Wigner-Eckart theorem of group theory (to be described later) to replace $\Sigma_i \xi(r_i)\, \mathbf{l}_i \cdot \mathbf{s}_i$ by $\lambda \mathbf{L} \cdot \mathbf{S}$, where λ is a term constant.

The inclusion of $\lambda \mathbf{L} \cdot \mathbf{S}$ in the ionic Hamiltonian removes the freedom enjoyed by the ion in respect to L and S separately. However, the total angular momentum $J(\mathbf{L}+\mathbf{S}=\mathbf{J})$ should remain unchanged under the combined rotation of spatial and spin coordinates of the ion. The projection of $\mathbf{J}$ on a conveniently defined z-axis is written $M_J (=J, J-1, \ldots, -J)$ and should be independent of how the axis is chosen. The $(2J+1)$ states with different M_J should be degenerate, and we can represent $L-S$ coupled states within a term by $|J, M_J\rangle$. The energies of the J multiplets within a term are obtained by evaluating the matrix elements $\langle L, M'_L, S, M'_S | \times$ $\times \lambda \mathbf{L} \cdot \mathbf{S} | L, M_L, S, M_S\rangle$. In our example of $4p^2$ configuration of Ge the introduction of spin-orbit coupling results in:

$$^1D\,(L = 2,\, S = 0;\, J = 2) \rightarrow {}^1D_2$$

$$^3P\,(L = 1,\, S = 1;\, J = 2, 1, 0) \rightarrow {}^3P_2,\, {}^3P_1,\, {}^3P_0$$

$$^1S\,(L = 0,\, S = 0;\, J = 0) \rightarrow {}^1S_0.$$

It is clear from the discussion of the exchange interaction potential that the states with the largest total spin will have minimum energy. However it is not apparent which states will have lowest energy when nonspherical potential and spin-orbit interaction are introduced. Hund's rule provides the answer for the ground multiplet – 'Choose maximum S and then maximum L consistent with Pauli principle; take $J=|L-S|$ if a shell is less than half filled and $J=L+S$ if the shell is more than half filled.' Schematically we can show the multiplet structure of Ge as in Figure 4.

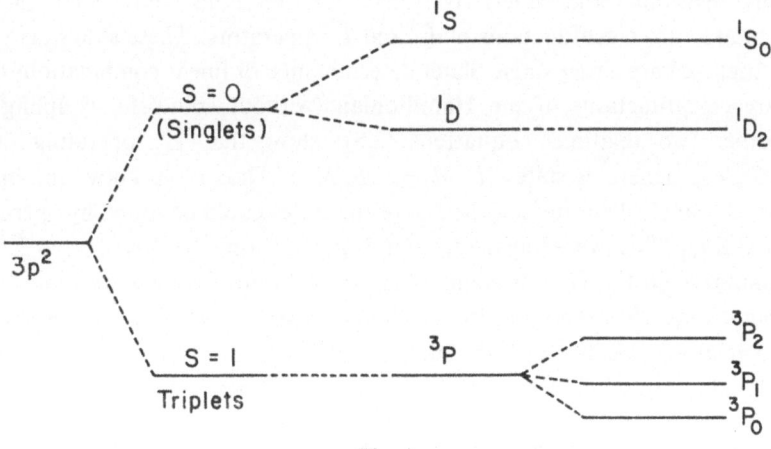

Fig. 4.

In heavy atoms the LS term description is not valid since v/c becomes large and magnetic forces become important compared to the electrostatic repulsion between the electrons. In these cases we first couple l_i and s_i for each electron to form j_i and then couple j_i's of all the electrons to form J, which will still give a valid description of the atom. It can be seen that the Hamiltonian, Equation (2.1), has singularities; when $r_{ij} \to 0$ the potential $\to \infty$. We should therefore, strictly speaking, include an infinite set of orthonormal functions to describe the state ψ of the ion. The electrons then have a finite probability of occurring in configurations other than the ground state and the matrix elements of H between the configurations are non-zero.

B. ANGULAR MOMENTUM OPERATORS

The orbital angular momentum operator is defined as

$$\tilde{l} = - i\hbar (\mathbf{r} \times \mathbf{V}).$$

We can write its x, y and z components in spherical coordinates and operate on spin-orbitals $u'_{nlyl}{}^{m_l}\chi(m) \equiv |nlm_l m_s\rangle$. The corresponding components of the total orbital angular momentum operator are obtained by summing the individual one-electron operators over all the electrons. When they operate on states $|L, M_L, S, M_S\rangle$, the results are:

$$\tilde{L}_z |L, M_L, S, M_S\rangle = \hbar M_L |L, M_L, S, M_S\rangle$$

$$\tilde{L}^2 |L, M_L, S, M_S\rangle = \hbar^2 L(L+1) |L, M_L, S, M_S\rangle$$

$$\tilde{L}_+ |L, M_L, S, M_S\rangle \equiv (\tilde{L}_x + i\tilde{L}_y) |L, M_L, S, M_S\rangle \tag{2.7}$$

$$= \hbar \sqrt{(L + M_L + 1)(L - M_L)} |L, M_L + 1, S, M_S\rangle$$

$$\tilde{L}_- |L, M_L, S, M_S\rangle \equiv (\tilde{L}_x - i\tilde{L}_y) |L, M_L, S, M_S\rangle$$

$$= \hbar \sqrt{(L - M_L + 1)(L + M_L)} |L, M_L - 1, S, M_S\rangle \tag{2.8}$$

where $M_L = \Sigma_i m_{l_i}, M_S = \Sigma_i m_{s_i}$. (For total spin angular momentum operators $\tilde{S}_z, \tilde{S}^2, \tilde{S}_\pm$, analogous relations hold true.) The first two relations show that the states $|L, M_L, S, M_S\rangle$ are eigenfunctions of $\tilde{L}_z$ and $\tilde{L}^2$ operators. These states, as we have already discussed are either single Slater determinants, or linear combinations of them which are eigenfunctions of the Hamiltonian (without spin-orbit coupling term).

The other two relations (Equations (2.8)) show that $\tilde{L}_\pm$ operating on state $|L, M_L, S, M_S\rangle$ generates states $|L, M_L \pm 1, S, M_S\rangle$. Thus if we know any one state of a term as a single determinant, the rest of the states could be found by operating on it by $\tilde{L}_\pm \equiv \Sigma_i l_{i\pm}$. The preceding description is possible only because $\tilde{L}_z$ and $\tilde{L}^2$ operators commute with the Hamiltonian. When they do not, we have to look for some other operator which commutes with the Hamiltonian and whose eigenfunctions could be found relatively easily. With the inclusion of the spin-orbit coupling term, $\lambda \mathbf{L} \cdot \mathbf{S}$ in the Hamiltonian we face such a situation. In the operator form, $(\mathbf{L} \cdot \mathbf{S})$ could be written as

$$(\mathbf{L} \cdot \mathbf{S})_{op} = \tfrac{1}{2}(\tilde{L}_+ \tilde{S}_- + \tilde{L}_- \tilde{S}_+) + \tilde{L}_z \tilde{S}_z. \tag{2.9}$$

Now

$$(\mathbf{L} \cdot \mathbf{S})_{\text{op}} |L, M_L, S, M_S\rangle =$$
$$= \tfrac{1}{2} \hbar^2 \left[(L + M_L + 1)(L - M_L)(S - M_S + 1)(S + M_S) \right]^{1/2} \times$$
$$\times |L, M_L + 1, S, M_S - 1\rangle + \tfrac{1}{2} \hbar^2 \left[(L - M_L + 1)(L + M_L) \times \right.$$
$$\times (S + M_S + 1)(S - M_S) \right]^{1/2} |L, M_L - 1, S, M_S + 1\rangle +$$
$$+ \hbar^2 M_L M_S |L, M_L, S, M_S\rangle. \tag{2.10}$$

Thus $|L, M_L, S, M_S\rangle$ is not an eigenfunction of $(\mathbf{L} \cdot \mathbf{S})_{\text{op}}$. However we notice that the sum of the z components of L and S in all the states in Equation (2.10) is constant, i.e. $M_L' + M_S' = M_L + M_S \equiv M_J$. But $(\mathbf{L} \cdot \mathbf{S})_{\text{op}}$ does not commute with the Hamiltonian. Therefore $|L, S, M_J\rangle$ are not correct eigenfunctions. The total angular momentum $\mathbf{J} = \mathbf{L} + \mathbf{S}$ commutes with the Hamiltonian. We should look for the eigenfunctions $|L, S, J, M_J\rangle$ such that

$$\bar{J}^2 |L, S, J, M_J\rangle = \hbar^2 J(J + 1) |L, S, J, M_J\rangle$$
$$\bar{J}_z |L, S, J, M_J\rangle = \hbar M_J |L, S, J, M_J\rangle. \tag{2.11}$$

For our purpose we only need to know that it is possible to obtain $|L, S, J, M_J\rangle$ eigenfunctions from $|L, M_L, S, M_S\rangle$ states using the group theoretical methods and that the result is

$$|L, S, J, M_J\rangle = \sum_{M_L, M_S} C_{M_L M_S; J M_J} |L, M_L, S, M_S\rangle. \tag{2.12}$$

$C_{M_L M_S; J M_J}$ are well known Clebsch-Gordon coefficients related to Wigner 3-j symbols by

$$C_{M_L M_S, J M_J} = (-1)^{S-L-M_J} \sum_{M_L, M_S} (2J + 1)^{1/2} \begin{pmatrix} L & S & J \\ M_L & M_S & -M_J \end{pmatrix} \tag{2.13}$$

C. IONS IN CRYSTAL POTENTIAL

Let us now consider the effect of crystal potential V on the free ion states. Three cases may arise: (1) *Weak crystal potential* – non-spherical part of the crystal potential, say V', is weaker than the spin-orbit interaction potential. The perturbation is considered on J multiplets $|L, S, J, M_J\rangle$. (2) *Intermediate crystal potential* – V' is stronger than the spin-orbit interaction potential but weaker than the interelectronic repulsion potential. The perturbation of V' on $L - S$ terms (states $|L, M_L, S, M_S\rangle$) is considered before the spin-orbit interaction. (3) *Strong crystal potential* – V' is stronger than the interelectronic repulsion. The term description is not useful then and we go back to single particle description of the atomic states and introduce perturbations in the order of decreasing magnitudes.

Conceptually the simplest procedure to obtain energy levels of the ion in crystal potential of any strength is to include all the perturbations at once in the Hartree-Fock Hamiltonian, to calculate the matrix elements of the resulting Hamiltonian between all determinental functions, and to diagonalise the matrix so formed. This procedure involves the diagonalisation of large matrices and makes it difficult to correlate various energy levels thus obtained with the free ion energy levels. For the intermediate and weak crystal potentials the use of 'operator equivalent' method of Stevens [5] makes the problem easier to handle and provides much of the relevant information adequately and elegantly.

Leaving aside the strong potential case our problem is essentially to evaluate the matrix elements of the type

$$\langle L, M'_L, S, M'_S \rangle \left(- e \sum_i V(r_i, \theta_i, \phi_i) \right) |L, M_L, S, M_S\rangle$$

or

$$\langle L, S, J, M'_J | \left(- e \sum_i V(r_i, \theta_i, \phi_i) \right) |L, S, J, M_J\rangle,$$

where the summation i is over all the electrons of the ion and $V(r_i, \theta_i, \phi_i)$ is the crystal potential at the ith electron. These matrix elements do not take into account the admixture of the $L-S$ terms, or of the J multiplets due to the crystal potential. The admixture of different $L-S$ terms and of different J-multiplets are possible but we shall assume that the separation between the consecutive terms or J-multiplets is large enough to ignore it. If desired, the matrix elements off diagonal in L, S or in J can be easily included with the extension of the method we shall outline here [6].

The spherically symmetric part of crystal potential V, as a first approximation, will only shift the energies of all the free ion states equally. Our interest here is in the study of the relative energies of the states, and we shall, therefore, not consider it.

D. METHOD OF OPERATOR EQUIVALENTS

This method makes use of the Wigner-Eckart theorem in a convenient form. Let us consider the matrix elements $\langle \alpha' J' M'_J | T_{kq} | \alpha J M_J \rangle$ where α, α' stand for any extra parameter (such as L, S) which might be necessary to specify the states $|J'M'_J\rangle$, $|J, M_J\rangle$, and T_{kq} is the tensor operator of rank k. The states $|J, M_J\rangle$ form a complete set of $2J+1$ basis functions and completely describe a state of quantum number J. The same is true for states $|J', M'_J\rangle$. The Wigner-Eckart theorem states that the dependence of the $2k+1$ components of the irreducible tensor T_{kq} on the projection quantum numbers M_J, M'_J, q is entirely determined by symmetry, e.g. by Clebsch-Gordon coefficients [7]:

$$\langle \alpha' J' M'_J | T_{kq} | \alpha J M_J \rangle = \langle \alpha' J' \| T_k \| \alpha J \rangle \, C(JkJ', M_J q M'_J). \tag{2.14}$$

Here $\langle \alpha' J' \| T_k \| \alpha J \rangle$ is called the reduced matrix element and is the only one to be calculated explicitly. Any suitable choice of $|\alpha' J' M'_J\rangle$, $|\alpha, J M_J\rangle$ states will give the reduced matrix element. Also we can calculate the matrix elements of any tensor operator equivalent to T_{kq} and the Wigner-Eckart theorem will enable us to equate the matrix elements of the new operator to those of T_{kq} within a single multiplicative constant. Stevens [5] noted that for diagonal matrix elements $\alpha' = \alpha$, $J' = J$, particularly useful 'operator equivalents' can be formed with angular momentum operators whose matrix elements are easy to calculate. As for example, the matrix elements of the angular momentum operator $3\tilde{J}_z^2 - J(J+1)$ are given by

$$\langle \alpha J M'_J | (3\tilde{J}_z^2 - J(J+1)) | \alpha J M_J \rangle = \delta_{M'_J M_J} \{3M_J^2 - J(J+1)\}.$$

Now consider an operator $\sum_i (3z_i^2 - r_i^2)$ with the summation extending over all the electrons of the ion. Within a manifold (α, J) the Wigner-Eckart theorem permits us to write

$$\sum_i (3z_i^2 - r_i^2) = a[3\tilde{J}_z^2 - J(J+1)] \tag{2.15}$$

where the constant a is obtained by calculating explicitly one matrix element of $\sum_i (3z_i^2 - r_i^2)$ over any one convenient state $|\alpha, J, M_{J_0}\rangle$:

$$a\{3M_{J_0}^2 - J(J+1)\} = \langle \alpha J M_{J_0}| \sum_i 3z_i^2 - r_i^2 |\alpha J M_{J_0}\rangle =$$

$$= \int \Psi^* \sum_i (3z_i^2 - r_i^2) \Psi \, d\tau. \tag{2.16}$$

The operator equivalents of the function $\sum_i r_i Z_{kq}(\theta_i, \phi_i)$ (Table II) occurring in the crystal potential could be obtained by going from x, y, z to $\tilde{J}_x, \tilde{J}_y, \tilde{J}_z$ respectively while always allowing for the noncommutation of $\tilde{J}_x, \tilde{J}_y, \tilde{J}_z$. This is done by replacing the products of x, y, z by an expression consisting of the average of all possible different combinations of the corresponding products of $\tilde{J}_x, \tilde{J}_y, \tilde{J}_z$, for example, by replacing $\sum_i x_i y_i$ by $\frac{1}{2}(\tilde{J}_x\tilde{J}_y + \tilde{J}_y\tilde{J}_x)$. This procedure gives us operators with the same transformation properties under rotation as the crystal potential. Some of the operator equivalents are given in Table IV. [4, 6]

TABLE IV

$\sum(3z^2 - r^2)$	$\rightarrow [3\tilde{J}_z^2 - J(J+1)]$	$\equiv O_{20}$
$\sum xz$	$\rightarrow \frac{1}{2}[\tilde{J}_x\tilde{J}_z + \tilde{J}_z\tilde{J}_x]$	$\equiv O_{21}{}^c$
$\sum(x^2 - y^2)$	$\rightarrow \frac{1}{2}[\tilde{J}_+^2 + \tilde{J}_-^2]$	$\equiv O_{22}{}^c$
$\sum yz$	$\rightarrow \frac{1}{2}[\tilde{J}_y\tilde{J}_z + \tilde{J}_z\tilde{J}_y]$	$\equiv O_{21}{}^s$
$\sum 2xy$	$\rightarrow [\tilde{J}_x\tilde{J}_y + \tilde{J}_y\tilde{J}_x]$	$\equiv O_{22}{}^s$
$\sum(35z^4 - 30r^2z^2 + 3r^4)$	$\rightarrow [35\tilde{J}_z^4 - 30J(J+1)\tilde{J}_z^2 + 25\tilde{J}_z^2$	
	$\quad - 6J(J+1) + 3J^2(J+1)^2]$	$\equiv O_{40}$
$\sum(x^4 - 6x^2y^2 + y^4)$	$\rightarrow \frac{1}{2}[\tilde{J}_+^4 + \tilde{J}_-^4]$	$\equiv O_{44}{}^c$
$\sum(231z^6 - 315z^4r^2 + 105z^2r^4 - 5r^6)$	$\rightarrow [231\tilde{J}_z^6 - 315J(J+1)\tilde{J}_z^4 + 735\tilde{J}_z^4$	
	$\quad + 105J^2(J+1)^2\tilde{J}_z^2 - 525J(J+1)\tilde{J}_z^2 + 294\tilde{J}_z^2$	
	$\quad - 5J^3(J+1)^3 + 40J^2(J+1)^2 - 60J(J+1)] \equiv O_{60}$	
$\sum(11z^2 - r^2)(x^4 - 6x^2y^2 + y^4)$	$\rightarrow \frac{1}{4}[11\tilde{J}_z^2 - J(J+1) - 38)(\tilde{J}_+^4 + \tilde{J}_-^4)$	
	$\quad + (\tilde{J}_+^4 + \tilde{J}_-^4)(11\tilde{J}_z^2 - J(J+1) - 38)]$	$\equiv O_{64}{}^c$
$\sum(x^6 - 15x^4y^2 + 15x^2y^4 - y^6)$	$\rightarrow \frac{1}{2}[\tilde{J}_+^6 + \tilde{J}_-^6]$	$\equiv O_{66}{}^c$
$\tilde{J}_x = \frac{1}{2}[\tilde{J}_+ + \tilde{J}_-];\ \tilde{J}_y = -i/2[\tilde{J}_+ - \tilde{J}_-]$		

We shall now see as an example how the matrix elements of $\sum(3z^2 - r^2)$ between the states $|L, M_L, S, M_S\rangle$ (or $|L, S, J, M_J\rangle$) are calculated. From Equation (2.16) it is clear that the constant a is simply the reduced matrix element of Wigner-Eckart theorem (Equation (2.14)) and that if the electrons giving rise to the ionic states are equivalent a could be factorized in two parts – a radial part and an angular part:

$$\langle L, M_L'| \sum_i (3z_i^2 - r_i^2) |L, M_L\rangle = \langle L\| \alpha \|L\rangle \langle r^2\rangle \times$$

$$\times \langle L, M_L'| 3\tilde{L}_z^2 - L(L+1) |L\ M_L\rangle. \tag{2.17}$$

Here $\langle r^2 \rangle = \int r^2 u'^2 \, dr$ (if we had a function $\Sigma_i f(r_i) \, (3z_i^2 - r_i^2)$ instead of $\Sigma_i (3z_i^2 - r_i^2)$ we would have had the radial part of the matrix element as $\langle f(r) \, r^2 \rangle$). $\langle L \| \alpha \| L \rangle$ is calculated as follows. First we calculate the reduced matrix element $\langle l \| \alpha \| l \rangle$ for a single electron with angular momentum l.

$$\langle l, l | 3z^2 - r^2 | l, l \rangle = \langle r^2 \rangle \int Y_l^{l*} \left(\frac{3z^2 - r^2}{r^2} \right) Y_l^l \, d\Omega =$$

$$= \langle r^2 \rangle \, 2 \sqrt{\frac{4\pi}{5}} \int Y_l^{l*} Y_2^0 Y_l^l \, d\Omega =$$

$$= \langle r^2 \rangle \frac{-2l}{2l + 3}. \tag{2.18}$$

But at the same time we can write

$$\langle l, l | 3z^2 - r^2 | l, l \rangle = \langle l \| \alpha \| l \rangle \langle r^2 \rangle \langle ll | O_{20} | ll \rangle =$$

$$= \langle l \| \alpha \| l \rangle \langle r^2 \rangle \{3l^2 - l(l+1)\}. \tag{2.19}$$

Comparison of Equation (2.18), and Equation (2.19) gives

$$\langle l \| \alpha \| l \rangle = - \frac{2}{(2l + 3)(2l - 1)}. \tag{2.20}$$

Next we calculate the matrix element $\langle L \| \alpha \| L \rangle$. Consider a ground state term (L, S) belonging to configuration with t equivalent electrons of orbital angular momentum l. Then according to Hund's rule $M_L = L$ and $M_S = S$ and the state is a single Slater determinant. If the shell giving rise to the term is not more than half filled $[t \leqslant (2l + 1)]$ then $t = 2S$ and the Slater determinant in terms of the m_l values of the electrons is $(l, l-1, l-2, \ldots, l-t+1)$.
Now

$$\langle LL | \sum 3z^2 - r^2 | LL \rangle = \sum_{m_l = l}^{l - t + 1} \langle lm_l | 3z^2 - r^2 | lm_l \rangle =$$

$$= \langle l \| \alpha \| l \rangle \langle r^2 \rangle \sum_{m_l} \langle lm_l | 3l_z^2 - l(l+1) | lm_l \rangle =$$

$$= \langle l \| \alpha \| l \rangle \langle r^2 \rangle [3 \{l^2 + (l-1)^2 +$$

$$+ \cdots + (l - t + 1)^2\} - tl(l+1)] =$$

$$= \langle l \| \alpha \| l \rangle \langle r^2 \rangle [L(2l + 1 - 4S)]. \tag{2.21}$$

Also

$$\langle LL | \sum 3z^2 - r^2 | LL \rangle = \langle L \| \alpha \| L \rangle \langle r^2 \rangle \langle LL | 3\tilde{L}_z^2 - L(L+1) | LL \rangle =$$

$$= \langle L \| \alpha \| L \rangle \langle r^2 \rangle \{3L^2 - L(L+1)\}. \tag{2.22}$$

Comparing Equations (2.21) and (2.22) ($-$ sign is good if the shell is more than half filled)

$$\langle L \| \alpha \| L \rangle = (\pm) \frac{2l + 1 - 4S}{2L - 1} \langle l \| \alpha \| l \rangle =$$

$$= (\pm) \frac{2(2l + 1 - 4S)}{(2l - 1)(2l + 3)(2L - 1)}.$$

Abragam and Bleaney [6] have discussed the method of calculating the reduced matrix elements in detail and have compiled tables of these matrix elements. They have also tabulated the matrix elements of the O_{kq} operators.

E. ENERGY LEVELS OF METAL-IONS IN CRYSTAL POTENTIAL

We shall now discuss the procedure of obtaining the energy levels of a rare earth ion in the weak crystal potential of D_{3h} symmetry. We rewrite the Equation (1.11) for D_{3h} symmetry defining the crystal potential Hamiltonian as

$$H_{CP} = -e \sum_i V_{D_{3h}} = \sum_i [A_2^0(3z_i^2 - r_i^2) + A_4^0(35z_i^4 - 30r_i^2 z_i^2 + 3r_i^4) +$$
$$+ A_6^0(231z_i^6 - 315r_i^2 z_i^4 + 105r_i^4 z_i^2 - 5r_i^6) +$$
$$+ A_6^6(x_i^6 - 15x_i^4 y_i^2 + 15x_i^2 y_i^4 - y_i^6)], \qquad (2.24)$$

where

$$A_2^0 = -e\gamma_{20} \frac{1}{4}\left(\frac{5}{\pi}\right)^{1/2}, \quad A_4^0 = -e\gamma_{40} \frac{3}{16}\left(\frac{1}{\pi}\right)^{1/2}$$

$$A_6^0 = -e\gamma_{60} \frac{1}{32}\left(\frac{13}{\pi}\right)^{1/2}, \quad A_6^6 = -e\gamma_{66} \frac{231}{64}\left(\frac{26}{231\pi}\right)^{1/2}.$$

The matrix elements of H_{CP} between the multiplets $|J, M_J\rangle$, using the operator equivalents are:

$$H_{M'_J M_J}^{CP} = A_2^0 \langle r^2\rangle \langle J\|\alpha\|J\rangle \langle J, M'_J| O_{20}|J, M_J\rangle +$$
$$+ A_4^0 \langle r^4\rangle \langle J\|\beta\|J\rangle \langle JM'_J| O_{40}|JM_J\rangle +$$
$$+ A_6^0 \langle r^6\rangle \langle J\|\gamma\|J\rangle \langle J, M'_J| O_{60}\|JM_J\rangle +$$
$$+ A_6^6 \langle r^6\rangle \langle J\|\gamma\|J\rangle \langle JM'_J| O_{66}^c|JM_J\rangle. \qquad (2.25)$$

Here $\langle J\|\alpha\|J\rangle$, $\langle J\|\beta\|J\rangle$ and $\langle J\|\gamma\|J\rangle$ are the reduced matrix elements corresponding to the operator O_{2q}, O_{4q} and O_{6q} respectively. The matrix elements $H_{M'_J M_J}^{CP}$ could be explicitly obtained in terms of $A_k^q \langle r^k\rangle$. If $A_k^q \langle r^k\rangle$ are known, the diagonalization of the matrix $(H_{M'_J M_J}^{CP})$ immediately gives the energy levels of the ion in crystal potential. For the diagonalization purpose we define the ion-states in the crystal potential as

$$\psi_v = \sum_{M_J} (a_{M_J}^v) |J, M_J\rangle \qquad (2.26)$$

with normalization condition $\Sigma_{M_J}(a_{M_J}^v)^2 = 1$. We thus get the coefficient $a_{M_J}^v$ and hence the eigenfunctions ψ_v.

Alternatively if the ion-energy levels in crystal are known from experiments $A_k^q \langle r^k\rangle$ and ψ_v can be computed. The crystal potential parameters, A_k^q, are difficult to calculate satisfactorily and in most cases are obtained from the experiments. There is some modification of the crystal potential at any site within an ion due to the polarization of the closed electronic shells (Sternheimer effect) caused by the quadrupole and higher moments of the potential. This effect is taken into account by replacing $A_k^q \langle r^k\rangle$ with $A_k^q \langle r^k\rangle (1 - \lambda_{nl}^{(k)})$ for the electrons in (nl) shell. What one measures experimentally is, therefore, $A_k^q \langle r^k \rangle (1 - \lambda_{nl}^{(k)})$. We shall later discuss this subject in some detail.

For the iron group ions the crystal potential is of intermediate strength and the matrix elements of the crystal potential Hamiltonian are evaluated between a term states $|L, M_L\rangle$ in the same way as for the J-multiplets in weak crystal potential. The matrix is diagonalized and the new energy levels and eigenfunctions are obtained (without LS coupling). If the spin-orbit interaction is also important, then $-\lambda \, L \cdot S$ and H_{CP} can be considered together to calculate the matrix elements between the states $|L, M_L, S, M_S\rangle$ and the matrix can be diagonalized by defining new states $\psi_v = \Sigma_{M_L, M_S} a_{M_L, M_S}^v |L, M_L, S, M_S\rangle$. For the iron group ions the crystal potential symmetry is generally high, and the multiplet splitting and their structure in the symmetry of the potential are known from the group theory [3, 8, 9].

F. NUCLEAR AND CORE-HOLE ENERGY LEVELS IN CRYSTAL POTENTIAL

There are mainly two ways in which the crystal potential interacts with the nuclear moments or with the core holes (we shall first discuss the core-hole states and later see how our expressions are simplified for the nuclear states). First is the same interaction as on the ionic states (discussed earlier); second is the interaction of the core holes with the electrons in unfilled shells distorted by the crystal potential.

1. Unfilled Shell Interaction

The interaction potential energy between the two systems of point charges, e_i at $\mathbf{R}_i$ and e_j at $\mathbf{r}_j$, is

$$H_{\text{int}} = \sum_i \sum_j \frac{e_i e_j}{|\mathbf{R}_i - \mathbf{r}_j|} = \sum_i \sum_j e_i e_j \sum_{k=0}^{\infty} \frac{r_<^k}{r_>^{k+1}} P_k(\cos \omega) =$$

$$= \sum_i \sum_j e_i e_j \sum_{k=0}^{\infty} \frac{4\pi}{2k+1} \frac{r_<^k}{r_>^{k+1}} \sum_{q=0}^{k} \sum_{\alpha=c,s} Z_{kq}^\alpha(\Theta_i \Phi_i) Z_{kq}^\alpha(\theta_j \phi_j) =$$

$$= \sum_i \sum_j e_i e_j \sum_{k=0}^{\infty} \frac{4\pi}{2k+1} \left[\frac{R_i^k}{r_j^{k+1}} \theta(r_j - R_i) + \right.$$

$$\left. + \frac{r_j^k}{R_i^{k+1}} \theta(R_i - r_j) \right] \sum_{q=0}^{k} \sum_{\alpha} Z_{kq}^\alpha(\Theta_i \Phi_i) Z_{kq}^\alpha(\theta_j \phi_j)$$

where we have introduced the function $\theta(r - R) = 1$ for $r - R > 0$ and $\theta(r - R) = 0$ for $r - R < 0$.

We define

$$A_{kq}^\alpha = \sum_i e_i R_i^k Z_{kq}^\alpha(\Theta_i \Phi_i) \sum_j e_j r_j^{-(k+1)} \theta(r_j - R_i) Z_{kq}^\alpha(\theta_j \phi_j)$$

$$B_{kq}^\alpha = \sum_i e_i R_i^{-(k+1)} Z_{kq}^\alpha(\Theta_i \Phi_i) \sum_j e_j r_j^k \theta(R_i - r_j) Z_{kq}^\alpha(\theta_j \phi_j)$$

and thus write

$$H_{int} = \sum_{k=0}^{\infty} \frac{4\pi}{2k+1} \sum_{q=0}^{k} \sum_{\alpha=c,s} [A_{kq}^{\alpha} + B_{kq}^{\alpha}]. \tag{2.27}$$

Let subscript i denote the electrons in a core shell (n, l) with a vacancy (subscripted 'h' as hole) such that $L_h = l$, $S_h = \frac{1}{2}$, $J_h = l \pm \frac{1}{2}$. Let the subscript j denote all electrons of the ion (i.e. the electrons outside the core shells of the ion) giving rise to the ground term LS (or ground multiplet J). For single core-hole states we have only one term $(l, \frac{1}{2})$. (We shall assume that the core-hole does not mix with the ionic configuration without the hole). In most physical situations the core-holes have $l \leqslant 2$. The matrix of H_{int} between the core-hole states will be non-zero for even $k \leqslant 4$. We shall again ignore $k = 0$ term since it does not give rise to the splitting of the energy levels.

From the Tables II and IV it is seen that the tesseral harmonics Z_{kq}^{α} can be written directly in the operator form. We define new operators Q_{kq}^{α} such that they are related to O_{kq}^{α} through the constant factors in the tesseral harmonics, e.g.

$$Q_{22}^{c} = \frac{\sqrt{15}}{2} \frac{1}{\sqrt{n}} O_{22}^{c}; \qquad Q_{40} = \frac{3\sqrt{2}}{16} \frac{1}{\sqrt{2\pi}} O_{40}.$$

We can now use the Wigner-Eckart theorem (Equation (2.14)) and, within the manifold of the ion term (LS), write ($e_j = -e$ for all electrons).

$$A_{kq}^{\alpha}(L, S) = \sum_{i} e_i R_i^k Z_{kq}^{\alpha}(\Theta_i, \Phi_i)(-e) \times$$
$$\times \langle L \| a_k \| L \rangle \langle r^{-(k+1)} \theta(r - R_i) \rangle Q_{kq}^{\alpha}(L)$$
$$B_{kq}^{\alpha}(L, S) = \sum_{i} e_i R_i^{-(k+1)} Z_{kq}^{\alpha}(\Theta_i \Phi_i)(-e) \times$$
$$\times \langle L \| a_k \| L \rangle \langle r^k \theta(R_i - r) \rangle Q_{kq}^{\alpha}(L)$$

where the reduced matrix elements $\langle L \| a_2 \| L \rangle = \langle L \| \alpha \| L \rangle \langle L \| a_4 \| L \rangle = \langle L \| \beta \| L \rangle$ and so on. Further, within a J_h manifold of the core-hole with $e_i = +e$,

$$A_{kq}^{\alpha}(L, S, J_h) = -e^2 \langle J_h \| a_k \| J_h \rangle \langle L \| a_k \| L \rangle \times$$
$$\times \langle R^k \langle r^{-(k+1)} \theta(r - R) \rangle \rangle Q_{kq}^{\alpha}(L) Q_{kq}^{\alpha}(J_h)$$
$$B_{kq}^{\alpha}(L, S, J_h) = -e^2 \langle J_h \| a_k \| J_h \rangle \langle L \| a_k \| L \rangle \times$$
$$\times \langle R^{-(k+1)} \langle r^k \theta(R - r) \rangle \rangle Q_{kq}^{\alpha}(L) Q_{kq}^{\alpha}(J_h)$$

and therefore,

$$H_{int}(L, S, J_h) = \sum_{k} \frac{4\pi}{2k+1} \sum_{q=0}^{k} \sum_{\alpha=c,s} \times$$
$$\times (-e^2) \langle J_h \| a_k \| J_h \rangle \langle L \| a_k \| L \rangle Q_{kq}^{\alpha}(L) Q_{kq}^{\alpha}(J_h) \times$$
$$\times \langle R^k \langle r^{-(k+1)} \theta(r - R) \rangle + R^{-(k+1)} \langle r^k \theta(R - r) \rangle \rangle. \tag{2.28}$$

The most important assumption made to obtain a Hamiltonian with the separated angular momentum operators for core-hole states and ion-states is that each angular momentum J_h and L is separately a good quantum number. The last factor in Equation (2.28) could explicitly be written as

$$\langle R^k \langle r^{-(k+1)} \theta(r-R)\rangle + R^{-(k+1)} \langle r^k \theta(R-r)\rangle\rangle =$$

$$= \int_0^\gamma \int \frac{r_<}{r_>^{k+1}} u_h'^2(R) u_v'^2(r) \, dR \, dr \equiv F^k(u_h(R), u_v(r)) \qquad (2.29)$$

where $u_h'(R)$ is the radial part of the core-hole (electron) wave function u_h and $u_v'(r)$ is the radial part of the valence-electron wave function u_v.

We have not considered the exchange interaction here. The wave functions u_h and u_v overlap. We ultimately calculate the matrix elements of the type $\langle a(1) \, b(2) \, |e^2/r_{12}| \times \times a(1) \, b(2)\rangle$. We should therefore, rigorously speaking, include the exchange interaction by writing the matrix elements as $\langle a(1) \, b(2) \, |e^2/r_{12}(1-P_{12})| \, a(1) \, b(2)\rangle$ where P_{12} is an operator which permutes a and b with respect to the coordinates 1 and 2. One can write P_{12} in terms of tensor operators separable in 1 and 2 coordinates and then use Wigner-Eckart theorem to get the correct $H_{int}(L, S, J_h)$, but for simpliciy we shall be satisfied here without the exchange term.

For core p-holes, $l=1$ and therefore only $k=2$ term in Equation (2.28) will have non-zero matrix elements (apart from the uninteresting $k=0$ term). We get

$$H_{int}^p(L, S, J_h) = \frac{-4\pi}{5} e^2 \sum_{q=0}^{2} \sum_\alpha \langle J_h\| \, a_2 \, \|J_h\rangle \times$$

$$\times \langle L\| \, a_2 \, \|L\rangle \, F^2(u_h', u_v') \, Q_{2q}^\alpha(J_h) \, Q_{2q}^\alpha(L). \qquad (2.30)$$

In order to obtain the nuclear quadrupole interaction Hamiltonian from Equation (2.28) we imagine the nucleus to be a squeezed multiple core hole structure. For the purpose electrons R_i's (now nucleon coordinates) could be taken zero and F^k integral breaks into two integral

$$\int_0^\infty R^k u_n'^2(R) \, dR = \langle R^k\rangle_n \quad (n \text{ for nucleus})$$

$$\int_0^\infty r^{-(k+1)} u_v'^2(r) \, dr = \langle r^{-(k+1)}\rangle.$$

Further in this case, J_h manifold will be the manifold of nuclear spin I. The quadrupole moment of a nucleus with spin I is defined as

$$eQ = \langle II| \sum_i^{nucleons} e_i(3Z_i^2 - R_i^2) |II\rangle =$$

$$= e \langle I\| \, a_2 \, \|I\rangle \langle R^2\rangle \langle II| \, 3I_z^2 - I(I+1) |II\rangle =$$

$$= e \langle I\| \, a_2 \, \|I\rangle \langle R^2\rangle \{I(2I-1)\}.$$

or

$$\langle I\| \, a_2 \, \|I\rangle \langle R^2\rangle = Q/I(2I-1).$$

Thus for nuclear quadrupole interaction we need consider only $k=2$ term in Equation (2.28) and write

$$H_{int}^{Q}(L, S, I) = \frac{4\pi}{5} \sum_{q=0}^{2} \sum_{\alpha} \left[\frac{-e^2 Q \langle r^{-3} \rangle}{I(2I-1)} \langle L \| \alpha \| L \rangle \, Q_{2q}^{\alpha}(I) \, Q_{2q}^{\alpha}(L) \right].$$

We substitute explicit expressions of $Q_{2q}^{\alpha}(I)$ and $Q_{2q}^{\alpha}(L)$ and after some manipulations get the well known form of H_{int}^{Q}:

$$H_{int}^{Q}(L, S, I) = -\frac{e^2 Q \langle r^{-3} \rangle}{I(2I-1)} \langle L \| \alpha \| L \rangle \left[\tfrac{1}{4} \{3L_z^2 - L(L+1)\} \{3I_z^2 - \right.$$
$$- I(I+1)\} + \tfrac{3}{8} \{ (L_z L_+ + L_+ L_z)(I_z I_- + I_- I_z) +$$
$$\left. + (L_z L_- + L_- L_z)(I_z I_+ + I_+ I_z) \} + \tfrac{3}{8} (L_+^2 I_-^2 + L_-^2 I_+^2) \right].$$
$$(2.31)$$

2. Direct Crystal Potential Interaction

The expressions for crystal potential (Equations (1.10) and (1.11)) and H_{CP} (Equation (2.24)), suggest that we can write a general form of crystal potential Hamiltonian for a manifold J_h (I for nucleus) as

$$H_{CP}(J_h) = -\sum_{k=0}^{\infty} \sum_{q=0}^{k} \sum_{\alpha} A_k^{q\,(\alpha)} \langle R^k \rangle \langle J_h \| a_2 \| J_h \rangle O_{kq}^{\alpha}(J_h) \qquad (2.32)$$

(negative sign indicates a hole). For core p-holes ($k=2$)

$$H_{CP}^{p}(J_h) = - \langle J_h \| a_2 \| J_h \rangle \langle R^2 \rangle \sum_{q=0}^{2} \sum_{\alpha} A_2^{q\,(\alpha)} O_{2q}^{\alpha}(J_h). \qquad (2.33)$$

The term with $q=1$ and $\alpha=s$ will be absent due to the symmetry of crystal potentials, and therefore

$$H_{CP}^{p}(J_h) = - \langle J_h \| \alpha \| J_h \rangle \langle R^2 \rangle \left[A_2^0 (3J_{h_z}^2 - \right.$$
$$\left. - J_h(J_h + 1)) + \tfrac{1}{2} A_2^{2\,(c)} (J_{h+}^2 + J_{h-}^2) \right]. \qquad (2.34)$$

For nucleus with quadrupole moment Q and spin I,

$$H_{CP}^{Q}(I) = - \frac{Q}{I(2I-1)} \left[A_2^0 (3I_z^2 - I(I+1)) + \tfrac{1}{2} A_2^{2\,(c)} (I_+^2 + I_-^2) \right]. \qquad (2.35)$$

We derived the expressions for interaction Hamiltonians assuming implicitly that the electrons in the closed shells are passive. This is not true, and the closed shells, through their polarization, do affect the electrostatic interactions. This effect, as Sternheimer [10] first showed, can be accounted for by linear shielding (antishielding) multiplicative factors $(1 - R^{(k)})$ and $(1 - \lambda^{(k)})$ in H_{int} and H_{CP} respectively. The Sternheimer parameters R and λ are suffixed 0 for the nuclear site and nl for the nl-core-hole site. We shall later discuss the Sternheimer effect in some detail.

3. *Crystal Potential Splitting of Core $p_{3/2}$ Levels of* Tm^{3+} *in Thulium Ethyl Sulphate*

The symmetry of the crystal potential at Tm^{3+} ions in thulium ethyl sulphate (TmES) is C_{3h} and therefore for the core p-hole states we need consider only $q=0$ in H^p_{CP} (Equation (2.33)). It is not obvious that only $q=0$ term need to be considered also for H^p_{int} (Equation (2.30)). However, it can be proved that the crystal potential splits the degenerate ion-energy levels such that the symmetry of the charge distribution of the ion is *not* lower than that of the crystal potential. For Tm^{3+} ion, with $4f^{12}$ configuration, the spin-orbit interaction is stronger than the crystal potential. We should, therefore, consider J manifold instead of LS manifold. Thus we write

$$H^p_{int}(J, J_h) = \frac{4\pi}{5} (-e^2) \langle J_h \| \alpha \| J_h \rangle \langle J \| \alpha \| J \rangle F^2 \left(u'_p(R) \, u'_{4f}(r) \right) \times$$

$$\times \sqrt{\frac{5}{4\pi} \frac{1}{2}} \, O_{20}(J_h) \times \sqrt{\frac{5}{4\pi} \frac{1}{2}} \, O_{20}(J) =$$

$$= -\frac{e^2}{4} \langle J_h \| \alpha \| J_h \rangle \langle J \| \alpha \| J \rangle F^2 (u'_p u'_{4f}) \times$$

$$\times \{3J^2_z - J(J+1)\} \{3J^2_{h_z} - J_h(J_h+1)\}.$$

and

$$H^p_{CP}(J_h) = - A^0_2 \langle R^2 \rangle \langle J_h \| \alpha \| J_h \rangle \{3J^2_{h_z} - J_h(J_h+1)\}.$$

Adding the two,

$$H^p(J, J_h) = - \left[A^0_2 \langle R^2 \rangle + \frac{e^2}{4} \langle J \| \alpha \| J \rangle F^2 (u'_p u'_{4f}) \{3J^2_z - J(J+1)\} \right] \times$$

$$\times \langle J_h \| \alpha \| J_h \rangle \{3J^2_{h_z} - J_h(J_h+1)\}. \tag{2.36}$$

The contribution of crystal potential ion state ψ_v (Equation (2.26)) with energy E_v is $\langle v | H^p(J, J_h) | v \rangle$. Experimentally one observes contribution from various ions which have the Maxwell-Boltzmann energy distribution. We assume that only those electronic states ψ_v, which belong to the ground J-multiplet, are populated at temperatures T of interest. The contribution of all ψ_v's is obtained by replacing $3J^2_z - J(J+1)$ in Equation (2.36) by

$$\langle 3J^2_z - J(J+1) \rangle_T = \sum_{v=1}^{2J+1} \langle v | 3J^2_z - J(J+1) | v \rangle \times$$

$$\times \exp(-E_v/k_B T) / \sum_{v=1}^{2J+1} \exp(-E_v/k_B T) \tag{2.37}$$

where k_B is the Boltzmann constant.

The J_h manifolds for single p-hole are $J_h = \frac{3}{2}$, $M_{J_h} = \pm\frac{3}{2}$, $\pm\frac{1}{2}$ and $J_h = \frac{1}{2}$, $M_{J_h} = \pm\frac{1}{2}$. The Hamiltonian H^p is diagonal in J_h as well as in M_{J_h} and, therefore, the matrix elements $\langle J_h M_{J_h} | H^p | J_h M_{J_h} \rangle$ are the energies and $| J_h, M_{J_h} \rangle$ the eigenstates of the Hamiltonian H^p. The states are degenerate in $| J_h, \pm M_{J_h} \rangle$. Therefore the multiplet $J = \frac{1}{2}$ remains unsplit whereas the multiplet $J_h = \frac{3}{2}(p_{3/2})$ splits into two levels,

$|\frac{3}{2}, \pm\frac{3}{2}\rangle$ and $|\frac{3}{2}, \pm\frac{1}{2}\rangle$. The splitting of $np_{3/2}$ multiplet is then given by

$$\Delta E(np_{3/2}) = A_2^0 \langle r^2 \rangle_{np} + \frac{e^2}{4} \langle J \| \alpha \| J \rangle F^2(u'_{np} u'_{4f}) \times$$

$$\times \langle 3J_z^2 - J(J+1) \rangle_T] \langle \tfrac{3}{2} \| \alpha \| \tfrac{3}{2} \rangle \times$$

$$\times \{ \langle \tfrac{3}{2}, \pm \tfrac{3}{2} | 3J_z^2 - J(J+1) | \tfrac{3}{2}, \pm \tfrac{3}{2} \rangle -$$

$$- \langle \tfrac{3}{2}, \pm \tfrac{1}{2} | 3J_z^2 - J(J+1) | \tfrac{3}{2}, \pm \tfrac{1}{2} \rangle \}. \qquad (2.38)$$

The inclusion of the Sternheimer effect gives finally

$$\Delta E(np_{3/2}) = [A_2^0 \langle r^2 \rangle_{np} (1 - \lambda_{np}^{(2)}) + \frac{e^2}{4} \langle J \| \alpha \| J \rangle F^2(np, 4f) \times$$

$$\times (1 - R_{np}^{(2)}) \langle 3J_z^2 - J(J+1) \rangle_T] \times$$

$$\times \langle \tfrac{3}{2} \| \alpha \| \tfrac{3}{2} \rangle \{ \langle \tfrac{3}{2}, \pm \tfrac{3}{2} | 3J_z^2 - J(J+1) \times$$

$$\times | \tfrac{3}{2}, \pm \tfrac{3}{2} \rangle - \langle \tfrac{3}{2}, \pm \tfrac{1}{2} | 3J_z^2 - J(J+1) | \tfrac{3}{2}, \pm \tfrac{1}{2} \rangle \}. \qquad (2.39)$$

Wong and Richman [11], made optical obsorption measurements on TmES and obtained $C_2^0 = 129.8$, $C_4^0 = -71.0$, $C_6^0 = -28.6$ and $C_6^6 = 432.8$ in units of Cm^{-1}, where $C_k^q = A_k^q \langle r^k \rangle_{4f} \times (1 - \lambda_{4f}^{(k)})$. These crystal potential parameters were used to calculate the matrix elements $H_{M'_J, M_J}^{CP}$ (Equation (2.25)) for the ground multiplet $^3H_6$ and the matrix was diagonalized to get E_ν's and ψ_ν's (Equation (2.26)). Then $\langle \nu | 3J_z^2 - J(J+1) | \nu \rangle$ was calculated [12] and the statistical average $\langle 3J_z^2 - J(J+1) \rangle_T$ obtained (Equation (2.37)). The reduced matrix elements $\langle J \| \alpha \| J \rangle = \frac{1}{99}$ ($J = 6$, $4f^{12}$ configuration) and $= \frac{2}{15}$ ($J = \frac{3}{2}$, np^{1+} configuration). The factor in curly bracket (Equation (2.39)) is $\{3 - (-3)\} = 6$. $F^2(np, 4f)$ and $\langle r^2 \rangle_{np}$ were evaluated along with the Sternheimer parameters $\lambda_{np}^{(2)}$, $\lambda_{4f}^{(2)}$ and $R_{np}^{(2)}$ [13, 14] and are given in Table V.

TABLE V

in atomic units

Site	$\langle r^2 \rangle$	$F^2(np, 4f)$	$\lambda_{nl}^{(2)}$	$R_{np}^{(2)}$
$2p$	0.0076	0.089	1.388	0.194
$3p$	0.0600	0.474	-0.381	0.171
$4p$	0.3397	0.689	0.812	0.172
$5p$	2.7272	0.222	0.181	0.142
$4f$	0.7062	N.A.	0.601	N.A.

The calculated $\Delta E(np_{3/2})$ are plotted in Figure 5.

The contributions to ΔE (Equation (2.39)) are from the crystal potential term $(A_2^0 \langle r^2 \rangle_{np} (1 - \lambda_{np}^{(2)}))$ and from the potential due to the ion-states. The former contribution is opposite in sign to the later and is less than 2% to the total splitting except 0.102 eV for the $5p_{3/2}$ level. The result for $5p_{3/2}$ level cannot be taken seriously since $5p$ electrons constitute the outermost shell in Tm^{3+} ions and are rather loosely bound with the ion. The effect of neighbouring ions on them is therefore complicated. The

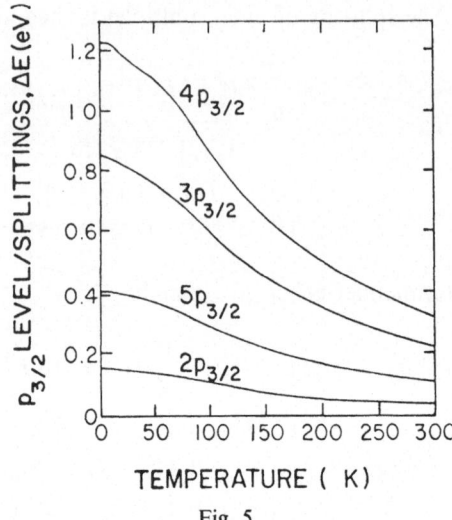

Fig. 5.

Mössbauer experiments on nuclear quadrupole interaction measure the splitting of nuclear energy levels (spin $I \geqslant \frac{3}{2}$) where it is known that the two contributions are comparable ($\lambda_0 \sim -100$). The splittings $\Delta E(p_{3/2})$, therefore, are direct measure of the state of valence electrons of the ions and are insensitive (directly) to the ligand ions. (Of course any change in ligand ion states or in their geometric positions will be reflected in $\Delta E(p_{3/2})$ through the valence electrons).

The measurements of $\Delta E(p_{3/2})$ are not limited to the compounds with Mössbauer nuclei. TmES is a weak field case and we may expect larger splittings in transition metal and heavy metal complexes.

Novakov and Hollander [15, 16] have reported measurements on the splitting of the atomic core $p_{3/2}$ levels of Th, U, Pu, Au, Pt in their compounds by using photoelectron and internal conversion spectroscopy. They observed splittings of the order of 1 eV for Th, Au and Pt in their compounds, and of more than 10 eV for U in Uranyl acetate and for Pu. We may expect that the technique of high resolution X-ray emission spectroscopy should also be able to measure the splitting of the inner shell vacancy energy levels.

3. Sternheimer Effect

A perturbing potential, whether originating within an ion or outside it, polarizes the spherically symmetric complete shells of electrons. The complete shells, having thus lost their spherical symmetry, contribute destructively (shielding) or constructively (antishielding) to the perturbing potential at any site. The reduction or enhancement of the perturbing potential, the so called Sternheimer effect, could be accounted for in terms of some parameters. It is difficult to predict whether a parameter, when calculated, would shield or antishield the perturbation. We prefer the following convention for nomenclature. When the origin of the perturbation is within an ion (e.g., valence electrons, nuclear quadrupole moment) we call the appropriate parameter as

Sternheimer R-parameter, and when the origin of perturbation is outside the ion (e.g. ligands) we call the appropriate parameter as Sternheimer λ-parameter. We attach subscripts to denote the site at which the perturbation is considered (e.g. nl for the nl-shell electrons, 0 for nucleus) and superscripts to denote the tensorial order of perturbation (e.g., 2 for quadrupole potential, 4 for hexa-decapole potential). We shall drop the superscript from the parameters when we consider only the quadrupole perturbations. The parameters R and λ are different for different ions. The parameter R depends on the wave functions of the valence electrons which may change appreciably while forming a compound, and further from one compound to another.

We shall give here [13, 17], the calculation of $\lambda^{(2)}$-parameter. We can show schematically the Sternheimer effect as in Figure 6.

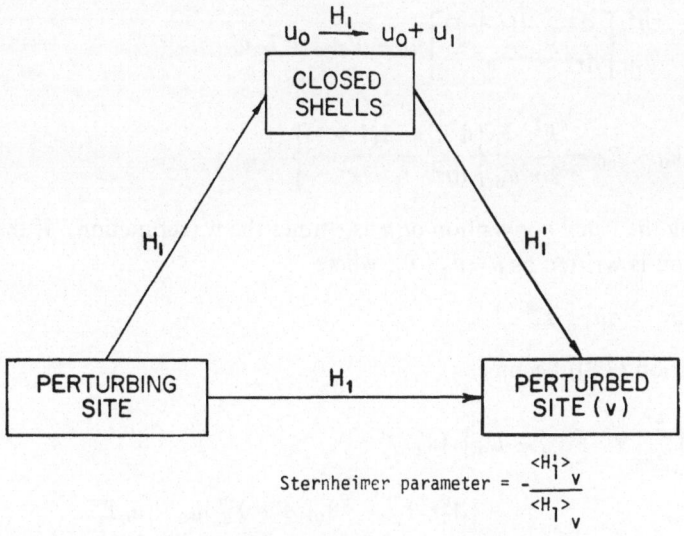

Fig. 6.

We consider the perturbation Hamiltonian at (r, θ, ϕ)

$$H_1 = A_2^0 r^2 Y_2^0 (\theta, \phi). \tag{3.1}$$

The closed shell electrons will be represented by the wave function $u_0(nlm_l) = u'_{onl}(r)\, Y_l^{m_l}(\theta, \phi)$ and the perturbed site electrons by $v(n_e l_e m_{l_e})$. Let an electron with wave function ψ_0 satisfy wave equation

$$H_0 \psi_0 = E_0 \psi_0, \tag{3.2}$$

and the perturbed equation

$$(H_0 + H_1)(\psi_0 + \psi_1) = (E_0 + E_1)(\psi_0 + \psi_1)$$

after the introduction of perturbation H_1 where $\psi_0 + \psi_1$ is the perturbed wave function and $E_0 + E_1$ is the perturbed energy. Then

$$(H_0 - E_0)\,\psi_1 = -(H_1 - E_1)\,\psi_0, \tag{3.3}$$

and the perturbation energy is $E_1 = \langle \psi_0| H_1 |\psi_0 \rangle$. The normalization of the perturbed wave function $\langle \psi_0 + \psi_1 | \psi_0 + \psi_1 \rangle = 1$ gives the orthogonality condition $\langle \psi_0 | \psi_1 \rangle = 0$.

The perturbation energy of the electron at the perturbed site with wave function $v = v'_{n_e l_e} Y_{l_e}^{m_e}$ is

$$
\begin{aligned}
E_1 &= \langle v| H_1 |v \rangle. \\
&= A_2^0 (\langle v'| r^2 |v' \rangle \langle Y_{l_e}^{m_e}| Y_2^0 |Y_{l_e}^{m_e} \rangle.
\end{aligned}
\tag{3.4}
$$

For the closed shell electrons we may write Equation (3.2) as

$$
\left(-\frac{\hbar^2}{2m} \nabla^2 + V_0 \right) u'_0 Y_l^m = E_0 u'_0 Y_l^m
$$

or

$$
-\frac{\hbar^2}{2m} \left[\frac{d^2}{dr^2} - \frac{l(l+1)}{r^2} \right] u'_0 + V_0 u'_0 = E_0 u'_0
$$

or

$$
V_0 - E_0 = \frac{\hbar^2}{2m} \frac{1}{u'_0} \left\{ \frac{d^2 u'_0}{dr^2} - \frac{l(l+1)}{r^2} \right\}
\tag{3.5}
$$

(we are using the usual convention of u as r times the wavefunction). If the perturbed wave function is written as $u = u_0 + u_1$, where

$$
u_1 = A u'_{1 n l'} Y_{l'}^{m'}.
\tag{3.6}
$$

Then (Equation (3.3)) becomes

$$
\left(\frac{\hbar^2}{2m} \nabla^2 + V_0 - E_0 \right) A u'_1 Y_{l'}^{m'}
$$

$$
= - \{A_2^0 r^2 Y_2^0 - \langle u_0| A_2^0 r^2 Y_2^0 |u_0 \rangle \} u'_0 Y_l^m
$$

or

$$
\left[-\frac{\hbar^2}{2m} \left(\frac{d^2}{dr^2} - \frac{l'(l'+1)}{r^2} \right) + V_0 - E_0 \right] A u'_1 Y_{l'}^{m'}
$$

$$
= - A_2^0 \{r^2 Y_2^0 Y_l^m - \langle u'_0| r^2 |u'_0 \rangle \langle Y_l^m| Y_2^0 |Y_l^m \rangle Y_l^m \} u'_0
$$

$$
= - A_2^0 \left\{ r^2 \sum_{L = |l-2|}^{l+2} \sqrt{\frac{5}{4\pi}} C^{(2)} (l, m; L, m) Y_L^m \right.
$$

$$
\left. - \langle r^2 \rangle \sqrt{\frac{5}{4\pi}} C^{(2)} (l, m; l, m) Y_l^m \right\}
$$

$$
= - A_2^0 \{r^2 - \langle r^2 \rangle \delta (L, l) \} \times
$$

$$
\times \sum_{L = (l-2)}^{l+2} \sqrt{\frac{5}{4\pi}} C^{(2)} (l, m; L, m) Y_L^m u'_0
\tag{3.7}
$$

where

$$
C^{(k)} (l', m'; l, m + q) = \sqrt{\frac{4\pi}{2k+1}} \sum_q \langle Y_{l'}^{m'}| Y_k^q |Y_l^m \rangle
\tag{3.8}
$$

and is non-zero for $m' = m+q$ and $k = l'+l, l'+l-2, \dots |l'-l|$. Both sides of Equation (3.7) should have same angular dependence; all allowed L, m on the right hand side should also be present on the left hand side. Therefore for each allowed L, m value we have a solution of Equation (3.7):

$$
\left[-\frac{d^2}{dr^2} + \frac{l'(l'+1)}{r^2} + \right.
$$

$$
\left. + \frac{2m}{\hbar^2} (V_0 - E_0) \right] u_1'(l') = -[r^2 - \langle r^2 \rangle \delta(l', l)] u_0'(l) \tag{3.9}
$$

$$
Y_{l'}^{m'} = \sqrt{\frac{5}{4\pi}} C^{(2)} (l, m; L, m) Y_L^m \big|_{L=l+2, l, |l-2|} \tag{3.10}
$$

$$
A = A_2^0/(\hbar^2/2m). \tag{3.11}
$$

The radial equation for $u_1'(l')$ (Equation (3.9) is solved numerically by substituting $(V_0 - E_0)$ from Equation (3.5) and writing it as a difference equation:

$$
u_1'(r+\delta) = u_1'(r) \left[2 + \delta^2 \left\{ \frac{l'(l'+1) - l(l+1)}{r^2} \right\} + \right.
$$

$$
\left. + \frac{u_0'(r+\delta) - 2u_0'(r) + u_0'(r-\delta)}{u_0'(r)} \right] -
$$

$$
- u_1'(r-\delta) + u_0'(r) \delta^2 (r^2 - \langle r^2 \rangle \delta(l', l)). \tag{3.12}
$$

The perturbed charge distribution of the core electron is given by $u = u_0 + u_1$. The interaction between perturbed site electron $v(1)$ at $\mathbf{r}_1$ and the charge distribution of a core electron at $\mathbf{r}_2$ (including exchange) is

$$
\langle u(2) | \frac{e^2}{r_{12}} (1 - P_{12}) | u(2) \rangle =
$$

$$
= \langle u_0(2) | \frac{e^2(1 - P_{12})}{r_{12}} | u_0(2) \rangle + \langle u_1(2) | \frac{e^2(1 - P_{12})}{r_{12}} | u_0(2) \rangle
$$

$$
+ \langle u_0(2) | \frac{e^2(1 - P_{12})}{r_{12}} | u_1(2) \rangle + \langle u_1(2) | \frac{e^2(1 - P_{12})}{r^2} | u_1(2) \rangle.
$$

The first term summed over all the electrons in a closed shell is zero (since u_0 is an unperturbed wave function). The last term is of third order and is neglected. The second and third terms are of interest to us;

$$
H_1'(l') = 2 \langle u_{1l'}(2) | \frac{e^2(1 - P_{12})}{r_{12}} | u_0(2) \rangle. \tag{3.13}
$$

The perturbation energy E'_1 of $v(1)$ due to H'_1 is

$$E'_1(l') = \langle v| H'_1 |v\rangle =$$

$$= 2\left[\langle v(1)\, u_1(2) \left| \frac{e^2}{r_{12}} \right| v(1)\, u_0(2)\rangle - \right.$$

$$\left. - \langle v(1)\, u_1(2) \left| \frac{e^2}{r_{12}} \right| v(2)\, u_0(1)\rangle \right] =$$

$$= E'_{1l'}(\text{Direct}) - E'_{1l'}(\text{Exchange}). \tag{3.14}$$

$$E'_{1l'}(\text{Direct}) = 2e^2 A \sum_{k=0}^{\infty} \frac{4\pi}{2k+1} \times$$

$$\times \left[\langle v'(1)\, u'_{1l'}(2) \left| \frac{r_<^k}{r_>^{k+1}} \right| v'(1)\, u'_0(2) \times \right.$$

$$\times \sum_{q=-k}^{k} \langle Y_{l_e}^{m_e}(1)\, Y_{l'}^{m'}(2)|\, Y_k^{q*}(1)\, Y_k^q(2)|Y_{l_e}^{m_e}(1)\, Y_l^m(2)\rangle \left. \right] =$$

$$= 2A_2^0 \frac{2me^2}{\hbar^2} \sum_{k=0}^{\infty} [H^k(v;u_{1l'})\sum_{q=-k}^{k} \langle Y_{l_e}^{m_e}|\, Y_k^{-q}|Y_{l_e}^{m_e}\rangle \times$$

$$\times \langle Y_{l'}^{m'}|\, Y_k^q|Y_l^m\rangle] = \frac{2me^2}{\hbar^2}\, 2A_2^0 \sum_{k=0}^{\infty} \left[H^k(v, u_{1L}) \times \right.$$

$$\times C^{(k)}(l_e m_e, l_e m_e)\sqrt{\frac{5}{4\pi}}\, C^{(2)}(lm, Lm)\, C^{(k)}(Lm, lm) \left. \right].$$

$$H^{(k)}(v, u_1) = \langle v'(1)\, u'_1(2) \left| \frac{r_<^k}{r_>^{k+1}} \right| v'(1)\, u'_0(2)\rangle.$$

From Equation (3.10) $L = l+2, l, |l-2|;\ k(L=l+2) = l+2+l, l+l, \dots, 2;\ k(L=l) = l+l, l+l-2, \dots, 0;\ k(L=|l-2|) = |l-2|+l, \dots |(|l-2|-l)|$. It turns out that $k=2$ term is the most important for E'_1 (Direct) ($k \neq 2$ terms are zero when summed over all electrons in a closed shell).

$$E'_{1L}(\text{Direct}) = \frac{2me^2}{\hbar^2}\, 2A_2^0 H^{(2)}(v, u_{1L}) \times$$

$$\times C^{(2)}(l_e m_e, l_e m_e)\sqrt{\frac{5}{4\pi}}\, [C^{(2)}(Lm, lm)]^2 \tag{3.15}$$

$$E'_{1l'}(\text{Exchange}) = \frac{2me^2}{\hbar^2}\, 2A_2^0 \sum_{k=0}^{\infty} \frac{4\pi}{2k+1}\, Z^{(k)}(v, u_{1l'}) \times$$

$$\times \sum_{q=-k}^{k} \langle Y_{l_e}^{m_e}|\, Y_k^{-q}|Y_l^m\rangle \langle Y_{l'}^{m'}|\, Y_k^q|Y_{l_e}^{m_e}\rangle] =$$

$$= \frac{2me^2}{\hbar^2}\, 2A_2^0 \sum_{k=0}^{\infty} Z^{(k)}(v, u_{1L})\, C^{(k)}(l_e m_e; l, m-q) \times$$

$$\times \sqrt{\frac{5}{4\pi}}\, C^2(lm, Lm)\, C^{(k)}(Lm; l_e, m_e+q) \tag{3.16}$$

where

$$Z^{(k)}(v, u_1) = \langle v'(1) u_1'(2) \left| \frac{r_<^k}{r_>^{k+1}} \right| v'(2) u_1'(1) \rangle .$$

It turns out that most of the contribution to E_1' (Exchange) comes from the term having minimum allowed value of k; C-coefficients decide the selection rule for k. Total E_1 is obtained by summing over all electrons in a closed shell:

$$E_{1L}' = \sqrt{\frac{5}{4\pi}} \frac{2me^2}{\hbar^2} 2A_2^0 \Bigg[H^{(2)}(v, u_{1L}) \times$$

$$\times C^{(2)}(l_e m_e, l_e m_e) \sum_{m=-l}^{l} 2\{C^{(2)}(Lm, lm)\}^2 -$$

$$- Z^{(k)}(v, u_{1L}) \sum_{m=-l}^{l} \{C^{(k)}(l_e m_e; l, m-q) \times$$

$$\times C^{(2)}(lm, Lm) C^{(k)}(Lm; l_e, m_e+q)\} \Bigg]. \tag{3.17}$$

(The exchange term is summed only on the electrons of the same spin.)

The Sternheimer parameter, defined as $\lambda = -E_1'/E_1$, is

$$\lambda_{n_e l_e}(nl \to L) = - E_{1L}'/E_1 \tag{3.18}$$

(Equation (3.17) gives E_{1L}' and Equation (3.4) gives E). For the direct part of E_{1L}' a more compact form of λ is obtainable:

$$\lambda_{n_e l_e}(nl \to L, \text{Direct}) = \frac{8me^2}{\hbar^2} \sum_{m=-l}^{l} \{C^{(2)}(Lm, lm)\}^2 H^{(2)}(v, u_{1L})/\langle r^2 \rangle_{n_e l_e}. \tag{3.19}$$

Turning now to nuclear quadrupole moment we observe that the exchange part of E_1' has no meaning and that the nucleus as seen by the electrons has zero radius. Then

$$F^{(2)}(\text{nucl}, u_{1L}) = \int_0^\infty \Bigg[r_1^{-3} \int_0^{r_1} r_2^2 u_1' u_0' \, dr_2 +$$

$$+ r_1^2 \int_{r_1}^\infty r_2^{-3} u_1' u_0' \, dr_2 \Bigg] v_{\text{nucl}}'^2 \, dr_1 =$$

$$= \langle r^2 \rangle_{\text{nucl}} \int_0^\infty r_2^{-3} u_1' u_0' \, dr_2$$

and

$$\lambda_0(nl \to L) = \frac{8me^2}{\hbar^2} \sum_{m=-l}^{l} [C^{(2)}(Lm, lm)]^2 \int_0^\infty r^{-3} u_1' u_0' \, dr. \tag{3.20}$$

TABLE VI

Perturbation	λ_0	λ_{2p}	λ_{3p}	λ_{3d}	λ_{4p}	λ_{4d}	λ_{4f}	λ_{5p}
$5p \to p$	-64.487	-1.394	-0.677	-1.166	-0.238	-0.325	-0.189	0.079
$5p \to f$	0.573	1.497	-0.703	-0.513	0.568	0.558	0.491	0.148
$5s \to d$	0.221	1.889	0.910	1.018	0.350	0.360	0.236	-0.017
$4d \to s$	-0.013	-0.890	-0.358	-0.419	0.011	-0.041	-0.027	-0.001
$4d \to d$	-2.080	-0.449	-0.155	-0.098	-0.006	0.033	-0.006	0.000
$4d \to g$	0.184	0.191	0.188	0.193	0.072	0.074	0.039	0.004
$4p \to p$	-6.411	-0.105	-0.036	-0.082	0.020	-0.003	-0.002	0.001
$4p \to f$	0.142	0.105	0.199	0.191	0.048	0.064	0.033	0.003
$4s \to d$	0.379	0.470	0.215	0.245	-0.008	0.035	0.024	0.002
$3d \to s$	-0.014	-0.195	0.007	-0.042	-0.002	-0.002	-0.001	0.000
$3d \to d$	-0.239	-0.035	-0.004	0.017	0.000	0.001	0.000	0.000
$3d \to g$	0.073	0.057	0.016	0.024	0.002	0.001	0.001	0.000
$3p \to p$	-1.154	-0.010	0.012	-0.003	0.001	0.000	0.000	0.000
$3p \to f$	0.065	0.074	0.017	0.032	0.002	0.002	0.001	0.000
$3s \to d$	0.036	0.170	-0.007	0.042	0.003	0.003	0.001	0.000
$2p \to p$	-0.200	0.011	0.000	0.000	0.000	0.000	0.000	0.000
$2p \to f$	0.030	0.005	0.000	0.001	0.000	0.000	0.000	0.000
$2s \to d$	0.022	0.000	0.000	0.001	0.000	0.000	0.000	0.000
$1s \to d$	0.010	0.000	0.000	0.000	0.000	0.000	0.000	0.000
Total	-72.863	1.388	-0.381	-0.559	0.812	0.753	0.601	0.181
$\langle r^2 \rangle$	0.0076	0.0076	0.0600	0.0473	0.3397	0.3848	0.7062	2.7272
$(1-\lambda)\langle r^2 \rangle$	-0.003	-0.003	0.083	0.074	0.064	0.095	0.282	2.233

Equations (3.19) and (3.20) are summed over all allowed L and then over all closed electron shells (nl) to obtain final value of the Sternheimer parameters.

We give in Table VI the Sternheimer λ-parameters for Tm^{3+} ion [13]. There is no definite trend seen in the value of λ-parameters as we go from the outer shells to the inner shells. In general the contribution of $nl \to l$ and $nl \to l-2$ type perturbations of the closed shells is antishielding while that of $nl \to l+2$ type perturbation is shielding. We also see that the magnitude of the contribution decreases with the decreasing principal quantum number n. This is because the inner shell electrons are more tightly bound to the ion than the outer shell electrons and are difficult to be polarized by the crystal potential.

Various other approaches have been used to calculate the Sternheimer parameters. Principal among them are the analytic perturbation method of Das and Bersohn [18] and Unrestricted Hartree-Fock method of Freeman and Watson [19]. A discussion of the different perturbation methods employed has been given by Allen [20], and Dalgarno [21]. Recently the Sternheimer effect related to nuclear quadrupole coupling has been discussed by Lucken [22] and Armstrong [23]. Gupta *et al.* [13] and Gupta and Sen [14, 24, 25] have reported calculation of the Sternheimer parameters for the core electron sites. Sternheimer [17] has given extensive references of the work on this subject.

Acknowledgement

The authors wish to thank Dr B. S. Bhakar of this Department for many helpful discussions. Acknowledgements are also due to the National Research Council of Canada for financial support of this work. The authors are grateful to Miss Lynda Jones for her valuable contributions in typing the manuscript with great patience.

References

1. Jackson, J. D., *Classical Electrodynamics*, John Wiley & Sons, Inc., New York, 1962.
2. Prather, J. L., 'Atomic Energy Levels in Crystals', NBS Monographs No. 19, Washington, 1961.
3. Griffith, J. S., *The Theory of Transition-Metal Ions*, Cambridge University Press, London, 1961.
4. Hutchings, M. T., *Solid State Phys.* **16**, 227 (1964).
5. Stevens, K. W. H., *Proc. Phys. Soc.* **65**, 209 (1952).
6. Abragam, A. and Bleaney, B., *Electron Paramagnetic Resonance of Transition Ions*, Clarendon Press, Oxford, 1970.
7. Rose, M. E., *Elementary Theory of Angular Momentum*, John Wiley & Sons, Inc., New York, 1957.
8. Stevenson, R., *Multiplet Structure of Atoms and Molecules*, W. B. Saunders Company, Philadelphia, 1965.
9. Ballhausen, C. J., *Introduction to Ligand Field Theory*, McGraw-Hill Book Company, Inc. New York, 1962.
10. Sternheimer, R. M., *Phys. Rev.* **80**, 102 (1950).
11. Wong, E. Y. and Richman, I., *J. Chem. Phys.* **34**, 1182 (1961).
12. Barnes, R. G., Mössbauer, R. L., Kankeleit, E., and Poindexter, J. M., *Phys. Rev.* **136**, A175 (1964).
13. Gupta, R. P., Rao, B. K., and Sen, S. K., *Phys. Rev.* **A3**, 545 (1971).
14. Gupta, R. P. and Sen, S. K., *Phys. Rev. Letters* **28**, 1311 (1972).
15. Novakov, T. and Hollander, J. M., *Phys. Rev. Letters* **21**, 1133 (1968).

16. Novakov, T. and Hollander, J. M., *Bull. Amer. Phys. Soc.* **10**, 597 (1969).
17. Sternheimer, R. M., *Phys. Rev.* **146**, 140 (1966) (and references therein).
18. Das, T. P. and Bersohn, R., *Phys. Rev.* **102**, 833 (1956).
19. Freeman, A. J. and Watson, R. E., *Magnetism* (ed. by G. T. Rado and H. Suhl), Academic Press, New York, Vol. IIA, p. 167 (1965).
20. Allen, L. C., *Phys. Rev.* **118**, 167 (1960).
21. Dalgarno, A., *Adv. Phys. (Phil. Mag. Suppl.)* **11**, 281 (1962).
22. Lucken, E. A. C., *Nuclear Quadrupole Coupling Constants*, Academic Press, London, 1969.
23. Armstrong, L., Jr., *Theory of Hyperfine Structure of Free Atoms*, Wiley-Interscience, New York, 1971.
24. Gupta, R. P. and Sen, S. K., *Phys. Rev.* **A7**, 850 (1973).
25. Gupta, R. P. and Sen, S. K., *Phys. Rev.* **A** (Sept. 1973).

MOLECULAR PHOTOELECTRON SPECTROSCOPY

E. LINDHOLM

Physics Department, Royal Institute of Technology, Stockholm, Sweden

1. Historical Development

To ionize an electron in a molecule the 'ionization energy' of the electron must be supplied. As there are many electrons in each molecule there are many ionization energies, in a hydrocarbon from about 10 eV to about 300 eV.

The electron in the molecule moves in an orbital and this is characterized by the 'orbital energy'. The orbital energy cannot be measured but it can easily be estimated as it does not differ much from the ionization energy.

The ionization energies of a molecule are thus of fundamental importance for the description of the molecule, but before the photoelectron spectroscopy was introduced, they were nearly unknown.

Important studies of ionization potentials were performed around 1935 by Price by use of Rydberg series. Such a series converges towards the ionization potential, and for many molecules the first ionization potential was determined in this way.

Later, electron impact was used to study the ionization potentials, and when mass spectrometry started, mass spectrometers were used for this purpose. In a mass spectrometer the positive ions are observed, and when the kinetic energy of the incident electron is diminished below a certain limit (for example 10 eV) no positive ions are observed. This gives the ionization potential (as 10 eV in the example mentioned).

In this way the first ionization potentials of most molecules were determined, in good agreement with the Rydberg values in the few cases when they were available. The higher ionization potentials were, however, nearly completely unknown.

The main way to study the higher IP's was band spectroscopy. Spectroscopic studies of for instance N_2^+ had given the energy differences between the ground state and the

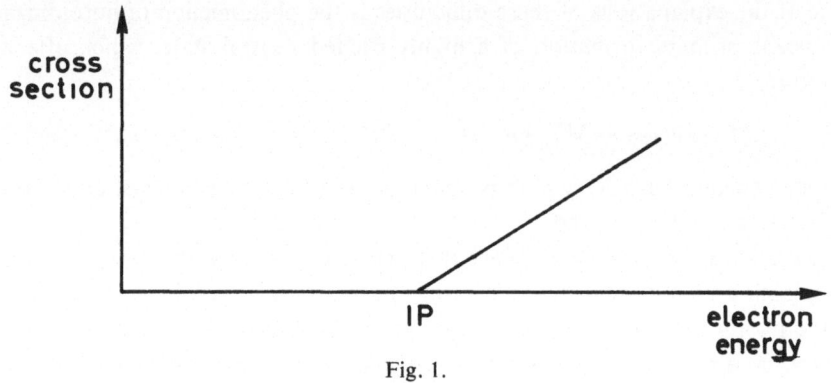

Fig. 1.

excited states of N_2^+, and from the lowest IP of N_2 the higher IP's could be calculated. The diatomic molecules (N_2, O_2, CO) were partly known in this respect, but of the triatomic molecules only CO_2 was partly known from spectroscopic studies of CO_2^+ (this situation refers to 1954).

This situation was very unfortunate for our understanding of the molecules, especially the chemical bonding. Mulliken had introduced the concept of molecular orbitals a long time ago, but the orbitals could be observed and studied only indirectly.

Very many efforts were therefore made to study the higher IP's. It was known that using electron impact the cross section for ionization increases linearly from the ionization potential (Figure 1) and one expected therefore that the increased probability for ionization above a higher IP would be observed as a break in the cross-section curve. (Figure 2). Usually, no such break could be seen although different experimental methods were used to control the supplied energy (electrostatic analysis of the incident electrons, retarding-potential-difference analysis of the incident electrons, photons instead of electrons).

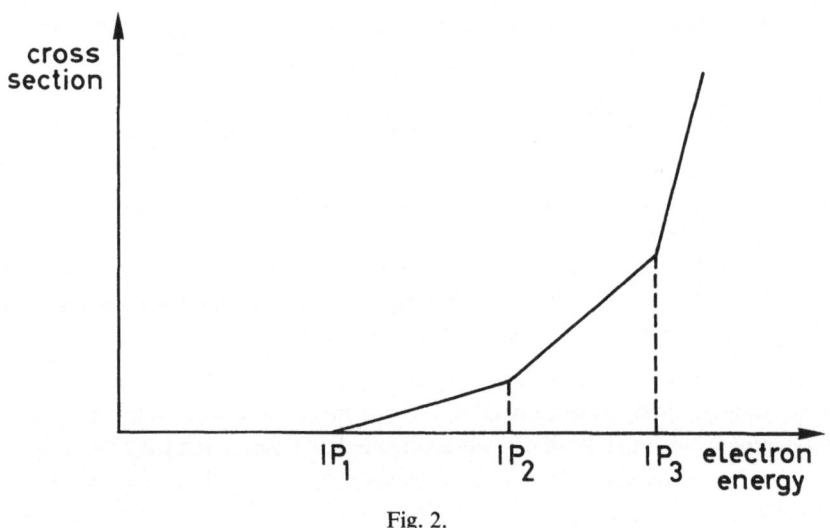

Fig. 2.

One of the explanations of these difficulties is the phenomenon of autoionization. This means primary formation of a highly excited neutral state, which afterwards autoionizes.

$$M + \text{energy} \rightarrow M^* \rightarrow M^+ + e.$$

This phenomenon has a very high probability, and the direct process, giving breaks, can therefore not be observed.

In 1962 Turner at Imperial College in London built a new apparatus for study of the ionization of molecules by measuring the kinetic energy of the photoelectrons. For the photons from the light source he chose the helium resonance line at 584 Å. To avoid a window Turner invented an ingenious differential pumping system. To

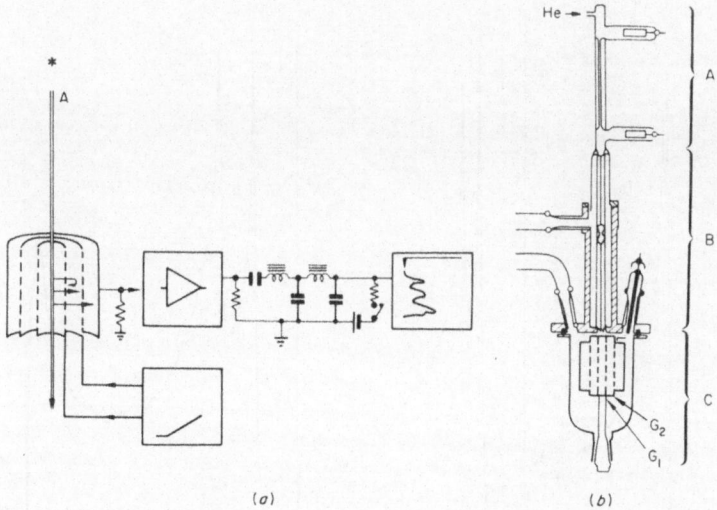

Fig. 3. Helium resonance photoelectron spectrometer of Al-Joboury and Turner. (a) General arrangement; (b) cross-section of light source A, collimator section B and target chamber C. From Turner *et al.*: *Molecular Photoelectron Spectroscopy*, Copyright © 1970 John Wiley & Sons Ltd. By Permission of John Wiley & Sons Ltd.

analyze the ejected photoelectrons he used a retarding grid together with a differentiating circuit [6] (Figure 3).

The process in the photoelectron spectrometer is:

$$h\nu + M = M^+ + e$$
21.21 eV IP   kin. energy
584 Å

During the ionization process M^+ may be formed in an excited electronic state or vibrations may be excited. The kinetic energy of the ejected electron is therefore:

21.21 eV – the total energy absorbed by the molecule.

The first test with the new apparatus on the noble gases gave remarkable results (Figure 4) [16]. The two doublet states of the ions gave sharp peaks from which the energy split could be determined with high accuracy. This proved that by use of the new method higher ionization potential could be observed and measured for the first time.

The success of Turner's apparatus depends upon several factors.

Autoionization processes cannot contribute.

The 584 Å line is extremely intense and extremely narrow. The signals are therefore rather large even at high resolution.

The differential pumping enables work to be performed without a window. As a window makes studies impossible above about 11 eV, it is impossible to test the method with for instance the noble gases. With a window the study is thus limited to only

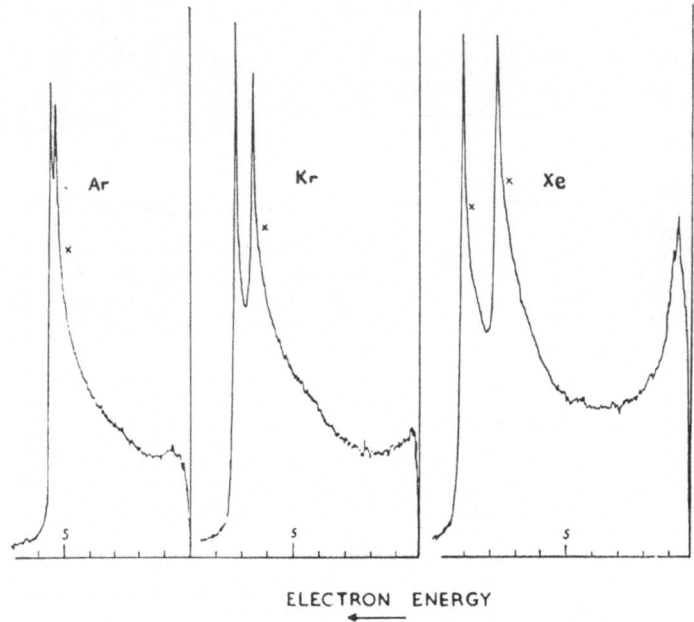

Fig. 4.   Photoelectron spectra of noble gases.

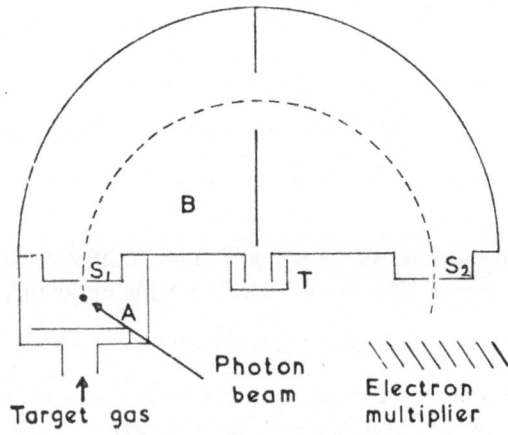

Fig. 5.   A 180° magnetic deflection photoelectron velocity analyzer (schematic, see text).

the first part of the spectrum of large organic molecules, as small molecules usually have IP's larger than 11 eV.

At the same time as a large number of molecules were measured and interpreted by use of the new method, Turner started construction of better instruments.

The magnetic instrument (Figure 5) [17] used a magnetic field from two Helmholz' coils and gave beautiful spectra of a number of small molecules.

Finally, the electrostatic instrument [18] gave a half-width of about 20 meV. (Figure 6).

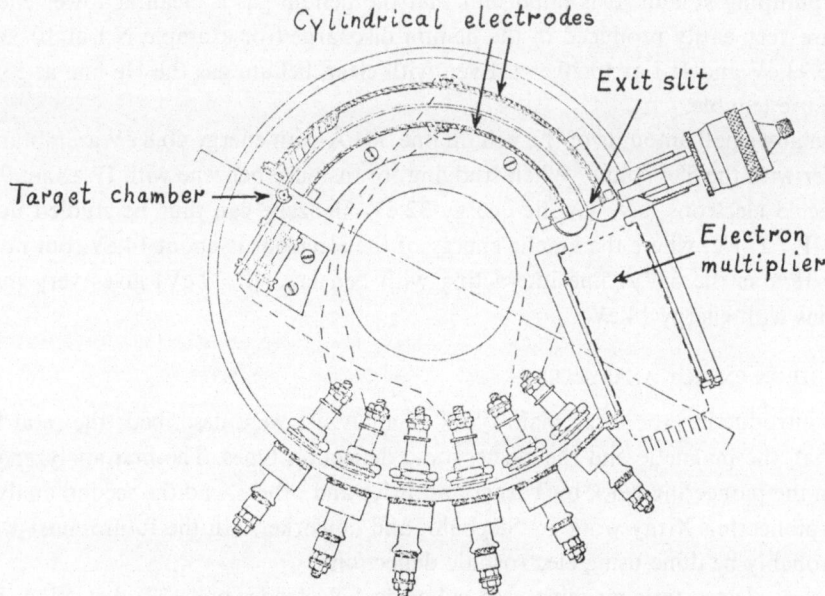

Fig. 6.   Turner's electrostatic photoelectron spectrometer. [6, 18]

Besides the work by Turner and his many coworkers important work has been performed by other groups.

In Leningrad, Vilessov, Kurbatov and Terenin studied photoelectrons from substituted benzenes already 1961. The vacuum monochromator enabled them to use different wavelengths of the incident photons but the window restricted the work to below about 11 eV. The diffuse curves obtained from these molecules did not indicate any special importance of the method used. After this preliminary work no more papers seem to have been published from this group (Review: Terenin and Vilessov, *Adv. Photochemistry* **2**, 385, 1964).

In Uppsala Siegbahn's group had studied mainly X-ray photoelectron spectra, but recently also used the 584 Å helium line.

Much important work has been done at King's College in London by Price's group.

Other fundamental work has been performed in many laboratories. Systematic studies of organic compounds have been performed especially in Heilbronner's group in Basel. In Physics Department of the Royal Institute of Technology in Stockholm different kinds of molecules have been measured by Åsbrink in an apparatus with spherical electrodes.

2. Instruments

A. LIGHT SOURCES

Most work has been performed by use of the helium 584 Å line with energy 21.21 eV. It can be produced by a dc discharge or a microwave discharge in flowing pure helium gas. The light is then collimated by a capillary, which also serves as part of the differential pumping system. It is important that the helium gas is clean as lower energy lines are very easily produced in the helium discharge (for example N I at 10.33 eV and 10.93 eV and H I at 10.20 eV). Even with clean helium gas the He line at 537 Å may cause trouble.

Normally small amounts of the helium line 304 Å with energy 40.8 eV are obtained together with the 584 Å line. When studying for instance benzene with IP about 9 eV the ejected electrons have kinetic energy 32 eV. Benzene can thus be studied up to about IP = 27 eV, where the kinetic energy of the electrons is about 14 eV, but not at higher IP's as the 537 Å line interacting with benzene (IP = 9 eV) gives very many electrons with energy 14 eV.

B. ELECTRON ENERGY ANALYZERS

In the introduction the three main types of analyzers were described: the retarding potential, the magnetic and the electrostatic deflection types. The first analyzer was used in the pioneering work by Turner and Price and others, and the second analyzer in the pioneering X-ray work by Siegbahn and coworkers. In the future most work will probably be done using electrostatic deflection.

Turner's electrostatic machine used cylindrical electrodes and 127° deflection. It is also possible to use spherical electrodes and 180° deflection (Simpson, the Stockholm instrument) or parallel plates (Eland and Danby) or cylindrical mirror electrodes (Berkowitz).

It is probably of minor importance which geometry is used. It is possible that the spherical type gives the best transmission. On the other hand, the cylindrical electrodes are probably cheapest.

Of major importance is instead the shielding of stray magnetic fields, including the earth's field, by use of shielding material or Helmholz coils. Further, the electric circuits must give extremely stable voltages.

To avoid surface charges all surfaces ought to be painted with colloidal graphite in alcohol [19].

C. DETECTING SYSTEMS

Usually an electron multiplier is placed after the exit slit of the analyzer. Often a Chaneltron is used. With a ratemeter the spectrum can be recorded directly.

For high-precision work the spectrum can be measured at one point during for instance 20 s, and then the number of electrons is recorded. Next point is then studied during the same interval, and so on.

D. VACUUM CHAMBER

For studies of molecules in the gas phase no especially good vacuum is necessary. However, as large gas quantities leak from the target chamber and from the light source, high pumping speed is desirable.

It is recommended to make the vacuum chamber and pump from non-magnetic material.

3. Diatomic Molecules: Photoelectron Spectra
and Potential Energy Diagrams

A. HYDROGEN

The photoelectron spectrum of H_2 has been measured by Åsbrink [20] (Figure 7) with high resolution using the 584 Å helium line. A progression of vibrational bands is obtained with $v' = 0$ at 15.45 eV (outside the picture) and $v' = 18$ at 18.08 eV. Maximum intensity occurs at $v' = 3$. The half-width of the bands is about 20 meV because of the thermal motion of the H_2 molecules.

In the figure the band $v' = 0$ is outside the measurement. 3 means $J'' = 3 \rightarrow J' = 3$. The transitions $0 \rightarrow 0$, $1 \rightarrow 1$ and $2 \rightarrow 2$ are not resolved.

To interpret this spectrum the Franck-Condon principle is used. The ionization process is very rapid so that during the ionization the distance between the heavy nuclei does not change. In a potential energy diagram the ionization process will therefore take place along a vertical line from the minimum of the potential energy curve for the ground state. In quantum theory the extension of the wavefunctions broadens the vertical line to the Frank-Condon region (shaded portion in Figure 8).

The potential energy curve for H_2^+ has been calculated by Cohen et al. [21]. From their tables the diagrams in Figures 9a and b have been drawn. On one of them the Franck-Condon line at $R = 0.74$ Å has been drawn. It can be seen that it cuts the potential energy curve at $v' = 3$, explaining the intensity maximum. It can further be seen that the number of photoelectron bands agrees with the calculated number of vibrational levels.

In Åsbrink's [20] study of H_2 also the rotational structure of the spectrum could be resolved. Neon-light was used instead of the helium line. The photon energy was thus 16.85 eV, and when the $v' = 5$ band at 16.64 eV was studied, the energy of the ejected electrons was only 0.2 eV. As the half-width decreases with the energy of the photoelectrons, the study could be performed with a halfwidth between 4 and 7 meV. The $v' = 5$ band is shown below. This seems to be the only case where resolution of rotational structure in photoelectron spectroscopy has been successful.

B. NITROGEN OXIDE

The photoelectron spectrum of NO consists of a short progression between 9 and 10 eV, a complicated system of bands between 15.5 and 20 eV, and some bands at 22 and 23 eV.

From the photoelectron spectrum the potential energy curves for NO^+ could be determined. From the maximum in a progression the internuclear distance could easily be estimated.

The potential energy curves of NO and NO^+ had earlier been drawn by Gilmore [24] from spectroscopic studies, but very incompletely compared with the new diagram.

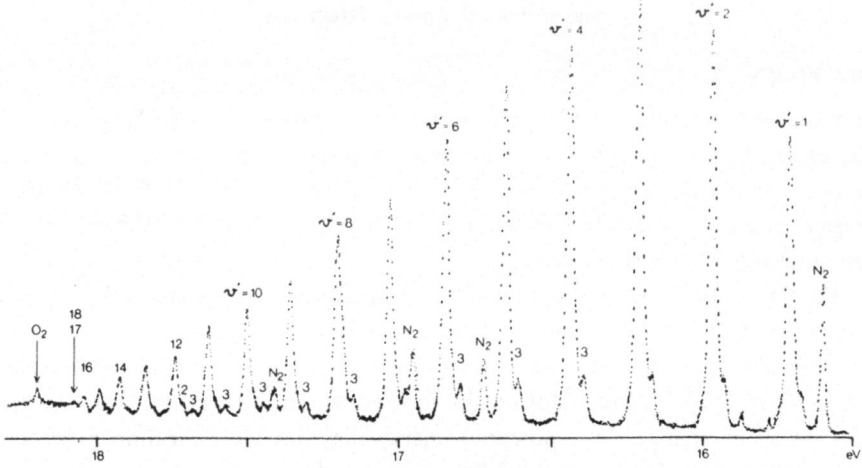

Fig. 7.   H₂-spectrum from [20].

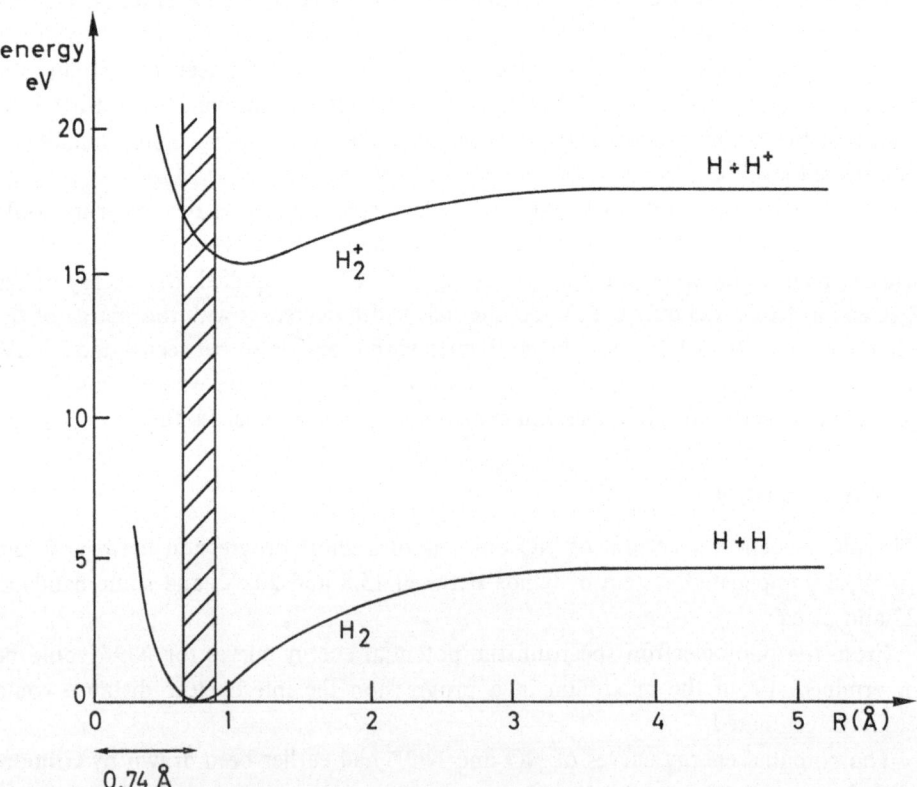

Fig. 8.   Franck-Condon region for H₂ and H₂⁺.

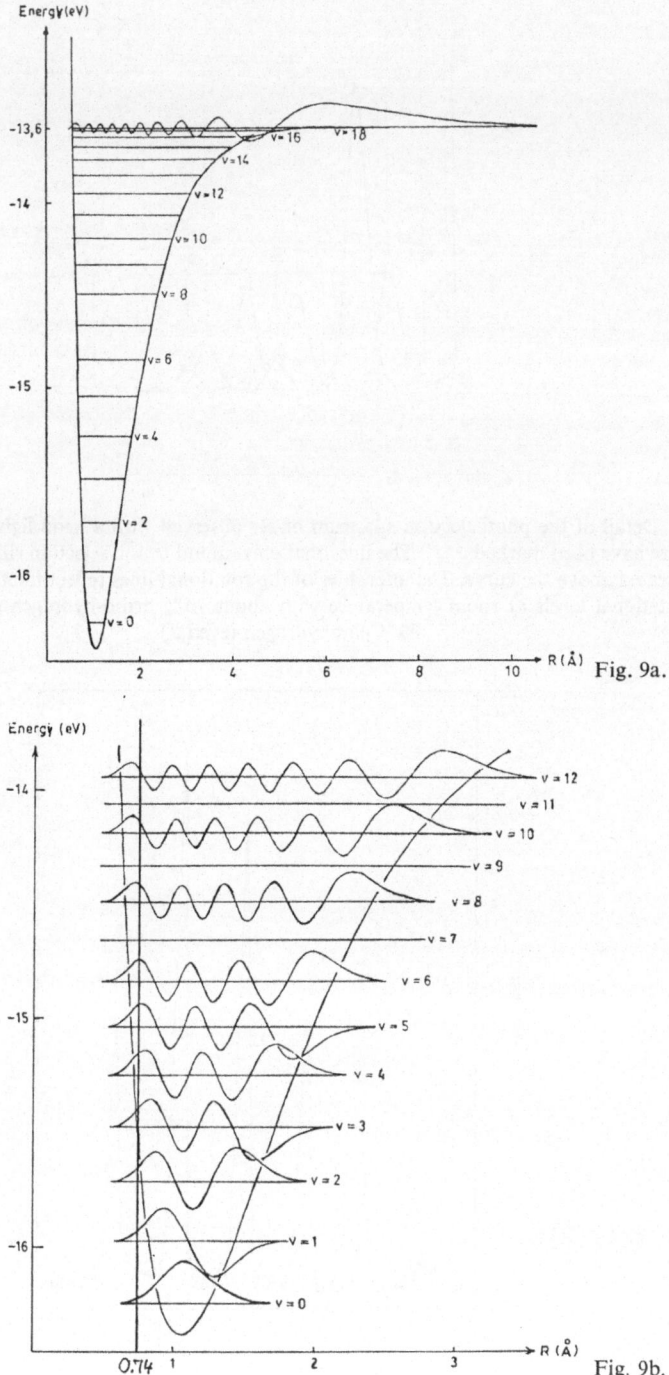

Figs. 9a–b.   The potential energy curve for H_2^+ with vibrational energy levels. 19 such levels exist. The wavefunctions are also shown for $v' = 18$ and for several levels between $v' = 0$ and $v' = 12$.

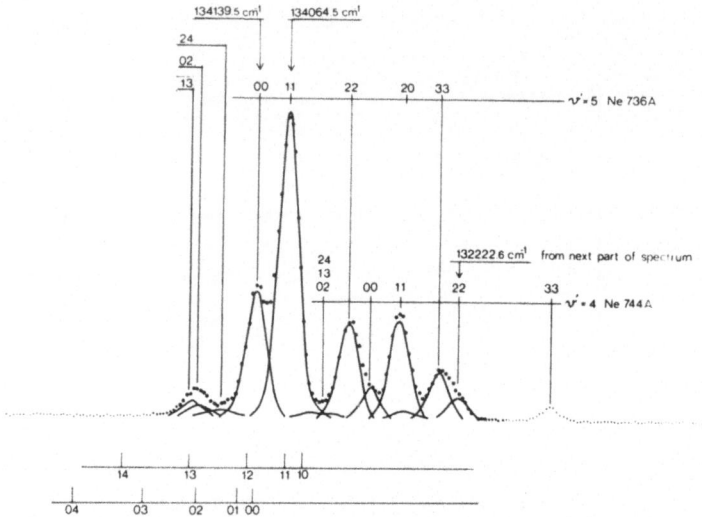

Fig. 10. Detail of the photoelectron spectrum of H_2 observed with a neon light source. The rotational lines have been marked $J''J'$. The lines that correspond to the selection rule $\Delta J = 0, \pm 2$ have been indicated above the curve. The intensities of the rotational lines reflect the thermal distribution of the rotational levels at room temperature with about 70% ortho-hydrogen (odd J) and about 30% parahydrogen (even J).

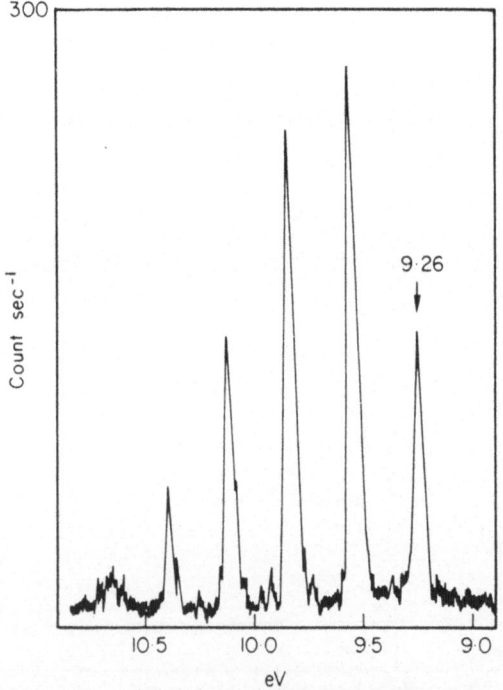

Fig. 11 Photoelectron spectrum of NO between 9.0 and 10.5 eV[6]. From Turner *et al.: Molecular Photoelectron Spectroscopy*. Copyright © 1970 John Wiley & Sons Ltd. By permission of John Wiley & Sons Ltd.

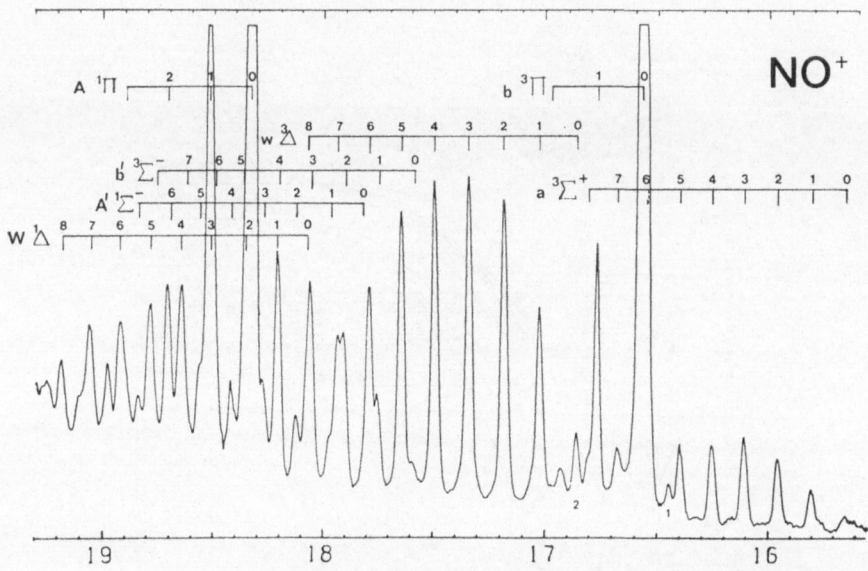

Fig. 12. Photoelectron spectrum of NO using the helium 584 Å line between 15.5 and 19 eV from Edqvist *et al.* [22]. Peak 1 is due to 537 Å and peak 2 partly to N₂.

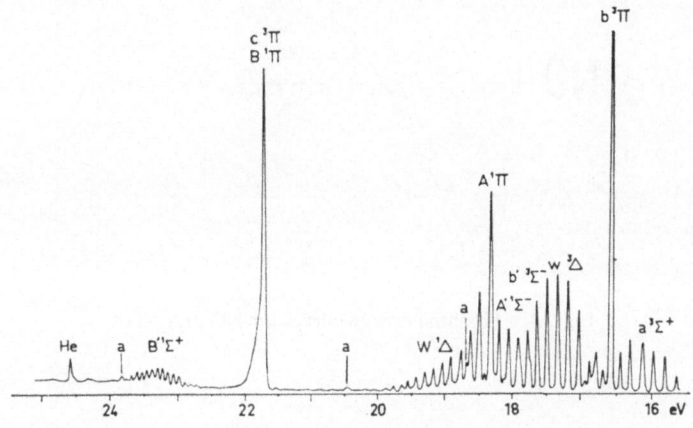

Fig. 13. Photoelectron spectrum of NO using the He 304 Å line. Three small peaks marked with *a* are due to the He 320 Å line. Recording time was 75 h. [23].

To explain the complicated photoelectron spectrum of NO its electron configuration must be discussed. It is

$$1\sigma^2 \quad 2\sigma^2 \quad 3\sigma^2 \quad 4\sigma^2 \quad 5\sigma^2 \quad 1\pi^4 \quad 2\pi$$

$$560 \quad 425 \quad 40 \quad \sim 22 \quad \sim 17 \quad \sim 17 \quad 9 \qquad eV$$

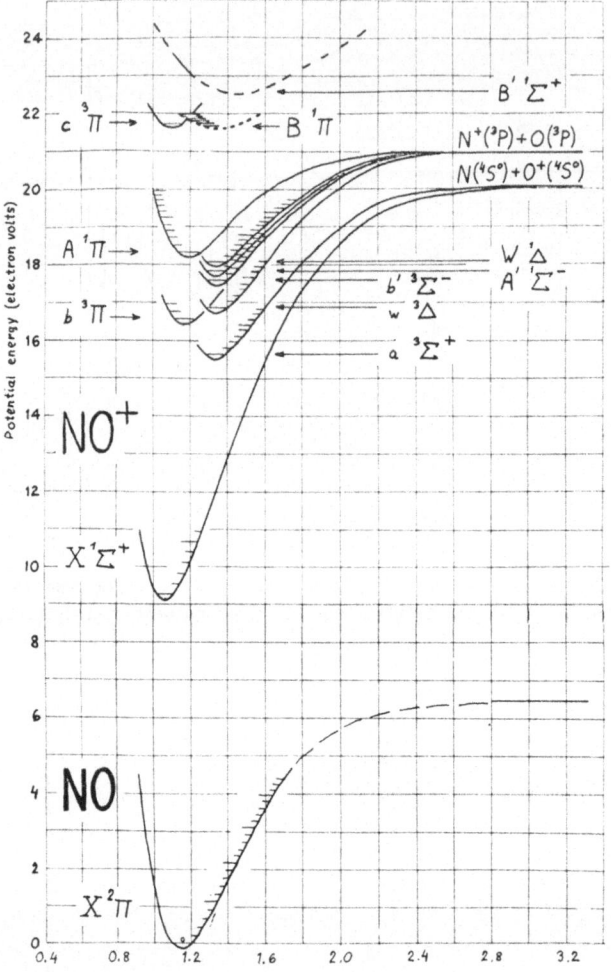

Fig. 14. Potential energy curves for NO and NO⁺.

Below the orbital the approximate ionization potential is given
    The orbitals can easily be described approximately:

> 1σ is $1s$ on oxygen
> 2σ is $1s$ on nitrogen
> 3σ is $2s + 2s$
> 4σ is $2s - 2s$
> 5σ is the σ bond $2pz - 2pz$ (in fact, nonbonding)
> 1π is the bonding π orbital
> 2π is the antibonding π orbital

As NO has one unpaired 2π electron, the molecule has the orbital angular momentum $L=1$ and is $X\,^2\Pi$.

If during the ionization the 2π electron is ionized, $L=0$ and the lowest state of NO^+ is $X\,^1\Sigma^+$. If during the ionization one of the four 1π electrons is ionized, sevral states are possible, as now both the 2π and the unpaired 1π electron have $1=1$. The resulting L is either 2 (Δ state) or 0 (Σ^+ or Σ^-). The spins of the two electrons produce singlets and triplets. This explains the states $^3\Sigma^+$, $^3\Delta$, $^3\Sigma^-$, $^1\Sigma^-$, $^1\Delta$, and $^1\Sigma^+$. If during the ionization one of the 5σ electrons is ionized $b\,^3\Pi$ or $A\,^1\Pi$ are produced, and 4σ gives in the same way $c\,^3\Pi$ or $B\,^1\Pi$.

For identification of the different states theoretical calculations, comparison with N_2 and the higher intensities of triplets and Δ states compared with singlets and Σ states were used.

C. OXYGEN

The photoelectron spectrum of O_2 is given below.

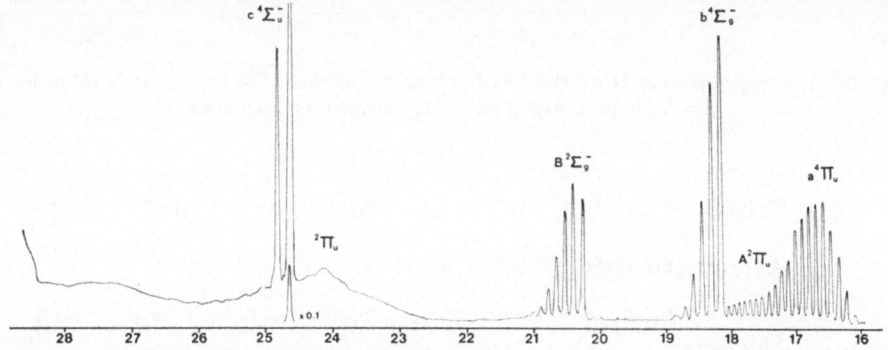

Figs. 15a–b. (a) Photoelectron spectrum of O₂ between 16 eV and 28 eV. [25] and (b) between 12 eV and 13 eV [16]. From Turner *et al.*: *Molecular Photoelectron Spectroscopy.* Copyright © 1970 John Wiley & Sons Ltd. By Permission of John Wiley & Sons Ltd.

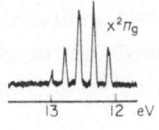

Fig. 15b.

This photoelectron spectrum is in excellent agreement with the potential energy curves of O_2^+ constructed by Gilmore and reproduced below. Some modifications are, however, necessary in Gilmore's diagram. The $A\,^2\Pi_u$ curve ought to be moved upwards a little, and a repulsive $^2\Pi_u$ curve ought to be inserted at 24 eV.

To explain the different states of O_2^+ the electron configuration must be discussed. The orbitals are the same as in NO and have already been described.

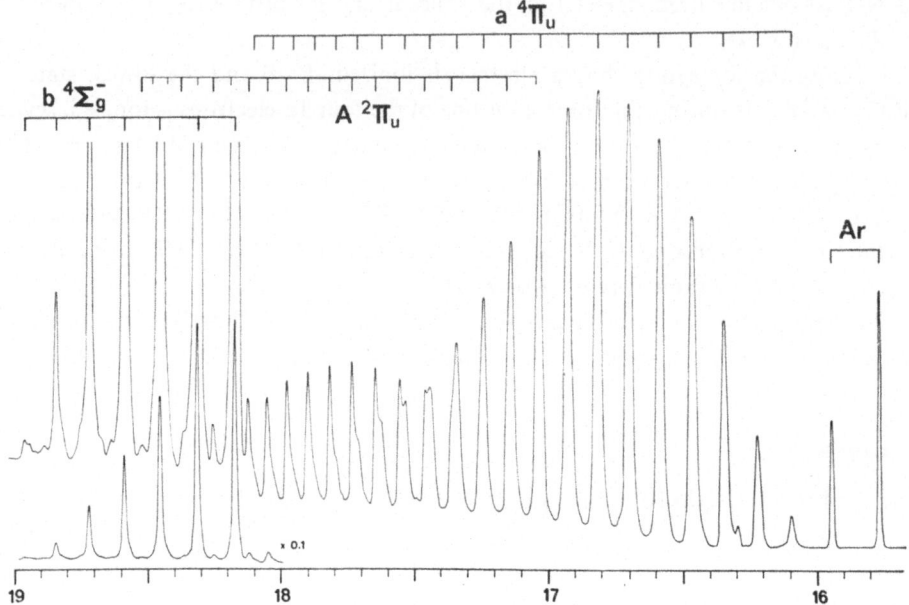

Fig. 16.   The region between 16 eV and 19 eV at higher resolution. The peak at 16.30 eV is due to
537 Å photons. Argon peaks are included for calibration.

Orbital	$1\sigma_g^2$	$1\sigma_u^2$	$2\sigma_g^2$	$2\sigma_u^2$	$3\sigma_g^2$	$1\pi_u^4$	$1\pi_g^2$
Approximate IP	561	560	43				
Doublets { State				$^2\Sigma_u^-$	$B^2\Sigma_g^-$	$A^2\Pi_u$	$X^2\Pi_g$
IP				?	20.30	17.05	12.07
Quartets { State				$c^4\Sigma_u^-$	$b^4\Sigma_g^-$	$a^4\Pi_u$	
IP				24.57	18.17	16.10	

In the ground state of O_2 the two antibonding $1\pi_g$ electrons have antiparallel
angular momentum but parallel spins, and therefore the ground state is $X\,^3\Sigma$. After
ionization of an inner electron, one doublet and one quartet state is therefore obtained.

D. NITROGEN

The previous examples, NO and O_2, are very complicated as the have unpaired π
electrons which interact with the ionized core of the molecule. Normally, in for
instance organic molecules, only paired electrons are present, and the photoelectron
spectrum is simpler. As an example of such a simple molecule, N_2 will be chosen.

The photoelectron spectrum of N_2 observed by Åsbrink by use of the 304 Å helium
line is shown below (Figure 18).

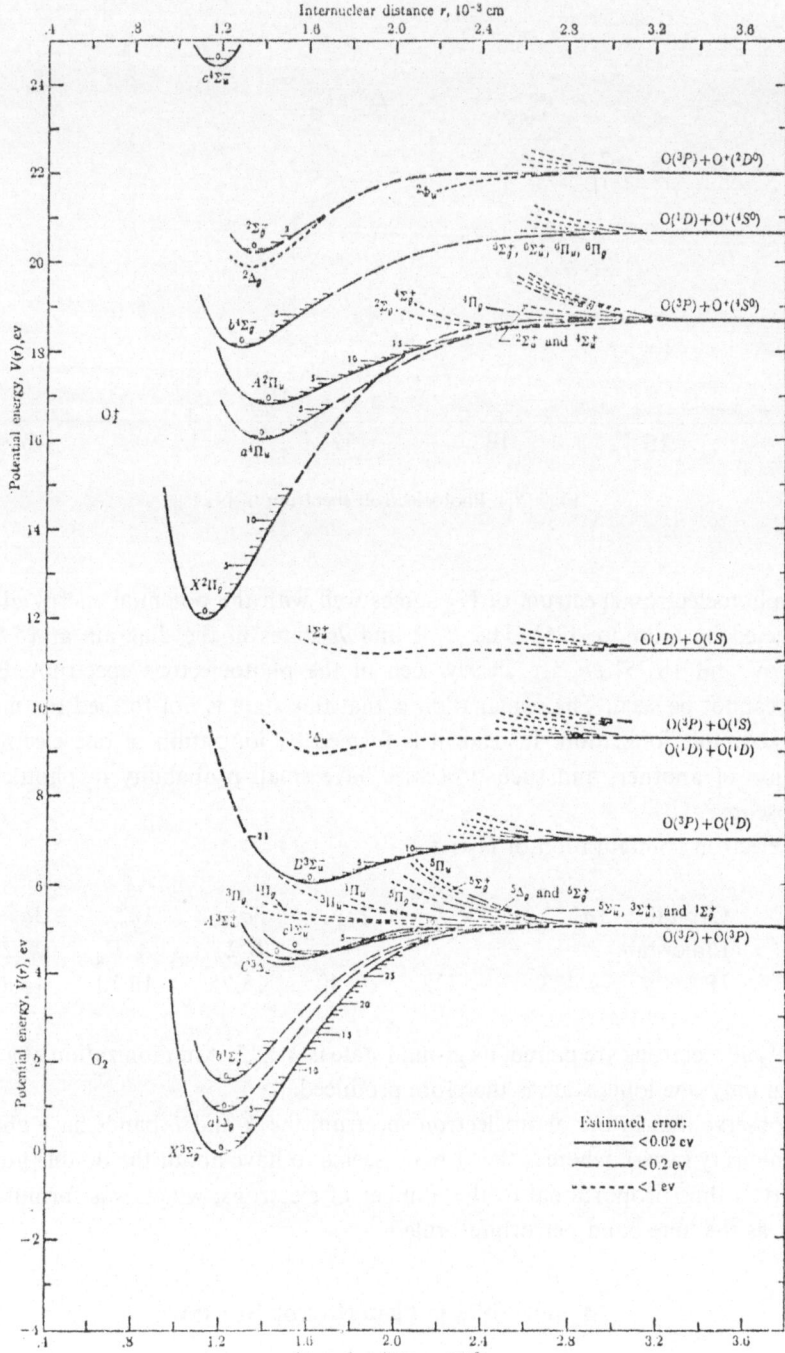

Fig. 17. Potential curves for the O_2, and O_2^+ molecules based upon spectroscopic and other experimental data according to Gilmore, [24].

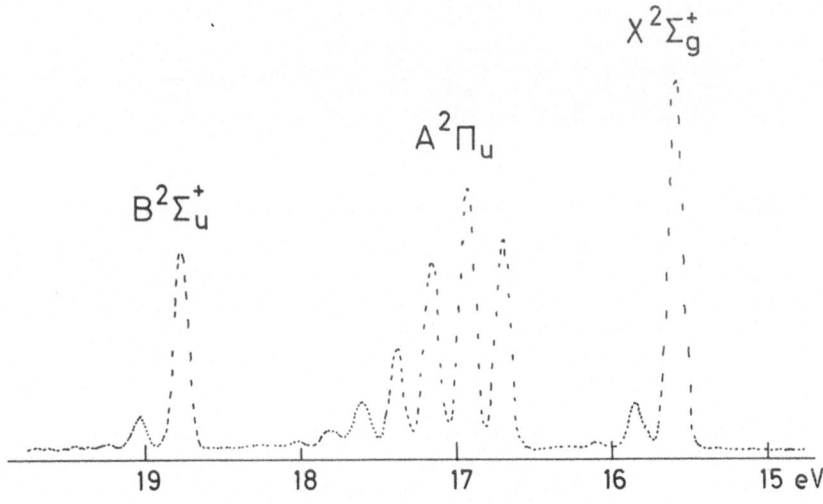

Fig. 18.   Photoelectron spectrum of N_2.

The photoelectron spectrum of N_2 agrees well with the potential energy diagram, constructed by Gilmore [26]. The X, A and B states in the diagram at 15.581 eV, 16.698 eV and 18.751 eV are clearly seen in the photoelectron spectrum. But the $C\,^2\Sigma_u^+$ cannot be seen. The explanation is that this state is not formed the molecule N_2 by a simple ionization. Instead, it is formed by ionization of one electron and excitation of another, and such processes have small probability in photoelectron spectroscopy.

The electron configuration of N_2 is:

Orbital	$1\sigma_g^2$	$1\sigma_u^2$	$2\sigma_g^2$	$2\sigma_u^2$	$1\pi_u^4$	$3\sigma_g^2$
Ionic state				$B^2\Sigma_u^+$	$A^2\Pi_u$	$X^2\Sigma_g^+$
IP	~ 423	~ 422	~ 40	18.75	16.70	15.88

As in N_2 all electrons are paired, its ground state is $X\,^1\Sigma_g^+$. After ionization of a certain electron only one ionic state is therefore produced.

We observe that in the photoelectron spectrum the X and B bands have about the same intensity (area), whereas the A band seems to have about the double area. The intensity is thus proportional to the number of electrons, which was formulated by Turner as the 'one band per orbital' rule.

4. Intensities in Photoelectron Spectra

The intensity in the photoelectron process is a complicated function of the wavelength of the incident light, the nature of the molecular orbital, and the experimental angles.

It is known that if the valence bands of a molecule are studied using X-ray photons

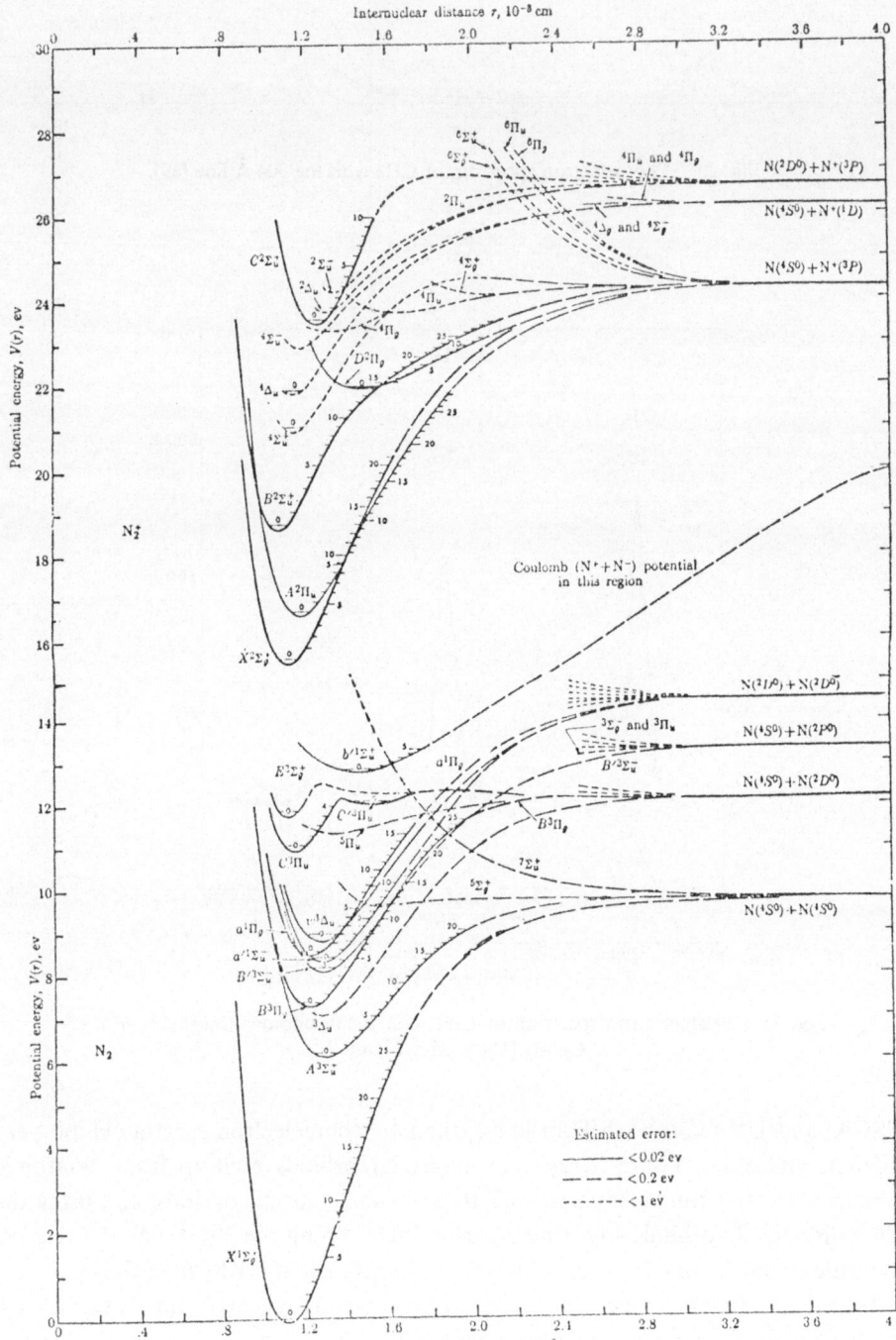

Fig. 19. Potential curves for the N_2, and N_2^+ molecules based upon spectroscopic and other experimental data [26].

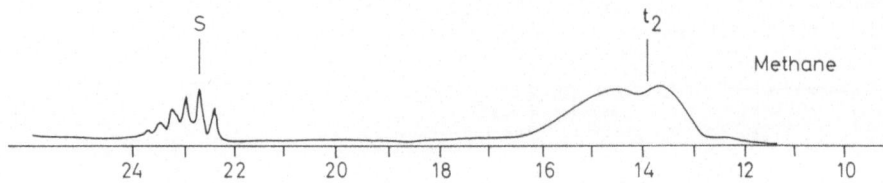

Fig. 20.   Photoelectron spectrum of CH₄ with the 304 Å line [39].

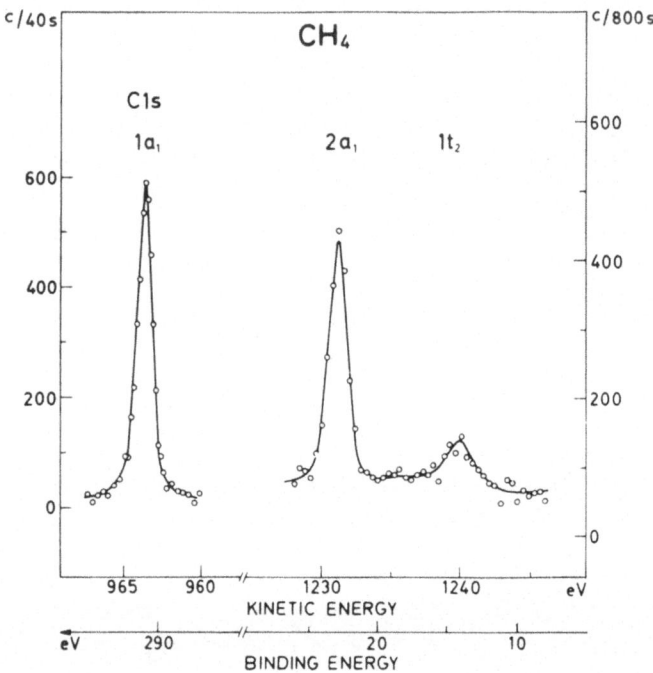

Fig. 21.   Photoelectron spectrum of CH₄ with X-ray photons (Siegbahn *et al.*;
Gelius [15] p. 320, Figure 4.

(ESCA) and UV photons (helium lines), the two photoelectron spectra exhibit very different intensities. Using X-rays the molecular orbitals built-up from 2*s* atomic orbitals are very strong compared with those from 2*p* atomic orbitals, and using the 584 Å line the 2*p* orbitals sometimes predominate. Using the 304 Å line the relative intensities from 2*s* and 2*p* seem to be approximately equal for hydrocarbons.

As an illustration two spectra of methane are shown in Figures 20 and 21. With 304 Å the 2*p* band 1t₂ is about three times stronger than the 2*s* band 2a₁, which is what one expects from the number of electrons in the two orbitals. With the X-ray the 2*s* band predominates.

The energy dependence for neon has been studied systematically by Krause [15].

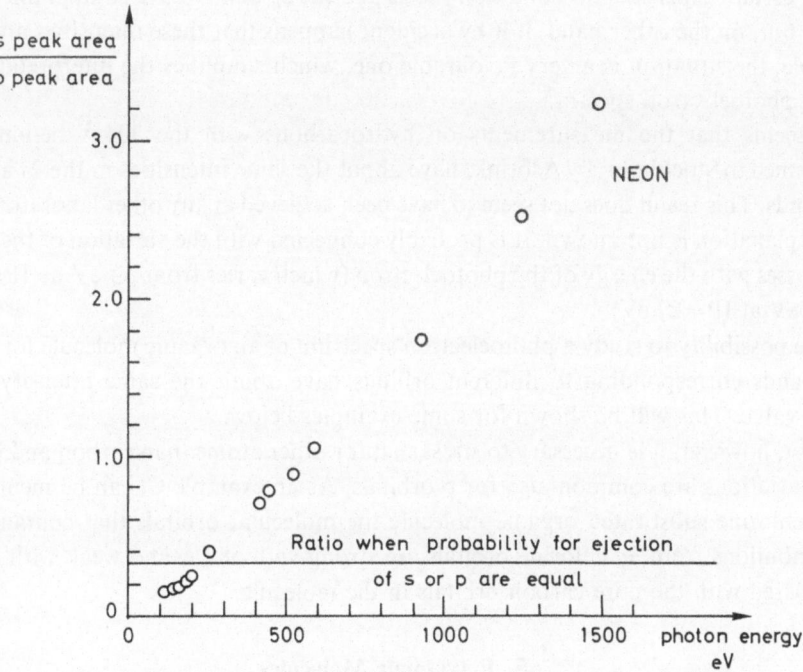

Fig. 22. Ratio of cross sections for s and p in neon after Krause. ([15] after data from Figure 2, p. 263.)

To cover the energy gap between the helium lines and the normal X-ray lines (MgKα at 1253 eV) Krause used X-ray lines from rare earths and other metals. His results for neon are presented in the diagram below.

If the probability for ejection of a photoelectron is the same for s and p electrons, the $2p$ band in neon should be three times as intense as the $2s$ band, as there are six $2p$ electrons but only two $2s$ electrons. In the diagram this case is indicated by a horizontal line at the ratio 0.33. From the diagram follows that at low photon energy the ratio is smaller and at high photon energy it is larger.

Further experimental and theoretical studies have been performed by Price and coworkers [15] and the Uppsala group (see for instance Gelius [15]).

The intensity in the photoelectron spectra depends also upon the experimental angles. Usually, the experiments have been performed with an angle 90° between the direction of the incident photon and the direction of the ejected electron, and it was once believed, that this geometry would give the highest intensity. It has, however, appeared that the angular intensity distribution depends upon the molecular orbital, and efforts have been made to determine the nature of a molecular orbital from the angular dependence (see for instance Carlson *et al.* [15]).

The dependence of the cross sections on the photon energy and the experimental angles imply that the situations is complicated. One has no possibility to state that under certain experimental condition the $2s$ and the $2p$ bands have comparable intensities, but, on the other hand, if it by accident happens that these intensities are comparable, the situation is a very favourable one, which simplifies the interpretation of many photoelectron spectra.

It seems that the measurements on hydrocarbons with the 304 Å helium line, performed in Stockholm by Åsbrink, have about the same intensities in the $2s$ and the $2p$ bands. This result does not seem to have been achieved at any other laboratory, and the explanation is not known. It is probably connected with the variation of the intensity losses with the energy of the photoelectron (which varies from 32 eV at IP=9 eV to 14 eV at IP=27 eV).

The possibility to study a photoelectron spectrum of an organic molecule for which the bands corresponding to different orbitals have about the same intensity, is of great value. This will be shown for some examples below.

First, however, it is necessary to stress that for other atoms than carbon and hydrogen deviations are common also for p orbitals. As an example Cl can be mentioned. In a chlorine substituted organic molecule the molecular orbitals that contain large contributions from $3p$ chlorine orbitals are strong with 584 Å and weak with 304 Å, compared with the pure carbon orbitals in the molecule.

5. Polyatomic Molecules

A. BENZENE

The photoelectron spectrum of benzene, obtained by use of the 304 Å line [27] is shown in Figure 23 together with pictures of the molecular orbitals. This interpretation was given by Jonsson and Lindholm [28].

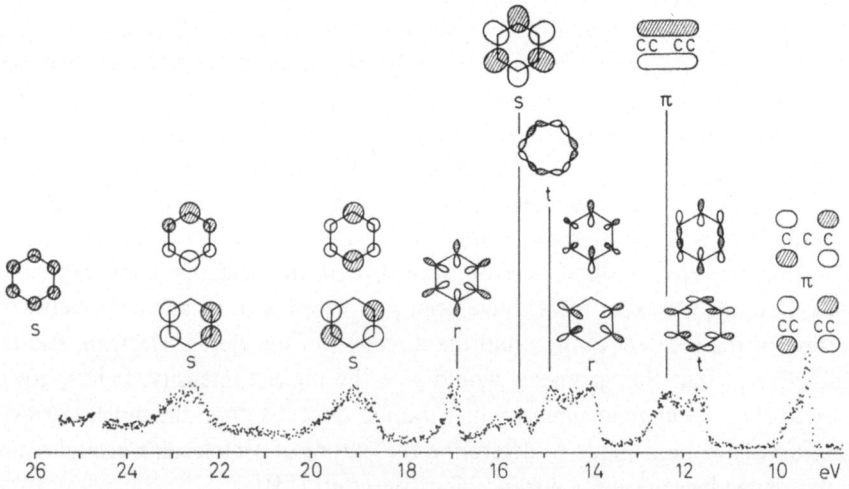

Fig. 23.   Photoelectron spectrum and molecular orbitals of benzene from [27] and [28].

Quantum chemical calculations by Hoffmann [29] give complete information about the molecular orbitals of benzene regarding structure, symmetry and approximate energies. Two orbitals of s-type with low orbital energies (high IP's) are degenerate and must therefore correspond to the bands at 19 eV and 23 eV, whose areas seem to be twice as large as the band at 17 eV. Having interpreted the bands at 19 eV and 23 eV, we known from the calculations, that there must be 7 bands with lower IP's, three of them degenerate. Every interpretation of the photoelectron spectrum of benzene must therefore imply a distribution of the orbitals which corresponds to the distribution of bands in the spectrum.

Some time ago it was discussed whether not one or two more orbitals should be put in the low energy part of the spectrum. The consequences of such an interpretation should have been that some bands in the high energy part should have been without interpretation.

B. NAPHTHALENE

The photoelectron spectrum of naphthalene has recently been compared with the orbital energies from two theoretical calculations: an ab initio calculation and a new calculation called SPINDO [37] (Figure 24).

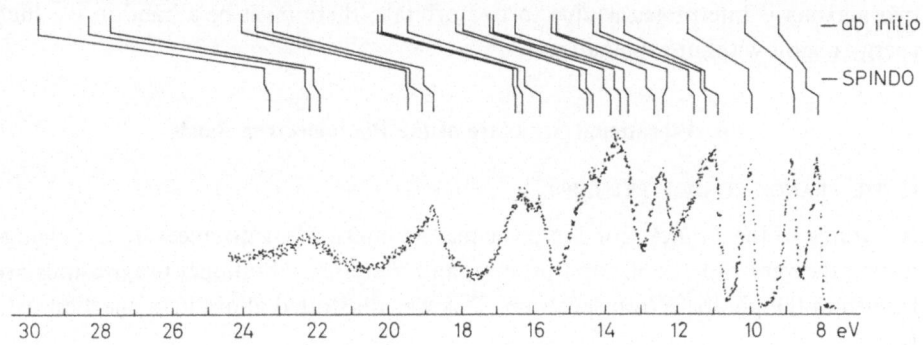

Fig. 24.   Photoelectron spectrum of naphthalene.

It can be seen that the three broad bands at 23 eV, 19 eV and 16 eV have about the same area. The SPINDO calculation interprets them as due to three orbitals each, which seems acceptable. It should be stressed that this satisfactory result was obtained

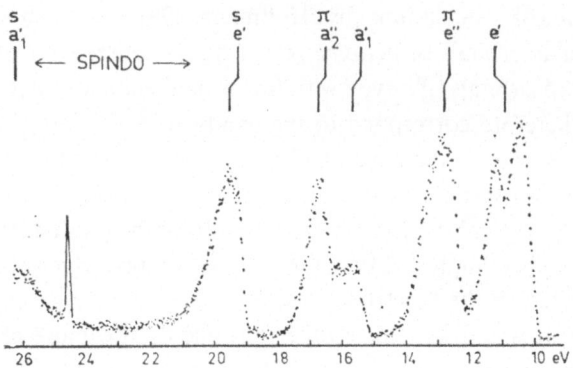

Fig. 25. Photoelectron spectrum of cyclopropane and comparison with SPINDO calculations.

although the nature of these three bands is different. The first two are of 2s-type but the last is mainly of 2p-type.

C. CYCLOPROPANE

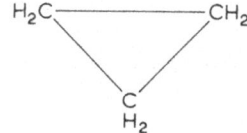

In the photoelectron spectrum of cyclopropane the first two bands may be interpreted either as due to two orbitals or to one orbital which then must be split. The 304 Å spectrum was therefore studied.

The areas of the different bands correspond well to the degeneracies of the orbitals (*e* means degenerate, *a* non-degenerate), given by the calculation. If the first band with two maxima is interpreted as due to two orbitals, there must be a band in the high energy region without explanation.

6. Vibrational Structure of the Photoelectron Bands

A. THE FRANCK-CONDON PRINCIPLE

According to the Franck-Condon principle the internuclear distances do not change during the very rapid ionization process, and therefore vibrational progressions are formed in the photoelectron spectrum. This was illustrated above in connection with H_2.

If the ionized electron is nonbonding, the potential energy curve of the ionized state will have its minimum at the same internuclear distance as for the neutral molecule. The photoelectron spectrum will then consist of one vibrational band.

In NO the 5σ electron is nonbonding and therefore the $b\ ^3\Pi$ state at 16.562 eV and the $A\ ^1\Pi$ state at 18.319 eV consist mainly of one peak each.

It is thus possible to see directly from the photoelectron spectrum whether the ionized electron is bonding or nonbonding.

B. THE VIBRATIONAL ENERGIES

The vibrational energy can easily be measured on a photoelectron spectrum. It is the energy difference between two consecutive peaks in a vibrational progression.

If the ionized electron is strongly bonding, we expect that after the ionization the bond will be weaker and therefore the vibrational energy lower than before the ionization. This is the case for hydrogen where for H_2^+ the vibrational energy is 2191 cm^{-1} (0.272 eV) and for H_2 4280 cm^{-1}.

If the ionized electron is antibonding, the vibrational energy will be larger in the photoelectron spectrum than in the molecule. As an example NO^+ $X^1\Sigma^+$ with 2345 cm^{-1} can be mentioned. NO in its ground state has 1890 cm^{-1}.

We find thus that a strongly bonding electron upon ionization gives a progression with the following two features: (a) the progression is long which means a large distance between the adiabatic and vertical energy; (b) the vibrational energy is low compared with that of the neutral molecule.

Turner has studied the correlation between the length of the progression and the change of the vibrational energy for many small molecules. A useful diagram is given in for instance [6], p. 11. It is not certain, however, whether the method can be used for larger molecules. It has been stated that for some six-membered rings (pyridine, pyrazine, pyrimidine, pyridazine) the vibrational energies do not change much although the progressions are very long [30].

C. STRUCTURELESS BANDS

Most of the bands of a large molecule have no clear vibrational structure. There are two explanations for this:

(a) The large number of vibrational bands overlap so that an apparent continuum is obtained. The vibrational bands themselves consist further of a large number of rotational transitions.

(b) Immediate dissociation of the molecular ion takes place after a transition to a repulsive state, in which the vibrational energy is not quantized.

First, two cases will be discussed for which the explanation is certain.

In the photoelectron spectrum of O_2 (see above) there is a broad feature around 24 eV, which must be due to a transition to a repulsive state. In O_2^+ there is only one vibrational mode, and therefore this is the only possible explanation.

The photoelectron spectrum of CH_4 consists of a broad band between 12.5 eV and 16 eV. It is known from mass spectrometry that the CH_4^+ ion is stable up to 14.2 eV. No dissociation takes therefore place between 12.5 and 14.2 eV, and the nearly structureless band must be due to overlap of many vibrational bands with different vibrational energies. In the first part (12.5 eV–13.0 eV) the number of bands is smaller and here the different vibrations can be distinguished although owing to the Franck-Condon principle the intensity is very low.

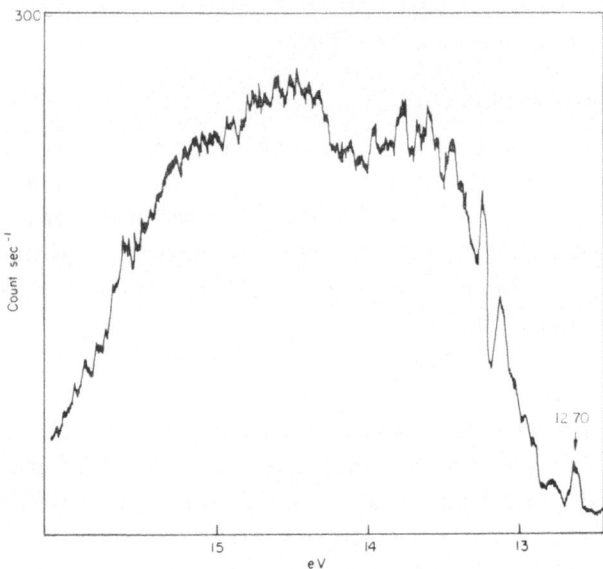

Fig. 26. Photoelectron spectrum of methane [6]. From Turner *et al.*: *Molecular Photoelectron Spectroscopy*. Copyright © 1970 John Wiley & Sons Ltd. By permission of John Wiley & Sons Ltd.

Many other examples could be given. The benzene ion does not dissociate until above 13.8 eV but the band between 12 eV and 13 eV has no vibrational structure. The cyclopropane ion does not dissociate until above 12 eV but the band between 10 eV and 12 eV is nearly without structure.

The explanation is less certain, however, when we discuss diffuse bands above the dissociation limit of a molecule. It has often been assumed that the explanation is the short lifetime of the ion, but in fact very little is known about the lifetimes of the ionic states. An observation in a mass spectrometer that the ion dissociates has normally no connection at all with our problem. In the mass spectrometer the ions are studied after 10^{-6} s, which is a very long time in photoelectron spectroscopy.

Instead, we will discuss some examples. O_2^+ dissociates into $O+O^+$ at about 19 eV but the progressions at 20 eV and 24 eV consist of narrow vibrational bands. In methane, benzene and cyclopropane one finds good vibrational progressions at much higher energy than the dissociation limits.

There seems therefore to be rather little correlation between the dissociation of the molecular ion and the diffuseness of the photoelectron bands, and we will therefore assume, that in large molecules the normal reason for the structureless bands is overlap of many vibrations.

In the case of degenerate orbitals the vibrational structure is especially complicated. According to the Jahn-Teller theorem the geometry changes. For example, the equilateral triangle changes its form into one of the following isoscele triangles during the ionization of cyclopropane [31].

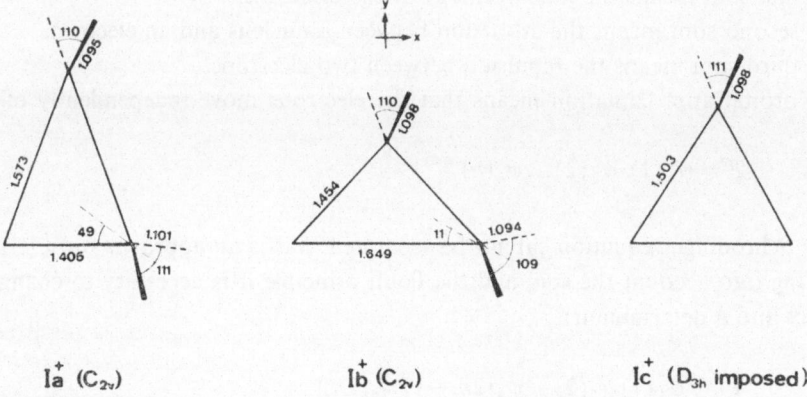

$$\overset{+}{Ia} \; (C_{2v}) \qquad\qquad \overset{+}{Ib} \; (C_{2v}) \qquad\qquad \overset{+}{Ic} \; (D_{3h} \; imposed)$$

Fig. 27.   Geometrical parameters of $C_3H_6^+$ for the energy minima of C_{2v} symmetry (I_a^+ and I_b^+) and the local minimum of D_{3h} symmetry (I_c^+). (Bond length in Å, angles in degrees.)

This explains the double maxima in the first band of cyclopropane, but the complicated vibrational structure explains also the apparent diffuseness of the band. Jahn-Teller effect seems to be especially important in the photoelectron spectrum of methane [32].

7. Theoretical Calculation of Orbital Energies

The most general method for interpretation of a photoelectron spectrum consists of a comparison of the ionization energies in the spectrum with the orbital energies from a quantum-mechanical molecular orbital calculation. This is important also as the other properties of a molecule follow from the quantum-mechanical description.

The basis for this comparison is the assumption that to a first approximation an observed ionization potential equals the negative of the orbital energy of the orbital. This was suggested already 1933 by Koopmans [33]. Koopmans' theorem assumes that when the photoelectron is ejected, the orbitals and energies of the remaining electrons are unchanged. The reorganization energy is thus neglected, but this neglect is partly compensated by the neglect of the correlation energy. One should therefore not expect to obtain perfect agreement between experiment and theory.

We will now give a very brief description of the possibilities to treat a molecule by use of approximate quantum-mechanical methods. As this topic has been treated in detail by Pople and Beveridge [38], we will follow the presentation in this book.

If the nuclei of the molecule are assumed to be fixed, the hamiltonian is

$$\mathscr{H} = -\sum_p \tfrac{1}{2}\nabla_p^2 - \sum_A \sum_p \frac{Z_A}{r_{A_p}} + \sum_{p<q} \frac{1}{r_{pq}}. \tag{1}$$

Here the electrons are denoted p and the nuclei A and the nuclear charge Z_A. The distance between nucleus A and electron p is denoted r_{A_p}.

The first sum means the kinetic energy of the electrons.

The second sum means the attraction between a nucleus and an electron.

The third sum means the repulsion between two electrons.

The orbital approximation means that the electrons move independently of each other:

$$\Psi = \psi_1(1) \cdot \psi_2(2) \ldots \psi_n(n). \tag{2}$$

As the Schrödinger equation cannot be separated, this is an approximation.

Taking into account the spin and the Pauli principle it is necessary to change the product into a determinant:

$$\Psi = |\psi_1(1)\bar{\psi}_1(2) \ldots \psi_n(2n-1)\bar{\psi}_n(2n)|. \tag{3}$$

Here ψ_1 is the first molecular orbital and (2) means the second electron. Positive spin direction is marked ψ and negative $\bar{\psi}$.

LCAO means that the molecular orbital ψ is approximated as a sum of atomic orbitals ϕ (for example carbon 2s or carbon 2p or hydrogen 1s) from all atoms in the molecule

$$\psi = c_1\phi_1 + c_2\phi_2 + c_3\phi_3 + \cdots + c_\mu\phi_\mu + \cdots \tag{4}$$

LCAO is thus an approximation and the situation is still worse if simplified atomic orbitals are used (Slater type orbitals STO).

The variational method gives us a possibility to calculate the constants c_μ for the ground state of the molecule. The constants c_μ are varied until the calculated energy

$$\langle \Psi | \mathscr{H} | \Psi \rangle \tag{5}$$

is as low as possible. The theory is tedious (p. 32–45 in Pople's book [38]), as the wavefunction is a complicated determinant and as $\mathscr{H}$ consists of three sums. We must also remember that there are many different molecular orbitals and change therefore the formula above into

$$\psi_i = c_{1i}\phi_1 + c_{2i}\phi_2 + c_{3i}\phi_3 + \cdots + c_{\mu i}\phi_\mu + \cdots. \tag{6}$$

The result of all this is that the constants c_{vi} are determined by the following equations (Roothaans' equations):

$$\sum_v (F_{\mu v} - \varepsilon_i S_{\mu v}) c_{vi} = 0 \tag{7}$$

where F is the Fock matrix

$$F_{\mu v} = H_{\mu v} + \sum_{\lambda\sigma} P_{\lambda\sigma} [(\mu v | \lambda\sigma) - \tfrac{1}{2}(\mu\lambda | v\sigma)]. \tag{8}$$

Here the different quantities are defined in (9), (11), (12) and (13).

$$H_{\mu\nu} = \int \phi_\mu(1) \, H^{core} \phi_\nu(1) \, d\tau_1 \tag{9}$$

where ϕ_μ is an atomic orbital (e.g. carbon $2s$) and where the integration is performed for the first electron (1), and

$$H^{core} = -\nabla_p^2 - \sum_A \frac{Z_A}{r_{pA}}. \tag{10}$$

This means that an electron moves in the field of only the nuclei. A diagonal element $H_{\mu\mu}$ is connected with the energy of the atom in question. An off-diagonal element $H_{\mu\nu}$ is connected with the chemical bond between the atoms (if μ and ν belong to different atoms).

$$(\mu\nu \,|\, \lambda\sigma) = \int\int' \phi_\mu(1) \cdot \phi_\nu(1) \, \frac{1}{r_{12}} \, \phi_\lambda(2) \, \phi_\sigma(2) \, d\tau_1 \, d\tau_2. \tag{11}$$

If μ and λ belong to different atoms, the special case $(\mu\mu \,|\, \lambda\lambda)$ can easily be understood as the electron repulsion between the two electrons.

$$S_{\mu\nu} = \int \phi_\mu(1) \, \phi_\nu(1) \, d\tau_1. \tag{12}$$

This is called the overlap integral and is easy to visualize.

ε_i is the orbital energy of the molecular orbital. (For the first orbital of benzene it is about -9.3 eV.)

$$P_{\lambda\sigma} = 2 \cdot \sum_i^{occ.} c_{\lambda i} c_{\sigma i} \tag{13}$$

where the summation extends over all occupied molecular orbitals. This means that $P_{\lambda\sigma}$ can be obtained only after the calculation has been completed. It is called 'Density matrix' or 'Charge density and bond order matrix'.

To solve Roothaans's equations the determinant is equated to zero

$$|F_{\mu\nu} - \varepsilon_i S_{\mu\nu}| = 0. \tag{14}$$

The Hartree-Fock determinant is illustrated below for a planar molecule with at least two carbon atoms and at least two hydrogens. In the determinant only six atomic orbitals are included. As $2s$ and $2p$ are orthogonal, $S_{12}=0$. The x-axis is in the plane of the molecule and z is perpendicular to it.

μ	$\nu = 1$ $\quad C_1(2s)$	$\nu = 2$ $\quad C_1(x)$	$\nu = 3$ $\quad C_1(z)$
1 $C_1(2s)$	$-S_{11}\varepsilon + H_{11} + \sum P_{\lambda\sigma}[(11\mid\lambda\sigma) - \tfrac{1}{2}(1\lambda\mid 1\sigma)],$	$H_{12} + \sum P_{\lambda\sigma}[(12\mid\lambda\sigma) - \tfrac{1}{2}(1\lambda\mid 2\sigma)],$	$H_{13} + \sum P_{\lambda\sigma}[(13\mid\lambda$
2 $C_1(x)$	$H_{21} + \sum P_{\lambda\sigma}[(21\mid\lambda\sigma) - \tfrac{1}{2}(2\lambda\mid 1\sigma)],$	$-S_{22}\varepsilon + H_{22} + \sum P_{\lambda\sigma}[(22\mid\lambda\sigma) - \tfrac{1}{2}(2\lambda\mid 2\sigma)],$	$H_{23} + \sum P_{\lambda\sigma}[(23\mid\lambda$
3 $C_1(z)$	$H_{31} + \sum P_{\lambda\sigma}[(31\mid\lambda\sigma) - \tfrac{1}{2}(3\lambda\mid 1\sigma)],$	$H_{32} + \sum P_{\lambda\sigma}[(32\mid\lambda\sigma) - \tfrac{1}{2}(3\lambda\mid 2\sigma)],$	$-S_{33}\varepsilon + H_{33} + \sum P_{\lambda\sigma}[(33\mid\lambda$
4 $C_2(2s)$	$-S_{41}\varepsilon + H_{41} + \sum P_{\lambda\sigma}[(41\mid\lambda\sigma) - \tfrac{1}{2}(4\lambda\mid 1\sigma)],$	$-S_{42}\varepsilon + H_{42} + \sum P_{\lambda\sigma}[(42\mid 2\sigma) - \tfrac{1}{2}(4\lambda\mid 2\sigma)],$	$H_{43} + \sum P_{\lambda\sigma}[(43\mid\lambda$
5 $H_1(1s)$	$-S_{51}\varepsilon + H_{51} + \sum P_{\lambda\sigma}[(51\mid\lambda\sigma) - \tfrac{1}{2}(5\lambda\mid 1\sigma)],$	$-S_{52}\varepsilon + H_{52} + \sum P_{\lambda\sigma}[(52\mid\lambda\sigma) - \tfrac{1}{2}(5\lambda\mid 2\sigma)],$	$H_{53} + \sum P_{\lambda\sigma}[(53\mid\lambda$
6 $H_2(1s)$	$-S_{61}\varepsilon + H_{61} + \sum P_{\lambda\sigma}[(61\mid 2\sigma) - \tfrac{1}{2}(6\lambda\mid 1\sigma)],$	$-S_{62}\varepsilon + H_{62} + \sum P_{\lambda\sigma}[(62\mid\lambda\sigma) - \tfrac{1}{2}(6\lambda\mid 2\sigma)],$	$H_{63} + \sum P_{\lambda\sigma}[(63\mid\lambda$

$$(15)$$

The Hartree-Fock determinant has the same number of rows and columns as the number of atomic orbitals (μ). It is an equation in ε giving as many solutions ε_i as the number of atomic orbitals. Afterwards, another calculation gives $c_{\nu i}$ for each ε_i and the 'Eigenvalues with eigenvectors in columns' is obtained and perhaps printed. The 'First order density matrix' $P_{\lambda\sigma}$ can now be calculated and perhaps printed.

As $P_{\lambda\sigma}$ now is known, a better determinant can be used as the starting point of a second calculation. This is repeated until 'Self-consistency' is obtained (SCF).

Ab initio methods mean that in the Hartree-Fock determinant all integrals are computed exactly.

8. Approximate Calculations of Orbital Energies

The *Extended Hückel Method* [29] is obtained if the Hartree-Fock determinant is simplified in the following way.

Instead of the diagonal element

$$- S_{\mu\mu}\varepsilon + H_{\mu\mu} + \sum P_{\lambda\sigma}[...] \tag{16}$$

we write

$$- S_{\mu\mu}\varepsilon + \overline{H_{\mu\mu}} \tag{17}$$

with

$$\overline{H_{\mu\mu}} = I_{\mu} \text{ (the ionization energy)} \tag{18}$$

Instead of the off-diagonal element

$$- S_{\mu\nu}\varepsilon + H_{\mu\nu} + \sum P_{\lambda\sigma}[...] \tag{19}$$

we write

$$- S_{\mu\nu}\varepsilon + \overline{H_{\mu\nu}} \quad \text{with} \quad \overline{H_{\mu\nu}} = \tfrac{1}{2}\cdot 1.75\cdot S_{\mu\nu}\cdot[I_{\mu} + I_{\nu}]. \tag{20}$$

The following ionization potentials are used: for carbon $2s$ 19 eV, for carbon $2p$ 11 eV, for hydrogen 13.6 eV.

The Extended Hückel determinant

$$\begin{vmatrix} - S_{11}\varepsilon + \overline{H_{11}}, & \overline{H_{12}}, \\ \overline{H_{21}}, & - S_{22}\varepsilon + \overline{H_{22}}, \\ \overline{H_{31}}, & \overline{H_{32}}, \\ - S_{41}\varepsilon + \overline{H_{41}}, & - S_{42}\varepsilon + \overline{H_{42}}, \\ - S_{51}\varepsilon + \overline{H_{51}}, & - S_{52}\varepsilon + \overline{H_{52}}, \\ - S_{61}\varepsilon + \overline{H_{61}}, & - S_{62}\varepsilon + \overline{H_{62}}, \end{vmatrix} \tag{21}$$

can be solved directly and no iteration is necessary.

The expression for $\overline{H_{\mu\nu}}$ is called the 'Mulliken approximation'.

In the *CNDO method* by Pople *et al.* [34] the integrals in the Hartree-Fock determinant are treated and approximated in detail.

The first assumption is that orbitals on different atoms do not overlap. It follows that $S_{\mu\nu}=0$ if μ and ν on different atoms. Further, most of the integrals $(\mu\nu \mid \lambda\sigma)$ are zero. Only the electron repulsion integrals $(\mu\mu \mid \lambda\lambda)$ are left.

The second assumption concerns $H_{\mu\mu}$. The electron in the atomic orbital ϕ_{μ} around the nucleus A experiences attraction from A ($U_{\mu\mu}$) but also from the other nuclei B (V_{AB}). The energy is thus

$$H_{\mu\mu} = U_{\mu\mu} - \sum_B V_{AB}. \tag{22}$$

Here $U_{\mu\mu}$ is sometimes taken as -49 eV.

The third assumption concerns $H_{\mu\nu}$. The Mulliken approximation is used also in CNDO.

The *INDO method* by Pople *et al.* [35] and also Dixon (1967) differs not much from CNDO. Only a few more integrals $(\mu\nu \mid \lambda\sigma)$ are included.

The *MINDO method* by Baird *et al.* [40] differs not much from INDO. The integrals were determined to get the best possible agreement with the heat of formation and geometry for a few organic molecules.

The *SPINDO method* [37] differs not much from MINDO. The integrals have been determined to get the best possible agreement between the orbital energies and the ionization potentials for a few organic molecules (benzene, methane, ethane, ethylene). The success can be judged from the picture in Figure 28.

The interesting feature of the SPINDO method is that it has appeared to be necessary to have one formula for $H_{\mu\nu}$ for each kind of interaction. As $H_{\mu\nu}$ is connected with the chemical bonding, it might be considered as natural to have different expressions for the $2s$–$2s$ bond, the $2p\sigma$–$2p\sigma$ bond, the $2p\pi$–$2p\pi$ bond, and so on.

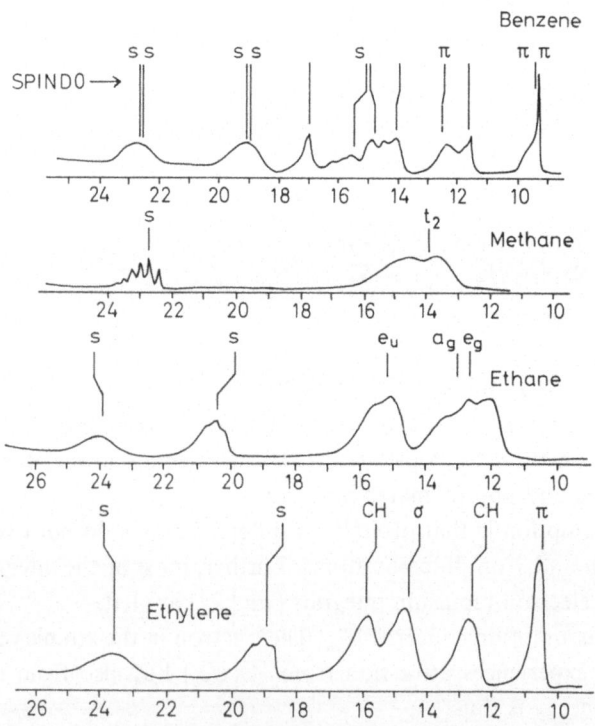

Fig. 28.   Photoelectron spectra of some hydrocarbons, obtained using the 304 Å helium line, with interpretations according to SPINDO/-calculations.

9. The Study of Large Molecules with Molecular Photoelectron Spectroscopy

As further examples of the methods described above we will now discuss the study of two substituted benzenes.

A. TOLUENE [41]

The photoelectron spectrum of toluene with the 304 Å helium line is shown below after measurements in the Stockholm spectrometer. The spectrum is rather diffuse with only weak vibrational structure, and the interpretation of the electronic structure of toluene directly from this curve has therefore proved to be difficult.

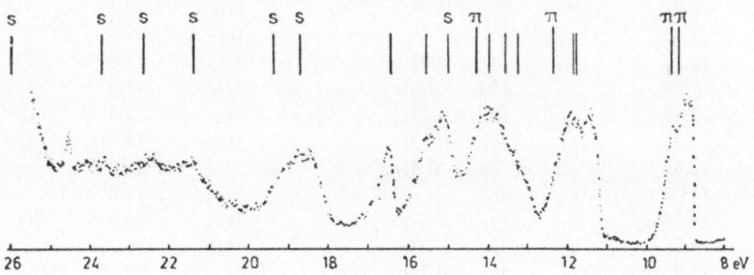

Fig. 29. The photoelectron spectrum of toluene with the 304 Å helium line.

The SPINDO/1 calculation was performed with the C–C distance 1.397 Å in the ring and 1.510 Å to the methyl group, and the negative orbital energies and the approximate forms of the molecular orbitals are given in Table I.

In presenting the electronic structure of a large molecule in a short table several problems appear. Presentation of the species can be very informative in a molecule such as benzene, but in a molecule with low symmetry, such as toluene, there are 13 orbitals of symmetry a_1 and there seems to be no reason to give them separate numbers 1...13.

Very often the molecular orbitals have been presented in form of schematic pictures. This method is practical when only π orbitals are discussed, but is clearly impossible for presentation of all orbitals of a polysubstituted benzene.

In the first study of benzene by Jonsson and Lindholm [28] the approximate forms of the molecular orbitals were presented in a small table by use of the following abbreviations: s, r, t, z mean carbon $2s$, $2p$ in radial direction, $2p$ in tangential direction and $2p$ perpendicular to the plane of the benzene ring. Further, h means hydrogen $1s$ and rh_i means $r_i + h_i$. Using these abbreviations the molecular orbitals of benzene were divided into four types: s, r, t, and π.

To extend this method to a monosubstituted benzene such as toluene we will treat it as a composite molecule RS, where S is the substituent. This means that we will present a certain molecular orbital Ψ_i as

$$\Psi_i = a \cdot \Psi_{R_i} + b \cdot \Psi_{S_i}$$

TABLE I

Approximate forms and orbital energies (in eV) of the molecular orbitals of toluene from a SPINDO/1 calculation, presented as a sum of the benzene orbital and the methyl orbital. For comparison, the orbitals and the vertical ionization potentials of benzene are also included. The methyl carbon is denoted C_7 and the methyl hydrogens have been omitted. Interaction diagrams are included.

Toluene			Benzene		
Orbital energy (negative)	Type	Approximate form	Orbital	IP (vert)	Interaction (see text and diagrams below)
9.24	π	$2b_1 - 0.09z_7$	$1e_{1g}$	9.3	
9.38	π	$1a_2$	$1e_{1g}$	9.3	
11.81	t	$7b_2 - 0.07t_7$	$3e_{2g}$	11.8	
11.89	t	$11a_1 + 0.35r_7$	$3e_{2g}$	11.8	
12.19	π	$1b_1 - 0.19z_7$	$1a_{2u}$	12.5	(a)
13.31	t	$5b_2 - 0.53t_7$	$1b_{2u}$	14.9	(b)
13.59	r	$10a_1 + 0.41r_7$	$3e_{1u}$	14.0	
14.04	r	$6b_2$	$3e_{1u}$	14.0	
14.34	π	$1b_1 + 0.61z_7$	$1a_{2u}$	12.5	(a)
15.05	s	$9a_1 + 0.23r_7$	$2b_{1u}$	15.5	
15.57	t	$5b_2 + 0.28t_7$	$1b_{2u}$	14.9	(b)
16.46	r	$8a_1 + 0.21r_7$	$3a_{1g}$	17.0	
18.71	s	$7a_1 \; 0.19s_7$	$2e_{2g}$	19.2	
19.35	s	$4b_2$	$2e_{2g}$	19.2	
21.23	s	$6a_1 - 0.49s_7$	$2e_{1u}$	22.7	(c)
22.72	s	$3b_2$	$2e_{1u}$	22.7	
23.72	s	$6a_1 + 0.52s_7$	$2e_{1u}$	22.7	(c)
25.97	s	$5a_1 + 0.20s_7$	$2a_{1g}$	25.8	

where a and b are constants to be obtained from the calculation, supposed that Ψ_R and Ψ_S are normalized. To abbreviate the presentation, however, a will be omitted and b will be taken directly from the eigenvectors in the SPINDO print-out.

In the case of a monosubstituted benzene the ring orbitals Ψ_R have usually C_{2v} symmetry. The benzene molecular orbitals from the study by Jonsson and Lindholm will therefore be given in C_{2v} symmetry. For large b orbital Ψ_R is sometimes badly distorted from its form in benzene, but this will be neglected, as then Ψ_R plays a minor role in Ψ.

In the substituent S the radial and tangential directions are determined by the ring. We can therefore classify the orbitals of the substituent as s, r, t and z. The hydrogens of the methyl group will normally be neglected, as their contribution to the molecular orbital normally follows from the coefficient of the carbon orbital.

Finally, the sign in the formula will be determined by the following convention: in

$$\Psi \approx \Psi_R + b \cdot \Psi_S$$

the sign + means positive overlap (bonding) between S and the neighbouring carbon atom, whereas the sign − means an antibonding overlap. In this way the interaction between R and S is sufficiently illustrated.

Owing to the diffuseness of the photoelectron spectrum the observed ionization potentials are not well defined and are therefore not given in the table. Instead, the orbital energies are plotted in the figure.

The distribution of the orbital energies in the figure corresponds well to the different maxima of the photoelectron spectrum. The areas of the first five bands in the spectrum are roughly proportional to the number of orbitals: 2, 3, 4, 2 and 1. The good agreement can be taken as a support for the SPINDO/1 calculation.

In the table the interaction between the ring and the substituent is beautifully illustrated by the forms and energies of the orbitals. In case (a) in the table, the interaction takes place between the benzene $1b_1$ orbital ($1a_{2u}$ in D_{6h}) at 12.5 eV and the z_7 orbital in methyl at about 13.8 eV (cf. the SPINDO orbital energy of methane), giving one orbital at 12.19 eV and one, localized mainly on CH_3, at 14.34 eV. In case (b) the benzene $5b_2$ orbital ($1b_{2u}$) at 14.9 eV interacts with the methyl t_7 orbital, and in case (c) the benzene $6a_1$ orbital ($2e_{1u}$) at 22.7 eV with the methyl s_7 orbital at 22.5 eV Usually, only the first type of interaction has been studied (hyperconjugation).

B. STYRENE [43]

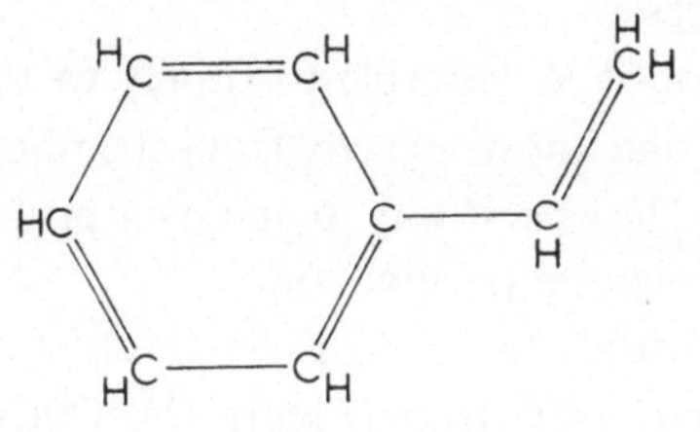

In styrene, $C_6H_5-CH=CH_2$, the C_2H_3 group can rotate around the C–C bond, and therefore two conformations are possible: the planar case and the perpendicular case. As no studies by electron diffraction seem to have been performed, the conformation of the molecule in the gas phase is unknown.

As molecular structures have been determined by molecular photoelectron spectroscopy in only few cases it is of interest to see whether this can be done for styrene.

The SPINDO calculations were performed for both the planar and the perpendicular conformation. The orbital energies are compared in Figure 30 with the photoelectron spectrum, observed using the 304 Å helium line. It can be seen that the distribution of the orbital energies for the planar case corresponds well to the maxima of the curve. This supports the method, and as the agreement is less good for the

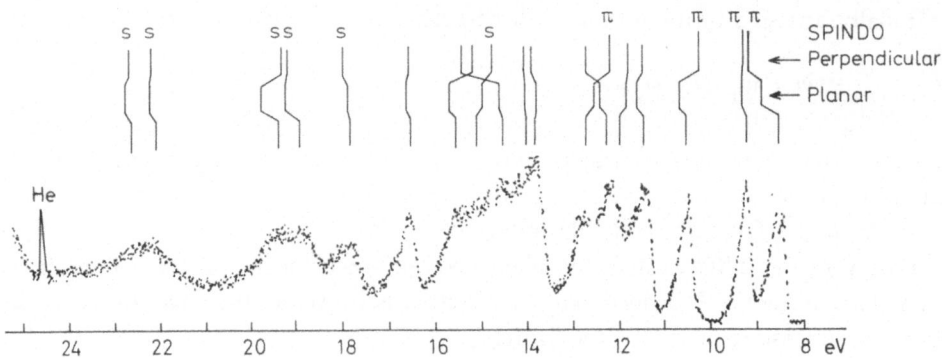

Fig. 30. The photoelectron spectrum of styrene and comparison with SPINDO calculations

perpendicular conformation, the molecule is probably more or less planar. This follows especially from the large energy split between the first two π bands.

10. The Dissociation of the Molecular Ion

After the ionization of a molecule dissociation often takes place, giving fragment ions, which can easily be studied experimentally in a mass spectrometer. The theoretical treatment is, however, less satisfactory which is regrettable since the unimolecular dissociation reaction is perhaps the simplest chemical reaction. The results from photoelectron spectrometry offer now better possibilities to understand the mass spectrometric fragmentation processes.

To study the dissociation it is desirable to supply to the molecule a well defined amount of energy so that one knows exactly from the photoelectron spectrum which electron has been ionized. The easiest way to do this is probably to ionize the molecule by use of charge exchange with a positive ion.

As an example of such a study the breakdown graph of cyclopropane [42] is shown together with its photoelectron spectrum (Figure 31). This picture is very informative. It shows that after ionization of an e' electron the molecular ion $C_3H_6^+$ is stable and does not dissociate. But after ionization of an e'' electron, the molecular ion is unstable and dissociates by loss of H or H_2. The fragment ions $C_3H_5^+$ or $C_3H_4^+$ are thus formed. This is not unexpected as the e'' electron can be described as a π electron in cyclopropane or as a C–H bonding electron in the CH_2 group. Further, after ionization of an a_1' electron or an a_2'' electron the dissociation mostly gives the fragment ion $C_3H_3^+$ after loss of $H_2 + H$.

The dissociation of the molecular ion does therefore not seem to be governed by statistical laws as it was believed earlier. Instead, the ejection of the ionized electron determines the fragmentation processes. But the detailed theoretical interpretation of mass spectra from the photoelectron spectrogram belongs to the future.

For some further examples reference is made to a recent review [36].

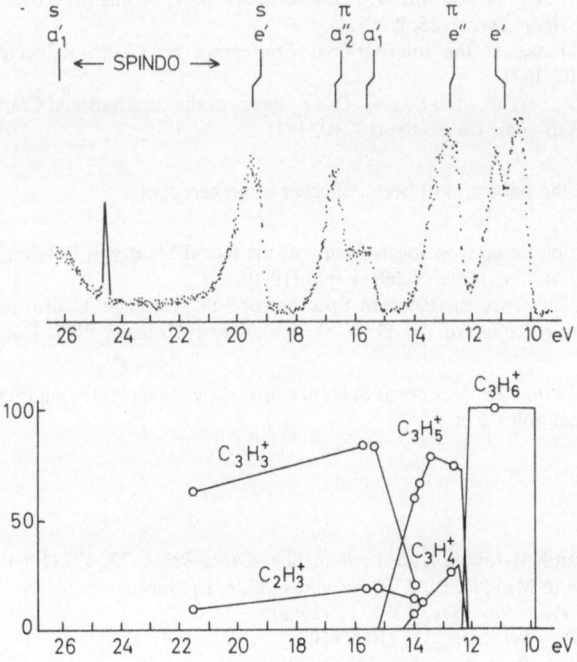

Fig. 31.   Photoelectron spectrum of cyclopropane and break down graph of the cyclopropane ion.

Acknowledgements

I wish to express my gratitude to my coworkers for their cooperation: L. Åsbrink for experimental photoelectron work and also theoretical work, C. Fridh for theoretical work, L. E. Selin and O. Edqvist for the initial construction of the photoelectron spectrometer, I. Szabo and B. Ö Jonsson for mass-spectrometric work, and M. Svantesson for typing.

We are all indebted to Prof. R. Manne, Bergen, for discussions and help.

References

A. *Some Review Articles*

1. Turner, D. W., in *Physical Methods in Advanced Inorganic Chemistry* (ed. by H. A. O. Hill and P. Day), p. 74, Wiley Interscience, New York, 1968.
2. Turner, D. W., *Adv. Mass Spectrom.* **4**, 755 (1968).
3. Price, W. C., in *Molecular Spectroscopy* (ed. by P. Hepple), p. 221, Institute of Petroleum, London, 1968.
4. Berry, R. S., *Ann. Rev. Phys. Chem.* **20**, 357 (1969).
5. Turner, D. W., *Ann. Rev. Phys. Chem.* **21**, 107 (1970).
6. Turner, D. W., Baker, C., Baker, A. D., and Brundle, C. R., *Molecular Photoelectron Spectroscopy*, Wiley, New York, 1970.
7. Turner, D. W., *Phil. Trans. Roy. Soc. London* **A268**, 7 (1970).
8. Betteridge, D. and Baker, A. D., *Anal. Chem.* **42**, 43A (1970).
9. Baker, A. D., *Acc. Chem. Res.* **3**, 17 (1970).
10. Brundle, C. R. and Robin, M. B., in *Determination of Organic Structures by Physical Methods,*

Vol. 3 (ed. by F. C. Nachod and J. J. Zuckerman), p. 1, Academic Press, New York, 1971.
11. Brundle, C. R., *Appl. Spectr.* **25**, 8 (1971).
12. Carlson, T. A., Paper at the International Conference on Electron Spectroscopy, Asilomar, Calif., Sept. 7–10, 1971.
13. Price, W. C., Potts, A. W., and Streets, D. G., Paper at the International Conference on Electron Spectroscopy, Asilomar, Calif., Sept. 7–10, 1971.

At two conferences the papers have been collected in proceedings:

14. A discussion on photoelectron spectroscopy at the Royal Society in London, 27 February 1969, in *Phil. Trans. Roy. Soc. Lond.* **A268**, 1–175 (1970).
15. International Conference on Electron Spectroscopy at Asilomar, California, 7–10 Sept. 1971, in *Electron Spectroscopy* (ed. by D. A. Shirley), North-Holland Publ. Co, Amsterdam, 1972.

This survey on 'Molecular Photoelectron Spectroscopy' depends especially on [6] and [10] and on the results from the Stockholm group.

B. *Specific References*

16. Turner, D. W. and Al-Joboury, M. I., *Bull. Soc. Chim. Belges* **73**, 428 (1964).
17. Turner, D. W. and May, D. P., *J. Chem. Phys.* **45**, 471 (1966).
18. Turner, D. W., *Proc. Roy. Soc.* **A307**, 15 (1968).
19. Lindholm, E., *Rev. Sci. Instr.* **31**, 210 (1960).
20. Åsbrink, L., *Chem. Phys. Letters* **7**, 549 (1970).
21. Cohen, S. H., Hiskes, J. R., Judd, D. L., and Riddell, R. J., UCRL-8802 and UCRL-8871, 1959.
22. Edqvist, O., Lindholm, E., Selin, L. E., Sjögren, H., and Åsbrink, L., *Arkiv Fysik* **40**, 439 (1970).
23. Edqvist, O., Åsbrink, L., and Lindholm, E., *Z. Naturf.* **26A**, 1407 (1971).
24. Gilmore, F., *J. Quant. Spectr. Radiative Transfer* **5**, 369 (1965).
25. Edqvist, O., Lindholm, E., Selin, L. E., and Åsbrink, L., *Physica Scripta* **1**, 25 (1970).
26. Gilmore, F., in *Potential Energy Curves for* N_2, NO, O_2 *and Corresponding Ions*, RM-4024, Santa Monica, Calif., The RAND Corporation, 1964.
27. Åsbrink, L., Edqvist, O., Lindholm, E., and Selin, L. E., *Chem. Phys. Letters* **5**, 192 (1970).
28. Jonsson, B. Ö. and Lindholm, E., *Arkiv Fysik* **39**, 65 (1969).
29. Hoffmann, R., *J. Chem. Phys.* **39**, 1397 (1963).
30. Fridh, C., Åsbrink, L., Jonsson, B. Ö., and Lindholm, E., *Int. J. Mass. Spectr. Ion Phys.* **8**, 101 (1972).
31. Haselbach, E., *Chem. Phys. Letters* **7**, 428 (1970).
32. Rabalais, J. W., Bergmark, T., Werme, L. D., Karlsson, L., and Siegbahn, K., *Physica Scripta* **3**, 13 (1971).
33. Koopmans, J., *Physica* **1**, 104 (1933).
34. Pople, J. A., Santry, D. P., and Segal, G. A., *J. Chem. Phys.* **43**, S129 (1965).
    Pople, J. A. and Segal, G. A., *J. Chem. Phys.* **43**, S136 (1965).
35. Pople, J. A., Beveridge, D. L., and Dobosch, P. A., *J. Chem. Phys.* **47**, 2026 (1967).
36. Lindholm, E., in *Ion-molecule reaction* (ed. by J. Franklin), Plenum Press, New York, 1972.
37. Fridh, C., Åsbrink, L., and Lindholm, E., *Chem. Phys. Letters* **15**, 282 (1972).
38. Pople, J. A. and Beveridge, D. L., *Approximate Molecular Orbital Theory*, McGraw-Hill, New York, 1970.
39. Åsbrink, L. unpublished work.
40. Baird, N. C. and Dewar, M. J. S., *J. Chem. Phys.* **50**, 1262 (1969).
    Bodor, N., Dewar, M. J. S., and Worley, S. D., *J. Am. Chem. Soc.* **92**, 19 (1970).
    Dewar, M. J. S. and Haselbach, E., *J. Am. Chem. Soc.* **92**, 590 (1970).
41. Åsbrink, L., Fridh, C., and Lindholm, E., *Chem. Phys. Letters* **15**, 567 (1972).
42. Jonsson, B. Ö., private communication.
43. Lindholm, E., Fridh, C., and Åsbrink, L., *Disc. Faraday Soc.* **54**, 127 (1972).

AUGER ELECTRON SPECTROSCOPY FOR
SURFACE ANALYSIS

J. C. TRACY

Bell Laboratories, Murray Hill, N. J. 07974, U.S.A.

1. Introduction

Auger electron spectroscopy has emerged as one of the most important experimental techniques for the characterization and study of solid surfaces. The purpose of these notes is to attempt to elucidate the important parameters involved in Auger electron spectroscopy, particularly as it applies to surface studies. We will consider and evaluate the various methods of obtaining Auger electron spectra and consider by means of specific example, some of the more important applications of the technique. A complete discussion of this entire field is beyond the scope of these notes. The most notable omissions are discussions of quantitative surface analysis and the influence of chemical environment on the observed spectra.

The extremely rapid growth of Auger electron spectroscopy over the last five years is the result of the increased importance of surface studies in a variety of areas. The recent popularity also stems in part from the relative ease of the technique, the commercial availability of high performance electron spectrometers, and the present high level of ultrahigh vacuum (UHV) technology. In addition to the obvious applications to basic surface physics research, Auger electron spectroscopy is finding uses in semiconductor technology, metallurgy, and catalysis.

Auger electron spectroscopy has made its biggest impact as a tool for the qualitative analysis of solid surfaces. Hence the technique has been used more for studying the 'chemistry' of solid surfaces than as a high energy spectroscopy applicable to solid state physics. This results primarily from the relative complexity of the basic physics of the Auger process as compared for example with ESCA or UV photoemission. While these complexities make it somewhat difficult to determine energy level structures using Auger spectroscopy, they do not in general render the spectra so complex as to impede the qualitative analysis. Indeed, the ultrahigh sensitivity achieved by the use of electron impact excitation and the high transmission of modern electron energy analyzers make the technique almost unrivaled in terms of surface sensitivity.

While studying the particle tracks in a cloud chamber irradiated with broad band (white) X-rays, Pierre Auger discovered in 1925 that not only photoelectrons were ejected from bombarded atoms but also electrons of constant energy irrespective of the X-ray excitation energy. He correctly interpreted these electrons as resulting from the electron reorganization in the photoexcited atom.

In 1953, Lander [155] observed Auger electron peaks in the secondary electron energy distributions from a variety of materials. He pointed out the applicability of

W. Dekeyser et al. (eds.), Electron Emission Spectroscopy, 295–372. All Rights Reserved
Copyright © 1973 by D. Reidel Publishing Company, Dordrecht-Holland

Auger electron spectroscopy for surface analysis and further suggested that the technique provided a complement to soft X-ray emission for the determination of energy band densities of states.

It seems that with a few exceptions Lander's work went virtually unnoticed until 1968 when Harris [119, 125] reported his extensive studies of the Auger spectra of metals. In retrospect it is difficult to understand this large gap in the literature but it most probably resulted from the lack of UHV technology, at least during the 1950's. Harris' most important contribution was the recognition that because of their sharp energy structure, the Auger peaks could be easily observed even in the presence of the large secondary background by electronically differentiating the measured energy distribution. Harris used a velocity (dispersion) analyzer for measuring the energy distribution and obtained the derivative by the well known methods of synchronous detection. In turn, Weber and Peria [297] showed the differentiated secondary electron energy distributions could easily be obtained using retarding field analyzers such as those used widely for low energy electron diffraction (LEED) studies. This marked the beginning of the real upswing in Auger electron spectroscopy for surface analysis. In 1969, Palmberg et al. [207] demonstrated the applicability of the cylindrical mirror electrostatic analyzer for Auger studies of solid surfaces. The large improvement in performance offered by this instrument relative to the previously used analyzers has further increased activity in the field.

We have included a complete Auger electron spectroscopy bibliography prepared in March 1972 by D. T. Hawkins of the Bell Telephone Laboratories Library Staff and have used its index for reference numbers.

2. Secondary Electron Energy Distributions

A. BASICS

The basic observable quantity which we will be dealing with is the energy distribution of electrons leaving a solid surface during bombardment by a primary beam of electrons. These secondary electrons arise from a variety of interactions of the primary beam with the solid and these will be considered in some detail. Of particular interest of course are the Auger electrons which have energies characteristic of the atoms on the surface under bombardment. In this section we will restrict ourselves exclusively to electron excitation and will take up at a later point the cases of X-ray and high energy ion excitation of low energy Auger electrons.

The secondary electron energy distribution we will call $N(E, E_p)$ where E is the energy of the electrons leaving the surface of the solid and E_p is the kinetic energy of the impinging electron. E and E_p are referenced to zero at the Fermi level of the target.

In Figure 1 energy distributions from a surface of GaP are shown for two different primary energies. The most conspicuous feature of these energy distributions is the elastic peak at the high energy end $(E=E_p)$. The remainder of the distribution is relatively featureless on the scales presented with the exception of a few small peaks at intermediate energies and a peak at the extreme low energy end near $E=0$. It must

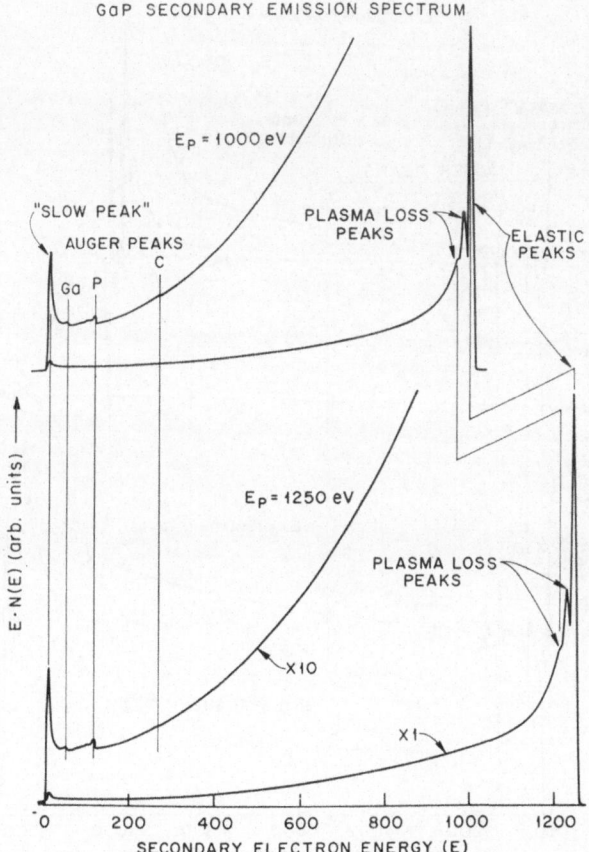

Fig. 1.   Secondary electron energy distributions from a slightly contaminated GaP surface taken with two different primary energies. Note that the elastic and plasma loss peaks whift with the primary energy but the Auger and 'slow' peaks do not.

be pointed out that these curves were obtained using a particular energy analyzer arrangement which measures $E \cdot N(E)$ and hence strongly attenuates the features in the low energy end of $N(E)$.

Of course the peaks we are most interested in for the present discussion are those small intermediate peaks which arise from Auger electrons emitted from the surface. In order to observe them more easily however it is clear that a convenient means must be found for suppressing the large background on which they appear. As noted in the introduction, Harris [119] first pointed out that this could readily be done by electronically differentiating the energy distribution, thereby obtaining $dN(E, E_P)/dE$ and suppressing the background. Figure 2 shows the result of applying this procedure to the energy distributions of Figure 1. Clearly a great deal of detail is brought out by this technique. We will return to the details of the experimental methods in 5B, but will first discuss the characteristics of the structure found in the (differentiated) energy distributions.

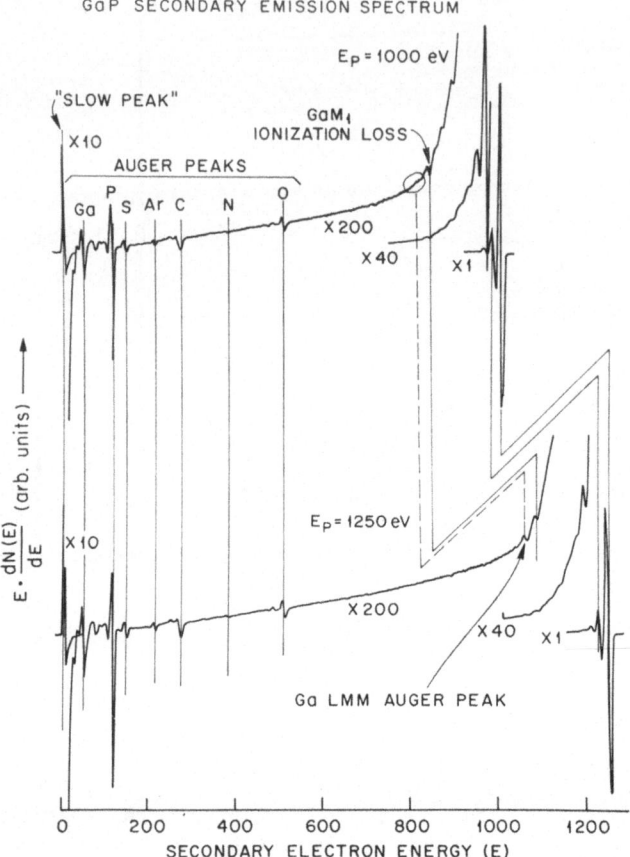

Fig. 2.   Differential secondary electron energy distributions from same surface as in Figure 1. Note how the suppression of the background enables the fine structure to be more easily observed. Note also that the Ga M_1 ionization loss peak moves with E_p while the Ga LMM Auger peak does not.

B. CHARACTERISTIC STRUCTURE

The structure found in electron energy distributions falls into two well defined categories. First there are those peaks whose energy position is independent of the primary energy E_p. Such structure is the result of secondary processes, that is mechanisms whereby the emitted electrons have no 'knowledge' of the excitation energy. The Auger electrons and the 'true secondaries' fall into this category. With reference to Figure 2. it is now easy to see which peaks are Auger peaks; just those whose position in energy is not changed on increasing E_p. In addition one sees that the low energy 'true secondary' or 'slow' peak does not change its energy position either. Now it should be noted that while the energy position is not affected by changes in E_p, the *magnitude* of particular structure will invariably change with variations in E_p. This results from changes in the excitation strength and will be considered in detail in the next section.

The second class of characteristic features are those which move in energy as the primary energy E_p is varied. These peaks appear at an energy $E = E_p - E'$ where E' is an energy characteristic of the mechanism producing the peak. For most cases to be considered here it is independent of E_p. The most conspicuous structure with this characteristic is of course the elastic peak for which $E' = 0$. There are also a variety of 'loss' peaks occurring in the energy distribution and these will be considered below.

There are in addition to these two classes of structure a number of other features. These include the broad high energy background of secondaries which while featureless is important with respect to noise considerations. Also there are twostep processes involving for example both an Auger electron and a discrete energy loss.

1. *Auger Electrons*

It is clear from Figures 1 and 2 that the Auger electrons constitute only a very small fraction of the total secondary electron current. Fortunately, as we have seen, their relatively sharp energy structure allows them to be accentuated by electronic manipulation.

Auger electrons are emitted when an atom, excited by the primary beam, relaxes to a lower lying energy state. The primary beam produces a vacancy or 'hole' in an inner level thus forming a highly excited ion. The Coulomb repulsion between two higher lying electrons may give rise to a radiationless deexcitation of the excited ion producing an ejected (Auger) electron and leaving a doubly ionized ion of lower energy. Thus, an electron in a higher level moves 'down' into the core hole and gives its energy to a second outer lying electron which is emitted. This deexcitation process is shown schematically in Figure 3.

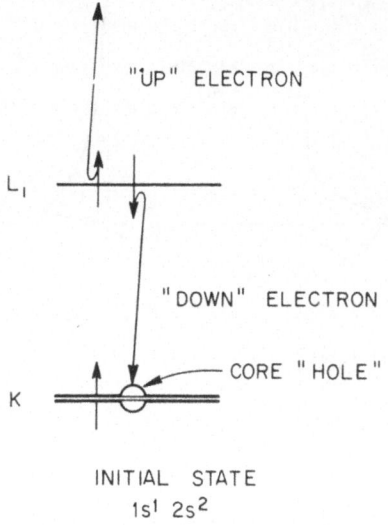

Fig. 3.   Auger deexcitation process for Be^+ ($1s^1, 2s^2$).

Inner shell ionization for primary energies of 3 keV occurs in times less than 10^{-16} s. As we will discuss in more detail later, the lifetime of a core hole is typically greater than 10^{-15} s and more than 10 times the characteristic ionization time. It is thus clear why the Auger electron energy distribution is independent of the primary energy.

Hence Auger electrons are characteristic of the excited atoms from which they are emitted and thus provide us with information on the chemical makeup of solid surface. We will consider the Auger process in more detail in a later section and will go on to show some of the many uses of the technique.

2. 'Slow' Peak

The electrons found at the low energy end of the secondary electron energy distribution constitute the majority of the total current leaving the sample during electron bombardment. They are usually considered as those electrons with less than 100 eV of kinetic energy. These 'slows' result from pair production cascade processes in which the primary beam excites electrons from the valence band which in turn may have enough energy to excite other valence band electrons and so on. The net effect of these electron–electron interactions is to build up an internal (to the solid) energy distribution of slow electrons which has its peak below the vacuum level as shown in Figure 4. Of all these internally excited electrons, some will have sufficient energy and momenta to escape into the vacuum and be measured. Thus the observed 'slow' peak will be the product of an internal energy distribution and an escape function which is

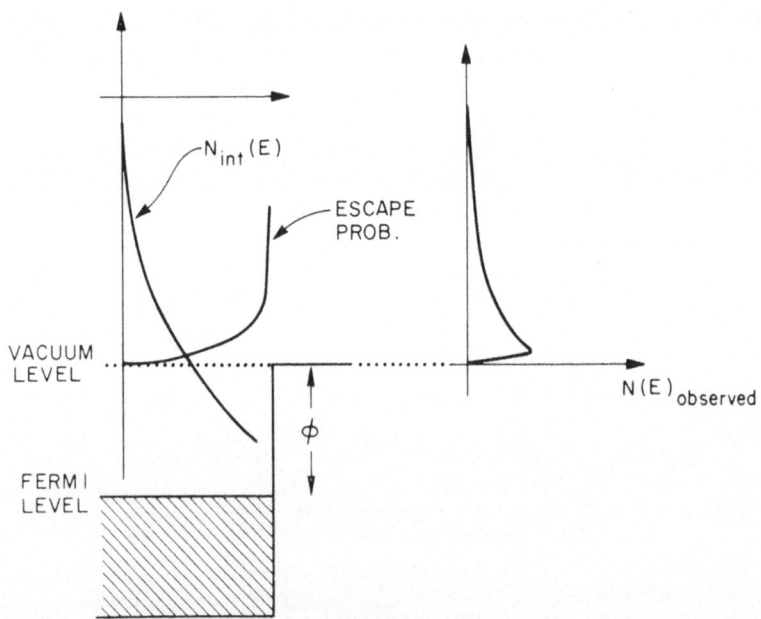

Fig. 4. Schematic representation of the origin of the 'slow' peak. $N_{int}(E)$ is the distribution in energy of hot electrons in the solid. Φ is the work function.

dependent on the surface barrier height and the angular distribution of the internally excited electrons. The electron–electron scattering is so strong at these low energies that a given electron will have experienced numerous interactions before escaping and will therefore lose any memory of the primary energy.

3. *Elastic Peak*

The elastic peak occurs at the highest energy in the secondary electron energy distribution. It is made up of those primary electrons which have been turned around by collisions with ion cores without significant energy loss (at least within the resolution of the energy analyzer). For measurement of $N(E, E_P)$ which we will consider the elastic peak will contain primary electrons which will not only be turned around but also scattered off phonons ($E' < 0.5$ eV).

The elastically scattered electrons are of course studied extensively in LEED (low energy electron diffraction) and we will not consider them in any further detail. In Auger studies the elastic peak is of considerable experimental value as it provides an ideal calibration of the energy scale as well as a monitor of analyzer performance with respect to alignment, resolution, and transmission.

4. *Discrete Energy Loss Peaks*

For our present purpose it will suffice to consider two particular types of 'loss' peaks, the plasma loss and the ionization loss.

A. *Plasma loss peaks.* As a primary electron moves through the solid it will perturb the potential which normally influences the valence band electrons. One of the most important responses to this change in potential is the collective oscillation of the valence electrons at the so-called plasma-frequency. This frequency corresponds to the point at which the dielectric function is a minimum and the system responds with a polarization equal but opposite to the applied field. The primary electron can be thought of as producing an impulse field on the valence electrons and thus in effect subjects them to the full spectrum of excitation frequencies. Because the valence electrons respond most strongly at the plasma frequency and the resulting collective mode is moderately long lived relative to the primary electron transit time, the primary electrons may lose an amount of energy equal to the plasmon energy $\hbar\omega_P = \hbar(4\pi n e^2/m_e)^{1/2}$ where n is the electron density, e is the electron charge and m_e the electron mass.

Plasma loss peaks are clearly observed in Figures 1 and 2 for GaP near the elastic peak. Note that in addition to a single loss peak near $E' = 17$ eV there are at least three additional loss peaks at $2E'$, $3E'$, and $4E'$ each with decreased amplitude. These are plural losses and correspond to the excitation of more than one plasmon during the time the primary electron interacts with the solid.

Figure 5 shows plasma loss peaks observed from Ga metal and anodized GaAs for $E_P = 1068$ eV. Ga metal displays at least five plasmon loss peaks with the one-plasmon peak $\sim 25\%$ of the elastic peak. On the other hand anodized GaAs, believed to be

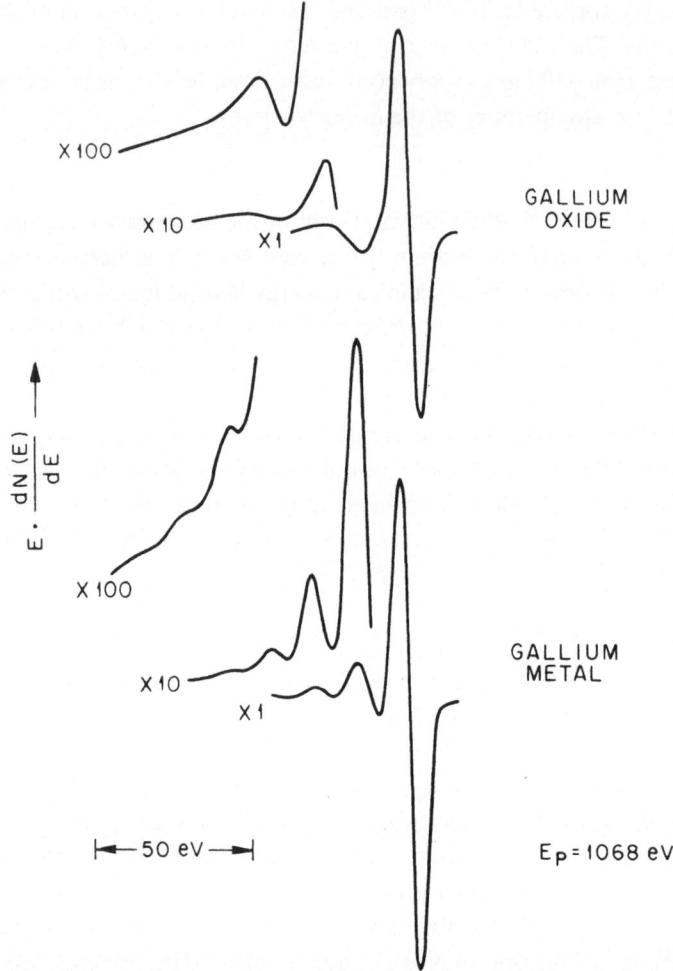

Fig. 5.   Energy loss structure in the vicinity of the elastic peak for Ga metal and
Ga oxide (anodized GaAs).

much more ionic and hence having its valence electrons more localized, reveals at
most three plasmon peaks with the one-plasmon loss being only 10% of the elastic
peak. This comparison reveals that the plasmons are, if you will, 'more well defined'
in the case of a free electron gas (Ga metal) than for a localized electron gas (anodized
GaAs). The difference may also be considered by noting that the effect of the primary
electron on the potential experienced by the valence electrons is greater where the
lattice potential is weaker (Ga metal). It should be noted here that the loss structure
in the anodized GaAs case may not even be associated with plasmons but rather result
from interband or localized one electron excitations.

Plasmon loss peaks are clearly very strong and as might be anticipated they often
appear as low energy satellites on Auger peaks.

B. *Ionization loss peaks.* Ionization loss peaks arise when the primary electron causes ionization of a core level (as might precede an Auger deexcitation) and leaves the solid without further inelastic interactions. The features in the energy distribution appear at $E = E_P - E_B$ where E_B is the binding energy of the level which has been ionized. Figure 6 shows an energy level diagram appropriate to this process. For $E_P \gg E_B$ the final state of the system following ionization will contain an excited ion of

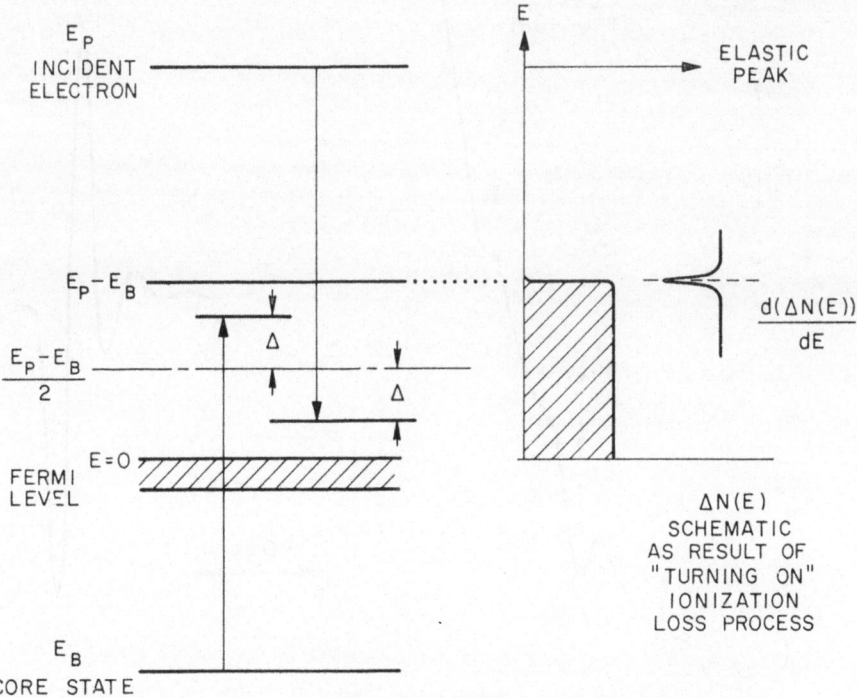

Fig. 6.   Energy level diagram for contribution of ionization loss process to measured energy distribution.

energy E_B and *two* electrons the sum of whose kinetic energy is $E_P - E_B$. Now of course this energy may be partitioned in any way between the two electrons. At the extreme one electron will be at the Fermi energy (energy equal zero) and the second electron may leave the crystal with kinetic energy $E_P - E_B$. No electrons of energy greater than $E_P - E_B$ could result from this ionization. This process which takes electrons from the elastic peak and puts them in the interval between 0 and $E_P - E_B$ is expected to cause a decrease of $N(E)$ as E is increased above $E_P - E_B$. As shown in Figure 7 this is indeed the case. While there is initially a small increase in $N(E)$, there follows with increasing E a rather large decrease in $N(E)$.

Further discussions of these type of experiments are given in [91, 88, 90, 92, 18 and 28].

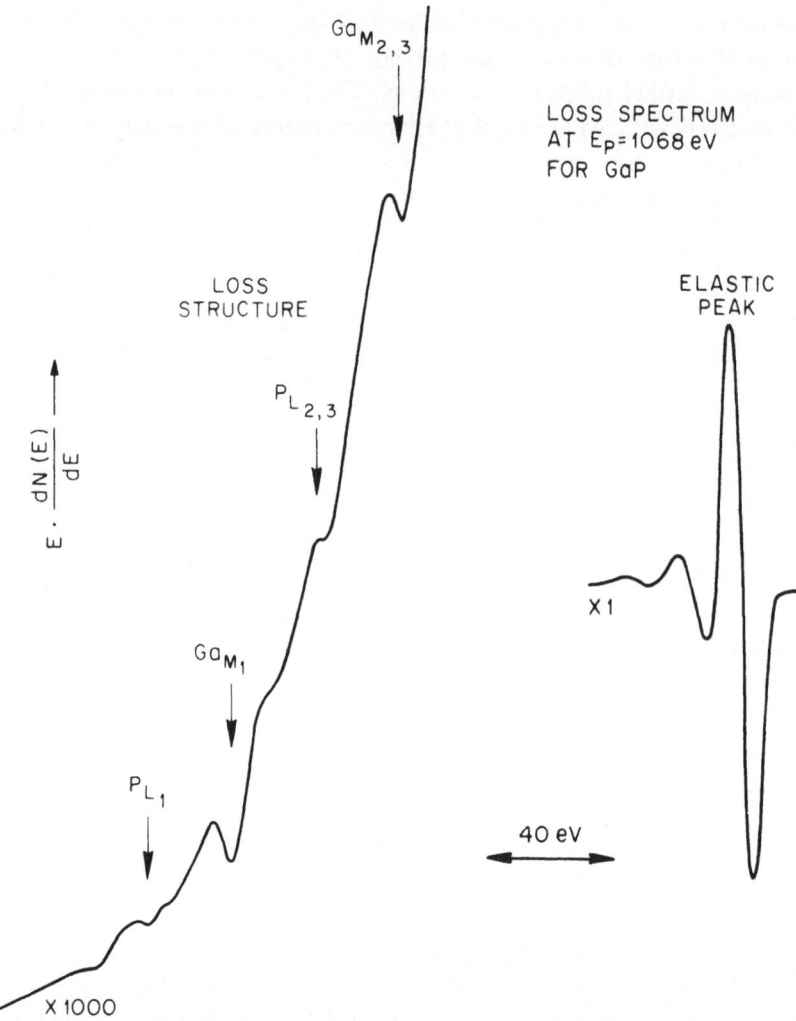

Fig. 7. Ionization loss structure from GaP in the range $100 < E' < 200$ eV.

C. ELECTRON TRANSPORT AND SURFACE SENSITIVITY

The surface sensitivity of the energy distribution with all of its characteristic features derives from the short escape depth $\lambda_e(E)$ of electrons in solids in the energy range of interest below 2 keV. $\lambda_e(E)$ is the electron path length for which the probability is $1/e$ of the electron suffering an inelastic scattering energy loss greater than several electron volts. Such scattering losses will 'remove' Auger electrons from the peak and diminish the observed signal. Hence the observation of any process, be in Auger, elastic scattering, or inelastic scattering, which depends for its specification on a discrete peak (or discontinuity) in the energy distribution, will be strongly influenced by the short mean free path for electron–electron scattering. Thus in the case of Auger electron spectroscopy where the depth of penetration of the primary beam is in general much greater than the escape depth of the Auger electrons, the sampled depth will be roughly equal to $\lambda_e(E)$. Likewise in LEED where the elastic wave field is rapidly attenuated by electron–electron scattering, the effective depth of structural analysis is approximately equal to $\lambda_e(E)$.

The relatively few measurement of the escape depth of electrons in the energy range between 25 and 2000 eV are plotted in Figure 8. Palmberg and Rhodin [210] measured the change in Auger signal from a Au substrate and a Ag overlayer as the Ag overlayer was built up layer by layer. As expected the probability of escape of an Auger electron

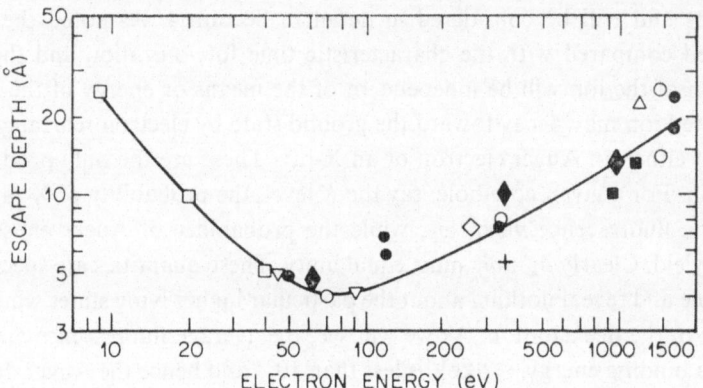

Fig. 8. Escape depth $\lambda_e(E)$ for various materials. All measurements except those of Jacobi et al. [139] were made by overlayer attenuation.

$\square$ – Ag   Eastman (1972) Phys. Elec. Conf., Albuquerque, N.M.
$\triangle$ – Fe   46 eV; Ridgeway and Haneman, *Surf. Sci.* **24**, 451 (1971).
$\triangle$ – Ni   92 eV; Ridgeway and Haneman, *Surf. Sci.* **26**, 683 (1971).
● – Mo, W   Tarng and Wehner (1972) Phys. Elec. Conf., Albuquerque, N.M.
○ – Ag   69 355 eV; Palmberg and Rhodin, *J.A.P.* **39**, 2425 (1968).
◇ – C   Jacobi and Holzman, *Surf. Sci.* **26**, 54 (1971).
■ – C   Stemhardt et al., ICES Asilomar (1971).
$\triangle$ – Au   Baer et al., *Sol. State Commun.* **8**, 1479 (1970).
○ – 1450 eV; W (8.9 Å); WO₃(18.3 Å)   Carlson and McGuire (preprint).
◆ – Be   Seah, M. P., *Surf. Sci.* **32**, 703 (1972).
+ – Cu   Seah, M. P., *ibid.*

formed in the substrate was found to behave as e^{-d/λ_e} where d is the thickness of the overlayer. On the other hand the signal strength from the Auger electron originating in the Ag overlayer was found to vary as $1 - e^{-d/\lambda_e}$. Because the Auger electrons from the Au and Ag had different energies, they were found to have different mean free paths. Ridgeway and Haneman have performed similar experiments (229) for Fe on Si while Baer et al. [311] using X-ray excited electrons have measured mean free paths at 1.2 keV in gold. Using an alternative method Jacobi and Hölze [139] have observed a mean free path of 7.5 Å for 262 eV electrons in carbon.

The data plotted in Figure 8 clearly show that despite wide variation in Z, the mean free path for electrons in the energy range of interest is extremely short. While this clearly demonstrates the surface sensitivity of *any* spectroscopy which relies on electron transport in this energy range, the observability of submonolayer quantities of adsorbed atoms which we will discuss in some detail later is perhaps even more convincing for the Auger case.

3. The Auger Process

A. INTRODUCTION

The starting point for our discussion will be the excitation of an atom to a highly excited ion containing a vacancy or 'hole' in a core level. This is most efficiently done by electrons and will be considered in detail in Section 4. As noted the core hole is long lived compared with the characteristic time for ionization and the resulting deexcitation of the ion will be independent of the means or energy of the excitation.

The excited ion may decay toward the ground state by electron rearrangement and emission of either an Auger electron or an X-ray. These are the only products of the deexcitation. For a given core hole, say the K level, the probability of X-ray emission is called the fluorescence yield, ω_K, while the probability of Auger emission is a_K, the Auger yield. Clearly $a_K + \omega_K$ must equal unity. These quantities are specific only to the core hole and reveal nothing about the particular higher lying states which actually participate in the deexcitation. As we will see, the average fluorescence yield for any shell with a binding energy $\lesssim 2$ keV is less than 10% and hence the Auger deexcitation mechanism dominates.

The Auger process takes an atom in an excited singly ionized initial state and converts it to a doubly ionized final state which is generally but not always an excited state. Thus the initial vacancy can be thought of as moving up to a higher lying level and releasing the energy to the second 'Auger' electron. For these cases in which the final state is not the ground state of the doubly ionized atom, the system will continue to deexcite by X-ray or Auger processes until (in the absence of externally supplied electrons) it reaches some multiply ionized ground state.

Let us consider the simplest example of an Auger process as found in a Be atom (Figure 9). In panel (a) the ground state $(1s^2, 2s^2)$ of neutral Be is shown. We will consider this as the reference state having zero energy. In (b) the initial state prior to Auger deexcitation is shown with the configuration $Be^+ (1s_u, 2s^2)$. The energy of this state is just the K shell binding energy E_K. Now the Auger mechanism is the only means of deexcitation in this case because dipole selection rules prohibit X-ray emission from transitions involving states of the same angular momentum such as $2s \rightarrow 1s$. The Coulomb repulsion between the two $2s$ electrons in the presence of the $1s$ hole drives the Auger transition shown in (b) with the result that the final state of the system is the ground state of doubly ionized $Be^{++} (1s^2, 2s^0)$. The energy of this particular final state is well known from experiment and is equal to the sum of the first and second ionization potentials of Be which we will denote $E_{L_1}(Be^0)$ and $E_{L_1}(Be^+)$ respectively. Thus the energy E_A of the emitted Auger electron is given by

$$E_A = E_K - E_{L_1}(Be^0) - E_{L_1}(Be^+) \tag{3.1}$$

and the ejected electron is denoted a KL_1L_1 Auger electron.

For this particularly simple example the fluorescence yield $\omega_K = 0$ while $a_K = 1$. This is of course only the case for isolated Be atoms. In Be metal the $2s(L_1)$ level broadens into a band which overlaps and interacts with the broadened $2p(L_{2,3})$ band (which is of course unoccupied in the free atom). This results in the valence band electrons having both $2s$ and $2p$ character to various degrees depending on their energy. This

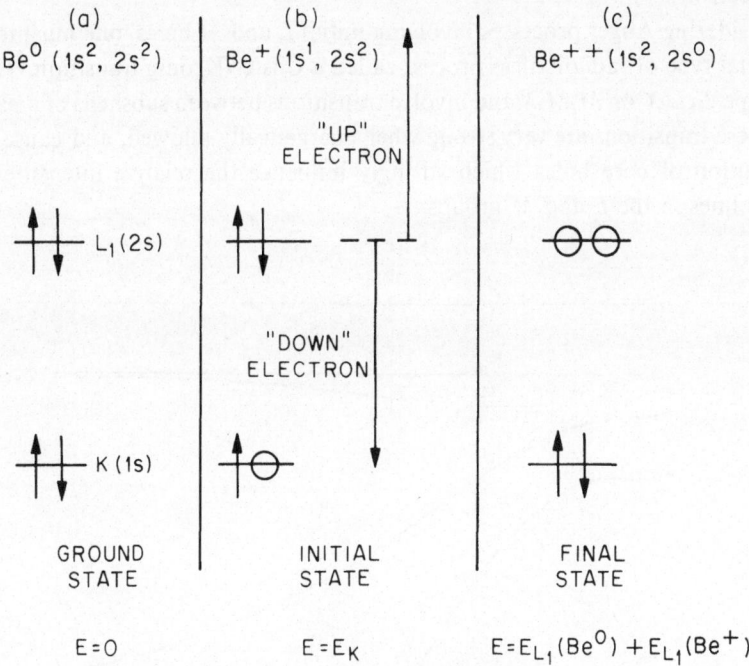

Fig. 9.   Auger process in a Be atom.

behavior makes the application of (3.1) somewhat uncertain because the final state is no longer clearly defined as $Be^{++}(1s^2, 2s^0)$. It should in principle be possible to turn the situation around and determine the energy distribution of final states from the measured KLL energy distribution. In the one electron approximation this would be simply related to the energy band density of states.

B. CLASSIFICATION OF AUGER SPECTRA

For most elements the situation is not nearly so straightforward as in Be. In general, a particular core hole can give rise to large numbers of different Auger transitions. These transitions are usually designated according to the three levels participating in the process. This is done with the aid of the X-ray level notation $K \to 1s$; $L_1 \to 2s$; $L_2 \to 2p^{1/2}$; $L_3 \to 2p^{3/2}$; $M_1 \to 3s$; $M_2 \to 3p^{1/2}$; $M_3 \to 3p^{3/2}$; $M_4 \to 3d^{3/2}$; $M_5 \to 3d^{5/2}$; etc. Consider as an example the Auger spectrum of Si with filled $1s$, $2s$, and $2p$ shells. If a vacancy in the K shell is filled by an electron from the L_1 shell and the energy is given to an electron in the L_2 shell, the resulting Auger electron is designated KL_1L_2. This process is indistinguishable from the exchange process KL_2L_1 because the initial and final state are identical. The convention will be to write the subshells in order of increasing index. Any electron, originating from an Auger process involving a K vacancy is called a K Auger electron and if only electrons in the L shell are involved in the deexcitation of a K ionized atom, the group of ejected electrons are designated KLL Auger electrons.

In considering Auger processes involving initial L and M holes, one must take note of a special type of radiationless process called a Coster-Kronig transition. These are of the type L_kL_jX or M_iM_jX and involve transitions between subshells of a particular shell. These transitions are very strong when energetically allowed, and cause a rapid redistribution of core holes which strongly influence the relative intensities of the observed lines in the L and M groups.

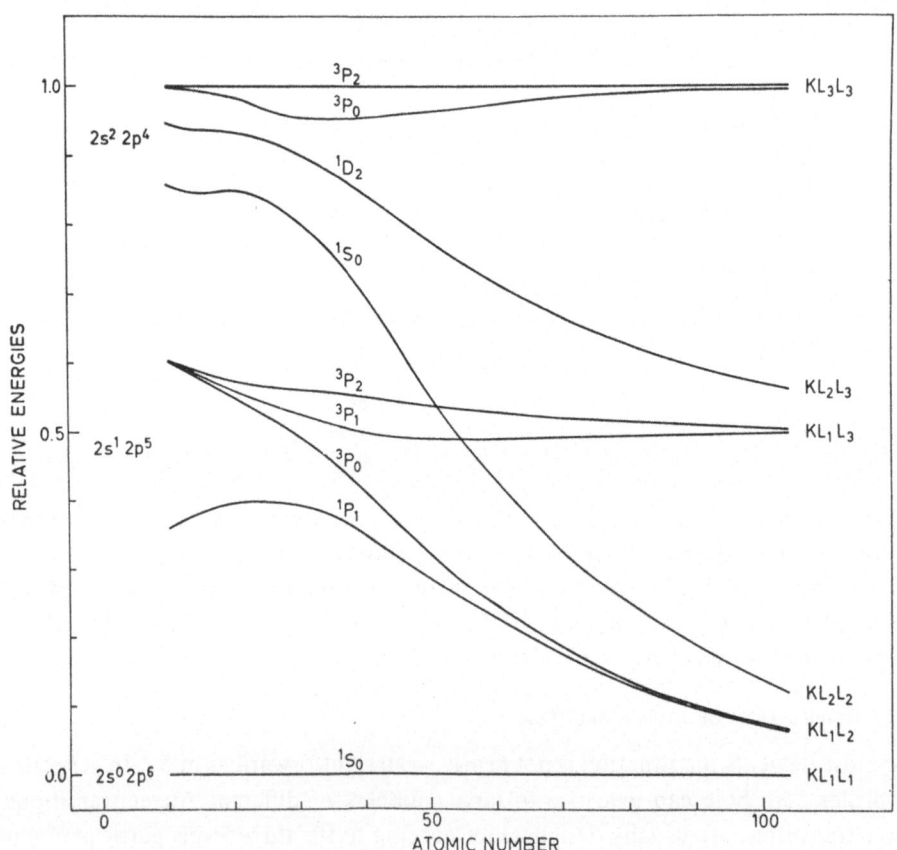

Fig. 10. Relative line positions in the KLL Auger group as a function of atomic number Z. The energy difference between the lines of highest and lowest energy ranges from 55 eV at $Z = 10$ to 17 keV at $Z = 104$. (From Siegbahn et al., ESCA; Atomic, Molecular, and Solid State Structure Studied by Means of Electron Spectroscopy, Almqvist and Wiksells Boktryckeri AB Uppsala, 1967.)

The X-ray level designations are only strictly correct when spin-orbit interactions are dominant and hence the final states are describable in terms of j–j coupling. For the energy range of interest in surface Auger work, this is not the situation at all, in fact, the final states are best described according to LS (Russell-Saunders) coupling. This is because for states with binding energies less than several keV, the spin-orbit interaction is much weaker than the Coulomb interaction. This is seen for example in Figure 10. Here the relative energies of various KLL Auger electrons are plotted as a function of Z. At high Z the spin-orbit coupling is so much stronger than the Coulomb interactions that the two holes can be considered to be in states of definite l, j; these being $0, \frac{1}{2}$; $1, \frac{1}{2}$; and $1, \frac{3}{2}$ for L_1, L_2, and L_3 respectively. The two holes then couple their j's giving rise to six possible lines. At low Z however, spin-orbit interactions are

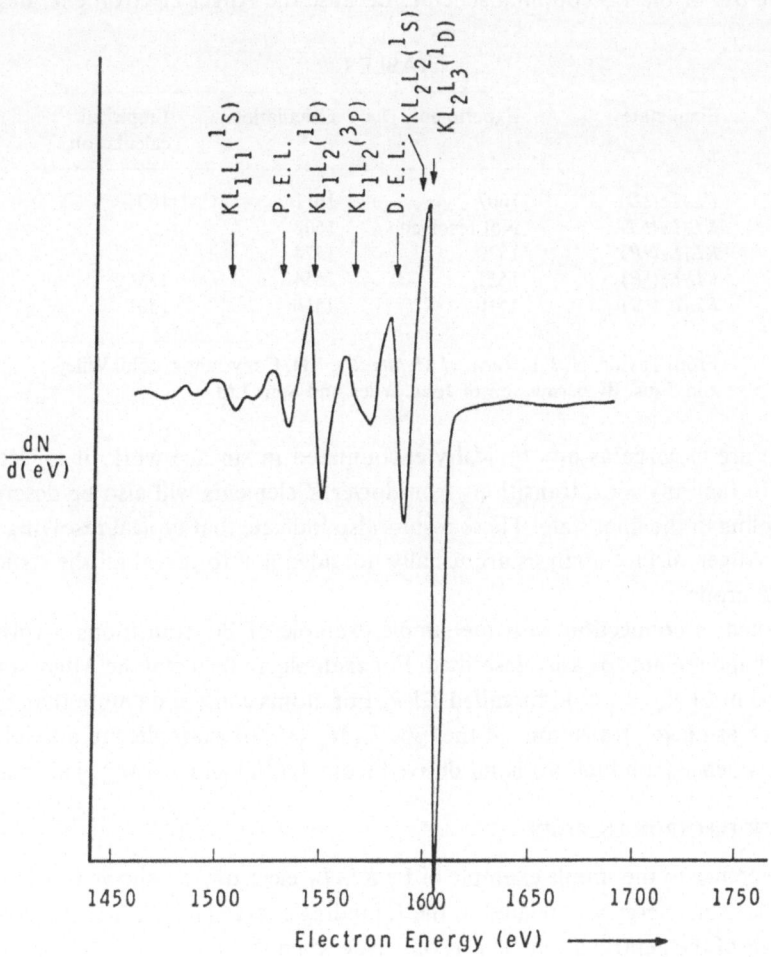

Fig. 11. Derivative of KLL Auger spectrum for Si obtained with a retarding field analyser. (From N. J. Taylor in *Tech. of Metals Res.* **VII** Copyright © J. Wiley and Sons (to be published).) By permission of John Wiley and Sons Ltd.

not important and the two hole state can no longer be described by the l, j's of the single hole states. The Coulomb interaction requires a description of the two hole state in terms of the coupling of the total L and S of the single hole states. In this case (Russell-Saunders coupling) the spectroscopic notation indicated in Figure 10 becomes appropriate. For intermediate Z a more complicated intermediate coupling scheme is required but it is clear from Figure 10 that there is a definite correspondence between the two extreme cases. The convention is then usually adopted in high resolution work that the X-ray notation is used followed by the spectroscopic notation in parentheses.

Figure 11 shows the *KLL* Auger spectrum of Si taken by Taylor using a retarding field analyzer (Section 5B) with a resolution of about 10 eV. This represents a typical spectrum which might be observed in Auger spectroscopy for surface analysis. Table I shows the reasonable agreement between theory and experiment in this case and confirms the use of the *LS* coupling scheme. Because the Auger electron energies in this

TABLE I

Final state	Experiment	Calculation	Empirical calculation
$KL_2L_3(^1D)$	1607	1611	1604
$KL_2L_2(^1S)$	Not resolved	1602	
$KL_1L_2(^3P)$	1570	1574	
$KL_1L_2(^1P)$	1551	1554	1554
$KL_1L_1(^1S)$	1510	1516	1501

From Taylor, N. J. in *Tech. of Metals Res.* **VII**. Copyright © John Wiley and Sons. By permission of John Wiley and Sons Ltd.

example are as large as any typically encountered in surface work, it is clear from Figure 10 that any *KLL* transitions from lower Z elements will also be described by *LS* coupling in the final state. These results also indicate that typical resolving powers used in Auger surface analysis are usually not adequate to reveal all the structure in the *KLL* group.

As noted in connection with the simple example of Be, transitions involving the valence band are not so easily described. For example, in Be metal the Auger transition discussed in Figure 9 would be called *KVV*. For atoms containing more than one core level such as nickel, transitions of the type $L_3M_{2,3}V$, for example, are possible where V is the valence (conduction) band derived from $4s(N_1)$ and $3d(M_{4,5})$ atomic levels.

C. AUGER ELECTRON ENERGIES

With reference to the simple example of Be KL_1L_1 electron, we showed (3.1) that the Auger electron energy was related to the K binding energy and the first two ionization potentials of the neutral atom. This is only true when the final state is the ground state of the doubly ionized atom. In general the final state is an excited doubly charged ion and its energy is not so easily determined. Consider the energy of KL_1L_2 Auger electron from Si and let us for the sake of this example neglect the coupling of the holes in the final state. Now in all cases

$$E_{KL_1L_2} = E_{\text{initial}} - E_{\text{final}}. \tag{3.2}$$

E_{initial} is given by the K core hole binding energy which in this case is 1839 eV. For E_{final} a first estimate is simply the sum of the one electron binding energies $E_{L_1} + E_{L_2} = 248$ eV. This gives a value for the Si KL_1L_2 Auger electron energy of 1591 eV which is 40 eV too large. The discrepancy arises because of the approximation used in determining the final state energy. Once an L_1 electron is removed, the binding energy of all the other electrons is increased as a result of a decrease in the screening of the nuclear charge or a decrease in the electron–electron repulsion, however you wish to consider it.

Burhop suggested that as an approximation to the binding energy of an electron in a singly charged ion, one can use binding energy of the equivalent electron in the atom of next higher atomic number. For our Si KL_1L_2 example this yields an Auger energy of 1551e V or 1558 eV depending on whether we consider the direct KL_1L_2 or exchange KL_2L_1 transition. Chung and Jenkins (48) pointed out that because these are equivalent one should in fact take their average which is then 1554.5 eV (see Table I) and in satisfactory agreement with experiment.

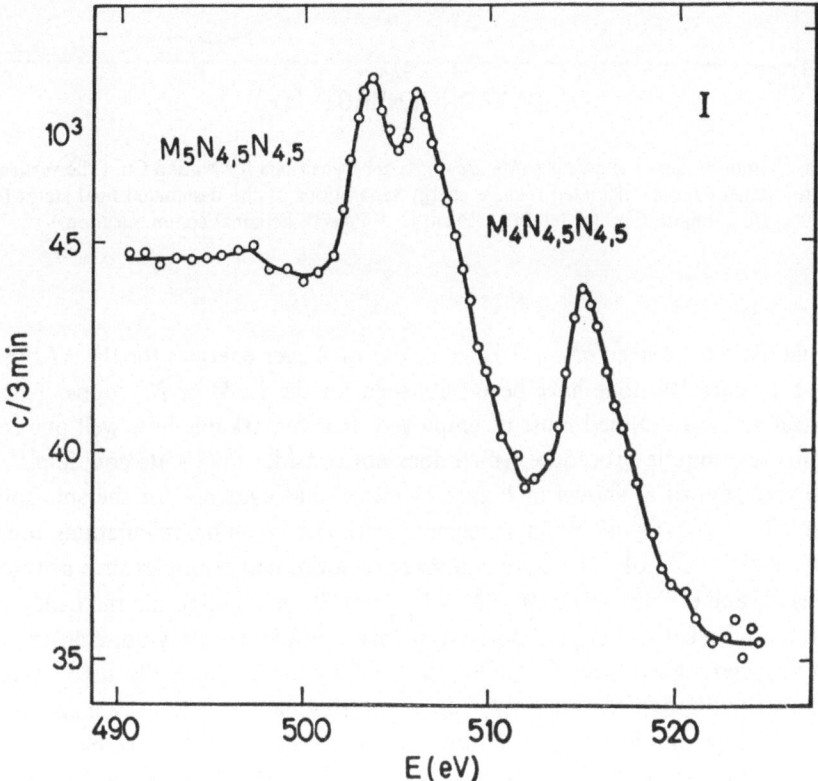

Fig. 12a.   MNN Auger electron spectrum of I from evaporated NaI [4].

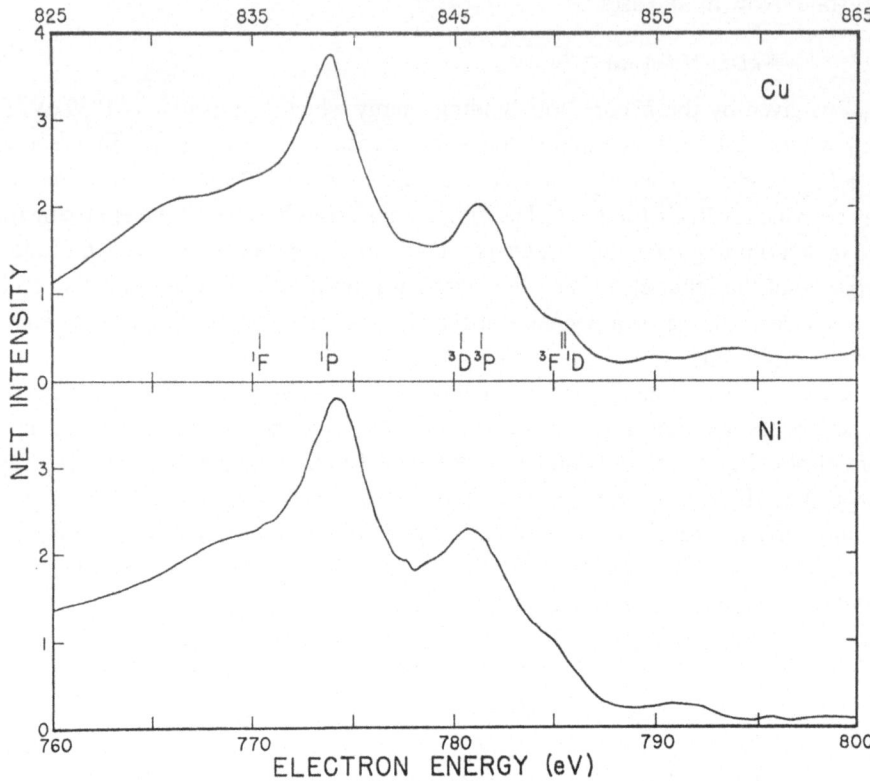

Fig. 12b.   Smooth curves showing $L_3 M_{2,3} M_{4,5}$ Auger transitions for Ni and Cu. (The vertical bars in the top panel denote calculated *relative* energy separations of the designated final states for the Cu configuration $3s^2\, 3p^5\, 3d^9$.) (From C. J. Powell, personal communication.)

While there have been *ab initio* calculations of Auger energies for the *KLL* group, Table I, (Figure 10) none have been published for the *L*, *M* or *N* groups. Thus, the empirical method outlined must be employed. It is remarkable how well one can do with this very simple procedure which does not consider final state coupling.

The spectrum of Si shown in Figure 11 clearly has evidence for the spin-splitting of the $KL_1 L_2$ line by 19 eV in agreement with the *ab initio* calculations found in Siegbahn's ESCA book. There are a number of additional examples such as shown in Figures 12a and b. Data for the $M_{4,5} N_{4,5} N_{4,5}$ transition in iodine and the $L_3 M_{2,3} M_{4,5}$ transition in nickel and copper both show this multiplet splitting quite clearly. Both of these spectra were recorded at higher resolution than normally used in surface Auger work although the Ni $L_3 M_{2,3} M_{4,5}$ splitting has been seen by many workers using lower resolution. This structure is of relatively little importance in surface analysis and provides additional difficulties in making measurements of energy band densities of states using Auger spectroscopy.

D. AUGER ELECTRON INTENSITIES

The intensity of a given Auger line is just the rate of production of the particular core hole involved times the ratio of the specific transition probability to the total Auger transition probability involving that core hole. These two terms can be conveniently separated. We will not be concerned at this point with the total Auger yield for a given group of transitions (*KLL*, etc) but rather with the distribution of intensities within each group.

1. *Vacancy Distribution and Coster-Kronig Processes*

It is generally believed that electron impact produces core holes in the various sub-shells of a particular shell in rough proportion to their occupation number. Thus for *L* shell ionization the initial distribution is $1:1:2$ for $L_1:L_2:L_3$. There are three things which can happen to the L_1 and L_2 holes. They can undergo X-ray or Auger deexcitation by combining with electrons in a higher lying shell or they can move to a higher

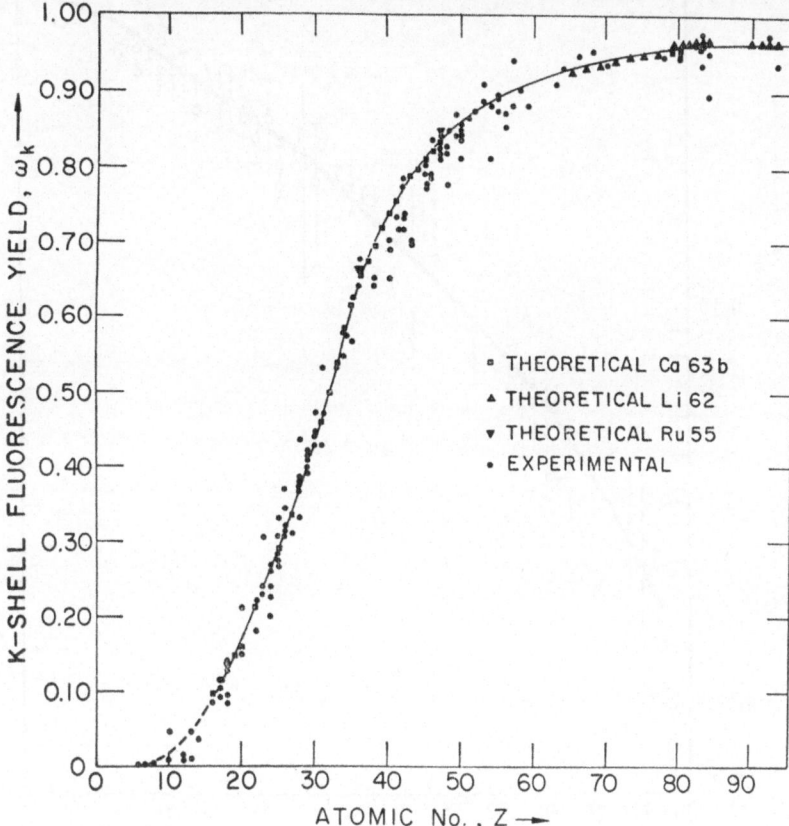

Fig. 13a. The *K*-shell fluorescence yields shown as a function of atomic number. The closed circles represent experimental values. The solid curve is the 'best fit' theoretical curve. (From Fink *et al.*, *Rev. Mod. Phys.* **38**, 513 (1966).)

lying subshell $L_1 \rightarrow L_2$ or L_3 and $L_2 \rightarrow L_3$ releasing energy to a weakly bound outer electron. These Coster-Kronig processes are extremely rapid where they are energetically favorable. McGuire [175] calculated for Ti that the L_1 lifetime is 20 times shorter than the L_2 or L_3 lifetimes and the Coster-Kronig probability for filling an L_1 hole from L_2 or L_3 is 90%. Thus as fast as L_1 holes are formed, they are shifted to L_2 and L_3 holes in the ratio of about 1:2 (a result mostly of the multiplicity of L_2 and L_3). Hence the vacancy distribution appropriate for comparison with experiment is probably more like 1:15:25 for $L_1:L_2:L_3$. A similar situation exists for the M shell except that here there are five subshlels to contend with. McGuire [312] has calculated

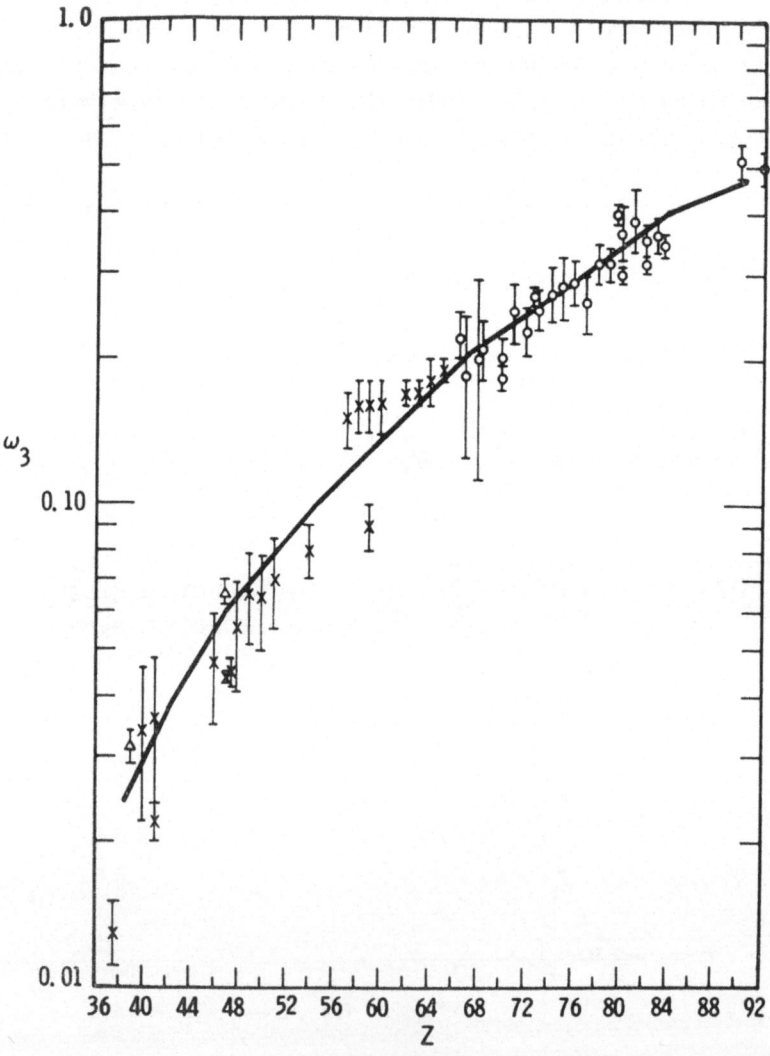

Fig. 13b.   Fluorescence yield $\omega_3(\omega_{L_3})$. Solid curve connects values calculated by McGuire [178]. The data points are experimental values. [178]

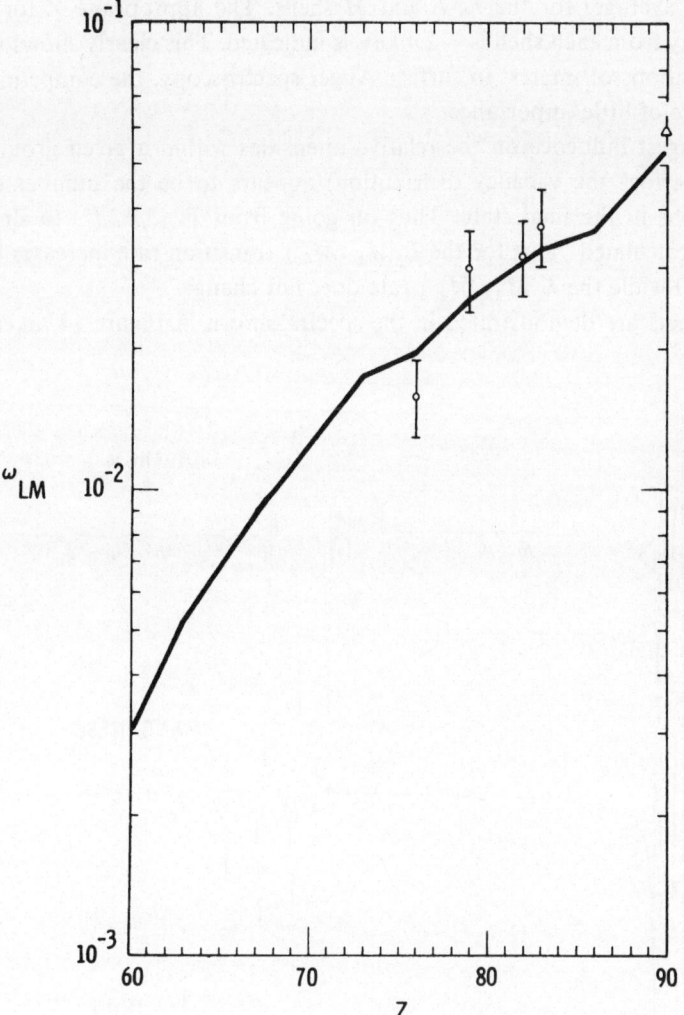

Fig. 13c. Mean *M*-shell fluorescence yield ω_{LM} vs *Z*. (From McGuire, *Phys. Rev.* **A5**, 1043 (1972).)

that for Ag as an example an initial hole distribution $M_1:M_{2,3}:M_{4,5}$ of $1:6:10$ becomes $1:15:280$ by the action of Coster-Kronig processes.

2. *Auger Transition Rates*

There have recently been published a large number of calculations of Auger transition rates [44, 61, 150, 175, 176, 177, 178]. While a number of these have included data for individual transitions, these should be used cautionsly in the low energy range as they were computed using *j–j* coupling in the final state. In spite of this, they are completely satisfactory in determining the total Auger rate for a given group (which is independent of the coupling scheme used) as well as the fluorescence yield.

Figures 13a, b, and c shows experimental and calculated fluorescence yields (or an appropriate average) for the K, L and M shells. The appropriate Z for which the Auger energy from each shell is ~ 2.4 keV is indicated. This clearly shows that for the Auger transitions of interest in surface Auger spectroscopy, the competing radiative processes are of little importance.

The strongest influence on the relative intensities within a given group of Auger electrons (besides the vacancy distribution) appears to be the number of electron configurations in the final state. Thus on going from Ti $(3p^63d^2)$ to Zn $(3p^63d^{10})$ McGuire's calculated value for the $L_3M_{4,5}M_{4,5}$ transition rate increases by a factor of almost 10 while the $L_3M_{2,3}M_{2,3}$ rate does not change.

These effects are demonstrated in the spectra shown in Figure 14 taken from the

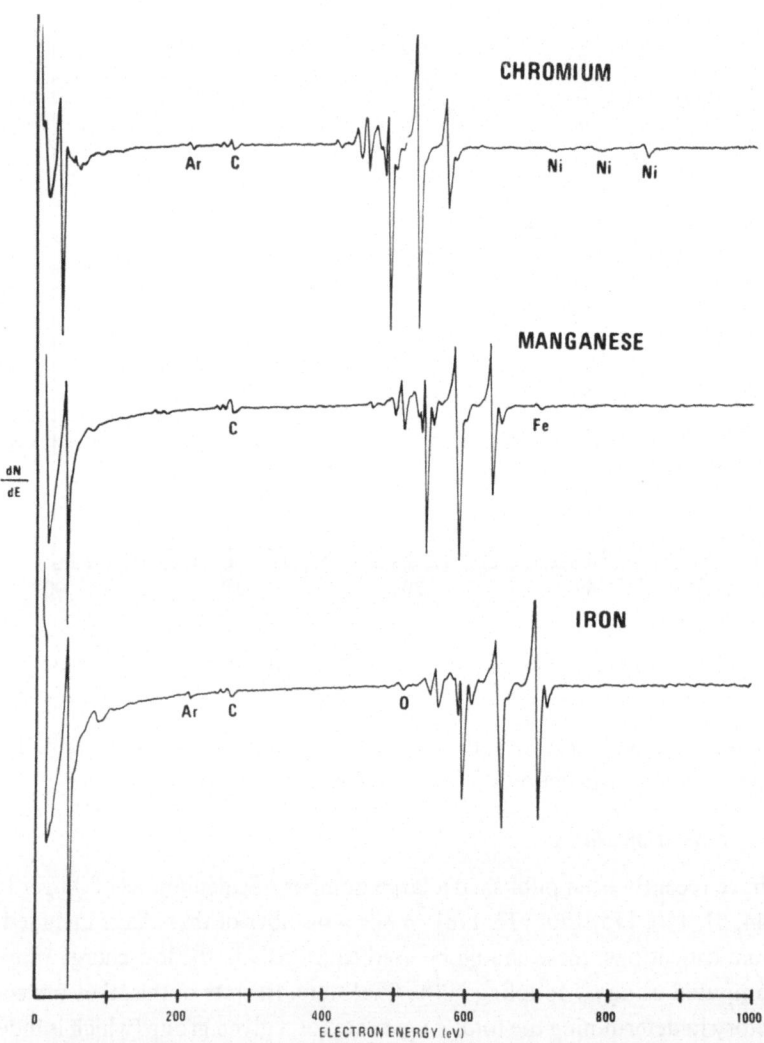

Fig. 14. Auger spectra from Cr, Mn, and Fe. $E_p = 3$ keV, $I_p = 50 \,\mu A$. (From P. W. Palmberg in *Electron Spectroscopy* (ed. by D. A. Shirley), North Holland (1972

Handbook of Auger Electron Spectroscopy written by the staff at Physical Electronics Ind. As the d-electrons are added in crossing from Ca to Cu the relative strength of the transitions involving these $M_{4,5}$ levels increases dramatically.

4. Electron Excitation of Auger Spectra

A. SOME RECAPITULATION

We have seen that the Auger process is purely secondary in nature, that is to say the peak shapes (energy distributions) are independent of the mode or energy of excitation provided of course that enough energy is available to produce a core hole. The importance of understanding excitation processes rests primarily with the hope of obtaining some quantitative information from the resulting Auger spectra concerning the amount of a particular element present on the surface.

The strength of a particular Auger 'line' will be directly proportional to the rate of ionization of the appropriate core state. As we have seen in the previous section the fluorescence yield can be considered zero for all core holes whose binding energy is less than 2 keV. Thus the total Auger current (within the sample and neglecting any losses to plasmons, etc.) is *equal* to the ionization rate times the electronic charge. This condition, appropriate only for low energy transitions, is of considerable value in determining experimental ionization cross sections.

Despite the title of this section we will first describe the three means used to date for excitation of Auger spectra from surfaces and point out their relative advantages and disadvantages. Next we will discuss in more considerable detail the studies of electron induced ionization which have yielded information concerning ionization cross sections, backscattering contributions, and the dependence on angle of incidence.

B. MEANS OF EXCITATION

1. *Electron Induced Ionization*

The most common means of exciting Auger spectra is with a primary electron beam as shown for example in Figure 2. The reasons for this common usage are simple. High intensity electron beam with currents on the order of 100 μA at energies from several hundred eV up to 5 keV are easily obtained. These beams may be readily focussed electrostatically and positioned by simple deflection plates at various points on the target.

The cross section for ionization will be seen to be quite large, often exceeding 10^{-19} cm^2 and is reasonably independent of primary energy over a wide range from 2–6 times the binding energy associated with the particular core hole being ionized.

This combination of high intensity primary beams coupled with large cross sections leads quite naturally to high sensitivity and therefore high speed measurement. On the other hand, electron excitation does give rise to a large secondary background upon which the Auger peaks are found. The elimination of this background has been shown in Section 2A to be achievable but as will be demonstrated in the following section, it is the energy dependence of this background which provides the present ultimate limit on the sensitivity of electron excited Auger spectroscopy in determining surface impurities. Electron excitation provides one additional complication, particularly with respect to quantitative analysis of surface impurities. The large number of high energy 'rediffused' secondaries constitute a backscattered current capable of producing secondary ionization. Thus the measured Auger current will not be so simply related to the cross section for ionization in these cases.

2. *X-Ray Induced Ionization*

While there has been considerable work on X-ray induced Auger spectroscopy by the

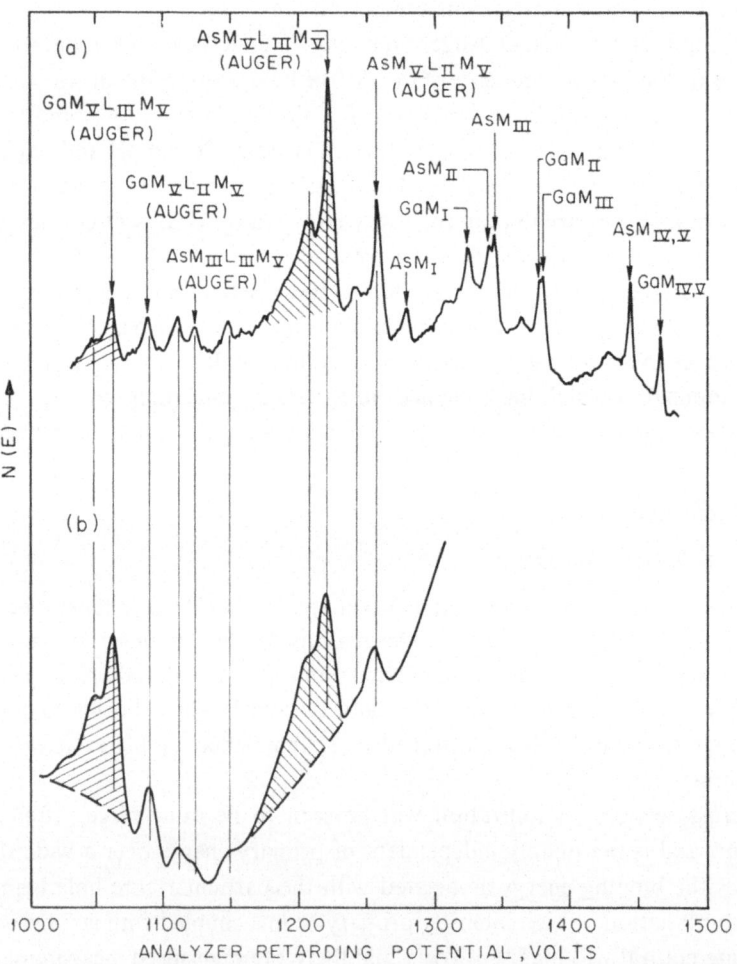

Fig. 15. (a) ESCA spectrum of vacuum cleaved GaAs. (From Smith, D. L. *et al.*, *J. Appl. Phys.* **43**, 2624 (1972).) (b) Electron excited Auger spectrum of clean GaAs. Reconstructed from d$N(E)$/dE spectrum to suppress background.

group at Uppsala on materials maintained in technical vacuum, there has been extremely little such work on clean surfaces in ultrahigh vacuum.

Recently Smith and Huchital [313] have reported Auger spectra from cleaved GaAs obtained using AlK_α X-rays for excitation. Their results are shown in Figure 15 along with results using electron excitation. The figure shows clearly the advantages and disadvantages of X-ray excitation. First the entire scan took slightly more than 20 min to run. This is to be compared with less than 5 s to obtain the data of Figure 15b which has comparable signal-to-noise ratio (and slightly poorer resolution) using a 50 μÅ, 3 keV primary electron beam with roughly comparable energy analyzer. Note that the X-ray induced measurement is of $N(E)$ and not $dN(E)/dE$. This is possible because using X-rays the strong 'rediffused' primary electron background is absent. The Auger contribution to $N(E)$ for the electron excited case was obtained by first differentiating electronically to suppress the background and then reintegrating.

The X-ray excitation has a particularly desirable advantage in that the interactions of the X-ray beam with the surface under bombardment, which may tend to degrade it, are much weaker than in the electron excited situation. Where highly sensitive surfaces such as photocathodes are under study, this may be a particularly important advantage.

There is one other very interesting feature of this comparison. The results reveal the same Auger line positions as for the X-ray excited case but the relative magnitudes of the Ga and As structure are very different. Now this might be explained by differences in stoichiometry between the two surfaces. This is considered unlikely for the following reason. The $dN(E)/dE$ spectrum from which the $N(E)$ of Figure 15b was obtained was identical with the Auger spectrum obtained by Uebbing and Taylor (282) from a cleaved GaAs surface like the one used to obtain the X-ray induced spectra of Figure 15a. It is rather suggested that the observed differences arise from the difference in excitation ratios of Ga_{L_3}/As_{L_3} for the two cases. The binding energy of Ga_{L_3} is 1115 eV while for As_{L_3} it is 1323 eV. For 3 keV electron excitation the ionization cross sections for these two levels will be roughly comparable. However for AlK_α excitation at $\hbar\omega = 1486$ eV the ionization of the As_{L_3} level may be expected to be considerably more probable than for the Ga_{L_3} owing to its proximity to the X-ray line.

3. *Proton Induced Ionization*

Musket and Bauer [314] have recently demonstrated that high energy (350 keV) protons can be used to produce core holes in the surface region within the escape depth of low energy Auger electrons. They found that the Auger yield was comparable to that obtained using 3 keV electron excitation. In addition they suggest that the straight line trajectories of the protons eliminate the backscattering correction which may be necessary for the electron excitation case. This would naturally lead to more straightforward quantitative analysis although the escape depths of the Auger electrons would still have to be determined. The method has the disadvantage that a high energy accelerator is needed and the beam currents are low ($\sim 1 \mu$Å).

C. ELECTRON INDUCED IONIZATION (Some Details)

1. *Auger Yield Measurements*

The dependence of the strength of an observed Auger peak on the primary electron energy E_P is known as the Auger yield. Such a measurement can be related to the ionization cross section only under the most ideal conditions. Figure 16 shows a number of MNN Auger yield curves obtained for elements in the second long period. Quantitative comparisons should not be made from this data as the amount of the particular element present on the surface was not well established. The qualitative

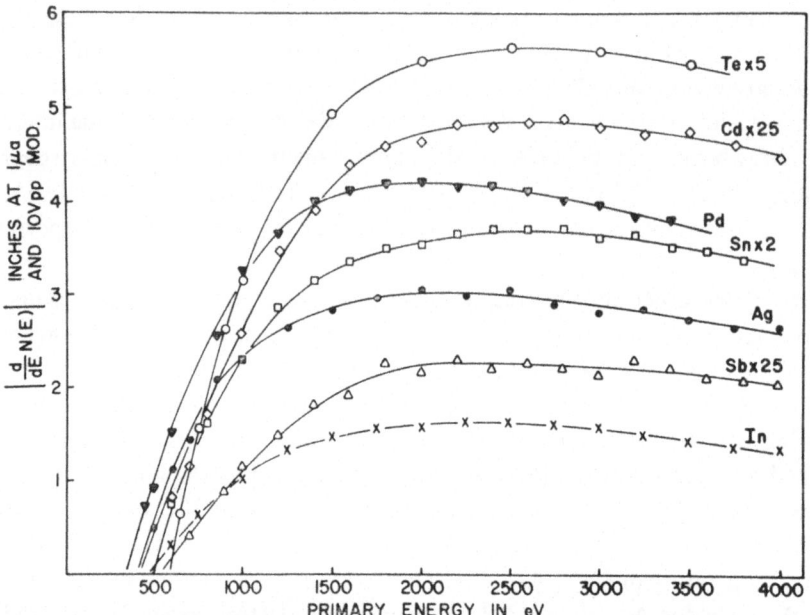

Fig. 16.   Auger yield curves for Pd–Te. (From Oswald [199].)

result that the Auger yield rises abruptly above the threshold for core hole production to a maximum value at about 3–6 times the threshold energy appears to hold for most materials studied to date. These results for the energy of maximum cross section do not in general agree with theoretically determined values which tend to peak at lower energies closer to threshold and then fall off more quickly with increased energy. The experimentally observed flatness of the Auger yield curves at higher energies is probably due to backscattering influences, particularly at grazing incidence excitation.

Under some conditions structure may be observed in Auger yield curves. Figure 17a shows the $P\,L_{2,3}M_{2,3}M_{2,3}$ Auger yield in GaP as a function of primary energy. Figure 17a shows the yield versus reduced energy $E_P/E_{L_{2,3}}$, where $E_{L_{2,3}}$ is the phosphorus $L_{2,3}$ binding energy (132 eV). The curve appears basically similar to the results shown in Figure 16 with the notable exception of a kink in the yield at $E_P = 189$ eV which is just the phosphorus L_1 binding energy. Figure 17b shows the yield data in a narrow region close to the L_1 edge. Note that the yield *increases* sharply above the kink indicating that more $L_{2,3}M_{2,3}M_{2,3}$ Auger electrons are being produced.

This observation can be explained with the aid of Figure 18. In the region $132 < E_P < 189$ eV only $L_{2,3}$ holes can be produced and the yield rises as expected. For $E_P > 189$ eV both $L_{2,3}$ and L_1 holes can be produced. It is known from the calculations of McGuire [178] that L_1 holes are shifted to the $L_{2,3}$ level by Coster-Kronig processes like $L_1 L_{2,3} M_{2,3}$ at a thirty times faster rate than they give rise to $L_1 M_{2,3} M_{2,3}$ Auger transitions. Thus the L_1 threshold will appear in the $L_{2,3}$ yield curve. This demonstrates quite clearly a role of Coster-Kronig processes which will have to be considered in detail for example when attempting to determine the L subshell ionization cross section.

2. *Ionization Cross Sections and Backscattering Factors*

The measurement of electron ionization cross sections is made difficult by backscattering problems, concentration determination, and a knowledge of the Auger electron escape. Despite these problems a few workers have succeeded in obtaining reliable cross section data.

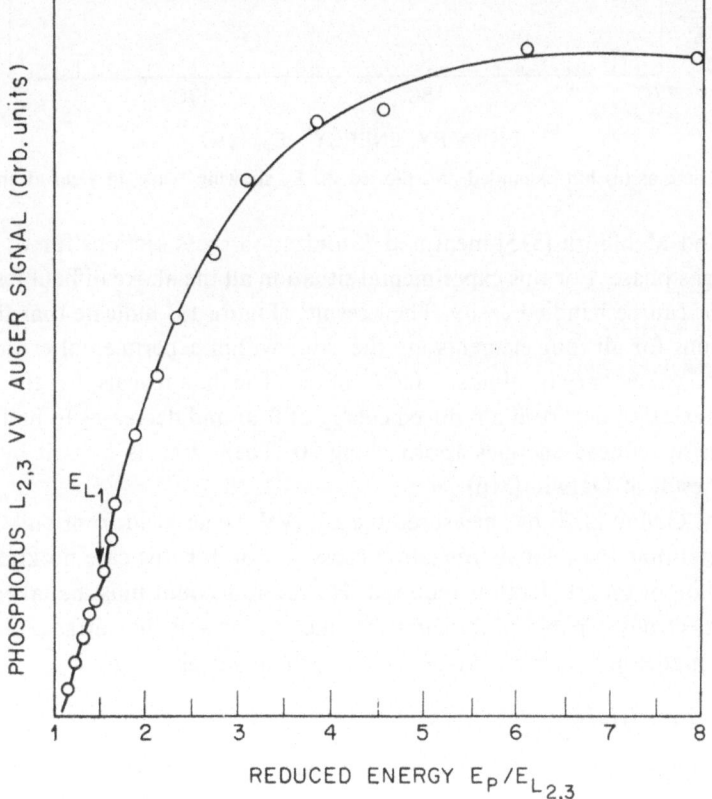

Fig. 17a. $PL_{2,3}$ VV Auger yield vs. reduced energy.

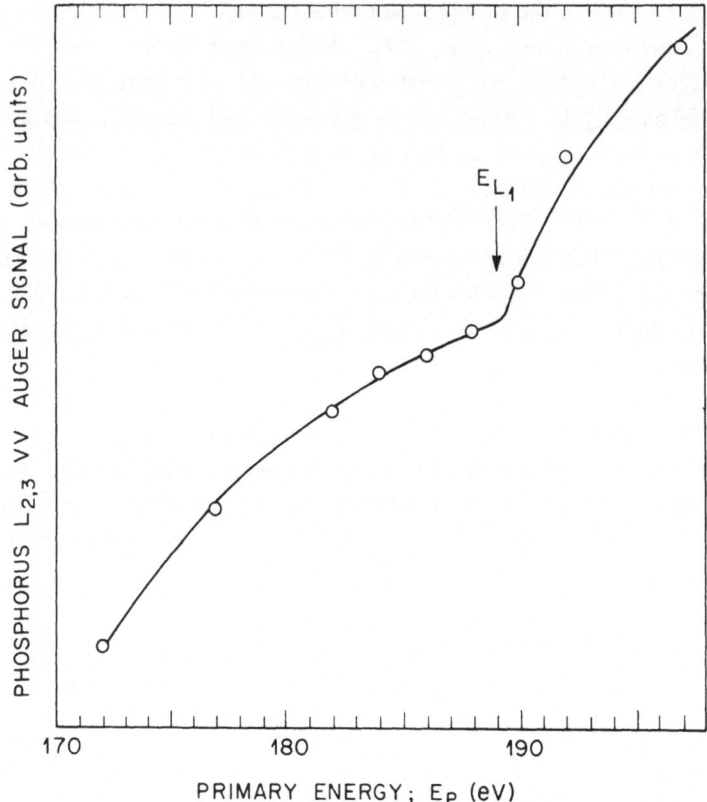

Fig. 17b. Same as (a) but expanded and plotted vs. E_p showing 'kink' in yield at the $E_p = E_{L_1}$.

Glupe and Mehlhorn [315] measured K ionization cross sections for C, N, O and Ne in the gas phase. For this experimental situation all the above difficulties either do not exist or can be handled easily. Their results (Figure 19) indicate that the relative cross sections for all four elements are the same within experimental error although the absolute values vary by almost a factor of ten. The data reveals that the maximum in the cross section occurs at a reduced energy of four and decreases to half the maximum value at reduced energies approaching 20. Their data is best fit by the semiempirical result of Darwin [316].

Recently, Gallon [317] has measured the $L_{2,3}$ VV Auger yield from bulk Si (Figure 20). To determine the relative ionization cross section for this case no knowledge of concentration or escape depth is required. However, account must be taken of backscattered electrons capable of ionizing the silicon atoms at the surface. Gallon constructed a model in which the Auger signal current was given by

$$I_A = C I_P (\varphi + \alpha \beta). \tag{4.1}$$

C is a constant, I_P is the primary beam current, φ is the energy dependent cross section, α is the ionization cross section integrated over the range of energies present in the backscattered current, and β is a geometrical term. The two terms in (4.1)

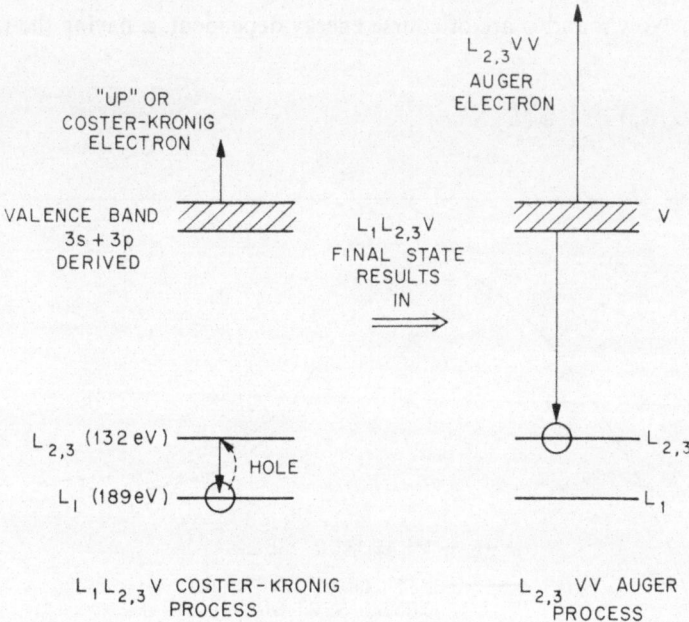

Fig. 18.   Energy diagram showing proposed mechanism for appearance of kink at E_{L_1} in $L_{2,3}$ VV yield curve.

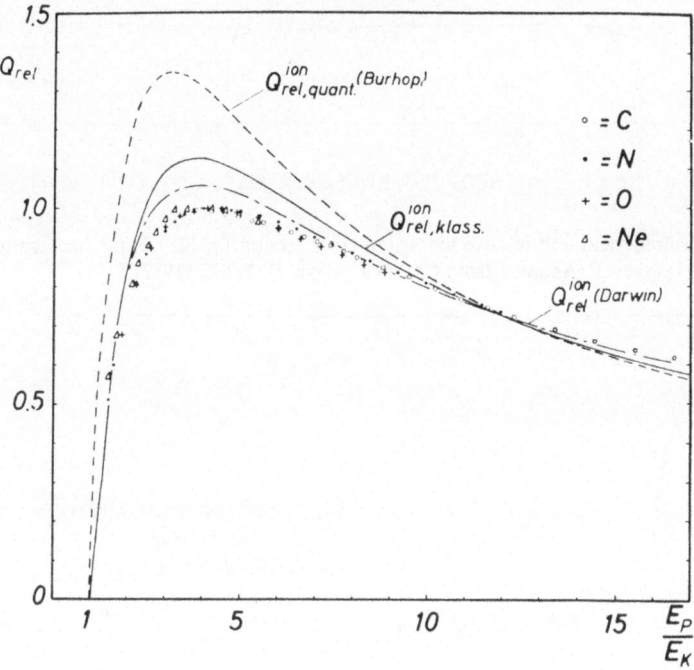

Fig. 19.   Relative experimental ionization cross sections of the K shells of C, N, O, and Ne as a function of reduced energy E_P/E_K. The curves are for various theoretical relative cross section. The data and theories have all been normalized at $E_P/E_K = 12$. (From Glupe and Mehlhorn, *Phys. Letters* **25A**, 274 (1967).)

represent the primary and secondary (backscatter) contribution to the ionization rate respectively. Now α and φ are of course energy dependent, α having the form

$$\alpha(E_P) = \int_{E_A}^{E_P} \varphi(E) N(E) \, dE. \qquad (4.2$$

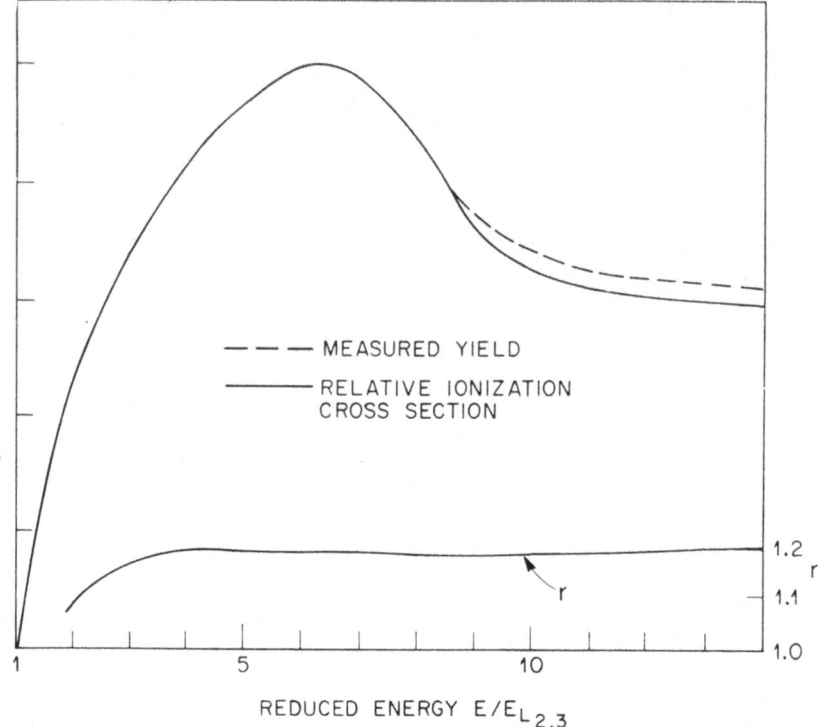

Fig. 20. Measured yield and relative ionization cross section for Si. r is the backscattering factor. (Adapted from Gallon J., *Phys. D.* **5**, 822 (1972).)

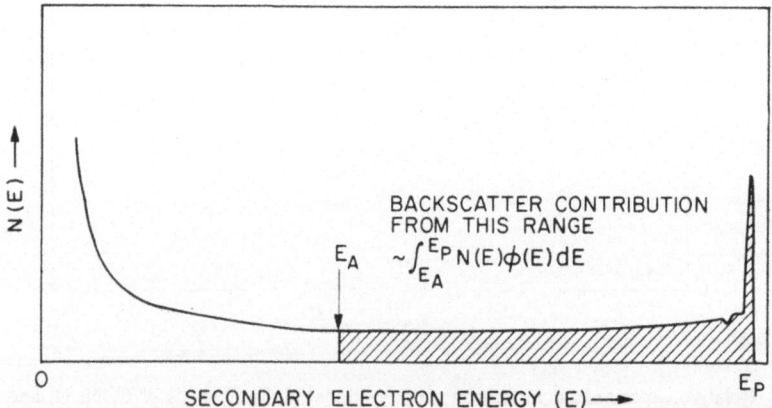

Fig. 21. Schematic diagram of the backscattering contribution for a given ionization threshold E_A.

E_A is the energy of the Auger threshold. This relation is shown schematically in Figure 21.

Gallon measured relative values of $I_A(E_P)$ as well as $N(E)$ for various values of E_P. He then proceeded as follows. As a first approximation he assumed that $I_A(E_P) \sim \varphi(P)$ and thereby neglected the backscattering. He then used this value of $\varphi(E_P)$ to calculate α from (4.2) and the measured energy distribution above threshold. From the value of α he then recalculated $\varphi(E_P)$ from (4.1). He repeated the procedure until successive iteration gave the same values for $\alpha(E_P)$ and $\varphi(E_P)$. He states that only three or four iterations were required to obtain 0.1% convergence which indicates that the back-scattering contribution was in fact small. Figure 20 shows the resulting cross section which is seen to deviate only slightly from the measured yield. Also shown is the scattering factor r which is equal to $1 + \beta\alpha/\varphi$ and reflects the amount of ionization produced by the backscattered electrons.

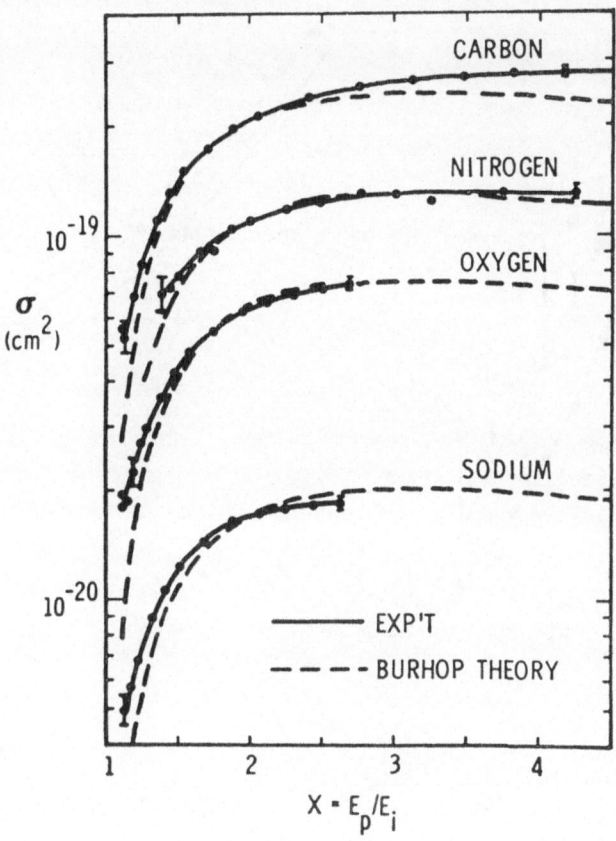

Fig. 22. Measured total K-shell ionization cross sections (dot and solid lines) and cross sections computed from the first Born approximation (dashed lines) vs. reduced energy. The measured cross sections are adjusted in absolute value to cross the theoretical curves at $X = 2$. Error bars are given for the extremes of each measured curve. (From Gerlach, R. L. and Ducharme, A., Surf. Sci. **32**, 329 (1972).)

These results lead one to the conclusion that for Si, at least up to reduced energies of 5 or 10, the Auger yield is a reasonably good approximation to the relative electron ionization cross section.

Neave *et al.* (NFJ) [318] have made measurements similar to those of Gallon on bulk Si and found similar Auger yields. In attempting to determine the backscattering contribution they used the theoretical value of $\varphi(E_P)$ obtained from the Worthington and Tomlin modification of Born approximation as discussed by Bishop and Riviere (22). Note that by doing this Equation (4.1) becomes completely specified (assuming $N(E)$ has been measured). The results however do not agree with the experimentally observed values for I_A. Following Bishop and Riviere [22], NFJ introduced a constant

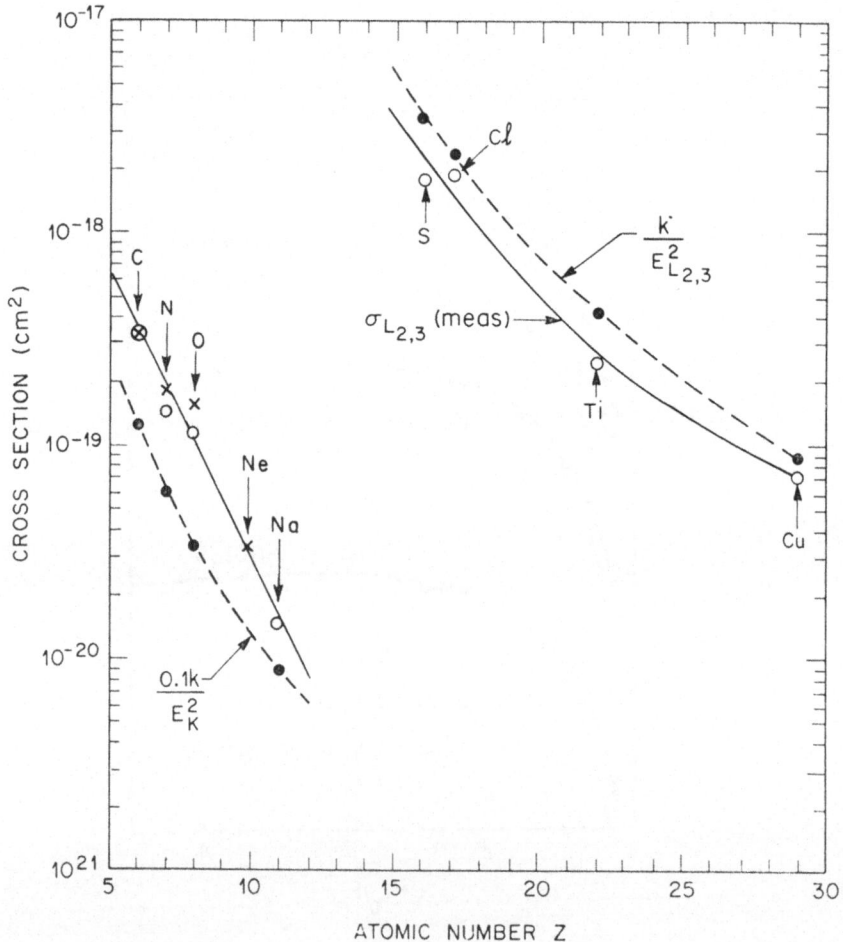

Fig. 23. Ionization cross section vs. Z at a reduced energy of four. Note these are the measured absolute values with no adjustment to theory as in Figure 22 (+) from Glupe and Mehlhorn, *Phys. Letters* **25A**, 274 (1967) and (0) from Gerlach and DuCharme, *Surf. Sci.* **32**, 329 (1972) and personal communication.

enhancement factor k in the backscattering term of (4.1) to change it to

$$I_A = CI_P(\varphi + k\alpha\beta).\tag{4.3}$$

The data was found to agree with (4.3) only for very large values of k. They were therefore led to the conclusion that even though the backscattered current was very small (<0.1 of I_P) it was producing most of the Auger electrons.

On considering the results of Gallon it is clear that the problem with the analysis presented by NFJ is their assumption that the cross section was accurately known. Gallon, by using an iterative procedure to solve the integral Equation (4.1) completely, obtained not only the cross section but also the backscattered contribution. Gallon's results clearly show that the cross section not only has its maximum at much higher reduced energy than the NFJ computed value, but also that it falls off much more slowly above the maximum.

Gerlach and DuCharme [312] have recently made some measurements of absolute ionization cross sections for C, N, O and Na adsorbed on a $W[100]$ single crystal surface. This method allows reasonable estimates of the concentration to be made (although these coverage determinations represent the largest error in the measurements) and avoids escape depth complications because the layers under excitation are only a single atom thick. They have computed the backscattering contribution using a method similar to that of Gallon and observed that it had less than a 25% effect on the C ionization cross section at reduced energies below 4.

The resulting absolute cross sections are shown in Figure 22. The agreement with the results of Glupe and Mehlhorn is within experimental error both in relative and absolute terms.

The absolute cross sections at a reduced energy of four are shown in Figure 23 as a function of Z. Also on this graph are plotted the relative values for $1/E_i^2$ which is seen to scale correctly with the measurements.

3. Dependence of Yield on Angle of Incidence

Palmberg [203] investigated the effect of varying the angle of incidence on the Auger yield from overlayers and substrates. His results (Figure 24) indicate that the entire Auger spectrum is very much enhanced when the primary electron beam is incident on the sample at angles near grazing incidence. This may be understood as follows with reference to Figure 25. Here we represent the overlayer schematically as a thin slab on an infinite substrate. As the angle of incidence φ is decreased, the primary electrons spend more time traversing the overlayer and thus have an increased probability of producing ionization in the region from which Auger electrons can escape. In addition, the Auger signal from the S and C impurities is stronger relative to the substrate Pd as grazing incidence is approached. This observation is not fully understood and is not always observed (see NFJ below). While the qualitative effect of relative overall enhancement is observed at all three angles, the overall strength of the spectrum is attenuated at very grazing incidence ($7°$) probably because of surface roughness.

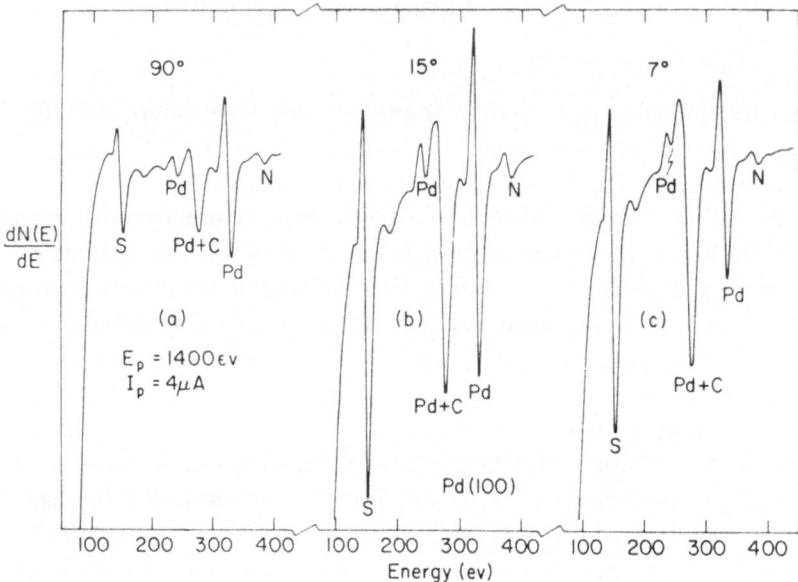

Fig. 24.   Auger spectra from Pd [100] surface for incident beam angles of (a) 90°, (b) 15° and (c) 7°. Incident beam energy and current were the same for all spectra. (From Palmberg [203].)

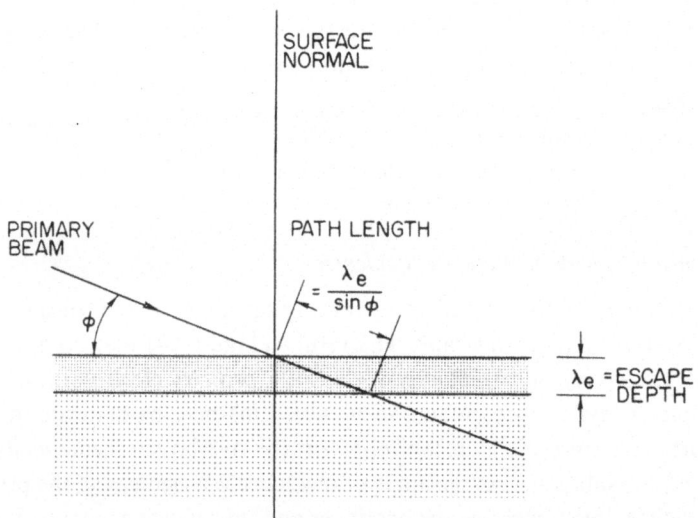

Fig. 25.   Pictorial description of increased ionization within the escape depth λ_e as the angle of incidence φ is decreased.

   The angle of incidence is seen to also have a marked influence on the Auger yield curves as shown in Figure 26 from Neaves et al. [318]. Note that while at normal incidence the Auger yield for both Si and the C overlayer show a well defined maximum, the curves are much flatter at the maximum for grazing incidence. Indeed, the

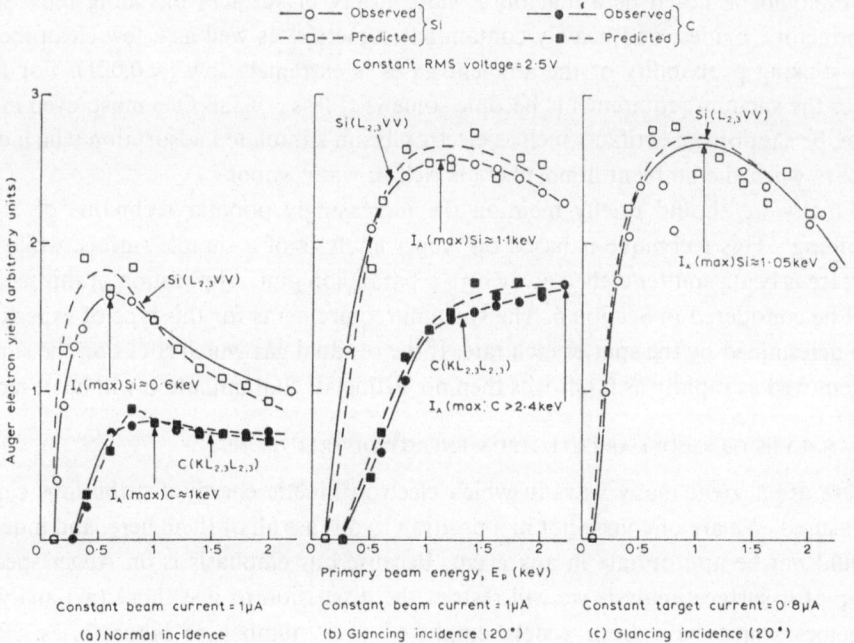

Fig. 26. Auger current as a function of primary beam energy at normal and glancing incidence with constant beam current (a) and (b) and constant target current (c). (From Neave, Foxon, and Joyce, *Surf. Sci.* **29**, 411 (1972).)

C Auger yield showed no maximum at all. This result is simply a consequence of the larger number of high energy backscattered electrons which are present at glancing incidence.

5. Experimental Considerations

A. VACUUM REQUIREMENTS; SOME GENERAL REMARKS

We have seen in Section 2C that Auger electron spectroscopy is very surface sensitive as a result of the short escape depths for Auger electrons with energies below several keV. In the next section we will go on to see just how sensitive the technique is to adsorbed layers.

While this characteristic surface sensitivity is extremely useful for a wide class of problems, at the same time it places extremely stiff requirements on the vacuum environment in which the experiments must be carried out. More specifically, if a clean metal surface is exposed to an ambient pressure of 1×10^{-6} torr (mm of Hg) of an active gas (CO_2, H_2, H_2O, O_2), it will become covered with adsorbed molecules in a time on the order of several *seconds*. As we have seen in Section 4B2, this would hardly be enough time to carry out a useful experiment. Typically then for work on clean metal surfaces, background pressures in the 10^{-10} torr range are required to maintain the surface *status quo*. Fortunately, ultrahigh vacuum technology has advanced sufficiently over the past decade that such an environment can be provided routinely.

It should be noted here that on a wide variety of surfaces including most semi-conductors, oxides, and heavily contaminated metals, as well as a few clean metals, the sticking probability of the ambient gases is extremely low (<0.001). For these cases the vacuum requirements become somewhat less crucial. One must, even in this case, be cautious of artifacts such as electron beam stimulated adsorption which often occurs when the ambient atmosphere is rich in water vapor.

Finally we should briefly mention the increasingly popular technique of 'depth profiling'. This technique is based on Auger analysis of a sample surface while that surface is being sputter etched away by a separate ion gun. Application of this method will be considered in Section 6. The vacuum requirements for this type of experiment are determined by the sputter etch rate. If the residual gas which sticks on the surface is removed as rapidly as it adsorbs then no 'artificial' contamination will be observed.

B. ANALYSIS OF SECONDARY ELECTRON ENERGY DISTRIBUTIONS

There are a great many ways in which electron kinetic energy distributions can be measured. We are of course not in a position to discuss all of them here, and indeed it would not be appropriate in any event. Because our emphasis is on Auger spectroscopy for surface analysis we will restrict the discussion to just those two analyzing schemes which have been widely employed in a number of laboratories. These analyzers have several features in common; they are electrostatic, compact, constructed so as to withstand the rigors of the ultrahigh vacuum environment, and provide easy sample access. They share these features because these are requirements of an analyzer which is compatible with instrumentation for other measurements besides Auger analysis of the surface being investigated. These other techniques may include LEED, work function measurement, photoemission, ion neutralization, etc.

1. *Retarding Field Analyzers*

As noted in the introduction, the popularity of LEED for studying single crystal surfaces grew rapidly through the 1960's. The basis for this measurement is a spherical retarding field analyzer as shown in Figure 27. An electron beam impinges normal to the surface and the secondary electrons move radially outward in the field free region between the sample and the #1 grid. Now in LEED we seek the angular distribution of elastically scattered electrons and therefore we must filter out of these radially drifting secondaries all those which have an energy less than E_p. This is done by applying a negative bias on the #2 (suppressor) grid which is maintained at a potential E close to E_p.

Only the elastically scattered electrons will have sufficient energy to 'climb over' the retarding barrier and be post accelerated by 5 keV to the fluorescent screen. Figure 28 shows a potential diagram for this situation. The purpose of grid #3 is to minimize field penetration of the collector potential into the retarding region.

It should be clear that if we measure the current I_c to the collector as a function of the retarding voltage E, it will be related to the energy distribution by

$$I_c(E) = I_P \int_E^{E_P} N(E)\,dE \tag{5.1}$$

where the energy distribution is normalized so that

$$\int_0^{E_P} N(E)\,dE = \delta. \tag{5.2}$$

δ is the secondary yield. Thus for E more negative than E_P all secondaries will be cut off and $I_c = 0$ while for E positive with respect to the target $I_c = \delta I_P$ (assuming of course that all secondaries emitted from the target are collected).

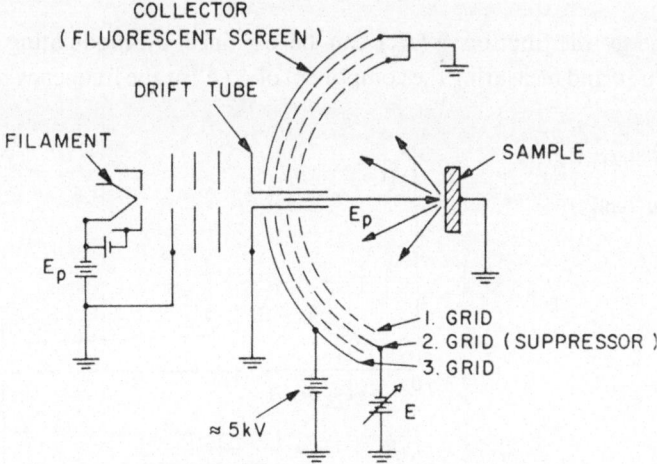

Fig. 27.   Schematic of LEED apparatus.

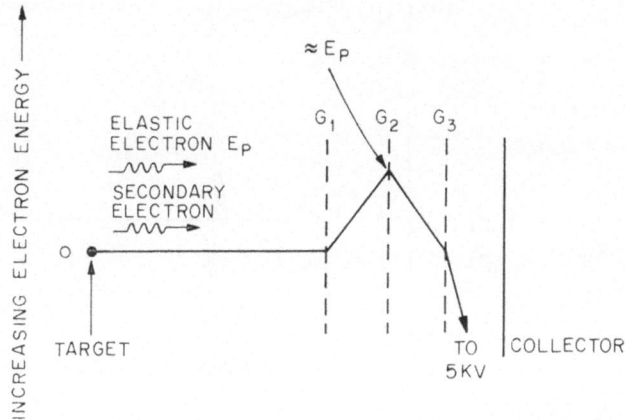

Fig. 28.   Potential energy diagram showing means by which a grid system energy analyses secondary electron energy distribution.

In 1967 Palmberg [206] and Scheibner and Tharp [235] independently recognized that the energy distribution $N(E)$ could be obtained from the measured $I_c(E)$ by electronic differentiation. Consider the current $I_c(E_0)$ at some retarding energy E_0. The change in I_c on moving to a slightly different energy is given by expanding $I_c(E)$ in a Taylor series about E_0. Thus

$$I_c(E) = I_c(E_0) + \frac{dI_c(E)}{dE}\bigg|_{E=E_0} (E - E_0) + \frac{d^2 I_c(E)}{dE^2}\bigg|_{E=E_0} \frac{(E - E_0)^2}{2!} + \cdots.$$

(5.3)

Substituting from (5.1) we get

$$I_c(E) = I_c(E_0) - I_P N(E_0)(E - E_0) - I_P \frac{dN(E)}{dE}\bigg|_{E=E_0} \frac{(E - E_0)^2}{2!} \cdots.$$

(5.4)

Now the energy distribution $N(E_0)$ can be obtained by modulating E such that $E - E_0 = k \sin \omega t$ and measuring the component of $I_c(E)$ at the frequency ω (Figure 29).

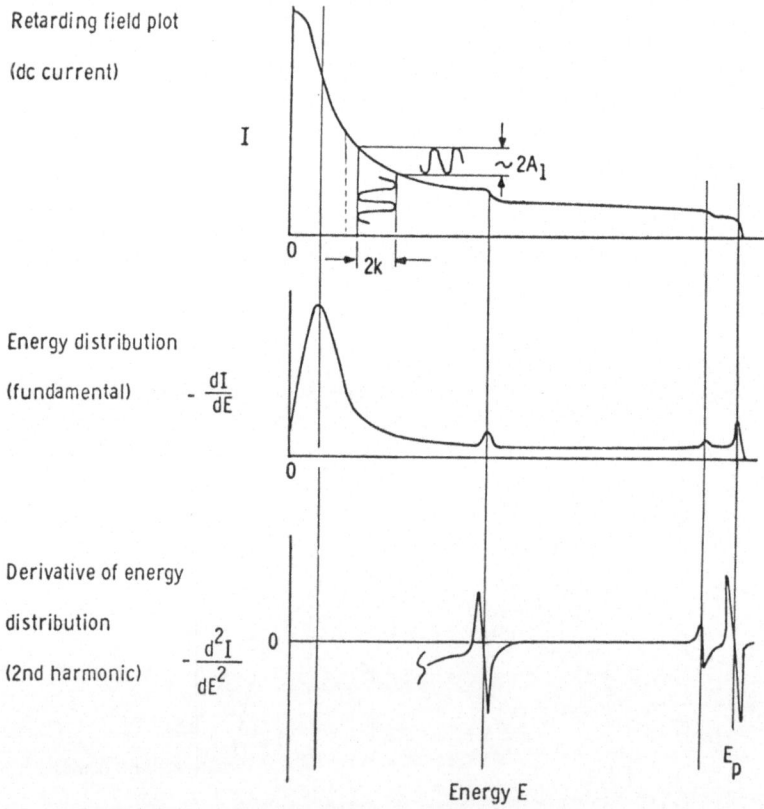

Fig. 29. Schematic diagram showing the collector currents in the retarding field analyser in different modes of operations E_p is the primary beam energy. (From Taylor, J. in *Techniques of Metals Research* **VII** (Copyright © J. Wiley and Sons) (to be published).). By permission of John Wiley and Sons Ltd.

This component will be $-kI_PN(E_0)$ for sufficiently small modulation voltages. In this way Palmberg [206] and Scheibner and Tharp [235] were able to obtain $N(E)$ using the LEED system. Furthermore Scheibner and Tharp actually observed Auger peaks although they appeared on a large and rapidly varying background.

Harris (119) recognized that $dN(E)/dE$ revealed the Auger peaks while surpressing the background. He used a dispersive analyzer which as we will show later measures $N(E)$ directly. He then obtained $dN(E)/dE$ by modulating the analyzing energy and determining the component of the current output at the modulation frequency ω.

Based on Harris' success at extracting Auger peaks from $dN(E)/dE$, Weber and Peria [297] recognized that the third term in (5.4) provided a simple means to obtain $dN(E)/dE$ spectra using a LEED retarding field analyzer. For sinusoidal modulation of the retarding potential this term becomes

$$I_c(2\omega) = -1/4k^2 I_P \frac{dN(E)}{dE}\Big|_{E=E_0} (\cos 2\omega t). \tag{5.5}$$

Hence by measuring the component of $I_c(E)$ at 2ω the $dN(E)/dE$ spectrum is easily obtained.

It was with this step that the explosion in Auger spectroscopy took place, for it was now possible for many workers using existing LEED systems to convert them in a straightforward manner to Auger spectrometers.

Let us now consider the operation of the LEED/retarding field analyzer as an Auger spectrometer (Figure 30). The scheme shown includes a double retarding grid

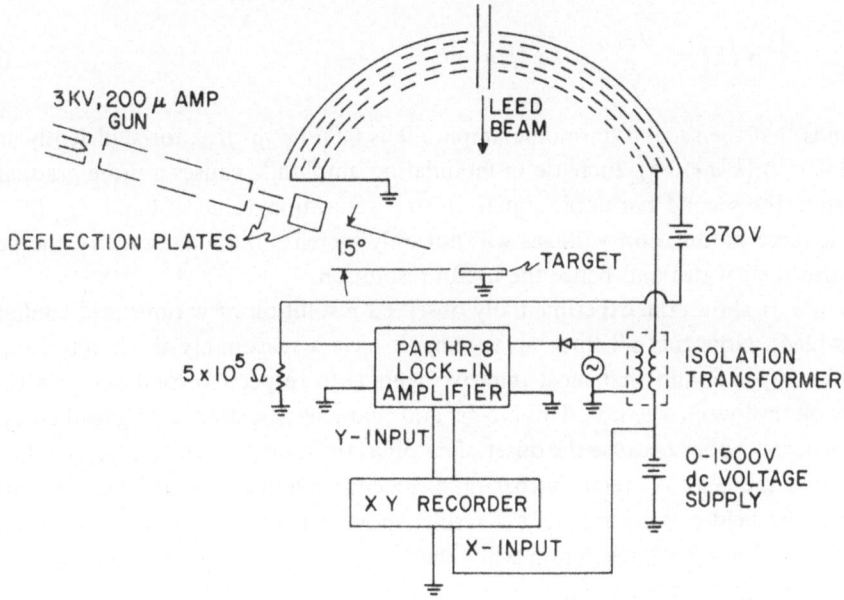

Fig. 30.   Scheme to obtain Auger spectra using 4-grid retarding field analyser.
(From Palmberg [203].)

first used by Palmberg (203) to improve the resolution. The retarding potential E is applied to grids #2 and #3 by a power supply capable of being swept from 0 to E_P. An voltage $k \sin \omega t$ is superimposed on top of E by means of an isolation transformer driven by an oscillator. Depending on the detection circuitry the frequency ω may range between 100 Hz and 5 kHz. The oscillator voltage is also fed through a diode into the reference channel of a synchronous (lock-in) detector. The diode half wave rectifies the oscillator signal and provides a 2ω signal at the reference input. The synchronous detector essentially performs the integral of the input signal times $\sin \omega_{ref} t$. More specifically for this case where $\omega_{ref} = 2\omega$

$$E_{out}^{lock\text{-}in} = C_1 \frac{1}{T} \int_0^T I_c(t) \sin(2\omega t + \varphi) \, dt \tag{5.6}$$

where φ is an adjustable phase angle and T is the time constant. For T sufficiently long compared with $1/2\omega$ this integral will be zero for all frequency components present in $I_c(t)$ except $I_c(2\omega)$. Because this 2ω component is $\sim \cos 2\omega t$, φ must be $\pi/2$. For these conditions then

$$E_{out} = \frac{C_1}{4} k^2 I_P \left. \frac{dN(E)}{dE} \right|_{E=E_0} \tag{5.7}$$

which is the desired result. Thus by measuring E_{out} as a function of E we obtain the spectrum of $dN(E)/dE$.

Taylor [264] has considered the performance of modulated LEED retarding field analyzers in considerable detail. For a Gaussian peak in $N(E)$ given by

$$N(E) = \frac{I_{tot}}{\sigma \sqrt{2\pi}} e - E^2/2\sigma^2 \tag{5.8}$$

he finds that the second harmonic amplitude is $0.06 \, (k^2/\sigma^2) I_{tot}$ for sufficiently small modulation $(k/\sigma < 0.5)$. Increase in modulation amplitude causes a more gradual increase in the second harmonic signal up to a saturation value of $0.051 \, I_{tot}$ [22]. Of course large modulation voltages will not only increase the magnitude of the peaks but also their width and hence the useful resolution.

Figure 31 shows the experimentally observed resolution of various grid configurations [264]. Note that all three arrangements have a reasonably sharp cutoff on the high energy side and that most improvements with respect to total peak width are made on the low energy side. This is a general characteristic of retarding field analyzers which comes about because the onset of cutoff as the retarding voltage approaches the energy of the peak is much sharper than the completion of cutoff which is strongly effected by field penetration. These results show that the 'three grid' configuration of Figure 27 has 2.5% resolution while the addition of a 'fourth grid' or the tying of two grids together for the cutoff electrode improves this to 0.5%.

We will return to noise considerations in 5B3 after a discussion of dispersive analyzers.

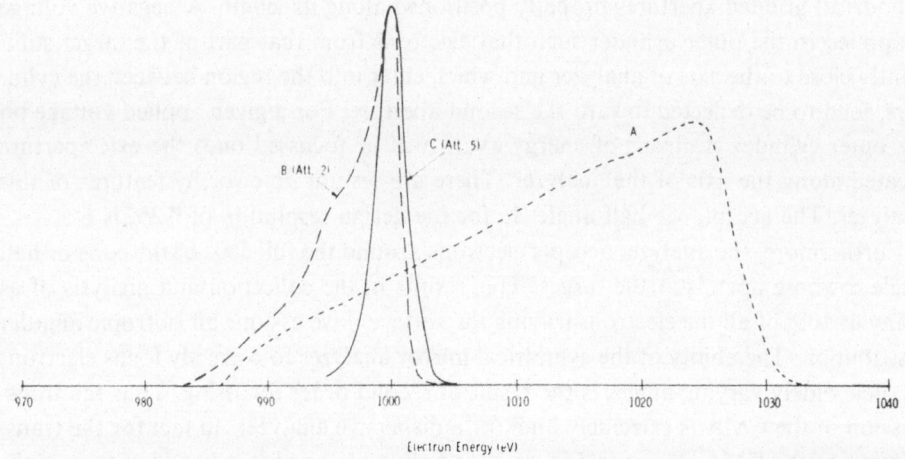

Fig. 31. Effect of grid arrangement on resolution. For a single retarding grid, instrumental line width is 2.4 %, for B double retarding grid 0.5 % and for C double retarding grid plus modified first grid < 0.3 %. (From Taylor [264].)

2. *Cylindrical Mirror Analyzer* (CMA)

The cylindrical mirror analyzer, first described by Blauth [320], appears to be the best analyzer for Auger spectroscopy of surfaces both from a constructional and performance standpoint. It has received limited attention in several gas phase studies until its rapid growth as the standard for Auger surface studies.

Palmberg *et al.* [207] were the first to use a cylindrical mirror analyzer for Auger electron spectroscopy of surfaces. Their arrangement is shown in Figure 32. The analyzer consists of two coaxial cylinders, the inner of which is grounded and has two

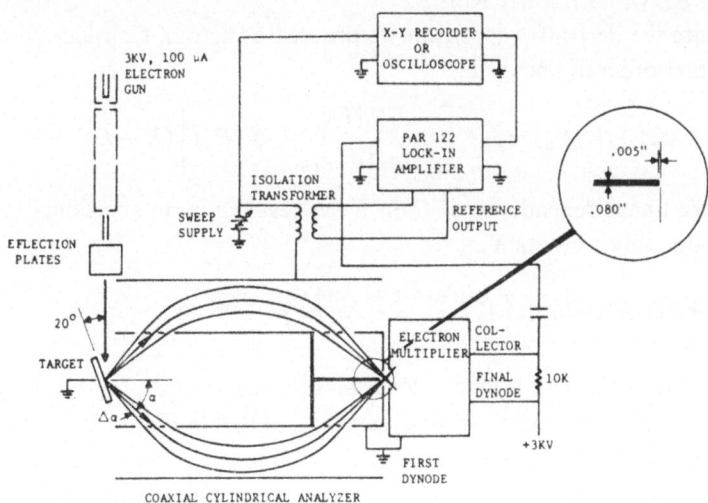

Fig. 32. Scheme for measuring the Auger spectrum with coaxial cylinder analyser. (From [207].)

cylindrical gridded apertures properly positioned along its length. A negative voltage is applied to the outer cylinder such that electrons from that part of the target sufficiently close to the axis of analyzer and which enter into the region between the cylinders, tend to be deflected toward the second aperture. For a given applied voltage on the outer cylinder electrons of energy eV_{pass} will be focussed onto the exit aperture located along the axis of the analyzer. There are several noteworthy features of this analyzer. The acceptance half angle $\Delta\alpha$ for the design resolution of 0.3% is 6°.

Furthermore, the analyzer accepts electrons around the full 360° of the cone of half angle α whose apex is at the target. This results in the collection and analysis of as many as 10% of all the electrons leaving the surface if we assume an isotropic angular distribution. The ability of the cylindrical mirror analyzer to correctly focus electrons of such widely varying angles is the result of second order focussing. Thus the transmission of the CMA is extremely high for a dispersive analyzer. In fact for the transmission of the CMA shown in Figure 32 is only a factor of two less than that of the retarding field analyzer of Figure 30. Other dispersion analyzers such as the 127° analyzer used by Harris in his pioneering work have several orders of magnitude lower transmission [264]. While the transmission of the CMA is comparable with retarding field analyzers, it has the advantage that it measures only those electrons at the pass energy. The retarding field analyzer collects *all* electrons greater than the cutoff energy. We will see below that these higher energy electrons contain no information about the energy distribution at the cutoff energy but do contribute a great deal of noise. Thus the signal noise ratio of the CMA is superior by as much as a factor of 100.

The current collected at the exit aperture of the CMA is given by

$$I_c = I_P \cdot K \cdot E \cdot N(E) \tag{5.9}$$

where K is a geometry factor. This linear dependence on the pass energy results from the constant resolution $\Delta E/E$ of the analyzer. Thus the sampled width of the energy distribution ΔE varies linearly with E.

To measure the derivative spectrum we proceed as before. Consider an expansion of (5.9) to first order at energy E_0

$$I_c(E) - I_c(E_0) = I_P K \left[E_0 \cdot \frac{dN(E)}{dE}\bigg|_{E_0} + N(E_0) \right] (E - E_0) + \cdots. \tag{5.10}$$

Note that the linear dependence of I_c on E has resulted in an additional term. For a $k \sin\omega t$ modulation we obtain

$$I_c(\omega, E) = kI_P K \left(E \frac{dN(E)}{dE} + N(E) \right). \tag{5.11}$$

Figure 32 shows the details of the electronics for measuring $I_c(\omega, E)$. There are two important differences between this arrangement and that of Figure 30 for the retarding field analyzer. An electron multiplier is used as the preamplifier for the CMA because it is the lowest noise–widest bandwidth low current amplifier available. The electrons passing through the exit aperture impinge on the first dynode which is biased near ground. Thus the multiplier collector (anode) is at a high positive *dc* potential. We must therefore capacitively couple the $I_c(\omega)$ back to near ground potential where it is applied to the synchronous detector input. The arrangement shown permits modulation frequencies up to 50 kHz to be used, thereby allowing high speed scans (>10000 eV s^{-1}) and oscilloscope displays as shown in Figure 33.

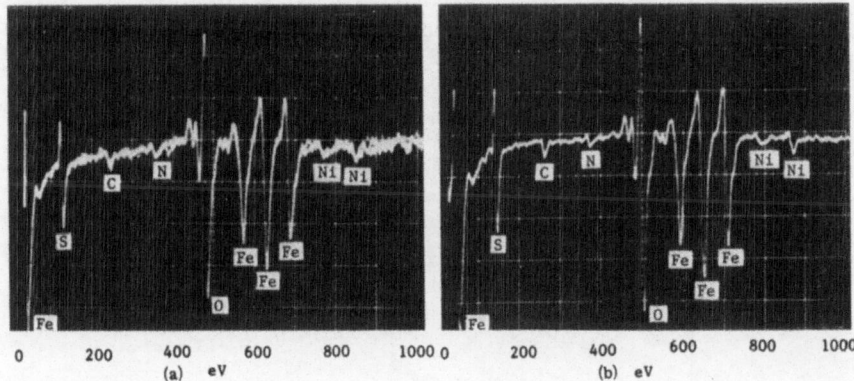

Fig. 33.   Oscilloscope trace of Auger spectrum taken at (a) 20000 V s^{-1}, 2.5 Hz bandwidth and (b) 2000 V s^{-1}, 0.25 kHz bandwidth. (From [207].)

If slow scan rates (<100 eV s^{-1}) are used and hence lower modulation frequencies (<500 Hz), there is no need for the multiplier and a simple load resistor arrangement may be employed as in Figure 30.

There are two noteworthy advantages of the CMA which should be pointed out at this time. When an electron multiplier is used the current measured is that given by (5.9) times a factor $g(E)$, the multiplier gain. g is energy dependent because the gain of the first dynode is determined by the energy E of the impinging electron. The gain of the successive stages is of course independent of E. Furthermore, the coefficient of the first harmonic component in (5.11) is $[E\,dN(E)/dE + N(E)]$. At low energy (<200 eV) $dN(E)/dE$ is negative while $N(E)$ is of course positive. This cancellation effect on the slow peak plus the attenuation of the whole spectrum at low energy resulting from the factor E and the low value of $g(E)$ at low energy, leads to an observed spectrum which is remarkably flat down to as low as 25 eV. This is a considerable convenience when recording data when compared with measurements using retarding field analyzers (see Figure 34). Thus while the gain for the grid system must be changed over as much as a factor of 300 over the whole energy range, the CMA measures the data using only a single gain.

The second advantage to be noted here is the adaptability of the CMA to accommodate an electron gun mounted within the inner cylinder which produces an incident beam along the analyzer axis. This enormously eases the alignment difficulties as the only remaining variable is the position of the sample along the axis. Thus the entire analyzer including electron gun can be mounted on a single flange.

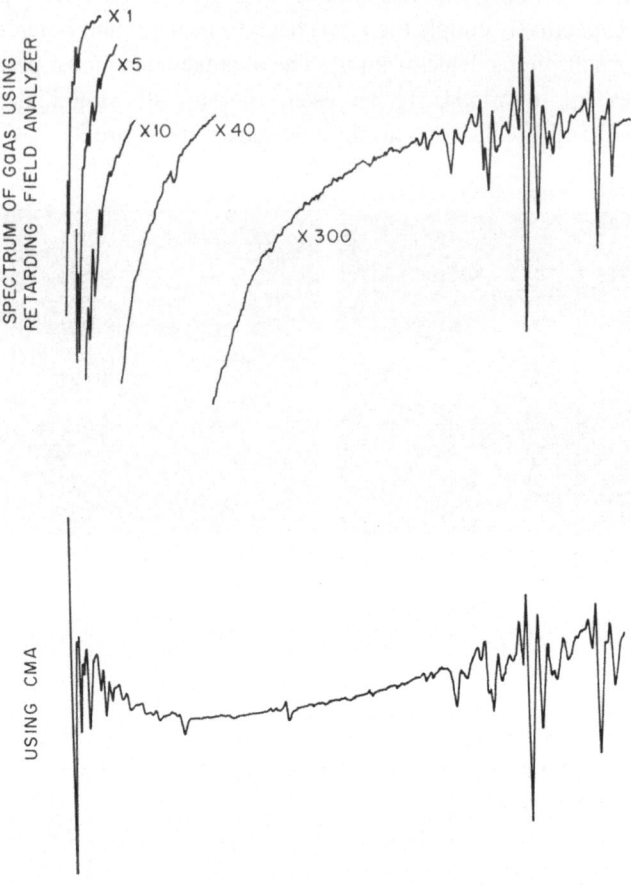

Fig. 34.   Comparative Auger spectra taken with retarding grid system and cylindrical mirror analyser.

3. *Sensitivity and Noise Considerations*

We have noted that the transmission in the retarding field and cylindrical mirror analyzers is comparable for a resolution of 0.5–1% which is adequate for most Auger surface analysis. Hence the Auger current measured at the respective collectors will be comparable. On the other hand, the noise will be quite different in the two cases. Consider in this regard an energy distribution $N(E)$ with an Auger peak at E_A. For the retarding field analyzer the total collector current at $E = E_A$ is

$$I_C^{RFA}(E_A) = C_1 I_P \int_{E_A}^{E_P} N(E)\,dE. \tag{5.12}$$

On the other hand for the CMA the corresponding quantity is

$$I_C^{CMA}(E_A) = C_2 I_P \int_{E_A}^{E_A + \Delta E} N(E)\,dE = C_2 I_P\, \Delta E N(E) \tag{5.13}$$

where C_1 and C_2 are constants which account for the transmission, resolution, etc. and are about equal.

The mean square shot noise associated with a current I_c is

$$\overline{i^2_{\text{noise}}} = 2I_c e\, \Delta f \tag{5.14}$$

where e is the electronic charge and Δf is the noise equivalent bandwidth. This expression is no more than an expression of the uncertainty $\sqrt{n}$ in counting n random events. Dividing both sides by $(2e\Delta f)^2$ we obtain

$$\frac{\overline{i^2_{\text{noise}}}}{(2e\,\Delta f)^2} = \frac{I_c}{2e\,\Delta f}. \tag{5.15}$$

$1/2\Delta f$ can be considered as an effective measurement time, that is a period over which particles are 'counted'. Naturally, the longer this time (the smaller Δf), the less uncertainty there will be in the measurement. Also I_c/e and $(\overline{i^2_{\text{noise}}}/e^2)^{1/2}$ are just electrons per second. Hence the mean square uncertainty in the electrons measured ('counted') in a time $1/2\Delta f$ is just the total number and the rms (root mean square) number is the square root of the total number 'counted'.

Consider now a simple estimate for the ratio of $I_C^{\text{RFA}}/I_C^{\text{CMA}}(E_A)$ from (5.12) and (5.13) assuming $C_1 = C_2$. If we assume $N(E)$ is constant from E_A to E_P at the value $N(E_A)$ then the result is simply

$$\frac{I_C^{\text{RFA}}}{I_C^{\text{CMA}}} = \frac{E_P - E_A}{\Delta E}. \tag{5.16}$$

E_P is generally set equal to $> 5\, E_A$ for efficient ionization while $\Delta E = 5 \times 10^{-3}\, E_A$ for a cylindrical analyzer. Thus

$$\frac{I_C^{\text{RFA}}}{I_C^{\text{CMA}}} \cong 10^3.$$

This is of course a low estimate because $N(E)$ in general increases quite substantially above E_A, particularly in the vicinity of the elastic peak. Thus this ratio may under some conditions be as high as 10^4.

The shot noise $\sim \sqrt{I_C}$ and because we have estimated that the signals are comparable from the two analyzers, we may estimate that the signal-to-noise ratio (S/N) will be from 10 to 100 better for the CMA. There are at least two important consesequences of this improved S/N. First because $S/N \sim (1/\Delta f)^{1/2}$, a spectrum with a given S/N may be obtained as much as 10^4 times faster with the CMA than a retarding field analyzer. This permits real time oscilloscope displays as shown in Figure 33. This is of considerable value in checking out spatial variation of impurities on a surface, as well as studying adsorption kinetics, catalytic reactions, or surface segregation. On the other hand if electron beam interactions are a problem, the beam current may be reduced by 10^4 (making the effective interaction rate lower by 10^4 times) and the measurement will only require 10^2 more time to obtain the same S/N. Thus the electron beam effects would be reduced to 1% of their influence at the high beam currents.

While it is clear from the above that for a given excitation condition the CMA is enormously superior to the retarding field analyzers, it should be pointed out that their respective operating conditions with respect to excitation are not the same.

The retarding field analyzer being limited by the number of electrons with energy above cutoff, is best operated with the excitation energy as low as possible consistent with efficient ionization. The CMA on the other hand, being limited by the magnitude of $N(E)$ at the Auger peak and to an even greater extent by the curvature with respect to energy in $N(E)$, is best operated at a very high excitation energy. The total high energy secondary current is reasonably independent of primary energy, decreasing at a given energy as E_P increases. This also leads to a flattening of $N(E)$. Both these conditions are favorable to CMA performance.

4. Data Manipulation

A. *Separation of inelastic and secondary features.* There are instances when it is desirable to separate the Auger spectrum from the loss spectrum and vice versa. For example when measuring Auger yields near threshold as in Figure 29, the plasma loss peaks may obscure the weak Auger peaks.

Gerlach *et al.* [91] pointed out that because the Auger peaks (as well as any other secondary structure) are independent of primary energy while the characteristic loss structure are not, these features may be separated by a clever change in the modulation scheme. If the sample potential is modulated while the analyzer and primary beam are maintained fixed, then to a good approximation the energy of an electron being analyzed which has simply suffered a characteristic loss will be uninfluenced by the modulation. That is, it did not interact with the target in such a way as to 'know' the location of the Fermi level. On the other hand, an Auger electron ejected from the target will have its energy modulated relative to the analyzer potential.

We may consider this also in the following way. The current measured at the output of the analyzer is proportional to $N(E, E_P)$. Loss features have an energy dependence like $N_{loss}(E - E_P - E')$ where both E and E_P are measured relative to the sample Fermi level. Thus by modulating the sample relative to both the analyzer (at E) and the primary beam (at E_P), the modulation term cancels out and the loss features do not appear. For Auger structure however $N(E, E_P)$ has a form like $N(E)$ and modulation of the sample potential causes Auger peaks to be observed.

To excite only loss peaks one modulates just the primary energy E_P. This clearly has only a weak influence on the Auger peaks and thus they are not observed for this condition. These situations are indicated in Figure 35.

B. *Deconvolution techniques.* Mularie and Rusch [188] have shown that by measuring the characteristic loss spectrum for primary energies equal to a particular Auger peak, it is possible to decide whether low energy structure on that Auger peak is the result of

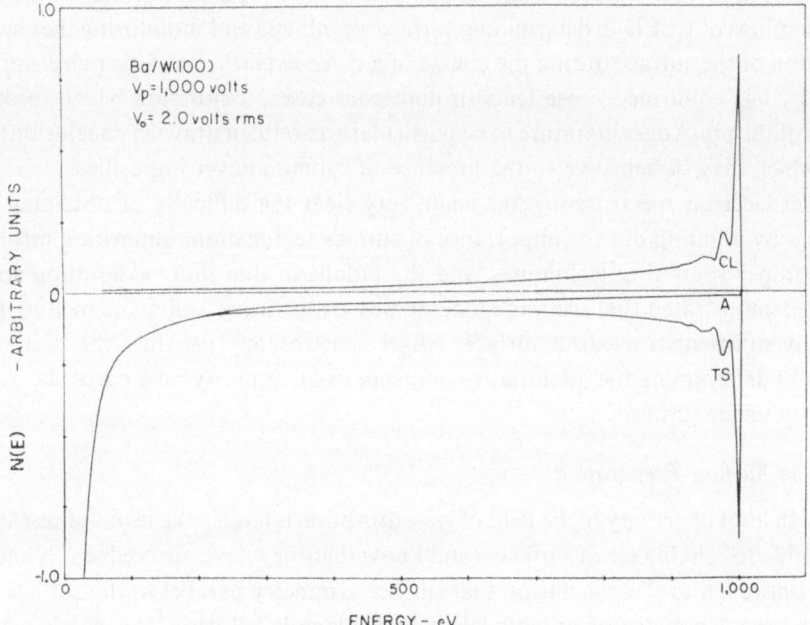

Fig. 35. Energy distribution curves for the three modes of operation. A – Auger, modulate only
target; CL – characteristic loss, modulate only E_p; TS – total spectrum, modulate
analyser energy (From [091].)

characteristic losses suffered by the Auger electrons on leaving the solid. If the position and magnitude of the characteristic loss peaks is similar to the low energy structure on the Auger peak then these satellites may be identified with the loss structure
produced by the Auger electron.

Mularie and Peria [187] have in turn used this idea to deconvolute measured spectra
using the characteristic loss measured at the Auger peak energy. While this method
has been quite successful, it must be noted that an Auger electron and a primary
electron will not necessarily be equivalent even though they have the same energy. This
results from the fact that the final doubly ionized state of the atom emitting the Auger
electron may be an excited one involving plasmons created as a result of the sudden
change in potential. The Auger electron would thus not really suffer a loss in the same
sense as a primary electron. This difference in mechanism would certainly be expected
to change at least the relative amplitudes of the loss-like structure.

6. Auger Spectroscopy for Surface Analysis

A. BASIC SURFACE PHYSICS STUDIES

The most widespread application of Auger electron spectroscopy has been in the area
of fundamental surface studies. This is to be expected in view of the course of development of LEED systems into retarding field Auger spectrometers. Workers involved in

single crystal surface research using LEED during the 1960's were acutely aware of the shortcomings of LEED in determining surface cleanliness and monitoring the chemical condition of the surface during the course of a given experiment. Auger electron spectroscopy has confirmed those fears in numerous cases, a situation which forces the reader of the pre-Auger literature to be particularly careful in drawing conclusions from data which may be sensitive to the presence of submonolayer impurities.

Auger electron spectroscopy has made very clear the difficulty of obtaining clean surfaces by pointing out the importance of surface segregation, impurities introduced by improper sputtering techniques, and the pitfalls in thin film evaporation studies. It also demonstrated that heating alone, to any temperature below the melting point, is rarely sufficient to clean a surface. Auger spectroscopy provides the user with a means of determining the quantitative influence of an impurity on a particular surface property under scrutiny.

1. *Clean Surface Preparation*

The high level of activity in the field of gas adsorption is largely the result of researchers being able to reliably clean surfaces (and know that they have succeeded). Numerous studies have led to the conclusion that surface symmetry parallel to the surface is the same as the symmetry of an equivalent bulk plane in all cases for metals with the exception of the [100] surfaces of Ir, Pt, and Au. These surfaces have approximately hexagonal symmetry presumably brought on by a rearrangement of the outer most layer. At the time this observation was first made, there was considerable controversy regarding the cleanliness of these surfaces, but subsequent Auger analysis has confirmed the clean surface rearrangement idea. Suggestions have been made that other metal surfaces also rearrange but these claims were dispelled by careful cleaning procedures using Auger spectroscopy as a surface monitor.

The most common method used to obtain a clean metal surface has been to sputter with inert gas ions (usually Ar^+) at room temperature and then to anneal the damage and flash off the imbedded Ar by heating the sample in vacuum. This process of sputtering and annealing is then repeated until the surface is clean. This cycling procedure will in time leach out all impurities which tend to surface segregate. Each sputtering will clean the immediate surface layer and the annealing will replenish it by providing the bulk impurities (which tend to surface segregate) with the mobility required to reach the surface by diffusion. Provided the sample is never heated above the final annealing temperature the surface will in general not be contaminated by impurities diffusing out.

Because a segregated layer of impurities sets up a concentration gradient opposite to the direction of the driving force and tends to saturate at a single layer thickness, each annealing cycle will only bring on the order of one layer of impurity to the surface. Thus to extract 10 ppm of segregatable impurity from a 1 mm (5×10^6 layer) thick sample may require more than 50 such cycles.

These problems may be overcome by sputtering at an elevated temperature. For the removal of endothermically dissolved impurities (increased solubility with increased temperature), the temperature of the target must be kept low enough to maintain the driving force for segregation but high enough to maintain the necessary mobility. On the other hand, for exothermically dissolved impurities the sputtering temperature may be as high as experimentally feasible.

Figures 36 and 37 show the usefulness of Auger spectroscopy in determining the cleaning procedure. Figure 36 shows the C surface concentration decreases with increasing temperature (endothermic). As the cleaning treatment proceeds and C is depleted from the bulk there is a temperature above which no C is present on the surface. Finally after leaching essentially all of the C out, the Auger signal is zero at any elevated temperature. This is in contrast to the case of oxygen which dissolves exothermically in Ni. Thus while the 0 signal is apparent at high temperatures, it is absent at RT.

In addition to helping to clean the surface of these major contaminants arising chiefly from the bulk, Auger spectroscopy helps avoid pitfalls in the sputtering process

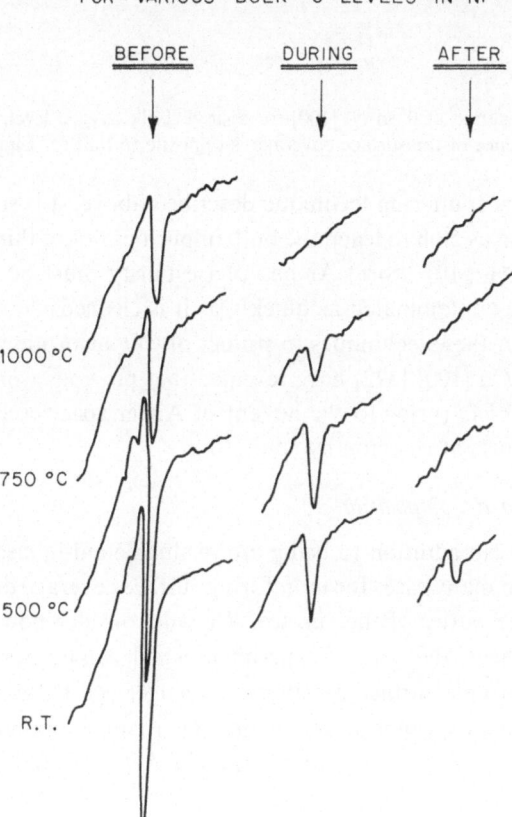

CARBON AUGER SIGNAL VS TEMPERATURE
FOR VARIOUS BULK C LEVELS IN Ni :

Fig. 36. Surface segregation of C on Ni [100]. Before, during, and after refer to the leaching out of bulk C. Thus the bulk C level is decreasing on going from left to right.

OXYGEN AUGER SIGNAL
VS   TEMPERATURE
FOR   FIXED   BULK   O LEVEL   IN   Ni

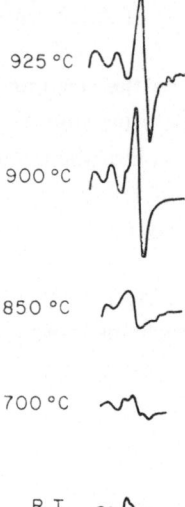

925 °C

900 °C

850 °C

700 °C

R.T.

Fig. 37.   Surface segregation of 0 on Ni [100] for a single bulk oxygen level. Note the temperature
dependence of the surface coverage is opposite to that for Figure 36.

itself. Even using the sputtering technique described above, it is sometimes necessary
to sputter for as long as 50 h to leach the bulk impurities out of thin crystals (0.1 mm).
The high pressure $(1 \times 10^{-4}$ torr) Ar gas phase purity must be maintained or the
surface will become contaminated as quickly as it is cleaned.

The application of these techniques to studies of CO adsorption on Pd [100] [278],
Ni [100] [321] and Cu [100] [322] have revealed that previous work on these systems
which used only LEED (prior to the advent of Auger spectroscopy) was greatly in
error despite all efforts at cleaning by sputtering.

2. *Monitoring Surface Composition*

Auger spectroscopy, in addition to being an invaluable aid in clean surface preparation, is also useful in many cases for monitoring surface coverage during adsorption as
well as checking the purity of the surface. We will consider now several illustrative
examples which indicate the types of experiment which can be done. These results all
demonstrate the extreme surface sensitivity of Auger spectroscopy by its ability to
detect small fractions of a monolayer of the adsorbate. It should be indicated that
barring peak overlap, Auger sensitivity using a CMA is typically 0.1% of a monolayer
for most elements.

A. Adsorption. Palmberg (200) has used Auger spectroscopy in conjunction with
LEED and work function measurements to study the adsorption of Xe on Pd [100].

This system is ideally suited for a Auger study because the electron beam interactions with the adsorbed layer are very weak. One might think that in view of the weak Xe–Pd bond the interaction would be sufficient to desorb it easily. The reason for the weak interaction is that the Xe is physically adsorbed, i.e. held by dispersion (van der Waals) forces. Unlike normal molecular bonds, these van der Waals bonds have no higher lying antibonding state into which they may be excited by the electron beam. Thus, whereas even a simple molecule like CO adsorbed on a surface is rapidly decomposed by an electron beam, the Xe layer is hardly influenced except of course by the indirect effect of sample heating.

Figure 38 shows the basic result of Palmberg's experiment. The Xe Auger peak height, work function change, and LEED overlayer intensity are plotted as a function

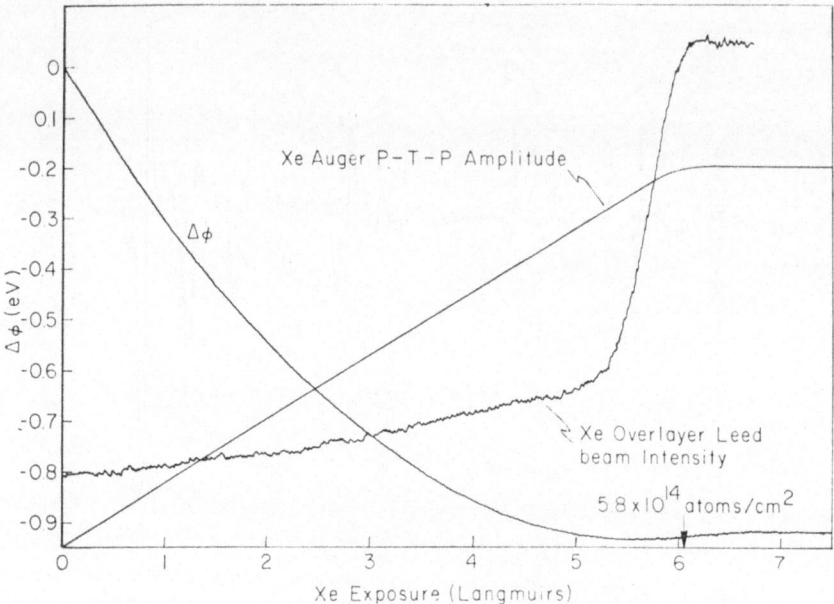

Fig. 38. Work function ($\triangle \Phi$), Xe – Auger peak-to-peak amplitude, and Xe overlayer LEED beam intensity versus Xe exposure at 77 K. (From [200].)

of exposure to Xe gas at 77 K. These results show several interesting features. The linear increase in the Xe Auger signal with exposure implies a constant near unity sticking probability up to $6L$ exposure followed by an abrupt drop to zero at the saturation. At saturation the LEED intensity arising from a hexagonal Xe overlayer is seen to rise abruptly to constant value. These observations lead to a model based on a random lattice gas of Xe atoms at coverages of less than 90% of saturation followed by an abrupt ordering as the remaining 10% are adsorbed. The use of Auger spectroscopy allows the specific determination of coverage (provided one calibration point is known as from LEED in this case) so that the $\Delta \varphi$-coverage relation is established. This data is shown to fit a topping model for simple dipole–dipole repulsion very well.

These experiments represent an elegant example of the power of using as many various experimental techniques as possible in the same system.

B. *Thin film deposition.* As another example of the utility of Auger spectroscopy in basic surface studies, consider the deposition of Cs on Ni as a means to lower the work function for doing band structure studies at low photon energies. The Cs was provided by a zeolite source consisting of a Pt strip coated with Cs-zeolite with an appropriate extractor grid. The source provides a beam of Cs^+ at a current of $100 \, nÅ \, cm^{-2}$ which forms a monolayer on the sample in 5–10 min. The system pressure during the Cs^+ deposition did not rise above 7×10^{-10} torr. However, the Auger spectrum after deposition of a monolayer of Cs on an initially clean Ni surface revealed (in addition to Cs) large amounts of S, C, and Pd (Figure 39). The low system

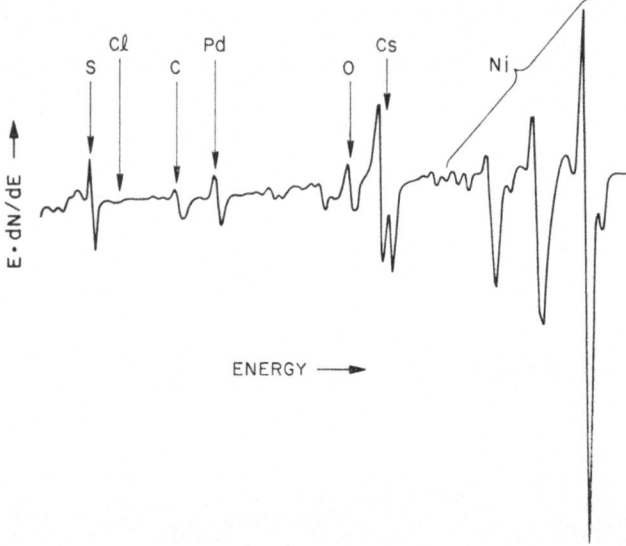

Fig. 39.   Auger spectrum of cesiated-Ni containing various impurities.

pressure during the deposition implies that these impurities must have come by direct transport from the source to the target. Thus, a low pressure during deposition is not a sufficient condition for obtaining a clean film. The Pd probably came from evaporation from the commercial grade Pt foil used for the source. After repeated use of source the Pd impurity flux decreased but reappeared when a new piece of Pt foil was installed.

The experiments conducted on cesiated nickel revealed that the presence of sulphur caused an anomolous peak to appear in the photoemitted energy distribution which was attributed to a S–Cs complex.

C. *Catalysis studies.* Bonzel has recently reported some studies [323] of the oxidation of sulfur on a Cu [110] surface to produce SO_2 in the gas phase. Using a combination of LEED, mass spectroscopy and Auger spectroscopy he was able to make detailed

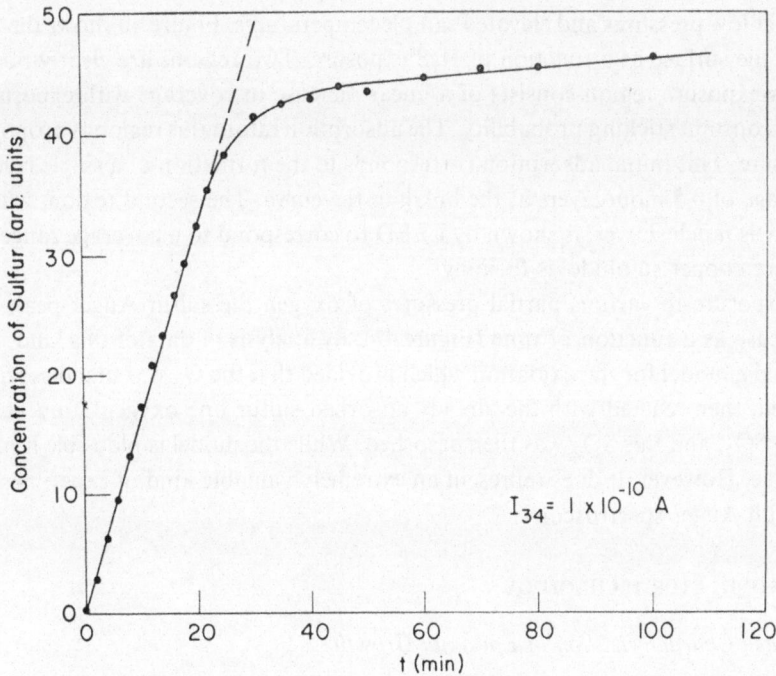

Fig. 40. Concentration of adsorbed sulfur as a function of time at room temperature and constant partial pressure of H_2S. (From Bonzel, *Surf. Sci.* **27**, 387 (1971).)

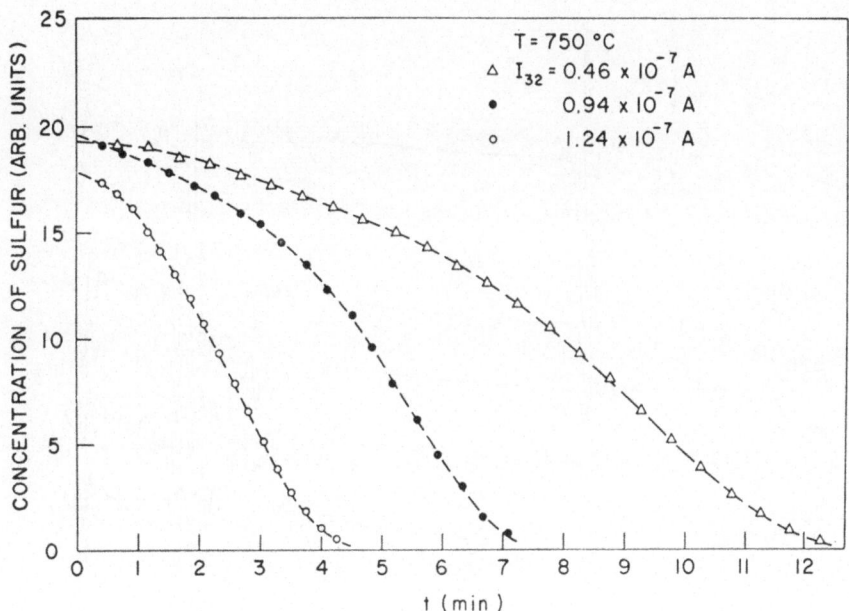

Fig. 41. Concentration of adsorbed sulfur as a function of reaction time for constant temperature and partial pressure of O_2, $T = 750\,°C$. I_{32} is a direct measure of the partial pressure of oxygen. (From Bonzel, *ibid.*)

measurements of the adsorption of S (from H_2S) and its subsequent removal be oxygen at low pressures and elevated sample temperatures. Figure 40 shows the uptake of S on the surface as a function of H_2S exposure. Two regions are clearly observed. The low exposure region consists of a linear increase in coverage with exposure and hence a constant sticking probability. The adsorption rate in this region is proportional to pressure. This initial adsorption corresponds to the formation of a single layer with a coverage of 0.5 monolayers at the break in the curve. The second region, where the S uptake is much slower, is shown by LEED to correspond to a coverage range where a 'surface copper sulphide' is forming.

On exposure to various partial pressures of oxygen the sulfur Auger peak is seen to decrease as a function of time (Figure 41). By analysis of data of this kind, Bonzel suggested a model for the oxidation which provided that the O_2 was first dissociatively adsorbed, then reacted with the already adsorbed sulfur one oxygen atom at a time to form SO_2, and this SO_2 was then desorbed. While the model is plausible it may not be unique. However, it does represent an extremely valuable kind of experiment to be done with Auger spectroscopy.

B. SEMICONDUCTOR TECHNOLOGY

1. *Surface Characterization in Epitaxial Growth*

The feasibility of growing GaAs semiconductor layer structures from molecular beams

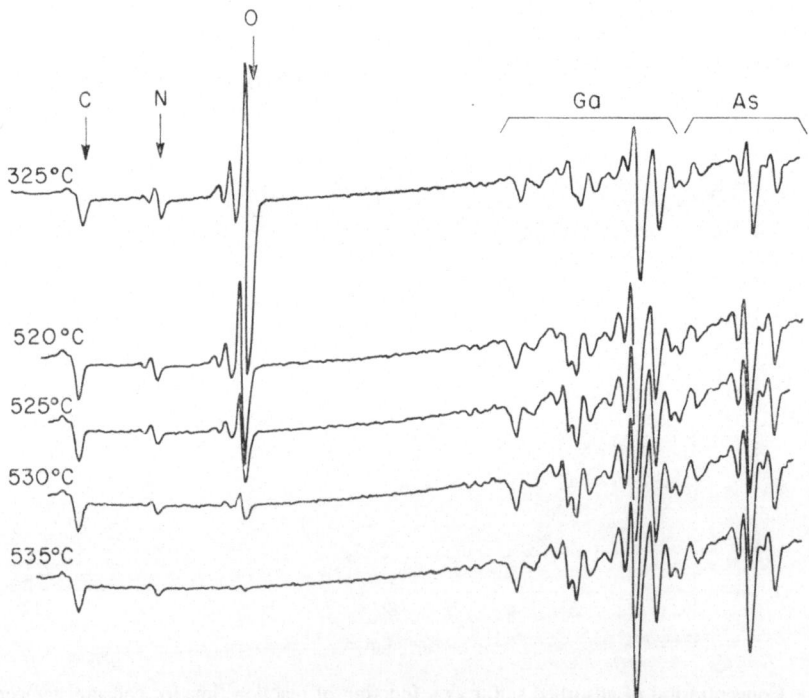

Fig. 42.   Auger spectra of passivated GaAs during heating in vacuum.

of constituent Ga and As atoms has been amply demonstrated. One of the more important considerations in achieving ideal growth is the production of a clean initial surface. It is desirable to do this without sputtering if possible. It has been found that by passivating the clean surface with H_2O after preparation in a suitable etchant, the as introduced sample surface contains an acceptably small amount of carbon with no other impurities (besides the oxide). When such a surface is heated in vacuum the Auger spectrum changes as indicated in Figure 42. The oxygen peak disappears very abruptly at around 525 °C and the weak N peak at 535 °C. The oxygen leaves the surface as Ga_2O as determined by mass spectrometer observation. However the carbon remains unchanged up to the decomposition (in vacuum) temperature of 627 °C. Thus, Auger electron spectroscopy has helped to determine the most favorable preparation conditions for the sample such that it may be cleaned by heat alone.

2. *Composition (Depth) Profiling*

Figure 43 shows the experimental arrangement used extensively by Palmberg [324]

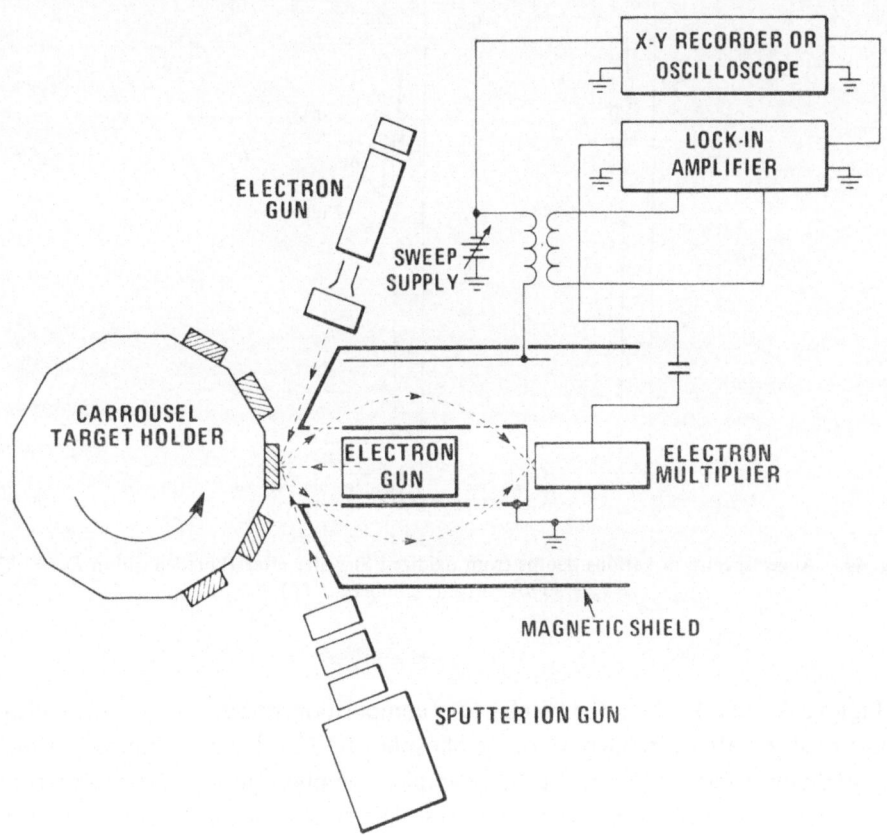

Fig. 43. Schematic arrangement for use of the cylindrical-mirror analyser as an Auger electron spectrometer. (From Palmberg in *Electron Spectroscopy* (ed. by D. A. Shirley), North Holland 1972.)

for determining the composition of thin films as a function of depth. The technique
has been applied extensively to semiconductor problems and analysis.

Profiles are obtained by sputtering the sample with a 1 mm^2 beam of Ar$^+$ while
simultaneously recording the Auger spectrum from a small area 0.01 mm^2 in the
center of the sputtered crater. This insures a uniform depth across the analyzed region.
Sputter etch rates greater than 100 Å min^{-1} are easily achieved.

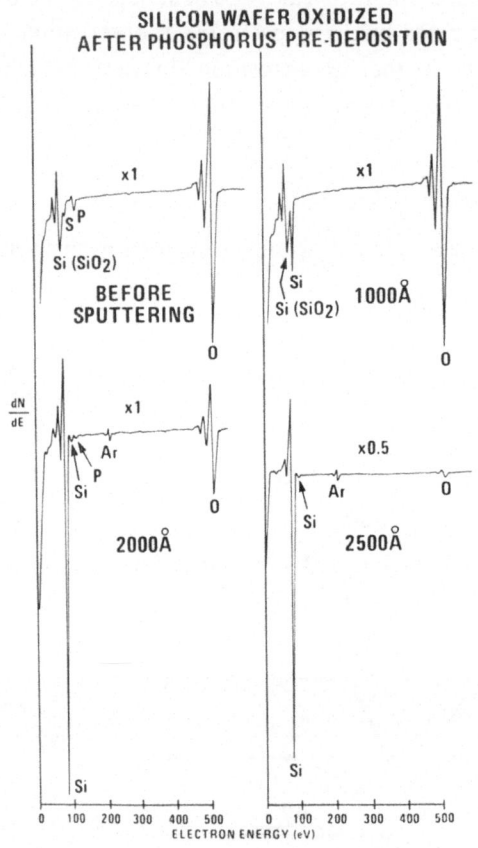

Fig. 44. Auger spectra at various depths from oxidized Si wafer after P predeposition $E_p = 3$ kV;
$I_p = 50$ μÅ. (From Palmberg, *ibid.*)

Figures 44 and 45 show the results of a composition profile analysis of a silicon
wafer oxidized after a predeposition of phosphorus. The depth resolution is several
hundred angstroms for this particular example. The phosphorus is seen to segregate
at the oxide-air and oxide-silicon interfaces. The silicon–silicon dioxide interface is
not only characterized by a change in oxygen concentration, but also a strong chemical
shift of the silicon peak.

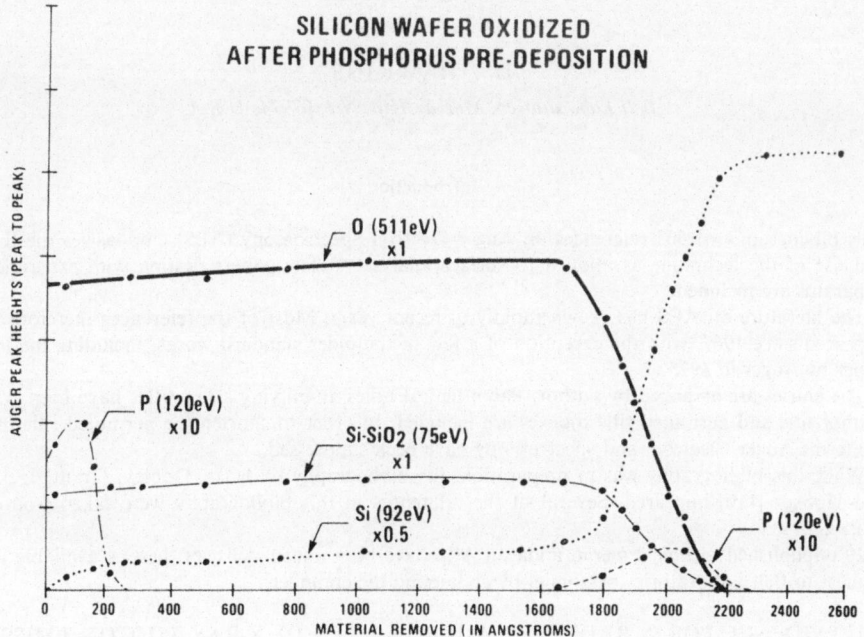

Fig. 45. Composition profiles for Si, P, and O on oxidized Si wafer after *P* predeposition. (From Palmberg, *ibid.*)

BIBLIOGRAPHY

D. T. HAWKINS
Bell Laboratories, Murray Hill, N.J. 07974, U.S.A.

Introduction

This bibliography of 305 references on Auger Electron Spectroscopy (AES) emphasizes the theory and use of the technique as applied to surface analysis. A few papers dealing with experimental apparatus are included.

The literature on AES has grown rapidly in recent years. Most of the references therefore have appeared since 1967, with the exception of a few of the older standard works, including the initial paper by Auger in 1925.

The entries are arranged by author. Parenthetical notes amplifying a few titles have been added. Author-title and permuted title indexes are included. In order to shorten the permuted title index, the terms Auger, electron, and spectroscopy have been suppressed.

While this bibliography was in preparation, the bibliography by Haas, Dooley, Grant, Jackson, and Hooker [110] appeared. Several of the references in this bibliography were taken from that source.

20 unpublished technical memoranda on AES have been issued. A list of these is available upon request to Bell Laboratories personnel by contacting the compiler.

1. EVIDENCE FOR A RADIATIVE AUGER EFFECT IN X-RAY PHOTON EMISSION
   ABERG T
   PHYS REV LETT 22: 1346–8 (1969)
2. AUGER EFFECT IN THE MULTIPLE IONIZATION OF MANGANESE AND
   CADMIUM BY ELECTRON IMPACT.
   ABOUAF R
   J PHYS (PARIS) 31: 277–83 (1970) (IN FRENCH)
3. MULTIPLE IONIZATION AND AUGER TRANSITIONS IN INDIUM AND SILVER
   BY ELECTRON IMPACT.
   ABOUAF R
   J PHYS (PARIS) 32: 603–8 (1971) (IN FRENCH)
4. HIGH RESOLUTION MNN AUGER SPECTRA OF SILVER, CADMIUM, INDIUM,
   ANTIMONY, TELLURIUM AND IODINE.
   AKSELA S
   Z PHYS 244: 268–74 (1971)
5. HIGH RESOLUTION LMM AUGER SPECTRA OF LOW ENERGY FROM SOLID
   SURFACES.
   AKSELA S + PESSA M + KARRAS M
   Z PHYS 237: 381–7 (1970)
6. KLL AUGER SPECTRUM OF FLUORINE. (IN FLUORIDE SALTS)
   ALBRIDGE RG + HAMRIN K + JOHANSSON G + FAHLMAN A
   Z PHYS 209: 419–27 (1968) (IN GERMAN)
7. BAND STRUCTURE OF SILICON BY CHARACTERISTIC AUGER SPECTRUM
   ANALYSIS.
   AMELIO GF
   SURFACE SCI 22: 301–18 (1970)
8. AUGER ELECTRON SPECTROSCOPY AND SECONDARY EMISSION IN SEMI-
   CONDUCTORS. (SILICON, GERMANIUM, GRAPHITE, ELECTRON DIFFUSION
   THEORY)
   AMELIO, GF
   GEORGIA INST TECHNOL, THESIS 1968
9. TRUE SECONDARY ELECTRON ENERGY DISTRIBUTIONS.

AMELIO GF + SCHEIBNER EJ
P 11-1 To 11-15 OF STRUCTURE AND CHEMISTRY OF SOLID SURFACES
SOMORJAI GA (ED.) NY WILEY 1969. (541.375/B51) 140258

10. AUGER SPECTROSCOPY OF GRAPHITE SINGLE CRYSTALS WITH LOW-
ENERGY ELECTRONS.
AMELIO GF + SCHEIBNER EJ
SURFACE SCI 11: 242-54 (1968)

11. CLEAN TELLURIUM SURFACES STUDIED BY LEED.
ANDERSSON S + ANDERSSON D + MARKLUND I
SURFACE SCI 12: 284-98 (1968)

12. DEVELOPMENT AND APPLICATION OF AN "AUGER ELECTRON SPECTRO-
METER" FOR HIGH TEMPERATURE ADSORPTION STUDIES.
ARAMATI VS
MASS INST TECHNOL, MS THESIS 1971

13. ON THE COMPOUND PHOTOELECTRIC EFFECT.
AUGER P
J PHYS RADIUM 6: 205-8 (1925)

14. KLL AUGER ELECTRON SPECTRA OF THULIUM AND LUTETIUM.
BABENKOV MI + BOBYKIN BV
BULL ACAD SCI USSR, PHYS SER 32: 1840-3 (1968) (IN ENGLISH)

15. KLL AUGER ELECTRONS OF CADMIUM AND ANTIMONY.
BABENKOV MI + BOBYKIN BV + NORGORODOV AF + TASKARIN BT
BULL ACAD SCI URRS, PHYS SER 33: 1205-9 (1969) (IN ENGLISH)

16. NATURE OF ANNEALED SEMICONDUCTOR SURFACES. (AUGER)
BAUER E
PHYS LETT A26: 530-1 (1968)

17. COMMENTS ON "ANGULAR DEPENDENCIES IN ELECTRON-EXCITED AUGER
EMISSION".
BISHOP HE + RIVIERE JC
SURFACE SCI 17: 446-7 (1969)

18. CHARACTERISTIC IONIZATION LOSSES OBSERVED IN AUGER EMISSION
SPECTROSCOPY.
BISHOP HE + RIVIERE JC
APPL PHYS LETT 16: 21-3 (1970)

19. AUGER SPECTROSCOPY OF TITANIUM.
BISHOP HE + RIVIERE JC
SURFACE SCI 24: 1-17 (1971)

20. SURFACE ABSORPTION BY POLYPEPTIDE FILMS EXAMINED BY AUGER
EMISSION SPECTROSCOPY.
BISHOP HE + RIVIERE JC
J COLLOID INTERFACE SCI 33: 272-7 (1970)

21. SEGREGATION OF GOLD TO SILICON (111) SURFACE OBSERVED BY AUGER
EMISSION SPECTROSCOPY AND BY LEED.
BISHOP HE + RIVIERE JC
J PHYS D 2: 1635-42 (1969)

22. ESTIMATES OF EFFICIENCIES OF PRODUCTION AND DETECTION OF
ELECTRON-EXCITED AUGER EMISSION.
BISHOP HE + RIVIERE JC
J APPL PHYS 40: 1740-4 (1969)

23. SURFACE SEGREGATION IN BORON DOPED IRON OBSERVED BY AUGER
EMISSION SPECTROSCOPY.
BISHOP HE + RIVIERE JC
ACTA MET 18: 813-17 (1970)

24. AUGER SPECTROSCOPY OF SILICON.
BISHOP HE + RIVIERE JC + TAYLOR NJ
SURFACE SCI 17: 462-5 (1969)

25. SEGREGATION OF CARBON TO (100) SURFACE OF NICKEL.

BLAKELY JM + KIM JS + POTTER HC
J APPL PHYS 41: 2693–7 (1970)

26. AUGER ELECTRON SPECTROSCOPY OF A SULFUR- OXYGEN SURFACE
    REACTION ON A COPPER (110) CRYSTAL.
    BONZEL HP
    SURFACE SCI 27(3): 387–410 (1971)

27. DETECTION OF IMPURITIES ON COPPER SURFACES BY LEED AUGER
    ELECTRON SPECTROSCOPY AND THEIR INFLUENCE ON SURFACE SELF
    DIFFUSION OF COPPER.
    BONZEL HP + GJOSTEIN NA
    J METALS 21(3): A88 (1969)

28. ELECTRON SPECTROSCOPY OF SURFACES.
    BRUNDLE CR
    SURFACE SCI 27(3): 681–5 (1971)

29. THE AUGER EFFECT AND OTHER RADIATIONLESS TRANSITIONS.
    BURHOP EHS
    CAMBRIDGE UNIVERSITY PRESS, 1952. (539.76/B95) 110732 188PP

30. AUGER ELECTRON STUDIES OF SURFACES: URANIUM DIOXIDE, URANIUM,
    GRAPHITE, 300 SERIES STAINLESS STEEL, AND NIOBIUM.
    CAMPBELL BD
    LOS ALAMOS SCI LABS 1969. REPT NO LA–4010

31. AUGER ELECTRON STUDIES OF URANIUM DIOXIDE SURFACES.
    CAMPBELL BD + ELLIS WP
    J CHEM PHYS 52: 330 3–4 (1970)

32. IDENTIFICATION OF HIGH ENERGY LINES IN THE KLL AUGER SPECTRUM
    OF NITROGEN.
    CARLSON TA + MODDEMAN WE + PULLEN BP + KRAUSE MO
    CHEM PHYS LETT 5: 390–2 (1970)

33. HIGH RESOLUTION ELECTRON SPECTROMETER FOR PHOTOELECTRON AND
    AUGER ELECTRON STUDIES.
    CARLSON TA + PULLEN BP + MODDEMAN WE + KRAUSE MO + WARD FW
    ACS ANN MEET 1969, 158TH PROC ABSTR PAPERS NO PHYS-179

34. STUDY OF MUSCOVITE AND SILICON BY AUGER ELECTRON SPECTRO-
    SCOPY.
    CARRIERE B + DEVILLE JP + GOLDZSTAUB S
    COMPT REND 271: 796–8 (1970) (IN FRENCH)

35. AUGER ELECTRON SPECTROSCOPY STUDY OF SOME OXYGENATED SILICON
    COMPOUNDS.
    CARRIERE B + DEVILLE JP + GOLDZSTAUB S
    COMPT REND B272: 951–4 (1971) (IN FRENCH)

36. CONTAMINANTS ON CHEMICALLY ETCHED SILICON SURFACES LEED-
    AUGER METHOD.
    CHANG CC
    SURFACE SCI 23: 283–98 (1970)

37. AUGER ELECTRON SPECTROSCOPY FOR CHEMICAL ANALYSIS.
    CHANG CC
    ACS ANN MEET 1970, 160TH PROC ABSTR PAPERS NO PHYS-7

38. AUGER ELECTRON SPECTROSCOPY. (REVIEW ARTICLE)
    CHANG CC
    SURFACE SCI 25: 53–79 (1971)

39. SILICON ON SAPPHIRE EPITAXY BY VACUUM SUBLIMATION – LEED-
    AUGER STUDIES AND ELECTRONIC PROPERTIES OF FILMS.
    CHANG CC
    J VACUUM SCI TECHNOL 8: 500–11 (1971).

40. CARBON CONTAMINATION OF SILICON (111) SURFACES.
    CHARIG JM + SKINNER DK
    SURFACE SCI 15: 277–85 (1969)

41. AUGER ELECTRON SPECTROSCOPY OF NICKEL DEPOSITS ON THE SILICON (111) SURFACE.
    CHARIG JM + SKINNER DK
    SURFACE SCI 19: 283–90 (1970)
42. LEED, AUGER, AND WORK FUNCTION STUDIES OF CLEAN AND SODIUM-COVERED SURFACES OF GALLIUM ARSENIDE.
    CHEN JM
    SURFACE SCI 25: 305–14 (1971)
43. DITUNGSTEN CARBIDE OVERLAYER ON TUNGSTEN (112).
    CHEN JM + PAPGEORGOPOULOS CA
    SURFACE SCI 20: 195–200 (1970)
44. THEORETICAL L(2) – AND L(3) – SUBSHELL FLUORESCENCE YIELDS AND L(2)-L(3)X COSTER-KRONIG TRANSITION PROBABILITIES.
    CHEN MH + CRASEMANN B + KOSTROUN VO
    PHYS REV A 4: 1–7 (1971)
45. LEED AND SURFACE POTENTIAL STUDY OF CARBON MONOXIDE AND XENON ADSORBED ON COPPER (100).
    CHESTERS MA + PRITCHARD J
    SURFACE SCI 28: 460–8 (1971)
46. CHARACTERISTIC ENERGIES IN SECONDARY ELECTRON SPECTRA FROM SILICON (111) SURFACES.
    CHUNG MF + JENKINS LH
    SURFACE SCI 26: 649–63 (1971)
47. LLM VERSUS LMM AUGER TRANSITIONS FOR THE LIGHT ELEMENTS.
    CHUNG MF + JENKINS LH
    SURFACE SCI 28: 637–44 (1971)
48. AUGER ELECTRON SPECTROSCOPY OF THE OUTER SHELL ELECTRONS.
    CHUNG MF + JENKINS LH
    SURFACE SCI 22: 479–85 (1970)
49. AUGER ELECTRON EMISSION SPECTRA: A SIMPLE TREATMENT OF ENERGIES AND INTENSITIES.
    CLARKE TA + MASON R + RANDACCIO L + THOMAS JM
    J CHEM SOC LONDON, PART A, INORG, PHYS, THEOR 9: 1156–60 (1971)
50. APPLICATION OF AUGER ELECTRON SPECTROSCOPY TO REACTOR MATERIALS RESEARCH.
    CLAUSING RE
    J VACUUM SCI TECHNOL 7: 5124 (1970)
51. KLL AUGER SPECTRUM OF CHLORINE. (IN CARBON CHLORINE (4))
    CLEFF B + MEHLHORN W
    Z PHYS 219: 311–24 (1969) (IN GERMAN)
52. CHEMICAL SHIFTS IN THE AUGER SPECTRA FROM OXIDIZED CHROMIUM AND VANADIUM.
    COAD JP
    PHYS LETT A35: 185–6 (1971)
53. LMM AUGER SPECTRA OF SOME TRANSITION METALS OF THE FIRST SERIES.
    COAD JP
    Z PHYS 244: 19–30 (1971)
54. ELECTRON-BEAM ASSISTED ADSORPTION ON SILICON (III) SURFACE.
    COAD JP + BISHOP HE + RIVIERE JC
    SURFACE SCI 21: 253–64 (1970)
55. COMBINATION OF AUGER SPECTROSCOPY AND CHARACTERISTIC LOSS SPECTROSCOPY FOR THE ELEMENTS VANADIUM TO COBALT.
    COAD JP + RIVIERE JC
    PHYS STATUS SOLIDI A 7: 571–5 (1971)
56. AUGER SPECTROSCOPY OF CARBON ON NICKEL.
    COAD JP + RIVIERE JC

SURFACE SCI 25: 609–24 (1971)

57. STUDY OF MULBERRY SURFACES BY AUGER AND IEE SPECTROSCOPY.
COLMENARES CA
UNIV CALIF, LAWRENCE RADIATION LAB 1970. REPT UCID-15599

58. AUGER ELECTRON SPECTROSCOPY OF GRAPHITE FIBER SURFACES.
CONNELL GL
NATURE 230: 377 (1971)

59. AUGER ELECTRON SPECTROSCOPY OF LUNAR MATERIAL.
CONNELL GL + GUPTA YP
2ND NASA LUNAR SCIENCE CONF (1971) PROC

60. AUGER ELECTRON SPECTROSCOPY.
CONNELL GL + GUPTA YP
MATER RES STAND 11(1): 8–13 (1971)

61. AUGER AND COSTER-KRONIG TRANSITION PROBABILITIES TO THE
ATOMIC 2S STATE AND THEORETICAL L(1) FLUORESCENCE YIELDS.
CRASEMANN B · CHEN MH + KOSTROUN VO
PHYS REV A 4: 2161–3 (1971)

62. COMMENTS ON "AUGER SPECTROSCOPY OF CARBON ON NICKEL" BY
JP COAD AND JC RIVIERE.
DALMAI-IMELIK G · BERTOLINI JC + ROUSSEAU J
SURFACE SCI 27: 379 (1971)

63. ADSORBATE EFFECTS IN ELECTRON EJECTION BY RARE GAS METASTABLE
ATOMS.
DELCHAR TA + MACLENNAN DA + LANDERS AM
J CHEM PHYS 50: 1779–87 (1969)

64. SPECTROSCOPY OF AUGER ELECTRONS.
REVIEW, 24 REFERENCES
DEVILLE JP
REV PHYS APPL, SUPPL J PHYS 3: 351–5 (1968) (IN FRENCH)

65. MUSCOVITE CLEAVAGE STUDIED BY AUGER ELECTRON SPECTROSCOPY.
DEVILLE JP + GOLDZSTAUB S
COMPT REND B268: 629–30 (1969) (IN FRENCH)

66. AUGER ELECTRON SPECTROSCOPY STUDIES OF REFRACTORY METAL
SURFACES.
DOOLEY GJ + GRANT JT + HAAS TW
BULL AMER PHYS SOC 14: 948 (1969)

67. CHEMICAL EFFECTS ON KLL AUGER ELECTRON SPECTRUM FROM
OXYGEN.
DOOLEY GJ + GRANT JT + HAAS TW
BULL AMER PHYS SOC 15: 1507 (1970)

68. BEHAVIOR OF REFRACTORY METAL SURFACES IN ULTRAHIGH VACUUM
AS OBSERVED BY LEED AND AUGER ELECTRON SPECTROSCOPY.
DOOLEY GJ + HAAS TW
J VACUUM SCI TECHNOL 7: S90–100 (1970)

69. AUGER ELECTRON SPECTROSCOPY: METALLURGICAL APPLICATIONS.
DOOLEY GJ + HAAS TW
J METALS 22(11): 17–24 (1970)

70. LEED STUDY OF MOLYBDENUM (100) SURFACE.
DOOLEY GJ + HAAS TW
BULL AMER PHYS SOC 14: 270 (1969)

71. COMPARATIVE STUDY OF LEED INTENSITY DATA.
DOOLEY GJ + HAAS TW
BULL AMER PHYS SOC 15: 632 (1970)

72. CHEMISORPTION ON SINGLE CRYSTAL MOLYBDENUM (112) SURFACES.
DOOLEY GJ + HAAS TW
J VACUUM SCI TECHNOL 7: 49–52 (1970)

73. SOME PROPERTIES OF THE RHENIUM (0001) SURFACE.

DOOLEY GJ + HAAS TW
SURFACE SCI 19: 1–8 (1970)

74. FURTHER STUDIES OF GAS ADSORPTION ON MOLYBDENUM (100) SURFACE.
DOOLEY GJ + HAAS TW
J CHEM PHYS 52: 461–2 (1970)

75. AUGER ELECTRON ANALYSIS OF ELECTROPOLISHED HIGH-PURITY
ALUMINIUM.
DUNN CG + HARRIS LA
J ELECTROCHEM SOC 117: 81–2 (1970)

76. SECONDARY-ELECTRON ENERGY DISTRIBUTION STUDIES OF URANIUM
DIOXIDE SURFACES.
ELLIS WP + CAMPBELL BD
J APPL PHYS 41: 1858–61 (1970)

77. ADSORPTION ON SINGLE CRYSTAL SURFACES OF COPPER-NICKEL
ALLOYS. I.
ERTL G + KUPPERS J
SURFACE SCI 24: 104–24 (1971) (IN GERMAN)

78. THE INTERACTION BETWEEN CHLORINE AND THE (100) SURFACE OF GOLD.
FEDAK DG + FLORIC JV + ROBERTSON WD
PP 74-1 TO 74-18 OF STRUCTURE AND CHEMISTRY OF SOLID SURFACES,
SOMORJAI GA (ED). NY, WILEY, 1969. (541.375/B51) 140258

79. AUGER SPECTROSCOPY AND LEED STUDY OF EQUILIBRIUM SURFACE
SEGREGATION IN COPPER-ALUMINIUM ALLOYS.
FERRANTE J
ACTA MET 19: 743–8 (1971)

80. PHASE TRANSFORMATIONS OF SILICON (111) SURFACE.
FLORIO VJ + ROBERTSON WD
SURFACE SCI 22: 459–64 (1970)

81. CHLORINE REACTIONS ON THE SILICON (III) SURFACE.
FLORIO JV + ROBERTSON WD
SURFACE SCI 18: 398–427 (1970)

82. CHEMICAL EFFECTS ON THE AUGER ELECTRON SPECTRA OF BERYLLIUM.
FORTNER RJ + MUSKET RG
SURFACE SCI 28: 339–43 (1971)

83. SIMPLE MODEL FOR DEPENDENCE OF AUGER INTENSITIES ON SPECIMEN
THICKNESS.
GALLON TE
SURFACE SCI 17: 486–9 (1969)

84. (100) SURFACES OF ALKALI HALIDES PART-1: AIR AND VACUUM CLEANED
SURFACES.
GALLON TE + HIGGINBOTHAM IG + PRUTTON M + TOKUTAKA H
SURFACE SCI 21: 224–32 (1970)

85. GROWTH OF SILVER ON POTASSIUM CHLORIDE OBSERVED BY LEED AND
AUGER EMISSION SPECTROSCOPY.
GALLON TE + HIGGINBOTHAM IG + PURTTON M + TOKUTAKA H
THIN SOLID FILMS 2: 369–73 (1968)

86. IMPROVED APPARATUS FOR MEASUREMENT OF AUGER ELECTRON
SPECTRA.
GALLON TE + HIGGINBOTHAM IG + PRUTTON M
J PHYS E 2: 894–6 (1969)

87. LOW ENERGY AUGER EMISSION FROM LITHIUM FLUORIDE.
GALLON TE + MATTHEW JAD
PHYS STATUS SOLIDI 41: 343–51 (1970)

88. IONIZATION SPECTROSCOPY OF CONTAMINATED METAL SURFACES.
GERLACH RL
J VACUUM SCI TECHNOL 8: 599–604 (1971)

89. ELECTRON BINDING ENERGIES OF BARIUM FROM THE SECONDARY

ELECTRON YIELD SPECTRUM.
GERLACH RL
SURFACE SCI 28: 648–50 (1971)

90. DIFFERENTIAL CROSS SECTIONS FOR K-SHELL IONIZATION OF SURFACE
ATOMS BY ELECTRON IMPACT.
GERLACH RL + DUCHARME AR
PHYS REV LETT 27: 290–2 (1971)

91. IONIZATION SPECTROSCOPY OF SURFACES.
GERLACH RL + HOUSTON JE + PARK RL
APPL PHYS LETT 16: 179–81 (1970)

92. IONIZATION SPECTROMETER FOR ELEMENTAL ANALYSIS OF SURFACES.
GERLACH RL + TIPPLING DW
REV SCI INSTRUM 42: 151–4 (1971)

93. MODERN TECHNIQUES FOR SURFACE STUDIES.
GJOSTEIN NA + BONZEL HP + CHAUKA NG
RES DEVELOP 21(10): 24–30 (1970)

94. DETECTION OF IMPURITIES ON A SILICON (111) SURFACE BY AUGER-
LEED ANALYSIS.
GOFF RF + JACOBSON RL
BULL AMER PHYS SOC 13: 944 (1968)

95. SENSITIVITY VARIATIONS IN BAYARD- ALPERT GAUGES CAUSED BY
AUGER EMISSION AT COLLECTOR.
COPALARAMAN CP + ARMSTRONG RA + REDHEAD PA
J VACUUM SCI TECHNOL 6: 910 (1969)

96. LEED STUDY OF IRIDIUM (100) SURFACE.
GRANT JT
SURFACE SCI 18: 228–38 (1969)

97. STUDY OF IRIDIUM (100) SURFACE USING LEED AND AUGER ELECTRON
SPECTROSCOPY.
GRANT JT
BULL AMER PHYS SOC 14: 794 (1969)

98. STUDIES ON THE IRIDIUM (111) SURFACE USING LEED AND AUGER
ELECTRON SPECTROSCOPY.
GRANT JT
SURFACE SCI 25: 451–6 (1971)

99. STRUCTURE OF THE PLATINUM (100) SURFACE.
GRANT JT + HAAS TW
SURFACE SCI 18: 457–61 (1969)

100. AUGER STUDIES OF (111) SILICON SURFACES.
(ABSTRACT)
GRANT JT + HAAS TW
J VACUUM SCI TECHNOL 6: 903 (1969)

101. AUGER ELECTRON SPECTROSCOPY STUDIES OF CARBON OVERLAYERS ON
METAL SURFACES.
GRANT JT + HAAS TW
SURFACE SCI 24: 332–4 (1971)

102. AUGER ELECTRON SPECTROSCOPY OF SILICON.
GRANT JT + HAAS TW
SURFACE SCI 23: 347–62 (1970)

103. LEED STUDY OF PLATINUM (100) SURFACE.
GRANT JT + HAAS TW
BULL AMER PHYS SOC 14: 948 (1969)

104. IDENTIFICATION OF THE FORM OF CARBON AT A SILICON (100) SURFACE
USING AUGER ELECTRON SPECTROSCOPY.
GRANT JT + HAAS TW
PHYS LETT A33: 386–7 (1970)

105. COMBINED LEED AND AUGER ELECTRON SPECTROSCOPY STUDIES OF

SILICON, GERMANIUM, GALLIUM ARSENIDE, AND INDIUM ANTIMONIDE SURFACES.
GRANT JT + HAAS TW
J VACUUM SCI TECHNOL 8(1): 94–7 (1971)

106. STUDY OF RUTHENIUM (0001) AND RHODIUM (111) SURFACES USING LEED AND AUGER ELECTRON SPECTROSCOPY.
GRANT JT + HAAS TW
SURFACE SCI 21: 76–85 (1970)

107. STUDY OF INDIUM ARSENIDE (111) AND (–1–1–1) SURFACES USING LEED AND AUGER ELECTRON SPECTROSCOPY.
GRANT JT + HAAS TW
SURFACE SCI 26(2): 669–76 (1971)

108. AUGER STUDIES OF CLEAVED SILICON (111) SURFACES.
GRANT JT + HAAS TW
J VACUUM SCI TECHNOL 7: 77–9 (1970)

109. NATURE OF SILICON (111) SURFACES.
GRANT JT + HAAS TW
APPL PHYS LETT 15: 140–1 (1969)

110. A BIBLIOGRAPHY OF LOW ENERGY ELECTRON DIFFRACTION AND AUGER ELECTRON SPECTROSCOPY.
(766 REFERENCES)
HAAS TW + DOOLEY GJ + GRANT JT + JACKSON AG + HOOKER MP
PROGRESS IN SURFACE SCIENCE, VOL. 1(2). PERGAMON, 1971

111. CHEMICAL SHIFTS IN AUGER ELECTRON SPECTROSCOPY FROM INITIAL OXIDATION OF TANTALUM (110).
HAAS TW + GRANT JT
PHYS LETT A30: 272 (1969)

112. CHEMICAL EFFECTS ON THE KLL AUGER ELECTRON SPECTRUM FROM SURFACE CARBON.
HAAS TW + GRANT JT
APPL PHYS LETT 16: 172–3 (1970)

113. CHEMICAL SHIFTS IN AUGER ELECTRON SPECTROSCOPY.
HAAS TW + GRANT JT + DOOLEY GJ
BULL AMER PHYS SOC 14: 948 (1969)

114. AUGER ELECTRON SPECTROSCOPY OF TRANSITION METALS.
HAAS TW + GRANT JT + DOOLEY GJ
PHYS REV B 1: 1449–59 (1970)

115. AUGER ELECTRON SPECTROSCOPY OF SOME REFRACTORY METALS.
HAAS TW + GRANT JT + DOOLEY GJ
J VACUUM SCI TECHNOL 6: 903 (1969)

116. SOME PROBLEMS IN ANALYSIS OF AUGER ELECTRON SPECTRA.
HAAS TW + GRANT JT + DOOLEY GJ
J VACUUM SCI TECHNOL 7: 43–5 (1970)

117. SOFT X-RAY APPEARANCE POTENTIAL SPECTROSCOPY IN A DISPLAY LEED SYSTEM.
HAAS TW + THOMAS S + DOOLEY GJ
SURFACE SCI 28: 645–7 (1971)

118. ANGULAR DEPENDENCIES OF ELECTRON-EXCITED AUGER EMISSION.
HARRIS LA
SURFACE SCI 15: 77–93 (1969)

119. ANALYSIS OF MATERIALS BY ELECTRON-EXCITED AUGER ELECTRONS.
HARRIS LA
J APPL PHYS 39: 1419–27 (1968)

120. CARBON EVAPORATION FROM A THORIUM DISPENSER CATHODE OB-SERVED BY AUGER ELECTRON EMISSION.
HARRIS LA
J APPL PHYS 39: 4862 (1968)

121. AUGER ELECTRON SPECTROSCOPY FOR SURFACE ANALYSIS.
     HARRIS LA
     J METALS 20(12): A 16 (1968)
122. AUGER ELECTRON EMISSION ANALYSIS.
      (GOOD "LAYMAN'S" INTRODUCTORY ARTICLE.)
     HARRIS LA
     ANAL CHEM A40(14): 24–34 (1968)
123. SECONDARY ELECTRON SPECTROSCOPY.
     HARRIS LA
     IND RES 10: 52 (1968)
124. REPLY TO COMMENTS OF HE BISHOP AND JC RIVIERE ON "ANGULAR
     DEPENDENCIES IN ELECTRON-EXCITED AUGER EMISSION."
     HARRIS LA
     SURFACE SCI 17: 448–9 (1969)
125. SOME OBSERVATIONS OF SURFACE SEGREGATION BY AUGER ELECTRON
     EMISSION.
     HARRIS LA
     J APPL PHYS 39: 1428–31 (1968)
126. SURFACE ANALYSIS BY AUGER ELECTRON SPECTROSCOPY.
     HARRIS LA
     J ELECTROCHEM SOC 115: C250 (1968)
127. SECONDARY ELECTRON SPECTROSCOPY. (SURFACE ANALYSIS OF MATE-
     RIALS BY ELECTRON-EXCITED AUGER ELECTRON EMISSION)
     HARRIS LA
     APPL SPECTROSC 22: 372 (1968)
128. AUGER ELECTRON EMISSION IN THE ENERGY SPECTRA OF SECONDARY
     ELECTRONS FROM MOLYBDENUM AND TUNGSTEN.
     HARROWER GA
     PHYS REV 102: 340–7 (1956)
129. IN-PROCESS CONTROL TECHNIQUES FOR COMPLEX SEMICONDUCTOR
     STRUCTURES. TASK-II. APPLICATION OF SECONDARY ELECTRON
     SPECTROSCOPY (AUGER ELECTRON ANALYSIS) TO SURFACE CONTROL
     PROGRAM I THE MANUFACTURING OF SILICON DEVICES.
     HARTMANN DI + HARRIS LA + AFFLECK JH
     GENERAL ELECTRIC CO. SEMICONDUCTOR PROCUCT DEPT.
     PART I   REPT NO AD 845596 (1968)
     PART II   REPT NO AD 845597 (1968)
     PART III REPT NO AD 847775 (1969)
     PARTIV   REPT NO AD 852179 (1969)
     PART V   REPT NO AD 856920 (1969)
     PART VI   REPT NO AD 861921 (1969)
130. METHOD FOR PACKET OF WAVES IN AUGER ELECTRON-ELECTRON PRO-
     CESSES.
     HAYMANN P
     COMPT REND B272: 1029–32 (1971)
131. EMISSION OF AUGER ELECTRONS BY ATOMS IN ATOMS IN A METALLIC
     TARGET SUBJECTED TO AN IONIC BOMBARDMENT.
     HENNEQUIN JF
     J PHYS (PARIS) 29: 1053–65 (1968) (IN FRENCH)
132. LOW ENERGY PHOTO- AUGER ELECTRON SPECTROSCOPY.
     HENKE BL
     APPL SPECTROSC 22:372 (1968)
133. AUGER EXCITATION BY INTERNAL SECONDARY ELECTRONS.
     HOUSTON JE + PARK RL
     APPL PHYS LETT 14: 358–60 (1969)
134. CROSS CORRELATION TECHNIQUES IN AUGER SPECTROSCOPY.
     HOUSTON JE · PARK RL

BULL AMER PHYS SOC 14: 793 (1969)

135. HIGH SENSITIVITY ELECTRON SPECTROMETER.
HUCITAL DA + RIDGEN JD
APPL PHYS LET 16: 348–51 (1970)

136. EPITAXY OF ULTRATHIN METAL FILMS ON BODY CENTERED CUBIC
SUBSTRATES USING LEED- AUGER TECHNIQUES. (ABSTRACT)
JACKSON AG

137. J VACUUM SCI TECHNOL 8: 23 (1971)
AUGER -LEED INVESTIGATION OF THE DEPOSITION OF ALUMINIUM ONTO
THE MOLYBDENUM (110) SURFACE.
JACKSON AG + HOOKER MP

138. SURFACE SCI 28: 373–94 (1971)
AUGER -LEED INVESTIGATION OF TIN ON MOLYBDENUM (100).
JACKSON AG + HOOKER MP

139. SURFACE SCI 27: 197–210 (1971)
AUGER ELECTRON EMISSION OF THIN CARBON FOILS IN REFLECTION AND
TRANSMISSION.
JACOBI K + HÖLZL J
SURFACE SCI 26: 54 (1971) (IN GERMAN)

140. AUGER AND OTHER CHARACTERISTIC ENERGIES IN SECONDARY.
ELECTRON SPECTRA FROM ALUMINIUM SURFACES.
JENKINS LH + CHUNG MF
SURFACE SCI 28: 409–22 (1971)

141. ENERGY SPECTRUM OF BACK-SCATTERED ELECTRONS AND CHARAC-
TERISTIC LOSS AND GAIN PHENOMENA OF COPPER (111).
JENKINS LH + CHUNG MF
SURFACE SCI 26: 151–64 (1971)

142. LEED AND AUGER INVESTIGATIONS OF COPPER (111) SURFACE.
JENKINS LH + CHUNG MF
SURFACE SCI 24: 125–39 (1971)

143. LOW-ENERGY ELECTRON DIFFRACTION, CIRCA 1968.
(REVIEW OF LEED AND AUGER.)
JONA F
HELV PHYS ACTA 41: 960–4 (1968)

144. AUGER SPECTROSCOPIC ANALYSIS OF BISMUTH SEGREGATED TO GRAIN
BOUNDARIES IN COPPER.
JOSHI A + STEIN DF
J INST METALS 99: 178–81 (1971)

145. INTERANGULAR BRITTLENESS STUDIES IN TUNGSTEN USING AUGER
SPECTROSCOPY.
JOSHI A + STEIN DF
MET TRANS 1: 2543–6 (1970)

146. SILICON- OXYGEN INTERACTIONS USING AUGER ELECTRON SPECTRO-
SCOPY.
JOYCE BA + NEAVE JH
SURFACE SCI 27(3): 499–515 (1971)

147. THE INFLUENCE OF SUBSTRATE CONDITIONS ON THE NUCLEATION AND
GROWTH OF EPITAXIAL SILICON FILMS.
JOYCE BA + NEAVE JH + WATTS BE
SURFACE SCI 15: 1–13 (1969)

148. THE ADSORPTION OF CARBON MONOXIDE ON COPPER (001) LEED AND
AUGER EMISSION STUDIES.
JOYNER RW + MCKEE CS + ROBERTS MW
SURFACE SCI 26: 303–9 (1971)

149. THE INTERACTION OF HYDROGEN SULFIDE WITH COPPER (001).
JOYNER RW + MCKEE CS + ROBERTS MW
SURFACE SCI 27: 279–85 (1971)

150. ATOMIC RADIATION TRANSITION PROBABILITIES TO THE 1S STATE AND THEORETICAL K- SHELL FLUORESCENCE YIELDS.
KOSTROUN VO + CHEN MH + CRASEMANN B
PHYS REV A 3: 533–45 (1971)

151. OBSERVATIONS OF BETA- SILICON CARBIDE FORMATION ON RECON- STRUCTED SILICON SURFACES.
KRAUSE MO
PHYS STATUS SOLIDI A3: 899–906 (1970)

152. SECONDARY EMISSION AND CONTAMINATION OF METAL SURFACES.
KULOV SK + SHERTNEV LG
INSTRUM EXP TECH 4: 917–8 (1967)

153. ADSORPTION OF OXYGEN ON MOLYBDENUM (111): EFFECT OF TRACE IMPURITIES.
LAMBERT RM + LINNETT JW + SCHWARZ JA
SURFACE SCI 26: 572–86 (1971)

154. LEED- AUGER INVESTIGATION OF A STABLE CARBIDE OVERLAYER ON A PLATINUM (111) SURFACE.
LAMBERT RM + WEINBERG WH + COMRIE CM + LINNETT JW
SURFACE SCI 27: 653–8 (1971)

155. AUGER PEAKS IN THE ENERGY SPECTRA OF SECONDARY ELECTRONS FROM VARIOUS MATERIALS.
LANDER JJ
PHYS REV 91: 1382–7 (1953)

156. LEED- AUGER ANALYSIS OF THE BERYLLIUM (0001) SURFACE.
LE JEUNE EJ
J VACUUM SCI TECHNOL 8(1): 9 (1971)

157. CORRELATION OF ELECTRONIC; LEED, AND AUGER DIAGNOSTICS ON ZINC OXIDE SURFACES.
LEVINE JD + WILLIS A + BOTTOMS WR + MARK P
SURFACE SCI 29: 144–64 (1972)

158. KLL AUGER SPECTRUM OF MANGANESE.
LIU YY + ALBRIDGE RG
NUCL PHYS A92: 139–44 (1967)

159. IMPURITIES, INTERFACES, AND BRITTLE FRACTURE.
(GIVES SHORT REVIEW OF AUGER ELECTRON SPECTROSCOPY AS APPLIED TO INTERFACES.)
LOW JR
TRANS MET SOC AIME 245: 2481–94 (1969)

160. POTENTIAL MAPPING USING AUGER ELECTRON SPECTROSCOPY.
MACDONALD NC
ANN SCANNING ELECTRON MICROSCOPE SYMP 1970, 3RD PROC

161. AUGER ELECTRON SPECTROSCOPY IN SCANNING ELECTRON MICROSCOPY: POTENTIAL MEASUREMENTS.
MACDONALD NC
APPL PHYS LETT 16: 76–80 (1970)

162. AUGER ELECTRON SPECTROSCOPY FOR SCANNING ELECTRON MICRO- SCOPY.
MACDONALD NC
P89–96 OF ANN SCANNING ELECTRON MICROSCOPE SYMP 1971, 4TH PROC
PHYS ABSTR 74: 52003 (1971)

163. MICROSCOPIC AUGER ELECTRON ANALYSIS OF FRACTURE SURFACES.
MACDONALD NC + MARCUS HL + PALMBERG PW
ANN SCANNING ELECTRON MICROSCOPE SYM 1970, 3RD PROC

164. AUGER EJECTION OF ELECTRONS FROM TUNGSTEN BY OXYGEN CHEMISORPTION.
MACLENNAN DA
BULL AMER PHYS SOC 13: 197 (1968)

165. ROLE OF WORK FUNCTION IN ELECTRON EJECTION BY METASTABLE
ATOMS: HELIUM AND ARGON ON (111) AND (110) TUNGSTEN. (AUGER)
MACLENNAN DA + DELCHAR TA
J CHEM PHYS 50: 1772–8 (1969)

166. WORK FUNCTION EFFECTS ON AUGER ELECTRON EJECTION BY NOBLE
GAS METASTABLE ATOMS.
MACLENNAN DA + DELCHAR TA
BULL AMER PHYS SOC 13: 197 (1968)

167. LOW ENERGY ELECTRON DIFFRACTION STUDY OF THE POLAR (111)
SURFACES OF GALLIUM ARSENIDE AND GALLIUM ANTIMONIDE.
MACRAE AU
SURFACE SCI 4: 247–64 (1966)

168. AN ELECTRON DIFFRACTION STUDY OF CESIUM ADSORPTION ON
TUNGSTEN.
MACRAE AU + MULLER K + LANDER JJ + MORRISON J
SURFACE SCI 15: 483–97 (1969)

169. ELECTRONIC AND LATTICE STRUCTURE OF CESIUM FILMS ADSORBED ON
TUNGSTEN.
MACRAE AU + MULLER K + LANDER JJ + MORRISON J + PHILLIPS JC
PHYS REV LETT 22: 1048–51 (1969)

170. SCATTERING OF LOW-ENERGY ELECTRONS FROM A COPPER (111) SURFACE.
MARKLUND I + ANDERSSON S + MARTINSON J
ARKYV FTS 37: 127–39 (1967)

171. FRACTURE SURFACE ANALYSIS OF TEMPER EMBRITTLED STEEL BY
AUGER ELECTRON SPECTROSCOPY.
MARCUS HL + PALMBERG PW
J METALS 211(3): A96 (1969)

172. AUGER FRACTURE SURFACE ANALYSIS OF A TEMPER EMBRITTLED
3340- STAINLESS STEEL.
MARCUS HL + PALMBERG PW
TRANS MET SOC AIME 245: 1664–6 (1969)

173. AUGER ELECTRON FINE PROFILES IN IONIC CRYSTALS.
MATTHEW JAD
PHYS LETT 32A: 261–2 (1970)

173. A TEMPERATURE DEPENDENT CONTRIBUTION TO AUGER ELECTRON
ENERGY DISTRIBUTIONS.
MATTHEW JAD
SURFACE SCI 20: 183–6 (1970)

175. L- SHELL AUGER AND COSTER-KRONIG ELECTRON SPECTRA.
MCGUIRE EJ
PHYS REV A 3: 1801–10 (1971)

176. K- SHELL AUGER TRANSITION RATES AND FLUORESCENT YIELDS FOR
ELEMENTS ARGON TO XENON.
MCGUIRE EJ
PHYS REV A 21: 273–8 (1970)

177. K- SHELL AUGER TRANSITION RATES AND FLUORESCENT YIELDS FOR
ELEMENTS BERYLLIUM TO ARGON.
MCGUIRE EJ
PHYS REV 185: 1–6 (1969)

178. ATOMIC L- SHELL COSTER-KRONIG, AUGER, AND RADIATIVE RATES AND
FLUORESCENCE YIELDS FOR SODIUM- THORIUM.
MCGUIRE EJ
PHYS REV A 3: 587–94 (1971)

179. LEED AND AUGER ELECTRON SPECTROSCOPY.
(REVIEW ARTICLE.)
MCKEE CS + ROBERTS MW
CHEM BRIT 6: 106–10 (1970)

180. ENERGY WIDTHS OF ROENTGEN LEVELS BY AUGER ELECTRON SPECTRO-
SCOPY.
MEHLHORN W + STAHLHERM D + VERBEEK H
Z NATURFORSCH A23: 287–94 (1968)

181. INELASTIC INTERACTIONS OF SLOW ELECTRONS WITH ABSORBED
PARTICLES.
(SHORT SECTION ON AUGER SPECTROSCOPY)
MENZEL D
ANGEW CHEM INTERNAT ED 9: 255–66 (1970)

182. DETERMINATION OF KLL AUGER SPECTRA OF NITROGEN, OXYGEN,
CARBON DIOXIDE, NITRIC OXIDE, WATER, AND CARBON MONOXIDE.
MODDEMAN WE + CARLSON TA + KRAUSE MO + PULLEN BP + BULL
WE + SCHWEITZ G
J CHEM PHYS 55: 2317 (1971)

183. AUGER SPECTRA OF SIMPLE MOLECULES.
MODDEMAN WE + CARLSON TA + PULLEN BP + KRAUSE MO
ACS ANN MEET 1969, 158TH PROC ABSTR PAPER NO PHYS-178

184. APPLICATION OF TRIPLE GRID LEED SYSTEM TO AUGER SPECTRUM
ANALYSES.
MORRISON J + LANDER JJ
J VACUUM SCI TECHNOL 6: 338–42 (1969)

185. THE ADSORPTION OF IONIC SALTS ON A TUNGSTEN (100) SURFACE.
MORRISON J + LANDER JJ
SURFACE SCI 18: 428–30 (1969)

186. EPITAXIAL GROWTH OF COPPER ON (110) SURFACE OF A TUNGSTEN
SINGLE CRYSTAL STUDIED BY LEED, AUGER ELECTRON AND WORK
FUNCTION TECHNIQUES.
MOSS ARL + BLOTT BH
SURFACE SCI 17: 240–61 (1969)

187. DECONVOLUTION TECHNIQUES IN AUGER ELECTRON SPECTROSCOPY.
MULARIE WM + PERIA WT
J VACUUM SCI TECHNOL 8: 90 (1971)
SURFACE SCI 26: 125 (1971)

188. INELASTIC EFFECTS IN AUGER ELECTRON SPECTROSCOPY.
MULARIE WM + RUSCH TW
SURFACE SCI 19: 469–74 (1970)

189. CESIUM ADSORPTION ON TUNGSTEN STUDIED BY LEED AND SECONDARY
ELECTRON SPECTROSCOPY.
MULLER K
P1–5 OF SURFACE PHENOMENA OF THERMIONIC EMITTERS, 1969

190. AUGER ELECTRONS, INDICATORS IN ANALYSIS OF SOLID BODY SURFACES.
MULLER K
MIKROCHIM ACTA SUPPL 4: 1–9 (1969)

191. ROOM TEMPERATURE ADSORPTION OF OXYGEN ON TUNGSTEN SURFACES.
(REVIEW OF METHODS OF MEASUREMENT)
MUSKET RG
J LESS COMMON METALS 22: 175–91 (1970)

192. OBSERVATION AND INTERPRETATION OF THE AUGER ELECTRON
SPECTRUM FROM CLEAN BERYLLIUM.
MUSKET RG + FARTNER RJ
PHYS REV LETT 26: 80–2 (1971)

193. AUGER ELECTRON SPECTROSCOPY STUDY OF ELECTRON IMPACT
DESORPTION.
MUSKET RG + FERRANTE J
SURFACE SCI 21: 440–2 (1970)

194. AUGER ELECTRON SPECTROSCOPY STUDY OF OXYGEN ADSORPTION ON
TUNGSTEN (110).

MUSKET RG + FERRANTE J
J VACUUM SCI TECHNOL 7: 14–7 (1970)

195. AUGER ELECTRON EMISSION FROM GOLD DEPOSITED ON SILICON (111) SURFACE.
NARUSAWA T
JAP J APPL PHYS 10: 280–1 (1971)

196. SINGULARITIES IN AUGER EMISSION SPECTRUM OF METALS.
NATTA M + JOYES P
J PHYS CHEM SOLIDS 31: 447–52 (1970)

197. EFFECT OF THE POLARITY OF INDIUM ANTIMONIDE ON THE EXTERNAL PHOTOEFFECT IN THE X-RAY SPECTRAL REGION.
(EFFECT OF IMPURITIES ON THE AUGER SPECTRUM)
NIKOLAENYA AZ + NEKRASHEVICH IG + SEMERENKO VV
IZV VYSSH UCHEB ZAVED, FIZ 12(3): 68–73 (1969).
(FOR TRANSLATION, SEE SOV PHYS J)

198. AUGER SPECTROSCOPY ON COPPER- NICKEL ALLOY SURFACES RELATED TO CATALYSIS.
ONO M + TAKASU Y + NAKAYAMA K + YAMASHINA T
SURFACE SCI 26: 313–6 (1971)

199. AUGER SPECTROSCOPY FOR THE CHEMICAL ANALYSIS OF SURFACES.
OSWALD RC
UNIV OF MINNESOTA, ELECTRICAL ENG. DEPT. MS THESIS (1969)

200. PHYSICAL ADSORPTION OF XENON ON PALLADIUM (100).
PALMBERG PW
SURFACE SCI 25: 598–608 (1971)

201. CHEMICAL ANALYSIS OF PLATINUM (100) AND GOLD (100) SURFACES BY AUGER ELECTRON SPECTROSCOPY.
PALMBERG PW
P 29-1 TO 29-18 OF STRUCTURE AND CHEMISTRY OF SOLID SURFACES
SOMORJAI GA (ED) NY, JOHN WILEY 1969. (541.375/B51) 140258

202. TECHNIQUE AND APPLICATIONS OF AUGER ELECTRON SPECTROSCOPY.
PALMBERG PW
J VACUUM SCI TECHNOL 7: 76 (1970); 6: 903 (1969)

203. OPTIMIZATION OF AUGER ELECTRON SPECTROSCOPY IN LEED SYSTEMS.
PALMBERG PW
APPL PHYS LETT 13: 183–5 (1968)

204. AUGER ELECTRON SPECTROSCOPY IN LEED SYSTEMS.
PALMBERG PW
PP 29-1 TO 2918 OF STRUCTURE AND CHEMISTRY OF SOLID SURFACES,
SOMORJAI GA (ED), NY WILEY, 1969. (5419375/B51) 140258

205. STRUCTURE TRANSFORMATIONS ON CLEAVED AND ANNEALED GERMANIUM (111) SURFACES.
PALMBERG PW
SURFACE SCI 11: 153–8 (1958)

206. SECONDARY EMISSION STUDIES ON GERMANIUM AND SODIUM-COVERED GERMANIUM.
PALMBERG PW
J APPL PHYS 38: 2137–47 (1967)

207. HIGH SENSITIVITY AUGER ELECTRON SPECTROMETER.
PALMBERG PW + BORN GK + TRACY JC
APPL PHYS LETT 15: 254–5 (1969)

208. AUGER SPECTROSCOPIC ANALYSIS OF GRAIN BOUNDARY SEGREGATION.
PALMBERG PW + MARCUS HL
TRANS AMER SOC METALS 62: 1016–8 (1969)

209. ATOMIC ARRANGEMENT OF GOLD (100) AND RELATED METAL OVER-LAYER SURFACE STRUCTURES.
PALMBERG PW + RHODIN TN

J CHEM PHYS 49: 134–46 (1968)

210. AUGER ELECTRON SPECTROSCOPY OF FACE CENTERED CUBIC METAL SURFACES. (GOLD, SILVER, PALLADIUM, COPPER, NICKEL)
PALMBERG PW + RHODIN TN
J APPL PHYS 39: 2425–32 (1968)

211. ATOMIC ARRANGEMENT OF GOLD (100), GOLD-COVERED (100), AND SILVER-COVERED COPPER (100) SURFACES.
PALMBERG PW + RHODIN TN
BULL AMER PHYS SOC 13: 944 (1968)

212. SURFACE DISSOCIATION OF POTASSIUM CHLORIDE BY LOW ENERGY ELECTRON BOMBARDMENT.
PALMBERG PW + RHODIN TN
J PHYS CHEM SOLIDS 29: 1917–24 (1968)

213. CHARACTERIZATION OF CHEMISORPTION BY LEED.
PARK RL + HOUSTON JE
J METALS 21(3): A87 (1969)

214. QUANTITATIVE USE OF AUGER SPECTROSCOPY: STANDARDIZATION OF THE METHOD.
PERDEREAU M
SURFACE SCI 24: 239–47 (1971)

215. STUDY OF INTENSITIES OF AUGER LINES EXCITED BY ELECTRON BOMBARDMENT AND ALUMINIUM K- ALPHA X-RAY IRRADIATION.
PESSA M
J APPL PHYS 42: 5831–6 (1971)

216. NEW FINE STRUCTURE IN ELECTRON-EXCITED AUGER SPECTRA FROM SOLID SURFACES.
PESSA M + AKSELA S + KARRAS M
PHYS LETT A31: 382–3 (1970)

217. INTERACTION OF LOW-ENERGY ATMOSPHERIC IONS WITH CONTROLLED SURFACES. (AUGER NEUTRALIZATION AT (100) FACE OF TUNGSTEN, POLYCRYSTALLINE MOLYBDENUM.)
PIERCE RH + FRENCH JB
UNIV TORONTO REPT. AD682373 (1968) (CHEM ABS 71: 16846G (1969); REPT. AD675206 (1969) (CHEM ABS 70: 81182 (1969))

218. A CORRELATION OF AUGER SPECTROSCOPY, LEED AND WORK FUNCTION MEASUREMENTS FOR EPITAXIAL GROWTH OF THORIUM ON A TUNGSTEN (100) SUBSTRATE.
POLLARD JH
SURFACE SCI 20: 269–84 (1970)

219. SURFACE COMPOSITION OF MICA SUBSTRATES.
POPPA H + ELLIOT AG
SURFACE SCI 24: 149–63 (1971)

220. CHARACTERIZATION OF COPPER- GOLD ALLOY SINGLE SURFACES.
POTTER HC
CORNELL UNIV. PH.D THESIS (1970)

221. SPECTROSCOPY OF A METAL- GAS INTERFACE.
PRITCHARD J
ANN REP PROGR CHEM SECT A66: 65 (1969)

222. UHV EVAPORATOR FOR LEED AND AUGER EMISSION STUDIES.
PRUTTON M + TOKUTAKA H
THIN SOLID FILMS 3: 411–16 (1969)

223. IDENTIFICATION OF AUGER SPECTRA FROM ALUMINIUM.
QUINTO DT + ROBERTSON WD
SURFACE SCI 27(3): 645–8 (1971)

224. AUGER SPECTRA OF COPPER- NICKEL ALLOYS.
QUINTO DT + SUNDARAN VS + ROBERTSON WD
SURFACE SCI 28: 504–16 (1971)

225. SECONDARY (AUGER) ELECTRON SPECTROSCOPY.
RAMSEY JA
VACUUM 21: 115–19 (1971)
226. AUGER ELECTRON SPECTRUM OF OSMIUM AT ENERGIES UP TO 300 EV.
REDKIN VS + ZASHKVARA VV + KORSUNSKIIMI + TSVEIMAN EV
SOV PHYS SOLID STATE 13: 1269–70 (1971)
227. THE EFFECT OF TELLURIUM ON INTERGRANULAR COHESION OF IRON.
RELLICK JR + MCMAHON CJ + MARCUS HL + PALMBERG PW
MET TRANS 2: 1492–4 (1971)
228. AUGER SPECTRA AND LEED PATTERNS FROM NICKEL DEPOSITS ON
CLEAVED SILICON.
RIDGWAY JWT + HANEMAN D
SURFACE SCI 26: 683–7 (1971)
229. AUGER SPECTRA AND LEED PATTERNS FROM VACUUM CLEANED
SILICON CRYSTALS WITH CALIBRATED DEPOSITS OF IRON.
RIDGWAY JWT + HANEMAN D
SURFACE SCI 24: 451–8 (1971)
230. CHARACTERISTIC AUGER ELECTRON EMISSION AS A TOOL FOR THE
ANALYSIS OF SURFACE COMPOSITION.
RIVIERE JC
PHYS BULL 20: 85 (1969)
231. DIFFUSION OF SULFUR TOWARD THE (110) FACE OF NICKEL.
RIWAN R
SURFACE SCI 27: 267–72 (1971)
232. CHARACTERISTIC ENERGY LOSS AND AUGER ELECTRON SPECTROSCOPY
APPLIED TO THE STUDY OF ADSORPTION PHENOMENA AT METAL SUR-
FACES.
ROUSSEAU J + PRALIAND H
J CHIM PHYS 67: 1493–505 (1970) (IN FRENCH)
233. THEORETICAL STUDY OF THE AUGER EFFECT IN THE LIGHT TO MEDIUM
RANGE OF ATOMIC NUMBER.
RUBENSTEIN RA
UNIV OF ILLINOIS PHD THESIS (1955)
DISSERTATION ABST 15: 851 (1955)
234. EXPERIMENTAL OBSERVATION OF CHEMICAL SHIFTS IN AUGER SPECTRUM
FROM SURFACE LAYERS OF SILICON DIOXIDE DURING ELECTRON
BOMBARDMENT.
SALMERON M + BARC AM
SURFACE SCI 29: 300–2 (1972)
235. INELASTIC SCATTERING OF LOW ENERGY ELECTRONS FROM SURFACES.
SCHEIBNER EJ + THARP LN
SURFACE SCI 8: 247–65 (1967)
236. THEORY OF THE AUGER EFFECT IN III-V SEMICONDUCTORS.
SCHRENE D
Z NATURFORSCH A 24: 1752–9 (1969)
237. SLOW ELECTRON SCATTERING FROM METALS. PART-2: INELASTICALLY
SCATTERED PRIMARY ELECTRONS.
SEAH MP
SURFACE SCI 17: 161–80 (1969)
238. SLOW ELECTRON SCATTERING FROM METALS. PART-1: EMISSION OF
TRUE SECONDARY ELECTRONS.
SEAH MP
SURFACE SCI 17: 132–60 (1969)
239. FARADAY CUP LEED APPARATUS WITH FACILITY FOR INVESTIGATING
ENERGY AND ANGULAR DISTRIBUTIONS OF INELASTICALLY SCATTERED
OR PHOTOEMITTED ELECTRONS.
SEAH MP + FORTY AJ

J PHYS E3: 833–41 (1970)

240. SECONDARY EMISSION AND ELASTIC REFLECTION OF ELECTRONS FROM MONOCRYSTALS.
SHULMAN AR + KORABLEV VV + MOROZOV YA
IZV AKAD NAUK SSSR SER FIZ 35: 218 (1971) (IN RUSSIAN)

241. SECONDARY ELECTRON EMISSION OF SILICON DIOXIDE SINGLE CRYSTALS.
SHULMAN AR + KORABLEV VV + MOROZOV YA
SOV PHYS SOLID STATE 12: 519–20 (1970)

242. SECONDARY ELECTRON EMISSION OF MOLYBDENUM CRYSTALS.
SHUMAN AR + KORABLEVV VV + MOROZOV YA
SOV PHYS SOLID STATE 12: 586–9 (1970)

243. ENERGY SPECTRA OF INELASTICALLY SCATTERED AND AUGER ELECTRONS FROM SINGLE CRYSTALS.
SHULMAN AR + KORABLEV VV + MOROZOV YA
SOV PHYS SOLID STATE 12: 1487–8 (1970)

244. AUGER ELECTRON SPECTROSCOPY APPLIED TO SURFACE COMPOSITION PROBLEMS.
SICKAFUS EN
J METALS 21(3): A72 (1969)

245. SULFUR AND CARBON ON (110) SURFACE OF NICKEL.
SICKAFUS EN
SURFACE SCI 19: 181–97 (1970)

246. SECONDARY EMISSION ANALOG FOR IMPROVED AUGER SPECTROSCOPY WITH RETARDING POTENTIAL ANALYZERS.
SICKAFUS EN
REV SCI INSTRUM 42: 933–41 (1971)

247. LEED AND AUGER ELECTRON SPECTROSCOPY STUDY OF NICKEL (110) SURFACE EFFECTS DUE TO CARBON AND SULFUR.
SICKAFUS EN
BULL AMER PHYS SOC 14: 793 (1969)

248. A MULTICHANNEL MONITOR FOR REPETITIVE AUGER ELECTRON SPECTROSCOPY WITH APPLICATION TO SURFACE COMPOSITION CHANGES.
SICKAFUS EN + COLVIN AD
REV SCI INSTRUM 41: 1349–54 (1970)

249. AUGER ELECTRON SPECTROSCOPY AND INELASTIC ELECTRON SCATTERING IN THE STUDIES OF METAL SURFACES.
SIMMONS GW
J COLLOID INTERFACE SCI 34: 343–56 (1970)

250. ORDER- DISORDER PHENOMENA AT THE SURFACE OF ALPHA TITANIUM OXYGEN SOLID SOLUTIONS.
SIMMONS GW + SCHEIBNER EJ
J MATER 5: 933–49 (1970)

251. SURFACE MADELUNG POTENTIALS IN ELECTRON SPECTROSCOPY.
SLATER RR
SURFACE SCI 23: 403–8 (1970)

252. SURFACE ANALYSIS BY LOW-ENERGY ION REFLECTION.
SMITH DP
APPL SPECTROSC 25: 147 (1971)

253. ANALYSIS OF SURFACE COMPOSITION WITH LOW-ENERGY BACK-SCATTERED IONS.
SMITH DP
SURFACE SCI 25: 171–91 (1971)

254. SPUTTER CLEANING AND ETCHING OF CRYSTAL SURFACES (TITANIUM, TUNGSTEN, SILICON) MONITORED BY AUGER SPECTROSCOPY, ELLIPSO-METRY AND WORK FUNCTION CHANGE.
SMITH T
SURFACE SCI 27: 45–59 (1971)

255. INTENSITIES OF SPECTRA- AUGER AND PHOTOELECTRON.
STADNIKOV CG + NIKOLSHII AP
IZV AKAD NAUK SSSR, FIZ ESKA 35: 330 (1971) (IN RUSSIAN)

256. ENERGIES OF EXCITED STATES OF DOUBLY IONIZED MOLECULES BY
MEANS OF AUGER ELECTRON SPECTROSCOPY. PART-1: ELECTRONIC
STATES OF NITROGEN (2).
STAHLHERM D + CLEFF B + HILLIG H + MEHLHORN W
Z NATURFORSCH A24: 1728–33 (1969)

257. STUDY OF GRAIN BOUNDARY SEGREGATION USING AUGER ELECTRON
EMISSION SPECTROSCOPY.
STEIN DF
J METALS 21(3): A88 (1969)

258. STUDIES USING AUGER ELECTRON EMISSION SPECTROSCOPY ON TEMPER
EMBRITTLEMENT IN LOW ALLOY STEELS.
STEIN DF + JOSHI A + LAFORCE RP
TRANS AMER SOC METALS 62: 776–83 (1969)

259. STUDY OF GRAIN BOUNDARY SEGREGATION USING AUGER EMISSION
SPECTRA.
STEIN DF + RAMASUBRAMANIAN PV
UNIV OF MINNESOTA ANN TECH PROG REPT I. N69-26102 (1968)

260. AUGER ELECTRON SPECTROSCOPY OF METAL SURFACES.
STEIN DF + WEBER RE + PALMBERG PW
J METALS 23(2): 39–44 (1971)

261. AUGER SPECTROSCOPY OF BERYLLIUM.
SULEMAN M + PATTINSON EB
J PHYS F 1: 124–7 (1971)

262. OBSERVATION OF A PLASMON GAIN IN THE FINE STRUCTURE OF THE
ALUMINIUM AUGER SPECTRUM.
SULEMAN M + PATTINSON EB
J PHYS F 1: L21–4 (1971)

263. ALLOY SPUTTERING STUDIES WITH IN SITU AUGER ELECTRON SPECTRO-
SCOPY.
TARNG ML + WEHNER GK
J VACUUM SCI TECHNOL 8: 23 (1971)

264. RESOLUTION AND SENSITIVITY CONSIDERATIONS OF AN AUGER ELECTRON
SPECTROMETER BASED ON LEED DISPLAY OPTICS.
TAYLOR NJ
REV SCI INSTRUM 40: 792–804 (1969)

265. AUGER ELECTRON SPECTROMETER AS TOOL FOR SURFACE ANALYSIS.
(CONTAMINATION MONITOR.)
TAYLOR NJ
J VACUUM SCI TECHNOL 6: 241–5 (1969)

266. THIN REACTION LAYERS AND SURFACE STRUCTURE OF SILICON (111).
TAYLOR NJ
SURFACE SCI 15: 169–74 (1969).

267. TECHNIQUE OF AUGER ELECTRON SPECTROSCOPY IN SURFACE ANALYSIS.
TAYLOR NJ
IN TECHNIQUES OF METALS RESEARCH VOL-7, RF BUNSHAH (ED). NY
INTERSCIENCE 1971. (669.028/B94)

268. ROLE OF AUGER ELECTRON SPECTROSCOPY IN SURFACE ELEMENTAL
ANALYSIS.
TAYLOR NJ
VACUUM 19: 575–8 (1969)

269. REPLY TO COMMENTS OF HE BISHOP AND JC RIVIERE ON "AUGER
SPECTROSCOPY OF SILICON".
TAYLOR NJ
SURFACE SCI 17: 466–8 (1969)

270. ENERGY SPECTRA OF INELASTICALLY SCATTERED ELECTRONS AND
     LEED STUDIES OF TUNGSTEN.
     THARP LN + SCHEIBNER EJ
     J APPL PHYS 38: 3320–30 (1967)
271. AUGER SPECTROSCOPY STUDY OF THE ADSORPTION OF RUBIDIUM ON
     MOLYBDENUM (100).
     THOMAS S + HAAS TW
     SURFACE SCI 28: 632–6 (1971)
272. ELECTRON SPECTROSCOPY OF METAL BLACKS.
     THOMAS S + SULEMAN M + PATTINSON EB
     J PHYS D 3: L77–80 (1970)
273. K-, L-, AND M- AUGER AND L- COSTER-KRONIG SPECTRA OF PLATINUM.
     TOBUREN LH + ALBRIDGE RG
     NUCL PHYS A90: 529–44 (1967)
274. (100) SURFACES OF ALKALI HALIDES. II. ELECTRON STIMULATED DISSO-
     CIATION.
     TOKUTAKA H + PRUTTON M + HIGGINBOTHAM IG + GALLON TE
     SURFACE SCI 21: 233–40 (1970)
275. SURFACE CHEMICAL ANALYSIS BY AUGER ELECTRON SPECTROSCOPY
     AND APPEARANCE POTENTIAL SPECTROSCOPY: A COMPARISON.
     TRACY JC
     APPL PHYS LETT 19: 353–6 (1971)
276. AUGER ELECTRON SPECTROMETER PREAMPLIFIER.
     TRACY JC + BOHN GK
     REV SCI INSTRUM 41: 591–2 (1970)
277. THE KINETICS OF OXYGEN ADSORPTION ON THE (112) AND (110) PLANES
     OF TUNGSTEN.
     TRACY JC + BLAKELY JM
     SURFACE SCI 15: 257–76 (1969)
278. STRUCTURAL INFLUENCES ON ADSORBATE BINDING ENERGY. I. CARBON
     MONOXIDE ON (100) PALLADIUM.
     TRACY JC + PALMBERG PW
     J CHEM PHYS 51: 4852–62 (1969)
279. USE OF AUGER ELECTRON SPECTROSCOPY IN DETERMINING THE EFFECT
     OF CARBON AND OTHER SURFACE CONTAMINANTS ON GALLIUM
     ARSENIDE- CESIUM-OXYGEN PHOTOCATHODES.
     UEBBING JJ
     J APPL PHYS 41: 802–4 (1970)
280. AUGER ELECTRON SPECTROSCOPY OF CONTAMINATED GALLIUM AR-
     SENIDE SURFACES.
     UEBBING JJ + JAMES LW
     J VACUUM SCI TECHNOL 7: 81–3 (1970)
281. AUGER ELECTRON SPECTROSCOPY OF CLEAN GALLIUM ARSENIDE.
     UEBBING JJ + TAYLOR NJ
     J APPL PHYS 41: 804–8 (1970)
282. AUGER ELECTRON SPECTROSCOPY OF GALLIUM ARSENIDE PHOTO-
     SURFACES.
     UEBBING JJ + TAYLOR NJ
     BULL AMER PHYS SOC 14: 792 (1969)
283. ANGLE OF INCIDENCE EFFECTS IN AUGER ELECTRON EMISSION FROM
     CLEAN MOLYBDENUM.
     VANCE DW
     BULL AMER PHYS SOC 13: 947 (1968)
284. AUGER ELECTRON EMISSION FROM CLEAN AND CARBON- CONTAMINATED
     MOLYBDENUM BOMBARDED BY POSITIVE IONS.
     VANCE DW
     PHYS REV 164: 372–80 (1967)

285. AUGER ELECTRON EMISSION FROM CLEAN AND CARBON-CONTAMINATED
    MOLYBDENUM BOMBARDED BY POSITIVE IONS. PART-2: EFFECT OF ANGLE
    OF INCIDENCE. PART-3; EFFECT OF ELECTRONICALLY EXCITED IONS.
    VANCE DW
    PHYS REV 169: 252–72 (1968)

286. AUGER TRANSITIONS AND SHAPE OF X-RAY SPECTRUM.
    VEDRINSKII RV + KOLESNIKOV VV
    BULL ACAD SCI USSR, PHYS SER 31: 904–10 (1967)

287. OBSERVATION OF AUGER ELECTRONS IN THE ENERGY SPECTRUM OF
    SECONDARY ELECTRONS EMITTED BY COPPER UNDER BOMBARDMENT
    BY RARE GAS IONS.
    VIEL L + FAGOT B + COLOMBIE N
    COMPT REND B272:623–8 (1971)

288. AUGER EMISSION SPECTROSCOPY VANADIUM (2) OXYGEN (5) (010) AND
    VANADIUM (100) SURFACES.
    FIERMANS L + VENNIK J
    SURFACE SCI 24: 541–54 (1971)

289. AUGER ELECTRON SPECTROSCOPY MADE QUANTITATIVE BY ELLIPSO-
    METRIC CALIBRATION.
    VRAKKING JJ + MEYER F
    APPL PHYS LETT 18: 226–8 (1971)

290. NONRELATIVISTIC AUGER RATES, X-RAY RATES, AND FLUORESCENCE
    YIELDS FOR THE 2P SHELL.
    WALTERS DL + BALLA CP
    PHYS REV A 4: 2164–70 (1971)

291. NONRELATIVISTIC AUGER RATES, X-RAY RATES, AND FLUORESCENCE
    YIELDS FOR THE K SHELL.
    WALTERS DL + BALLA CP
    PHYS REV A 3: 1919–27 (1971)

292. Z DEPENDENCE OF THE KLL AUGER RATES.
    WALTERS DL + BALLA CP
    PHYS REV A 3: 519–20 (1971)

293. THIN FILM ANALYSIS BY AUGER ELECTRON SPECTROSCOPY.
    (SHORT REVIEW)
    WEBER RE
    SOLID STATE TECH 13(12): 49–53 (1970)

294. DETERMINATION OF SURFACE STRUCTURES USING LEED AND ENERGY
    ANALYSIS OF SCATTERED ELECTRONS.
    WEBER RE + JOHNSON AL
    J APPL PHYS 40: 314–8 (1969)

295. DETERMINATION OF SURFACE STRUCTURES BY LEED AND AUGER
    ELECTRON SPECTROSCOPY.
    WEBER RE + JOHNSON AL
    BULL AMER PHYS SOC 13: 945 (1968)

296. WORK FUNCTION AND STRUCTURAL STUDIES OF ALKALI- COVERED
    SEMICONDUCTORS.
    WEBER RE + PERIA WT
    SURFACE SCI 14: 13–38 (1969)

297. USE OF LEED APPARATUS FOR THE DETECTION AND IDENTIFICATION
    OF SURFACE CONTAMINANTS.
    WEBER RE + PERIA WT
    J APPL PHYS 38: 4355–8 (1967)

298. AUGER RECOMBINATION IN GALLIUM ARSENIDE.
    WEISBERG LR
    J APPL PHYS 39: 6096–8 (1968)

299. THEORY OF AUGER EJECTION OF ELECTRONS FROM METALS BY IONS.
    WENAAS EP + HOWSMAN AJ

P13-1 TO 13-22 OF STRUCTURE AND CHEMISTRY OF SOLID SURFACES.
SOMORJAI GA (ED). NY WILEY, 1969. (541.375/B51) 140258

300. SIZE EFFECT IN IONIC CHARGE RELAXATION FOLLOWING AUGER EFFECT.
WERTHEIM GK + GUGGENHEIM HJ + BUCHANAN DN
J CHEM PHYS 51: 1931–4 (1969)

301. AUGER ELECTRON ENERGIES (0–2000 EV) FOR ELEMENTS OF ATOMIC
NUMBER 5–103.
YASKO RN + WHITMOYER RD
J VACUUM SCI TECH 8: 733–7 (1971)

302. SOME PARAMETERS AFFECTING AUGER AND PHOTOELECTRON SPECTRO-
SCOPY AS AN ANALYTICAL TECHNIQUE.
YIN LI + ADLER I + LAMOTHE R
APPL SPECTROSC 23: 41–50 (1969)

303. X-RAY EXCITED LMM AUGER SPECTRA OF COPPER, NICKEL AND IRON.
YIN LI + YELLIN E + ADLER I
J APPL PHYS 42: 3595–600 (1971)

304. AUGER SPECTRA OF URANIUM.
ZENDER MJ
Z PHYS 218: 245–59 (1969)

305. A CARBON STRUCTURE ON THE RHENIUM (0001) SURFACE.
ZIMMER RS + ROBERTSON WD
SURFACE SCI 29: 230–6 (1972)

306. DIRECT COMPARISON OF AUGER ELECTRON SPECTROSCOPY WITH
APPEARANCE POTENTIAL SPECTROSCOPY.
MUSKET RG
J VAC SCI TECH 9, 603 (1972)

307. A QUANTITATIVE STUDY OF AUGER ELECTRON SIGNALS OF PHOSPHORUS
ON SILICON USING A QUARTZ MICROBALANCE.
LEVENSON LL + DAVIS LE + BRYSON III CE + MELLES JJ + KON WH
J VAC SCI TECH 9, 608 (1972)

308. AUGER ELECTRON SPECTROSCOPY STUDIES OF SPUTTER DEPOSITION AND
SPUTTER REMOVAL OF Mo FROM VARIOUS METAL SURFACES.
TARNG ML + WEHNER GK
JVST 9, 625 (1972)

309. CARBON MONOXIDE OXIDATION ON A Pt(110) SINGLE CRYSTAL SURFACE.
BONZEL HD + KU R
JVST 9, 663 (1972)

310. AUGER SPECTROSCOPY STUDY ON THE SURFACE COMPOSITION OF
COPPER-NICKEL ALLOYS AFTER ANNEALING AND SPUTTERING.
NAKAYAMA K + ONO M + SHIMIZU H
JVST 9, 749 (1972)

311. Baer, Y., Hedén, P. F., Hedman, J., Klasson, M., and Nordling, C., *Solid State Commun.* **8**, 1479 (1970).

312. McGuire, E. J., *Phys. Rev.* **A5**, 1043 (1972).

313. Smith, D. L. and Huchital, D. A., *J. Appl. Phys.* **43**, 2624 (1972).

314. Musket, R. G. and Bauer, E., *Appl. Phys. Letters* **20**, 455 (1972).

315. Glupe, G. and Mehlhorn, W., *Phys. Letters* **25A**, 244 (1967).

316. Drawin, H. W., *Z. Phys.* **164**, 513 (1961).

317. Gallon, T. E., *J. Phys.* **D5**, 822 (1972).

318. Neave, J. H., Foxon, C. T., and Joyce, B. H., *Surf. Sci.* **29**, 411 (1972).

319. Gerlach, R. L. and Ducharme, A. R., *Surf. Sci.*, in press.

320. Blauth, E., *Z. Phys.* **147**, 228 (1957).

321. Tracy, J. C., *J. Chem. Phys.* **56**, 2736 (1972).

322. Tracy, J. C., *J. Chem. Phys.* **56**, 2748 (1972).

323. Bonzel, H. P. *Surf. Sci.* **27**, 387 (1971).

324. Palmber, P. W., in *Electron Spectroscopy* (ed. by D. A. Shirley), North Holland, Amsterdam, 1972, p. 835.

CHEMICAL ASPECTS OF ESCA

D. T. CLARK

Dept. of Chemistry, University of Durham, England

Foreword

The material presented in this review is a transcript of the lectures given at the NATO summer school. The aim was to give a broad view of the chemical applications of ESCA.

Lecture 1. Introduction to ESCA, Applications to Chemical Analysis and Surface Studies. (Section 1)

Lecture 2. Applications of ESCA to Studies of Structure and Bonding of Organic Molecules. (Section 2)

Lecture 3. Applications of ESCA to Inorganic and Theoretical Chemistry. (Section 3)

Lecture 4. Application of ESCA to Studies of Structure and Bonding of Polymers. (Section 4)

The course aimed to give a broad view of the chemical applications of ESCA with examples largely selected from the research program in my own laboratories. I make no apologies for this bias since it can readily be justified. Thus many of the examples which are discussed are representative of work going on in many laboratories around the world. However ones own material for preparation of slides etc. is more readily to hand and anyway one always talks best about the work with which one is most familiar. In the particular case of the application of ESCA to the study of polymers the only work of substance available at the time the lectures were being prepared was our own.

1. Introduction to ESCA, Application to Chemical Analysis and Surface Studies

A. INTRODUCTION TO ESCA

In common with most other spectroscopic methods ESCA is a technique originally developed by physicists and now gradually being taken over by chemists to be developed to its full potential as a tool for investigating structure and bonding. The technique has largely been developed by Professor Kai Siegbahn and his collaborators at the University of Uppsala over the past 20 years or so and the early work has been extensively documented in the, by now well thumbed pages of, 'The first ESCA book'. [1] [2] It is within the last decade however that the potential of the technique has been revealed with the development of spectrometers with sufficient resolution and sensitivety. The field has been opened to chemists with the advent of commercially produced instruments (at the last count there were ten instrument manufacturers) and from this point of view and in terms of the work so far published the state of development of the method is akin to that of NMR in the late 1950's.

W. Dekeyser et al. (eds.), Electron Emission Spectroscopy, 373–507. All Rights Reserved
Copyright © 1973 by D. Reidel Publishing Company, Dordrecht-Holland

In terms of alphabet soup* we are rapidly approaching '57 varieties' and in addition to the name ESCA originally coined by Siegbahn the technique is also variously known as:

(1) X-ray Photoelectron Spectroscopy (XPS),

(2) High Energy Photoelectron Spectroscopy (HEPS),

(3) Induced Electron Emission Spectroscopy (IEES),

(4) Photoelectron Spectroscopy of the Inner Shell (PESIS)

The designation ESCA is however descriptive and aesthetically pleasing and is to be preferred.

As we shall see ESCA is an extremely powerful tool with wide ranging applicability and the principal advantages of the technique may be summarized as follows:

(1) The sample may be solid, liquid or gas (it is as easy to study a high molecular weight polymer as it is to study a gas) and the technique is essentially non-destructive. (Some samples do undergo X-ray damage but there are ways of circumventing the effects of this.) X-rays do however produce much less radiation damage than an electron beam as used in conventional Auger spectroscopy and in this respect ESCA has a definite advantage.

(2) The sample requirement is modest, in favourable cases 1 m gm solid, 0.1 μl of liquid or 0.5 cc of gas (at STP).

(3) The technique has high sensitivity and is independent of the spin properties of any nucleus and is applicable in principle to any element of the periodic table. H and He are exceptions being the only elements for which the core levels are also the valence levels.

(4) The information it gives is directly related to the electronic structure of a molecule and the theoretical interpretation is relatively straightforward.

(5) Information can be obtained on both the core *and* valence energy levels of molecules.

B. PROPERTIES OF CORE ORBITALS

A clearer understanding of ESCA as a technique is obtained with some knowledge of the properties of core orbitals. The material presented below provides a convenient introduction and is particularly apposite since it was from a research program involving non-empirical quantum mechanical calculations of cross sections through potential energy surfaces for simple reactions that this author proceeded to an experimental interest in ESCA as a technique.

Traditionally chemists have discussed the electronic structure of molecules, dealing only with the valence electrons and neglecting inner shell or core electrons. The reason for this being:

(1) core electrons are not explicitly involved in bonding (although most of the *total* energy of a molecule resides in the core electrons) and

(2) it is only in the past five years that sufficient computing capability has become

---

* A term first brought to the authors' attention by Professor C. B. Duke.

available to allow non-empirical quantum mechanical calculations on molecules in which the core electrons are explicitly considered.

It has become clear however that although core electrons are not involved in bonding, the core energy levels of a molecule encode a considerable amount of information concerning structure and bonding. This is illustrated by examples of work carried out at Durham over the past few years.

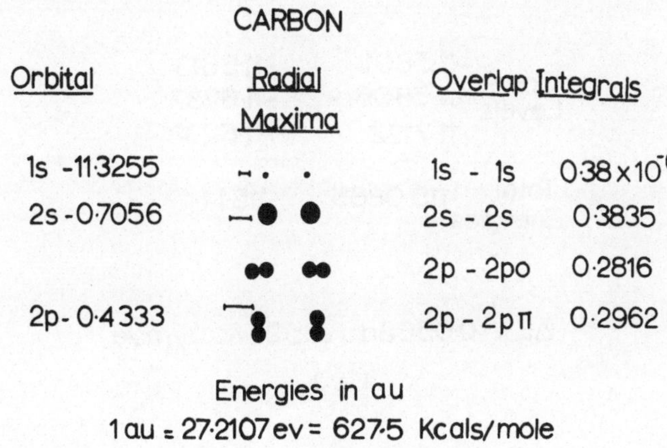

Fig. 1.

Figure 1 shows the orbital energies, total energy and radial maxima for the carbon atom calculated non-empirically in a Gaussian basis set.

$$1 \ AU = 2V = \frac{e^2}{\alpha_0 \times 1\,602 \times 10^{-12}}$$

$2 \times$ potential energy of the electron in the Hydrogen atom

$e = 4.8030 \times 10^{-10}$ esu = electronic charge

$\alpha_0 =$ Bohr radius $= 0.529\,17 \times 10^{-8}$ cm

$1 \ eV = 1.602 \times 10^{12}$ erg

Considering first the orbital energies, it is clear that the $1s$ (core) level is very much lower in energy than the $2s$ and $2p$ (valence) levels. From the radial maxima the $1s$ orbital is confined to a region in the immediate vicinity of the nucleus whereas the valence orbitals are much 'larger'. Since the core orbital is essentially localized around the nucleus overlaps with orbitals on adjacent atoms are negligible. Shown in Figure 1 are overlap integrals between orbitals on two adjacent carbon atoms with a bond length of 1.39 Å. The negligible value for the overlap integral involving the core orbitals is one reason why they are not involved in bonding.

It has tacitly been assumed by chemists in the past that in discussing the transformation of one molecule into another the energies of the inner shell or core electrons could be taken as constant and effectively ignored. A particularly interesting transformation which illustrates that this is not the case is the transformation of cyclo-

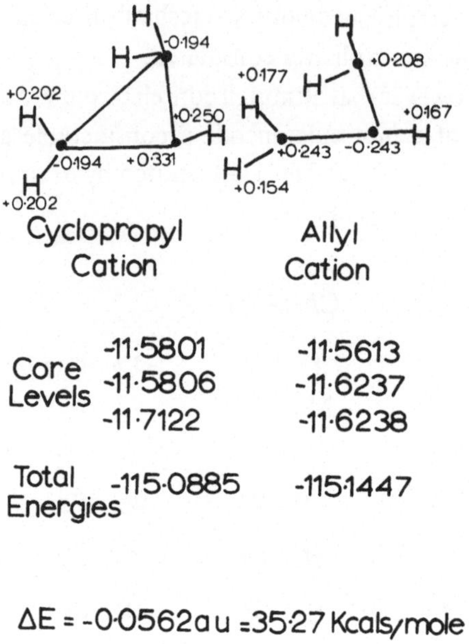

Cyclopropyl
Cation

Allyl
Cation

Core Levels	-11·5801	-11·5613
	-11·5806	-11·6237
	-11·7122	-11·6238

Total Energies	-115·0885	-115·1447

$$\Delta E = -0·0562\,\text{au} = 35·27\ \text{Kcals/mole}$$

Fig. 2.

propyl to allyl cation which occurs in a disrotatory fashion. The relevant energy levels are shown in Figure 2. As far as the carbon atoms are concerned there are fairly drastic charge migrations involved in this transformation. Thus C1 which carries a substantial positive charge in cyclopropyl cation becomes C2 with a substantial negative charge in allyl cation. As a result the C_{1s} orbital energy changes from -11.7122 AU to -11.5613 AU.

The change in energy of this particular core level is in fact almost three times the total energy change in the transformation.* The almost degenerate pair of C_{1s} orbitals for C2 and C3 in cyclopropyl cation change in energy by 0.044 AU in the transformation to allyl cation. Inspection of the charge distributions and core energy levels reveals that a more negative energy (i.e. increased binding energy)** is associated with an increased *positive* charge on an atom. The charge on a given atom is determined by the valence electron distribution, and the core levels of a molecule reflect this. Clearly although the core levels are not involved in bonding they are a sensitive function of the electronic environment about a given atom.

---

* It should be remembered that the total energy of a molecule may be expressed as

$$E = \sum_{i=a}^{h} \varepsilon_r - \sum_{\text{pairs } rs} (2J_{rs} - K_{rs}) + V_{nn}$$

where the ε_i are the occupied orbital energies, J_{rs} and K_{rs} are coulomb and exchange repulsion integrals and V_{nn} is the nuclear repulsion energy.

** Binding energy is defined here as the energy required to remove the electron in a given orbital to infinity and may be equated to the $-ve$ of the orbital energy (Koopmans' Theorem). The approximations involved in this are discussed elsewhere.

The different electronic environments about C1 and C2(C3) in cyclopropyl cation is therefore reflected in the 'shift' in C_{1s} binding energies of ~4.1 eV.

Core electrons are localized in space close to a nucleus and this is reflected in the fact that the binding energies are characteristic of a given element. Table I shows the computed orbital energies for acetamide and acetone.

TABLE I

Energy levels (in AU)

	$CH_3 CONH_2$		$CH_3 COCH_3$	
core	- 20.632	O_{1s}	- 20.721	O_{1s}
electrons	15.706	N_{1s}	- 11.507	
	- 11.521		- 11.424	C_{1}
	11.424	C_{1s}	- 11.424	
valence	1.394		- 1.451	
electrons	- 1.220		- 1.069	
	1.034		- 1.008	
	- 0.792		- 1.758	
	0.707		- 0.647	
	- 0.642		- 0.662	
	- 0.613		- 0.616	
	0.564		- 0.570	
	0.518		- 0.540	
	- 0.505		- 0.530	
	0.364		- 0.448	
	- 0.350		- 0.389	

The O_{1s} core levels are clearly distinguished from say the N_{1s} and C_{1s} levels by 135 eV and 240 eV respectively. Within a given molecule the C_{1s} levels also show the differing electronic environments about the respective atoms.

The characteristic identifying nature of core orbitals is emphasized on consideration of the approximate core binding energies for first and second row elements given in Table II. Clearly on the basis of their core binding energies it is an easy matter to distinguish say sulphur from chlorine.

TABLE II

Approximate core binding energies for 1st and 2nd row elements (in eV)

	Li	Be	B	C	N	O	F	Ne
$1s$	55	111	188	284	399	532	686	867

	Na	Mg	Al	Si	P	S	Cl	Ar
$1s$	1072	1305	1560	1839	2149	2472	2823	3203
$2s$	63	89	118	149	189	229	270	320
$2p_{1/2}$			74	100	136	165	202	247
	31	52						
$2p_{3/2}$			73	99	135	164	200	245

To summarize:

– Core orbitals are essentially localized on atoms, their energies are characteristic for a given element and are sensitive to the electronic environment of an atom.

C. INSTRUMENTATION

ESCA involves the measurement of binding energies of electrons in molecules, by determining the energies of electrons ejected by the interactions of a molecule with a monoenergetic beam of X-rays. In principle all electrons, from the core to the valence levels can be studied and in this respect the technique differs from uv photoelectron spectroscopy (UPS) in which only the lower energy valence levels can be studied.

Since extensive documentation concerning instrumentation is readily available elsewhere the minimum amount of information commensurate with achieving the objective of this review is given here.

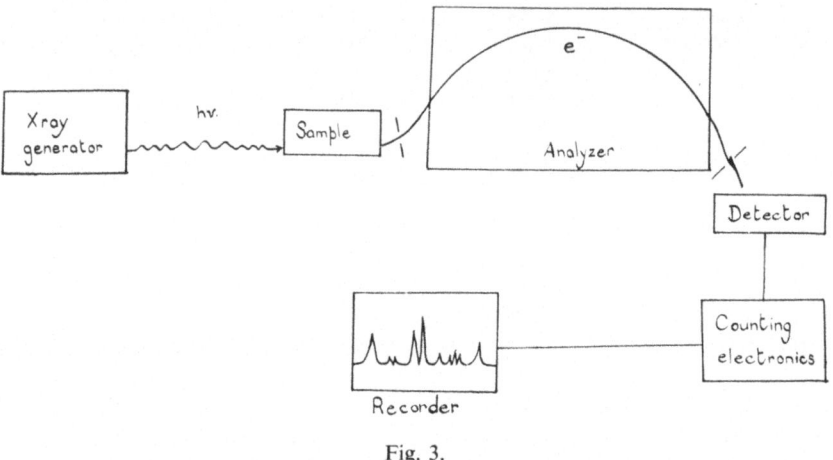

Fig. 3.

The essential features of an electron spectrometer are shown in Figure 3. The components are:

1. *An X-Ray Generator*

The most commonly used X-ray sources are $MgK\alpha_{1,2}$ and $AlK\alpha_{1,2}$ with photon energies (and linewidths) of 1253.7 eV (~ 0.7 eV) and 1486.6 eV (~ 1.0 eV)* respectively. (Harder X-ray sources (e.g. $CrK\alpha_1$ 5414.7 and $CuK\alpha_1$ 8047.8) have much larger inherent linewidths and thus give lower resolution. If this is not important however

---

* In the most recent designs a considerable effective reduction can be made in the apparent linewidth by dispersion compensation using bent quartz crystal monochromators. The loss in intensity can be compensated for by using rotating anodes which can be run at high power and by taking advantage of the focal plane in e.g. spherical electrostatic analyzers by employing multi-detector systems. In this way Siegbahn and co-workers have achieved *total* linewidths of ~ 0.5 eV for the C_{1s} levels in simple organic molecules. Another promising development which is only now being explored is to use synchrotron radiation.

the use of sources with widely differing photon energies can be very useful for studying escape depth dependence (see later).)

With $MgK\alpha_{1,2}$ and $AlK\alpha_{1,2}$ photon sources there is sufficient energy to study the $1s$ and valence levels of first row elements, the $2s$, $2p$ and valence levels of second row elements and so on. There is no particular virtue in studying the most tightly bound core levels of a given element (e.g. say the $1s$ level of gold), since this may well have a very large natural linewidth (see later) and the required higher energy photon source inevitably has a larger linewidth. Information concerning the valence electron distribution is encoded in all core levels (the information content may differ from level to level, see section on multiplet splittings), so that the core level selected for study depends on the cross section for photo-ionization, inherent linewidth, binding energy, photon energy and photon linewidth and the information required. For first row transition elements for example, the best levels for the study of 'shifts' in core binding energies are the $2p$ levels. Typical operating conditions for the X-ray generator would be (vac $\leqslant 4 \times 10^{-6}$ torr, 12 kV, 30 mA (for Mg target)).

2. *Sample Region*

The sample region of the spectrometer is usually separated from the X-ray tube by a thin metal window (usually $\frac{3}{10}$ thou Al or better Be) which ensures that electrons from the electron gun do not enter the sample region. Samples may be studied as solids and volatiles can be studied as thin films by employing a cryogenic probe, in which case typical operating pressures would be $\leqslant 4 \times 10^{-6}$ torr. With provision of differential pumping, samples may be studied directly in the gaseous state and preliminary experiments have been made on studying liquid beams, an area which has tremendous potential.

3. *Analyzer*

The analyzer must typically have a* resolution of something like one part in 10^4. In the precise energy analysis of the photoelectrons therefore two types of analyzers have mainly been used.

a. *Magnetic*. These are largely of the double focussing type and are generally made from brass or aluminium typically with a 30 cm radius. Double focussing is provided by an inhomogeneous magnetic field produced by a set of four cylindrical coils placed about the electron trajectory as shown in Figure 4. The chief advantage of a magnetic analyzer is relative ease of construction, the major disadvantage, the requirement to eliminate stray magnetic fields by employing necessarily bulky Helmholtz coils.

b. *Electrostatic*. Most are based on the hemispherical double focussing design first described by Purcell as long ago as 1938 (Figure 4). By employing a retarding field on the electrons before they enter the analyzer, the dimensions of the latter may be

---

* The main barrier to the development of the technique has been the design of analyzers of sufficiently high resolution and luminosity.

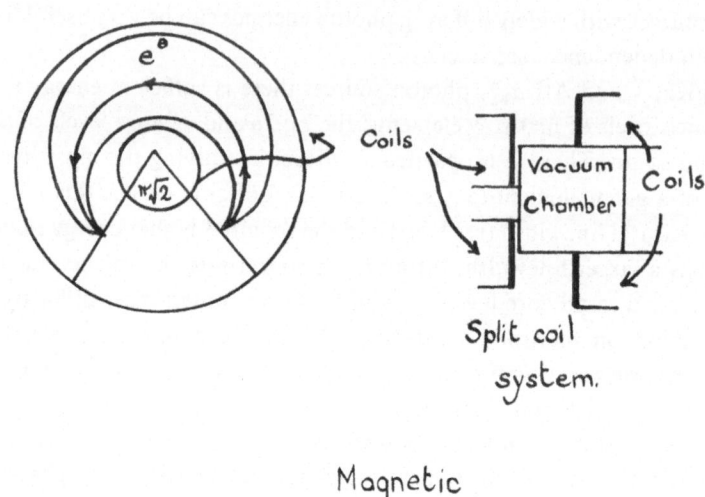

Magnetic

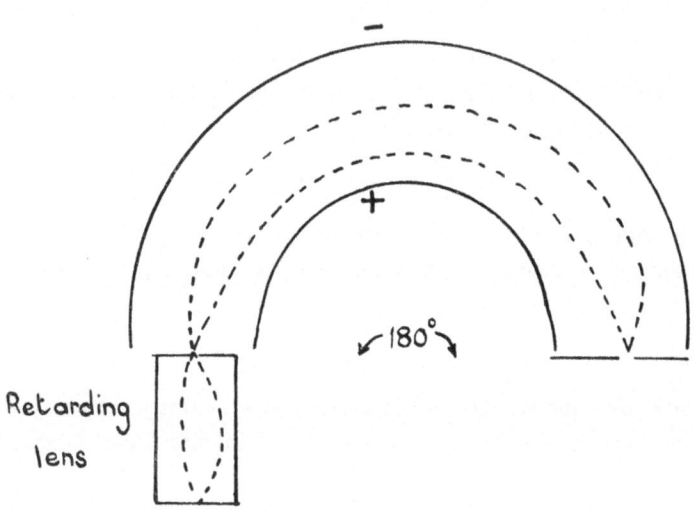

Electrostatic

Fig. 4.

considerably reduced and μ metal may also be used for screening. Although more difficult to construct therefore, spectrometers employing electrostatic analyzers can be made more compact. In fact the majority of the commercial instruments available are all of this basic design. Typical dimensions would be 25 cm mean diameter for the hemispheres which are usually constructed of stainless steel or gold plated aluminium or glass. Typical operating pressure in the analyzer $\leqslant 8 \times 10^{-7}$ torr.

4. Detector

The minute electron currents involved means that detection is via counting and most spectrometers employ channel electron multipliers. With most designs of double focussing analyzers their focal plane properties may be exploited by incorporating multichannel detectors which can give spectacular increases in the rate of data acquisition and this together with dispersion compensation is the most important development in commercial instruments in the short term. The output from the multiplier is then amplified and fed to counting electronics.

5. Scan and Readout

There are basically two ways of generating a spectrum either continuous or step scan. In the continuous mode of operation the field (either electric or magnetic) is increased continuously and the signal from the detector is continuously monitored by a rate

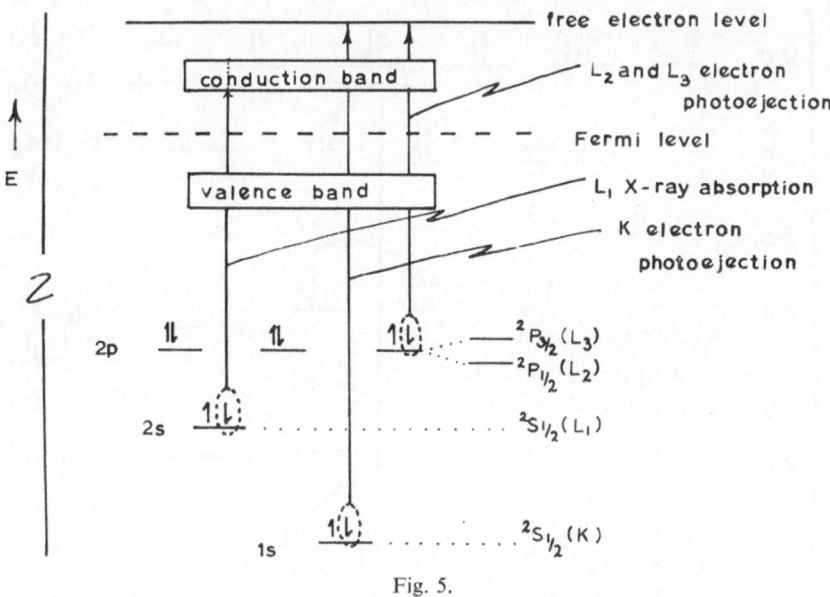

Fig. 5.

metei. If the signal to background and overall count rates are sufficiently high, this is the routine way to obtain a spectrum which may be plotted out directly on an XY recorder. Alternatively, the field may be incremented in small steps and at each setting either a fixed number of counts may be timed or a count can be made for a fixed length of time. Where signal to background is poor then this is the method of choice. It is advisable to have both wide and narrow scan facilities available, the former for carrying out preliminary searches and the latter for detailed study.

D. PROCESSES INVOLVED IN ESCA

The fundamental processes involved in ESCA are:

(1) *Photo ionization*

$$A + hv_1 \rightarrow (A^+)^* + e_1^-$$

(2) *Electronic relaxation* by either

(a) X-ray emission

$$(A^+)^* \rightarrow A^+ + hv_2$$

or   (b) Auger process

$$(A^+)^* \rightarrow A^{++} + e_2^-.$$

The relationship between the primary processes involved in photo ionization and X-ray absorption are outlined in Figure 5. Both relaxation processes and their relationship is depicted in Figure 6. The relative probabilities for electronic relaxation by X-ray emission and the Auger process depend on the atomic number of the element concerned and this is outlined in Figure 7a. In Figures 7b, c and d the similarity between the information gained from the related techniques is shown.

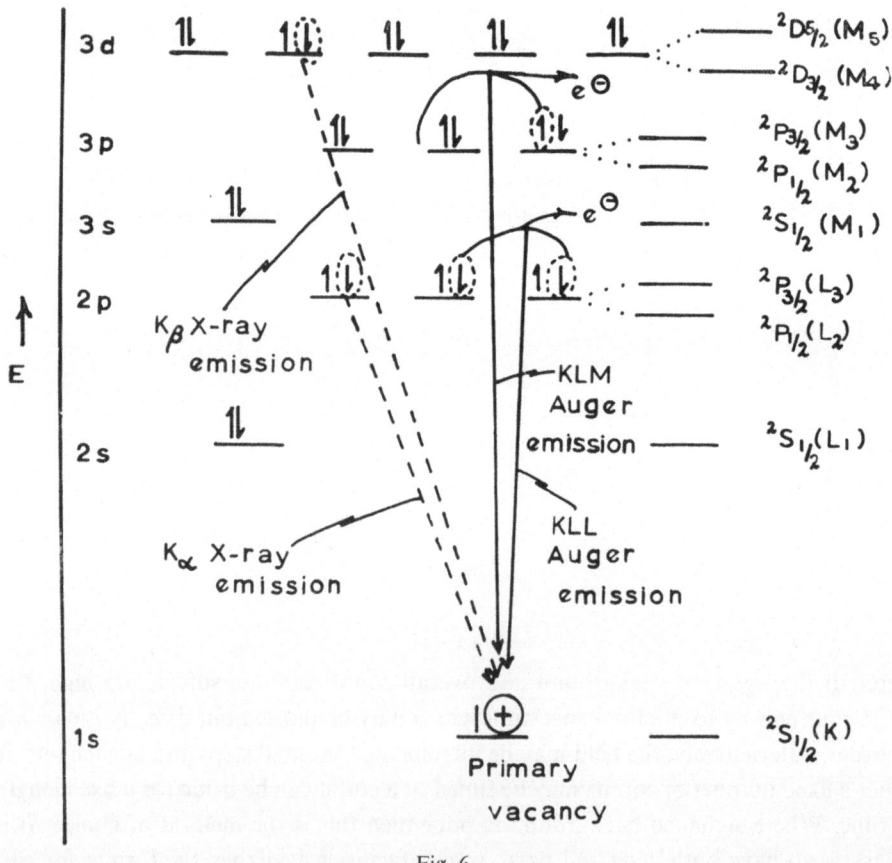

Fig. 6.

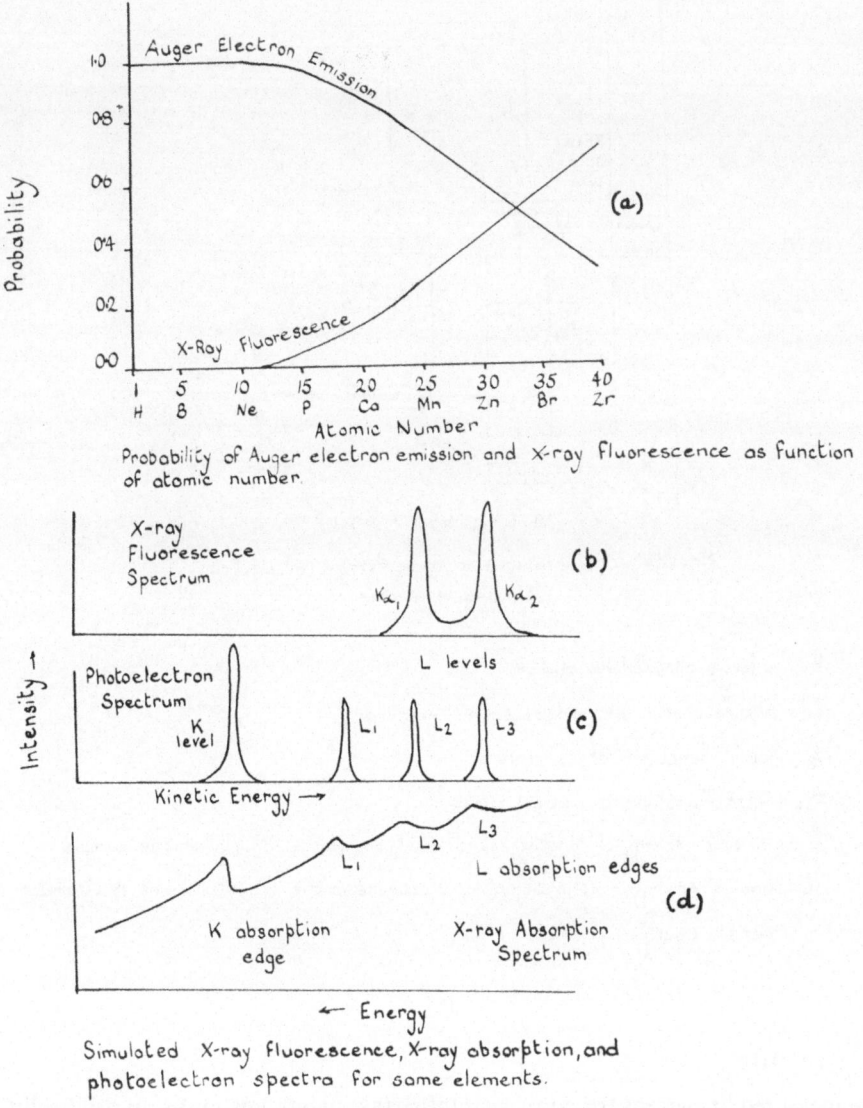

Probability of Auger electron emission and X-ray fluorescence as function of atomic number.

Simulated X-ray fluorescence, X-ray absorption, and photoelectron spectra for some elements.

Fig. 7.

Since the spectrometer is set up to detect and measure the energies of electrons expelled from the sample both the photoelectron and Auger spectra may be studied.

For photo ejection of a core electron in a solid sample of an insulator (in electrical contact with the spectrometer) the energy considerations in the measurements of the electron binding energies are shown in Figure 8. The reference is the Fermi Level and if the work function of the spectrometer is known then the absolute binding energy of a given level may be calculated. This will differ from the absolute binding energy defined with respect to the vacuum level by the work function of the sample and this is dealt with in some detail in a later section.

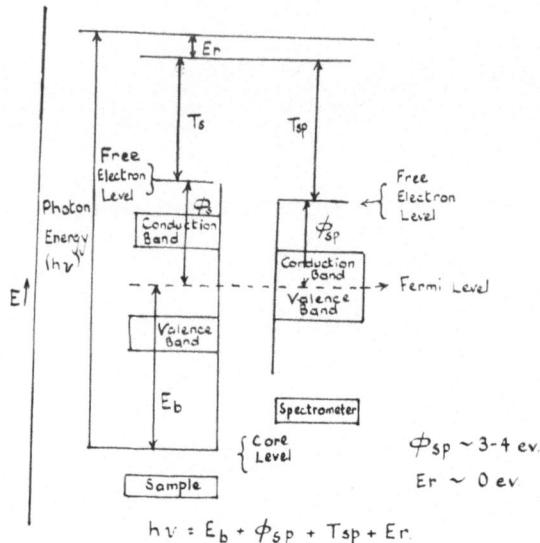

$$h\nu = E_b + \phi_{sp} + T_{sp} + E_r$$

Energy considerations in measurement of electron binding energies

$h\nu$ = energy of exciting photon

E_b = binding energy of photoejected electron

ϕ_s = work function of sample

ϕ_{sp} = work function of spectrometer

T_s = kinetic energy of photoejected electrons in the vicinity of the sample

T_{sp} = kinetic energy of photoejected electrons in the vicinity of the spectrometer

E_r = recoil energy

Fig. 8.

E. LINEWIDTHS

The measured linewidths for core levels (after taking into account spin orbit splittings if these are not resolved) may be expressed as

$$(\Delta E_m)^2 = (\Delta E_x)^2 + (\Delta E_s)^2 + (\Delta E_{Cl})^2$$

where

ΔE_m is the measured width at half height, the so called full width at half maximum (FWHM),

ΔE_x is the FWHM of the photon source

ΔE_s is the contribution to the FWHM due to the spectrometer (i.e. analyser)

ΔE_{Cl} is the natural width of the core level under investigation (For solids this includes solid state effects not directly associated with the lifetime of the hole state but rather with slightly differing binding energy due to differences in lattice environment.)

It has previously been pointed out (in subsection C) that $MgK\alpha_{1,2}$ and $AlK\alpha_{1,2}$ are the most useful photon source from the standpoint of keeping the contribution of ΔE_x to the total linewidth small. With well designed magnetic or electrostatic analysers the contribution ΔE_s can be reduced to negligible proportions so that the major limiting factors in terms of resolution are photon linewidths (which may be reduced by monochromatization) and the inherent width of the level itself. (For solids in which longer range interactions are important e.g. ionic lattices or hydrogen bonded covalent solids, solid state effects can contribute to the overall linewidths.)

Some examples of natural linewidths (ΔE_{Cl}) derived from X-ray spectroscopic studies are given in Table III. The uncertainty principle in the form

$$\Delta E \, \Delta t \geqslant \frac{h}{4\pi}$$

TABLE III
Approximate natural widths of some core levels (in eV) [20]

	Level	Atom							
		S	A	Ti	Mn	Cu	Mo	Ag	Au
Appr.	$1s$	0.35	0.5	0.8	1.05	1.5	5.0	7.5	54
natural widths	$2p_{3/2}$	0.10		0.25	0.35	0.5	1.7	2.2	4.4
Radiative widths	$1s$	0.04	0.07	0.2	0.33	0.65	3.6	6.0	50
Fluorescence yields	$1s$	0.1	0.14	0.22	0.31	0.43	0.72	0.8	0.93

shows that for a hole state lifetime of $\sim 6.6 \times 10^{-16}$ s the linewidth i.e. uncertainty in the energy of the state is ~ 1 eV.

It is evident from Table III that there are large variations in natural linewidths both for different levels of the same element and for the same levels of different elements. These reflect differences in lifetimes of the hole state. From Figures 6 and 7a it is clear that the lifetime is a composite of radiative (fluorescence) and non-radiative (Auger) contributions, the importance of the former increasing with atomic number (approximately as Z^4). This is clearly shown in Table III. This emphasises the fact that there is no particular virtue in studying the innermost core level as has been previously pointed out. For gold for example the 1s level has a halfwidth of ~ 54 eV so that even if a monochromatic X-ray source with the requisite photon energy were available any subtle chemical shift effects we might wish to investigate would be swamped.

Typically the lifetimes of the core hole states involved in ESCA are in the range $10^{-14} \sim 10^{-17}$ s emphasizing the extremely short time scales involved in ESCA compared with molecular vibration and the process may fairly be called *sudden* with respect to the nuclear (but not electronic) motions.

F. SIMPLE EXAMPLES ILLUSTRATING POINTS DISCUSSED IN A–E

Figure 9 shows an ESCA spectrum of gold foil and it is clear that there are sharp peaks located on sloping backgrounds. The sharp peaks arise from photo ionization of the core levels as indicated and since these are so characteristic of the appropriate element are readily identified. The tail to the low kinetic energy side of a given peak arises from inelastic scattering of the emitted photoelectrons and thus reduces the intensity of the 'elastic' or photoelectron peak. This will be discussed in some detail in a subsequent section. The spin orbit splitting of the gold $4p$ and $4d$ levels is clearly evident as is the large natural linewidths of the $4s$ and $4p$ levels. Peaks associated with carbon and oxygen can also be identified and are attributed to *surface* contamination by water and hydrocarbon material. Examination of the gold $4f$ levels and the conduction band under high resolution (Figure 10) reveals the spin orbital splitting in the former and fine structure in the latter. To the high kinetic energy side of the gold $4f$ levels satellite peaks are evident and these arise from the small $AlK\alpha_{3,4}$ contribution excited in the X-ray generator. The relative intensities and position of such satellites with respect to the main $AlK\alpha_{1,2}$ induced photoelectron peaks are well known and in more complex spectra the effect of the small percentage of $AlK\alpha_{3,4}$

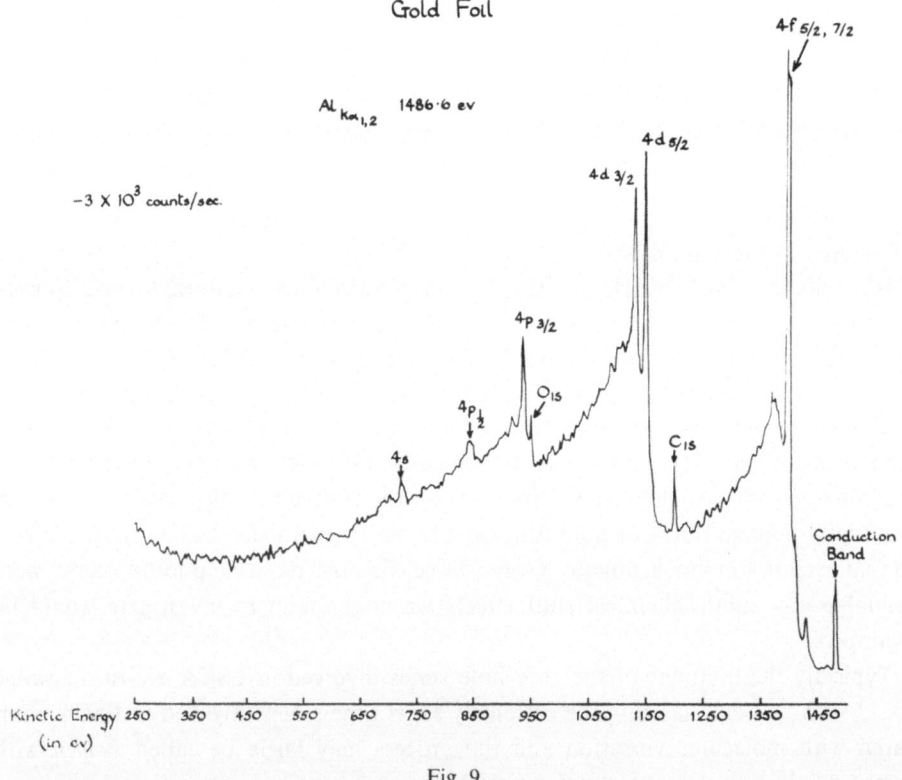

Fig. 9.

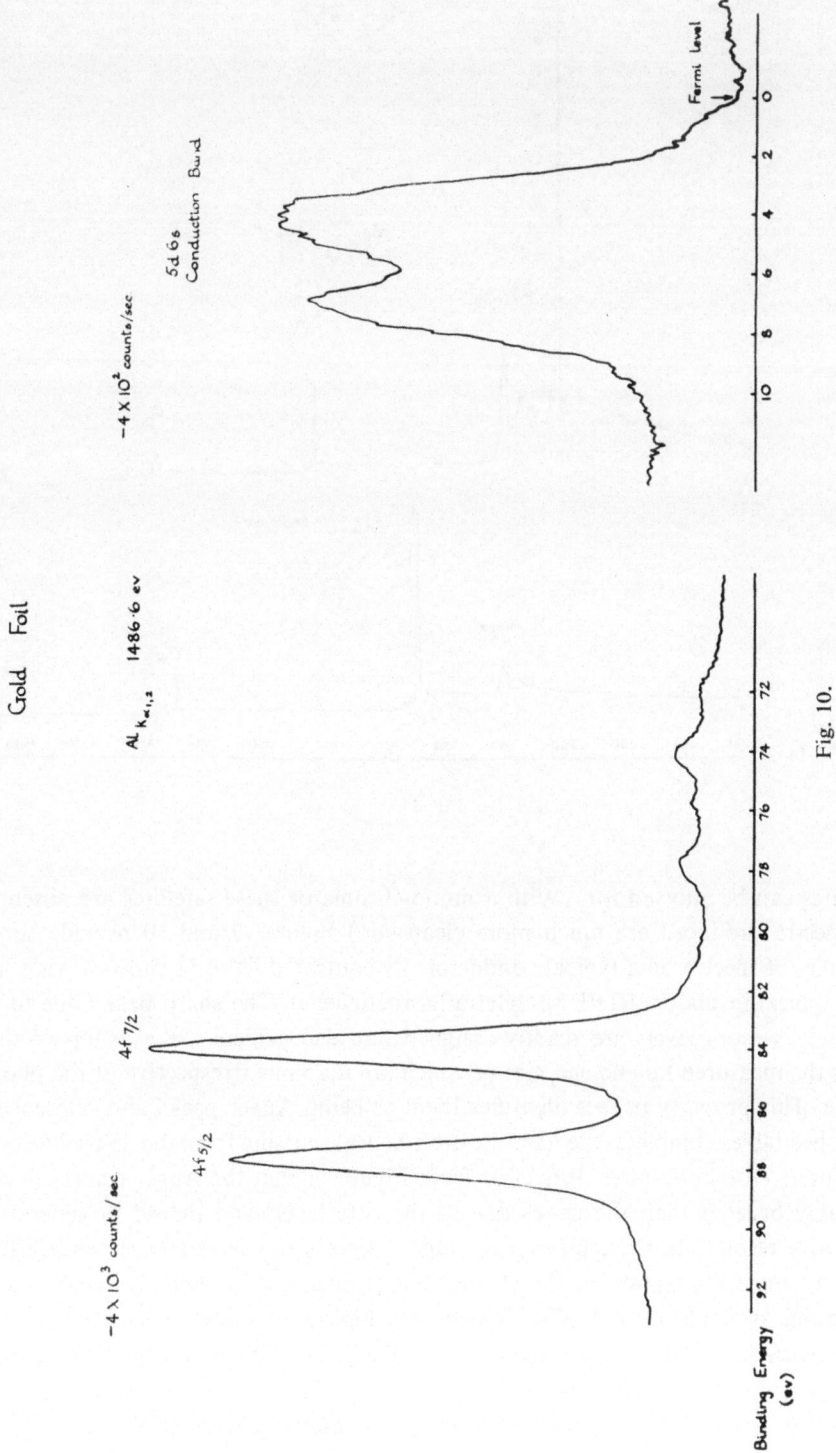

Fig. 10.

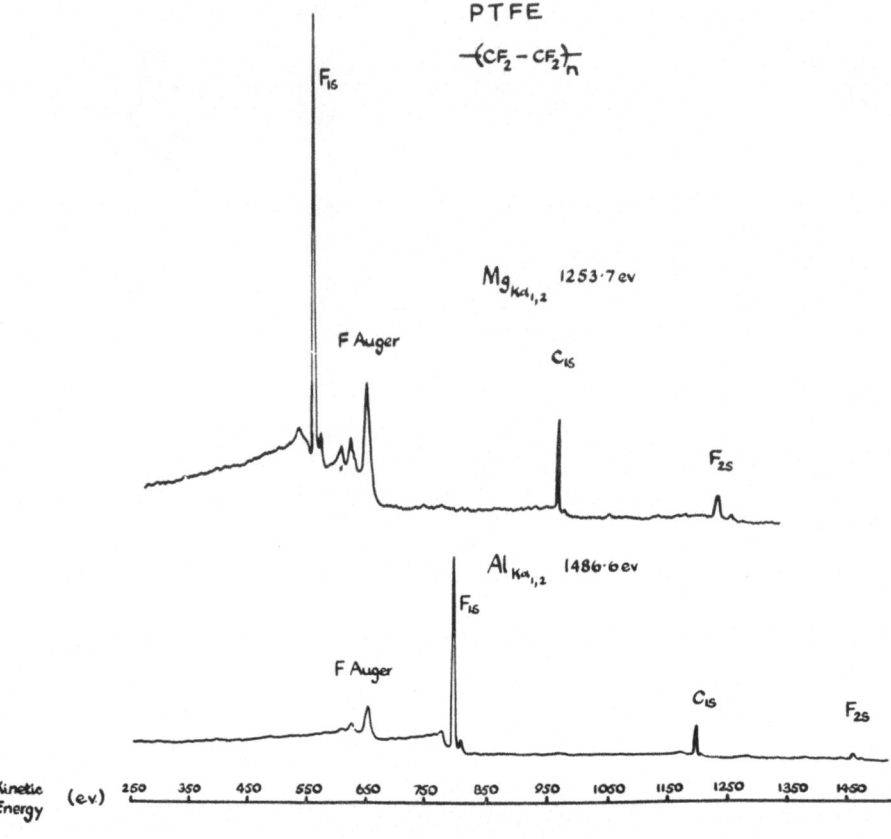

Fig. 11.

satellites can be allowed for. (With a mono-chromator these satellites are absent so the spectra produced are much more clean cut.) Figures 9 and 10 provide simple examples of spectra on a typical conductor. By contrast Figure 11 shows a wide scan for a typical insulator, PTFE (polytetrafluoroethylene). The sharp peaks due to the F_{1s} and C_{1s} core levels are readily assigned but also evident are a group of three peaks the measured kinetic energies of which are the same irrespective of the photon source. This property in fact identifies them as being Auger peaks and reference to published tables completes the identification as peaks arising from the de-excitation of fluorine 1s core hole states. It is clear from Figure 11 that the Auger peaks are considerably broader than the peaks due to the core levels and indeed in general the attainable resolution is much lower in Auger spectroscopy as compared with ESCA.

As a simple illustration of the chemical shift phenomenon and the wealth of information available from ESCA experiments Figure 12 shows spectra of the core and valence levels of a simple halocarbon CF_3CHCl_2. The previous examples have involved involatile samples whereas CF_3CHCl_2 is a volatile material which was studied as a thin film condensed on gold. The completely resolved C_{1s} levels shows

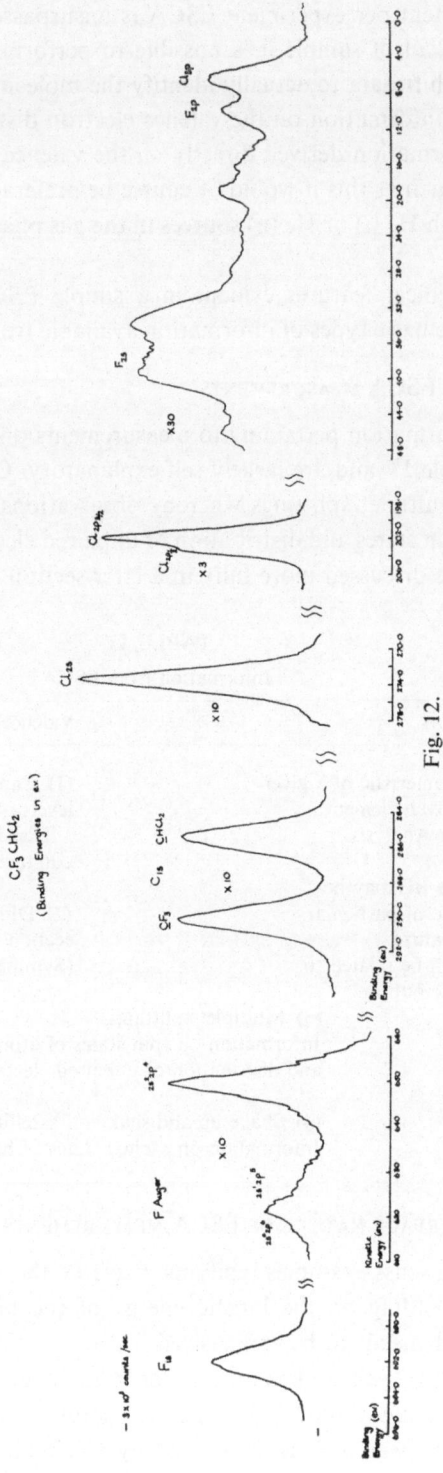

Fig. 12.

the great sensitivity of core binding energies to electronic environment. In terms of sheer information content per experiment ESCA is unsurpassed. Thus from one experiment involving 0.2 μl of sample it is possible to perform an elemental analysis and indeed from the shift data to actually identify the molecule. The chemical shifts can also give detailed information on the valence electron distribution which can be complemented by information derived directly on the valence energy levels. (With a low M.Wt. molecule such as this it would of course be preferable to study the higher lying valence levels with He (I) or He (II) sources in the gas phase to obtain improved resolution).

Having discussed typical features evident in a simple ESCA spectrum we now proceed to examine the basic types of information available from such measurements.

G. INFORMATION FROM ESCA MEASUREMENTS

The basic types of information pertaining to measurements on either valence or core levels are given in Table IV and are largely self explanatory. Of particular interest is the phenomenon of multiplet splittings whereby observations on core levels can be used to monitor the spin states and distribution of unpaired electrons in paramagnetic species and this will be discussed more fully in a later section.

TABLE IV

Information from ESCA

Core electrons	Valence electrons
(1) BE characteristic of a given level for a given element. Useful for Analysis.	(1) Can study valence energy levels of solids as well as gases. Densities of states, conduction bands of metals.
(2) Absolute BE may be characteristic of particular structural feature. 'Shifts' can be related to electron distribution.	(2) Differential changes in cross section with photon energy (Symmetries of orbitals)
(3) Multiplet splittings Information on spin states of atom and distribution of unpaired electrons.	
(4) Shake up and shake off satellites. Information on excited states of hole states.	

H. SURFACE OR SEMI-SURFACE NATURE OF ESCA MEASUREMENTS ON SOLIDS

Before proceeding to discuss examples typifying Table IV the surface or semi-surface nature of ESCA (depending on the kinetic energy of the photoemitted electrons) measurements on solids needs to be emphasized.

Depending on the core level studied and on the X-ray source the escape depths of photoelectrons contributing to the 'elastic' (i.e. no energy loss) peaks are typically in the range 0–100 Å. This was first demonstrated by Siegbahn and co-workers in an elegant series of experiments in which successive layers of α-iodo stearic acid were deposited on a probe and the $I_{3d_{5/2}}$ core level monitored. The results are shown schematically in Figure 13. For depths of 1,3 and 10 double layers, the observed signal intensities were in the ratio of $1:2:3:5$ respectively. This indicates that the maximum escape depth corresponds to something like 100 Å. It might be added that in the same study Siegbahn [1] obtained spectra of a surface layer containing one iodine atom per 10 Å^2, i.e. less than a monolayer. Therefore the sensitivity of the technique is such that submonolayers on surfaces may be studied. The potential for catalysis studies is therefore very great and it offers significant advantages over many other techniques since it is possible to monitor both the surface and the adsorbed species simultaneously.

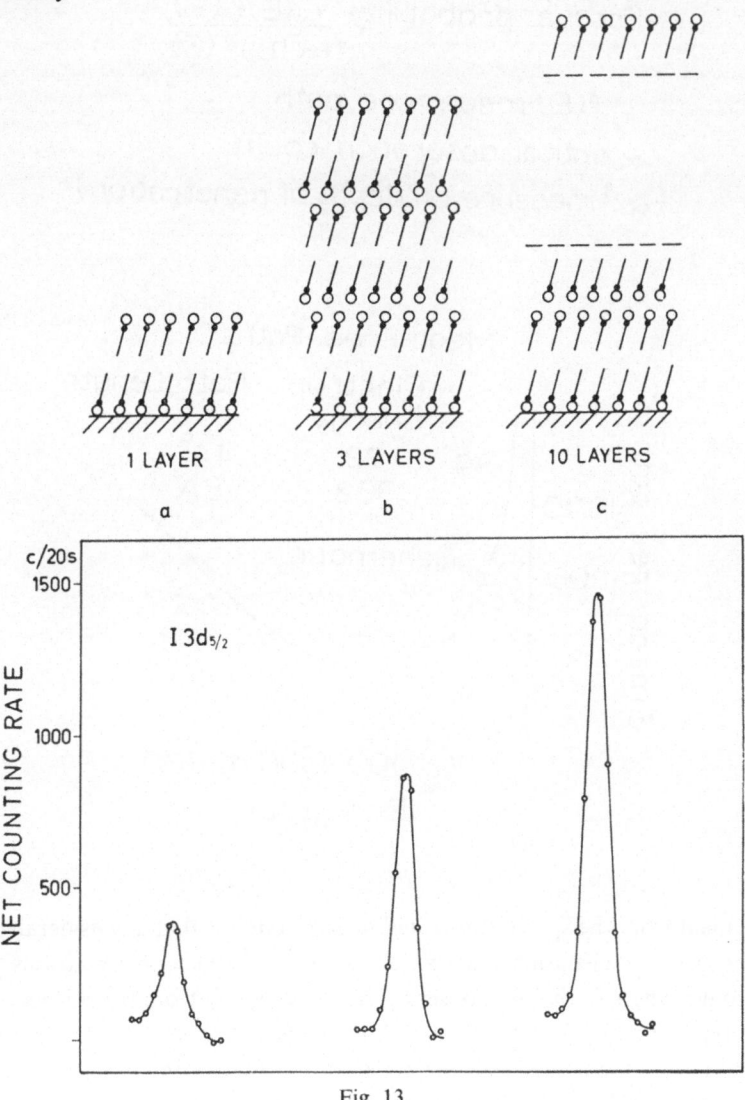

Fig. 13.

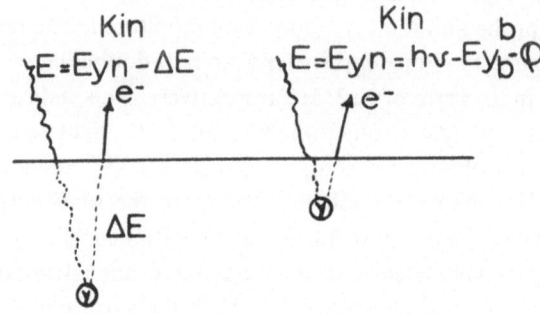

Escape probability $\dfrac{\alpha(E)\ell(E)}{1+\alpha(E)\ell(E)}$

$\ell(E)$ mean free path

α optical absorption coefft.

(α^{-1} measure of depth of penetration)

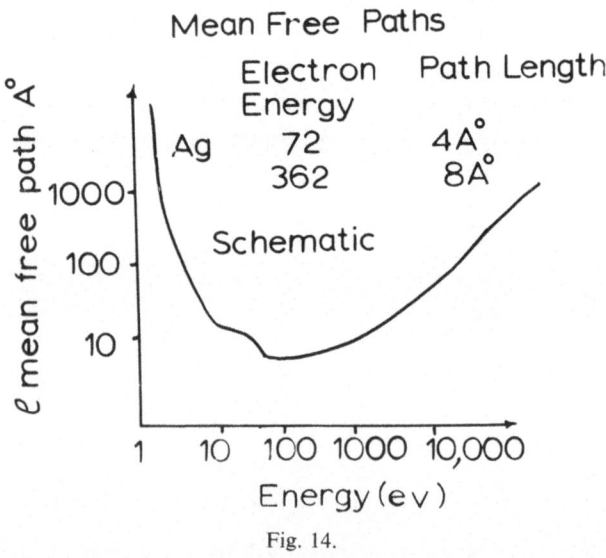

Fig. 14.

It is evident from the spectra presented in Subsection F that a considerable portion of the photoemitted electrons undergo scattering and lose energy by various processes (discussed elsewhere in this lecture series) before escaping from the surface. A typical (universal) plot of mean free path versus kinetic energy for electrons is given in Figure 14. (It should be emphasized that the penetration of the X-ray beam into the sample under typical conditions is $> 10^3$ Å so that it is the escape depth of the electrons which determines the intensity of the peak due to the photo-emitted electrons for a given intensity of irradiation*.)

TABLE V

Measured escape depths

Level	ke(eV)	$\lambda(\text{Å})$
C_{1s}	920	15
C_{1s}	1169	18
Ag_{3d}	362	8
W_{4f}	1455	13
$WO_{3\ 4f}$	1450	26
Au	1405	22

Some typical escape depths which have been measured (see later) are given in Table V. It is evident therefore that with $MgK\alpha_{1,2}$ or $AlK\alpha_{1,2}$ photon sources ESCA measurements on solids monitor the bulk, semi surface, or surface depending on the core level involved.

I. APPLICATIONS OF ESCA TO CHEMICAL ANALYSIS AND SURFACE STUDIES

As has already been pointed out, since the binding energies of core levels are characteristic for a given element the technique provides a valuable tool for routine chemical analysis and this is implicit in the name coined by Siegbahn. There are various levels of analysis which one might be interested in.

(1) *Qualitative analysis* (e.g. is a particular element present or not or direct identification of all the elements present in a sample etc., and their chemical form i.e. is sulphur present as S^{2-} or SO_4^{--} etc.).

(2) *Semi quantitative analysis* (e.g. approximate stoichiometries, differences in bulk and surface compositions etc.).

(3) *Quantitative analysis* (e.g. quantitative analysis of chemical compositions with accuracy comparable or approaching that of classical techniques).

Examples of each type of analysis are described below and at the present time it is routine in favourable cases to accomplish qualitative or semi quantitative analyses. For quantitative analysis however there are distinct difficulties to be dealt with which seem not to have been fully appreciated by some workers.

1. *Qualitative Analysis*

a. *Element mapping.* ESCA provides a very quick, convenient non-destructive means of 'mapping' the elements present in a sample especially when the modest sample requirement is taken into account. As a trivial example Figure 15 shows a wide scan ESCA spectrum of a simple halocarbon studied as a thin film condensed on gold. The elements present i.e. F, C, Cl, Br are readily identified by the characteristic binding energies of the relevant core levels. In fact with appropriate background information

---

* Enhancement of surface properties is possible by employing grazing incidence of the photon beam but very little work has been done in this direction.

$CF_2Br - CFClBr$

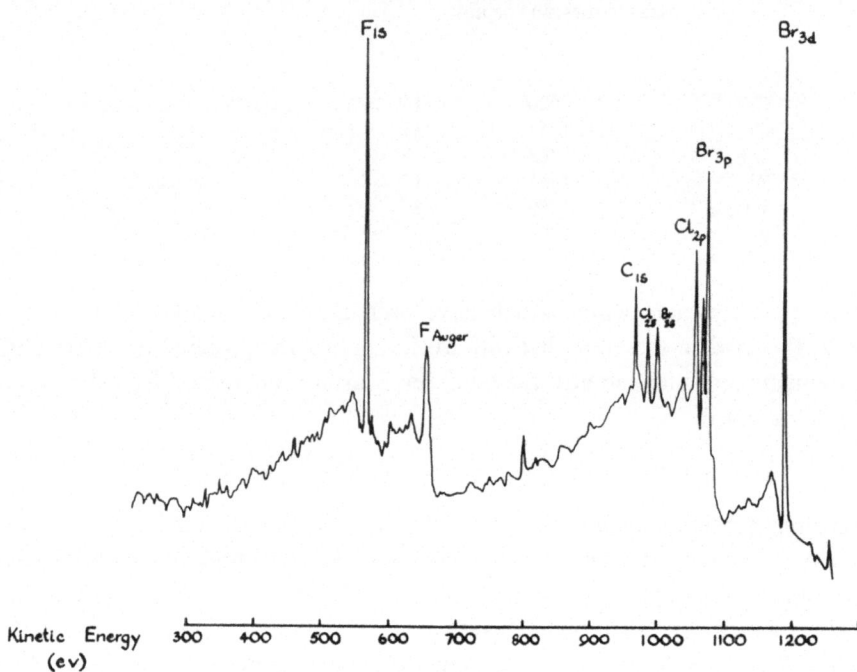

Fig. 15.

PTFE

$-(CF_2 - CF_2)_n$

pressed film

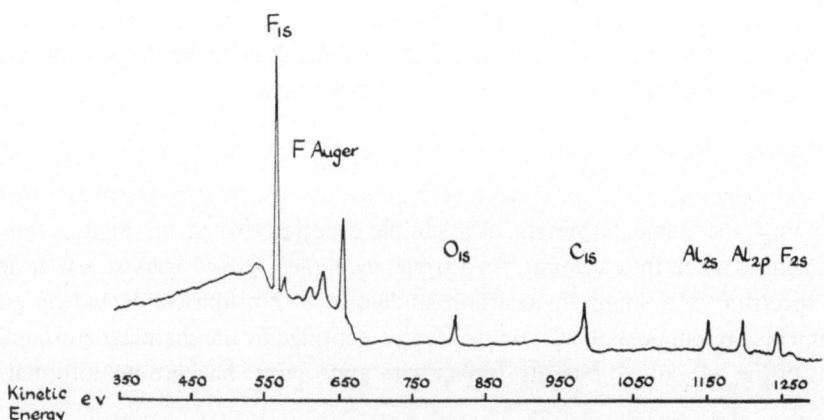

Fig. 16.

it is possible to demonstrate that the structure must be that indicated. Typically the spectrum in Figure 15 would take ~ 5 min to run on a modern commercial spectrometer and require $\sim 0.02 \, \mu l$ of sample.

It should be emphasized that no hard and fast rules can be laid down concerning the level at which certain elements can be detected by ESCA since so many factors are involved (see next section). It is probably fair to say however with regard to solids that for routine applications a lower limit of detection would be $\sim 0.1\%$ in favourable cases. Often therefore ESCA will not be a direct competitor with established analytical techniques for qualitative analysis for trace amounts of elements *not* specifically concentrated at the surface of the sample.

b. *Detection of surface coatings.* For chemists the chemical composition of the outermost < 100 Å of a solid sample is often of crucial importance (in catalysts for example). By conventional analytical techniques however (available in the average laboratory) it is the bulk composition of a sample which is *routinely* available. ESCA provides a convenient tool for monitoring the outermost < 100 Å of a sample and comparison with bulk analyses can then yield information of vital importance from both an academic and technological viewpoint.

As a simple example we have previously described the ESCA spectrum of PTFE, Figure 11. In this case the sample is conveniently studied as a thin film produced by a hot pressing technique. This involves pressing a powdered sample between sheets of 'clean' aluminium foil at the minimum temperature necessary for plastic flow to produce a film using a handpress. At much higher temperatures $\sim 320 \, ^\circ C$ films are produced at much lower pressures which are visually and *bulk* chemically identical to that produced at lower temperatures and higher pressures. The ESCA spectra of such films however are very revealing (Figure 16). It is clear from the appearance of peaks associated with the core levels of both oxygen and aluminium that on the high temperature pressing process a contaminant surface layer of alumina Al_2O_3 is deposited on the surface. The Thickness of the layer is almost certainly < 10 Å and would be undetectable by most spectroscopic techniques (including IR) although routinely evident from the ESCA experiments. Although this is a trivial example it can readily be appreciated that if surfaces are mechanically prepared the possibility of contamination is always present and is readily monitored by ESCA. From a technological viewpoint this is extremely importance since if one were interested in friction or lubrication for example the chemical composition of the surface of the sample is a prime consideration.

As a further technological application along these lines, Figures 17 and 18 show spectra of samples of muscovite mica concerning a problem presented to us by a heavy electrical engineering company. Muscovite mica composites have found extensive use in the heavy electrical engineering industry. The uppermost spectrum in Figure 18 shows a sample of muscovite mica itself providing a convenient element map (O, K, Al, Si). In use in high voltage equipment electrical breakdowns can occur and microscopic white dendritic growths are then apparent on the mica platelets. The questions

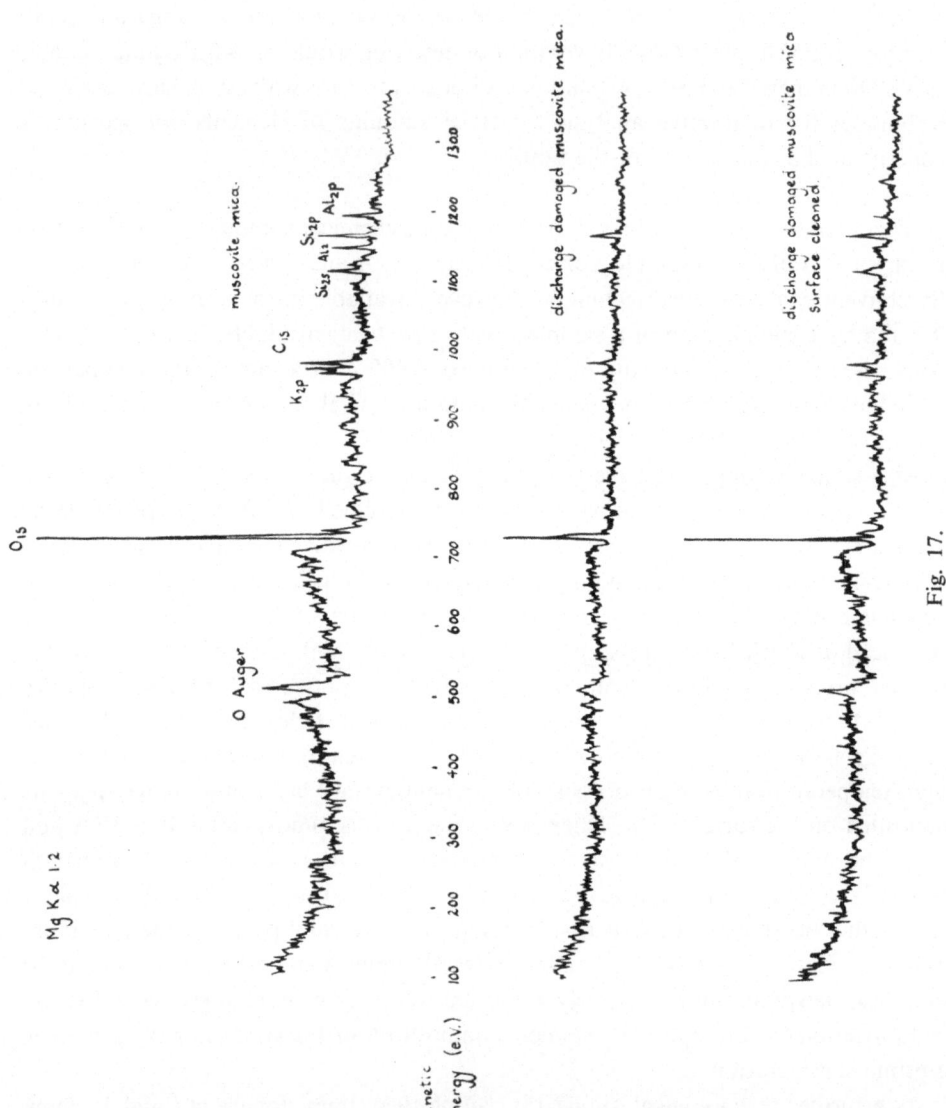

Fig. 17.

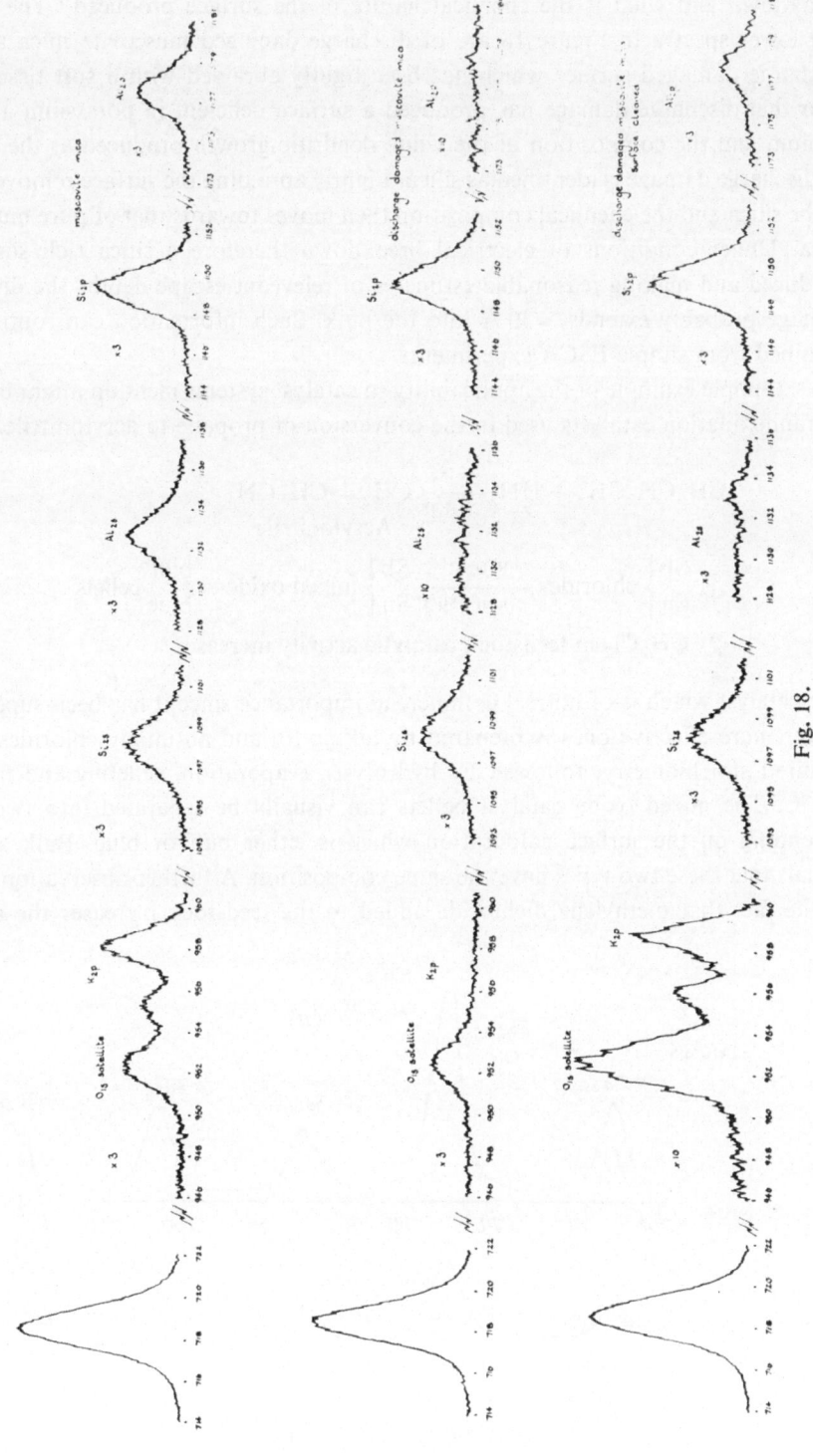

Fig. 18.

then arise what happens to the surface of the mica under conditions of high voltage breakdown and what is the chemical nature of the surface produced? The middle and lower spectra in Figure 18 are of discharge damaged muscovite mica and the discharge damaged surface which has been lightly abraided with a soft tissue. It is clear that discharge damage has produced a surface deficient in potassium and aluminium and the composition of the white dendritic growth produced at the surface by discharge damage is identified as silica. Lightly abraiding the surface removes some of the silica and the chemical composition then moves towards that of pure muscovite mica. Under conditions of electrical breakdown therefore a silica rich surface is produced and making reasonable estimates of releveant escape depths the discharge damage probably extends ~ 30 Å into the bulk. Such information can routinely be obtained from simple ESCA experiments.

As a simple example of the applicability to catalyst systems mention might be made of ammoxidation catalysts used in the conversion of propane to acrylonitrile.

$$CH_3CH_2CH_3 + HN_3 \xrightarrow[\text{Catalyst}]{\text{air}} CH_2 = CH\text{–}CN$$
$$\text{Acrylonitrile}$$

① $\left.\begin{array}{l} Sb \\ Sn \end{array}\right\}$ chlorides $\xrightarrow[\text{heat 750°C}]{H_2O}$ $\left.\begin{array}{l} Sb \\ Sn \end{array}\right\}$ mixed oxide → $\begin{array}{l} \text{buff} \\ \text{blue} \end{array}$ pellets

② CH_2Cl_2 in feedstock catalytic activity increases

The catalyst which is of no real commercial importance since it has been superceded by far more effective ones is prepared by taking tin and antimony chlorides in the required stoichiometry, followed by hydrolysis, evaporation, pelleting and firing at 750°C. The mixed oxide catalyst pellets can visually be separated into two types depending on the surface colouration which is either buff or blue. Bulk analysis reveals that these two types have the same composition. A further observation relates to the fact that methylene dichloride added to the feedstock increases the activity

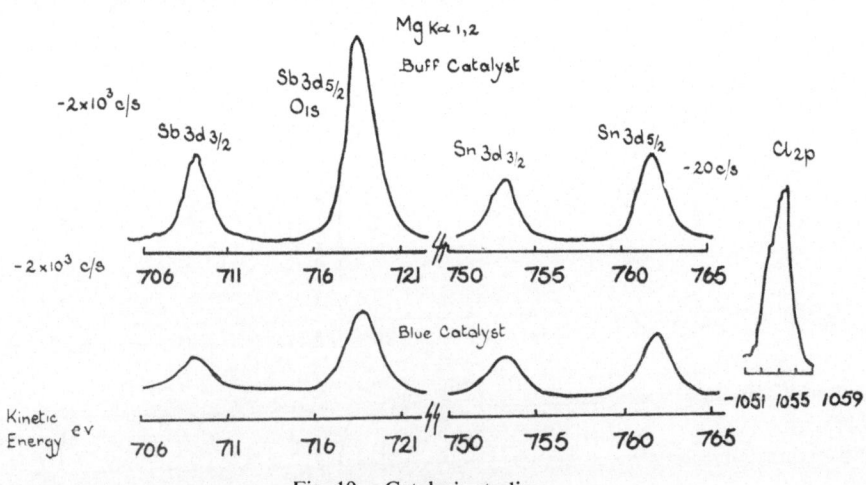

Fig. 19.   Catalysis studies.

of the catalyst. This suggests that oxidation may produce (e.g. $COCl_2$) which mya then incorporate chloride ion into the surface of the catalyst and that indeed the catalytic activity may arise from lattice sites with residual chloride at the surface remaining after hydrolysis. (Bulk analysis for chloride indicated $\sim 0.1\%$.) The feature of interest as far as ESCA is concerned are: (1) What are the differences between buff and blue coloured pellets and do the stoichiometries determined by ESCA correspond with the bulk. (2) Can chloride ion be detected and if so what is the stoichiometry with respect to the bulk.

The ESCA spectra are shown in Figure 19 and it is clear that the difference between buff and coloured pellets is in the surface stoichiometry of tin to antimony. The measurements also show that the surface stoichiometry is different from the bulk. Chloride ion is indeed readily detected and identified (by its characteristic binding energy) and with suitable calibration the approximate stoichiometry found for the surface of Sn/Cl is $\sim \frac{10}{1}$. This suggests that there is a factor of at least 100 difference in concentration of chloride at the surface compared with the bulk analysis and hence provides strong evidence that most of the residual chloride resides at the surface. This then provides a satisfactory account of the increased activity on adding methylene dichloride to the feed stock.

c. *Detection of surface impurities in analytically pure samples.* The great surface sensitivity is beginning to reveal information which previously was virtually unobtainable. In inorganic synthesis use is often made of metathetical reactions and advantage taken of the high lattice energy (and hence insolubility) of the desired product to produce a pure sample. As a simple example of the use of ESCA in the detection of surface contamination in such syntheses we might cite a recent investigation in our laboratory of the halogen bridged species

$$(\pi C_5 H_5) \, Fe \, (CO)_2 - X^+ - (CO)_2 \, Fe \, (\pi C_5 H_5) \, PF_6^- \quad X = Cl, \, Br, \, I$$

These were prepared by the following routes:

$$(1) \; (\pi C_5 H_5) \, Fe \, (CO)_2 \, X \xrightarrow{H_2 SO_4} [\, \pi C_5 H_5) \, Fe \, (CO)_2]_2 \, X^+$$

$$(1) \; H_2 O$$

$$(2) \; HPF_6$$

$$[(\pi C_5 H_5) \, Fe \, (CO)_2]_2 \, X^+ \, PF_6^-$$

$$(2) \; (\pi C_5 H_5) \, Fe \, (CO)_2 \, X \xrightarrow[\text{in } C_6 H_6]{AgPF_6} [(\pi C_5 H_5) \, Fe \, (CO)_2]_2 \, X^+ \, PF_6^-$$

Samples prepared by route 1 which were analytically (microanalysis, Mössbauer, IR, UV, NMR) pure were examined by ESCA. Strong signals were recorded for Fe, C, O and X but on the basis of our previous experience the signals for the F_{1s} and P_{2p} levels indicated that the surface stoichiometry did not correspond to that indicated. A search revealed the presence of HSO_4^- as counter ion (identified by the S_{2p} binding energy). Clearly in precipitating the hexafluorophosphate some bisulphate as counter ion had also been incorporated on the outermost few Å of surface. To obviate this

problem the salts were prepared by route 2. In this case however, there was little
difficulty in detecting silver halide on the surface of the samples! This emphasizes the
great strength of ESCA as an analytical tool. It is extremely sensitive to surface
contamination. The contamination in this case had gone undetected by both infrared
and Mössbauer spectroscopy as well as bulk analysis. On the other hand clearly the
technique can save a tremendous amount of wasted effort if for example the samples
had been of interest from the catalyst point of view since the catalytic activity (in
this case hypothetical) would almost certainly reside in the surface contamination
which is undetected by more familiar techniques.

d. *Atmospheric pollution.* In an increasingly environmentally aware public it is not
unnatural that enlightened scientists should see the virtue of applying ESCA to not
only esoteric academic problems but the more mundane and immediate problems of

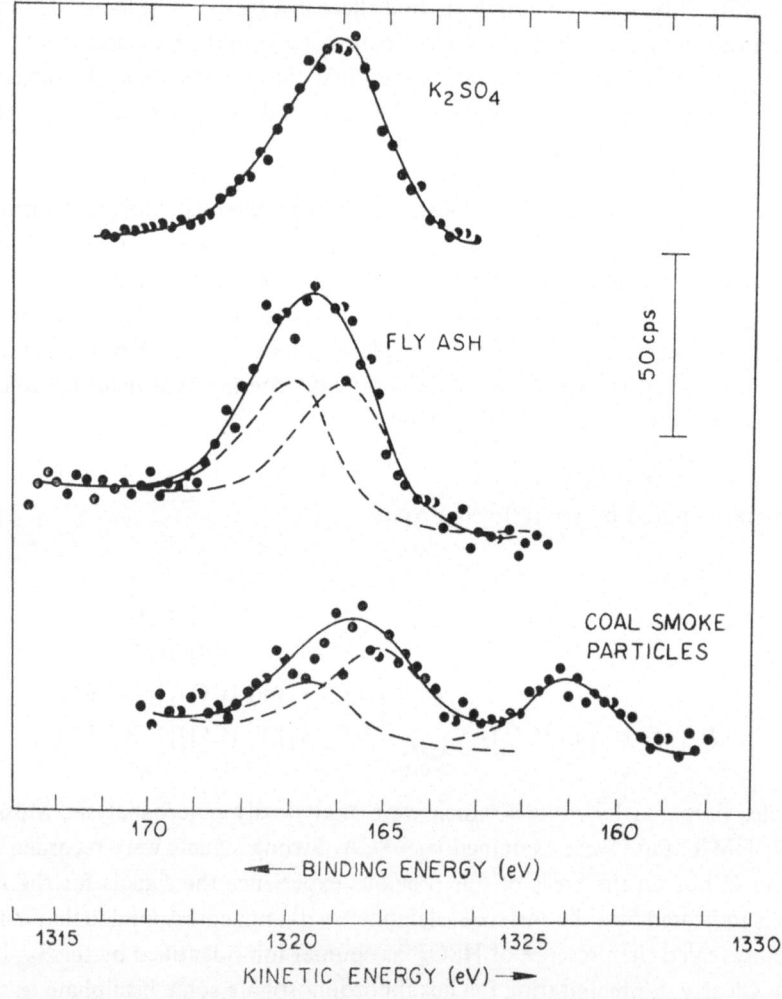

Fig. 20. Photoelectron spectra of Sulfur on various substrates.

atmospheric pollution. As an original example of qualitative analysis by ESCA the study of sulphur compounds adsorbed on smoke particles by Hulett and Carlson [5] and their co-workers deserves special mention.

It should be emphasized that conventionally air pollution monitoring and research has usually involved the study of gases and particulate matters as separate components. It is of course important to study the interaction between the two phases more particularly since it is known that 'aggressive' compounds are adsorbed on solid surfaces and in this form may well be more dangerous than in the gas phase.* For monitoring procedures to be more complete, the capability of *in situ* analysis of the surfaces of solid particles is needed, a task for which ESCA is ideally suited. Of particular interest is the sudy of sulphur dioxide adsorption on particulates a topic of investigation in several abatement schemes the efficiencies of which depend to some extent on the oxidation state of the sulphur after it is adsorbed.

Specimens of smoke particles from coal burned in a domestic fireplace were collected

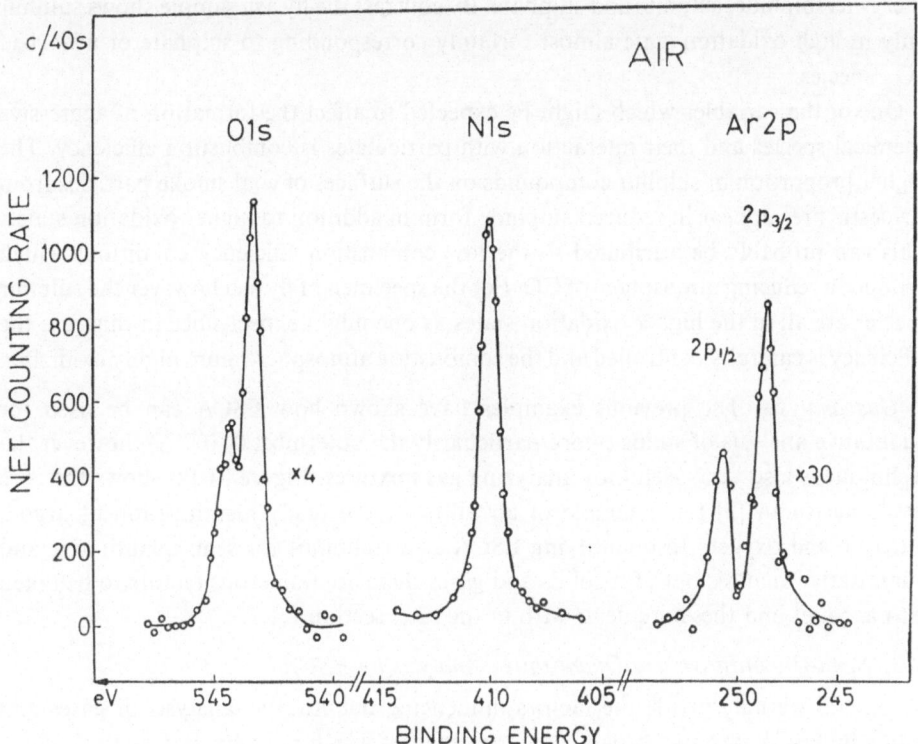

ESCA spectrum from air excited by MgKα radiation. The oxygen Is line shows spin splitting (paramagnetic molecule). Argon is detected through its 2p electrons and the spin-orbit splitting is well resolved.

Fig. 21.

* Minute traces of noxious gases inhaled into the lungs for example are rapidly exhaled. By contrast small particles once inhaled can become trapped in the lungs together with their adsorbed chemicals which may therefore be locally in quite high concentrations.

on filter paper, this process being facilitated by use of a vacuum cleaner! For comparison purposes fly ash (also collected on filter paper) emitted from a power station was also studied together with a reference sample of potassium sulphate. Bulk analysis indicated that the sulphur contents of the domestic smoke particles and fly ash samples were $\sim 0.95\%$ and 0.15% respectively. The ESCA spectra are shown in Figure 20. [5] The characteristic binding energy of the S_{2p} core levels makes identification of the element unambiguous. The intensities of the peaks by comparison with standard samples shows that the sulphur must be present mostly as layers adsorbed on the surfaces of the particles. This being the case it is perfectly feasible to study samples by ESCA corresponding to much lower bulk concentrations of sulphur (say in the range 10–100 ppm). Turning now to the chemical state of the adsorbed sulphur it is evident that for the coal smoke particles the S_{2p} levels appear as two distinct peaks. In fact employing standard linewidths it is possible to deconvolute the higher binding energy peak into two components. Comparison with binding energies for sulphate, sulphite and sulphide suggests the assignment of peaks in increasing binding energy as sulphide, sulphite and sulphate. By contrast the fly ash sample shows sulphur only in high oxidation state almost certainly corresponding to sulphate or adsorbed SO_3 species.

One of the variables which might be expected to affect the formation of aggressive chemical species and their interaction with particulates is combustion efficiency. The higher proportion of sulphur compounds on the surfaces of coal smoke particles from domestic fires appear in reduced sulphide form in addition to higher oxidation states. This can probably be attributed to the low combustion efficiency conditions which produce a reducing atmosphere of CO. For the specimen of fly ash however the sulphur species are all in the higher oxidation states as one might expect since in this case the efficiency is carefully controlled and the combustion atmosphere more highly oxidizing.

e. *Gas analysis*. The previous examples have shown how ESCA can be used for qualitative analysis of solids (more particularly the outermost 0–100 Å), however the technique is also very useful for analyzing gas mixtures. Figure 21 [2] shows the core levels measured [2] for a sample of air allowing the ready identification of argon, nitrogen and oxygen. In quantifying ESCA as a technique for semi-quantitative and quantitative analysis both for solids and gases there are numerous factors to be taken into account and these are dealt with in the next section.

2–3. *Semi-Quantitative and Quantitative Analysis by ESCA*

In general we may divide the factors influencing quantitative analysis of gases and solids by ESCA into two broad categories each of which must be dealt with if ESCA is to compete with classical methods of analysis on the same terms (viz. analysis os material representative of the bulk).

(a) Instrumental factors.

(b) Sample dependent factors.

Considering these in turn the major instrumental factors with regard to analysif of both solids and gases are:

a1. *Analyzer contributions to linewidths for peaks of different kinetic energy.* Inevitably quantitative measurements eventually hinge upon measuring peak areas (with due allowance for contributing factors which are sample dependent) for given core levels and comparison of these within the same sample or with a suitable reference. If comparison of areas is to be made it is important to ensure that the analyzer contribution to the total linewidths are the same for the core levels under consideration. Most of the recent spectrometer designs incorporate some form of retardation before the electrons enter the analyzer and in such designs a spectrum can be generated in one of two ways. Firstly, the kinetic energy range can be scanned by varying the retarding potential such that electrons brought to a focus at the multiplier always enter the analyzer with the same energy. This ensures that the analyzer contribution to the total linewidth is constant. On the other hand for other reasons it is often advantageous to generate spectra by scanning (keeping a constant ratio between the two) both the potential on the retarding lens and the potential on the plates (in a hemispherical electrostatic analyzer) in which case the absolute contribution of the analyzer to total linewidths for peaks of different kinetic energies is not constant so peak areas are not directly comparable.

a2. *Detector efficiency.* The total intensity of a given peak will depend on the number of electrons arriving at the detector (most usually a single channeltron) and their kinetic energies. Due allowance must therefore be made for the kinetic energy dependence.

a3. *Mono-chromaticity of X-ray source (satellites).* In discussing the core levels of gold (Subsection F) measured with an unmonochromatized $AlK\alpha_{1,2}$ X-ray source attention was drawn to the satellites due to the small percentage of $K\alpha_{3,4}$ radiation excited in the X-ray generator. The possibility that such satellites may overlap and interfere with the portion of a spectrum under consideration must be kept in mind and if necessary due allowance must be made for this factor. With the advent of commercial designs incorporating X-ray monochromators this potentially complicating feature with regard to quantitative analysis is absent.

b. *Sample dependent factors.* It is convenient to discuss gases and solids separately although there are some features in common.

Gases

(1) Sample pressures, consideration of shake up, shake off and multiplet effects.
(2) Overlapping peaks from different core levels etc.
(3) Cross sections for photoionization (hence sensitivity).

Since the prime interest of this school is in the solid state and since some of these factors are common to those detailed below further discussion of these points (which are largely self evident) will not be given here.

Solids

(1) Surface contamination from residual atmosphere in the spectrometer and/or

sample preparation. The surface sensitivity of ESCA is such that special care must be taken in sample preparation. With the advent of instruments operating under UHV conditions contamination of the surface from the extraneous atmosphere within the spectrometer may be obviated. With 'active' surfaces however with large sticking coefficients for hydrocarbon species or water present in an extraneous atmosphere for instruments operating in the range 10^{-6}–10^{-8} torr contamination can be a serious problem (Table VI). Surface contamination from relatively poor vacuum conditions can have its advantages however in providing useful 'contaminant' peaks (usually C_{1s}) that may be used for referencing the energy scale. If a fresh surface is being generated continuously by condensation from a directed beam of a vapourized sample,

TABLE VI

Surface contamination (hydrocarbon, H_2O etc.)

Pressure (torr)	Time for mono-layer coverage
High vac. 10^{-3}–10^{-6}	s
Very high vac. 10^{-6}–10^{-8}	min
Ultra high vac. $< 10^{-9}$	hr →

the minute partial pressures of the components of the extraneous atmosphere of a spectrometer operating in the range 10^{-6}–10^{-8} torr can effectively be ignored and surface contamination is very small. Under UHV conditions kinetically labile ligands from complexes for example may well be removed from surface layers at an appreciable rate in comparison with the overall time taken to study the core levels of the sample. The possibility of X-ray induced decomposition and/or isomerization in surface layers also needs to be kept in mind and is usually evidenced by change in physical appearance of the sample, by temperature and time dependance of spectra.

Contamination of the surface layers of a sample can of course arise in the preparation particularly for ionic compounds prepared by metathetical reactions. For ionic compounds in particular the surface may in any case not be representative of the stoichiometry of the bulk (Figure 22). (This of course very often represents the sort of information one wants to obtain from ESCA but needs to be borne in mind if the technique is being used to analyze the bulk phase.)

To summarize this section we may say that if semi-quantitative or quantitative analysis is to be made by ESCA we need to be sure that we are measuring initially what we think we should be and not something contaminated from one or more sources.

(2) Differing escape depths for levels with different KE's. Since in general for analyses our prime sources of information are the relative areas of the 'elastic' peaks, due allowance must be made for escape depth dependence. From Table I and Figure 41 it is clear that for soft X-rays the binding energies of the majority of the elements are such that mean free paths (escape depths) are typically < 30 Å. By studying different

Surface Contamination

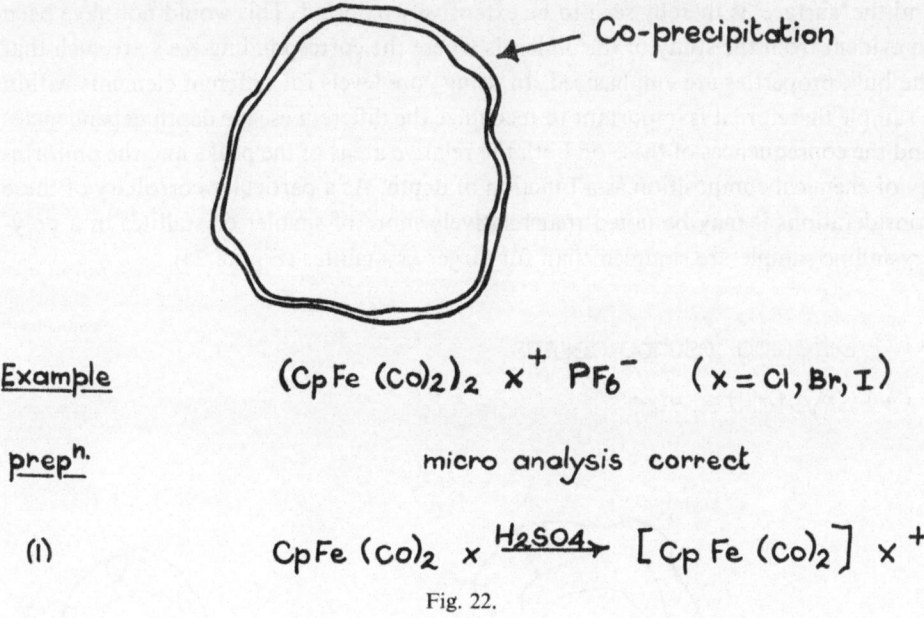

Co-precipitation

Example $(CpFe(CO)_2)_2\ X^+\ PF_6^-\quad (X = Cl, Br, I)$

prep^n.   micro analysis correct

(1) $CpFe(CO)_2\ X \xrightarrow{H_2SO_4} [CpFe(CO)_2]\ X^+$

Fig. 22.

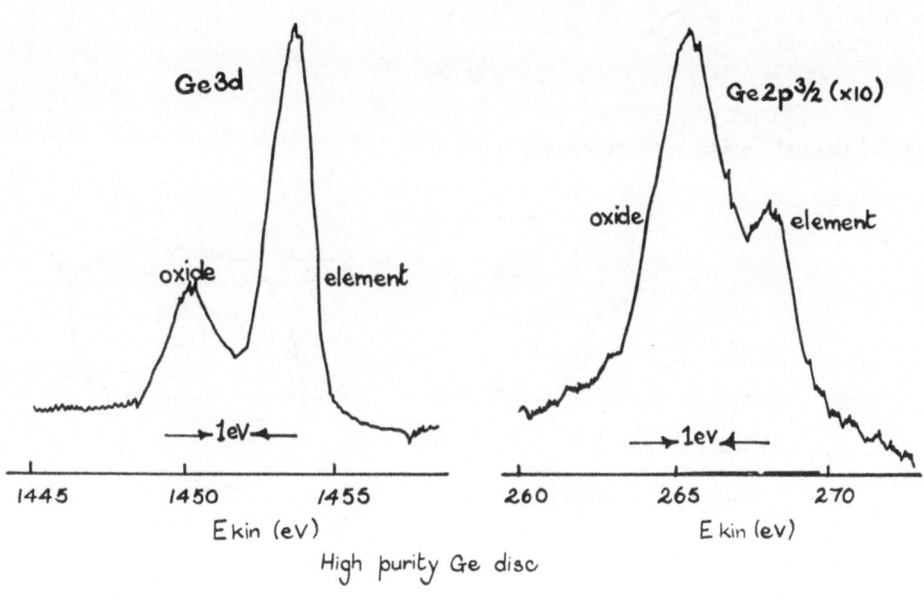

High purity Ge disc

Fig. 23.

core levels of the same element it is often possible to say something about chemical composition as a function of depth. A good example of this is provided by study of a high purity Ge disc (Figure 23). By studying the $2p$ levels of germanium corresponding to very low KE for the photo emitted electrons, the escape depth is small and the 'surface' is thereby seen to be extensively oxidized. This would not have been so evident from the study of the $3d$ levels where the corresponding Ke's are such that the bulk properties are emphasized. In using core levels for different elements within a sample therefore it is important to recognize the different escape depth dependencies and the consequences of these on both the relative areas of the peaks and the uniformity of chemical composition as a function of depth. As a particular corrollary of these considerations it may be noted that relatively more of smaller crystallites in a polycrystalline sample are sampled than for larger crystallites (Figure 24).

(2) Differing escape depths

(4) Crystallite size

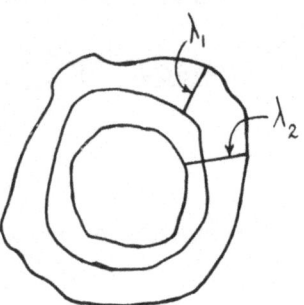 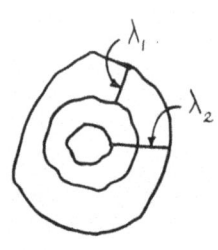

Fig. 24.

(v) Lattice site in solid

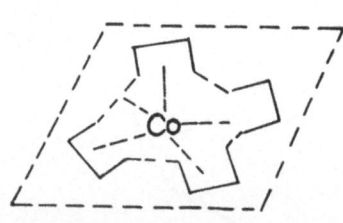

Vitamin B12

Co readily detected.

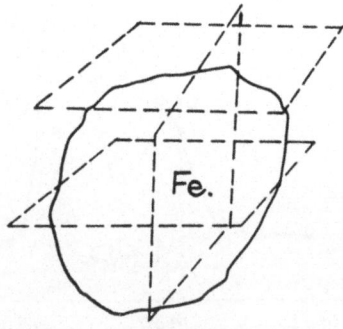

globular protein

Fe difficult to detect.

Fig. 25.

As a result of escape depth considerations account must also be taken of the differing lattice sites in a crystalline solid occupied by a given atom. For example the single cobalt atom in the relatively flat vitamin B12 system is readily detected whereas it is often difficult to detect iron in globular proteins. In the former the lattice sites

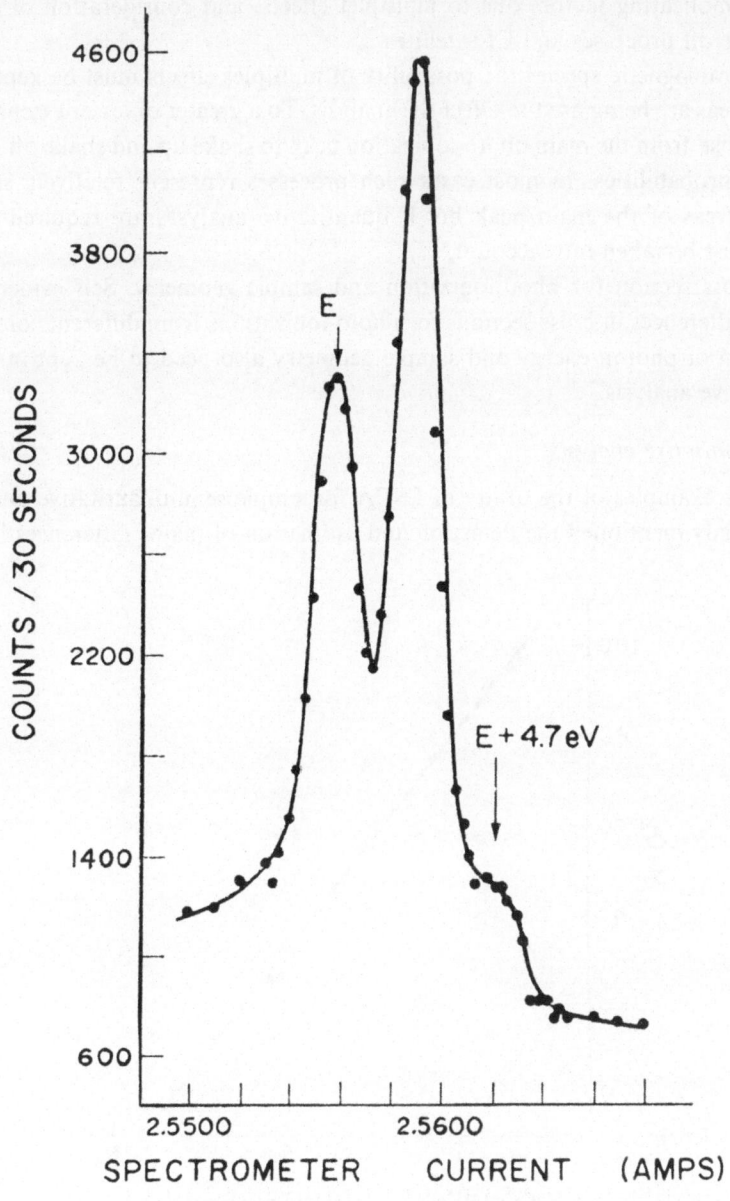

Molybdenum $(3d_{3/2}-3d_{5/2})$ electron
spectrum for a 25:75 $(MoO_3:MoO_2)$ mixture

Fig. 26.

for cobalt are relatively close to the surface so that despite the small escape depth for the electrons corresponding to photo emission from the $2p$ levels the 'elastic' peaks are readily observable. In the case of iron in a globular protein however the lattice site for iron may be such that the escape depth dependence corresponds to small elastic peaks difficult to detect against the background (Figure 25).

(3) Complicating factors due to multiplet effects, and consideration of shake up and shake off processes and CI satellites.

For paramagnetic species the possibility of multiplet effects must be kept in mind if peak areas are being used to effect the analysis. To a greater or lesser extent intensity may be 'lost' from the main photo-ionization peak to shake up and shake off processes of finite probabilities. In most cases such processes represent relatively small percentage areas of the main peak but if quantitative analyses are required then this factor must be taken into account.

(4) Cross section for photoionization and sample geometry. Self evident factors such as differences in cross sections for photo-ionizations from different core levels as a function of photon energy and sample geometry also need to be kept in mind for quantitative analysis.

Semi-quantitative analysis

As simple examples of the utility of ESCA for simple semi-quantitative analyses we have already mentioned the detection and estimation of major differences in surface

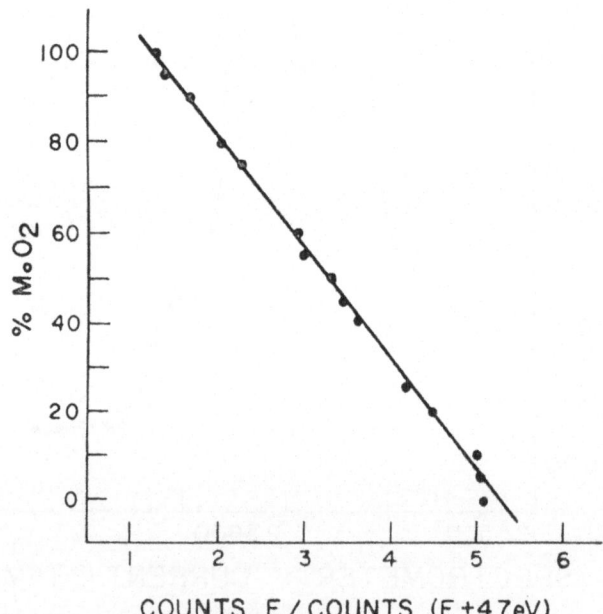

Calibration curve for quantitative analysis of
MoO_3–MoO_2 mixtures

Fig. 27.

and bulk compositions of catalyst samples and order of magnitude estimates of thickness of surface coatings. For covalent lattices where the surface and sub-surface might be expected to be representative of the bulk (in the absence of contamination) ESCA provides a rapid convenient non-destructive means of analysis of for example, the water of crystallization in samples.

Quantitative analysis

There have been few attempts to use ESCA in a truly quantitative way for analysis with an accuracy comparable with more conventional methods. The routine analysis of the composition of alloys was in fact one of the first applications of ESCA to analysis. [1] As a representative example of quantitative analysis of bulk compositions by ESCA, Figures 26–27 [6] and Table VII [6] gives data pertaining to the analysis

TABLE VII

Percentage of MoO_2 determined
for unknown mixtures of MoO_2 and MoO_3

Unknown	MoO_2, %	MoO_2 determined, %
A	51	53 ± 1
B	46	52 ± 6
C	47	50 ± 3
D	80	82 ± 2
E	20	23 ± 2

of oxides of molybdenum by Hercules and co-workers. [6] ESCA has a distinct advantage over the conventional 'wet' analysis which is time consuming. The accuracy represents that routinely available but could undoubtedly be improved upon with a careful consideration of the factors outlined above. Considerable work has been done on the analysis of polymers and will be discussed in a separate section. We have also previously indicated the large number of careful studies of escape depths.

2. Application of ESCA to Studies of Structure and Bonding in Organic Molecules

A. INTRODUCTION

The most extensively studied class of compounds are those of carbon and the application of ESCA to organic chemistry has contributed to both the experimental and theoretical development of the subject. As yet there have been relatively few applications of ESCA to the resolution of structural problems not readily amenable to solution by more conventional means but the subject is rapidly developing. As with any new spectroscopic technique initial development has been along two major lines. Firstly the systematic investigation of series of related molecules to build up a background of data which may allow empirical correlations to be drawn up. This approach follows that used in the development of most other forms of spectroscopy. Parallel to this systematic tabulation of data considerable effort has been expended in the development

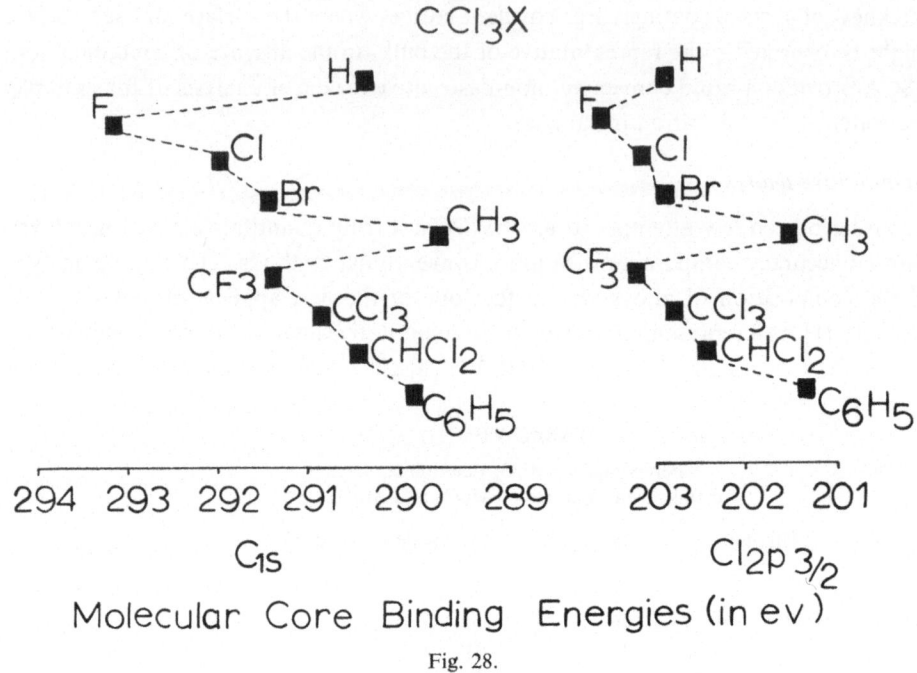

Fig. 28.

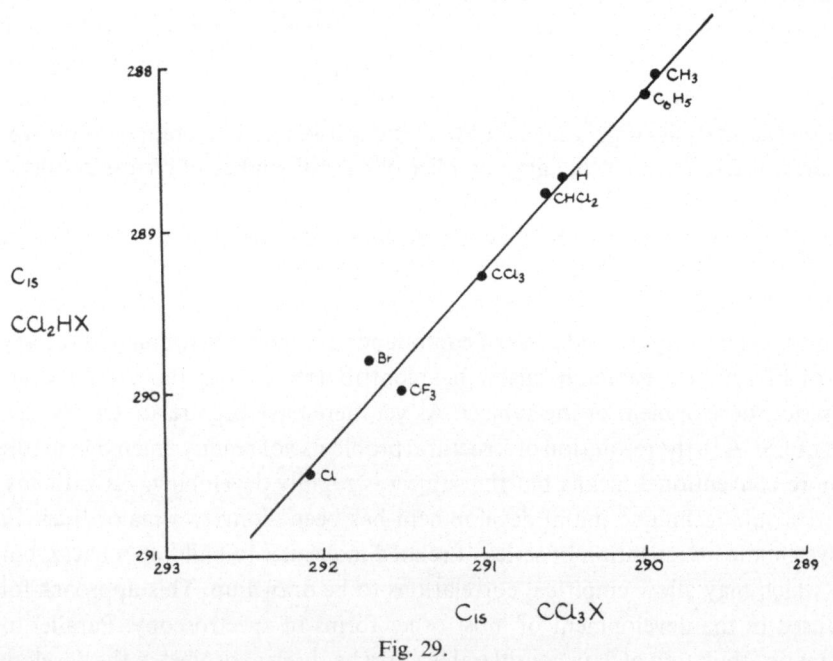

Fig. 29.

of theoretical models at various levels of sophistication to interpret the data. The close but inexact parallel between charge distributions and shifts in core binding energies was recognized early on and can be quantified as will be discussed later on. Systematic studies have been made of substituent effects on C_{1s} core binding energies and as representative examples of an area where ESCA is already fairly well developed we may consider the halocarbon field.

B. SUBSTITUENT EFFECTS

By studying series of closely related compounds as shown in Figure 28 for example substituent effects are clearly discernible and moreover follow 'intuitively' the order one might expect. Thus replacement of hydrogen by halogen increases the C_{1s} binding energies in the order $F > Cl > Br$ which follows the order one might naively expect for the electron density at carbon. As a bonus information may also be obtained on the effect of substituents on the chlorine core levels. By studying a related but different series of molecules and plotting the data, Figure 29, it becomes clear that for saturated systems at any rate substituent effects are highly characteristic; e.g. F and Cl increase C_{1s} binding energies by ~2.8 and 1.7 eV respectively. The short range nature of the factors contributing to shifts in binding energies in saturated systems may be readily demonstrated by studying pairs of compounds e.g. Table VIII. The 'secondary' shift of chlorine of 0.4–0.5 eV may be compared with the primary shift of ~1.7 eV.

TABLE VIII

Secondary shifts

Molecule	C_{1s} binding energy in eV	ΔC_{1s}
CCl$_2$ H–$\underline{C}$H$_3$	285.4	0.5
CCl$_3$–$\underline{C}$H$_3$	285.9	
CCl$_2$ H–$\underline{C}$HCl$_2$	288.8	0.5
CCl$_3$–$\underline{C}$HCl$_2$	289.3	
CHCl$_2$–$\underline{C}$Cl$_3$	290.6	0.4
CCl$_3$–$\underline{C}$Cl$_3$	291.0	
CHCl$_2$–$\underline{C}$F$_3$	294.8	0.4
CCl$_3$–$\underline{C}$F$_3$	295.2	

Detailed studies of aromatic and heterocyclic systems also allows 'information banks' to be built up on substituent effects in unsaturated systems. Studies of simple monosubstituted benzenes for example (Figure 30) show that small longer range effects are possible in unsaturated systems. In fluorobenzene detailed calculations show that the assignment of core levels is as shown with the ring carbon atoms following the order of charge distribution,

meta < ortho < para

which organic chemists *intuitively* feel is correct. (We will return to this point later on.) In studying all the fluorinated benzenes from mono through isomeric di, tri,

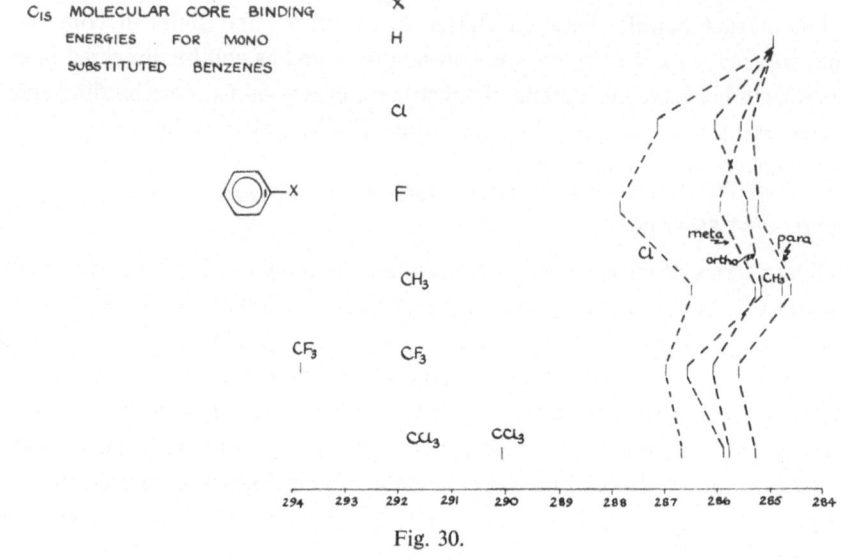

Fig. 30.

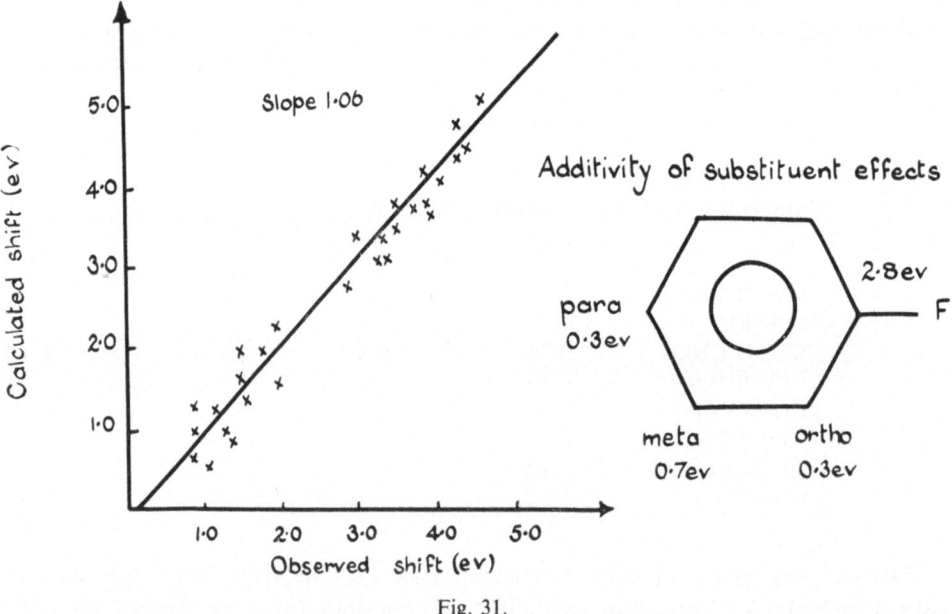

Fig. 31.

tetra, penta and perfluoro compounds we find the data can be nicely rationalized on the basis of a primary substituent effect of 2.8 eV for the carbon to which fluorine is attached, 0.3 eV for carbons ortho and para to the substituent and 0.7 eV for carbons meta to the substituent (Figure 31). With such studies as these, tables of substituent effects may be collected together which may be used empirically in structural studies.

Simple examples illustrative of this type of approach taken from our own research

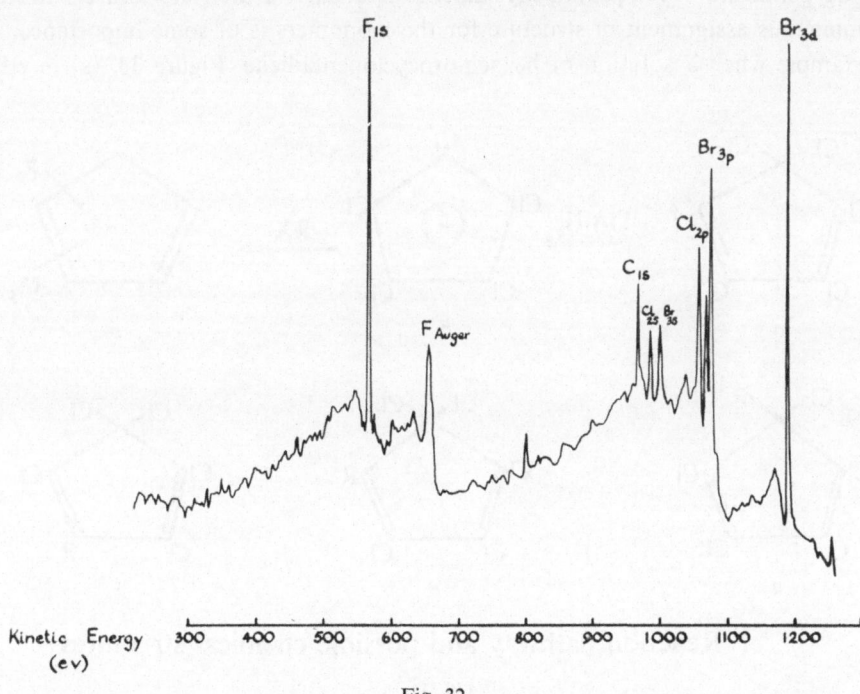

$CF_2Br - CFClBr$

Fig. 32.

interests are as follows. We have previously mentioned the utility of ESCA in element mapping. In the particular case of the simple fluorocarbon (Figure 32) the raw data shows the presence of F, C, Cl and Br. By careful calibration of peak areas etc. the empirical formula $C_2F_3ClBr_2$ may be demonstrated. The tables of substituent effects may then be used to distinguish between the possible isomers as shown below.

$C_2F_3Br_2Cl$

sample size 0.5 μl.

Formula from rel. intensities of core levels.

Isomers and predicted shifts in C_{1s} levels

(1) CF_3-CBr_2Cl

   ΔC_{1S} 4.2 eV

(2) $CF_2Br-CBrClF$

   ΔC_{1S} 1.0 eV

(3) $CF_2Cl-CBr_2F$

   ΔC_{1S} 2.0 eV

Observed 1.1 eV

$\therefore$ CF_2Br-CB_rClF

Although this is a trivial example it does give an indication of how ESCA can be used. A real example of ESCA used as a structural tool where more conventional spectroscopic techniques had not resolved ambiguities is the determination of the structure of allyl pentachlorocyclopentadiene. Cycloaddition polymerization involving pentachlorocyclopentadienyl alkenes is an active area of research and the unambiguous assignment of structure for the monomers is of some importance. As an example when a solution of hexachlorocyclopentadiene, Figure 33, (a) in ether

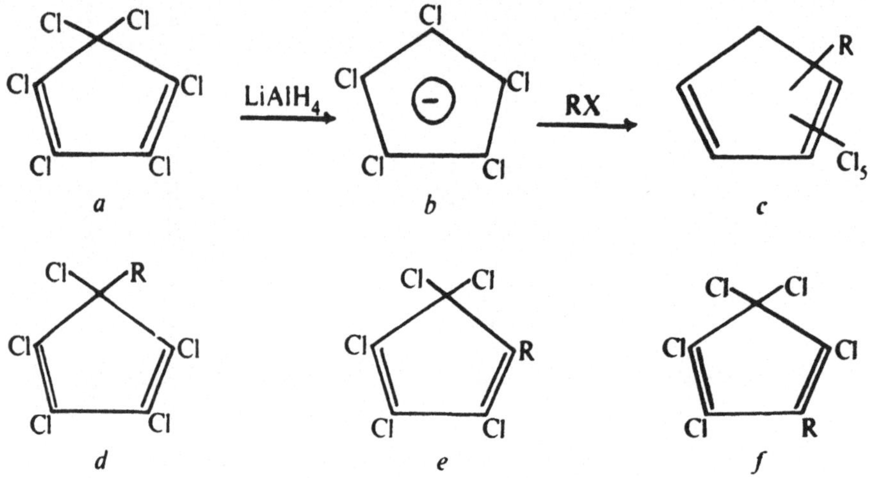

Reaction pathway and possible chemical structures

Fig. 33.

is reacted successively at -20–$30\,°C$ with two molar proportions of $LiAlH_4$ and an organic halide, the chief product is a substituted pentachlorocyclopentadiene (c) the proposed pathway being as shown. If this pathway is correct the initial product must be that with the substituent at C_5 (d) rather than at C_1 (e) or C_2 (f). Using conventional chemical and spectroscopic techniques, however, it has not been possible to give an unequivocal assignment. Reaction of (c) ($R=-CH_2-C_6H_5$) with cyclohexane and subsequent replacement of the chlorines of the Diels-Alder product by hydrogen, has ruled out (f), but failed to distinguish between (d) and (e). Using ESCA an unequivocal assignment may be made on the basis of two experiments each taking less than 45 min and requiring only $\sim0.2\,\mu l$ of sample. Figure 34 shows the C_{1s} spectra of (a) and (c) ($R=-CH_2-CH=CH_2$). In the C_{1s} spectrum of (a) there are two bands at binding energies 289.1 eV and 287.4 eV in the intensity ratio 1:4 corresponding to the dichloromethylene carbon (C_5) and four vinylic carbons respectively. The binding energies correspond quite closely to those predicted on the basis of our previous studies of aliphatic ($\underline{C}Cl_2$ in CH_2Cl_2, 288.7 eV) and unsaturated ($\underline{C}Cl$ in C_6H_5Cl, 287.1 eV) models. In the C_{1s} spectrum of (c) the dichloromethylene carbon level at 289.1 eV has disappeared and the spectrum shows basically two overlapping bands which may

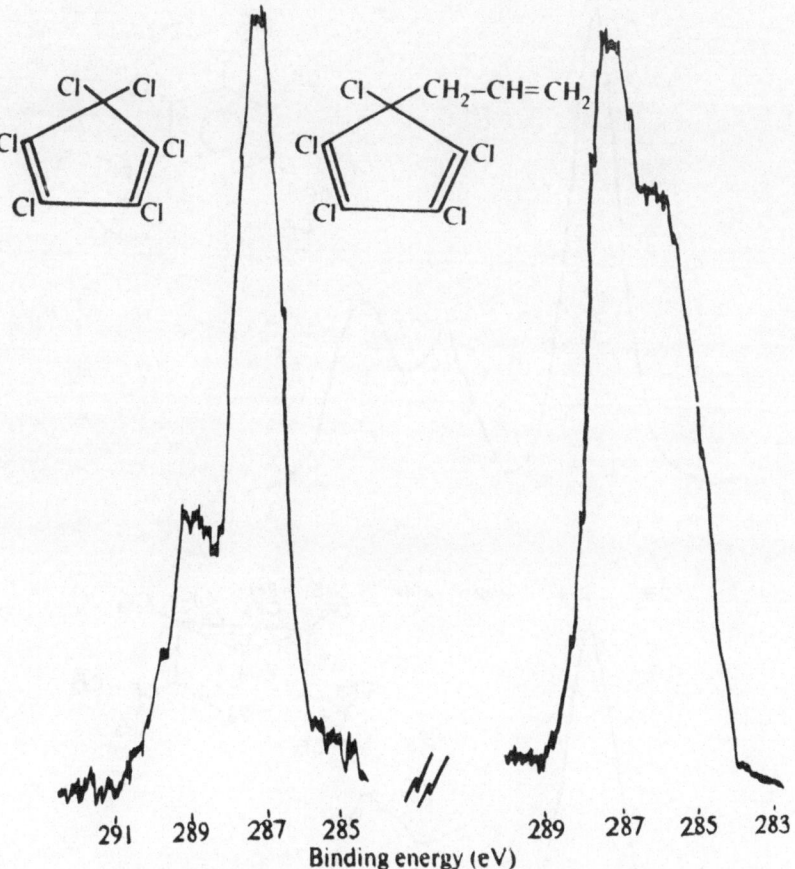

C_{1s} spectra of hexachlorocyclopentadiene and allyl-
pentachlorocyclopentadiene

Fig. 34.

be deconvoluted into area ratios 5:3. The band at 287.4 eV corresponds to a cyclo-
pentadiene residue in which each carbon has one chlorine attached, and the band at
285.9 eV corresponds to the three carbons of the allyl group, thus allowing the unambiguous assignment of structure as being (d).

The overall line profiles for the core levels of molecules may also be conveniently
used to recognize structurally isomeric species cf. Figure 35 and Figure 36. Having
recognized isomeric species the assignment of structures may be made by a consideration of the characteristic binding energies of the core levels and the intensities and
energies of satellite peaks due to shake up processes and if possible the finger print
nature of the low energy photoelectron spectrum of the valence energy levels. A
slightly more complex example of distinction between isomeric compounds on the
basis of molecular core binding energies is provided by the biochemically important
5 and 6 aza uracil ring systems (Figure 37). In this case the N_{1s} and C_{1s} spectra are

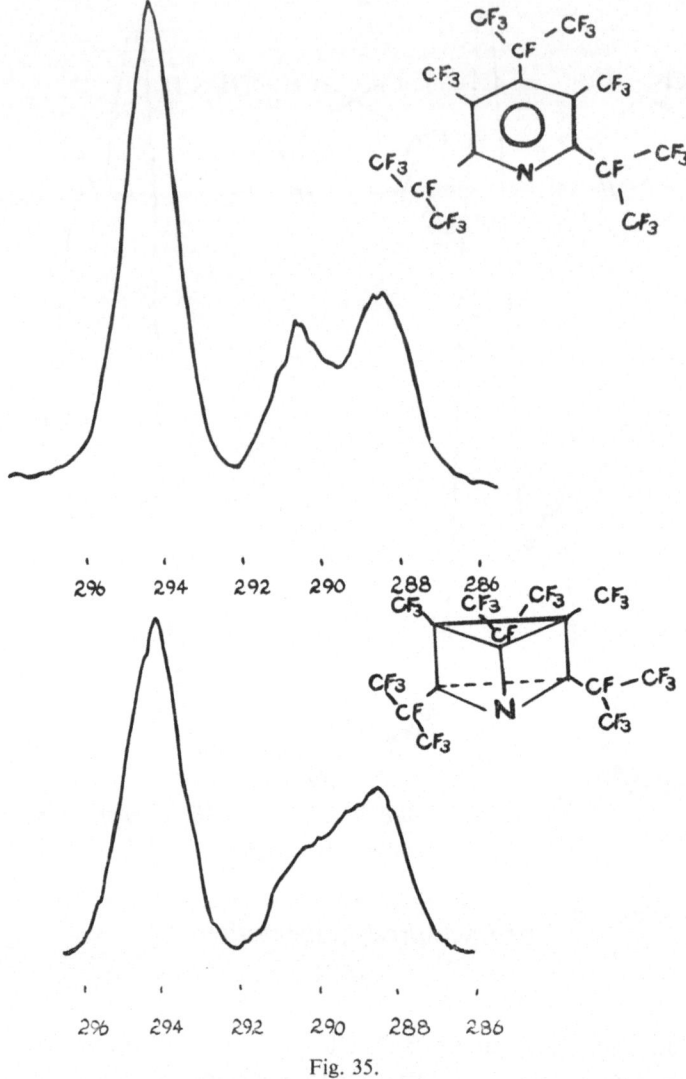

Fig. 35.

such as to allow ready identification of each isomer and comparison with the parent uracil also provides valuable information on the electronic structures of these im-important ring systems.

C. SAMPLE CHARGING, ENERGY REFERENCING AND LINE SHAPE ANALYSIS

It is apparent from a consideration of some of the data presented above that we often obtain incompletely resolved spectra. (The spectra presented in this section are almost exclusively obtained from thin films of the samples condensed onto a conducting backing and excited by $MgK\alpha_{1,2}$ radiation.) Under the experimental conditions routinely employed the dominant contribution to the linewidths arises from the in-

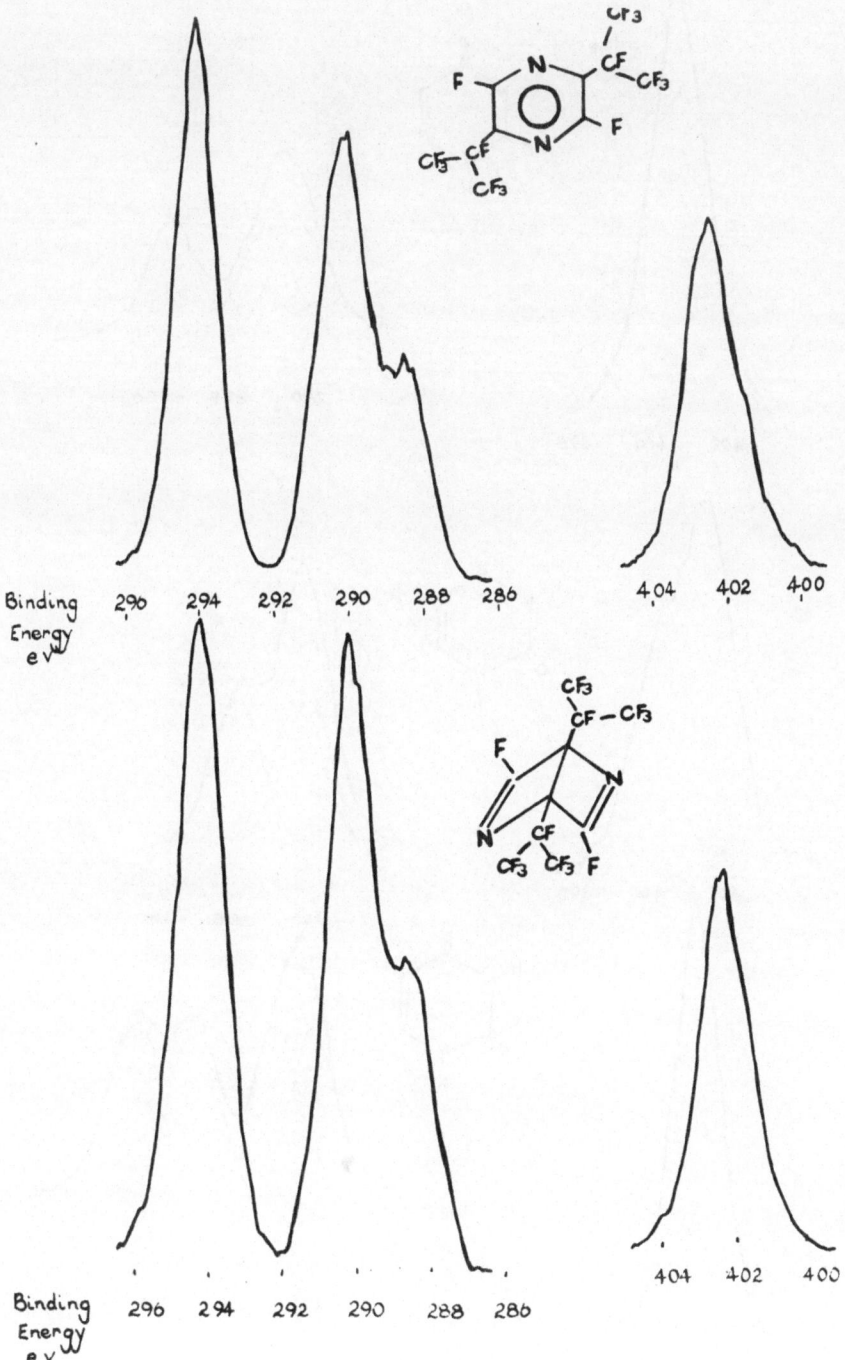

Fig. 36.

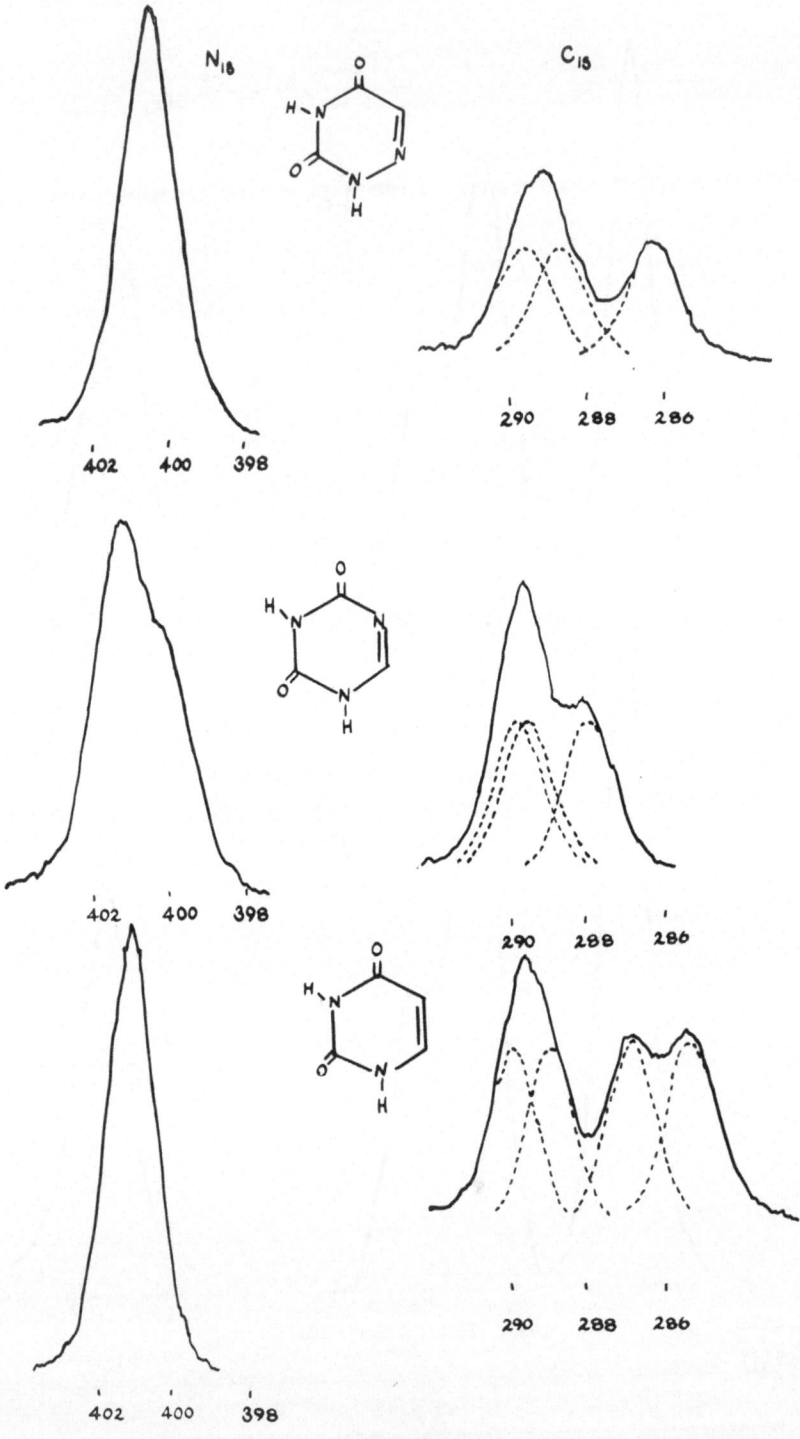

Fig. 37.

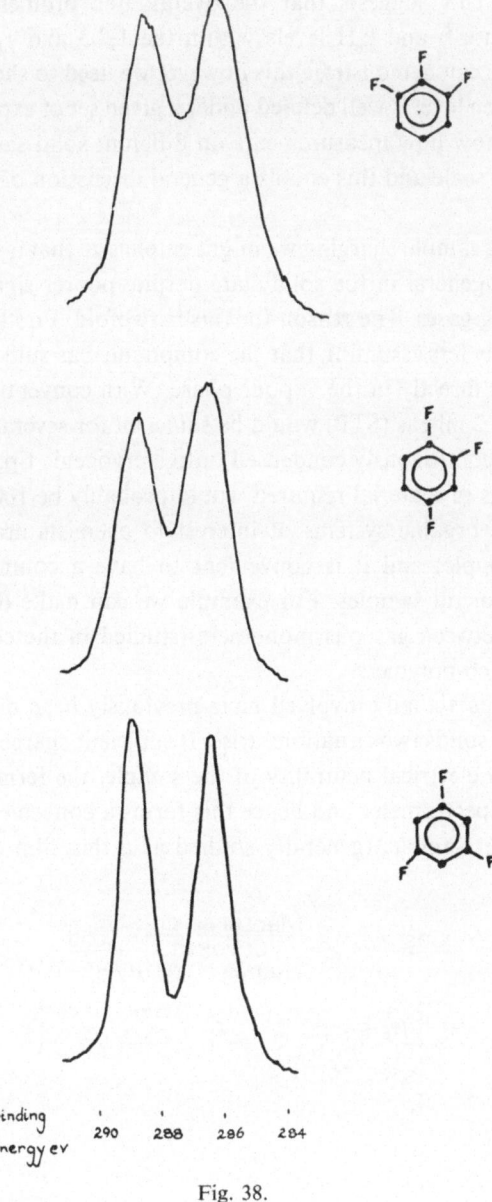

Binding
Energy ev   290   288   286   284

Fig. 38.

herent width of the MgKα$_{1,2}$ photon source ~0.8 eV). The question then arises can we extract useful information from the overall line profiles? For example, Figure 38 shows the C$_{1s}$ spectra for isomeric trifluorobenzenes. Clearly the C$_{1s}$ levels are broadly split into doublets the peaks at higher binding energy corresponding to carbon attached to fluorine the lower binding energy peaks from carbon attached to hydrogen. In the case of the symmetrically substituted 1,3,5,-trifluorobenzenes the peaks are much better resolved than for the less symmetrically substituted 1,2,3- and 1,2,4-trifluorobenzenes. This suggests that the overall line profiles might reflect slight differences between CF and CH levels within the 1,2,3 and 1,2,4 substituted compounds. Before one can demonstrate this however we need to show that the line shape and width for a given level is well defined under a given set of experimental conditions. We also need to show how measurements on different solid samples may be related to the same energy scale and this entails a general discussion of sample charging and line shape analysis.

Before discussing sample charging we might emphasize that it is convenient to study organic systems in general in the solid state despite poorer signal/background compared with studying gases. The reason for this is twofold. Firstly the sample requirement is considerably less assuming that the compound has sufficient vapour pressure for it to be studied directly in the vapour phase. With conventional instrumentation ∼0.2 μl liquid or 0.2 ml gas (STP) would be sufficient for several hours of experiment if the sample were continuously condensed onto a cryogenic tip. For direct gas phase studies the amounts of material required would probably be 100× greater. Secondly a large number of organic systems of interest to chemists are relatively involatile (polymers for example) and it is convenient to have a common energy reference (the Fermi level) for all samples. For example we can make direct comparisons of binding energies between gaseous monomers (studied in the condensed phase) and derived homo and co-polymers.

The measurements actually involved have previously been discussed cf. Figures 8 and 39. In studying solids two situations arise. If sufficient charge carriers are available to maintain overall electrical neutrality of the sample, the fermi level of the sample is the same as the spectrometer and hence this forms a convenient energy reference.* Thus as long as our sample, (generally studied as a thin film deposited on gold) is

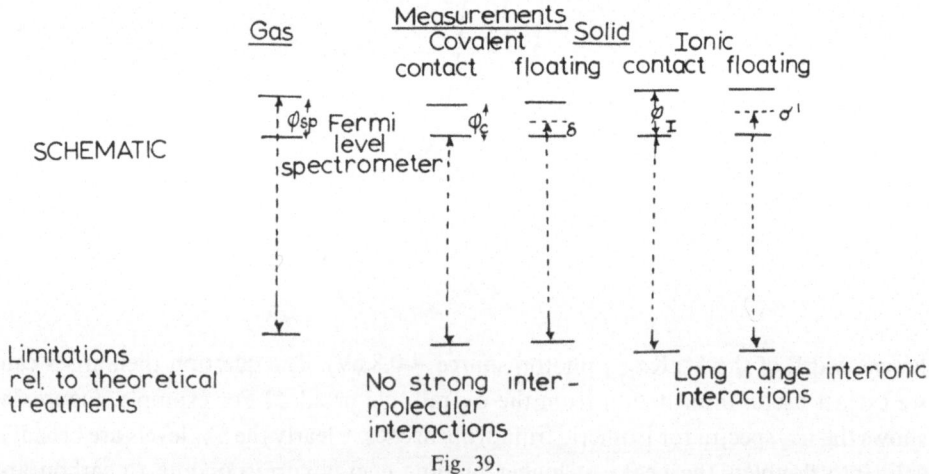

Fig. 39.

* Conventionally the Fermi level for an insulator is nominally located midway between the top of the valence (occupied) and bottom of the conduction (unoccupied) energy levels.

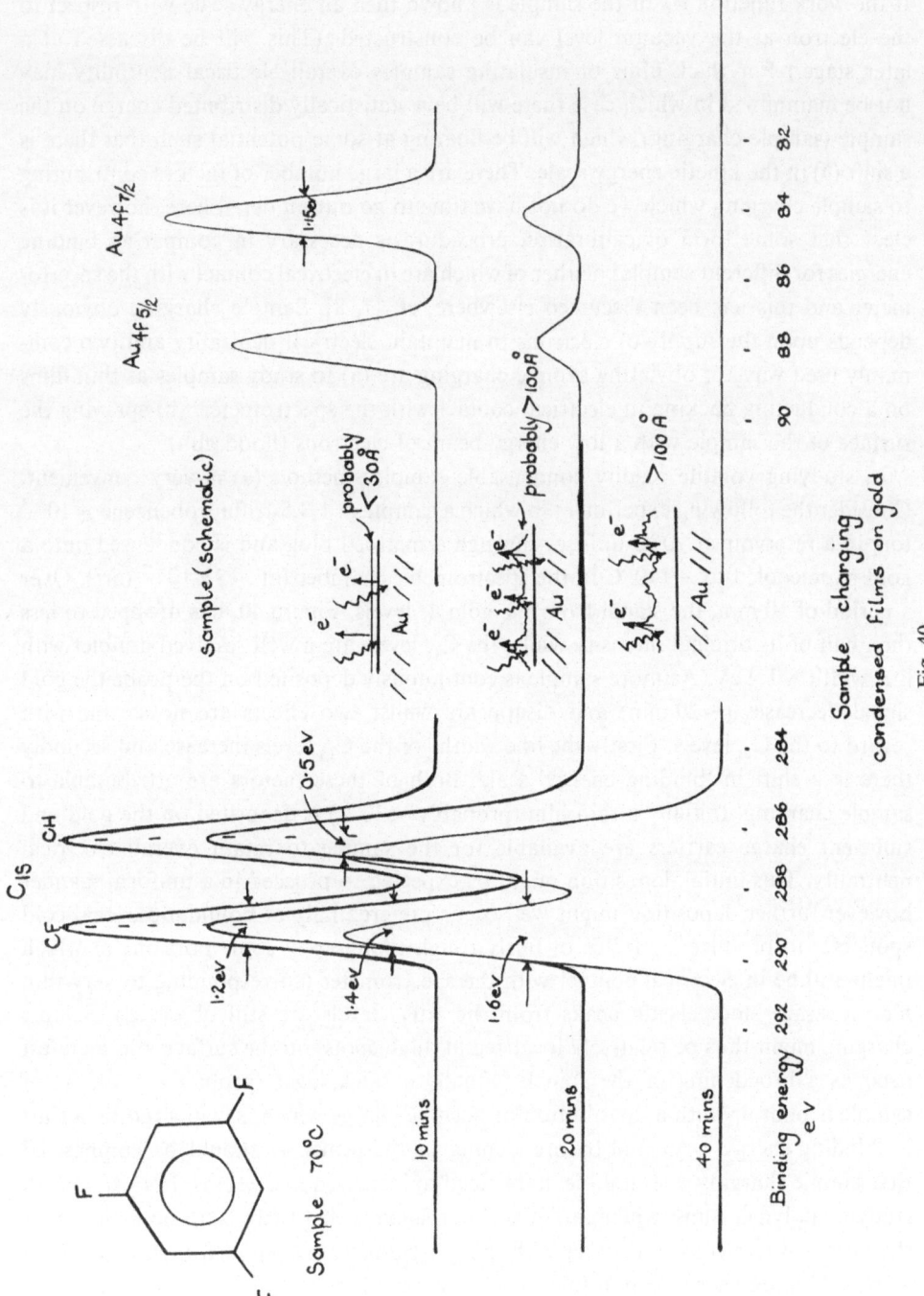

Fig. 40.

in electrical contact with the spectrometer energy referencing presents no problems. If the work function φ_C of the sample is known then an energy scale with respect to the electron at the vacuum level can be constructed. (This will be discussed at a later stage.) For thick films of insulating samples overall electrical neutrality may not be maintained in which case there will be a statistically distributed charge on the sample (sample charging) which will be floating at some potential such that there is a shift (δ) in the kinetic energy scale. There are a large number of factors contributing to sample charging which we do not have time to go into in detail here, however it is clear that some form of calibration procedure is necessary in comparing binding energies for different samples neither of which are in electrical contact with the spectrometer and this has been discussed elsewhere, cf. [7, 8]. Sample charging obviously depends upon the supply of electrons to maintain electrical neutrality and two commonly used ways of obviating sample charging are (a) to study samples as thin films on a conducting backing in electrical contact with the spectrometer. (b) spraying the surface of the sample with a low energy beam of electrons (flood gun).

In studying volatile readily condensable samples method (a) is very convenient. Consider the following experiment in which a sample of 1,3,5-trifluorobenzene $\sim 10^{-3}$ torr in a reservoir at 70 °C diffuses through a metrosil plug and is condensed onto a gold plate cooled to -110 °C in the spectrometer chamber (at $\sim 5 \times 10^{-7}$ torr). Over a period of 10 min, the signal from the gold $4f$ levels, Figure 40, has dropped to less than half of its original intensity whilst the C_{1s} levels are a well resolved doublet with line width ~ 1.2 eV. As more sample is continuously deposited on the probe the gold signal decreases (~ 20 min) and disappears whilst two effects are noticeable with regard to the C_{1s} levels. Firstly the line widths of the C_{1s} levels increase and secondly there is a shift in 'binding energy' scale. Both of these factors are attributable to sample charging. Initially a thin film (probably < 30 Å) is deposited on the gold and sufficient charge carriers are available for the sample to retain overall electrical neutrality. This initial deposition might be expected to proceed in a uniform manner however further deposition might well occur preferentially at boundaries, local cold spots etc. to produce a surface of fairly rough topography some portions of which might still be in electrical contact with the spectrometer (corresponding to very thin film coverage since elastic peaks from the Au_{4f} levels are still observed). Sample charging might thus be relatively localized at 'high spots' on the surface and manifest itself as a broadening of the signals. Finally a thick layer (probably > 100 Å) of sample is built up with a distribution of positive charge which is evidenced by a shift in 'binding energy' scale and by broadening of the peaks. It should be emphasized that sample charging does not inevitably lead to increased line widths. For example in studying polymer films, which are of such a thickness that there is no possibility that the sample is in electrical contact with the spectrometer (which is also clear from the shift in binding energy scale), line widths for core levels have been observed which are as narrow as those obtained from measurements on very thin films (< 30 Å) of monomers condensed onto gold. It appears that if the positive charge is *uniformly* distributed over the surface of the sample, line widths may not be broadened even if there is a shift in the 'binding energy' scale. One of the determining factors is obviously the surface topography which can result in non-uniform sample charging and hence line broadening. Repeat experiments show that initial measurements of C_{1s} binding energies are reproducible, corresponding to thin films in electrical contact with spectrometer, and this therefore suggests one way of energy referencing different samples.

In studying a closely related series of molecules such as the fluorobenzenes under closely controlled conditions (e.g. as thin films on gold), an estimate may be made of linewidths of individual levels. In the case of the symmetrical systems viz., benzene, 1,4-difluorobenzene, 1,3,5-trifluorobenzene, 1,2,4,5-tetrafluorobenzene and hexafluorobenzene in which only one or two types of chemically equivalent carbon are present (CF or C–H) the carbon 1s linewidths may be obtained by direct measurement. Under the experimental conditions employed both the linewidths (1.3 ± 0.1 eV) and line shapes are well defined and this is of some importance in attempting to extract additional data from the analysis of line shapes for more complex samples.

The basic philisophy behind such an analysis may be set out as follows: (It should be emphasized that such an analysis can be carried out in either an analogue or digital fashion. In general since one needs to have several variables under close control at any one time, analogue computer analysis is by far the most convenient.)

Steps in line shape analysis
(Analogue or Digital)
(Compounds of known structure)

(1) Study closely related series where no ambiguities arise and establish.
    (a) Line shape
    (b) Line width
(2) If (1) satisfied within narrow limits proceed to analyse line profiles with input.
    (a) Maximum number of peaks (i.e. different core levels) (can be established independently) i.e. use all available *chemical* data.
    (b) Peak areas (can be established independently).
(3) Fit line shape with peak position as *only* parameters.
    (a) Unique solution?
    (b) Several equally good solutions? (v. rare)
        (Reoptimize fit with line width allowed to float between limits previously set).
    (c) No solution? Check chemistry of system.
(4) Do any of the solutions make sense?
    (a) In terms of measurements actually made?
    (b) In terms of comparison with other data which did not depend on line shape analysis?
    (c) In terms of the chemistry of the system?
(5) If the answer to (4) is NO the 'deconvolution' is treated as spurious and rejected.
(6) If the answer to (4) is YES need to establish 'correct solution' if more than one obtained in (3) (some may be excluded on chemical grounds).

(7) Establish quantitative model which accurately reproduces shifts etc. in related systems where line shape analysis is not necessary.

(8) Apply Theoretical model for assignment.

(a) Very rare that this does not resolve any ambiguities.

(b) Results not compatible with theoretical model, reject deconvolution.

(9) Do the results fit in with data established without deconvolution?

(10) Are the results quantitatively described by the model?

(11) Check for self consistency.

(12) Alternative explanations of data.

(a) Drastically changed line widths (lifetime, lattice) and/or line shape (shake up?) evidence?

As an example we may consider 1,2,3-trifluorobenzene which has been discussed in detail elsewhere. Figure 41 ahows the analogue curve resolution of the C_{1s} levels, (a) experimental C_{1s} spectrum, (b) experimental envelope (a) normalized to a horizontal baseline showing component peaks giving optimum fit to available data, (c) optimum fit to experimental envelope obtained with the following linewidths:

$$(----) = \text{CH levels 1.3 eV, CF levels 1.3 eV};$$
$$(——) = \text{CH levels 1.3 eV, CF levels 1.2 eV};$$
$$(....) = \text{CH levels 1.3 eV, CF levels 1.4 eV}.$$

The latter illustrates the effect of quite small changes in linewidths to the overall envelope. With currently available instrumentation overall linewidths are such that line

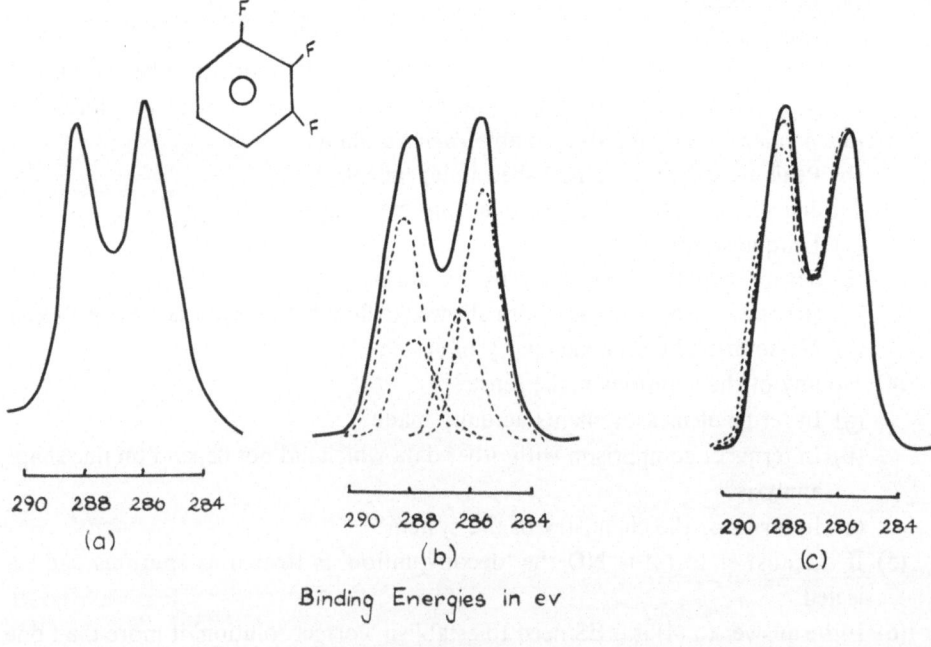

(a)

290 288 286 284

(b)

290 288 286 284

(c)

290 288 286 284

Binding Energies in ev

Fig. 41.

shape analysis is only feasible for relatively simple systems where there is sufficient information for unique solutions to be obtained. We might emphasize at this stage that a mathematically unique solution of a line shape analysis is never really feasible but this is of no real consequence since we are using all the information (chemical and physical) to select that particular mathematical solution which fits this information and provides a unique chemically meaningful solution. This crucial aspect of line shape analysis has been misunderstood in the past.

One of the first attempts to derive detailed information from line shape analysis was our own work on the biologically important molecules adenine, cytosine and thymine. [9] By studying large numbers of related systems to obtain data on line shapes and line widths it proved possible to extract valuable information on shifts in core binding energies for these molecules. By not fully understanding either the fundamental basis of either the chemistry or more importantly the physics of the measurements involved and of the analysis of the line shapes the claim has subsequently been made [10] that no useful information can be extracted from analysis of the line shapes. This is clearly not so and with due caution the analysis of line shapes can be very rewarding.

The need for line shape analysis arises from the unfavourable ratio of shift to linewidth ratio which is one of the major weaknesses of ESCA compared with say NMR or NQR. The dominant contribution to the linewidths with most commercial instrumentation is the inherent width of the unmonochromatised photon source. With improvements in instrumentation mentioned in the previous section, for simple systems the need for line shape analysis may well largely disappear and unambiguous

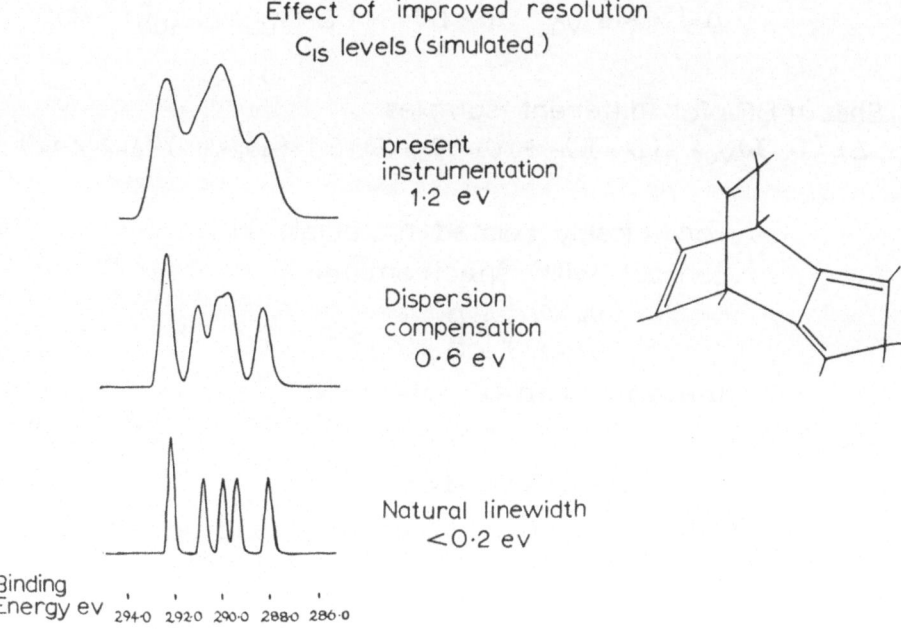

Effect of improved resolution
C_{1S} levels (simulated)

present
instrumentation
1·2 eV

Dispersion
compensation
0·6 eV

Natural linewidth
<0·2 eV

Binding
Energy eV 294·0 292·0 290·0 288·0 286·0

Fig. 42.

analysis of more complex systems become feasible. For example Figure 42 shows the C_{1s} spectrum of dodecafluorotricyclo [5.2.2.0^{2,6}] undeca-2,5,8-triene with a variety of linewidths. The analysis of the line shape, even for the linewidth (1.2 eV) appropriate to our present instrumentation, is straightforward into five basic components. With a well designed monochromator and either slit filtering or dispersion compensation a linewidth ~0.6 eV should be feasible and the improvement in resolution is quite dramatic. At the limit of the C_{1s} natural linewidth (probably <0.2 eV in the gas phase) the peaks would be completely resolved.

Relationship between energy levels of
gaseous and solid sample.
(Covalent, no strong intermolecular interactions)

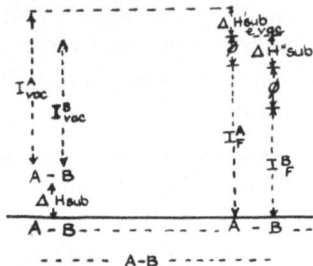

For organic molecules
Typical values
$I^{CIS} \sim 290$ e.v.
$\phi^{vac} \sim 5$ e.v.
$\Delta H_{sub} \sim 0.5$ e.v.

Shift in BE
$$\Delta = (I_A - I_B)_{vac} = (I_A - I_B)_F + (\Delta H'_{sub} = \Delta H_{\widetilde{sub}})$$
$$\overset{''}{O}$$

Shift in BE for different samples
$$\Delta = (I_A - I_x)_{vac} = (I_A - I_x)_F + (\phi_A - \phi_x) + (\delta_A - \delta_x) + (\Delta H_x - \Delta H_A) + (\Delta H'_A - \Delta H_x')$$

For closely related materials in
contact with spectrometer
$$(\phi_A - \phi_x) \approx 0$$
$$(\delta_A - \delta_x) = 0$$
$$(\Delta H_x - \Delta H_A) + (\Delta H_A' - \Delta H_x') \approx 0$$

$$(I_A - I_x)_{vac} \equiv (I_A - I_x)_F$$

∴ can understand shifts in terms of isolated molecules.
(excluding strong intermolecular interactions
eg. H bonding)

Fig. 43.

D. THE CHARGE POTENTIAL MODEL

In deconvoluting complex spectra ambiguities which may arise can often be resolved by appealing to model calculations. The theoretical models available for the interpretation of data have been discussed elsewhere in this review. In general the complexity of many of the organic systems studied renders a rigorous approach impossible. A theoretically valid but computationally inexpensive model at a slightly lower level of sophistication must therefore in general be used and the most successful of these is the charge potential model.

Before discussing the use of the charge potential model in the development of applications of ESCA to studies of structure and bonding it is worthwhile digressing to consider the fundamental relationship between theory and experiment. Theoretical calculations inevitably refer to isolated molecules in the gas phase with the vacuum level as energy reference. The measurements on the other hand refer to the condensed phase with the Fermi level as energy reference. The question then arises, under what circumstance will the direct comparison of the two be valid. For covalent solids in which there are no appreciable long range interactions (such as hydrogen bonding for example) the comparison is most readily seen from an appropriate Born cycle and the associated equations schematically presented in Figure 43. The terms are largely self explanatory and it can be seen that if the work function of the samples are closely similar and if there is no sample charging then the comparison between theoretical calculations on isolated molecules and measurements on thin films should be valid.

A non rigorous derivation of the equation relating binding energies to charge distributions may be made, starting from Koopmans' Theorem.

The crucial feature in the derivation of the charge potential equation is the constancy with varying electronic environments of many of the terms which arise (cf. Table IX).

The orbital energy of the core orbital ψ_r localized on atom m is given by Equation (1):

$$\varepsilon_r = \langle \psi_r | -\tfrac{1}{2}\nabla_1^2 - \sum_n \frac{Z_n}{r_{n1}} |\psi_r\rangle + \sum_{s=a}^{k} (2J_{rs} - K_{rs}) =$$

$$= \langle \psi_r | -\tfrac{1}{2}\nabla_1^2 - \frac{Z_m}{r_{m1}} |\psi_r\rangle + J_{rr} + \langle \psi_r | -$$

$$- \sum_{n \neq m} \frac{Z_n}{r_{n1}} |\psi_r\rangle + \sum_{s \neq r}^{k} (2J_{rs} - K_{rs}). \tag{1}$$

The terms $\langle \psi_r | -\tfrac{1}{2}\nabla_1^2 - (Z_m/r_{m1}) |\psi_r\rangle$ and J_{rr} (the Coulomb integral) are constants essentially independent of valence electron distribution. Grouping the constant terms into E_0 and neglecting the K_{rs} (exchange integrals, ψ_s orbitals on other atoms in molecule) in Equation (1), this transforms to Equation (2).

TABLE IX

C_{1s} levels

Molecule	Atom	q	$\langle \psi_r \mid -\frac{1}{2}\nabla_1^2 \mid \psi_r \rangle$ AU	$\langle 1/R \rangle$	Koopmans' shift
$H-C\equiv C-F$	$H-\underline{C}$	-0.061	16.0168	5.6559	0.33 eV
	$F-\underline{C}$	$+0.112$	16.0181	5.6563	2.96 eV
$H-C\equiv C-Cl$	$H-\underline{C}$	-0.136	16.0189	5.6515	0.24 eV
	$Cl-\underline{C}$	-0.078	16.0175	5.6515	1.60 eV

$$\varepsilon_r = E_0 + \langle \psi_r \mid - \sum \frac{Z_n}{r_{n1}} \mid \psi_r \rangle + \sum_{s \neq r}^{k} 2J_{rs}. \tag{2}$$

The interaction between $|\psi_r|^2$, localized on atom m, and the nuclei of other atoms may be approximated by an interaction between point charges $\sum_{n \neq m} (Z_n/r_{nm})$ and so Equation (2) may be transformed into Equation (3)

$$\varepsilon_r \approx E_0 + \sum_{n \neq m} - \frac{Z_n}{r_{nm}} + \sum_{s \neq r}^{k} 2J_{rs}. \tag{3}$$

The terms in the last summation including $\psi_j (\neq \psi_r)$, core orbitals on all atoms n, may be approximated by

$$J_{rj} = \langle \psi_r(1) \psi_j(2) \mid \frac{1}{r_{12}} \mid \psi_r(1) \psi_j(2) \rangle \approx \frac{1}{r_{nm}}.$$

The terms including the valence MO's ψ_i may be written in a LCAO approximation for ψ_i as:

$$J_{ri} = \sum_p c_{ip}^2 \langle \psi_r(1) \phi_p(2) \mid \frac{1}{r_{12}} \mid \psi_r(1) \phi_p(2) \rangle.$$

They may therefore be divided into two types of terms, whether
(i) ϕ_p is a valence AO on atom m, or
(ii) ϕ_p is a valence AO on atom n.
The contribution of the former type of terms to J_{ri} is

$$\sum c_{ip}^2 \langle \psi_r(1) \phi_p(2) \mid \frac{1}{r_{12}} \mid \psi_r(1) \phi_p(2) \rangle$$

where the sum is over all ϕ_p on atom m. The remaining terms in J_{ri} may be approximated as

$$\sum_n \sum_{\text{all } \phi_p \text{ on } n} c_{ip}^2 \frac{1}{r_{nm}}$$

so that Equation (3) becomes

$$\varepsilon_r = E_0 + \sum_{n \neq m} - \frac{Z_n}{r_{nm}} + \sum_{n \neq m} \frac{z_n}{r_{nm}} + \sum_{n \neq m} \sum_{\text{all } \phi_p \text{ on } n} \frac{2c_{ip}^2}{r_{nm}} +$$

$$+ \sum_{\text{all } \phi_p \text{ on } n} 2c_{ip}^2 \langle \psi_r(1) \phi_p(2) | \frac{1}{r_{12}} | \psi_r(1) \phi_p(2) \rangle$$

where z_n is the number of core electrons on n.

A diagrammatic representation of core and valence levels is given in Figure 44.

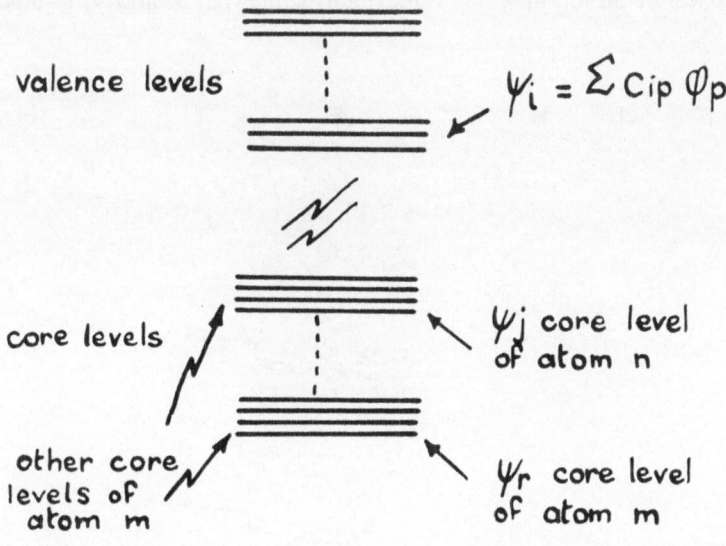

Fig. 44.

Collecting terms common to each atom and those containing valence orbitals ϕ_p on m, we obtain

$$\varepsilon_r = E_0 + \sum_{n \neq m} \left\{ \left(- \frac{Z_n}{r_{nm}} + \frac{z_n}{r_{nm}} \right) + \sum_{\text{all } \phi_p \text{ on } n} \frac{2c_{ip}^2}{r_{nm}} \right\} +$$

$$+ \sum_{\text{all } \phi_p \text{ on } n} 2c_{ip}^2 \langle \psi_r(1) \phi_p(2) | \frac{1}{r_{12}} | \psi_r(1) \phi_p(2) \rangle.$$

Remembering that

$$- Z_n + z_n + \sum_{\text{all } \phi_p \text{ on } n} 2c_{ip}^2$$

is the charge q_n on atom n, we get

$$\varepsilon_r = E_0 + \sum_{n \neq m} \frac{q_n}{r_{nm}} + \sum_{\text{all } \phi_p \text{ on } m} 2c_{ip}^2 \langle \psi_r(1) \phi_p(2) | \frac{1}{r_{12}} | \psi_r(1) \phi_p(2) \rangle.$$

Because $2c_{ip}^2$, the valence electron population on m, is proportional to the charge on m, q_m, we get finally

$$\varepsilon_r = E_0 + kq_m + \sum_{n \neq m} \frac{q_n}{r_{nm}}.$$

The derivation also shows that k is approximately the one centre coulomb repulsion integral between a core and valence electron on a given atom. The charge potential model is of particular importance to organic chemists since it relates charge densities (the theoretical definition of which is inevitably somewhat arbitrary) to binding ener-

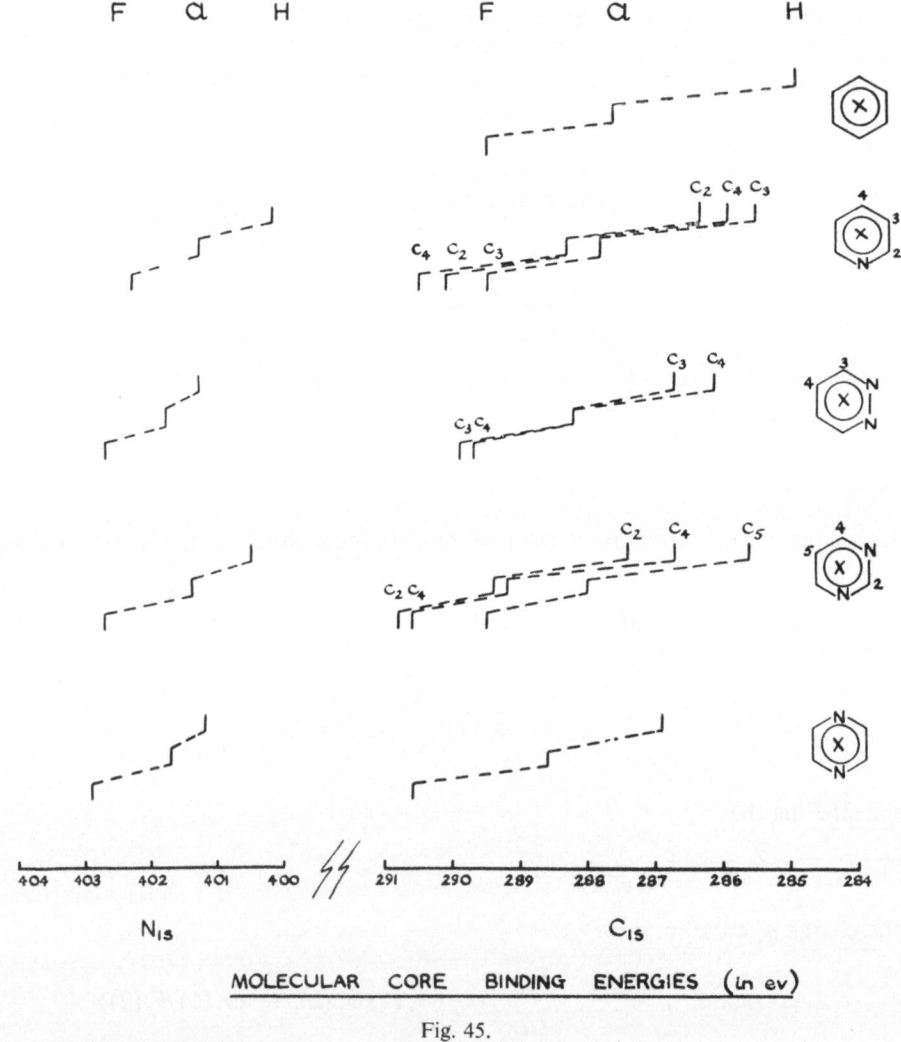

Fig. 45.

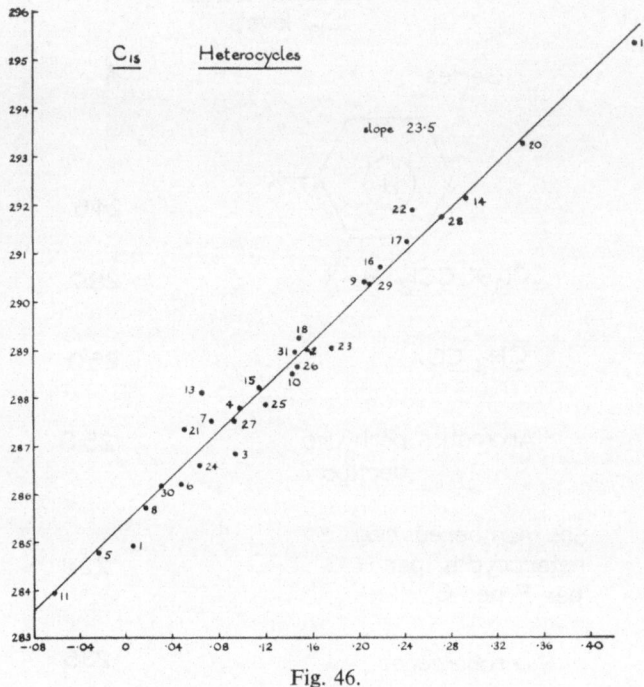

Fig. 46.

gies. The third term represents an intramolecular Madelung type potential arising from charges on other atoms in the molecule.

Using all valence electron CNDO/2 SCF MO calculations the quantitative discussion of data on quite complex molecules becomes feasible. For example the data in Figure 45 pertaining to C_{1s} levels for perhydro, perchloro and perfluoro aromatic and hetero aromatic compounds gives the charge potential correlation shown in Figure 46. The correlation is very good and from the slope and intercept, values of k and E_0 may be established. Since the charge potential model may be related to Koopmans' Theorem which neglects electronic relaxation of the valence electrons (on ejection of a core electron), it must suffer from some of the same deficiences. Since electronic relaxation will depend on the electronic structure of a molecule it is only for closely related series of molecules that one might expect Koopmans' Theorem to provide a quantitative interpretation of shifts in core binding energies. Similarly for the charge potential model we might expect different values of k and E_0 for different series of *closely related* compounds. This is indeed the case. Some values appropriate to series of organic molecules are given in Figure 47 where it is evident that with charges computed from CNDO/2 calculations the value of k clusters around 25.0 eV for C_{1s} levels. Similar correlations may be made for other levels e.g. F_{1s}, Cl_{2p}, N_{1s} etc.

Before proceeding to outline the central role of the charge potential model in the development of ESCA as a technique for studies of structure and bonding it is worth

<u>Charge</u>   <u>Potential</u> <u>Model</u>
C_{1S} levels

Series k

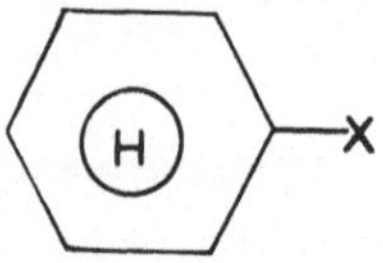

                   24·6

$\underline{C}Cl_3\text{-}X, \underline{C}Cl_2\ H\text{-}X$ 28·7

$\underline{C}H_3\ \underline{C}OX$ 25·0

Aromatics perhydro   25·0
perfluoro

Six membered ring
heterocycles per H   22·4
per F per Cl

Fluorobenzenes   23·5

Five membered ring   25·4
heterocycles

Fig. 47.

considering briefly the relationship between *shifts* in core binding energies and organic chemists intuitive ideas concerning 'charge distribution'. Since valence electron distributions in molecules are continuous functions the assignment of 'charges' to atoms within a molecule is somewhat arbitrary and depends on how the overlap density is partitioned between atoms. As such, theoretically calculated charge densities are only crude guides to the electron density about an atom. Nevertheless the idea of charge distributions in molecules is deeply rooted in chemists minds and within its limitations is a useful concept.

For example an organic chemist might infer from a study of the chemistry of the system the order of decreasing electron density (increasing charge) of C3(C5)> C4> > C2(C6) for pyridine. Similarly from studies of the chemistries of pyridine and perfluoropyridine it could be argued that the charge density at nitrogen is more positive in the case of the perfluoro compound. These charge distributions inferred by studying the chemistry of these molecules might be termed the organic chemists 'intuitive charge distributions'. (The charge distributions the organic chemist intuitively feels are correct from his knowledge of the chemistry of the systems.) In the particular case of the charge densities on nitrogen in pyridine and perfluoropyridine the line of argument used to deduce the 'intuitive charge distribution' might run as

follows. Perfluoropyridine is a much weaker base than pyridine therefore it is energetically unfavourable to bring a proton up to the nitrogen and the charge density
on nitrogen must therefore be much more positive for the perfluoro compound. It
is evident that what the organic chemist is calling 'charge density' is really a measure
of the potential experienced by the approaching proton. [In general the arguments
would refer to an approaching nucleophile (nucleophilic substitution) or electrophile
(electrophilic substitution) in inferring charge distributions from the chemistry of the
systems.] The potential however will depend on the charge distribution not only on
the atom concerned but also on the charges on the other atoms in the molecule. It is
clear from an inspection of the charge potential model involved in discussing shifts
in core binding energies that there are close similarities between the factors determining 'intuitive charge distributions' and shifts. This makes ESCA a particularly
valuable tool for organic chemists since shifts are closely related to his intuitive charge
distributions rather than *actual* charge distributions. From Figure 45 it is clear that
the C_{1s} and N_{1s} shifts parallel the 'intuitive charge distributions'. To highlight this
close parallel it is interesting to note that CNDO calculations show that the electron
density on nitrogen is actually higher (charge more negative) in perfluoropyridine
than in pyridine. The fact that the N_{1s} binding energy for the latter is much lower
arises from large contributions to the chemical shift from charges on the other atoms
in the two molecules.

E. INVERSION OF THE CHARGE POTENTIAL MODEL

The uses of the charge potential model in studies of structure and bonding in organic
molecules is illustrated in Figure 48. Starting on the LHS if we have geometries and
appropriate charge distributions (e.g. CNDO/2) we may use the experimental shifts
to obtain values of k and E^0 for a given level of a given element. We are now in a

USES OF THE CHARGE POTENTIAL MODEL

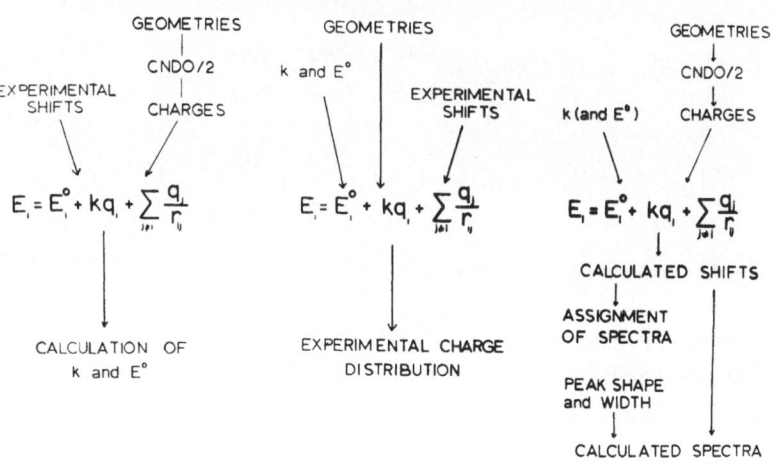

Fig. 48.

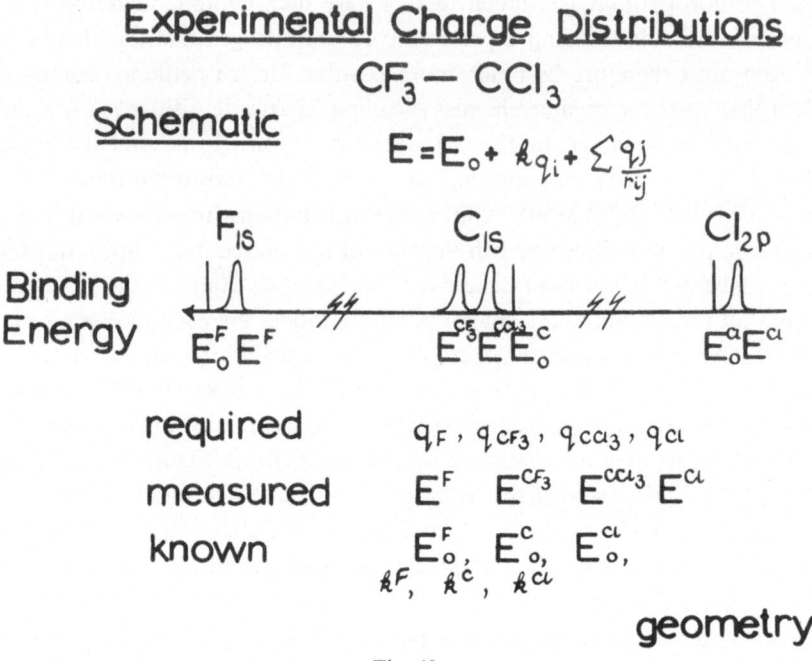

Fig. 49.

position to invert the charge potential model viz. by feeding in the geometry, appropriate values of k and E^0 (established from studies of related systems), the measured binding energies may be used to obtain 'experimental' charge distributions. This is outlined more fully for a simple halocarbon in Figures 49. Schematically the equations for F_{1s}, C_{1s} and Cl_{2p} levels are as follows:

$$\underline{F}_{1s} \quad E^F = E_0^F + k^F q_F + \frac{2q_F}{r_{C-F}} + \frac{3q_{Cl}}{r_{FCl}} + \frac{q_{CF_3}}{r_{F-CF_3}} + \frac{q_{CCl_3}}{r_{F-CCl_3}}$$

$$\vdots$$

$$\underline{C}_{1s} \quad E_{CF_3}^C = E_0^C + k q_{CF_3} + \frac{3q_F}{r_{C-F}} + \frac{3q_{Cl}}{r_{C-Cl}} + \frac{q_{CCl_3}}{r_{C-C}}$$

$$\vdots$$

$$\underline{Cl}_{2p} \quad E^{Cl} = E_0^{Cl} + k q_{Cl} + \frac{3q_F}{r_{Cl-F}} + \frac{2q_{Cl}}{r_{Cl-C_1}} + \frac{q_{CF_3}}{r_{Cl-C_1}} + \frac{q_{CCl_3}}{r_{Cl-C_2}}$$

$$\vdots$$

Also $3q_F + 3q_{Cl} + q_{CF_3} + q_{CCl_3} = 0$

and their relation is made clear.

Since the mathematical manipulations involved are trivial this method can be used to obtain charge distributions in complex systems for which direct computation would be either prohibitively expensive or impracticable. The method therefore has great potential. As a simple example Figure 50 shows the experimental and theoretical

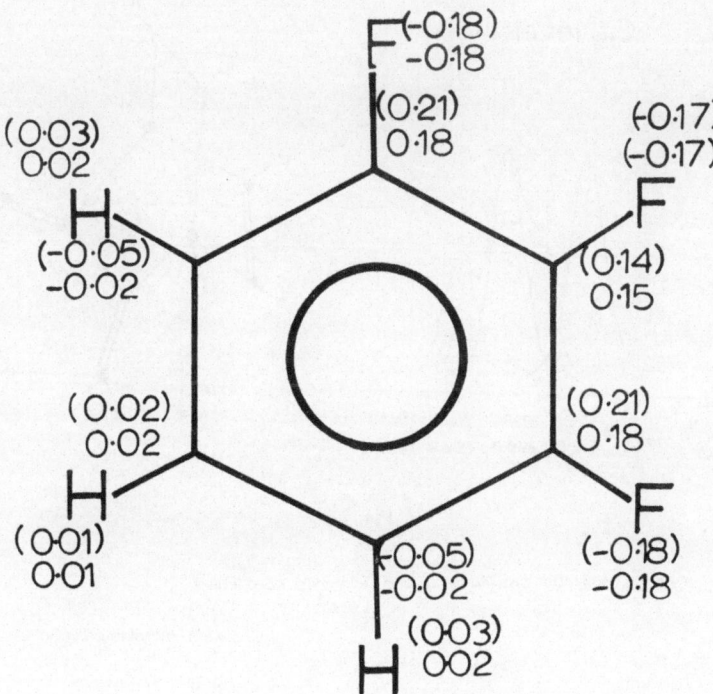

Fig. 50.

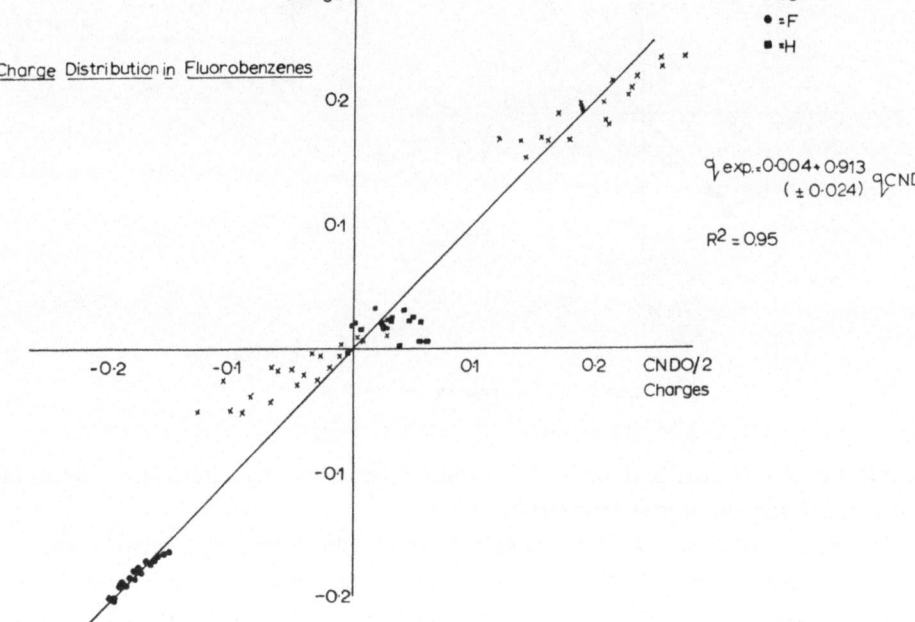

Charge Distribution in Fluorobenzenes

q exp. = 0·004 + 0·913 qCNDO
(± 0·024)

R^2 = 0·95

× = C
● = F
■ = H

Fig. 51.

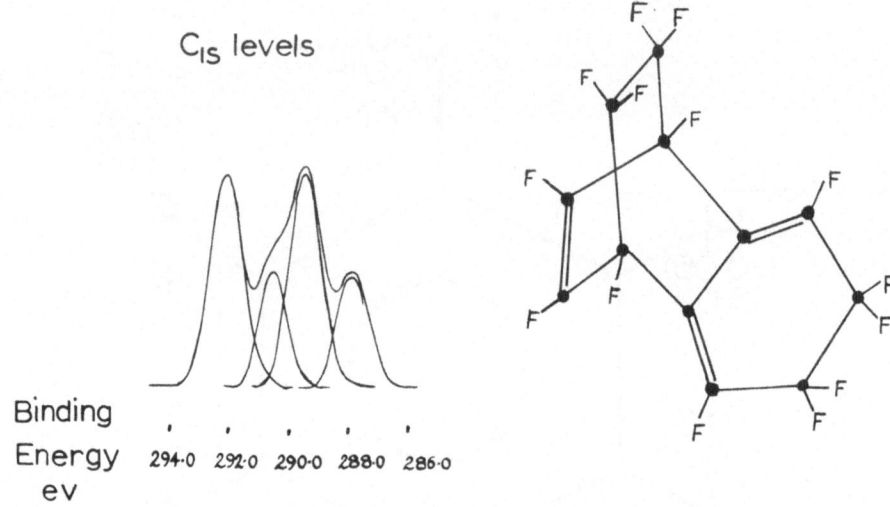

Fig. 52.

Charge distributions in Perfluoro Tricyclic compounds.

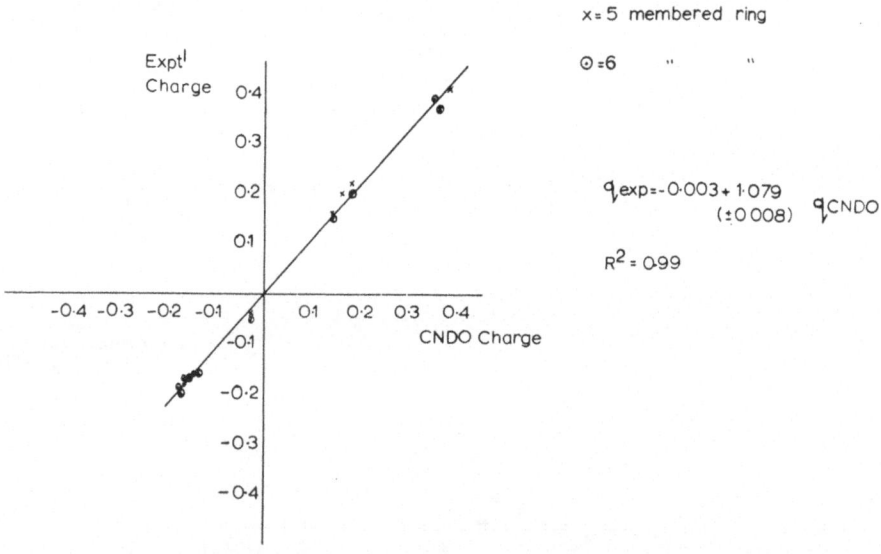

Fig. 53.

(CNDO/2) charge distributions for 1,2,3-trifluorobenzene. The agreement in terms of both magnitude and sign is impressive.

For the complete series of fluorinated benzenes the correlation between experimental and theoretical (CNDO/2) charge distributions is illustrated in Figure 51. The particular advantage of being able to derive charge distributions from shifts in core binding energies is nicely illustrated by studies of tetradecafluorotricyclo [6.2.2.0^{2,7}]

dodeca-2,6,9-triene, the C_{1s} spectrum of which is shown in Figure 52. CNDO/2 calculations on molecules of this sort of size require considerable computing facilities and are quite expensive to perform. The C_{1s} spectrum shows four peaks with area ratios, in order of decreasing binding energy, (4:2:4:2) and these are assigned to $\underline{C}F_2$, tertiary $\underline{C}F$, vinylic $\underline{C}F$ and bridgehead $\underline{C}$ carbons respectively. The derived charges for this and dodecafluorotricyclo $[5.2.2.0^{2,6}]$ undeca-2,5,8-triene mentioned previously are compared with those from direct calculation in Figure 53. For larger systems it is more convenient and practicable to obtain charge distributions from ESCA data than from direct calculations. A crude idea of the charge distribution is very often useful as a rule of thumb guide in discussing the chemistry of complex systems and also for preliminary assignment of other spectroscopic data (e.g. nmr) where the interpretation may not be as straightforward. Although we have illustrated the procedure for fluorocarbon systems where the shifts are relatively large it may be shown that detailed deconvolution of spectra is in general not required and with improvements in instrumentation this application of ESCA is likely to have considerable importance in many aspects of chemistry. Calculated charge distributions can also be usefully employed in investigating sample charging.

F. STRUCTURAL STUDIES

If appropriate values of k and E^0 are available theoretical charge distributions may be used for the assignment of spectra and if peak shapes and widths have been established then theoretically calculated spectra may be simulated. The possibility of using ESCA as a structural tool (particularly in cases where other spectroscopic techniques cannot resolve ambiguities) then becomes feasible. An interesting examples deriving from our own studies, which illustrates the logical sequence involved in such an analysis is outlined in Figures 54–60. The basic problem is set out in Figure 54 and concerns the question of the site of nucleophilic aromatic substitution in perfluoroindene, a topic of considerable theoretical and experimental importance in its own right.

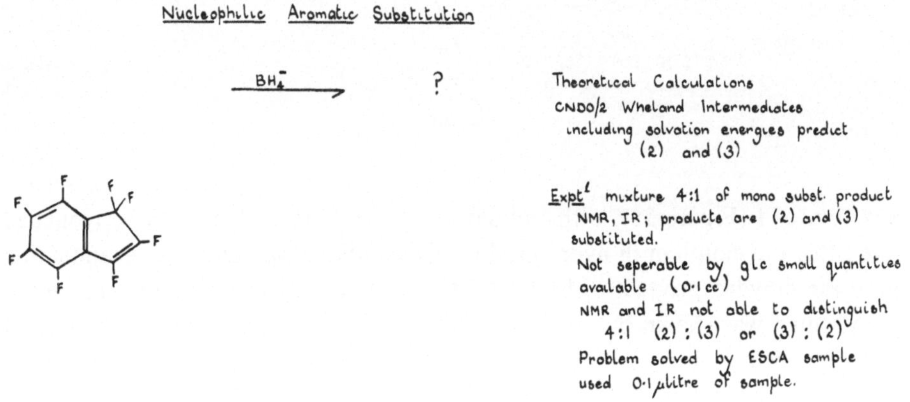

Fig. 54.

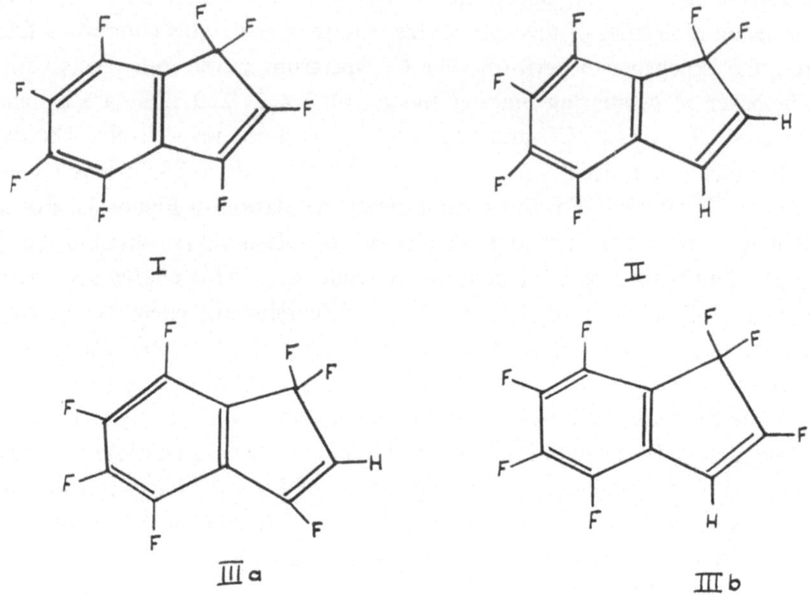

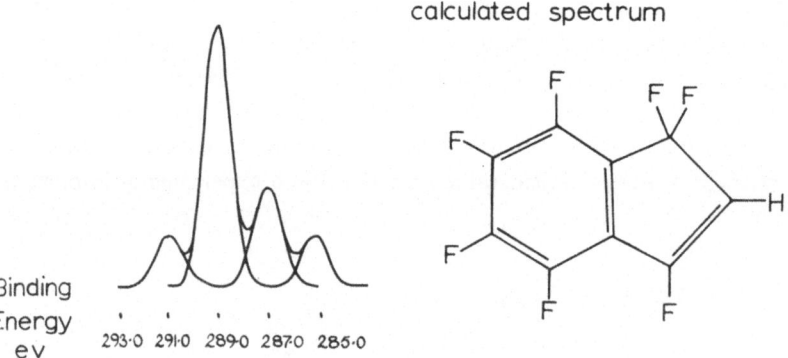

Fig. 55.

C_{1s} levels

Computer simulation of Theoretically
calculated spectrum

Binding
Energy
ev

293·0 291·0 289·0 287·0 285·0

Fig. 56.

Reaction of I (Figure 55), with sodium borohydride in diglyme can be regulated to give either the mono- or di-hydro product. Conventional spectroscopic examination showed the dihydro product to be 1,1,4,5,6,7-hexafluoroindene (II), and the mono-hydro product was shown by ^{1}H and ^{19}F nmr spectroscopy to be a 4:1 mixture of two isomers, which were, however, inseparable on available glc packings. Infra-red examination showed that the product consisted of a mixture of the 1,1,3,4,5,6,7- and the 1, 1, 2, 4, 5, 6, 7-heptafluoroindene (IIIa) and (IIIb), but identification of the

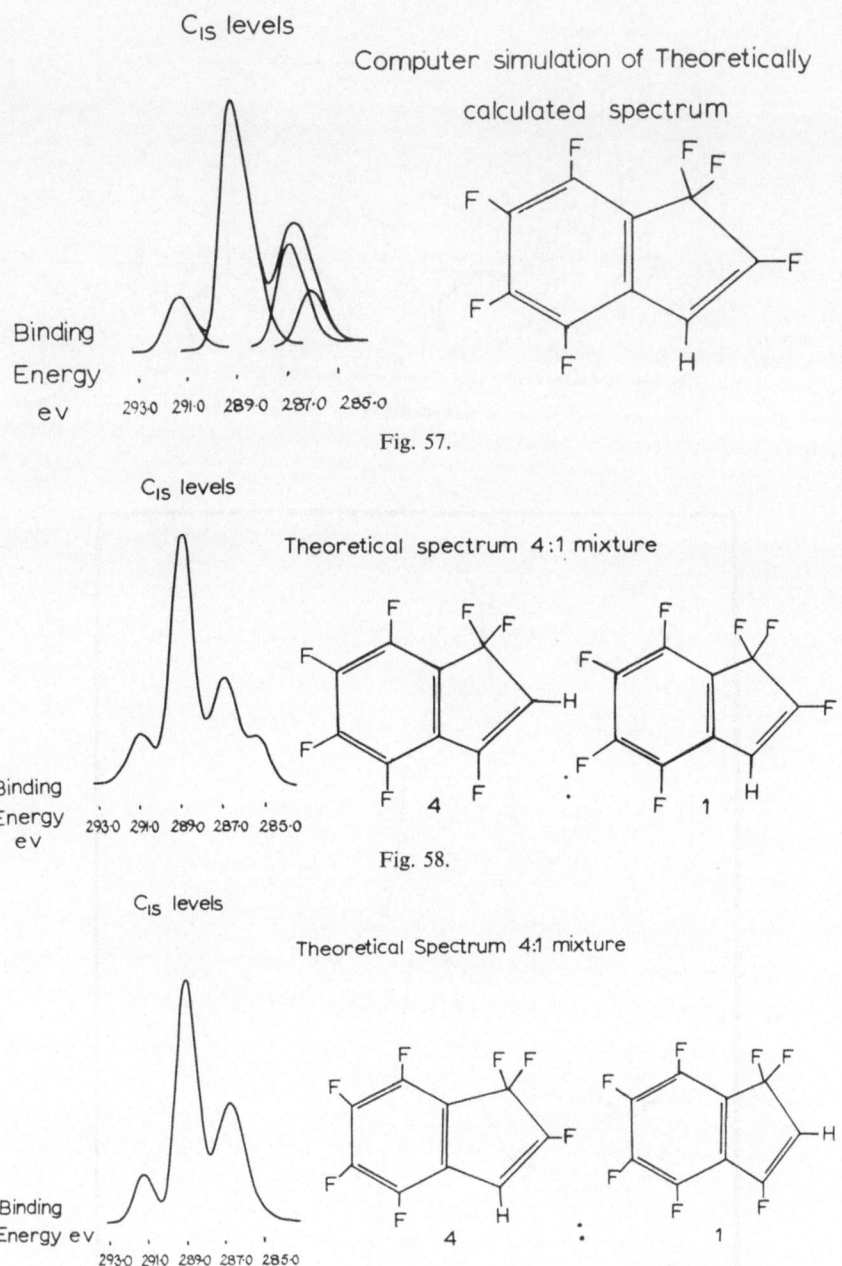

Fig. 57.

Fig. 58.

Fig. 59.

major component by ^{19}F nmr was not possible owing to the absence of the easily identified peri F–F coupling commonly found in other polycyclic fluoroaromatics.

The steps followed in elucidating the composition of this mixture were then as follows: first, CNDO/2 calculations were performed on (IIIa) and (IIIb) and the

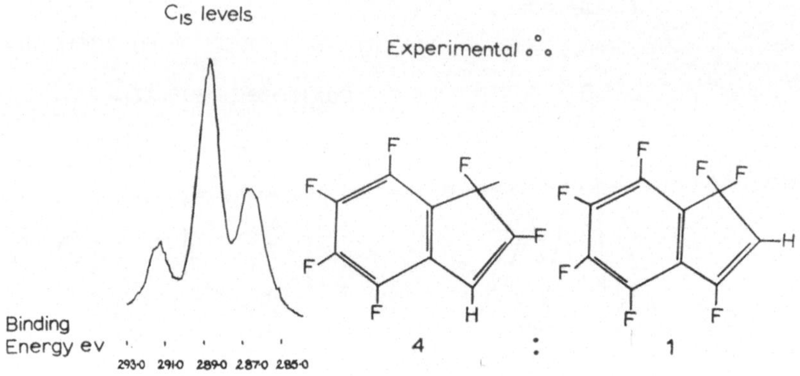

C$_{1S}$ levels

Experimental °°

Binding
Energy ev

293·0 291·0 289·0 287·0 285·0

4   :   1

Fig. 60.

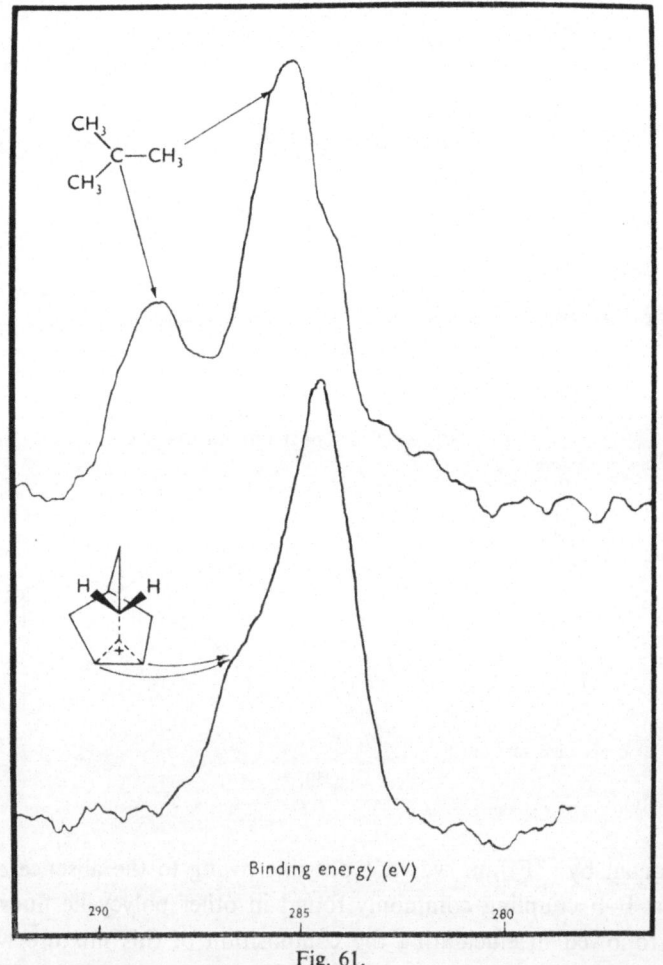

Binding energy (eV)

290   285   280

Fig. 61.

charge distributions so calculated were used, in conjunction with the charge potential model and previously established values of E_0 and k, to calculate theoretical C_{1s} binding energies. Using linewidths previously derived from a study of perfluoroindene (I), theoretical C_{1s} spectra were computed using a DuPont 310 Analogue Computer, and these are shown in Figures 56 and 57. The spectra corresponding to the 1:4 and 4:1 mixtures were then computed, and these are shown in Figures 58 and 59 compared with the experimental C_{1s} spectrum (Figure 60) for the mixture. The experimental spectrum matches exactly (when normalized to a horizontal baseline) the theoretical spectrum for the 1:4 mixture of (IIIa) and (IIIb) as regards both peak heights and peak positions. It can be seen that the theoretical 4:1 spectrum has a totally different appearance, with two distinct peaks at low binding energy, and therefore this completes the identification as a 1:4 mixture of (IIIa) and (IIIb).

Before leaving this thumbnail sketch of the sorts of application of ESCA which can be made in organic chemistry brief mention should be made of two important areas which are likely to be of considerable importance. The first concerns the application of ESCA to the study of carbonium ions (carbocations in general). The large differences in electronic environment about the carbon atoms in t-butyl cation for example (Figure 61), is directly manifest in the C_{1s} levels which gratifyingly reflect the organic chemists intuitive ideas concerning 'charge distribution' in this ion. A particularly important application pioneered by Olah and co-workers [11] is in the distinction between rapidly equilibrating classical and non-classical ions (e.g. norbornyl cation) where the ESCA time scale is such that clear distinctions can readily be made.

The second application concerns the study of hydrogen bonding. At first sight it might seem surprising that hydrogen bonding effects would be readily detectable by

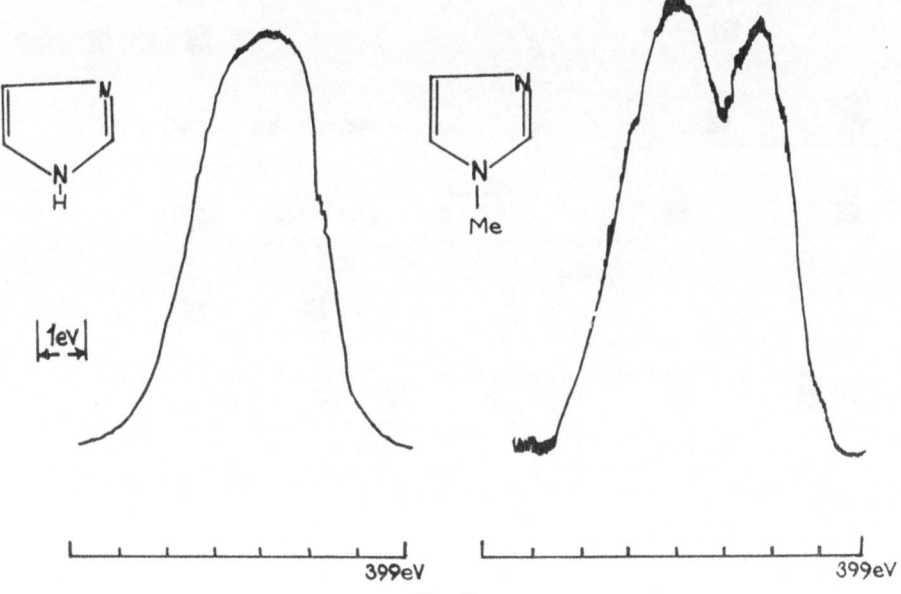

Fig. 62.

ESCA since the energies involved are so small (typically 5–10 Cals mole^{-1}). We have emphasized previously however that substantial shifts in core levels can occur on going from one system to another even when the energy difference between them is quite small. As an example we may consider the N_{1s} levels of imidazole (studied in the solid phase) and its N-methyl derivative. The spectra shown in Figure 62 (which were recorded some considerable time ago and could undoubtedly be improved upon) illustrate quite clearly the effect of hydrogen bonding in imidazole. In the solid state imidazole exists in a linearly hydrogen bonded chains viz., the net effect being that

$$--- N \overset{}{=\!\!=} N-H --- N \overset{}{=\!\!=} N-H ---$$

the electronic environment about the two nitrogens becomes more nearly the same· On N-methylation this is precluded and a much larger shift in N_{1s} binding energies is apparent. This is also evident for pyrazole. Figure 63 presents some interesting data for some other five-membered ring heterocyclic systems for comparison. It is clear that the NH nitrogen in both imidazole and pyrazole is at a higher binding energy. The reduction in shift between the nitrogens in these molecules due to hydrogen bonding is also clearly evident from a comparison with theoretically calculated shift from

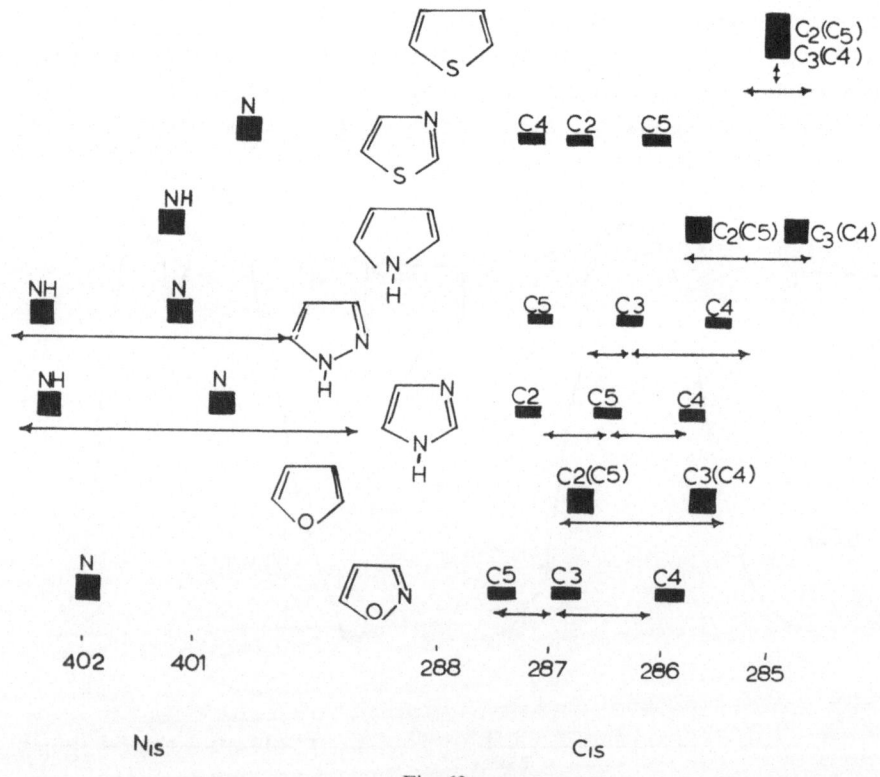

Fig. 63.

Koopmans' Theorem (indicated by the horizontal arrows). Whilst the shifts for the C_{1s} levels are well described those for the N_{1s} levels are considerably overestimated. The study of hydrogen bonding and of tautomeric equilibria in biologically important molecules such as the purine ring system is an active area of research.

3. Application of ESCA to Studies of Structure and Bonding in Inorganic Chemistry

A. INTROCUCTION

It is clear from Figure 64 [12] that extensive ESCA studies have been made for $\sim$ half of the periodic table to date and again we present only an outline of applications in inorganic chemistry which nonetheless illustrate the great potential of the technique. Before considering representative examples of the sort of information which can be

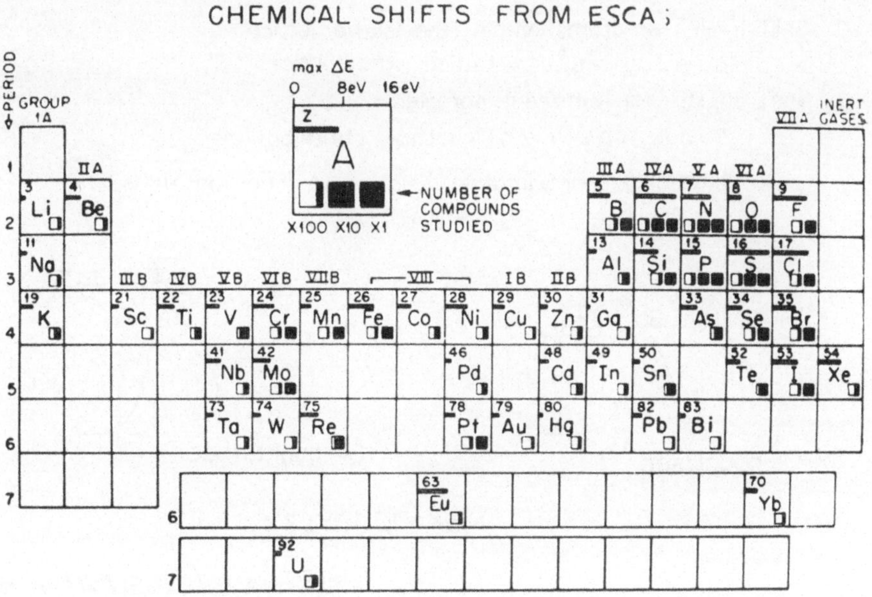

Fig. 64.

derived from such studies it is of interest to consider the relationship between the energy levels of isolated ions (which may readily be treated theoretically), and ions in a lattice. The relevant Born cycle is given in Figure 65. The lattice sublimation energy terms ΔH_{lat} and $\Delta H'_{lat}$ are now somewhat larger than in the case of the co-valent solids (Figure 43) and the binding energy measured with respect to the Fermi level of the sample is expected to depend on the lattice site. This is most readily appreciated if we generalize the charge potential model to ionic solids, when the third term takes the form of an interionic Madelung type potential. From this it is clear that we cannot expect to quantitatively understand shifts in binding energies for ions in a lattice by studying the energy levels of the isolated ion theoretically. An

Relationship between energy levels of
gaseous ion and ionic lattice

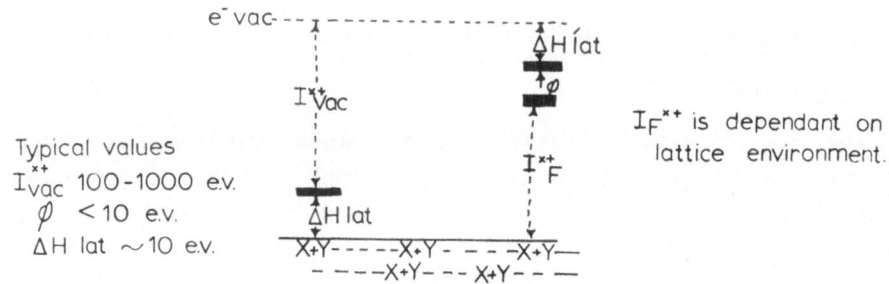

Typical values
I_{vac}^{x+} 100-1000 e.v.
$\emptyset$ < 10 e.v.
ΔH lat ~10 e.v.

I_F^{x+} is dependant on
lattice environment.

Shift in BE for atoms within the same lattice
$$\Delta = \Delta I_{vac} = \Delta I_F + \text{lattice contribution}$$
Shift in BE for different samples
$$\Delta = \Delta I_F + \Delta\emptyset + \Delta\sigma + \text{lattice contributions}$$
∴ shifts only approximately related to free ion shifts.

Fig. 65.

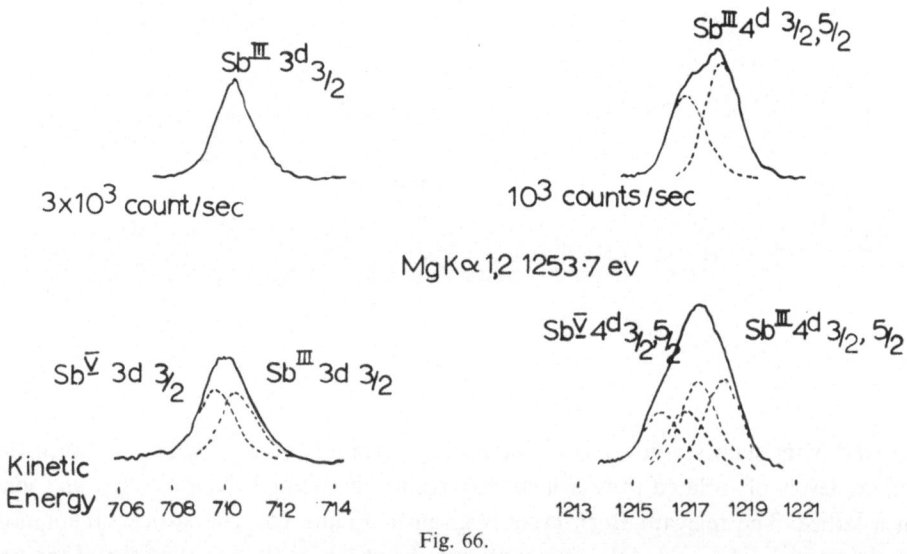

$Sb^{III} 3d_{3/2}$

3×10^3 count/sec

$Sb^{III} 4d\ 3/2, 5/2$

10^3 counts/sec

$Mg\,K\alpha\,1,2\ 1253\cdot7\ ev$

$Sb^{\underline{V}} 3d\ 3/2$ $Sb^{III} 3d\ 3/2$

Kinetic
Energy 706 708 710 712 714

$Sb^{\underline{V}} 4d\ 3/2, 5/2$ $Sb^{III} 4d\ 3/2, 5/2$

1213 1215 1217 1219 1221

Fig. 66.

interesting example which illustrates this rather nicely is provided by Sb_2O_4. The
lattice contains Sb(V) and Sb(III) ions and atomic SCF calculations would lead one
to expect ~12 eV difference in binding energies for the core levels of the isolated
ions. The spectra for the $3d$ and $4d$ core levels for Sb_2O_4 are shown in Figure 66. For
comparison spectra are also shown for Sb_2O_3 (which contains only a single oxidation

state Sb(III) of antimony) which may be used as a line shape and line width marker for analyzing the line shapes for Sb_2O_4. This shows that the difference in binding energy for the two oxidation states is ~ 1 eV; i.e. an order of magnitude smaller than that naively expected on the basis of free ion calculations. There are at least two reasons for this discrepancy. Firstly polarization of charge clouds probably means that description of the lattice in terms of discrete ions is not entirely correct. Secondly there is a substantial difference in lattice environment for the ions. This is shown in

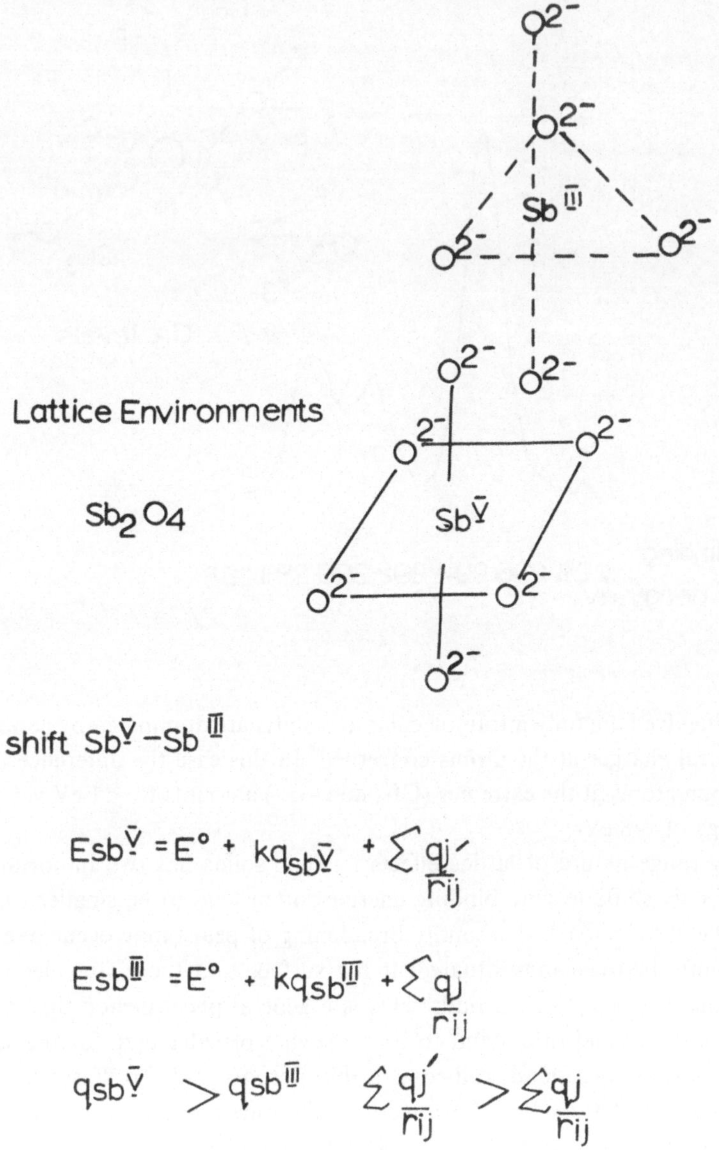

Lattice Environments

Sb_2O_4

shift $Sb^{\underline{V}} - Sb^{\underline{III}}$

$$Esb^{\underline{V}} = E° + kq_{sb}\underline{V} + \sum \frac{q_j'}{r_{ij}}$$

$$Esb^{\underline{III}} = E° + kq_{sb}\underline{III} + \sum \frac{q_j}{r_{ij}}$$

$$qsb^{\underline{V}} > qsb^{\underline{III}} \qquad \sum \frac{q_j'}{r_{ij}} > \sum \frac{q_j}{r_{ij}}$$

Fig. 67.

Figure 67. The Sb(III) is in a roughly trigonal bipyramidal site with five nearest neighbours whilst Sb(V) is in an octahedral site with six nearest neighbours. From the charge potential model, the potential provided by the nearest neighbours is more negative for Sb(V) than for Sb(III) thus acting in the opposite sense to the charge on the ion itself. On this basis the shift between Sb(III) and Sb(V) is much smaller than anticipated on the basis of free ion shifts. We may compare this situation with that for a typical organic molecule in the solid phase (e.g. Figure 68) where the

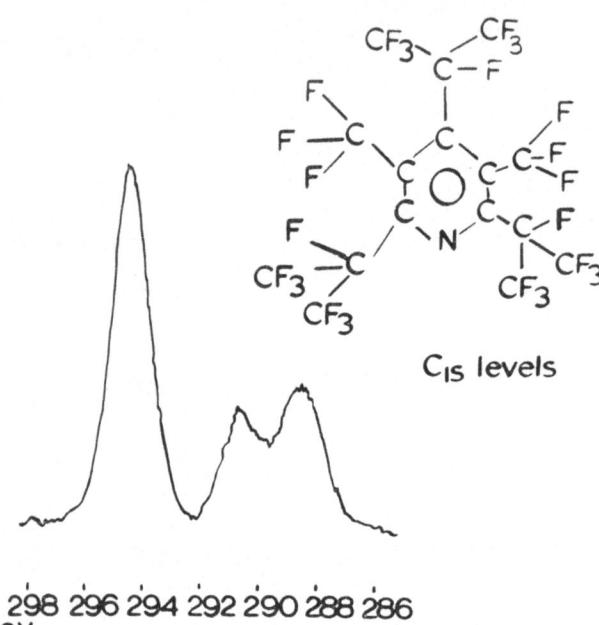

C_{1s} levels

Binding
Energy ev   298 296 294 292 290 288 286

Fig. 68.

potentials involved are only intramolecular and shifts are dominated by contributions from the local charges at the atoms concerned. In this case the difference in charge on the carbon atoms at the extremes ($\underline{C}F_3$ and $-\underline{C}-$) amounts to <1 eV yet the shifts span a range of ~8 eV.

The long range nature of lattice effects in ionic solids has two important consequences. Firstly shifts in core binding energies often tend to be smaller than might otherwise be anticipated and secondly broadening of peaks may occur due to small chemical shifts between ions situated at the surface and those (for electrons with sufficient kinetic energy) in the bulk. This is a general phenomenon that given core levels for an ionic solid often tend to be somewhat broader than for the same core level in a covalent compound studied as a thin film on gold. To illustrate this effect of lattice site, calculations have been made of chloride ion in an ionic lattice. Using a very large basis set of gaussian functions approaching the Hartree Fock limit for the chloride ion, shifts in core binding energies as a function of lattice environment may be calculated from the Fock eigenvalues as follows. By placing unit positive and negative charges about the ion, new one electron integrals (which are computationally inexpensive to compute) may be calculated and these together with the two electron integrals for the chloride ion can be used to perform new SCF calculations and derive new eigenvalues. Taking as energy reference an ion pair (Cl^- and a unit positive charge, lattice spacing appropriate to NaCl) the effect of extending the lattice in the plane and in planes above and below the chloride ion can be crudely simulated (Figure 69). This can in no way be regarded as a definitive study but does give some idea of the effect of lattice environment, and illustrates several important points. Firstly it is clear that although the absolute binding energies for the two core levels (Cl_{1s} and Cl_{2s}) are widely different ~ 2832 eV and 270 eV respectively the calculated shifts for both are the same. The effect of extending the lattice compared to the ion

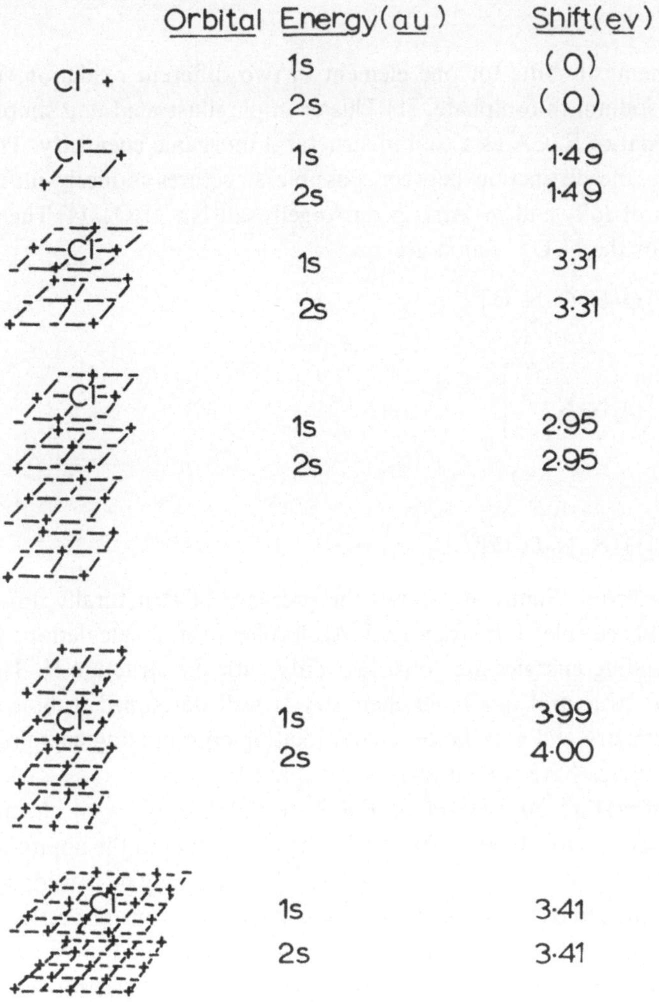

	Orbital	Energy (au)	Shift (ev)
$Cl^- +$	1s		(0)
	2s		(0)
	1s		1·49
	2s		1·49
	1s		3·31
	2s		3·31
	1s		2·95
	2s		2·95
	1s		3·99
	2s		4·00
	1s		3·41
	2s		3·41

Fig. 69.

pair is to increase the binding energies of the core levels. Comparing the results for the prototypes for ions at the surface (rows 3 and 6) the close similarity between the calculated shifts probably indicate that the major contributions to the lattice potential are adequately described since considerable extension of the lattice in both planes has such a small effect. (Computer limitations dictated the upper size of the lattice which could be investigated.) Considering the results for rows 4 and 5 the substantial difference between Cl^- at the surface and in the next layer down is apparent. These results suggest that a chemical shift range of ~ 1 eV could be accommodated by differences in lattice site. There are two points to be noted from this discussion. Firstly shifts in ionic solids often tend to be much smaller than one might naively expect and secondly core levels are often broadened, both effects being largely attributable to the lattice.

B. STRUCTURE AND BONDING IN INORGANIC SYSTEMS DERIVED FROM CHEMICAL SHIFT STUDIES

The first chemical shifts for one element in two different oxidation states were recorded for sodium thiosulphate. [1] This example illustrated in a simple manner the great potential of ESCA as a tool in structural inorganic chemistry. The application of ESCA to the distinction between possible structures is nicely illustrated by the recent work of Jolly and co-workers on Angelis salt $Na_2N_2O_3$. [4] The three possible structures for the $N_2O_3^{2-}$ anion are

$$[O-N-O-N-O]^{2-}$$

$$\left[O-N-N\begin{array}{c} \diagup O \\ \\ \diagdown O \end{array}\right]^{2-}$$

and

$$[O-N-N-O-O]^{2-}$$

The N_{1s} spectrum, Figure 70, shows the presence of structurally different nitrogen atoms and hence rules out structure I. Molecular orbital calculations show that the observed binding energies are consistent only with the structure II. The application of ESCA to structural inorganic chemistry is well developed in both nitrogen and sulphur chemistry and correlation charts relating core binding energies to particular functional groups have been drawn up (cf. [3, 4]).

As an interesting application of ESCA to sulphur nitrogen chemistry mention might be made of investigation of structure and bonding in the approximately planar heterocyclic cations $S_3N_2Cl^+$, $S_4N_3^+$ and $S_5N_5^+$. These may be considered as members of an inorganic Hückel series with 6π, 10π and 14π electrons respectively (Figure 71). The measured sulphur core binding energies are given in Figure 72 together with calculated (CNDO/2) total charges. For $S_3N_2Cl^+$ the local electronic charge distribution about the three sulphurs differ quite considerably and this is clearly reflected in

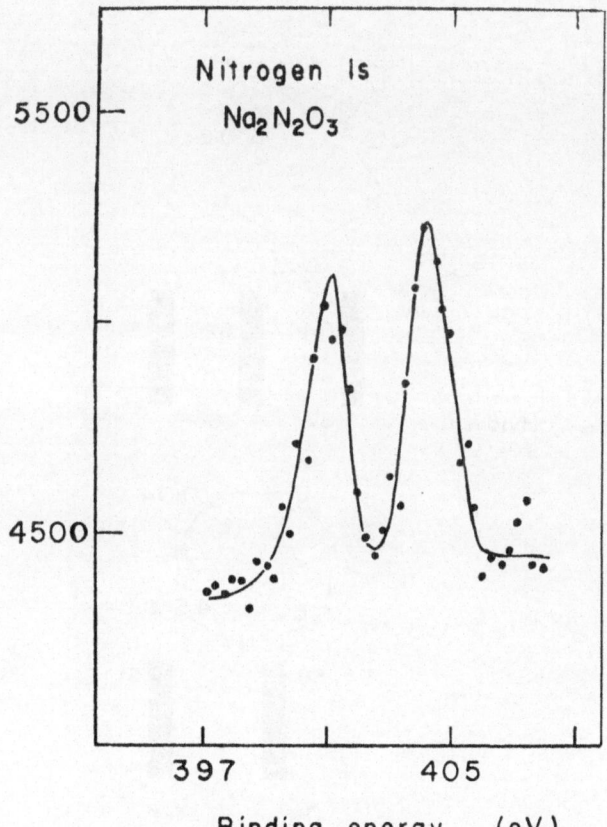

Nitrogen 1s spectrum of Angeli's salt, $Na_2N_2O_3$.

Fig. 70.

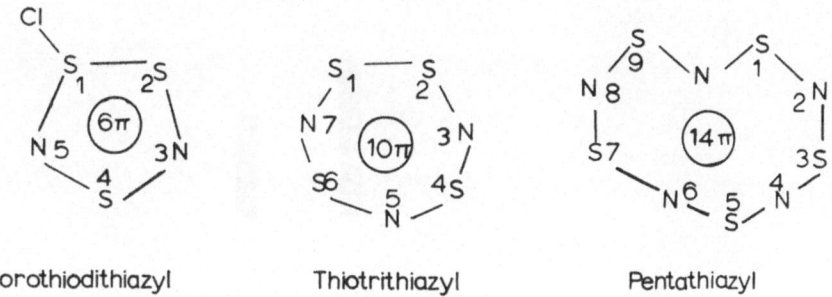

Chlorothiodithiazyl   Thiotrithiazyl   Pentathiazyl

S N Heterocyclic Cations

Fig. 71.

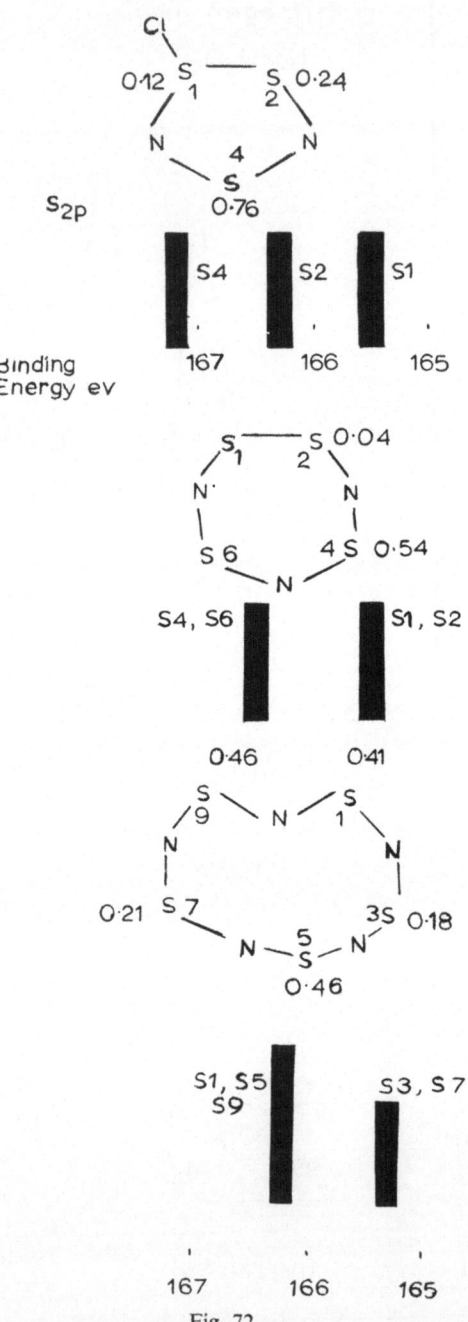

Fig. 72.

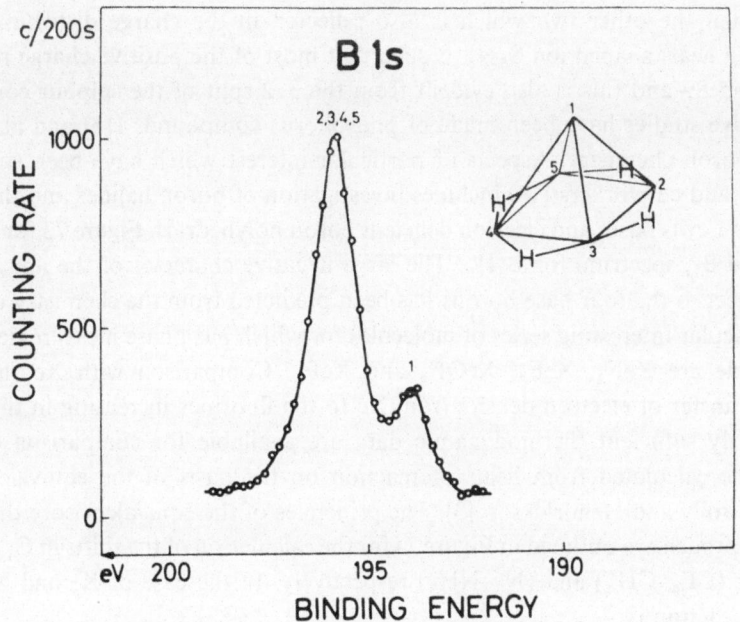

The B $1s$ spectrum of B_5H_9.

Fig. 73.

Equivalent core Thermodynamic Approach

Principle of equivalent cores (Jolly and Hendrickson 1970)
"When a core electron is removed from an atom in a molecule or ion, the valence electrons adjust as if the nuclear charge of the atom had increased by one unit."

$\cong$

* denotes core hole

$$CF_4 \longrightarrow CF_4^{*+1} + e^-(vac) \quad E_{CIS}$$
$$CF_4^{*+1} \longrightarrow NF_4^+ \qquad\qquad \delta$$

$$CF_4 \longrightarrow NF_4^+ + e^-(vac) \quad E_{CIS} + \delta$$
$$CH_4 \longrightarrow NH_4^+ + e^-(vac) \quad E_{CIS}' + \delta'$$

shift Δ_{CIS} $(CF_4 - CH_4) = (E_{CIS} - E_{CIS}') + (\delta - \delta') \quad (\delta - \delta') \approx 0$
$$= \Delta H \text{ for reaction}$$
$$CF_4 + NH_4^+ \longrightarrow CH_4 + NF_4^+$$

shift Δ N_{1S} $(N_2 - NH_3) = \Delta H$ for reaction
$$N_2 + OH_3^+ \longrightarrow NO^+ + NH_3$$

Limitation lack of Thermodynamic data.

Fig. 74.

the S_{2p} core levels. In $S_4N_3^+$ the sulphurs bonded to two nitrogens are at higher binding energy than the other two which is also reflected in the charge distribution. The interesting heart shaped ion $S_5N_5^+$ is such that most of the positive charge resides at S1, S5 and S9 and this is also evident from the 3:2 split of the sulphur core levels.

Extensive studies have been made of phosphorus compounds [13] and also in the field of boron chemistry. Aspects of particular interest which have been studied by Siegbahn and co-workers [14] includes investigation of boron halides and their reactions with Lewis bases and electron deficient boron polyhedrals. Figure 73 for example shows the B_{1s} spectrum for B_5H_9. The large negative character of the apical boron with respect to the four base borons has been predicted from the chemistry of B_5H_9.

A particular interesting series of molecules for which has phase measurements have been made are XeF_2, XeF_4, $XeOF_4$ and XeF_6. Comparison with Xe shows the overall transfer of electron density from Xe to the fluorines increasing in this series. Fortunately sufficient thermodynamic data are available for comparison of shifts with those calculated from heats of reaction on the basis of the equivalent cores model of Jolly and Hendrickson [3]. The principles of the equivalent core thermodynamic approach are outlined in Figure 74 for the calculation of the shifts in C_{1s} and N_{1s} levels for (CF_4-CH_4) and (N_2-NH_3) respectively. In the case of Xe and XeF_2 the relevant reaction is

$$XeF_2 + Cs^+ \rightarrow CsF_2^+ + Xe$$

Table X shows how well this ingenious approach accounts for shifts in core binding energies for some nitrogen, carbon and xenon compounds.

TABLE X

Shifts in core binding energies eV

Level	Compound	Shift exptl.	Equiv. core heat of reaction
N_{1s}	$(CH_3)_2NH$	-0.7	-0.7
	CH_3NH_2	-0.3	-0.4
	NH_3	(0)	(0)
	HCN	1.2	1.0
	$\underline{N}NO$	3.2	2.6
	N_2	4.4	3.5
	NO	5.5	4.4
	N_2F_4	6.8	6.3
	NO_2	7.3	6.8
C_{1s}	CH_4	(0)	(0)
	CO	5.4	4.1
	CO_2	6.8	6.9
	CF_4	11.0	12.3
$Xe_{3d5/2}$	Xe	(0)	(0)
	XeF_2	3.0	2.7
	XeF_4	5.5	5.4
	$XeOF_4$	7.0	6.3
	XeF_6	7.9	7.9

The agreement is impressive, but unfortunately the method suffers from the major deficiency that more often than not thermodynamic data are just not available.

Fortunately the reactions involved in interpreting shifts in core binding energies are for isodesmic processes. For such processes the number and type of bonds in reactants and products are the same and from the extensive studies of Pople [15] and of Snyder [16] it is known that the energetics of such processes are well described within the Hartree-Fock formalism since corelation energy corrections are very small. This being the case the relevant thermodynamic data may be generated from non-empirical LCAO MO SCF calculations. The equivalent core thermodynamic calculations of shifts for C_3O_2 for example shows that calculations of the total energies for C_3O_2, CO_2, NO_2^+, and $C_2NO_2^+$ enables shifts in C_{1s} levels between CO_2 and C_3O_2 to be computed.

OCCCO
Isodesmic Reactions
$OCCCO + ONO^+ \rightarrow OCCNO^+ + OCO$
$\rightarrow OCNCO^+ + OCO$

Obtain ΔC_{1s} C_3O_2 wrt CO_2
          ΔC_{1s} OCCCO wrt OCCCO

Shifts for both the C_{1s} and O_{1s} levels are shown in Table XI. The results for the equivalent core species (Table XI, Col. A) are in good agreement with experiment in both magnitude and sign. For comparison the results (Col. B) for calculations in which

TABLE XI

Experimental shifts of CO_2 and C_3O_2 compared with various calculated values

Molecule		Equiv. cores schift				Koopman's shift	Experimental
		A*		B**			
CO_2	C	(0)		(0)		(0)	(0)
	O	(0)		(0)		(0)	(0)
C_3O_2	OCCCO	− 1.6		− 4.3		− 0.9	− 2.6
			$\Delta = 3.5$		$\Delta = 6.6$	$\Delta = 5.8$	$\Delta = 3.4$
	OCCCO	− 5.1		− 10.9		− 6.7	− 6.0
	O	− 1.0		− 1.5		− 1.4	− 1.1

* A including electronic relaxation.
** B excluding electronic relaxation.

the effect of relaxation of the valence electrons on ionization has effectively been neglected, considerably over-estimates the shifts which are somewhat comparable to those predicted from Koopmans' theorem (which also neglects electronic relaxation). This indicates that the electronic relaxation is different for the two different types of carbon in C_3O_2. The approximations inherent in theoretical treatments of shifts within the Hartree Fock formalism are indicated in Figure 75. Since core levels are so localized, relativistic and correlation energy corrections to absolute core binding

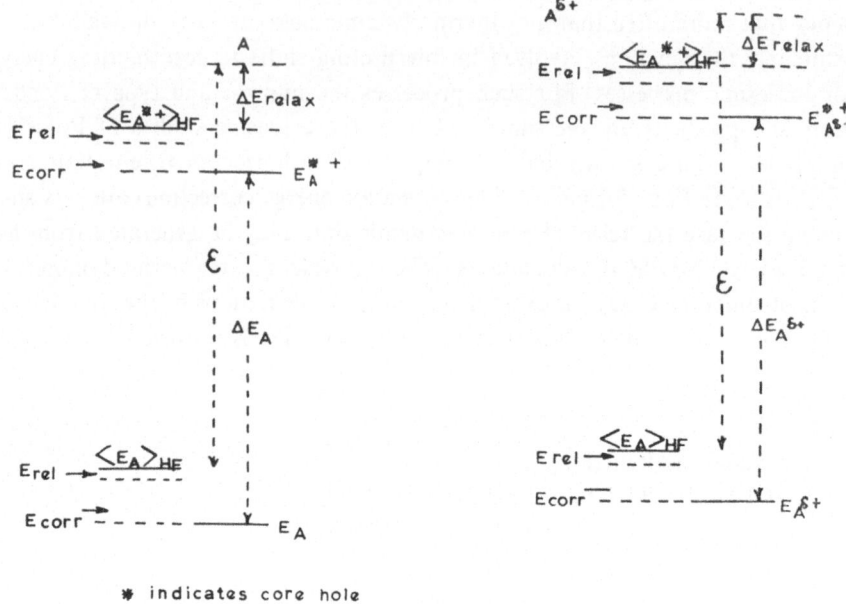

* indicates core hole

Fig. 75.

energies are essentially constant and play a negligible role in determining shifts in core levels. On the other hand the relaxation (or reorganisation) of electronic charge accompanying photoionization depends strongly upon the local electronic environment and can contribute significantly to shifts in core binding energies. The reorganization is predominantly associated with the valence electrons as is readily apparent from the data in Table XII. From the expectation values for r for the core and valence orbitals it is clear that for both core and valence ionizations it is the valence orbitals which undergo the most drastic reorganization. The neglect of this feature (i.e. Koopmans' Theorem) overestimates the $1s$ core binding energy for Ne by ~ 23 eV. Electronic re-organization can however be taken into account by carrying out hole state calculations or by equivalent core calculations. The importance of including this

TABLE XII

Reorganization energies in Ne^+

	Ne'S		$^2P_{2p}$ hole		$^2S_{2s}$ hole		$^2S_{1s}$ hole
$\langle r \rangle_{1s}$	0.1576		0.1576		0.1578		0.1545
$\langle r \rangle_{2s}$	0.8921		0.8603		0.8536		0.8171
$\langle r \rangle_{2p}$	0.9652		0.8759		0.8841		0.7993
		$\varDelta$		$\varDelta$		$\varDelta$	
Ion potl. (exptl.)	(0)		21.60	(0)	48.42	(0)	869.1
$^\varepsilon$HF	1.46		23.14	4.10	52.52	22.7	891.8
$\varDelta E_{HF}$	1.76		19.84	0.89	49.31	−0.5	868.6

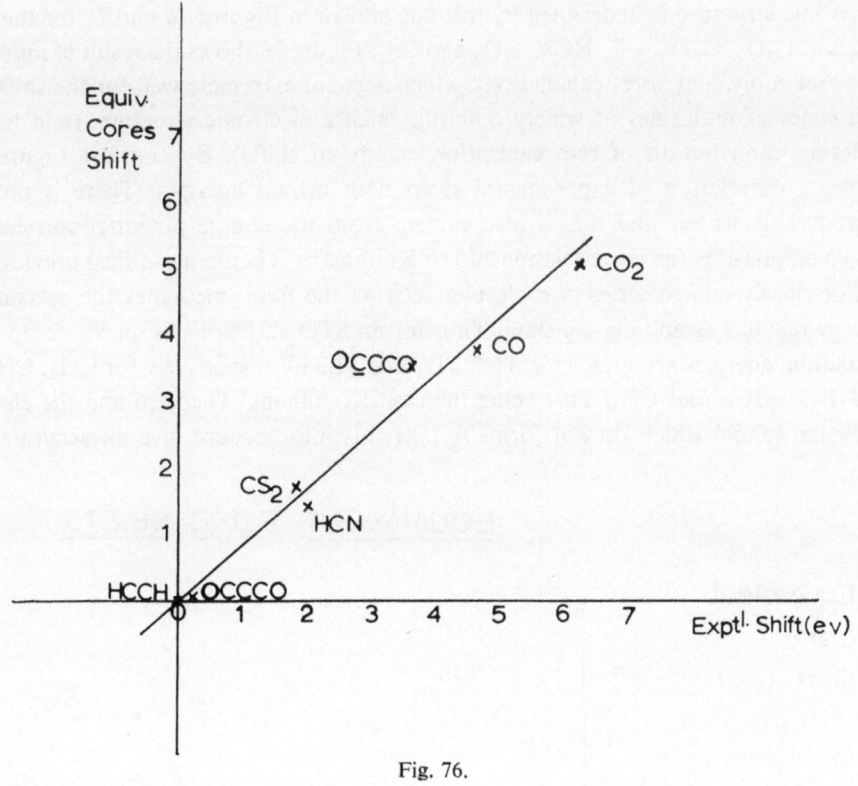

Fig. 76.

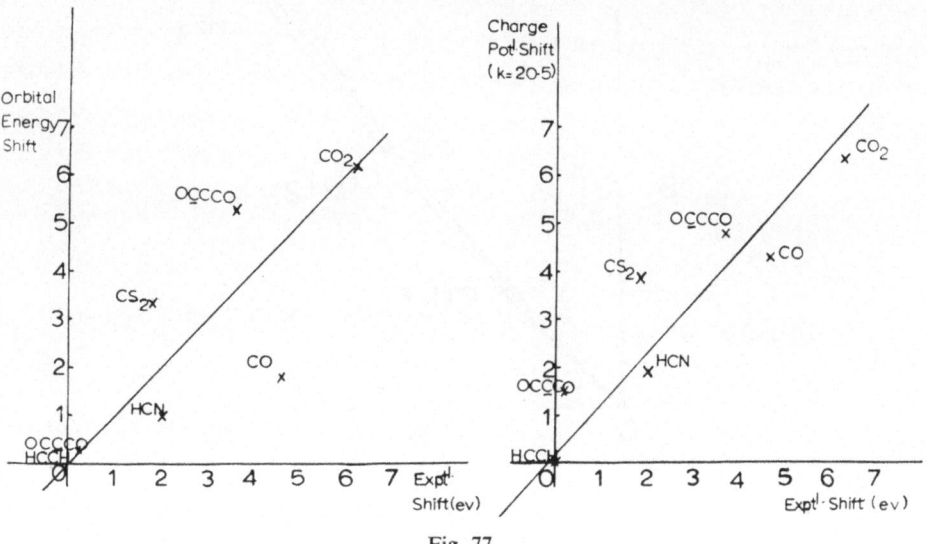

Fig. 77.

effect for a quantitative discussion of shifts for molecules of widely differing valence electronic structure is underlined by the data shown in Figures 76 and 77 for the C_{1s} levels of C_3O_2, C_2H_2, CO, HCN, CO_2 and CS_2. Figure 76 shows the result of minimal basis set equivalent cores calculations, which account extremely well for the shifts in this series of molecules of widely differing valence electronic structure, (and hence differing contribution of re-organization energy to shifts). By contrast Figure 77 shows a correlation of experimental shifts with orbital energies. There is now a considerable scatter and this is also evident from the charge potential correlation which emphasizes the close relationship to Koopmans' Theorem outlined previously.

For closely related series of molecules such as the fluoromethanes the relaxation energy remains essentially constant. Thus for an STO 4.31 G basis set the computed relaxation energies are 11.4, 11.2, 11.1, 11.1 and 11.3 eV respectively for CH_4, CH_3F, CH_2F_2, CHF_3, and CF_4. This being the case Koopmans' Theorem and the charge potential model which do not formally take this into account give an accurate re-

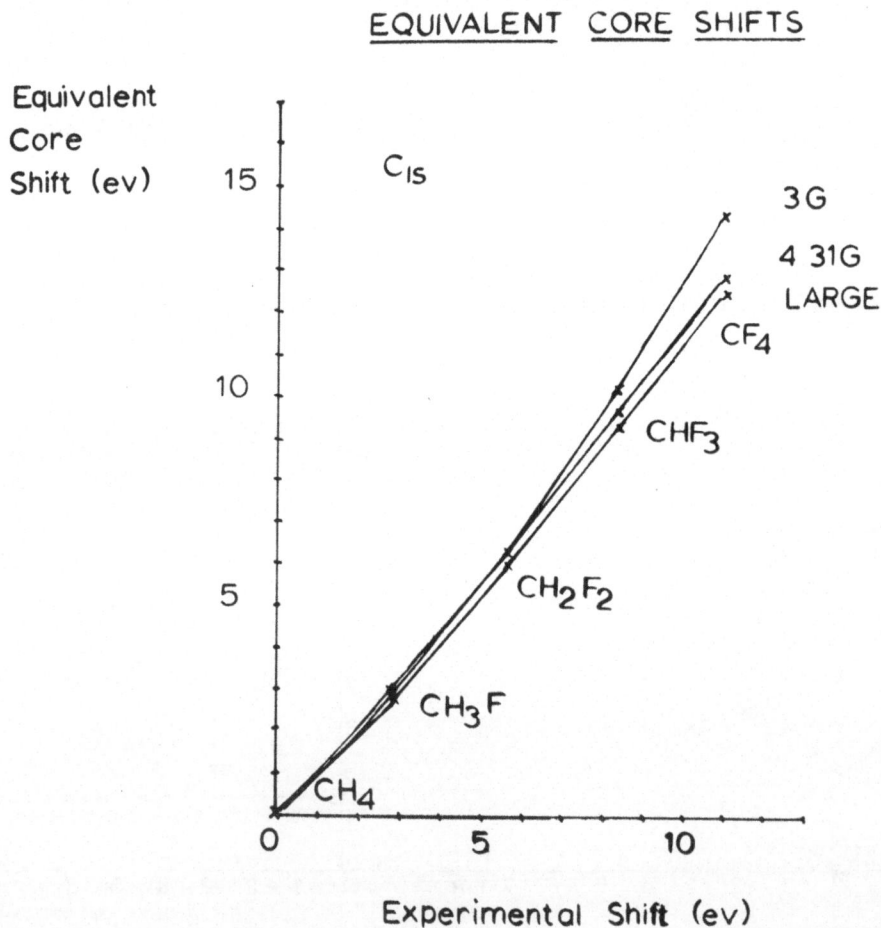

Fig. 78.

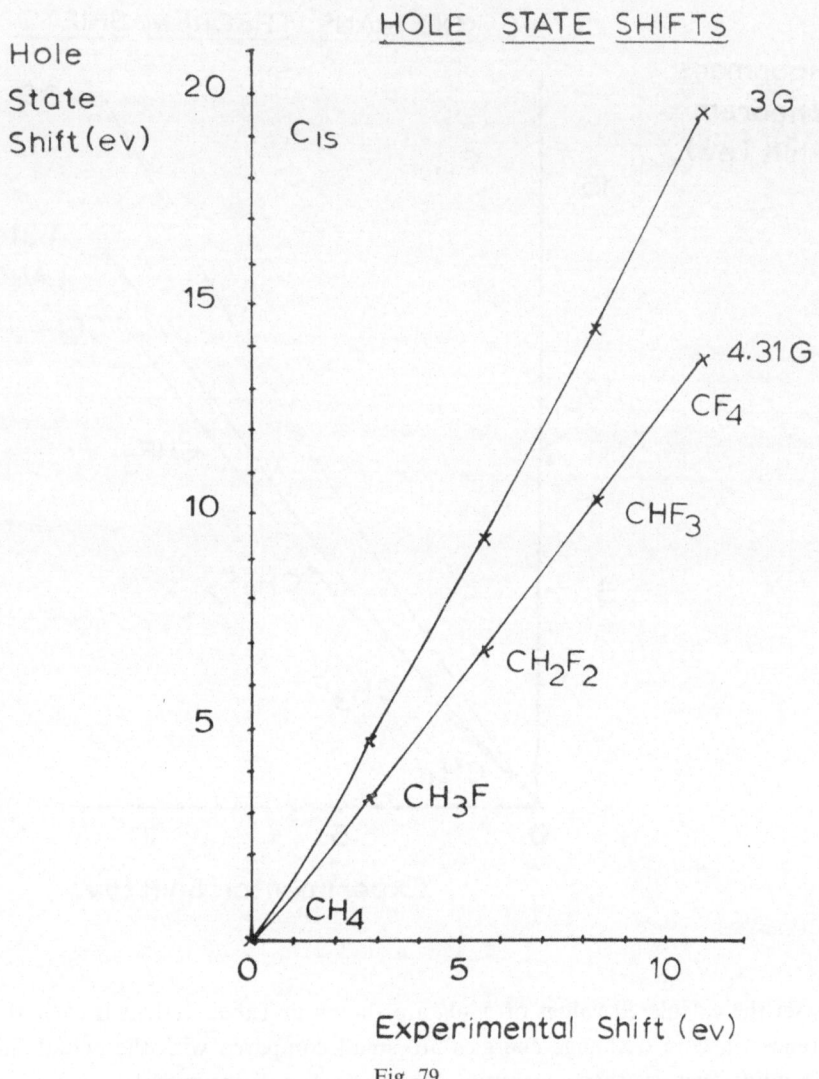

HOLE STATE SHIFTS

Hole State Shift (ev)

Experimental Shift (ev)

Fig. 79.

presentation of the shifts comparable with those obtained from equivalent cores and hole state calculations which do allow for relaxation. This is illustrated as a function of basis set employed in non-empirical LCAO MO SCF calculations in Figures 78–80. The clear advantage of the equivalent cores model is evident in that it is much less dependent on basis set size than either the hole state calculation or Koopmans' Theorem largely as a result of cancellation of errors. Since only calculations on closed shell species are involved the equivalent core calculations are computationally simpler and less expensive than hole state calculations. It is of interest to investigate for this series of molecules the differences in core exchange energies (δ's). For an STO 4.31 G

KOOPMANS' THEOREM SHIFTS

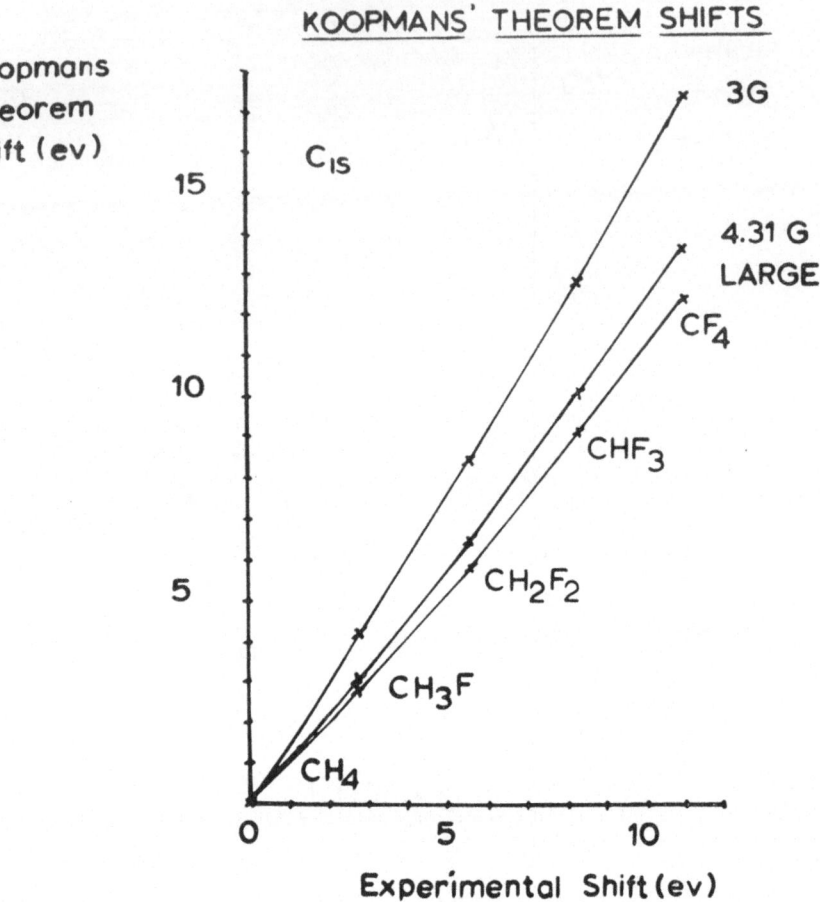

Fig. 80.

basis set the calculated values of $\delta_n - \delta_0$ are shown in Table XIII. It is clear that the differences in core exchange energies are small compared with the actual shifts in core binding energies thus accounting for the success of the model.

TABLE XIII

Calculated values of $\delta_n - \delta_0$

	n	$\delta_n - \delta_0$(eV)	Measured shifts in core binding energies
CH_4	0	(0.0)	(0)
CH_3F	1	-0.26	2.8
CH_2F_2	2	-0.49	5.6
CHF_3	3	-0.67	8.3
CF_4	4	-0.76	11.0

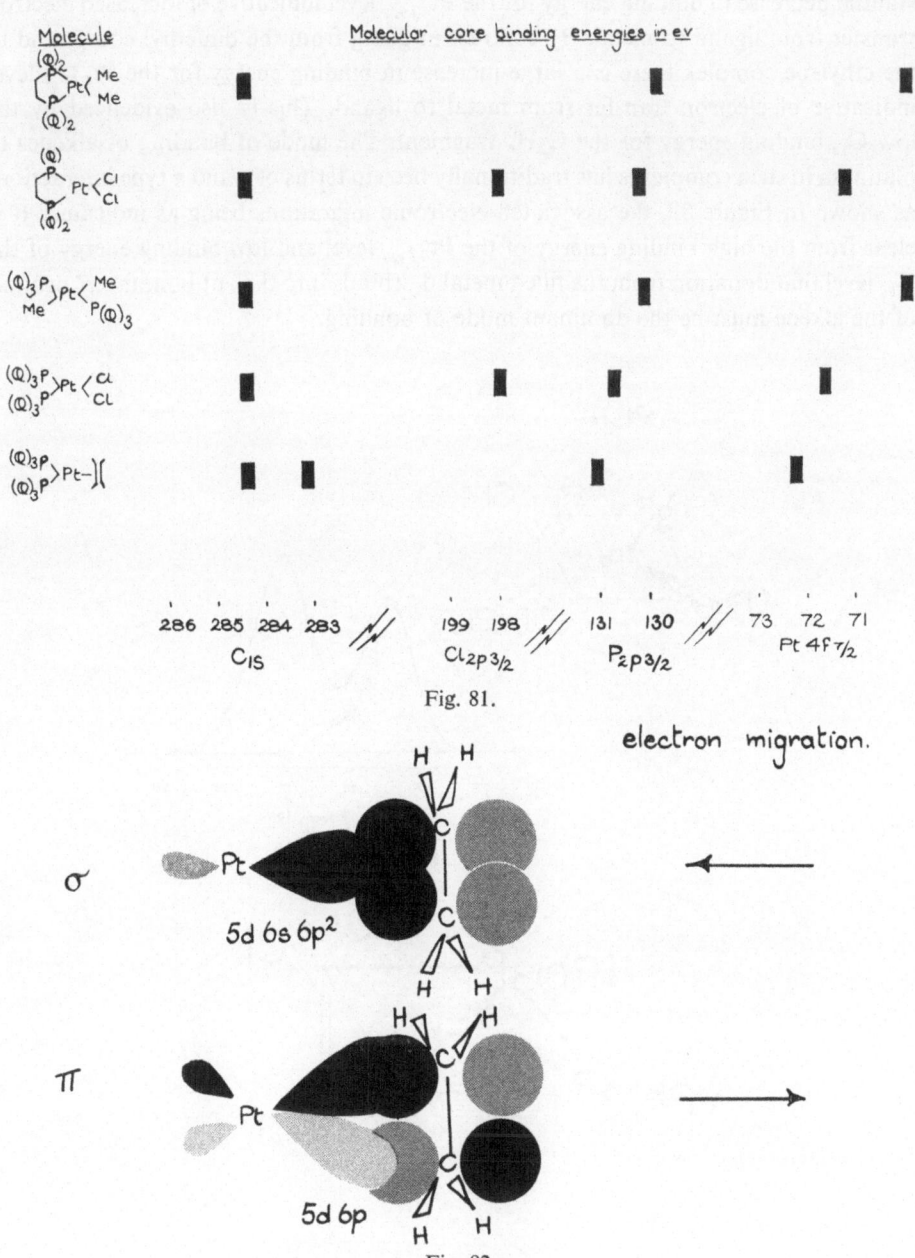

Molecule   Molecular core binding energies in ev

Fig. 81.

electron migration.

σ

5d 6s 6p²

π

5d 6p

Fig. 82.

286 285 284 283 // 199 198 // 131 130 // 73 72 71
C_{1s} $Cl_{2p\,3/2}$ $P_{2p\,3/2}$ Pt $4f\,7/2$

ESCA provides a powerful tool for investigating the mode of bonding of ligands in organometallic compounds, particularly neutral complexes where lattice effects are comparatively small. Figure 81 shows core binding energies for some platinum complexes which illustrate this rather effectively. Considering firstly the square planar dichloride, replacing chlorine by a strong σ donor such as methyl results in a substantial decrease in binding energy for the $Pt_{4f_{7/2}}$ level indicative of increased electron transfer from ligand to metal. By contrast in going from the dimethyl compound to the ethylene complex there is a large increase in binding energy for the $Pt_{4f_{7/2}}$ level indicative of electron transfer from metal to ligand. This is also evidenced by the low C_{1s} binding energy for the C_2H_4 fragment. The mode of bonding of alkenes to platinum in such complexes has traditionally been in terms of σ and π type interactions as shown in Figure 82, the associated electronic migrations being as indicated. It is clear from the high binding energy of the $Pt_{4f_{7/2}}$ level and low binding energy of the C_{1s} level that donation from the filled metal d orbitals into the antibonding π^* orbitals of the alkene must be the dominant mode of bonding.

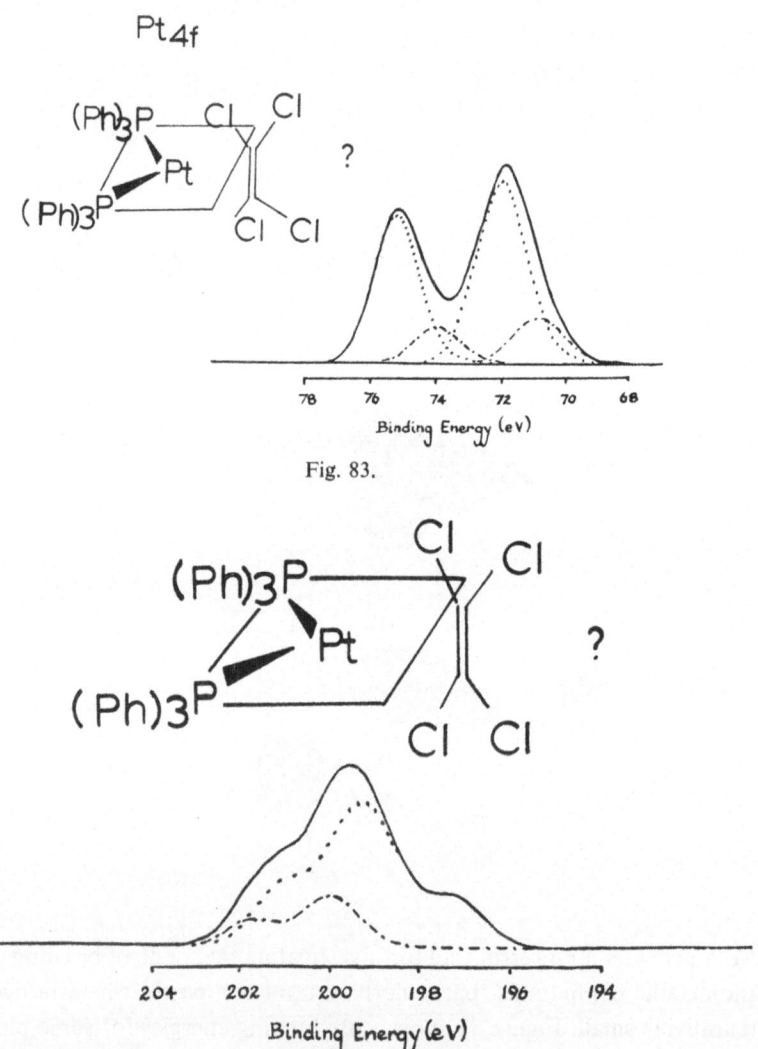

Fig. 83.

Fig. 84.

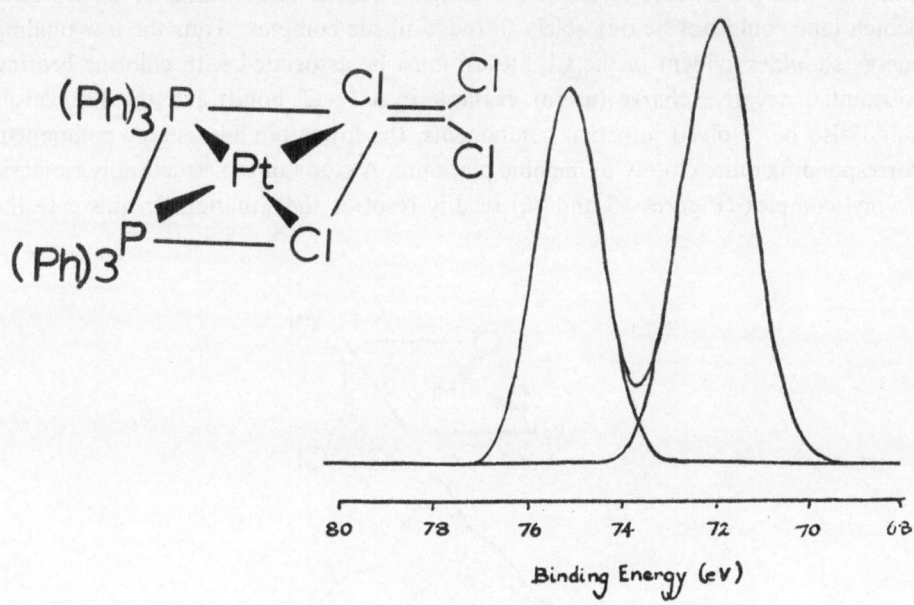

Fig. 85.

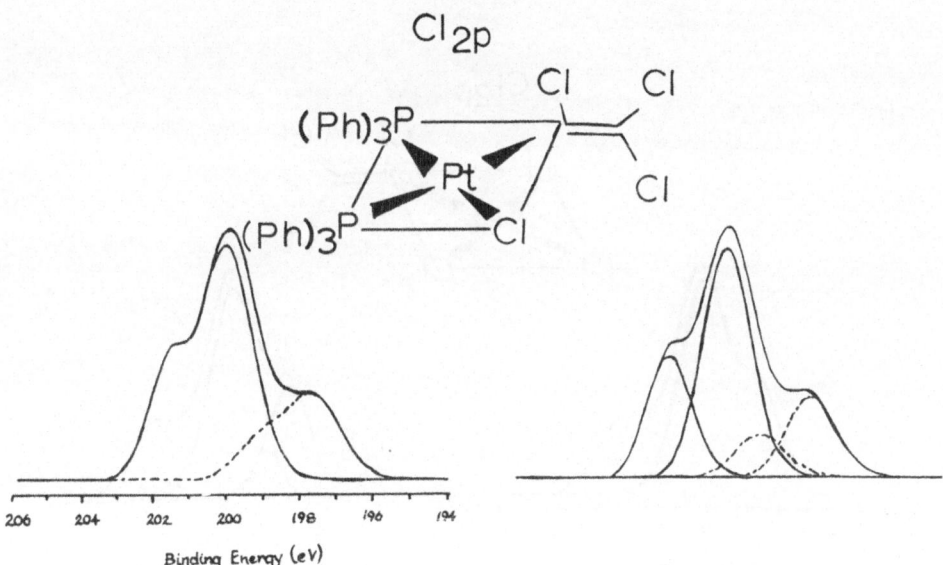

Fig. 86.

The application of ESCA to the distinction between structural isomers can be conveniently illustrated by reference to our attempts to study the analoguous π alkene complex involving tetrachloroethylene. The $Pt_{4f_{7/2}}$ and Cl_{2p} levels (Figures 83 and 84) indicated that the surface layers of the samples studied as thin films or dusted onto Scotch tape could not be due solely to the π alkene complex. Thus the low binding energy shoulder evident in the Cl_{2p} levels must be associated with chlorine bearing substantial negative charge (as for example in a Pt–Cl bond) and the Pt_{4f} levels could also be resolved into two components, the lower binding energy component corresponding quite closely to metallic platinum. A study of the structurally isomeric σ vinyl complex (Figures 85 and 86) readily resolves the situation. In this case the

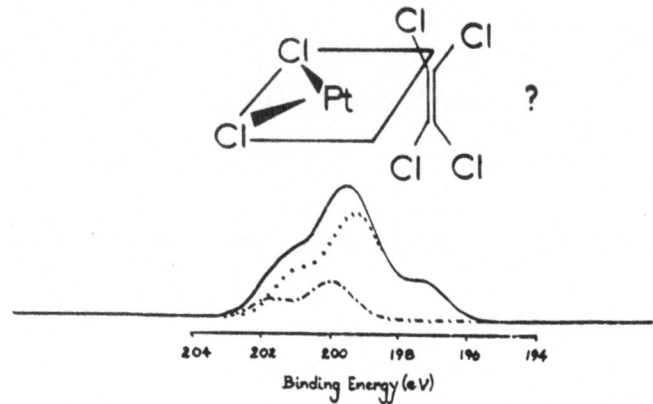

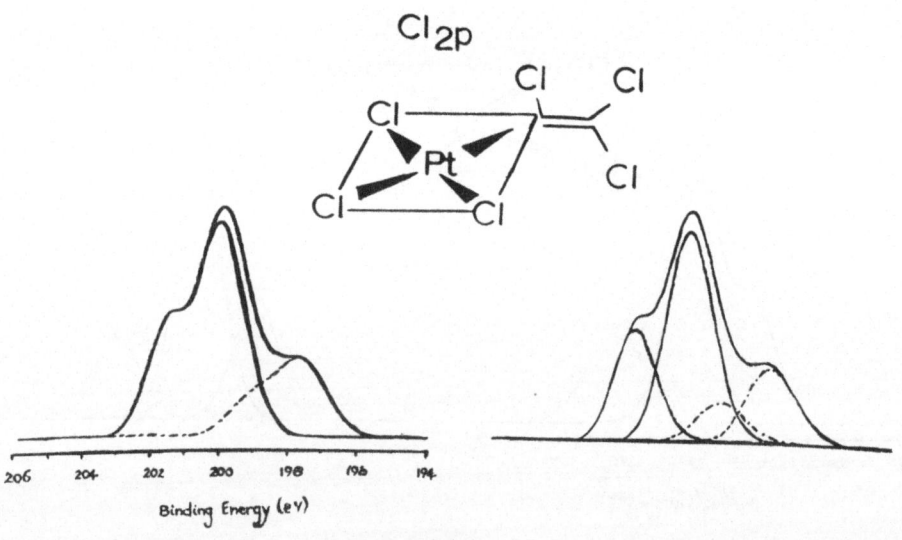

Fig. 87.

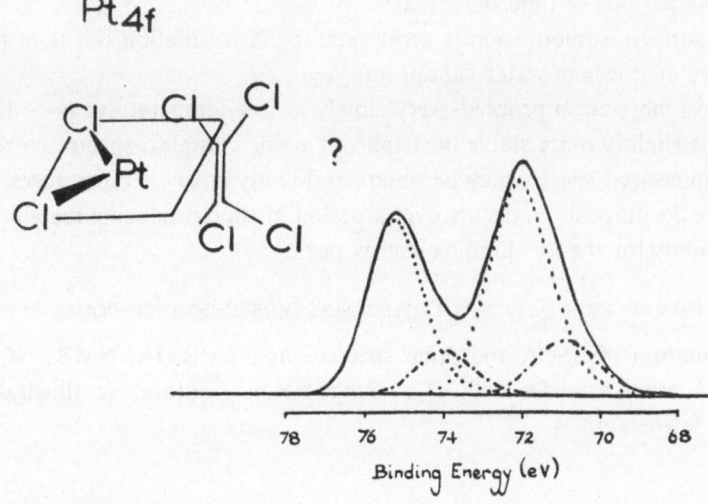

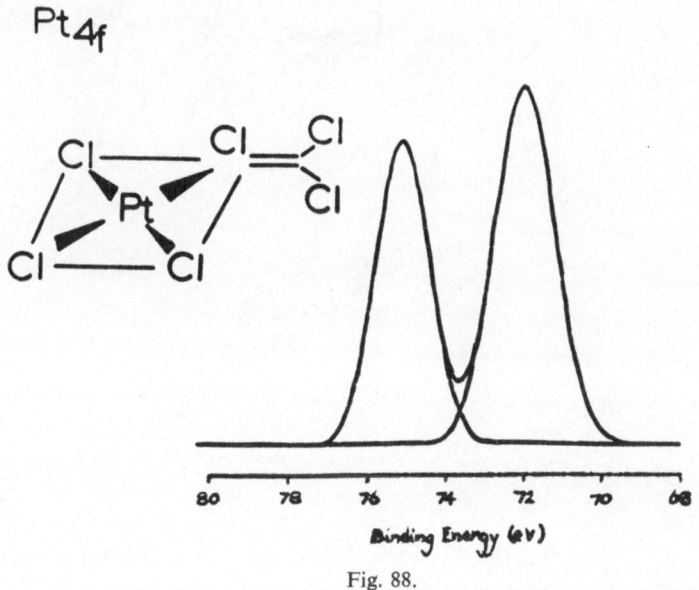

Fig. 88.

two types of chlorine C–Cl and Pt–Cl are readily distinguished. Comparison of the $Pt_{4f_{7/2}}$ and Cl_{2p} core levels of the two samples leaves little doubt that the surface layers of the π alkene complex have isomerized to the σ vinyl isomer. (Transmission IR studies confirm however that the bulk corresponds to the π alkene complex.) Using the line shape data for the σ vinyl isomer the residual Cl_{2p} signal actually arising from unisomerized π alkene complex may be detected (Figures 87 and 88).

Detailed studies over a temperature range with different X-ray beam intensities and over varying periods of time show that:

(1) The surface isomerization is accelerated by X-irradiation but does proceed in the presence of traces of water vapour anyway.

(2) The isomerization proceeds very slowly at low temperatures ($\sim -100\,°C$), indeed for the slightly more stable bis triphenyl arsine complex, spectra corresponding to the unisomerized species may be observed directly at low temperatures.

(3) Some decomposition occurs over a period of time producing metallic platinum which accounts for the low binding energy peak.

C. COMPARISON OF ESCA DATA WITH OTHER SPECTROSCOPIC TECHNIQUES

The combination of ESCA and other spectroscopic tools (IR, NMR, NQR, Mossbauer, etc.) often provides data of a complimentary nature. To illustrate this we consider a few examples.

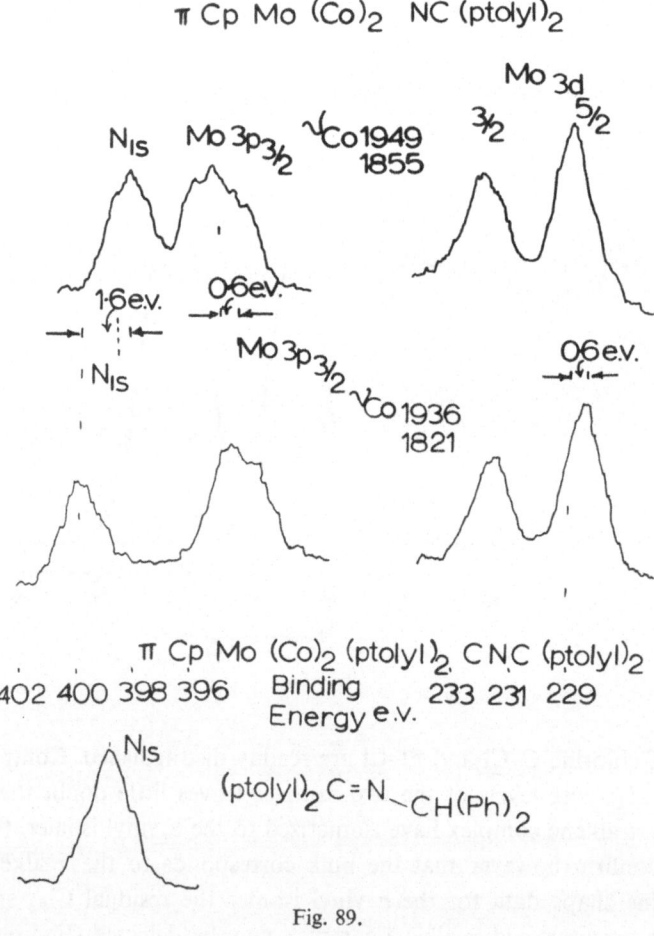

Fig. 89.

1. *Infrared Data*

An interesting application of ESCA to structure and bonding in transition metal complexes is the mode of bonding of ketimine and aza-allyl/allene ligands to transition metals. Some data pertaining to molybdenum complexes are given in Figures 89 and 90. An interesting feature of the spectra is the close proximity of the N_{1s} and $Mo_{3p3/2}$ levels. The latter core levels are less convenient for accurate study of shifts than the narrower more intense $3d_{5/2}$ levels, however, the spectra in this region give an immediate picture of the nitrogen and molybdenum binding energies and allow relative shifts to be measured without the errors imposed by corrections to reference levels. It is clear from Figure 89, for example, that in going from the ketimine (methylene amino) complex to the aza-allyl/allene complex that the increase in N_{1s} binding energy is accompanied by a decrease in the binding energy of the metal core levels. This suggests that the electron density increases at the metal atom and decreases at the nitrogen of the ligand. Without going into great detail it may be inferred that bonding of both ligands to the metal involve sigma and pi components. (The long range effect of the para substituents for example, Figure 90, is unambiguous evidence for pi component.)

It may readily be demonstrated by ESCA that in bonding CO to a transition metal in metal carbonyls there is overall electron transfer from metal to ligand. This demonstrates the dominance of the bonding interaction involving donation from filled d orbitals on the metal to antibonding π^* orbitals on the ligand. The decrease in IR

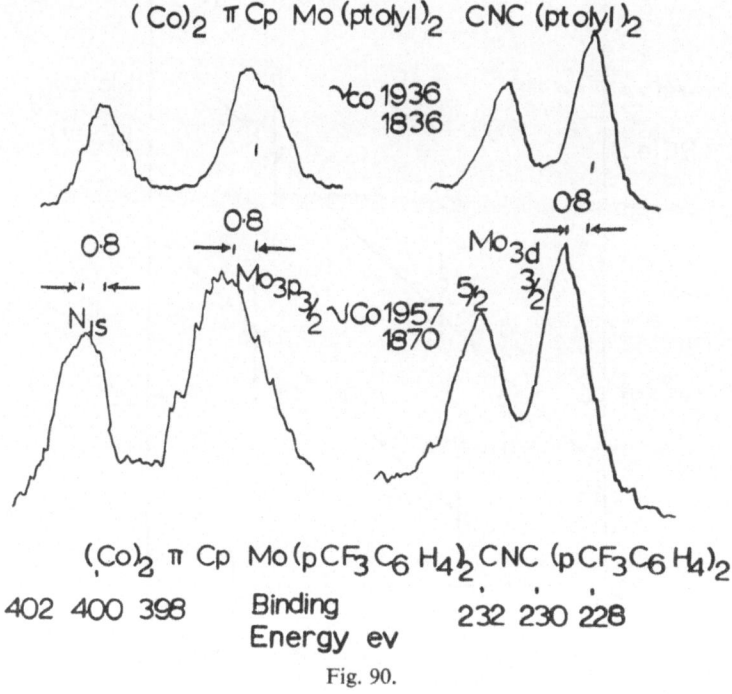

Fig. 90.

frequencies corresponding to CO stretching modes is consistent with this, although this data of itself is not sufficient to unambiguously distinguish between the relative magnitudes of the σ and π bonding components. The IR data therefore complements the ESCA data, and since the resolution available in the former technique is considerably higher than in the latter, serves as a more sensitive probe of the local electronic environment of the metal atom. Thus in Figure 89 it can be seen that the increased electron population on the metal in the aza-allyl/allene complex compared with the methylene amino complex evidenced by the decreased metal binding energies is accompanied by increased electron transfer from the metal to the carbonyl ligands evidenced by a decrease in carbonyl stretching frequencies. Replacing a para CH_3 by the overall electron withdrawing CF_3 (Figure 90), also decreases electron transfer from ligand to metal which is then relayed to the carbonyl ligands as is evidenced by *increased* stretching frequencies. Figure 91 shows a correlation between core binding energies for the metal and solid state carbonyl stretching frequencies (using the higher of the two ν_{CO} values). The parallel trends show that the correlation is independent of the metal indicative of the close similarity between these complexes of Mo and W.

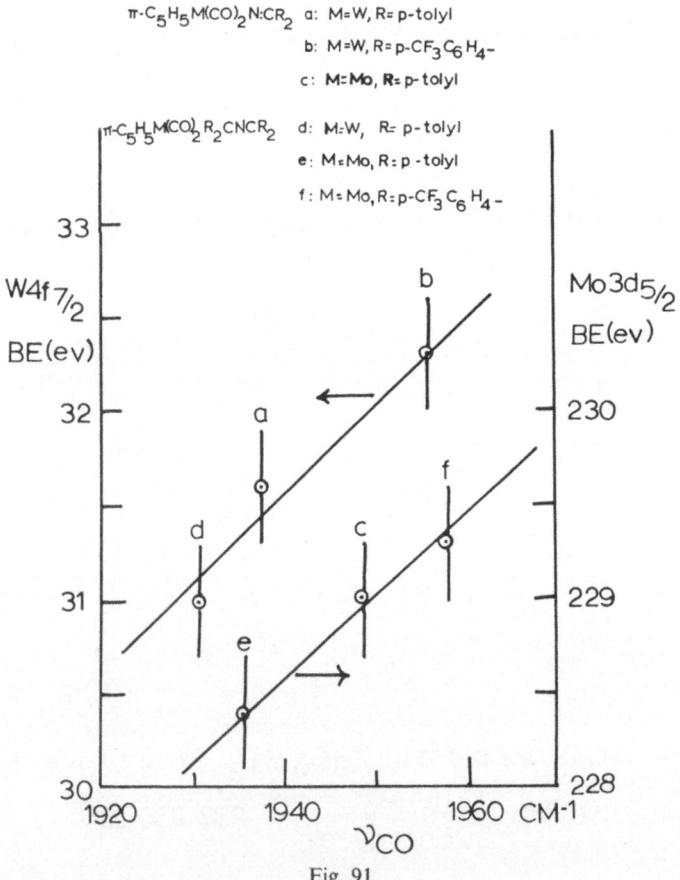

Fig. 91.

Since both the methylene amino and aza-allyl/allene complexes are used in the correlation the reasonable fit with limited data offers direct evidence for the strong stabilizing effect of the carbonyl group in complexes of metals in low formal oxidation states through dissipation of excess electron density on the central metal atom. There is no good theoretical reason why such correlations should be linear but since both measurements monitor to a lesser or greater extent the electronic environment of the metal atom a definite trend would be expected.

2. NMR Data

Molecular core binding energies reflect in some measure the local electronic environment about an atom and it is of interest therefore to consider other data which also depend on the wavefunction close to a given nucleus. We consider here NMR chemical shift data and in succeeding sections Mossbauer isomer shifts and NQR coupling constants. In the MO treatment of chemical shifts the dominant contributions to the screening constants for a given nucleus such as ^{13}C arises from the diamagnetic and paramagnetic contributions at the given nucleus (first two terms of the Ramsey equation, detailed below in Equation (1)).

$$\sigma_{A,\,av} = \sigma_{AA}^{d} + \sigma_{AA}^{p} + \sum_{B \neq A} \sigma_{AB} + \sigma_{A}^{ring} \tag{1}$$

$$\Delta\sigma_{A}^{d} = \frac{-\Delta E_{A}}{3cm^{2}} + \frac{e^{2}}{3mc^{2}}\Delta \sum_{B \neq A} \frac{Z_{B}}{R_{AB}} \tag{2}$$

$$\sigma_{p_{zz}} = \frac{-e^{2}\hbar^{2}\,\Delta}{2m^{2}c^{2}\,(\Delta E)} \{\langle r^{-3}\rangle_{2p}[(Q_{AA})_{zz} + (Q_{AB})_{zz}]\} \tag{3}$$

$$(Q_{AA})_{zz} = 2 - 2\,(P_{x}A_{x}A^{-1})\,(P_{y}A_{y}A^{-1}) + 2P_{x}A_{y}A^{2} \tag{4}$$

$$(Q_{AB})_{zz} = -\,2P_{x}A_{x}B\,P_{y}A_{y}B + 2P_{x}A_{y}B\,P_{y}A_{x}B. \tag{5}$$

The shift in core binding energies for a given nucleus can be related theoretically to the diamagnetic contribution to the chemical shift for that nucleus as is shown in Equation (2). The sign conventions are such that a more effective shielding corresponds to a positive shift in the diamagnetic screening constant, and a higher electron binding energy corresponds to a positive ESCA shift. According to the Pople LCAO–MO formalism the zz component of the paramagnetic screening tensor is given by the Equation (3), with corresponding equations for other components (4) and (5). $\sum_{B \neq A}$ is a summation over all atoms other than A, $\langle r^{-3}\rangle_{2p}$ is the mean inverse cube radius for carbon $2p$ orbitals, ΔE is an average electronic excitation energy and P_{AB} are elements of the charge density and bond order matrix in the MO theory of the unperturbed molecule. Both the factor $\langle r^{-3}\rangle_{2p}$ and the term $(Q_{AA})_{zz}$ depend primarily on the local electron density on the carbon atom and for a closely related series of molecules such as the fluoro- or chloro-methanes the average excitation energy ΔE, should remain essentially constant. In this situation it might be expected that the dominant contributions to the screening constants σ_d^{AA} and σ_p^{AA} should reflect the local electronic environment about the carbon atom and hence be related in some way to the corresponding molecular core binding energies. The data [14] in Figure 92 illustrate the validity of these simplistic arguments for data pertaining to the halomethanes. For the chloro- and fluoro-methanes the correlations are reasonably in agreement with these simple ideas. In general however no correlation is apparent (cf. data for bromo- and iodo-methanes and for boron Figure 93). It is clear therefore that the conditions under which direct correlations might exist are severely restricted. In this connection, it is worth noting also that in general the variation in the part of the NMR shift attributable to the diamagnetic screening constant which is directly related to the ESCA shift is generally one or two orders of magnitude smaller than the variation in total NMR chemical shift. The data also clearly illustrate the great advantage and disadvantage of ESCA compared with NMR as spectroscopic techniques. Firstly whereas the rationalization on a qualitative or quantitative basis of the ESCA data is straightforward, that for the NMR data is exceedingly difficult. Secondly taking the chloromethanes, whereas the C_{1s} shift to line width ratios in going from CH_4 to CCl_4 might typically be $\sim\frac{7}{1}$ in ESCA, the corresponding ^{13}C shifts to line width ratios would be typically $\sim 100/0.1$ illustrating the far superior resolution of NMR as a technique.

3. *Mössbauer Data*

Mössbauer isomer shifts may be related to the s electron density at the nucleus and as such provide data complementary to that obtainable from ESCA where shifts are often dominated by the total electron density at a given atom. An extremely interesting study [17] which reveal the great potential of complementary ESCA and Mössbauer studies for closely related series of compounds as listed in Table XIV was made by Adams *et al.* [17]. The correlation diagrams are shown in Figure 94. For octahedral

TABLE XIV

Mössbauer isomer shifts

S electron density at nucleus

Sn^{IV} Octahedral	Fe^{II} low spin
$Y_2 Sn(Ox)_2$	$[Fe^{II}(phen)_3] ClO_4$
$Y = Et, Ph, Cl, Br, I$	trans $Fe(isocy)_4 Cl$
	trans $Fe(isocy)_4 (SnCl_3)_2$
	$[Fe(SnCl_3)(isocy)_5] ClO_5$
	$Na_2Fe(CN)_5 NO. 2H_2O$
$\delta R/R + ve$	$\delta R/R - ve$
S electron density $\rightarrow$	S electron density $\rightarrow$
binding energy $\leftarrow$	binding energy $\rightarrow$

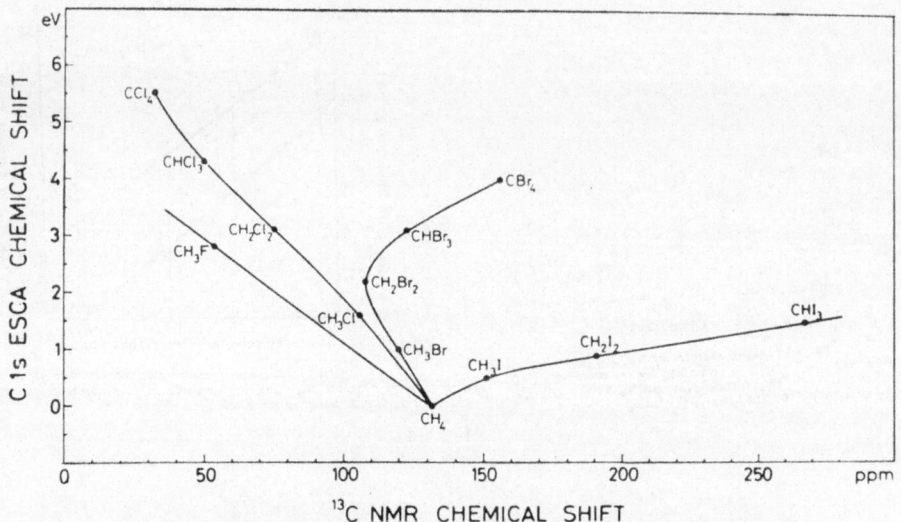

C 1s ESCA shifts versus ^{13}C NMR shifts for the halomethanes.

Fig. 92.

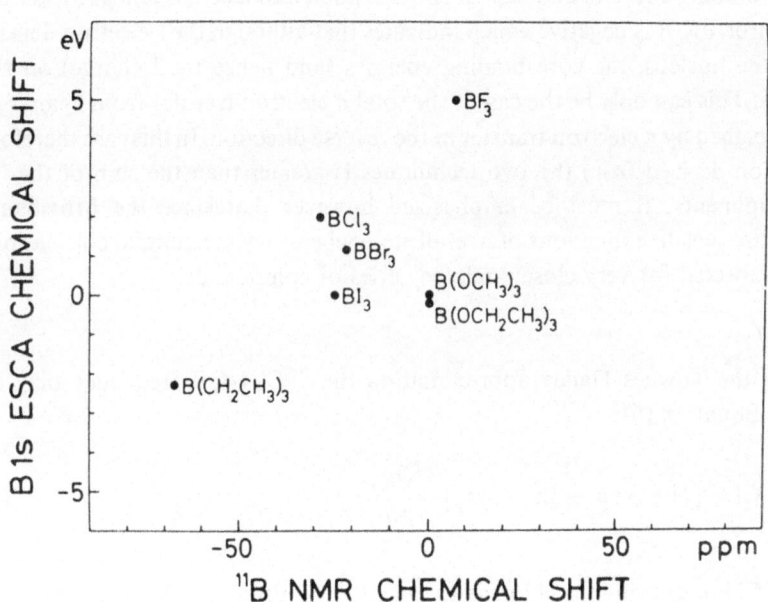

B 1s ESCA shifts versus ^{11}B NMR shifts for the BR_3 compounds

Fig. 93.

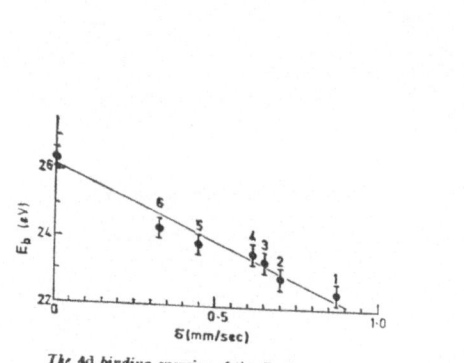

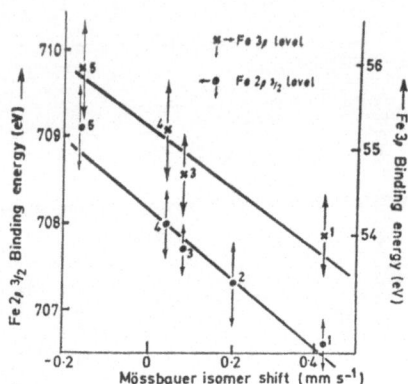

The 4d binding energies of tin, E_b (ev), against Mössbauer chemical shifts, δ (mm/s) for compounds (1–7). (1)
Et$_4$Sn(ox)$_2$; (2) Ph$_2$Sn(ox)$_2$; (3) SnBr$_4$.2oxH; (4) SnI$_2$(ox)$_2$;
(5) SnBr$_2$(ox)$_2$; (6) SnCl$_2$(ox)$_2$; (7) SnO$_2$.

Plot of Mössbauer isomer shift versus absolute binding
energy for Fe 2p$_{3/2}$ and Fe 3p levels. (The numbers refer to the
compounds listed in the Table.)

Fig. 94.

Sn(IV) complexes since $\delta R/R$ is positive for the tin nucleus increasing isomer shifts should correspond to increased s electron density at the nucleus. The correlation between isomer shift and 4d core binding energies for tin illustrated in Figure 94 shows that as might be expected for Sn(IV) the total electron density on tin (as deduced from the ESCA shifts) follows the s electron density (deduced from the Mössbauer isomer shifts). For the low spin Fe(II) complexes however the correlation between core binding energies and isomer shifts is good and also has a negative slope. However for iron $\delta R/R$ is negative which indicates that although the s electron density increases at the nucleus, the core binding energies (and hence total charge) on the iron increases. This can only be the case if the total σ electron transfer from ligands to metal is outweighed by π electron transfer in the reverse direction. In this case therefore the information derived from the two techniques is greater than the sum of the individual components. It must be emphasized however that since the Mössbauer isomer shifts are sensitive functions of overall stereochemistry meaningful correlations can only be expected for very closely related series of compounds.

4. NQR Data

According to the Townes Dailey approximation the ^{35}Cl NQR frequency may be expressed by Equation (6):

$$^{35}Cl = \left[(1-s)\,\sigma - \tfrac{1}{2}\pi_x - \tfrac{1}{2}\pi_y \right] \frac{e^2 Q q_{At}}{2h} \tag{6}$$

assuming M–Cl bonds using the chlorine $3p_z$ and 3_s orbitals in an hybrid with s degree of hybridization. σ is the covalent character of the σ-bond and π_x and π are the covalent character of the π-bonds. The charge configuration is depicted in Figure 95. For a closely related series of molecules involving a given type of M–Cl bond and

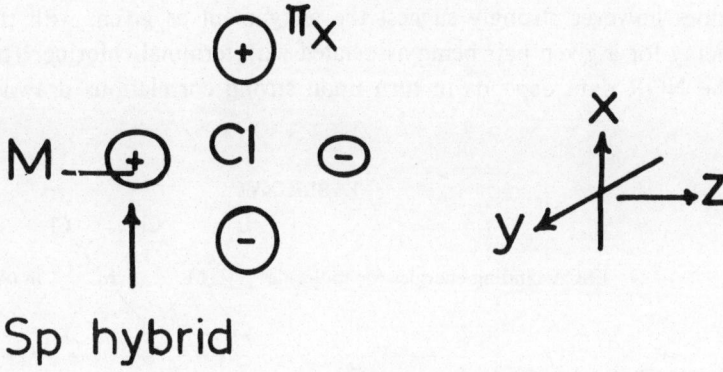

Sp hybrid

Fig. 95.

TABLE XV

ESCA binding energies and NQR frequencies for Cl in chloromethanes

	ESCA Binding energies (eV)		^{35}Cl NQR freq. (MHz)
	Cl_{2s}	$Cl_{2p_{3/2}}$	
CH_2Cl_2	272.8	201.3	36.233
$CHCl_3$	273.3	202.3	38.488
CCl_4	273.9	202.6	40.918

similar stereochemistry a case can be made for correlating NQR frequencies with the electron density at chlorine (as influenced by the population of the Cl_{3p_z} orbital). This being the case we might again expect in suitable cases for a correlation to be apparent between NQR and ESCA data. This is apparent for example for the chloromethanes where increasing binding energies correlated with increasing NQR frequency both relating to decreased electron population on chlorine (Table XV). This correlation implies that for closely related series of compounds the % s-character of the hybrid may be taken constant, the π-bonding contribution is insignificant and thus the ^{35}Cl NQR frequencies are directly proportional to the covalent bond character.

In studying square planar platinum and palladium complexes it is clear that both ESCA and NQR may be used to gain information on differences in electronic structure of isomeric cis and trans dichlorides. For bridged species both ESCA and NQR distinguish between terminal and bridging chlorine and indeed the NQR data is very useful as a guideline for assignment of the ESCA data. Thus for the halides listed in Table XVI the observation of two Cl_{2p} levels shows that there is a distinction between terminal and bridging halogen but does not allow an unambiguous assignment. The close correlation between the NQR and ESCA data for the square planar

dihalides does however strongly suggest the assignment as given, with the higher
binding energy for a given pair being associated with terminal chlorine. The assignment of the NQR data depends in turn upon strong correlations drawn between

TABLE XVI

ESCA binding energies for molecules (diagram of L, Cl, Cl bridging M, M, Cl, Cl, L) in eV

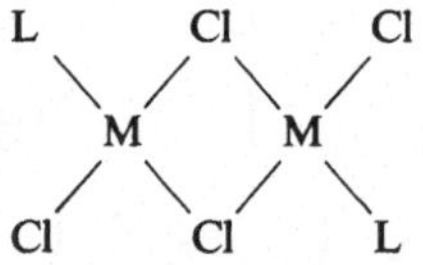

M = Pt	$Pt_{4f_{7/2}}$	$Cl_{2p_{3/2}}$	L
L = CH$_3$ – CH = CH$_2$	72.7	199.4 terminal	C_{1s} CH$_3$ 285.0 C = C 284.2
		198.4 bridging	
L = nPr$_3$P	71.9	199.1 terminal	C_{1s} 284.7 $P_{2p_{3/2}}$ 130.8
		198.1 bridging	
M = Pd	$Pd_{3d_{5/2}}$	$Cl_{2p_{3/2}}$	L
L = CH$_3$ – CH = CH$_2$	337.9	199.3 terminal	C_{1s} CH$_3$ 285.0 C = C 284.2
		198.3 bridging	
L = nBu$_3$P	337.7	198.7 terminal	C_{1s} 284.7 $P_{2p_{3/2}}$ 130.7
		197.7 bridging	

TABLE XVII

Correlation between NQR frequencies and bond lengths of M–C bond

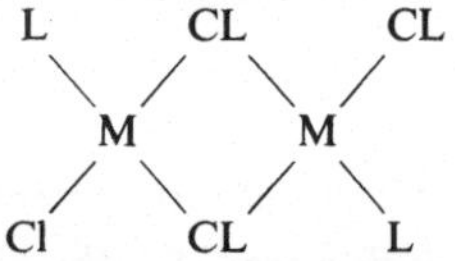

(diagram of L, CL, CL bridging M, M, Cl, CL, L)

		$\nu 35_{Cl}$ (MHz)		M–Cl (Å)	
M	L	Terminal Cl	Bridging Cl	Terminal	Bridging
Pt	CH$_3$–CH = CH$_2$	24.12	15.95	2.28	2.34
	nPr$_3$P	22.36	15.46	2.279	2.315
Pd	Bu$_3$P	19.50	12.77	–	–

NQR frequencies and bond lengths of M–Cl bonds (cf. Table XVII). For square
planar dihalides and bridged complexes the correlation between NQR frequencies
and Cl_{2p} binding energies for platinum complexes is shown in Figure 96. Although
the ESCA shifts span a small range and are therefore subject to relatively large experimental errors the correlation is clearly established.

Turning now to a slightly more complex situation it is of interest to investigate the
electronic structure of complexes of the form $[PtCl_3L]^- X^+$ (Figure 97) where the

trans influence of the ligand L may be apparent. The structure of and electron distribution in Zeise's salt ($L=C_2H_4$) has been debated for many years. The complex is

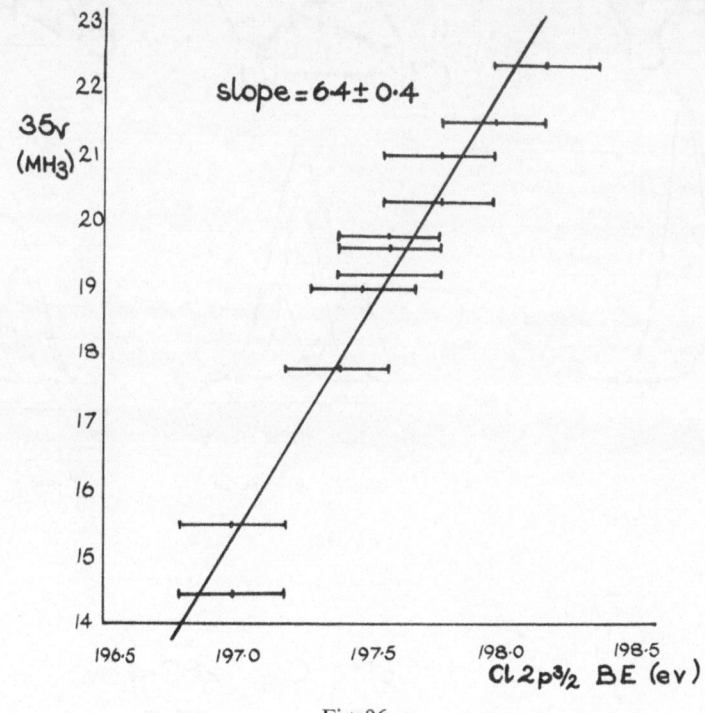

Fig. 96.

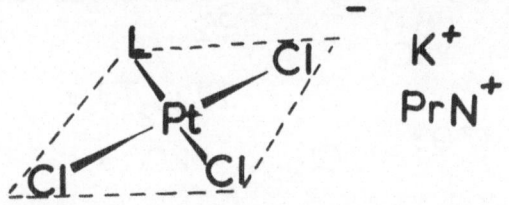

$L = C_2H_4,\ C_3H_6, Me_3\ P,\ Me_3As$

Fig. 97.

important since it demonstrates the high (kinetic) trans effect of the ethylene ligand. The trans influence which the ligand exerts in the ground state may be detected by NQR and for Zeise's salt three main signals are observed ca. 20.4, 20.2 and 16.0 MHz corresponding to cis and trans chlorines respectively. On the basis of the correlation

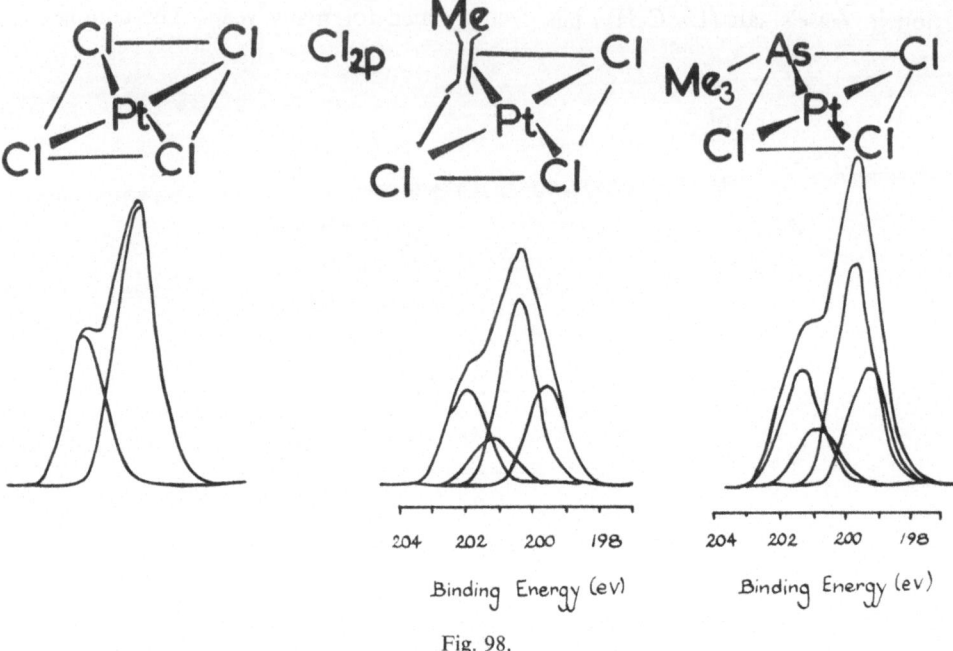

Cl
Cl
Pt
Cl
Cl
Me
Cl_{2p}
Cl
Pt
Cl
Cl
Me_3
As
Cl
Pt
Cl
Cl
204
202
200
198
Binding Energy (eV)
204
202
200
198
Binding Energy (ev)

Fig. 98.

L=C2H4 C1s 283·4 ev.
Pt4f7/2 74·5
L
Cl
Pt
Cl
Cl
35ν 20·4, 20·2 Cl2p3/2 200·4(2) 199·8(1)
16·0
L=Cl
Pt4f7/2 73·9
ν35 17·93
Cl2p3/2 200·3

Fig. 99.

Figure 96, we might therefore anticipate a difference in binding energy between cis and trans chlorines of ~0.6 eV which should manifest itself in a change in line shape relative to a standard (eg. K_2PtCl_4). The Cl_{2p} core levels for the standard and complexes ($L=C_3H_6$, Me_3As) are shown in Figure 98 and analysis of the overall line shapes yields the results shown in Figure 98. The trans influence of the ligands may clearly be seen and the line shape analysis is such that both for the propene and arsine ligands the trans chlorines have greatest electron density (hence lower binding energy).

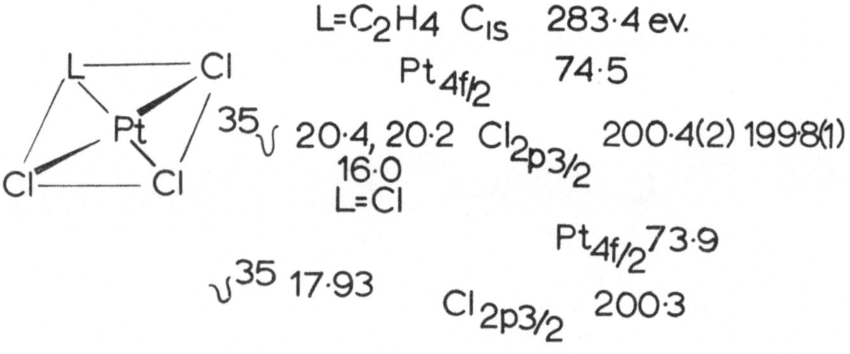

The data for K_2PtCl_4 and $K_2PtCl_3C_2H_4$ are given more fully in Figure 99. In summary it may be inferred that:

(a) The effect of replacing one Cl^- ligand in $[PtCl_4^{2-}]$ by a neutral ligand L to give $[LPtCl_3^-]$ is to split the remaining chlorine ligands into two differing types on the

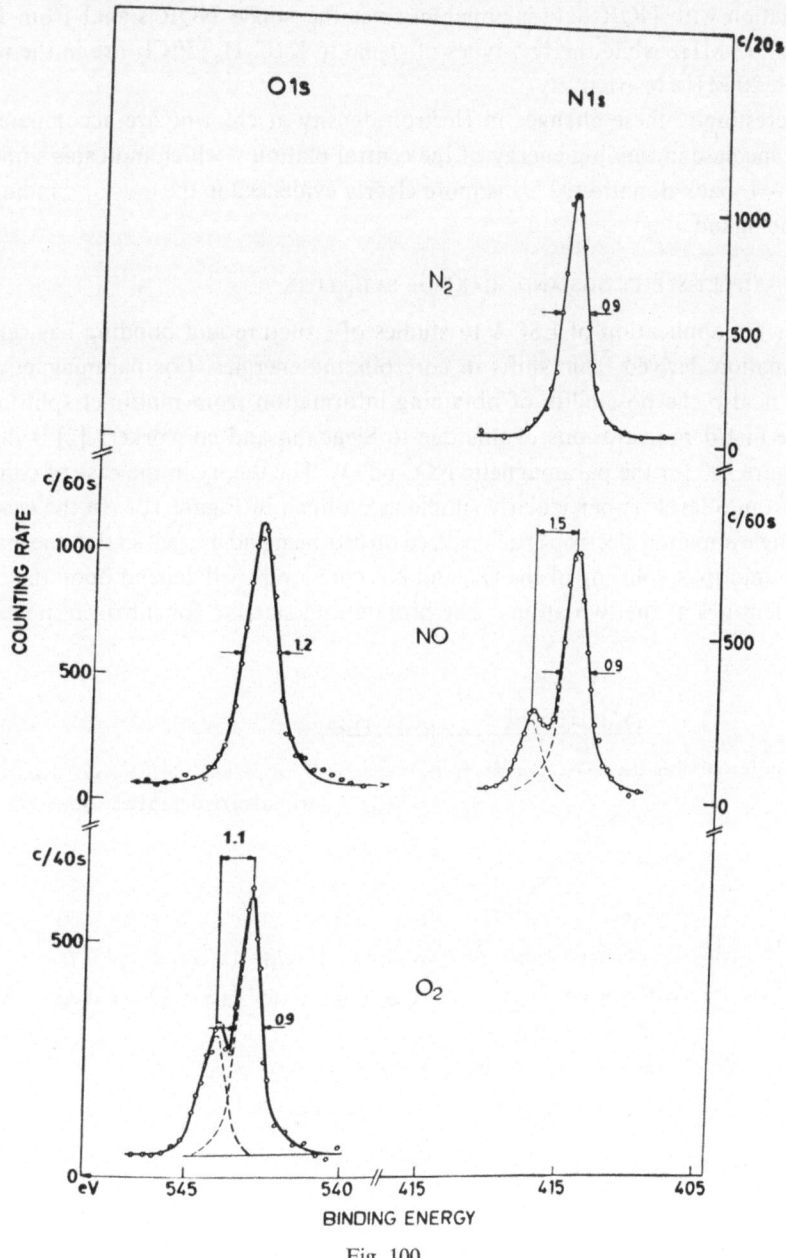

Fig. 100.

ratio 2:1 and these can therefore, be assigned to the cis and trans positions respectively. Hence ESCA detects a ground state influence of L on the remaining ligands.

(b) There is evidence that relative to $PtCl_4^{2-}$ the binding energy of the cis chlorines is raised (decrease in negative charge) and that of the trans-chlorine is lowered. This is in agreement with other ground state phenomena which relate to the trans-effect; increase in Pt–Cl bond length is accompanied by electron drift to chlorine. The correlation with NQR data is apparent since the single NQR signal from K_2PtCl_4 is at 17.93 MHz while the two types of signal in $K(C_2H_4)PtCl_3$ are in the region of 16 and 20 MHz respectively.

Interestingly, these changes in electron density at chlorine are accompanied by a slight increase in binding energy of the central platinum which indicates some degree of Pt → L back donation. This is more clearly evidenced in the low C_{1s} binding energy for the ligand.

D. MULTIPLET SPLITTINGS AND SHAKE-UP SATELLITES

So far the application of ESCA to studies of structure and bonding has centred on information derived from shifts in core binding energies. For paramagnetic species there is also the possibility of obtaining information from multiplet splittings. One of the first demonstrations of this due to Siegbahn and co-workers [2] is illustrated in Figure 100 for the paramagnetic NO and O_2. The theory in the case of core ionization from S levels is particularly simple as outlined in Figure 101. In the case of NO the single unpaired electron is delocalized on nitrogen and oxygen so that the magnitude of the multiplet splitting of the O_{1s} and N_{1s} core levels will depend upon the unpaired spin densities at the two atoms. The pronounced satellite for nitrogen indicates that

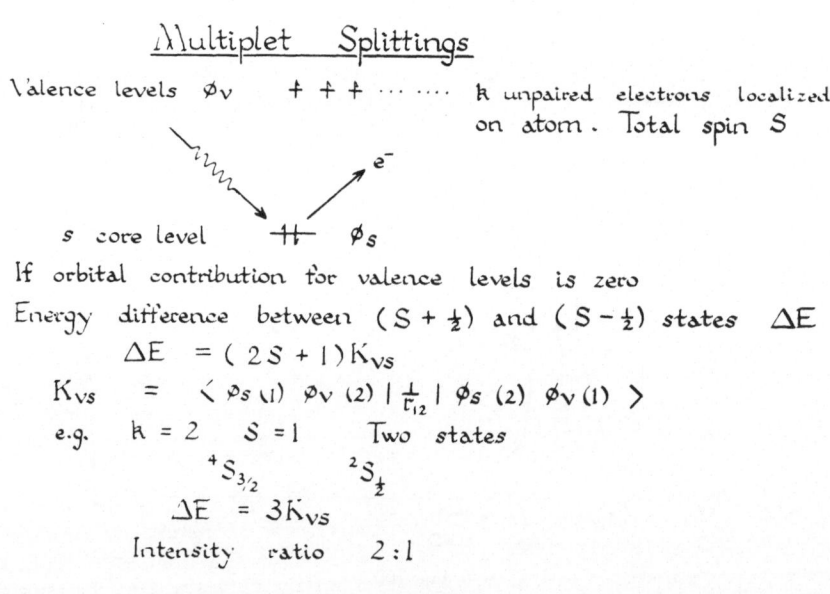

Fig. 101.

most of the unpaired spin density is on nitrogen and hence the magnitude of multiplet splittings can provide information comparable to that from ESR studies. The best developed application of multiplet effects is in transition metal chemistry originally pioneered by Fadley and co-workers. [18]

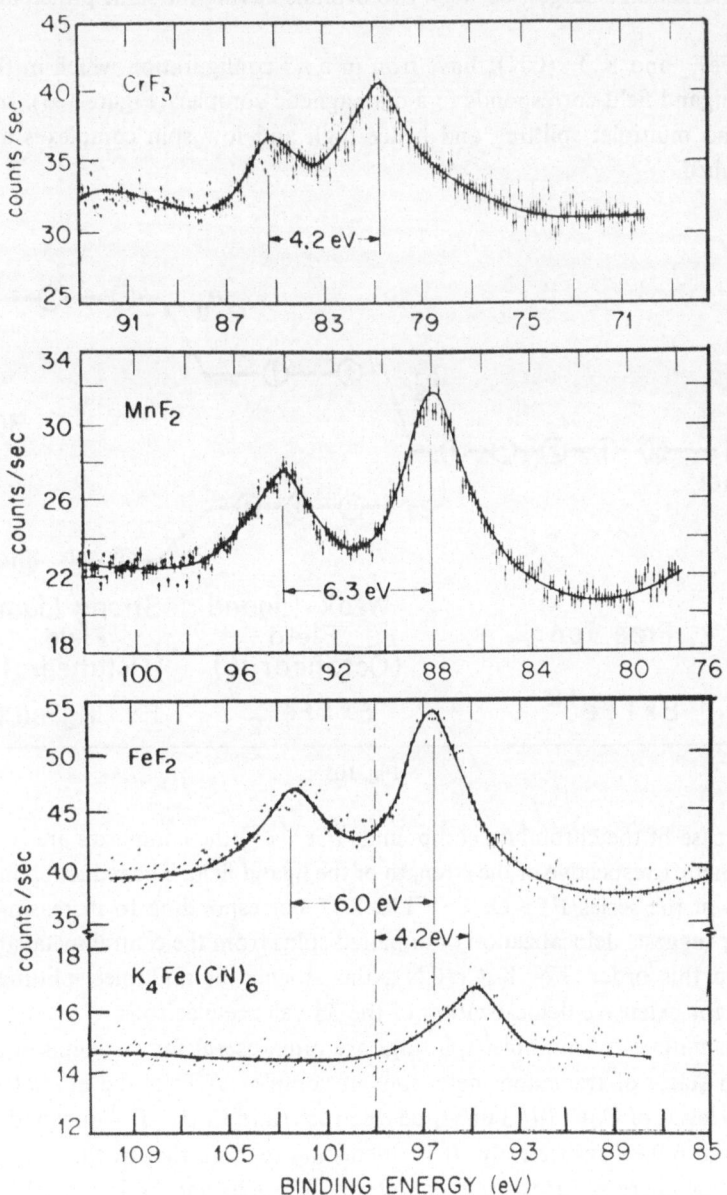

Photoelectron Spectra of 3s Shell for Metal Ion

Fig. 102.

Interesting examples of the application of ESCA in this field have been provided by Carlson and co-workers, [19] Figure 102. CrF_3 and MnF_2 are both paramagnetic with 3 and 5 unpaired electrons respectively and exhibit large multiplet splittings for the 3s shell. (The 3s level of first row transition elements is normally used for studying multiplet effects because of (1) the simplicity of the spectra, (2) the fact that the exchange interaction is largest between two orbitals having the same principal quantum number.)

Both FeF_2 and $K_4Fe(CN)_6$ have iron in a d^6 configuration which in the case of a strong ligand field corresponds to a diamagnetic complex (Figure 103). In this case there is no multiplet splitting and hence high and low spin complexes are readily distinguished.

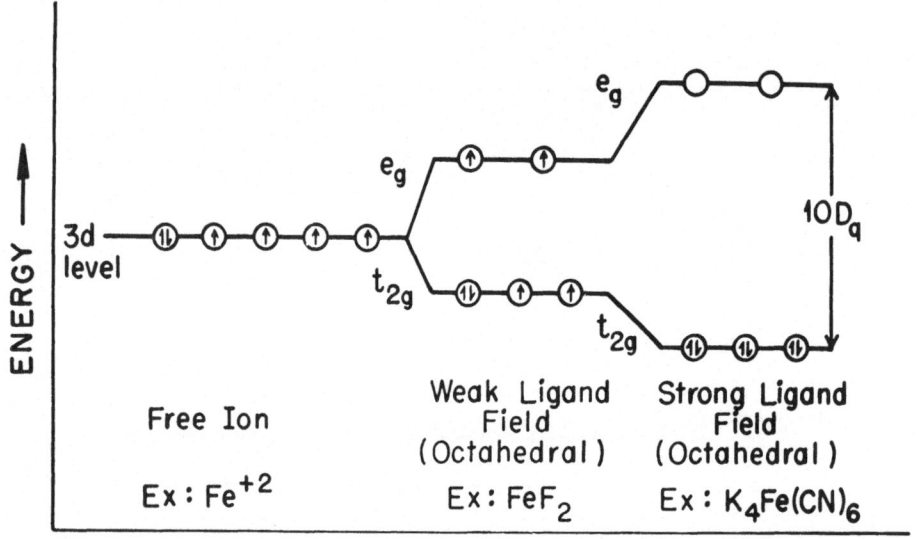

Fig. 103.

In the case of the chromium compounds ($Cr^{3+}d^3$) the complexes are (Figure 104) paramagnetic irrespective of the strength of the ligand field. The reduction in multiplet splittings in the series $F^- > O^{--} > Cl^- > S^{--}$ corresponding to increasing covalent character suggests delocalization of unpaired spins from the central metal atom to the ligands in this order. For $K_3Cr(CN)_6$ the absence of multiplet splittings is clear evidence for extensive delocalization of the 3d valence electrons.

The magnitudes of multiplet splittings are often useful for distinguishing different oxidation states of transition metal ions in complexes. Thus the multiplet splitting for the 3s levels of CoF_2 (d^7 3 unpaired electrons) and CoF_3 (d^6 4 unpaired electrons) are 5.0 and 6.0 eV respectively. It is interesting to note that in this case the metal core binding energies for the two complexes are such that the lower oxidation state corresponds to higher binding energy ($Co_{2p_{3/2}}$ 783.2 eV CoF_2 782.6 eV CoF_3). This illustrates the difficulty of correlating shifts in core binding energies measured for ionic solids with simple ideas concerning charge distributions. The reversal of the expected order of binding energies almost certainly arises to a large extent from lattice effects discussed briefly at the introduction to this section.

Although the limited scope of this review does not allow us to discuss them in detail, well developed satellite peaks to the low kinetic energy side of the main photo-

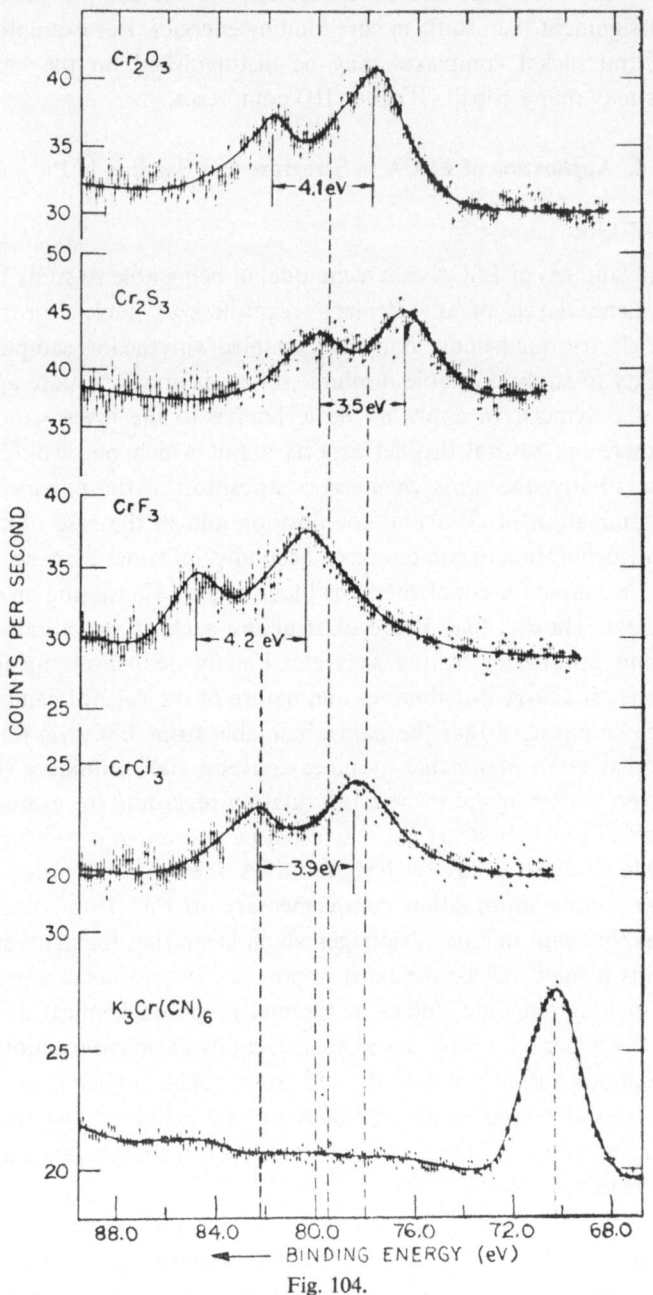

Fig. 104.

ionization peaks arising from shake-up processes are often well developed in transition metal compounds (e.g. in transition metal carbonyls there are often strong satellite peaks for the C_{1s} and O_{1s} core levels), and can yield valuable information on the electronic structure of the complexes. This is a rapidly developing area. The shake-up satellites for the central metal ion in a series of complexes often depends quite dramatically on the oxidation state and stereochemistry of the complex and often is more useful for assignment than shifts in core binding energies. For example square planar and tetrahedral nickel complexes may be distinguished on the basis of satellite structure as may many cobalt (II) and (III) complexes.

4. Application of ESCA to Structure and Bonding in Polymers

A. INTRODUCTION

The great advantages of ESCA as a technique, in being able to study in principle the core and valence levels of any element (regardless of nuclear properties such as magnetic or electric quadrupole moments), coupled with the low sample requirements, and the ability to study involatile insoluble solids, is nowhere more apposite than in the study of polymers. In applying the technique to the investigation of polymer structures there are several distinct aspects about which one would hope to gain information. Firstly, the gross chemical composition of the polymers. This would include determination of elemental composition and in the case of copolymers the percentage incorporation of comonomers. Secondly, information concerning the gross structure for example for copolymers the block and/or alternating or random nature of the linkages. Thirdly, finer detail of structure such as structural isomerism, the nature of end groups, branching sites etc. Finally deductions made from ESCA measurements, of charge distributions and nature of the valence bands for polymers. It should be emphasized that the data obtainable from ESCA is rather coarser in detail than that often obtainable by more conventional techniques (IR, NMR, inelastic neutron scattering etc.) where information regarding for example, conformational aspects of polymer structure, may often be inferred and which are in principle not amenable to direct study by ESCA. ESCA should be regarded as a powerful technique providing information complementary to that from other branches of spectroscopy, but with unique advantages which mean that for many studies of polymeric systems it may well be the most important. In particular aspects of polymer chemistry, such as dynamic studies of thermal or photochemical degradation, and in studies of polymeric films produced at surfaces by chemical reaction (e.g. fluorination or oxidation) the information derived from ESCA studies is not obtainable by other techniques. The application of ESCA to such problems has been pioneered in our laboratories, however in the space available we will concentrate mainly on aspects of the four headings detailed above.

The applications described below pertain to fluoropolymer systems since this has been our main research interest. Even without an underlying research interest in these systems there are two major reasons why an initial research program into the application of ESCA to polymers, should concentrate initially on fluoropolymer systems. Firstly the large shift in core levels induced by fluorine gives the most favourable situation for delineating the likely areas of applicability of ESCA in this field. Secondly, fluoropolymers are amongst the technologically most important systems of interest and since they are often insoluble are difficult to study by other spectroscopic techniques (e.g. ^{13}C, ^{1}H, NMR).

B. SAMPLE HANDLING

In general it is convenient to study polymers in the form of pressed films mounted on a suitable backing (e.g. gold). For elastomers it is possible to 'melt' a small amount of the sample and allow it to spread in the form of a thin film on the tip of a sampling probe or to slice a thin film from a larger sample. In preparing samples from powders it is often convenient to press films between sheets of clean aluminium foil at an appropriate temperature and pressure. There are two precautions to be taken in doing this

(1) The temperature and pressures used should be such that no decomposition or adhesion of surface contamination occurs.

(2) Since typically only the top $\sim 30 \, \text{Å}$ of the sample is studied by ESCA it is important to avoid chemical reaction at the surface during preparation. Thus pressing polyethylene films in air at the minimum temperature and pressure necessary results in considerable surface oxidation. This may be obviated by pressing in an inert atmosphere (e.g. N_2 or Ar). Surface contamination e.g. hydrocarbon etc., arising for example by inadvertant handling during processing can readily be removed by an appropriate solvent.

Of their very nature, most polymers of interest are extremely good insulators and as such in studying samples as thin films there is only a fortuitous possibility that the sample will be in electrical contact with the spectrometer i.e. sample charging occurs. Referencing of the energy scale is most readily accomplished by depositing a thin coating of a suitable reference (either hydrocarbon or gold) and monitoring the core levels of the sample and reference (cf. [8]).

An interesting observation is that for homopolymers, even in the presence of overall sample charging, with suitable preparation of sample, linewidths for given core levels are comparable with those for monomers cf. Figure 105. On this basis, in principle the range of applicability of ESCA in studying homopolymers is essentially the same as in studying simple monomers. With some preparation of thin films (by pressing or more usually with samples prepared by slicing from a larger sample or melted directly onto a probe tip) the less regular nature of the surface is often manifest in slight increase in linewidths arising from non-uniform distribution of charge on the sample. This is often of no consequence since it merely affects the overall resolution, however it may be obviated in many cases by modifying the way in which the film is produced.

C. HOMOPOLYMERS

As a first step in the ESCA examination of a polymer it is a straightforward matter

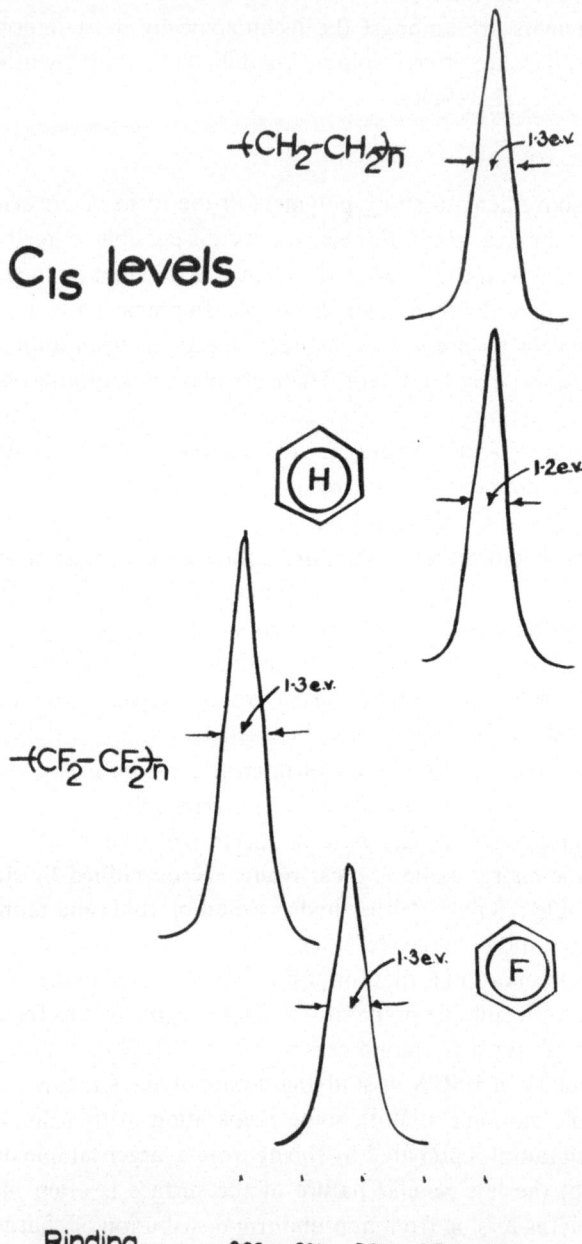

C$_{1S}$ levels

$+CH_2-CH_2+_n$ 1·3 e.v.

H

$+CF_2-CF_2+_n$ 1·3 e.v.

1·2 e.v.

1·3 e.v.   F

Binding
Energy ev   293 291   289 287 285

Fig. 105.

to establish its elemental composition, by examination of the relevant core levels together with appropriate calibration standards. In addition however, shifts in binding energies for a given level together with relative areas for different peaks, may be used to routinely acquire structural information, such as percentage incorporation of comonomers in copolymers (which may be checked by reference to the elemental composition). This particular application will be described in more detail later. It is convenient in developing the application of ESCA to polymer chemistry to follow the same general guidelines set out for application in organic chemistry. Namely to start by studying simple well characterized systems to build up banks of data from which trends may be discerned and comparisons drawn with simple monomers. Then develop a theoretical framework which will quantify the results and this then provides a strong basis for studying more complex systems. As a start in this direction the obvious choice for initial studies are simple homopolymers for which complications due to branching, end groups etc. are minimal. To this end we have undertaken detailed studies of polyethylene, polyvinyl fluoride, polytrifluoroethylene and polytetrafluoro-ethylene. The C_{1s} spectra for these polymers are shown in Figures 106–108. Typically measurements of all the core levels of a polymer take ~ 30 min whereas under the

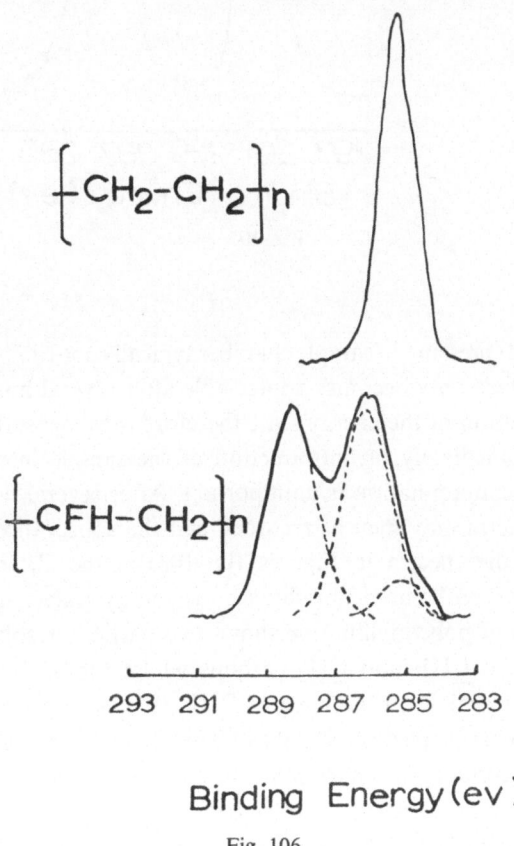

293 291   289 287 285 283

Binding Energy (ev)

Fig. 106.

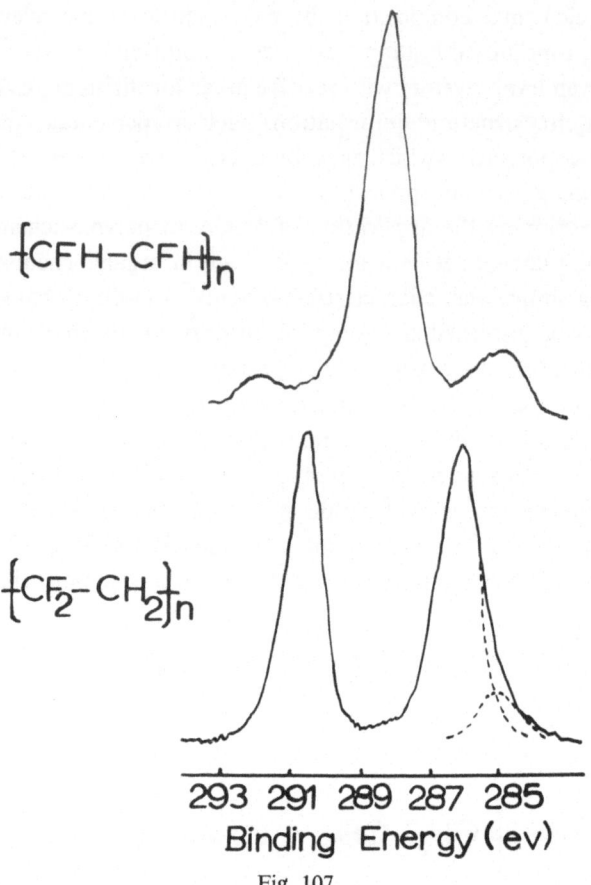

$\{CFH-CFH\}_n$

$\{CF_2-CH_2\}_n$

293 291 289 287 285
Binding Energy (ev)

Fig. 107.

conditions employed (pressure in sample chamber typically 5×10^{-7} torr) hydrocarbon build up on the surface only becomes appreciable after several hours. The technique employed for calibration of the energy scale therefore is to measure the core levels of a given polymer immediately on introduction of the sample into the spectrometer when hydrocarbon contamination is unimportant. After several hours in the sample chamber further spectra may then be recorded and the appearance of an extra peak, (clearly evident in the spectra in Figures 106–108) in the C_{1s} region at 285.0 eV binding energy may then be used to reference the energy scale.

The C_{1s} spectrum of polyvinylfluoride shows two partially resolved peaks of equal area corresponding to $\underline{C}HF$ and $\underline{C}H_2$ carbons whilst for polyvinylene fluoride in addition to the main peak corresponding to $\underline{C}HF$ carbons and hydrocarbon calibration peak there is a weak peak at 292.0 eV. This in fact corresponds to $\underline{C}F_2$ type carbon arising from contamination from the fluorocarbon soap $(H(CF_2)COO^-NH_4^+)$ used in the emulsion polymerization. For polyvinylidene fluoride well resolved peaks of equal area corresponding to $\underline{C}F_2$ and $\underline{C}H_2$ carbons are evident and for polytri-

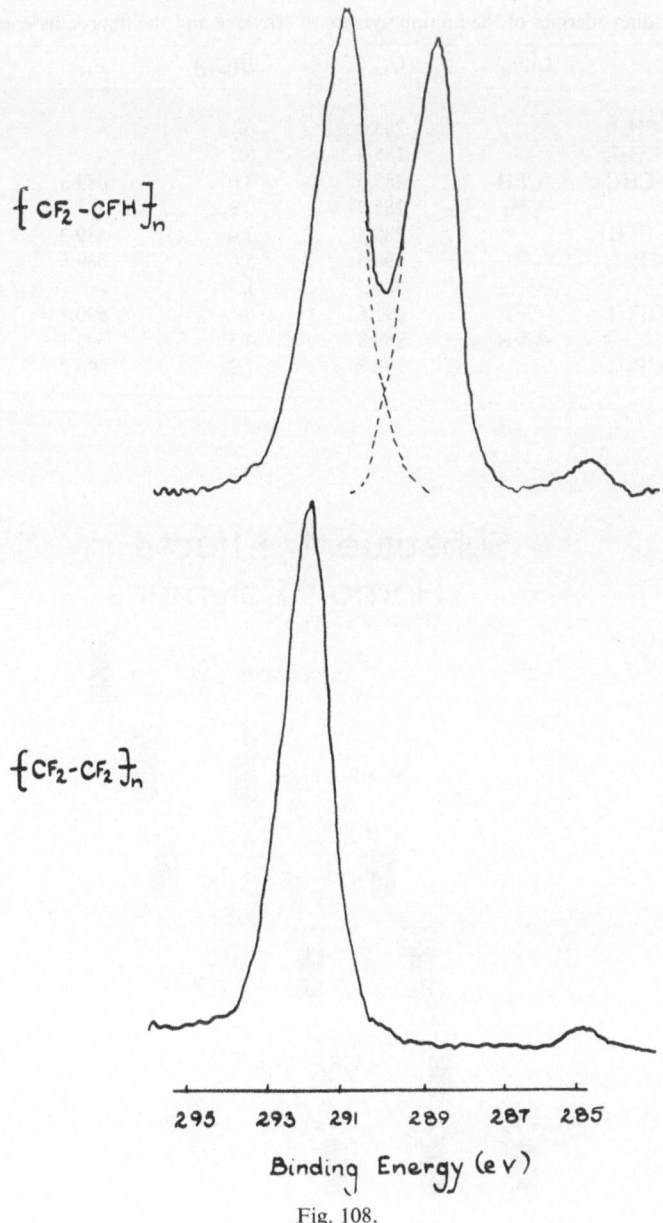

$\{ CF_2-CFH \}_n$

$\{ CF_2-CF_2 \}_n$

295   293   291   289   287   285

Binding Energy (ev)

Fig. 108.

fluoroethylene partially resolved peaks of equal area corresponding to $\underline{CF}_2$ and $\underline{CFH}$ carbons.

The data pertaining to these homopolymers are collected in Table XVIII and Figure 109 shows diagramatically the C_{1s} levels including data for polytrifluoro-chloroethylene and polyhexafluoropropene. The assignment of peaks arising from $\underline{CF}_3$, $\underline{CF}_2$, $\underline{CF}$ and $\underline{CH}$ structural units is readily apparent. By taking appropriate

TABLE XVIII

Binding energies of the homopolymers of ethylene and the fluoroethylenes

Nr	Polymer	Unit	C_{1s}	$\Delta(C_{1s})$	F_{1s}	$\Delta(F_{1s})$
Ia	$[CH_2-CH_2]_n$		285.0	(0)	–	–
Ib	$[CH_2-CH_2]_n$		285.0	(0)	–	–
II	$[CFH-CH_2]_n$	$-\underline{C}FH-$	288.0	3.0	689.3	(0)
		$-\underline{C}H_2-$	285.9	0.9	–	–
III	$[CFH-CFH]_n$		288.4	3.4	689.3	0.0
IV	$[CF_2-CH_2]_n$	$-\underline{C}F_2-$	290.8	5.8	689.6	0.3
		$-\underline{C}H_2-$	286.3	1.3	–	–
V	$[CF_2-CFH]_n$	$-\underline{C}F_2-$	291.6	6.6	690.1	0.8
		$-\underline{C}FH-$	289.3	4.3	690.1	0.8
VI	$[CF_2-CF_2]_n$		292.2	7.2	690.2	0.9

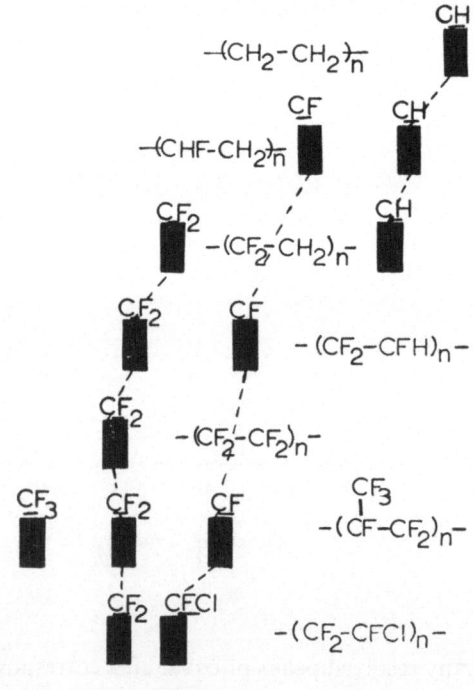

Fig. 109.

TABLE XIX

Primary substituent effects of replacing hydrogen by fluorine
in homopolymers

Polymer pairs	Shift in C_{1s} binding energy on replacing H by F (eV)
$(\underline{C}HFCH_2)_n$, $(\underline{C}H_2CH_2)_n$	3.0
$(\underline{C}F_2CH_2)_n$, $(\underline{C}HFCH_2)_n$	2.8
$(CHF\underline{C}HF)_n$, $(CHF\underline{C}H_2)_n$	2.5
$(CF_2\underline{C}HF)_n$, $(CF_2\underline{C}H_2)_n$	3.0
$(CF_2\underline{C}F_2)_n$, $(CF_2\underline{C}HF)_n$	2.9
$(\underline{C}F_2CHF)_n$, $(\underline{C}HFCHF)_n$	3.2
Average	2.9

TABLE XX

Secondary substituent effects of replacing hydrogen by fluorine
in homopolymers

Polymer pairs	Shift in C_{1s} binding energy on replacing H by F (per substituent) (eV)
$(CHF\underline{C}H_2)_n$, $(CH_2\underline{C}H_2)_n$	0.9
$(\underline{C}HFCHF)_n$, $(\underline{C}HFCH_2)_n$	0.4
$(CF_2\underline{C}H_2)_n$, $(CH_2\underline{C}H_2)_n$	0.7
$(\underline{C}F_2CHF)_n$, $(\underline{C}F_2CH_2)_n$	0.8
$(\underline{C}F_2CF_2)_n$, $(\underline{C}F_2CHF)_n$	0.6
$(CF_2\underline{C}HF)_n$, $(CHF\underline{C}HF)_n$	0.9
Average	0.7

pairs of polymers it is possible to investigate both primary and secondary effects of replacing hydrogen by fluorine. The results are given in Table XIX and XX. The average primary substituent effect of 2.9 eV is in excellent agreement with that discussed previously. The secondary effect of 0.7 eV is eminently reasonable by comparison with the value of 0.4 eV derived for chlorine as previously outlined. This emphasizes the characteristic nature of substituent effects. The rapid fall off in effect of the fluorine substituent provides a crude but immediate manifestation of the σ inductive effect exerted by fluorine. As the degree of fluorine substitution increases in going through the series, the fluorine substituents to some extent compete for the sigma electron drift from the carbon and hydrogen atoms and this is clearly evidenced by the increase in binding energy for the F_{1s} levels.

Quantitative Discussion of Results

The short range nature of substituent effects on core binding energies in saturated systems suggests that it may be feasible to quantitatively discuss the results in terms of calculations on simplified model systems which contain the essential structural features and accommodate all short range interactions. The success of the charge potential model coupled with CNDO SCF MO calculations of charge distributions in quantitative discussion of data for simple monomers presented earlier suggests that this is a feasible approach.

The model accounts quite nicely for the short range nature of substituent effects. For example Table XXI shows the calculated charges and Madelung potentials for a perfluoro alkane chain $CF_3CF_2CF_2$-R. Both of these quantities remain essentially the same at the CF_3 and adjacent CF_2 group as R is varied. The effect on calculated binding energies is shown in Table XXII. For both the CF_3 and adjacent CF_2 carbon the calculated relative binding energies are essentially independent of R.

TABLE XXI

CNDO/2 charges and Madelung potential for the series CF_3–CF_2–CF_2–R

CF₃		CF₂		CF₂		
qi	$\sum \frac{q_i}{r_{ij}}$	qi	$\sum \frac{q_i}{r_{ij}}$	qi	$\sum \frac{q_i}{r_{ij}}$	R
0.58	−4.68	0.29	0.47	0.58	−4.68	F
0.58	−4.69	0.32	−0.01	0.32	−0.01	CF_3
0.58	−4.68	0.32	−0.04	0.35	−0.49	C_2F_5
0.58	−4.70	0.31	−0.02	0.35	−0.52	nC_3F_7
0.58	−4.68	0.30	0.24	0.38	−1.14	iC_3F_7
0.58	−4.70	0.31	−0.04	0.34	−0.50	C_4F_9

TABLE XXII

Relative BE's calculated from potl. model
(CNDO/2 charges) in eV

CF₃	CF₂	CF₂	R
2.2	(0)	2.2	F
2.1	0.2	0.2	CF_3
2.0	0.2	0.4	C_2F_5
2.0	0.1	0.4	nC_3F_7
2.1	0.1	0.7	iC_3F_7
2.0	0.1	0.4	nC_4F_9

As our model for polyethylene and polyvinylene fluoride we have therefore taken our representative model unit as being the monomer linked to other monomer units and then appropriate end groups. In this way substituent effects over three carbon atoms are taken into account. (The calculations show that such long range effects are negligible so that the model incorporates all of the important short range interactions.) All valence electron SCF MO calculations have then been carried out on these models and the resulting charge distributions taken in conjunction with appropriate values for k's and E_0's used to calculate absolute binding energies for the core levels of the representative structural unit of polymer. The adequacy of the theoretical

TABLE XXIII

Comparison of calculated and observed C_{1s} binding energies for some polymer models

Polymer model	C_{1s} BE (eV)			
	calculated		observed	
I CH₃–CH₂–CH₂–CH₂–CH₂–CH₂–CH₂–CH₃	284.9		285.0	
CH₂F–CHF–CHF–CHF–CHF–CHF–CHF–CH₂F	288.6		288.4	
CF₃–CF₂–CF₂–CF₂–CF₂–CF₃	292.6		292.0	
(CHF–CH₂)ₙ				
II – regular:				
CFH₂–CH₂–CFH–CH₂–CFH–CH₂–CFH–CH₂–CFH–CH₃	288.1	285.4	288.0	285.9
– irregular:				
CFH₂–CH₂–CH₂–CFH–CFH–CH₂–CH₂–CFH–CFH–CH₃	288.0	285.6	–	
(CF₂–CH₂)ₙ				
III – regular:				
CF₂H–CH₂–CF₂–CH₂–CF₂–CH₂–CF₂–CH₂–CF₂–CH₃	291.0	286.1	290.8	286.3
– irregular:				
CF₂H–CH₂–CH₂–CF₂–CF₂–CH₂–CH₂–CF₂–CF₂–CH₃	291.0	286.3	–	–
(CF₂–CFH)ₙ				
IV – regular:				
CF₂H–CFH–CF₂–CFH–CF₂–CFH–CF₂–CFH–CF₂–CFH₂	291.8	289.3	291.6	289.3
– irregular:			–	–
CF₂H–CFH–CFH–CF₂–CF₂–CFH–CFH–CF₂–CF₂–CFH₂	291.7	289.4	–	–

treatment for polyethylene and polyvinylene fluoride is apparent from Table XXIII (I). The small discrepancy in the case of PTFE undoubtedly arises from the fact that computer limitations dictated that a smaller model system than optimum be studied.

For the unsymmetrical monomers polyvinyl fluoride, polyvinylidene fluoride and polytrifluoroethylene the possibility arises of structural isomerism by way of head to tail and/or head to head addition. In fact ^{19}F nmr studies show that structural isomerism does occur in both of the first two and indeed provides information on tacticity which is not available by ESCA studies. Since the C_{1s} spectra of the three polymers are relatively well resolved it is evident that information regarding structural isomerism if available must be encoded in the lineshapes and/or linewidths.

Theoretical calculations on suitable models incorporating the relevant structural features viz. head to tail and head to head linkages give the results shown in Table XXIII (II), (III), (IV). Considering firstly polyvinyl fluoride, the calculated binding energies for both types of structural arrangements are the same within experimental error and in fact in excellent agreement with the observed values. It seems clear therefore that for this particular polymer ESCA is unable to provide information on structural isomerism along the chain. (This contrasts with the situation to be described later concerning nitroso rubbers.) The same considerations apply to both polyvinylidene fluoride and polytrifluoroethylene, again the theoretical models are in excellent agreement with experiment.

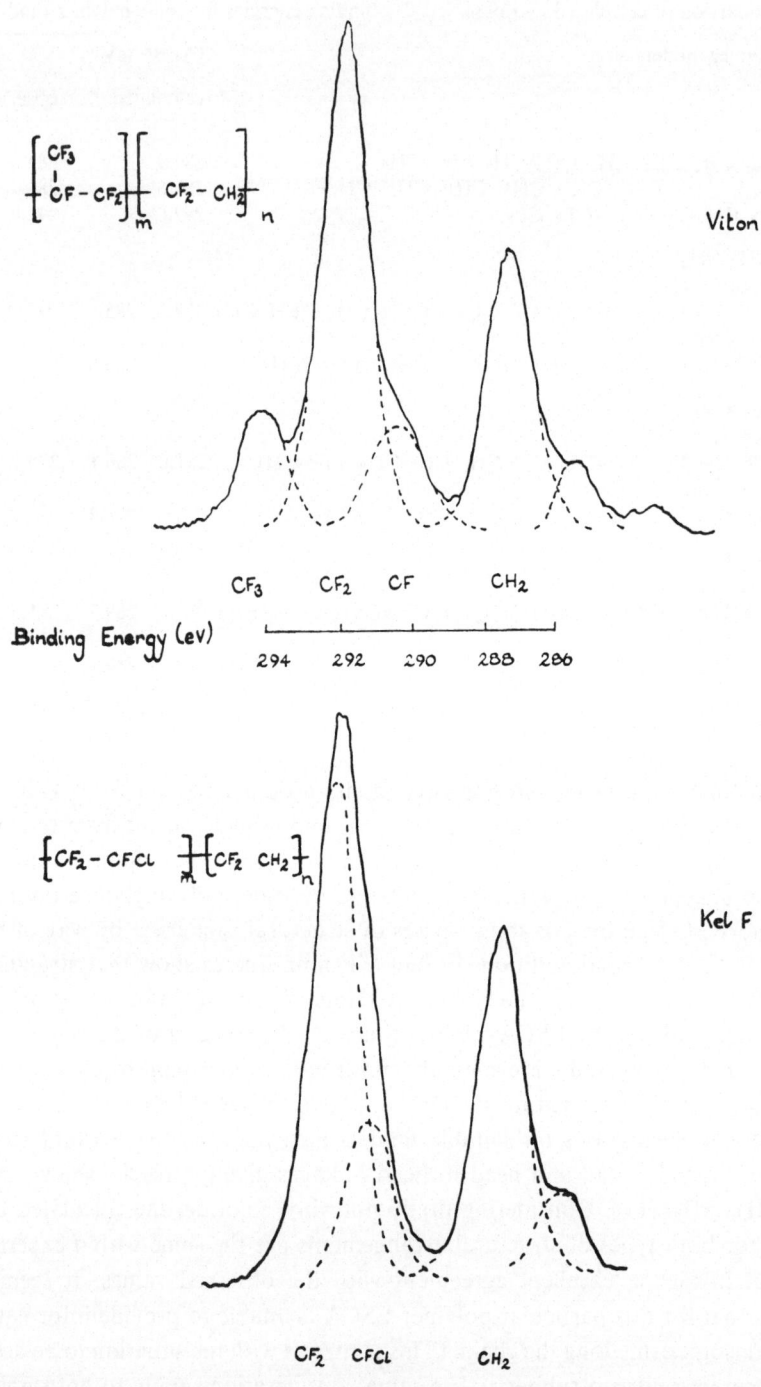

Fig. 110.

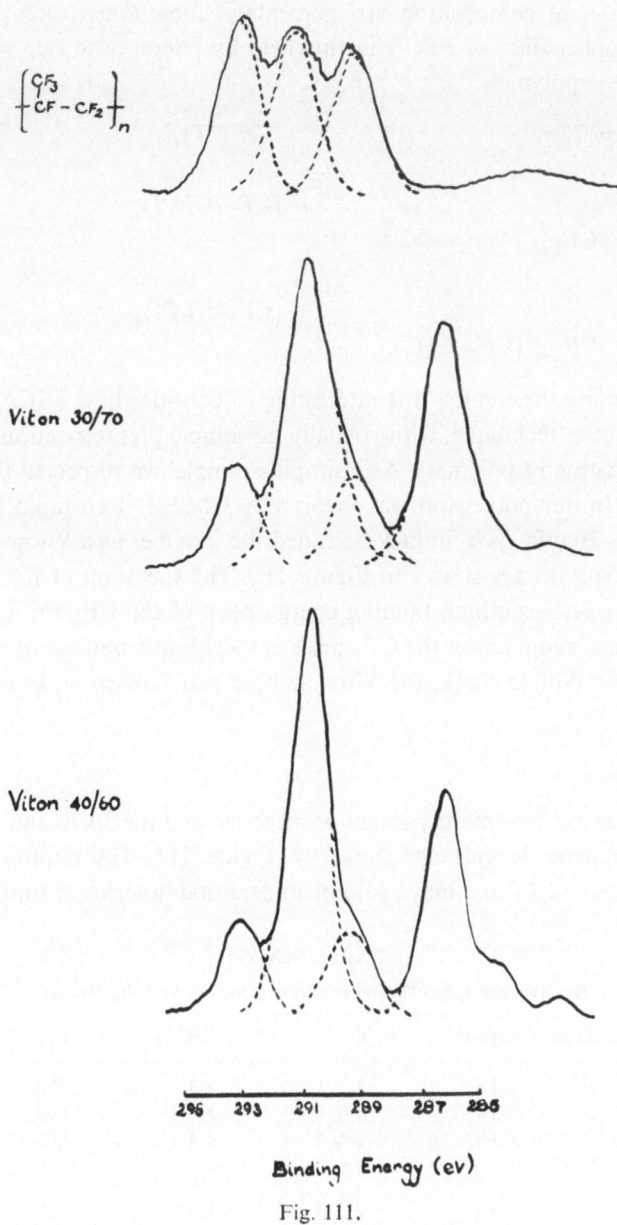

$$\left\{\begin{matrix}CF_3\\ CF-CF_2\end{matrix}\right\}_n$$

Viton 30/70

Viton 40/60

295   293   291   289   287   285

Binding Energy (ev)

Fig. 111.

D. COPOLYMERS

1. *Introduction*

It is evident from the previous sections that substituent effects on core binding energies
in polymers can be understood both qualitatively and quantitatively on the same
basis as those for simple monomeric systems. We are now in a position therefore to

proceed to more complex systems such as copolymers. The first feature of interest is the determination of composition viz. percentage comonomer incorporations. We illustrate the applicability of ESCA in this field by reference to our work on Viton and Kel F type copolymers.

Viton Type   *Kel F Type*

CF_3

| $(CF_2-CFCl)_n$

$(CF-CF_2)_n$

CF_3

| $(CF_2-CFCl)_n$

$(CF-CF_2)_m (CF_2-CH_2)_n$

Before discussing the results, it is interesting to consider how ESCA may be used as a non-destructive technique, with virtually no sample prepreparation necessary for routine identification of polymers. As a simple example, we suspected that one of the Kel F samples in our possession was incorrectly labelled. Two rapid ESCA experiments, taking ~ 20 min each, in fact identified the polymer as a Viton type polymer. The carbon 1s spectra are shown in Figure 110. The spectrum of the Viton sample exhibits the characteristic high binding energy peak of the CF_3 carbon at 294.3 eV and a pronounced shoulder on the CF_2 peak at 292 eV attributable to the CF carbon at 290.4 eV. As a double check, the Viton sample also showed no levels attributable to chlorine.

2. *Viton Polymers*

The C_{1s} levels for the parent polyhexafluoropropene and the 30/70 and 40/60 copolymers with vinylidene fluoride are shown in Figure 111. The binding energies are tabulated in Table XXIV and may again be understood in terms of simple substituent

TABLE XXIV

Binding energies of polyhexafluoropropene and the Vitons

No.	Polymer Type	Group	C_{1s}	$\Delta(C_{1s})$	F_{1s}	$\Delta(F_{1s})$
VIIa	40/60 Viton	CF_3	293.3	8.3	690.2	0.9
		CF_2	291.1	6.1	690.2	0.9
		CF	289.4	4.4	690.2	0.9
VIIb	30/70 Viton	CF_3	293.4	8.4	689.9	0.6
		CF_2	290.9	5.9	689.9	0.6
		CF	289.3	4.3	689.9	0.6
		CH_2	284.6	1.4	–	–
VIII	CF_3 | $[CF-CF_2]_n$	CF_3	293.7	8.7	690.2	0.9
		CF_2	291.8	6.8	690.2	0.9
		CF	289.8	4.8	690.2	0.9

effects. In obtaining the binding energies of the $\underline{CF}_2$ and $\underline{CF}$ carbon 1s levels a simple deconvolution is necessary. In any estimation of copolymer compositions, however, it is obviously desirable to avoid even such a minor complication. The procedure therefore is to measure the area of the $\underline{CF}_3$ peak, the total area of the $(\underline{CF}_2 + \underline{CF})$ peak and the area of the $\underline{CH}_2$ peak. The degree of incorporation of hexafluoropropene (HFP) in the copolymers may then be calculated from the percentage of the total area due to C_{1s} levels represented by each peak:

(a) The mole percent incorporation of HFP must be three times the area of the peak due to $\underline{CF}_3$, on the basis of the stoichiometry of the HFP unit.

(b) The area of the peak due to $\underline{CF}_2$ and $\underline{CF}$ (A) is made up of half the total C_{1s} peak area due to vinylene fluoride ($\frac{1}{2}VF_2$) and two thirds the total C_{1s} peak area due to HFP, i.e.

$$A = \tfrac{1}{2}VF_2 + \tfrac{2}{3}HFP \text{ and } 100 = VF_2 + HFP, \text{ by definition, hence \%}$$
$$HFP = 6(A\text{-}50).$$

(c) The area of the peak due to $\underline{CH}_2$ is half the total area due to CF_2 and hence the mole percent incorporation of HFP is given by the expression

$$100 - (2 \times \% \text{ Area due to } CH_2).$$

Using these three methods to determine the degree of incorporation of HFP gives an internal check on the reliability of the method, the results are as follows:

	% HFP incorporation		
	Method of calculation		
	(a)	(b)	(c)
Sample 40/60	39	42	40
Sample 30/70	33	30	32

The internal consistency is good to within 3% and the values obtained are in good agreement with those quoted for the copolymers investigated.

3. Kel-F Polymers

The C_{1s} spectrum of the parent polychlorotrifluoroethylene consists of a single, broad peak with a flattened top corresponding to overlapping lines from CF_2 and $CFCl$ carbons (Figure 112). The absolute binding energies and shifts (Table XXV) can again be neatly rationalized in terms of simple substituent effects. The C_{1s} spectra for the two copolymers differ considerably, the most noticeable feature being the high proportion of $-\underline{CH}_2$ units in the 30/70 copolymer coupled with the drastically reduced linewidth for the composite line at higher binding energy.

In estimating the compositions of these copolymers it is again desirable to avoid reliance on deconvoluted peak areas. Two methods are available: measurement of the total $(-\underline{CF}_2-) + (\underline{CF}Cl)$ peak area, and measurement of the $(\underline{CH}_2)$ peak area.

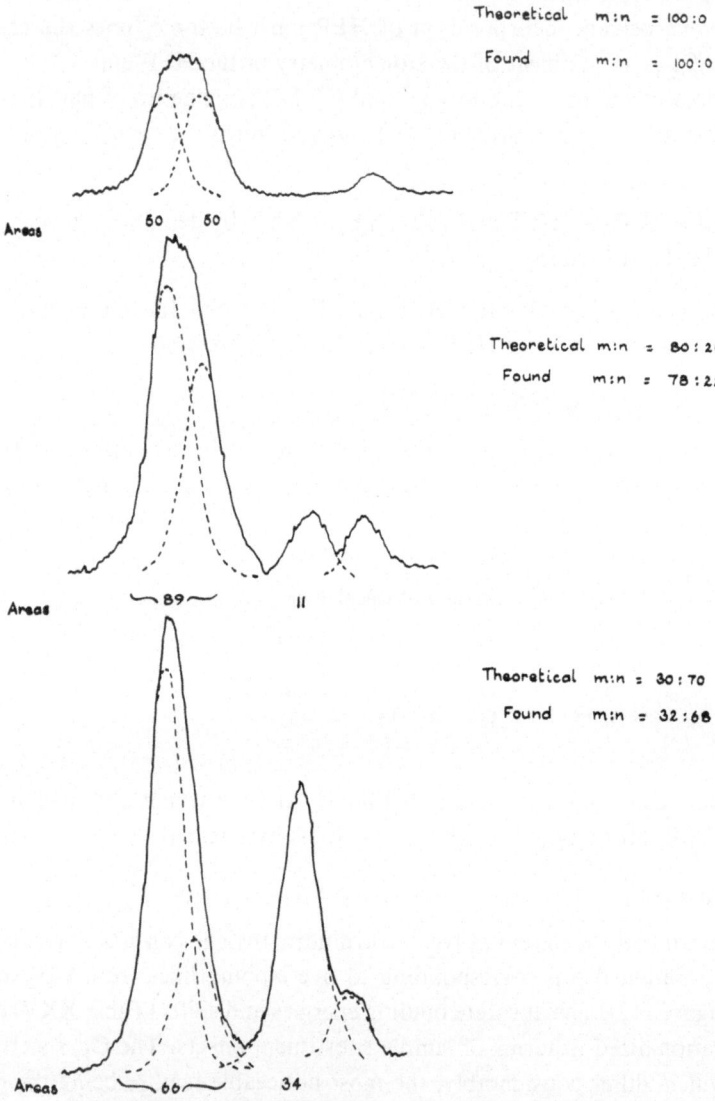

Fig. 112.

TABLE XXV

Binding energies of the Kel-F polymers

No.	Polymer Type	Group	C_{1s}	$\Delta(C_{1s})$	F_{1s}	$\Delta(F_{1s})$	$Cl_{2p3/2}$	Cl_{2s}
IXa	80/20 Kel-F	$-CF_2-$	291.7	6.7	690.3	1.0	–	–
		CFCl	290.5	5.5	690.3	1.0	201.4	272.8
		CH_2	286.8	1.8	–	–	–	–
IXb	30/70 Kel-F	CF_2	291.5	6.5	690.5	1.2	–	–
		CFCl	290.6	5.6	690.5	1.2	201.6	272.5
		CH_2	286.9	1.9	–	–	–	–
X	[CF_2–CFCl]	CF_2	291.9	6.9	690.8	1.5	–	–
		CFCl	290.8	5.8	690.8	1.5	201.1	272.2

Since the total $\underline{CF}_2$ content of each polymer is 50 mole %, the difference between the percentage of the total C_{1s} peak area attributable to ($\underline{CF}_2 + \underline{C}FCl$) and 50 gives the percentage of the total C_{1s} area due to $\underline{C}FCl$, twice this figure gives the amount of chlorotrifluoroethylene units in the polymer.

For the 80/20 and 30/70 copolymers this gives 78% and 32% respectively both being within 2% of the values of 80% and 30% based on elemental analysis. If the areas of the CH_2 peaks are used, the proportions of chlorotrifluoroethylene so obtained are again 78% and 32%. Thus both methods give exactly the same composition within 2% of the quoted values.

4. Nitroso Rubbers

We have thus far demonstrated how raw ESCA data may be used to obtain comonomer incorporations in copolymers. A more complex example is provided by nitroso rubbers.

Nitroso Rubbers

$$CF_3$$
$$|$$
$$[NO]\,[CF_2-CFX]$$
$$X = F,\ Cl,\ H$$

The C_{1s} levels for these three polymers are shown in Figure 113. The assignment of peaks is straightforward and again the shifts are understandable in terms of simple substituent effects (Figure 114). The 1:2 area ratio for $\underline{C}F_3$ with respect to $\underline{C}F_2$ carbons for the copolymer involving tetrafluoroethylene together with the relevant binding energies demonstrates the 1:1 alternating nature of this copolymer. Similar arguments apply to the other polymers. Having demonstrated the 1:1 alternating nature of these copolymers it is of interest to investigate the possibility of detecting structural isomerisms in these systems. It is clear for example that for the C_{1s} levels of the polymer involving trifluoroethylene although the three peaks are of equal area there are substantial differences in linewidth that are not apparent in the spectrum for the polymer formed from the symmetrical olefin, tetrafluoroethylene. To investigate the possibility that the line shapes may encode information concerning structural isomerisms therefore, calculations have been performed on model systems. It is clear from tables of binding energies that the binding energy of the $\underline{CF_3}$ carbons are virtually the same in all three polymers. Charge potential calculations on model systems (Table XXVI) also reproduce this feature.

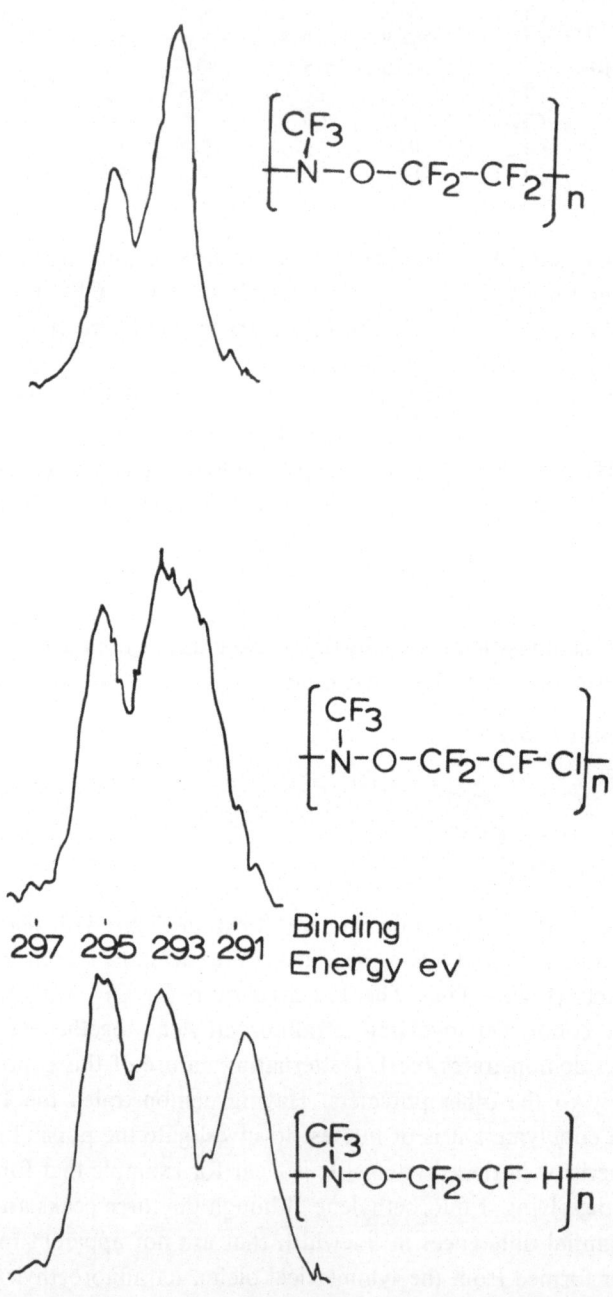

Fig. 113.

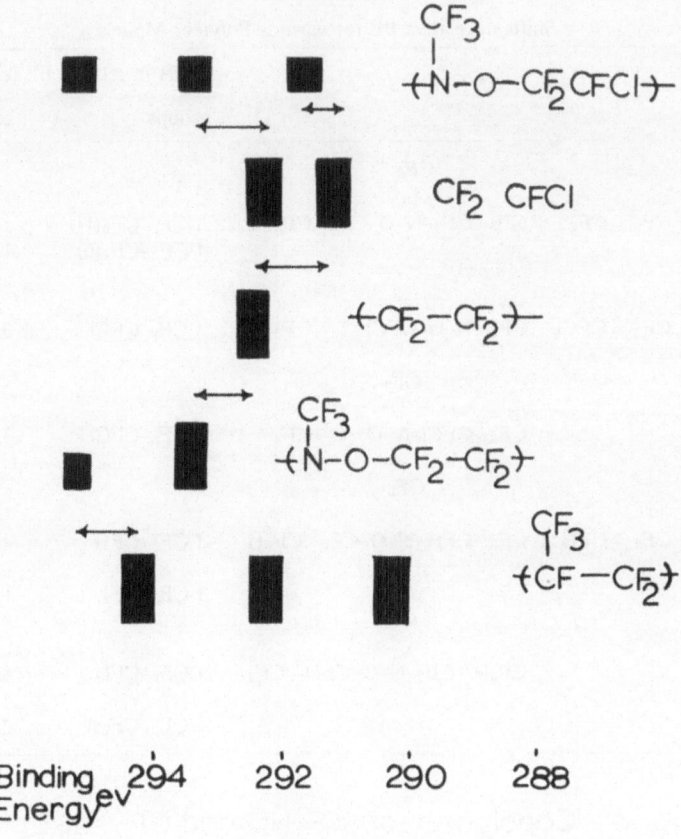

Fig. 114.

TABLE XXVI

Ei–Ei⁰ for $\underline{C}F_3$ from CNDO charges and
potential model in Polymer Models:

$$CF_3$$
$$|$$
$$R_1\text{–}N\text{–}O\text{–}R_2$$

R_1	Ei–Ei⁰	R_2
CF_3CF_2-	10.2	$-CF_2CF_3$
CF_2H-CF_2-	10.2	$-CFHCF_3$
$CF_3-CFCl-$	10.1	$-CF_3$
CF_3-	10.2	$-CFClCF_3$

The models for discussing structural isomerism are shown in Table XXVII together with relative binding energies with respect to $\underline{C}F_3$. Considering firstly the copolymer of tetrafluoroethylene, the model calculations suggest that $\underline{C}F_2$ attached to oxygen or nitrogen have closely similar binding energies thus accounting for the fact that the composite $\underline{C}F_2$ peak is only slightly broadened compared to that for the $\underline{C}F_3$ carbons.

TABLE XXVII

Shifts in relative BE for various Polymer Models

Copolymer		Shifts in relative BE (eV)		
		groups	calc.	obs.
CF₃NO and CF₂ = CF₂	CF_3 | CF_3 ⋮ $CF_2–N–O–CF_2$ ⋮ CF_3	Δ CF_3–CF_2(N)	1.7	1.8
		Δ CF_3–CF_2(0)	1.5	1.8
CF₃NO and CF₂ = CFCl	CF_3 | CF_3 ⋮ $N–O–CFCl$ ⋮ CF_3	Δ CF_3–$CFCl$	3.4	3.1
	CF_3 | CF_3 ⋮ $CFCl–N–O$ ⋮ CF_3	Δ CF_3–$CFCl$	3.6	3.1
CF₃NO and CF₂ = CFH	CF_3 | CF_3 ⋮ $CFH–N–O–CF_2$ ⋮ CF_2H	Δ CF_3–CFH	4.0	4.4
		Δ CF_3–CF_2	1.4	1.9
	CF_3 | CF_2H ⋮ $CF_2–N–O–CFH$ ⋮ CF_3	Δ CF_3–CFH	4.3	4.4
		Δ CF_3–CF_2	2.1	1.9

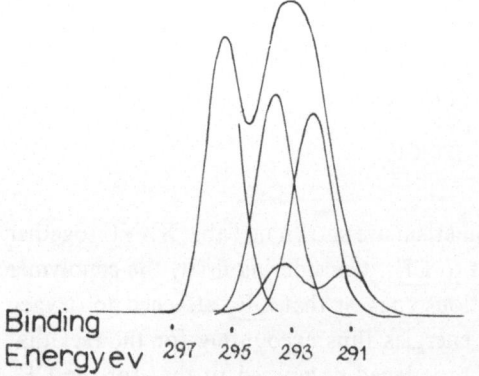
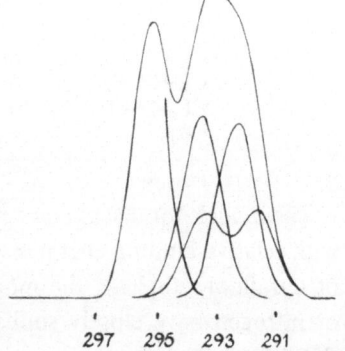

Copolymer of CF_3NO and $CF_2 = CFCl$

CF_3
\|
$-N-O-CF_2-CFCl-$

CF_3
\|
$-N-O-CFCl-CF_2-$

Binding
Energy ev   297   295   293   291   297   295   293   291

Fig. 115.

The results for the trifluoroethylene copolymer suggests that structural isomerism should indeed be detectable by ESCA. In view of this deconvolutions may be attempted for the trifluoro- and chlorotrifluoro-ethylene copolymers using five individual peaks with line shape and linewidth corresponding to that for the $\underline{CF_3}$ C_{1s} levels. The five peaks in each case corresponding to four $\underline{CF_2}$ and $\underline{C}FX$ carbons, (in two pairs corresponding to the two distinct modes of bonding) and only one for the CF_3 carbon.

No unique deconvolution is possible in the case of chlorotrifluoroethylene copolymers, however if two sets of two lines are used, with each member of each pair of equal intensity, a range of good deconvolutions are found, the extremes of the range being shown in Figure 115. The range of the relative areas of the two pairs is between 4:1 and 2:1 in the two cases. The average shifts in C_{1s} binding energies between $\underline{CF_3}$ and $\underline{C}FCl$ within this range are 3.0 eV and 3.8 eV compared with calculated values of 3.4 eV and 3.6 eV.

The measured ratios of 4:1 to 2:1 indicate a 20–33% contribution of structure

$$CF_3$$
$$|$$
$$(-NO-CFCl-CF_2)$$

in the polymer and this agrees with other evidence. The deconvolution for the polymer involving trifluoroethylene is less complicated and the results are shown in Figure 116. The experimental shifts in C_{1s} binding energies between $\underline{CF_3}$ and $\underline{CF_2}$ are 1.6 and 2.2 eV and for the shifts between $\underline{CF_3}$ and $\underline{C}FH$ 4.1 and 4.9 eV. The corresponding

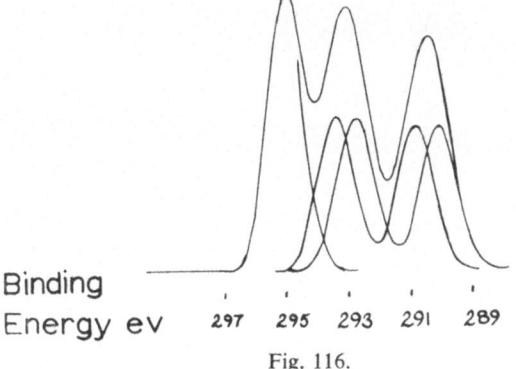

Fig. 116.

theoretical values from the model calculations are in good agreement ($\underline{CF}_3$ and $\underline{CF}_2$, 1.4 and 2.1 eV; $\underline{CF}_3$ and $\underline{CFH}$, 4.0 and 4.3 eV). This assignment would therefore appear to be perfectly reasonable. The area ratio between the two sets of pairs is 1:1, indicating a 50% contribution from

$$CF_3$$
$$|$$
$$(N-O-CFH-CF_2)$$

which is in excellent agreement with other studies.

For these particular copolymers therefore in addition to establishing their elemental composition and comonomer incorporations it is possible to say something about structural isomerisms. It should be evident from the data for the homopolymers discussed in the previous section (in particular the secondary shifts) that in copolymers involving simple fluorinated monomers it is a fairly straightforward matter to dis-

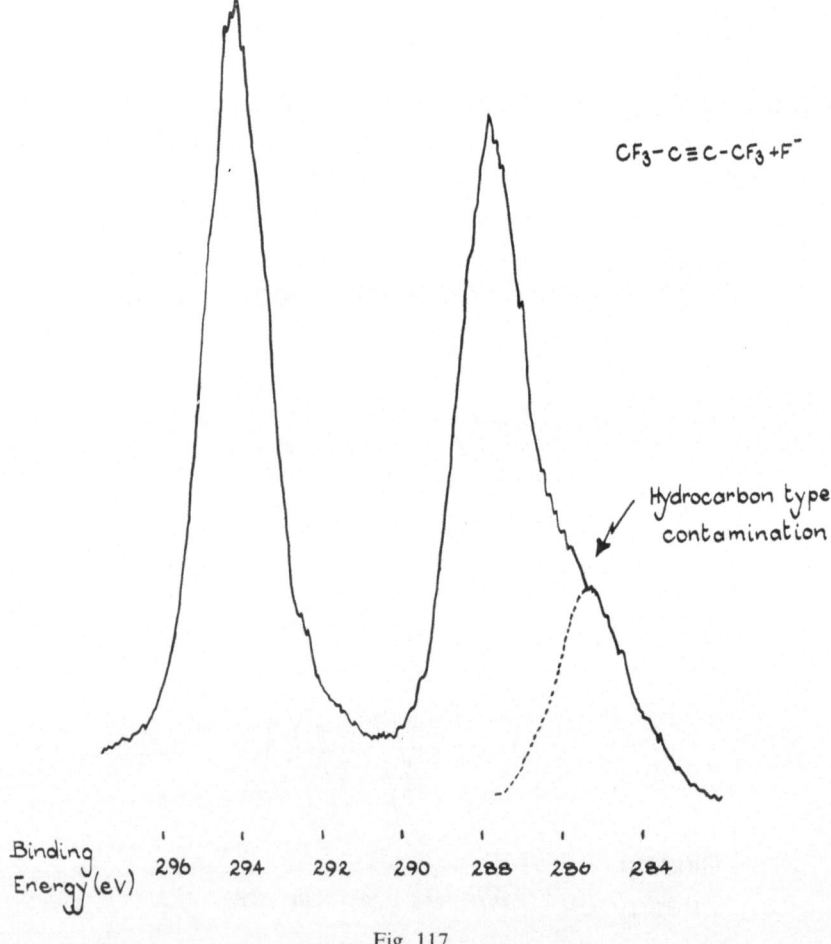

Fig. 117.

tinguish between copolymers which are essentially alternating, or blocked or random. This interesting aspect will not be discussed further here.

E. STRUCTURAL ELUCIDATION OF POLYMERS

With the background theoretical and experimental studies detailed in the previous sections it is now possible to study structure and bonding in polymer systems which have proved intractable by more conventional techniques. As an interesting example of this we may consider the identification of a polymer produced in substantial yield as a by-product in the fluoride ion initiated reaction of hexafluorobut-2-yne with fluorinated heterocyclic molecules. The reaction produces substantial amounts of an insoluble white polymer, with the carbon 1s spectrum shown in Figure 117. The hydrocarbon peaks at ~285 eV arises from solvent and/or the scotch tape backing. (It is difficult to remove the last traces of solvent used in the reaction, and it proved impossible to press a film of the sample.) The spectrum for the polymer is therefore extremely simple with just two peaks of approximately equal areas at 287.7 eV and 294.1 eV. From our studies of substituent effects (and since the polymer contains only C and F) the peak at high binding energy may be unambiguously assigned to $-\underline{C}F_3$

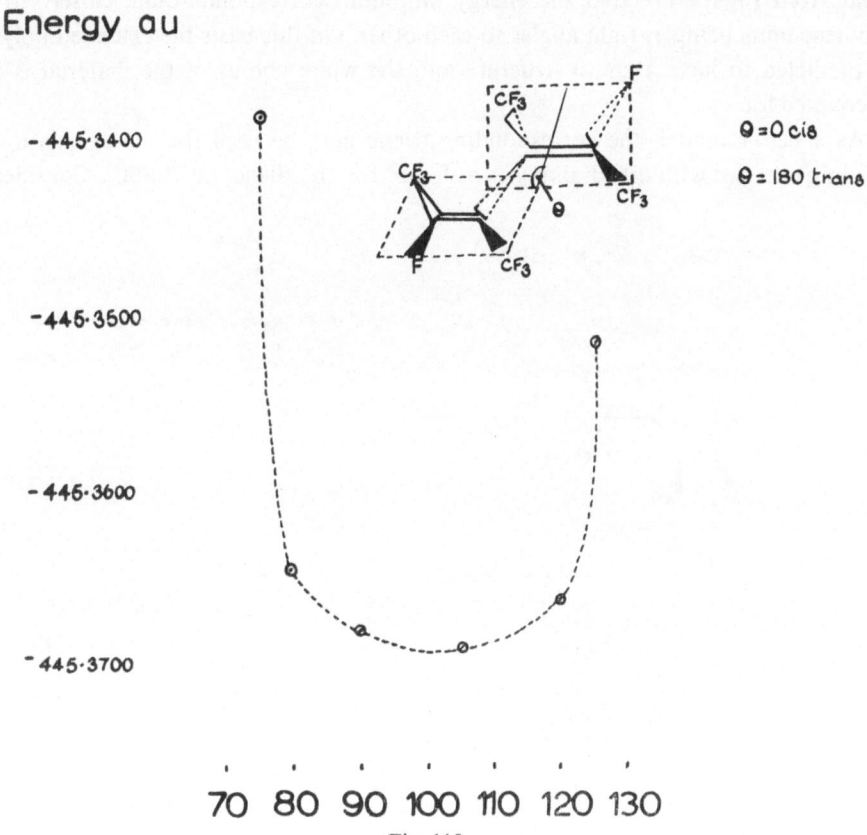

Energy au

Fig. 118.

groups. The assignment of the other peak is not as straightforward, however, the most likely structure for the polymer formed from fluoride catalysed polymerization would be a polyene of the form:

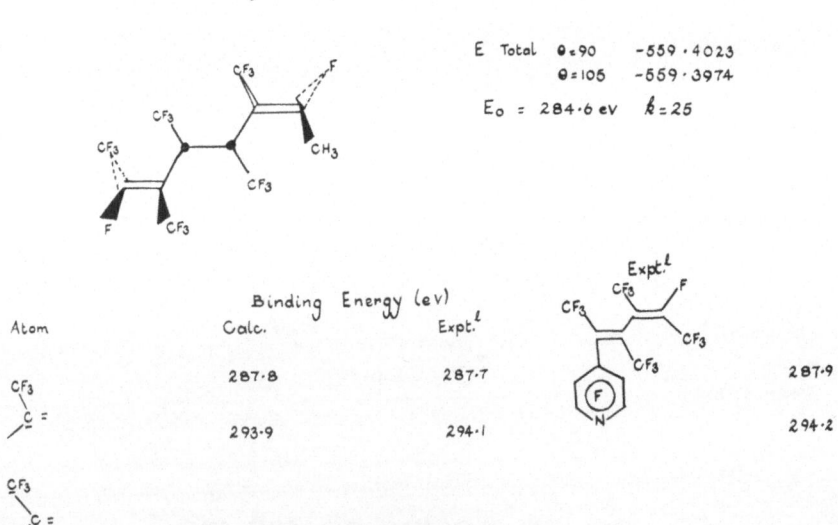

This however still leaves the question of the white colour of the polymer open. As confirmation of this structure a double check may be performed; firstly, a theoretical calculation of the relative binding energies of the two types of carbon, and, secondly experimental studies of model compounds.

Considering first the calculations on model systems it is of interest to investigate the likely stereochemistry of such a polyene. Considering the prototype diene for which the trans-configuration expected on chemical grounds has been assumed, it is clear from Figure 118 that the energy minimum corresponds quite closely to the ethylene units being at right angles to each other. On this basis the extended polymer is predicted to have a spiral structure and the white colour of the material is also accounted for.

As a better model, the corresponding triene may be used for the calculation of binding energies with dihedral angles as found for the diene, i.e. $\theta \approx 90°$. Considering

Model for charge potential Calculations.

E Total $\theta = 90$ $-559 \cdot 4023$
           $\theta = 105$ $-559 \cdot 3974$

$E_o = 284 \cdot 6$ ev $k = 25$

Atom	Binding Energy (ev)	
	Calc.	Expt.[l]
	287·8	287·7
	293·9	294·1

Expt.[l]

287·9

294·2

Fig. 119.

the central unit as characteristic of the polymer chain the calculated relative binding energies for the $\underline{CF}_3$ and the central $\diagdown C{=}C\diagup$ carbons are 293.9 and 287.8 eV respectively, in excellent agreement with the experimental values of 294.1 eV and 287.7 eV (Figure 119).

The C_{1s} spectrum for the substituted pyridine model compound yields the data given in Figure 120. This molecule contains both CF_3 groups and the polyene structure proposed for the polymer (*UV* spectral data confirms that in this system also the double bonds are twisted with respect to the ring). The assignment of the spectrum for this molecule (Figure 120) is made easier by comparison with the substituted vinyl pyridine Figure 121. The two peaks of equal intensity correspond to the $\underline{CF}_3$ carbons and the three carbons of the diene portion (the terminal carbon is at higher binding energies) and energy separation for this model are in excellent agreement with the model calculations and experimental data for the polymer thus completing the identification.

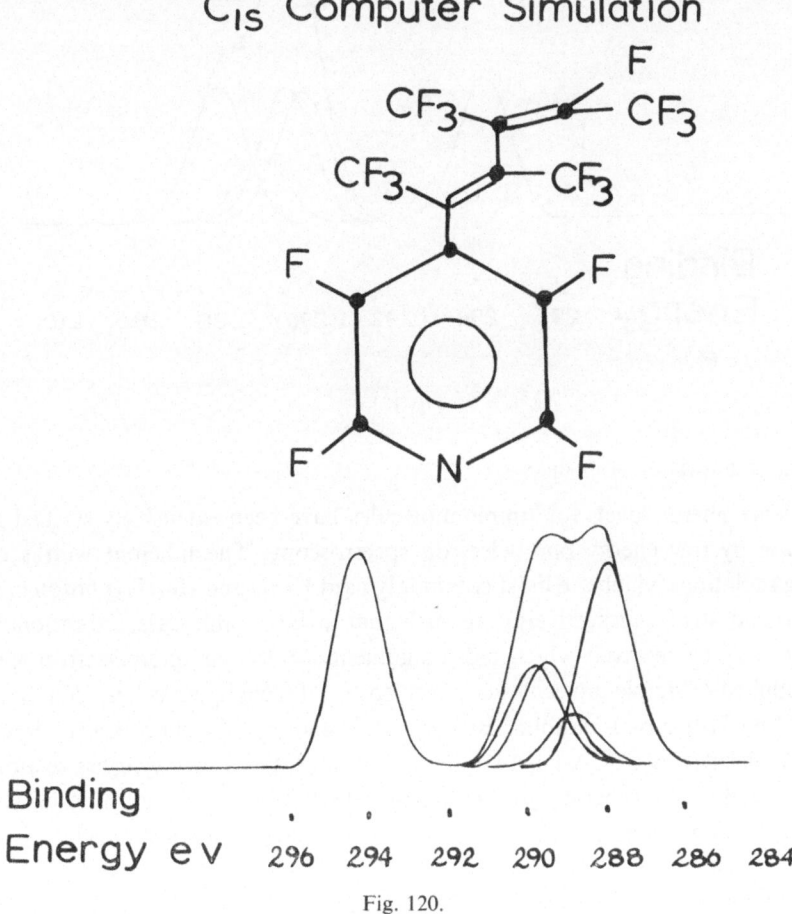

Fig. 120.

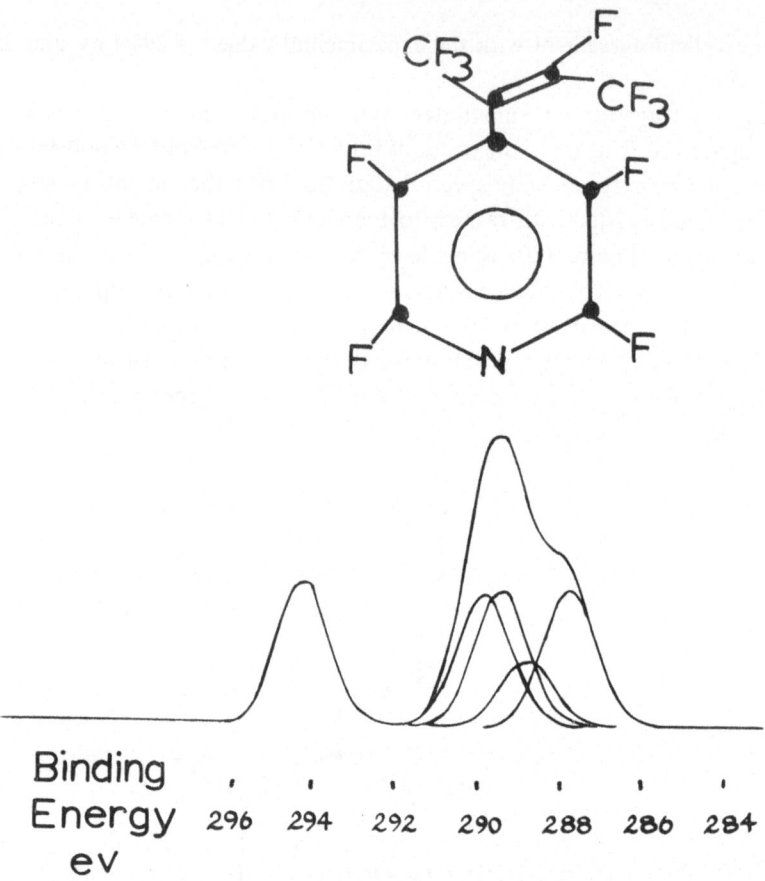

Fig. 121.

F. VALENCE BANDS OF POLYMERS

The valence energy levels for simple molecules have been extensively studied in the gas phase by low energy photoelectron spectroscopy. The inherent widths of the exciting radiations which are most commonly used He(I) and He(II) (photon energies ~ 20 eV and 40 eV) respectively) are such that in favourable cases vibrational progression may be resolved which aid assignment. Molecular photoelectron spectroscopy applied to simple molecules is a very powerful techniques as has been discussed in detail by Professor Lindholm. By contrast the study of valence energy levels for simple molecules by ESCA has two distinct disadvantages. Firstly cross sections for photoionization are generally lower than for the longer wavelength photon sources used in UPS and the resolution is much poorer, (viz. photon linewidths (He(I) ~ 5 meV MgK$\alpha_{1,2} \sim 800$ meV).

In studying involatile materials such as polymers however these disadvantages are considerably offset. Thus since there are so many vibrational modes possible, resolution becomes less of a problem since even with a He(I) source only broad unresolved bands would be obtained. Inspection of the escape depth versus kinetic energy curve discussed earlier reveals three major problems with using a low energy photon source which do not arise in ESCA examination of valence energy levels. Firstly with a low energy photon source not all of the valence energy levels may be studied, only the

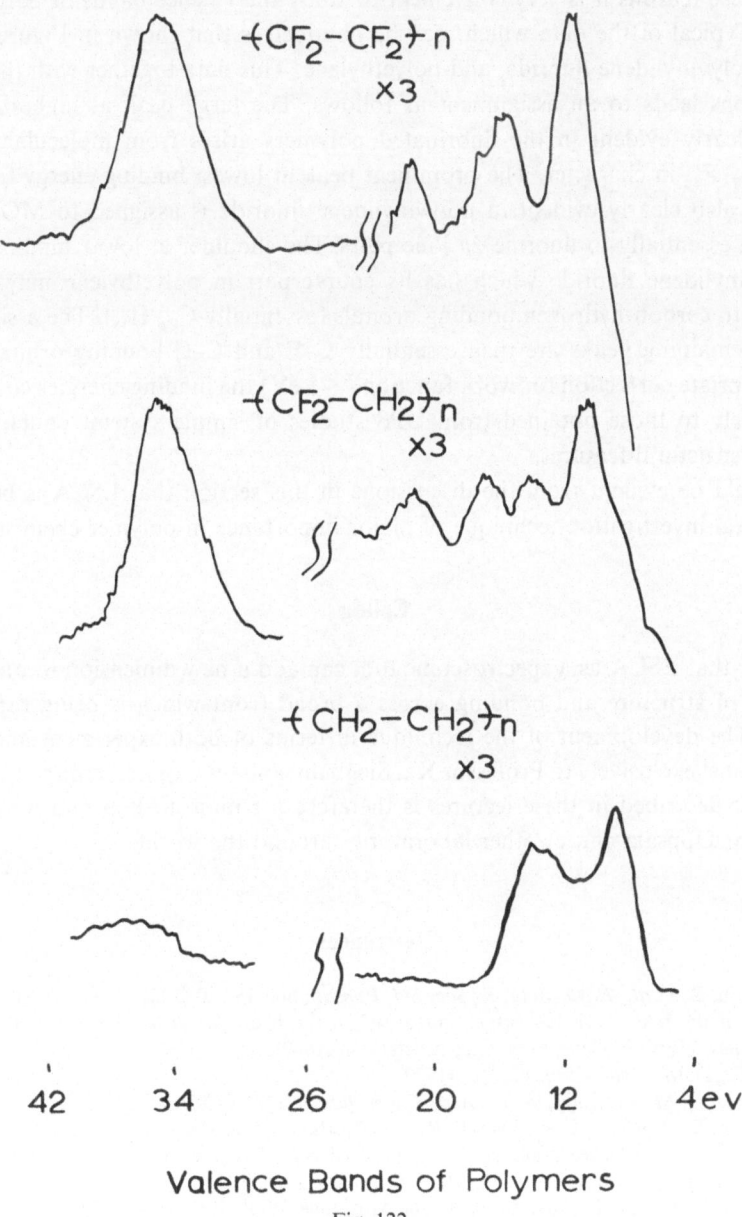

Valence Bands of Polymers

Fig. 122.

higher occupied levels. Secondly since the kinetic energy range for electrons will typically be in the range 0–20 eV (He I), 0–40 eV (He II) it is clear that this is a region of rapidly varying escape depth. Surface contamination is therefore very critical, much less so for an X-ray source. Thirdly in the absence of contamination there are still difficulties of interpreting the data because of marked differences in escape depths which do not arise in ESCA since the escape depth dependence is virtually constant across the valence band.

For these reasons it is very convenient to study the valence bands of polymers by ESCA. Typical of the data which may be obtained is that shown in Figure 122 for PTFE, polyvinylidene fluoride, and polyethylene. This data together with theoretical calculations leads to an assignment as follows. The large peak at highest binding energy clearly evident in the fluorinated polymers arises from molecular orbitals essentially F_{2s} in character. The prominent peak at lowest binding energy for PTFE which is also clearly evident in polyvinylidene fluoride is assigned to MO's corresponding essentially to fluorine $2p$ lone pairs. The shoulder at lower binding energy in polyvinylidene fluoride which has its counterpart in polyethylene may then be assigned to carbon-hydrogen bonding orbitals (essentially $C_{2p} H_{1s}$). The assignments for the remaining peaks are then essentially C–F and C–C bonding orbitals. With an appropriate correction for work function (~ 5 eV) the binding energies correspond quite nicely to those obtained from UPS studies of simple systems containing the essential structural features.

It should be evident from the discussions in this section that ESCA is becoming a structural investigation technique of major importance in polymer chemistry.

Epilog

It is clear that ESCA as a spectroscopic tool can add a new dimension to our understanding of structure and bonding across a broad front which is being rapidly extended. The development of the technique in terms of both experiment and theory is due almost exclusively to Professor Kai Siegbahn and his research group at Uppsala. The work described in these lectures is therefore a tribute to him and his research workers at Uppsala and at other laboratories around the world.

References

1. Siegbahn, K. *et al.*, *Nova. Acta. R. Soc. Sci. Uppsal.*, Ser. IV, 20 (1967).
2. Much of the later work has been summarised in Siegbahn, K. *et al.*, *ESCA Applied to Free Molecules*, North-Holland Publ. Co., Amsterdam (1969).
3. Hercules, D. M., *Anal. Chem.* **42**, 20A (1970).
4. Hollander, J. M. and Jolly, W. L., *Acts. Chem. Res.* **3**, 193 (1970).
5. Hulett, L. D., Carlson, T. A., Fish, B. R., and Durham, J. L., *Proceedings of the Symposium on Air Quality*, Los Angeles, California, U.S.A. April 1–2, 1971, Plenum Pub. Corp. (in press).
6. Swartz, W. E. and Hercules, D. M., *Anal. Chem.* **43**, 1774 (1971).
7. Clark, D. T., Kilcast, D., Adams, D. B., and Musgrave, W. K. R., *J. Electron Spectroscopy* **1**, 227 (1972).

8. Clark, D. T., Feast, W. J., Kilcast, D., and Musgrave, W. K. R., *J. Polymer Science A-1* **11**, 389 (1973).
9. Barber, M. and Clark, D. T., *J. Chem. Soc. D* (Chemical Communications) **23** (1970).
10. Brundle, C. R., *Appl. Spectroscopy* **25**, 8 (1971).
11. Olah, G. A., Mateescu, G. D., and Riemenschneider, J. L., *J. Amer. Chem. Soc.* **94**, 2529 (1972) and references therein.
12. Carlson, T. A., *General Survey of Electron Spectroscopy* (ed. by D. Shirley), Asilomar Conference California, U.S.A., September 1971, North-Holland (1972).
13. Pelavin, M. Hendrickson, D. N., Hollander, J. M., and Jolly, W. L., *J. Phys. Chem.* **74**, 1116 (1970).
14. Allison, D. A., Johansson, G., Allan, C. J., Gelius, U. Siegbahn, H. Allison, J., and Siegbahn, K., *J. Electron Spectroscopy* **1**, 269 (1972).
15. Ditchfield, R., Hehre, W. J., Pople, J. A., and Radom, L. *Chem. Phys. Letters* **5**, 13 (1970).
16. Snyder, L. C., Robert A. Welch Foundation Res., Bulletin No. 29, August 1971.
17. Adams, I., Thomas, J. M., Bancroft, G. M., Bulter, K. D., and Barber, M., *J. Chem. Soc. D* **751** (1972).
18. Fadley, C., this volume, p. 151.
19. Carver, J. C., Schweitzer, G. K., and Carlson, T. A., *J. Chem. Phys.* **57**, 973 (1972).
20. Parrat, L. G., *Rev. Mod. Phys.* **31**, 616 (1959).